Guide to the Taxonomic Literature of Vertebrates

Guide to the Taxonomic Literature of Vertebrates

RICHARD E. BLACKWELDER

The Iowa State University Press / Ames, Iowa

1972

RICHARD E. BLACKWELDER is Professor of Zoology, Southern Illinois University

Printed by The Iowa State University Press

First edition, 1972

International Standard Book Number: 0-8138-1630-0
Library of Congress Catalog Card Number: 70-39613

DEDICATED TO THE VERTEBRATE TAXONOMISTS OF THE WORLD AND THEIR LIBRARIANS

PREFACE

This book is intended to aid students to get a start into the literature of any group of vertebrates. It lists the known items published during the past 50 years or more which may be taxonomically useful themselves or may lead the student to the older literature.

The book is very carefully structured. Without an understanding of this structure, it is unlikely that anyone will succeed in putting it to maximum use. It also incorporates several novelties in details of presentation; these also need to be understood in advance. I urge the user to read the Introduction and to browse through the lists of works before he puts the book to the test of actual use to solve a problem of source literature.

Neither extreme accuracy of detail nor absolute completeness were goals in the preparation of this work. It is intended to provide clues for further search, and thus it is usually goal enough to try to start the student on the right track. Each book he examines from these lists will suggest other books he should also consult, giving the effect of a rolling snowball of increasing size and appropriateness.

It would be easy to get some false impressions of the purpose of parts of this work. An effort is made to state the real purposes and the limitations of each part, to prevent misunderstanding of the tool that is being offered to the user.

Taxonomist users will recognize that this is basically a summary (and extension) of the Zoological Record, to make it unnecessary for every student to search every page of each of the fifty volumes in his specialty. If he uses the time saved to produce good taxonomic work of his own, the compiler will be satisfied that his own time was not wasted.

Fossil vertebrates are covered herein in only two ways. Extinct families are listed in alphabetical order among the families of extant animals, but their literature is not shown. (Extinct classes are not listed at all as such.) A few general works on vertebrate paleontology are listed, because they do have great interest and value in the study of Recent vertebrates.

The extent to which the literature on fossils is relevant to animals living today is not always realized. The compiler would have preferred to treat them together. The decision was early made to deal with the fossils separately, and it was found to be impracticable to examine and report both literatures in the same way. The works on fossils are generally indexed in relation to stratigraphic data, rather than to family assignments or geographical occurrence. [Excellent references exist that list the works on vertebrate fossils, such as the bibliographies of O. P. Hay (1902 and 1929), A. S. Romer et al. (1962), C. L. Camp et al. (1940 —), and the Society of Vertebrate Paleontology.

Bibliographic aids are so few in most parts of zoology that no apology is necessary in presenting a new one. There are faults here, as in all such works, and the compiler merely hopes that the taxonomic approach and arrangement will be sufficiently useful to compensate for shortcomings.

R. E. Blackwelder

November 1, 1971

ACKNOWLEDGMENTS

Several groups of people were essential to the production of this book. Their help is gratefully acknowledged both by the compiler and in the name of the users of the book.

The first are the compilers and publishers of the Zoological Record. Although the purely revisionary works were usually found by other means, the subsidiary lists of works were largely from that indispensable taxonomic reference work. It is frequently cited herein, and it is believed to be even more useful to zoologists than sometimes realized. It has faults and contains a variety of errors, but it uniquely fills a vast need in recording literature not only taxonomically but also geographically and by subject.

Next are the librarians of natural history libraries across the country. Inasmuch as many persons were involved in these, it seems appropriate to list the libraries and express thanks to the staff of each. These include: the American Museum of Natural History in New York, where the basic compilation was made possible by the unexcelled availability of books and the unlimited cooperation of the library staff; the United States National Museum libraries, especially the four divisional libraries of Fishes, Reptiles and Amphibians, Birds, and Mammals; the Field Museum of Natural History in Chicago, where unusual sources of very recent titles were found; the California Academy of Sciences in San Francisco, where the main library is supplemented by the Department of Ornithology and Mammalogy, the Department of Ichthyology, and the library of the Steinhart Aquarium; the Biology Library at the University of California at Berkeley and the library of the Museum of Vertebrate Zoology; the Biology Library at the University of Illinois at Urbana; the six biological branches of the library of the University of Michigan at Ann Arbor; and the Science Library at Southern Illinois University at Carbondale. Special mention should be made of the separate Dean Memorial Library of Fishes at the American Museum of Natural History and the Exhibit Library of the Society of Systematic Zoology (now in the custody of the compiler).

Although the institutional libraries supplied the bulk of the data for this compilation, very great help was also given by specialists in the various groups at various museums. These taxonomists made their personal libraries available and helped in that most difficult task of finding the books too recent to be noted in the Zoological Record. To these colleagues sincere thanks are due. In fishes: W. I. Follett and E. S. Herald of the California Academy of Sciences; J. W. Atz at the American Museum of Natural History; B. B. Collette of the U. S. Bureau of Commercial Fisheries; and G. S. Myers of Stanford University. In reptiles and amphibians: J. L. Peters and G. R. Zug of the United States National Museum; and C. M. Bogert of the American Museum of Natural History. In birds: R. T. Orr of the California Academy of Sciences and G. W. Watson of the United States National Museum. In mammals: P. Hershkovitz of the Field Museum of Natural History; H. W. Setzer of the United States National Museum; and R. T. Orr of the California Academy of Sciences. In addition to these, and to the many helpful librarians, special thanks are due to R. L. Wenzel of the Field Museum of Natural History; to G. W. Wharton of Ohio State University; and to an unidentified reviewer for the Iowa State University Press.

CONTENTS

INTRODUCTION

This guide is intended to lead to answers to two sorts of questions asked by taxonomists or other persons needing taxonomic information. The first group includes such questions as the following:

What books on each group are available?
What inclusive papers have been published in serials?
How can one identify a serial title from an abbreviation?
What agency published this serial?
What else has this agency published?
Who was the man who wrote this work?
What else did he write?
Where is his collection deposited?
What other collections are in that institution?
Where can one find:

- Bibliographies?
- Descriptions of taxonomic methods?
- Identification aids, such as manuals and keys?
- Records of expeditions and voyages?
- Glossaries, lists of abbreviations, etc. ?
- Location of types and other specimens?

The second group of questions includes those concerned directly with taxonomic information:

What works have been published that are intended to permit the identification of species, either throughout the range of the group or in a specified lesser area?
What works list the species in a group, either worldwide or in a specified area, as checklists or synonymic catalogs?
What works deal taxonomically with the species of a region, as descriptive catalogs or faunal handbooks or revisionary studies?
What monographic works deal with the world species of a group, or the species of a major area, in comprehensive taxonomic manner?
For each family or group, what revisionary works cover all the species or all those in a single country?

Because these questions can be asked, and sometimes answered, at all taxonomic levels from species to kingdom, and because the relevant literature is indexed at a variety of these levels, it is difficult to arrange the answers to the taxonomic questions in any one manner that will serve all persons. The most important monograph of a family may be published in a larger work where it is combined with other families under a title showing only the order or the class, so that it is always necessary to look for relevant works at various levels.

In the case of the first class of questions referred to above, it is again necessary to search at various levels. A list of fish names would be listed in the section on fishes under the heading Nomenclators, but a nomenclator of all animals (such as Neave's) would be listed only under the general heading Animalia — Nomenclators, yet would be equally relevant to the enquirer concerned only with fishes or only with one family of them.

The section of Mammalia, therefore, must be used in conjunction with the Animalia section, and a subsection under Mammalia (order Rodentia or family Muridae, for example), must be used in conjunction with Animalia and Mammalia, and in the case of Muridae also in conjunction with Rodentia listings.

The plan adopted is to reduce the number of levels to a minimum; in the case of the fishes, to the artificial but useful "class" Pisces and the families. A revision of an order would be listed under Pisces or under the included families or both but usually not under the name of the order.

The citations to the literature in which the answers to these questions may be found are listed in varying form according to the requirements of each situation. Some are arranged geographically (faunal studies), some chronologically (classifications), and some merely by alphabetical sequence.

Titles of articles are frequently not given. The names of the periodicals are abbreviated according to the system long used by the U. S. National Museum and described briefly in TAXONOMY by R. E. Blackwelder (1967).

If the language is not clear from the title, there is a note stating the language. This includes all titles transcribed from non-Roman alphabets. If there is no note, the work is in the language of the title as given. Translations are indicated by annotation, if known to be such.

The purposes of the sections

The intent of the compiler as to the content and coverage of each section of this guide is indicated briefly below under the headings used in the text. In all of the lists, the limitations of each work, if not shown in the title, are indicated by parenthetical notes whenever such information was available. In general, however, it is not presumed to be appropriate to go beyond an explicit title, and it is left up to the user to decide whether Lozano y Rey, L. is the same as Lozano Rey, D. L.; whether different transliterations derive from one name (as Schmidt and Shmidt from Russian); and whether U. S. S. R., Soviet Union, and U. R. S. S., used at different periods, do in fact refer to the same geographical areas.

General books. Here are listed the most general books on the group, published in any part of the world and in any language. Included also are works on the entire group or a major part of it but restricted in subject to a single aspect, such as osteology. Many of these books are popular accounts, but it is intended to include only those that are taxonomically useful. Some books of doubtful general value are nevertheless included because they cover a unique area or may be useful in some special way.

Bibliographies, separate works. In most groups there have been published a few formal bibliographies, as separate works, sometimes of large size and extremely useful nature. In many cases these are arranged by author, necessitating a knowledge of the author involved to be readily used. When only the subject or region is known, recourse must be had to specialized bibliographies, where the subject is the key to the works. These generally fall under the next heading.

Bibliographies in other works. Here are listed what appear to be the major lists of References in large works. There is no end to these, and only the larger ones, those which appear to be comprehensive, are included. The student should examine the bibliography in every book he uses.

Institutional bibliographies. Infrequently, there are published lists of or indexes to the publications of a museum, society, or other organization. These are valuable sources of historical information on the papers, such as exact date of publication. Indexes to periodicals may be listed separately.

Biographies and personal bibliographies. These have been published for many taxonomists, especially in obituaries. They vary in completeness but offer the best means to identify a work when the author is known. It is not the purpose of the present work to list all the relevant biographical memoirs and personal lists of publications. Rather, it is intended to list the sources where these lists are available. Unfortunately these are almost non-existent. The references cited are therefore mostly examples of the type of bibliography that may be sought.

Periodical lists. Nearly all bibliographies and other references to periodical literature refer to the periodicals by abbreviations or shortened forms. Often these are so much abbreviated that they are difficult to recognize or no longer distinctive. There are more than 10,000 periodicals in which taxonomic papers have been published, and many of these have used various forms of their titles. The number of these and the variations make identification of some isolated abbreviations impossible without aids. There has never been, in recent times at least, a single list of all the journals in which taxonomic works have appeared. There are several extensive lists, however, and many on restricted groups. The most general are listed under Animalia, and the more restricted ones under the respective animal groups.

Directories. Directories are rare in most fields, at least as formal publications. Many persons have assembled lists of the current workers in their own field, and many societies have published lists in such an area as Paleontology or Parasitology. These are all listed, so far as known, but because they are soon out of date, and often merely mimeographed or otherwise informally duplicated, they are not available in most libraries, are not generally cited in bibliographic works, and are difficult to assemble.

Institutions. To identify a museum or other taxonomic organization, there are a few sources. These are usually of wide coverage and are listed under Animalia.

Museum lists. These are publications listing the species or actual specimens owned by that institution. They are useful as a means of finding material for revisionary work, for finding the location of specimens on which previous work was done, and apparently also for advertising the museum's holdings. These most frequently cite only types and other especially important specimens. In most cases they cover only one group of animals, and in a few cases they are annotated with much valuable taxonomic information.

Exploration and localities. Again, it is not intended that this section should be a bibliography of exploring expeditions or their reports but principally a list of sources in which to find such information. Some of the important reports are cited as examples.

Methods. General works on the taxonomic methods in each group are the prime interest here. As these are few, a variety of actual papers are also cited as examples of what is available. Only methods directly relevant to taxonomy are listed, but these include collecting, preservation, shipment, description, statistical analysis of variation, publication, illustration, naming, etc.

Glossaries. Many books contain lists of explanations of terms used in the group. A few separate glossaries and even dictionaries are available. General dictionaries are almost always unsatisfactory for defining such terms because they do not restrict themselves to a single definite meaning that permits the accuracy of communication that is necessary in descriptive taxonomy. Both separate glossaries and extensive ones in major books are listed, when known to the compiler.

Common-name lists. These include references to lists of the common names applied to species in each group. The few major lists for large areas, such as the United States or Europe, are elsewhere replaced by numerous shorter papers, often citing "the important" or "the economic" species only. Some recent examples of the latter are listed also.

Nomenclators. These are reference lists to published names at some particular level. The most inclusive ones (under Animalia) show all the names that have ever been used. The ones restricted to a special group may be useful to identify an unknown name. Nearly all are to generic names principally, although these also give some indirect help with family names.

Name studies. An important kind of taxonomic study is the analysis of the taxa and their names at some level in one group. The information given may be the history of the names, their synonymy, or their taxonomic application. Studies of the types of genera (genotypes) are a special form of these, necessary for the correct application of names to taxonomic groups. They are never numerous.

ARMADILLO WORLD • MAP NO. CO-325

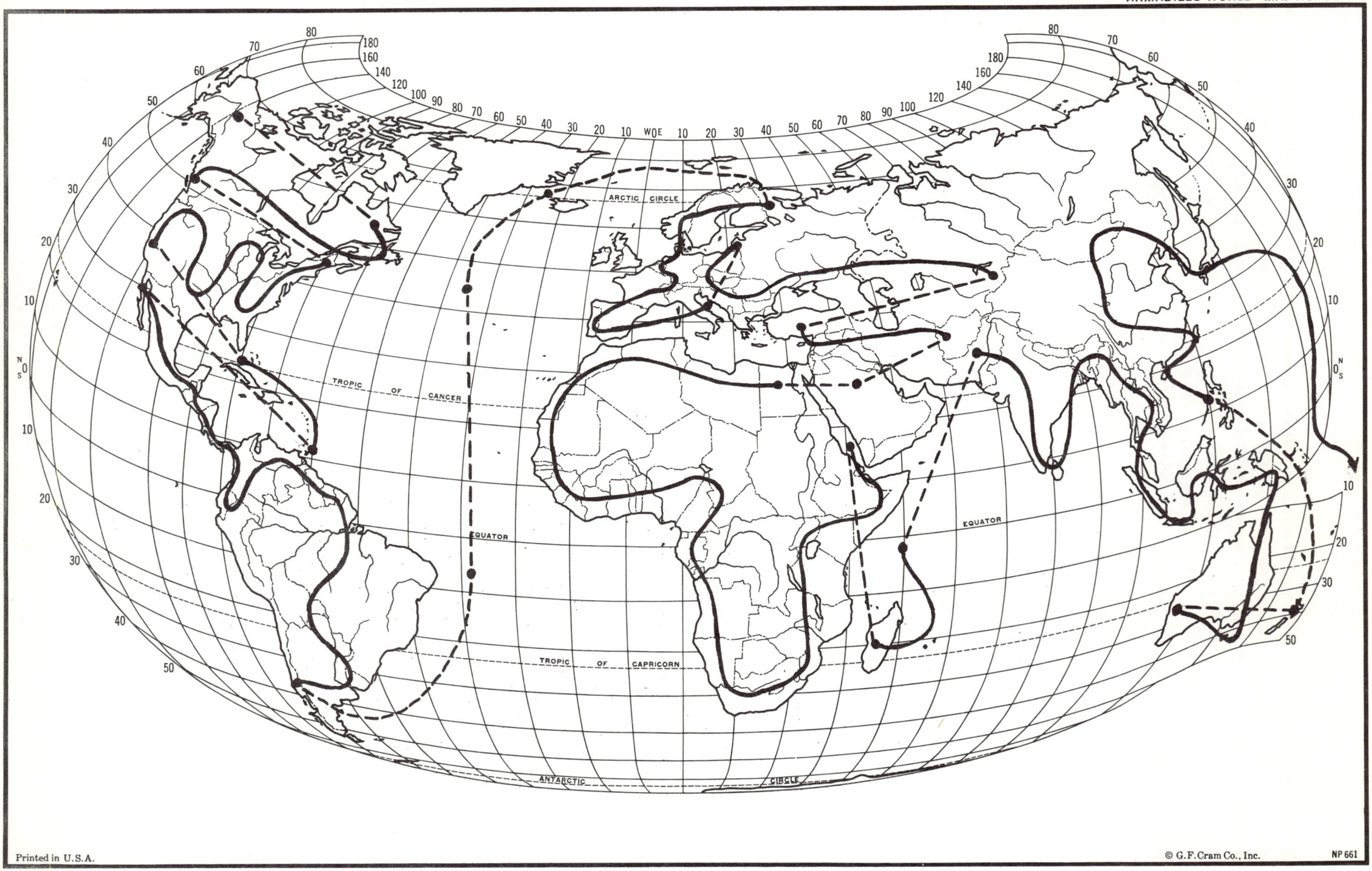

ARCTIC CIRCLE

TROPIC OF CANCER

EQUATOR

EQUATOR

TROPIC OF CAPRICORN

ANTARCTIC CIRCLE

Printed in U.S.A.

NP 661

Regional works. Among these are faunas, the large regional works, supposedly dealing with all the relevant species over a given large area. Included are those ostensibly dealing with the fauna of an entire country or group of countries, or, in the case of very large countries, with sections of that country. For example, in North America, the faunal works of various states are usually not listed, whereas those of groups of states usually are. Inclusion is primarily the result of decision that the study is truly taxonomic and detailed enough to give references to previous work in that area. The works are arranged geographically under regional subheads.

The faunal or regional lists (checklists) are frequently rather limited in geographical coverage, but they are one of the most persistent forms of taxonomic literature. In addition to lists covering whole countries or major regions, there are included here a few of more restricted coverage. They are interspersed with the faunas and identification works.

For identification, all monographs, revisions, and faunal studies will serve to some extent, but in addition there are many books and papers published solely to aid in this activity. These special ones are listed here, where they cover a region of some size and are not listed under other headings. They are usually indicated by the notation "(keys)", even if they do not actually use these devices.

Special lists. In most groups there are occasional studies based on non-taxonomic "groups". For example, a list of the fishes occurring in both fresh and salt water, the arthropods found in birds nests, the venomous animals, or the marine borers of all kinds. No careful search for these has been made, but those encountered are listed under the subject of the grouping.

Monographic, descriptive, and revisionary works. These works carry the responses to the second group of questions above, the taxonomic questions. Their collection was the principal goal of the preparation of this guide. These may individually cover a taxon of any size or rank or groups of taxa at any rank. Thus, a monograph may cover a species, or a family, or a class, or a phylum, and it becomes difficult to guess at which level help on a particular taxon should be sought. To a person unfamiliar with the details of the hierarchical classification of fishes, for example, a monograph listed under the heading "Sclerodermi" would not immediately be recognizable. It was therefore determined to reduce the number of levels essentially to two, the class as a whole and individual families, with the latter listed alphabetically. Major general revisionary works are therefore listed principally under the included families. A few revisions of lesser coverage, usually referring only to multi-family classifications, are listed separately in systematic order (as Group monographs).

Under the heading Classifications are listed principally those few works which present arrangement of the entire class or a major part of it.

Under Descriptive catalogs are cited those curious but often ponderous volumes, numerous in the productions of the late 19th century, that uniquely combine the three features of descriptive faunal revisions, museum lists, and catalogs of the known species. Some of these seem worth listing as modern sources of taxonomic information.

Under Synonymic catalogs are listed only those major taxonomic lists designed to show synonymy, literature, and often distribution of all species in the group. They are among the most useful taxonomic reference works but are lacking in many groups. They gradually merge into the more elaborate checklists on the one hand and into monographic studies on the other.

Family revisions. These make up the largest part of this guide. Under each family the works that treat of it monographically or faunally (regionally) are listed in a geographical sequence that is approximately this: North America southward to Patagonia, North Atlantic, Europe from Scandinavia to the Mediterranean and then the Black Sea and Russia, Africa, Madagascar, Red Sea and Indian Ocean, Indo-Malaya and Australasia, north to the Philippines and China and Japan, Oceania, and Antarctica. (See map). The papers cited have mostly been examined and determined to be substantially taxonomic revisions.

The manner of citation is noted at the beginning of the section, where there is a list of titles alphabetically by author that includes many of the comprehensive regional monographs or faunas.

The lists of families are themselves a unique feature of this book. Each one is the most complete alphabetical listing of the families of that class so far published. They include both extant and extinct families. All variant spellings found in recent literature are included. What appears to be synonymy in the listing is not expression of opinion as to actual zoological identity but merely notation of how references to these names are filed herein. For example:

Balistidae (incl. Aleuteridae, Aluteridae, Anacanthidae, Ballistidae, ...)
Ballistidae, as Balistidae

These do not claim synonymy of Aleuteridae and Anacanthidae with Balistidae, or certain seniority of Balistidae over Ballistidae, but merely announce that works using these names or spellings are herein grouped under Balistidae.

When a reference has used one of the parenthetically appended names instead of the alphabetized names, that name is listed in the annotation "(as Aluteridae)". Thus the actual name used in each reference is shown, and a taxonomist who wishes to may separate them out. Variant spellings are not so indicated, and the compiler is merely following some reviser in the choice of these. (The revisers are listed below.]

Sources of family lists

Each of the lists of families in a class has been gathered from a variety of sources. The works principally involved in each are these:

Cephalochordata. As there is no worldwide monograph, no synonymic catalog, and no other family name study known to me, these are simply the names that have been encountered in the Zoological Record and elsewhere. They are often treated as a single family.

Pisces. A very complete list of teleosts was published in 1966 by Greenwood et al. For the non-teleostean fishes the work of Berg (1940 and 1947) was used. In combining these two lists, also with that of Romer (1945, for fossils), some adjustments had to be made; any anomalies in the present list are not to be ascribed to these sources.

Amphibia. The list of Romer (Vertebrate Paleontology, 1945) is followed, with few additions from elsewhere.

Reptilia. The list of Romer (Vertebrate Paleontology, 1945) is followed, except for the usual additions from other literature.

Aves. None of the available lists of bird families (such as Wetmore, 1960, or the Peters checklist) give synonyms and variant spellings to any great extent. Therefore our basic list is from Romer, supplemented with the additions and variations found in the literature.

Mammalia. Two extensive lists that include both Recent and extinct forms were available: Simpson, 1945, and Romer, 1945. Oddly enough, these had to be so extensively supplemented (from Palmer, 1904) that the latter work became our basic source.

Details of presentation

Names of authors. Authors' names are alphabetized by the only simple system known to the compiler. All names are written exactly as spoken but alphabetized by the first capital letter (and all subsequent letters). For example:

Dollfus	DOLLFUS
van Emden	van EMDEN
de Haviland	de HAVILAND
Macfarlane	MACFARLANE
von Markham	von MARKHAM
McFarlane	McFARLANE
O'Reilly	O'REILLY
Petersen	PETERSEN
da Sylva	da SYLVA
Valloise	VALLOISE
Van der Stigchel	Van der STIGCHEL
Van Valen	Van VALEN
Wallingford	WALLINGFORD
van Walsingham	van WALSINGHAM

In most parts of the book, the authors' names are capitalized in full after the first capital, as shown in the second column above. When a second capital normally occurs in the name, as in LeBoeuf, the intervening letters are left small when the rest are capitalized, as LeBOEUF. Similarly in the name FitzPATRICK. Thus the reader will be able to tell that in McPHERSON the P is normally capitalized, whereas in MACPHERSON it is not.

Multiple authorship is not shown in full. It is assumed that credit is not being given but merely a clue to the filing of the work of that title. Associated authors will become known as soon as the work itself is consulted.

Geographic names. Taxonomy is international. Although North America is no doubt favored in this compilation, it is not deliberately so. Therefore anglicizing of geographic or political names has been avoided. It has been impossible to be consistent, however, because our sources were too diverse. Furthermore spellings change and authors of different nationalities use different forms. For example, there is a great temptation to cite one European country as Germany (without taking into account its present division), following the usage of the Zoological Record. But this country is Deutschland and is properly referred to only by that name. The French Allemagne or the Spanish Alemania are no better than Germany. The three latter translations are avoided, except in direct quotation of titles.

A few countries are so cosmopolitan that they themselves use several forms of the name, such as Nederland and Netherlands, Union of South Africa and Unie van Suid-Afrika, or Suomi and Finland. But in these it is not really correct to use Pays-Bas, Afrique del Sur, or Finlandia. In all countries that use the modern European alphabet, even if modified by a variety of diacritic marks, there seems to be no reason for using any one of the foreign spellings (or names). In the case of Slavic, Arabic, Chinese, or Japanese, persons in other countries are forced to transliterate. In the latter cases, translations are used if available, and, if not, the transliteration employed in our source is copied.

All titles are given exactly as on the original or as quoted by our source. The language is shown by the title (if given) or by annotation. If not so indicated, it may be presumed to be in English. (Some sources may have translated Russian or Japanese titles, especially, without specification; these will have been copied without indication of the original language.)

Citation. It was originally intended to employ herein a uniform style of citation for serials. Two factors made this impossible. First, the differing requirements in various parts of the manuscript made it desirable to vary the arrangement, using chronological, author, title, or geographic sequences, in appropriate places, with associated changes in punctuation and sequence of components. Second, the citations came from diverse sources and were usually already abbreviated. Standardization could have been achieved only by extended checking of these serials, a procedure for which time was not available.

The following general rules have been followed whenever possible:

1) Abbreviate only where confusion is not likely to result, assuming a worldwide approach.

2) Write out all proper names, when they can be identified, except for U.S.S.R., S.S.S.R., and U.R.S.S. (U.S.A. is not acceptable for United States of America, inasmuch as it is used for the Union of South Africa and for an organization in Japan.)
3) Write all words, abbreviations, and names in titles in the form of the original, if possible; otherwise in the form used in the source.
4) In transliterations, follow the system employed in the source. (This means that transliterations are not corrected into any one of the current systems. Thus, Terentiev and Terent'ev will stand together unaltered, as they were transliterated by the contemporary bibliographers in two different places.)

Diacritic marks. Among the more than two dozen languages in which these books are written, at least half employ diacritic marks in some manner to modify the letters of the alphabet. It was found to be impracticable to show these in most cases, because of certain features of arrangement. No instances were noted in which this would result in confusion, because this guide is intended to give clues for further search, rather than definitive and highly accurate bibliographic citations. Persons whose names are thus misspelled are asked to forgive the omission of the marks, noting that great care has been taken that names are correctly spelled and written in other respects.

Guide to the Taxonomic Literature of Vertebrates

ANIMALIA *including* VERTEBRATA

GENERAL BOOKS

[General works that refer primarily to single countries or regions are listed under those geographical places under Faunal Works]

(anonymous). Larousse encyclopedia of animal life. New York; 1-640. 1967.
ABEL, O. Die Stamme der Wirbeltiere. Berlin & Leipzig; 1-914. 1919.
ATWOOD, W. H. Introduction to vertebrate zoology. London; 1-512. 1940.
de BEER, G. R. The development of the vertebrate skull. Oxford; 1-552. 1937.
de BEER, G. R. Vertebrate zoology ..., 2nd ed. London; 1-435. 1951. (Comparative anatomy and embryology)
BLACKWELDER, R. E. Taxonomy .... New York; 1-698. 1967.
BLAIR, W. F. et al. Vertebrates of the United States, 2nd ed. New York; 1-616.
BOLK, L. et al. Handbuch der vergleichenden Anatomie der Wirbeltiere. Berlin; 6 volumes. 1931-38.
BRONN, H. G. Dr. H. G. Bronn's Klassen und Ordnungen des Thierreichs [wissenschaftlichdargestellt im Wort und Bild]. Leipzig & Heidelberg; 8 volumes. 1866-1919.
CALMAN, W. T. The classification of animals, an introduction to zoological taxonomy. New York; 1-54. 1949. (Also London, 1949)
CANSDALE, G. S. Animals of West Africa. London; 1-144. 1946.
COLBERT, E. H. Evolution of the vertebrates. New York; 1-479. 1955.
CURRY-LINDAHL, K. Djuren i Farg. Daggdjur,, Kraldjur, Groddjur, Tredje upplagan. Stockholm. 1963.
DARLINGTON, P. J., Jr. Zoogeography. The geographic distribution of animals. New York; 1-675; 1957.
DARLINGTON, P. J., Jr. Biogeography of the southern end of the world. Cambridge, Massachusetts; 1-236. 1965.
DUFRESNE, F. Alaska's animals and fishes. New York; 1-297. 1946.
FERRIS, G. F. The principles of systematic entomology. Stanford University Publ. Biol. Sci., 5, 103-269. 1928. (Also as separate publication, 1-169)
GIER, L. J. Principles of taxonomy. Liberty, Missouri; 1-94. 1965.
GRASSE, P. P. Traite de zoologie. Paris; 17 parts —. 1948 —
GRASSE, P. P. & DEVILLERS, C. Precis de sciences biologiques .... Zoologie. II. Vertebres. Paris; 1-1129. 1965.
GREGORY, W. K. Evolution emerging. New York; 2 volumes. 1951.
GUNTHER, K. & DECKERT, K. Wunderwelt der Tiefsee. Berlin; 1-240. 1950.
HARMER, S. F. & SHIPLEY, A. E. Cambridge natural history. London; 10 volumes. 1895-1909. (Reprinted 1958-1960)
HUXLEY, J. The new systematics. London; 1-583. 1940.
IHLE, J. E. W. Leerboek der Vergelijkende Ontleedkunde van de Vertebraten. Utrecht; 2 volumes. 1947.
JAEKEL, O. M. J. Die Wirbeltiere. Eine Ubersicht uber die fossilen und lebenden Formen. Berlin; 1-252. 1911
KASHKAROW, D. N. & STANCHINSKII, V. V. Textbook of zoology, vertebrates, 2nd ed. Moskva; 1-1024. 1940. (In Russian)
KEEN, A. M. & MULLER, S. W. Schenk & McMasters, Procedure in taxonomy, 2nd ed. Stanford, California; 1-92. 1948.
KEEN, A. M. & MULLER, S. W. Schenk & McMasters, Procedure in taxonomy, 3rd ed. Stanford, California; 1-149. 1956.
LANKESTER, R. A treatise on zoology. London; 8 volumes. 1900-1909. (Reprinted 1964)
LYDEKKER, R. The royal natural history. London; 6 volumes. 1893-1896.
MACNAE, W. A general account of the fauna and flora of mangrove swamps and forests in the Indo-West-Pacific region. Adv. Mar. Biol., 6, 73-270. 1968.
MARINELLI, W. & STRENGER, A. Vergleichende Anatomie und Morphologie der Wirbeltiere. Wien; several parts. 1954.
MARSHALL, A. J. Parker & Haswell, A textbook of zoology, volume 2, 7th ed. London & New York; 1-952. 1962.
MAYR, E. Systematics and the origin of species from the viewpoint of a zoologist. New York; 1-334. 1942.
MAYR, E. Principles of systematic zoology. New York; 1-428. 1969.
MAYR, E., LINSLEY, E. G., & USINGER, R. L. Methods and principles of systematic zoology. New York; 1-328. 1953.
NEWMANN, H. H. The phylum Chordata. New York; 1-477. 1939.
OGNEV, S. I. Zoology of the vertebrates. Moskva; 1-519. 1945. (In Russian)
ROMER, A. S. The vertebrate body, 2nd ed. Philadelphia; 1-644. 1955. (1st ed., 1949)
ROMER, A. S. The vertebrate story, 4th ed. Chicago; 1-437. 1959. (Prev. eds., 1933, 1939, 1941)

SAUNDERS, J. T. & MANTON, S. M. A manual of practical vertebrate morphology. Oxford; 1-255. 1949.
SCHENK, E. T. & McMASTERS, J. H. Procedure in taxonomy .... Stanford, California; 1-72. 1936.
SCHONMANN, R. Die Welt der Tiere. Wien; 1-654. 1949.
SIMPSON, G. G. The principles of classification and a classification of mammals. Bull. American Mus. Nat. Hist., 85, 1-350. 1945.
SIMPSON, G. G. Principles of animal taxonomy. New York; 1-247. 1961.
STANEK, V. J. Pictorial encyclopaedia of the animal kingdom. London; 1-614. 1962.
WALTER, H. E. Biology of the vertebrates, 3rd ed. New York; 1-882. 1939.
WALTER, H. E. & SAYLES, L. P. Biology of the vertebrates. New York; 1-875. 1949.
WEBB, J. E. & ELGOOD, J. H. Animal classification. Ibadan, Nigeria; 1-167. 1955.
WEICHERT, C. K. Anatomy of the chordates, 3rd ed. New York; 1-758. 1965.
YAPP, W. B. Vertebrates. Their structure and life. New York; 1-525. 1965.
YOUNG, J. Z. The life of vertebrates. Oxford; 1-767. 1950.

BIBLIOGRAPHIES

[Bibliographies on single groups or special subjects are listed under the group or subject. Because so many reference books are known primarily by title, these are arranged by title with the author indicated, if known.]

Animalium cavernarum catalogus, by B. WOLF. 's-Gravenhage, 3 volumes. 1934-38. (By regions & caves, II, 1-616; by authors, I, 1-108; by animal groups, III, 1-918)
Bibliographia zoologiae et geologiae. A general catalogue of all books, tracts, and memoirs on zoology and geology. London; 4 volumes. 1848-1854. By L. Agassiz.
Bibliographia zoologia, (anonymous). Zurich; 43 volumes. 1896-1934. (From Concilium Bibliographicum cards)
Bibliographic index. A cumulative bibliograph of bibliographies. 1937 —. New York; 8 volumes to 1967. 1945 —
Bibliotheca historico-naturalis ... 1700-1846, by W. Engelmann. Leipzig; 1-786. 1846.
Bibliotheca zoologica ... 1846-1860, by J. V. Carus & W. Engelmann. Leipzig; 2 volumes. 1861.
Bibliotheca zoologica II ... 1861-1880 ..., by O. Taschenberg. Leipzig; 8 volumes. 1887-1923.
Books of reference in zoology, chiefly bibliographical, by F. C. Sawyer. Journ. Soc. Biblio. Nat. Hist., 3, 72-91. 1955.
British Museum general catalogue of printed books. Photolithographic edition to 1955. London; 252 volumes. 1931 —
A catalogue of books represented by Library of Congress printed cards. Issued to July 31, 1941. Ann Arbor; 167 volumes. (And cumulative supplements to date)
A catalog of books represented by Library of Congress printed cards issued to July 31, 1942. Ann Arbor, Michigan; 167 volumes. 1942-46. (Supplement August 1, 1942 to December 31, 1947; 42 volumes. 1948)
Catalogue general des livres imprimes de la Bibliotheque Nationale. Auteurs. Paris; 189 volumes. 1897 —
A catalogue of papers concerning the dates of publication of natural history books. Third supplement. Journ. Soc. Biblio. Nat. Hist., 3, 165-174. 1957. By G. H. Goodwin, Jr.
A catalogue of papers concerning the dates of publication of natural history books, by G. H. Goodwin, Jr. et al. 4th supplement. Journ. Soc. Biblio. Nat. Hist., 4, 1-19. 1962.
Catalogue of scientific papers (1800-1863). London; 6 volumes. 1867-1872.
Catalogue of scientific papers 1864-1883. London; 6 volumes. 1867-1902. (Incl. supplement volume)
Catalogue of the books, manuscripts, maps and drawings in the British Museum (Natural History). London; 5 volumes and 3 supplement volumes. 1903-1940. (Reprinted 1964)
Catalogue of the printed books in the library of the British Museum. London; 58 volumes and 10 supplement volumes. 1881-1905. (Reprinted 1946-1950, Ann Arbor)
Dictionary catalog of the National Agricultural Library. New York; 68 volumes. 1967 —
Elements d'un guide bibliographique du naturaliste. Macan; 1-302. 1940. By F. Bouliere.
Guide to reference books, 7th ed., by C. M. Winchell. Chicago; 645 pp. 1951.
Guide to the literature of the zoological sciences, 6th ed., by R. C. Smith. Minneapolis, Minnesota; 1-232. (Prev. eds., 1942, 1945, 1952, 1955, 1958)
Index-catalogue of medical and veterinary zoology — authors, by C. W. Stiles & A. Hassell. United States Dep. Agric., Bureau Animal Industry, Bull. 39. 1902-1912.
Index-catalogue of medical and veterinary zoology. [Authors], by A. Hassell et al. Washington, D. C., 18 parts, 1-5711, and supplements. (Incorporates Stiles & Hassell, 1902-1912. (Lists parasites and hosts)
Index-catalogue of medical and veterinary zoology. [Authors], by M. A. Doss et al. Washington, D. C., 17 parts & continuation. 1953 — (Parasites and hosts)
International catalogue of scientific literature. N. Zoology. for 1904-1916. London. (Generally filed as part of Zoological Record, which it replaced temporarily)
An introduction to the literature of vertebrate zoology, by C. A. Wood. London; 1-643. 1931.
Journal of the Society for the Bibliography of Natural History. 1, 1936 —
Library of Congress and National Union Catalog author lists, 1942-1962. A master cumulation. Detroit; 152 volumes. 1969-1971. (And cumulative supplements)
The National Union Catalog. A cumulative author list representing Library of Congress printed cards. 1953-1957 —. 111 volumes.

Natural history index-guide. An index to 3, 365 books and periodicals in libraries. A guide to things natural in the field ..., 2nd ed. New York; 1-583. 1940. (By B. Altsheler)
Scientific, medical and technical books published in United States of America to 1956, 2nd ed., by R. R. Hawkins. Washington, D. C.; 1-1491. 1958.
Supplement zu C. O. Waterhouse's Index Zoologicus. Zool. Annalen, 2, 273-343, 1908; 6, 33-46, 1914. (By F. Poche)
Vertebrate paleontology, by A. S. Romer. Chicago; 1-687. 1945. (Of fossils)
Wildlife abstracts 1935-51, by N. Hotchkiss. Washington, D. C.; 1-434. 1954.
A world bibliography of bibliographies ..., 3rd ed., by T. Besterman. Geneve; 4 volumes. 1955. (2nd edition, London, 1947-49, 3 volumes)
Zoological Record 1863 —. London; 104 volumes & continuation. 1864 —.

Regional bibliographies

Arctic

COLLINS, H. B. Arctic bibliography. Washington, D. C., 6 volumes. 1953-56.

North America

MEISEL, M. A bibliography of American natural history. The pioneer century, 1867-1865. New York; 3 volumes. 1924-19. (Reprinted 1967)
WALKER, E. M. Bibliography of Canadian zoology (exclusive of entomology). Trans. Roy. Soc. Canada, 10, 201-215. 1917.
SHOUP, C. S. An annotated bibliography of the zoology of Tennessee. American Midl. Nat., 21, 583-635. 1939.
OSBORN, H. Bibliography of Ohio zoology. Ohio Biol. Surv. Bull., 4, (23), 351-410. 1930.
ISFORT, L. G. A partial bibliography of natural history in the Chicago region. American Midl. Nat., 42, 406-472. 1949.
BICK, G. H. A bibliography of the zoology of Louisiana. Proc. Louisiana Acad. Sci., 17, 5-48. 1954.

Latin America

KOERDELL, M. M. Bibliografia Mexicana de historia natural. III. Instrucciones a exploradores y colectores naturalistes. Rev. Soc. Mexicana Hist. Nat., 5, 273-280. 1944.
CURRA, R. A. Bibliografiae indice de los ciencias marinas de Venezuela. I. Biologias marina y pesquera. Lagena, 4, 2-62. 1964.
(anonymous). Bibliografia Brasileira de zoologia. Rio de Janeiro; 2 volumes. 1950-58. (By groups and by subjects)

Europe

SMART, J. Bibliography of key works for the identification of the British fauna and flora. Publ. Assoc. Study Syst. Rel. Gen. Biol., 1, 1-105. 1942.
SMART, J. & TAYLOR, G. Bibliography of key works for the identification of the British fauna and flora, 2nd ed. London; 1-126. 1953.
KERRICH, G. J. et al. Bibliography of key works for the identification of the British fauna and flora, 3rd ed. London; 1-186. 1967.
(anonymous). Bibliotheca zoologica Fenniae. Helsinki; 1-361. 1909.
LINDBERG, H. Bibliotheca zoologica Fenniae opera annorum 1901-1930. Acta Soc. Fauna Flora Fennica, 59, 1-438. 1937.
LINDBERG, H. Bibliotheca zoologica Fenniae opera annorum 1931-1940. Acta Soc. Fauna Flora Fennica, 70, 1-284. 1953.
JAKUBSKI, A. Bibliography of the Polish fauna. Prace Monogr. Kom. Fizy. Krakowie, 3, 1-470; 4, 1-384; 1927-28. (In Polish)
FELIKSIAK, S. et al. Bibliographical descriptive guide to the animals of Poland. Warszawa; 1-199. 1954. (In Polish)
PAX, F. Bibliografia zoologii Slaska Czesc. I - III. Wroclaw. 1930-57. (Also as Bibliography of the Silesian zoology; Bibliographie der Schlesischen Zoologie)
ALLODIATORIS, I. Bibliographie der zoologie in Karpatenbechen. Budapest; 1-574. 1966.
KANELLIS, A. & HADJISSARONOTOS, C. Bibliographia faunae Graecae. 1950-1960 und Nachtrage. Verh. Zool. -Bot. Ges. Wien, 100, 96-105. 1960.
(anonymous). Zoologiline Kirjandus 1945-1959. Tartu; 1-103. 1961. (In Estonian, Russian, German, and English)
(anonymous). Zoological bibliography 1945-1959. Tartu; 1-103. 1961. (In Russian)

Near East

NEU, W. & KUMMERLOWE, H. Bibliographie der zoologischen Arbeiten uder die Turkei und ihre Grenzgebiete. Leipzig; 1-62. 1939.
OREN, O. H. & STEINITZ, H. Regional bibliography of the Mediterranean coast of Israel and the adjacent Levant countries. Bull. Sea Fish. Res. Sta. Israel, 22, 1-32. 1957.

Africa & Indian Ocean

SEURAT, L. G. Exploration zoologique de l'Afrique de 1830 a 1930. Paris; 1-708. 1930.
GHIDINI, G. M. Materiali per una bibliografia zoologica dell'Africa Orientale Italiana. Riv. Biol. Coloniale, 2-6, many parts. 1939-1943.
PETERS, A. J. Bibliography of published works bearing on the natural history of the Seychelles and neighbouring archipelagos. Journ. Soc. Biblio. Nat. Hist., 3, 238-262. 1957.

Indo-Malaya

(anonymous). Bibliography of Indian Zoology .... In Indian Sci. Abstr. 1936, 1-66. 1939.
MATHEW, K & DE, M. L. A brief review and bibliography of Indian zoology. Bibliogr, Indian Zool., 5, 1-271. 1967.
ALAGARSWAMI, K. et al. Bibliography of the Indian Ocean 1900-1930. Bull. Cent. Mar. Fish. Res. Inst., 4, 1-5. 1968.
SUVATTI, C. Fauna of Thailand. Bangkok; 1-1100. 1950. (Chronological, 655-746; by phylum, 747-929)

Australia & New Zealand

FREED, D. Bibliography of New Zealand marine zoology 1769-1899. Mem. New Zealand Oceanogr. Inst., 16, 1-46. 1963. (Also as: Bull. Dep. Sci. Ind. Res. New Zealand, 148, 1-46.)

The Orient

ROMANOV, N. S. Annotated bibliography of Far Eastern aquatic fauna, flora and fisheries. 1923-1956. Moskva; 1-290. 1959. (In Russian)
ROMANOV, N. S. Annotated bibliography of Far Eastern aquatic fauna, flora and fisheries. Jerusalem; 1-391. 1966. (Translation of 1959)

Oceania

UTINOMI, H. Bibliographica Micronesica. Tokyo; 1-208. 1944.
FISHER, H. I. Utinomi's Bibliographica Micronesica. Chordate sections. Pacific Sci., 1, 129-150. 1947.

Abstracts

(serial). Archiv fur Naturgeschichte. Berlin; 1832-1922.
Bibliographia zoologica .... Berlin; 43 volumes. 1896-1934. (From Concilium Bibliographicum cards)
Biological Abstracts. Philadelphia; vol. 1, 1926 —.
Concilium bibliographicum. Zurich; 1896-1940. (3x5 cards)
International catalogue of scientific literature. N. Zoology. for 1904-1916. London; 12 volumes. (Generally filed as part of Zoological Record, which it replaced temporarily)
Zoological Record, 1863 —. London; 104 volumes and continuation.

Periodicals

Abbreviations of names of biological publications. Univ. Colorado Stud., 1 D, 177-191. 1941. By E. D. Crabb.
Abbreviations used in the Department of Agriculture for titles of publications. United States Dep. Agric. Misc. Publ., 337, 1-278. 1939. By C. Whitlock.
Annotated bibliography of Far Eastern aquatic fauna, flora and fisheries. 1923-1956. Moskva; 1-290. 1959. (Pp. 283-287; in Russian) By N. S. Romanov.
Annotated bibliography of Far Eastern aquatic fauna, flora and fisheries. Jerusalem; 1-391. (Translation of 1959. Pp. 377-384) By N. S. Romanov.
Bibliografia Brasileira de zoologia. Rio de Janeiro; 2 volumes. Pp. 17-21.
Bibliographia zoologiae et geologiae ..., vol. 1, 1-85. London; 1848. By L. Agassiz.
Bibliography of North American minor natural history serials in the University of Michigan libraries. Ann Arbor, Michigan; 1-197. 1954. By M. H. Underwood.
Bibliotheca zoologica. 1846-1860. 2 volumes. 1861. By J. V. Carus & W. Engelmann.
Bibliotheca zoologica II. 1861-1880 .... Leipzig; 8 volumes. By O. Taschenberg.
British Union-Catalogue of periodicals .... London; 4 volumes. 1955-58. By J. D. Stewart et al.
British Union-Catalogue of periodicals. Supplement to 1960. London. 1962. By J. D. Stewart et al.
A catalogue of scientific and technical periodicals (1665-1895), 2nd ed. Smithsonian Misc. Coll., 514, 1-1247. 1897. By H. C. Bolton.
Catalogue of scientific serials ... 1633-1876. Libr. Harvard Univ., Spec. Publ. 1, 1-358. 1879. (Reprinted, New York, 1965) By S. H. Scudder.
A check-list of periodical literature and publications of learned societies of interest to zoologists in the University of Michigan libraries. Circ. Mus. Zool. Univ. Michigan, 2, 1-83. By F. P. Allen.
Directory of Japanese scientific periodicals. (No reference available). 1964.
Guias de naturalistass sudamericanos. Buenos Aires; 1-138. (Latin American periodicals by country) By E. Martinez Fontes & J. J. Parodiz.

Index-catalogue of medical and veterinary zoology — Authors. United States Dep. Agric., Bureau Animal Industry Bull. 39, 36 parts. (Parasites and hosts) By C. W. Stiles & A. Hassell. 1902-1912.
Index-catalogue of medical and veterinary zoology [Authors]. Washington, D. C.; 18 parts and supplements. (Serials list) By A. Hassell et al. 1932-52.
Index-catalogue of medical and veterinary zoology [Authors]. Washington, D. C.; 17 parts and continuation. (Serials lists) By M. A. Doss et al. 1953 —.
International catalogue of scientific literature. List of journals.... London; 1-312. 1903. (Supplement, 1-68; 1904)
A list of abbreviations of the titles of biological journals. Selected from the "World list of scientific periodicals." London; 2nd ed., 1-32. 1954.
List of abbreviations with titles of the journals. Zool. Rec. for 1905, 1906, 1907, 1908, 1909, 1910, 1911, 1912, 1913.
List of abbreviations with titles of journals. Zool. Rec. for 1921. (With supplements for 1923, 1924, and 1925).
List of abbreviations with titles of journals. Zool. Rec. for 1927. (With supplements for 1928, 1929)
List of abbreviations with titles of periodical publications .... Zool. Rec. for 1930. (With supplement for 1931)
List of abbreviations with titles of periodical publications..... Zool. Rec. for 1934. (With supplement for 1935, 1936, 1937, 1938)
List of journals. International Catalogue of Scientific Literature. London; 1-312. 1903.
List of scientific and learned periodicals in the Netherlands. den Haag; 1-63. 1953.
List of serial publications in the library of the Department of Zoology British Museum (Natural History). London; 1-154. 1958.
List of serial publications in the library of the Department of Zoology, British Museum (Natural History) 2nd ed. London; 1-152. 1962. By F. C. Sawyer & S. FitzGerald.
List of serial publications in the libraries of the Departments of Zoology and Entomology in the British Museum (Natural History). London; 1-281. 1967.
List of the agricultural periodicals of the United States and Canada published during the century July 1810 to July 1910. Misc. Publ. United States Dep. Agric., 398, 1-190. 1941. By S. C. Stuntz & E. B. Hawks.
New serial titles 1950-1960. Supplement to the Union List of Serials Third Edition. Washington, D. C. 2 volumes. 1961 —.
New serial titles. A union list of serials commencing publication after December 31, 1949. 1961-1965 cumulation. Washington, D. C.; 2 volumes. 1966.
New serial titles. A union list of serials commencing publication after December 31, 1949. 1966-1969 cumulation. Washington, D. C.; 2 volumes. 1971.
Periodica zoologica. Abkurzungsverzeichnis .... Leipzig; 1-82. 1938. By C. Apstein & K. Wasikowski
Scientific and technical serial publications of the Soviet Union 1945-1960. Washington, D. C.; 1-347. 1963. By N. T. Zikeev.
A select list of British scientific periodicals: zoology. London; 1-55. 1963.
Union list of serials in libraries of the United States and Canada, 2nd ed. New York; 1-3065. 1943. And supplement. By W. Gregory.
Union list of serials in libraries of the United States and Canada, 3rd ed. New York; 5 volumes. 1965. By E. B. Titus.
Union list of technical periodicals, 3rd ed. New York; 1-285. 1947. By E. B. Bowerman.
Welche Zeitschriften sind fur ein zoologisches Institut am wichtigsten. Mem. Mus. Zool. Univ. Coimbra, 122, 1-57. 1941. By E. Matthes.
World list of periodicals for aquatic sciences and fisheries. FAO Fish Tech. Paper, 19, (1), 1-239; and suppl. 2, 1-28. 1962, 1964.
World list of scientific periodicals ... 1900-1960, 4th ed. Washington, D. C.; 3 volumes. By P. Brown & G. B. Stratton).
Zoological Record 1963 —. London; 104 volumes & continuation. Periodical lists in volumes 7-8, 43-50, 58-62, 64, 66-75.

Indexes

Index to the first 100 years 1866-1966 [of Journal of Anatomy]. Cambridge Univ. Press; 1-450. 1968. By D. Blake & R. E. M. Bowden.
List of new taxonomic names introduced in volume I to XVI of "Annales Zoologici". Ann. Zool. Polon., 16, 483-514. 1957.
Register zum Zoologischen Anzeiger. Leipzig; 8 volumes. 1889-1929. For vol. 1-75. By J. V. Carus & E. Korschelt.

Personal bibliographies

Various. MEISEL, M. A bibliography of American natural history. The Pioneer Century 1767-1865. New York; 3 volumes. 1924-29. (Reprinted 1967)
Deraniyagala, P. E. P. DERANIYAGALA, P. E. P. Bibliography from 1927-1952. Spol. Zeylanica, 26, 272-274. 1951.
Grant, C. GRANT, C. Bibliograph of Chapman Grant. San Diego; 1-16. 1952.
Hildebrand, S. F. SCHULTZ, L. P. & GREEN, A. S. Bibliography of Samuel F. Hildebrand. Copeia, 1950, 15-18. 1950.
Linnaeus, C. SOULSBY, B. H. A catalogue of the works of Linnaeus .... London; 1-246, 1-68. 1933.
(anon.) A catalogue of the work of Linnaeus .... Stockholm; 1-177. 1957.

Institutional bibliographies

United States. SKALLERUP, H. R. American state academy of science publications. Occ. Pap. Univ. Illinois Libr. School, 50, 1-20. 1957.

United States National Museum. (Anonymous). List of publications of the United States National Museum. Circ. United States Nat. Mus., 18, 1-12. 1881. (Anonymous). A list and index of the publications of the United States National Museum. United States Nat. Mus. Bull., 193, 1-306. 1947.

PERSONS AND INSTITUTIONS

Directories

Les biologistes belges au Congo. (Un bilan scientifique.) Bull. Acad. Belgique, Cl. Sci., (5), 47, (12), 1183-1218. 1961. By J. Lebrun.

Directory of hydrobiological laboratories and personnel of North America. Honolulu; 1-324. 1954. By R. W. Hiatt.

Directory of zoological taxonomists of the world. Carbondale, Illinois; 1-404. 1961. By R. E. Blackwelder & R. M. Blackwelder.

Guias de naturalistes sudamericanos. Buenos Aires; 1-138. 1949. By E. Martinez Fontez and J. J. Parodiz. (Includes all Latin America)

Guide du naturaliste dans le Midi de la France. 2 .... Neuchatel; 1-395. 1967. By H. Harant and D. Jarry.

A handlist of American naturalists, based on the Dictionary of American Biography. American Nat., 72, 534-546. 1938. By P. H. Oehser.

Handbook of zoologists of the Soviet Union. Moskva; 1-292. 1961. By R. E. F. Strelkov & K. Yurev.

Index ldes Zoologistes. Paris; 1-429. 1953.

Informator zoologiczny. Directory of Polish zoologists. Warszawa; 1-193. 1962.

List of Japanese zoologists partaking systematic zoology. Journ. Fac. Sci. Hokkaido Univ., ser. VI, (Zool.), 12, (1-2), suppl. pp. 1-33. 1954. By T. Uchida.

The naturalist's directory (international), 40th ed. South Orange, New Jersey; 1-181. 1968. (Prev. eds. by Samuel E. Cassino and successors)

A register of Hungarian zoologists. Allat. Kozlem., 47, 195-204. 1960. By L. Moczar (In Hungarian)

Systematic botanists and zoologists working in Africa. CCTA. London; 1-35. 1954.

Zoologi Sovetskogo soiuza. Moskva; 1-292. 1961.

The zoological Society of Japan. List of members. Zool. Mag. Tokyo, 76, 288-319. 1967.

Zoologisches Adressbuch. Namen under Adressen der Lebenden Zoologen, Anatomen, Physiologen und Zoopalaeontologen .... Berlin; 1-740. 1895.

Institutions

The world of learning, 12th ed. London; 1-1359. 1961-62.

A bibliography of American natural history. The pioneer century 1767-1865. New York; 3 volumes. 1924-29. Reprinted 1967. By M. Meisel.

Directory of hydrobiological laboratories and personnel in North America. Honolulu; 1-324. 1954. By R. W. Hiatt.

Museums directory of the United States and Canada. Washington, D. C.; 1-567. 1961.

Scientific and technical societies of the United States and Canada, 7th ed. Washington, D. C.; 1-413, 1-54. 1961.

Taxonomy .... New York; 1-698. 1967. By R. E. Blackwelder. (Pp. 43-48)

Guias de naturalistas sudamericanos. Buenos Aires; 1-138. 1949. By E. Martinez Fontez and J. J. Parodiz. (All Latin America, by country)

Directory of natural history and other field study societies in Great Britain .... London; 1-217. 1959.

Alphabetical list of Dutch zoological cabinets and menageries. Bijdr. Dierk., 27, 247-346. 1939. By H. Engel.

A list of scientific institutions in the Pacific Area. Pacific Sci., 2, 243-261. 1948. By O. A. Bushnell.

Museum lists

[See also museum lists under each class]

Naturhistorischen Museums in Basel.

REVILLIOD, P. Katalog der osteologischen Sammlung (rezente Abteilung) der Naturhistorischen Museums in Basel. Verh. Nat. Ges. Basel, 24, 184-227.

Museum type lists

[See also museum type lists under each class]

(list of lists)

BANFIELD, A. W. Une liste preliminaire de catalogues des specimens types en zoologie et paleontologie, 1967-1968. Bucuresti; 1-29, 1-30. 1967, 1968.

Brigham Young University

TANNER, W. W. A catalogue of the fish, amphibian and reptile types in the Brigham Young University Museum of Natural History. Great Basin Nat., 30, 219-226. 1970.

Stanford University

MAYER, W. V. Catalogue of type specimens in the Natural History Museum of Stanford University. Proc. California Zool. Club, 1, 29-32. 1949.

Instituto Miguel Lillo

HAYWARD, K. J. Lista de los tipos de vertebrados conservados in el instituto Miguel Lillo. Acta Zool. Lilloana, 19, 507-510. 1963.

University of Oslo

PETHON, P. List of type specimens of fishes, amphibians and reptiles in the Zoological Museum, University of Oslo. Rhizocrinus, 1, 1-17. 1969.

Museo de Mendoza

RUSCONI, C. Las piezas "tipos" del Museo de Mendoza. Rev. Mus. Hist. Nat. Mendoza, 7, 82-155. 1954.

Institute Francaise d'Afrique Noire

VILLIERS, A. Liste des types deposes au Museum National d'Histoire Naturelle par l'Institut Francaise d'Afrique Noire (7e liste). Bull. Mus. Hist. Nat., (2), 28, 495-499. 1957.

Collections of South & East Africa

(anonymous). A list of zoological and botanical types preserved in collections in south and East Africa. Vol. I, Zoology, part 1. Pretoria; 1-147. 1958.
(same). Part 2. Pp. 1-119. 1962.

Western Australian Museum

(anonymous). Type specimens in the West Australian Museum. Ann. Rep. Western Australian Mus., 1959-60, 1960-61, 1962-63. 1962-1964.

Auckland Museum

POWELL, A. W. B. Biological primary types in the Auckland Museum. Rec. Auckland Mus., 2, 239-259. 1941.
POWELL, A. W. B. Biological primary types in the Auckland Museum. Rec. Auckland Mus., 3, 403-409. 1949.

Collections

Where is the ________ collection? Cambridge; 1-149. 1940. By C. D. Sherborn.

FAUNAS & IDENTIFICATION

Arctic

WYNNE-EDWARDS, V. C. Fresh-water vertebrates of Arctic and Subarctic. Bull. Fish. Res. Board Canada, 94, 1-25. 1952.

North America

DRIVER, E. C. Name that animal .... Northampton, Massachusetts; 1-558. 1942. (Reprinted 1950)
EDMONDSON, W. T. et al. Ward & Whipple's Fresh-water biology, 2nd ed. New York; 1-1248. 1959.
MURIE, O. J. A field guide to animal tracks. Boston; 1-374. 1954.
PALMER, E. L. Fieldbook of natural history. New York; 1-664. 1949.
PRATT, H. S. Manual of land and fresh water vertebrate animals of the United States (exclusive of birds). Philadelphia; 1-422. 1923. (To families)
WARD, H. B. & WHIPPLE, G. C. Fresh-water biology. New York; 1-1111. 1918.
BLAIR, W. F. et al. Vertebrates of the United States. New York; 1-819. 1957.
BLAIR, W. F. et al. Vertebrates of the United States, 2nd ed. New York; 1-616. 1968.
JORDAN, D. S. Manual of the vertebrate animals of the northeastern United States inclusive of marine species, 13th ed. New York; 1-446. 1929.
BENTON, A. H. & STEWART, M. M. Key to the vertebrates of the northeastern states (excluding birds). Minneapolis, Minnesota; 1-57. 1965.
EDDY, S. & HODSON, A. C. Taxonomic keys to the common animals of the North Central states exclusive of the parasitic worms, insects and birds, 2nd ed, Minneapolis, Minnesota; 1-141. 1955.

Latin America

GODMAN, F. D. & SALVIN, O. Biologia Centrali-Americana..., Zoology. London; 215 pts. 1879-1915.

TITSCHACK, E. Beitrage zur Fauna Perus, Band 1. Jena; 1-405. 1951.
HUMMELINCK, P. W. Studies on the fauna of Curacao, Aruba, Bonaire and the Venezuelan islands. den Haag; 2 volumes. 1940.
ROHL, E. Fauna descriptiva de Venezuela, 3rd ed. Madrid; 1-516. 1956.
von IHERING, R. Dicionario dos animais do Brasil. Sao Paulo; 1-898. 1940.
MARCGRAVE, J. Historia natural do Brasil. Sao Paulo; 1-300. 1942.
von IHERING, R. Dicionario dos animais do Brasil. Sao Paulo; 1-790. 1968.

Europe

THIENEMANN, A. Verbreitungsgeschichte der Susswassertierwelt Europas .... Binnengewasser, 18, 1-809. 1950.
GERMAINE, L. A. P. La faune des lacs, des etangs et des marais de l'Europe occidentale. Encyclo. Prat. Natur., ed. 2, Tome XX, 1-549. 1957.
(anonymous). Guide to the British vertebrates. London; 1-122. 1910.
PARKER, H. W. List of British vertebrates. London; 1-66. 1935.
HINTON, M. A. C. et al. List of British vertebrates. London; 1-66. 1935.
CLEGG, J. The freshwater life of the British Isles. London; 1-351. 1952.
KEVAN, D. K. M. A practical key to the orders and suborders of soil and litter animals. In Soil Zool. London; 452-488. 1955.
LEUTSCHER, A. Tracks and signs of British animals. London; 1-252. 1960.
EALES, N. B. The littoral fauna of the British Isles. A handbook for collectors, 3rd ed. London; 1-306. 1961.
HOLME, N. A. The bottom fauna of the English channel. Journ. Mar. Biol. Assoc. U. K., 41, 397-461. 1961.
BRAESTRUP, F. W. et al. Vort lands dyreliv .... København; 3 volumes. 1949-50. (Danmark)
DEGERBOL, M. A list of Danish vertebrates. København; 1-180. 1950.
PFAFF, J. R. List of Danish vertebrates. København; 1-180. 1950.
GRIMPE, G. & WAGLER, A. E. Tierwelt der Nord- und Ostsee. Leipzig; many parts. 1925-1957.
van der BRINK, F. H. Lijst van Nederlandsche Vertebrata. 's-Gravenhage; 1-56. 1943.
BRAUER, A. Die susswasserfauna Deutschlands, eine Exkurzionsfauna .... Jena; 19 pts. 1909-12.
HASE, A. Fauna von Deutschland. (Reference not available) 1925.
DAHL, K. Tierwelt Deutschlands. Jena. 1928.
BROHMER, P. Tierbestimmungbuch ..., 2nd ed. Leipzig; 1-200. 1932.
BROHMER, P. et al. Fauna von Deutschland. Unserer heimischer Tierwelt, 4th ed. Leipzig; 1-561. 1932.
BROHMER, P. et al. Fauna von Deutschland. Leipzig; 1-584. 1944.
DODERLEIN, L. Bestimmungsbuch fur deutsche Land- und Susswassertiere. Wirbeltiere. Munchen; ed. 2, 1-304. 1955.
STRESEMAN, E. et al. Excursionsfauna von Deutschland. Wirbeltiere. Berlin; 1-340. 1955.
FATIO, V. Faune des Vertebres de la Suisse. Geneve; 6 volumes. 1969-1904.
CANTUEL, P. Faune des Vertebres du Massif Central de la France. Paris; 1-404; 1949.
de SEABRA, A. F. Catalogue systematique des Vertebres du Portugal. Bull. Soc. Port. Sci. Nat., 4, 91-114. 1910.
NOBRE, A. Fauna marinha de Portugal. I. Vertebrados (Mamiferos, reptis e peixes). Porto; 1-579. 1935.
TORTONESE, E. & LANZA, B. Piccola fauna Italiana. Pesci, Anfibi e rettile. Milano; 1-185. 1968.
FELIKSIAK, S. et al. Bibliographical descriptive guide to the animals of Poland. Warszawa; 1-199. 1954. (In Polish)
KOWALSKI, K. Keys for the identification of Polish vertebrates. V. Ssaki — Mammalia. Polska Akad. Nauk. Zakł. Zool. Syst. Krakowie, 5, 1-280.. 1964. (In Polish)
FERIANC, O. Fauna zoolenskeho okresu so zretel'om na stavovce. Prirod. Sbornik, 4, 37-76. 1949.
STEPANEK, O. Klic nasich obratluvcu. Praha; 1-250. 1950. (Key to Czech vertebrates; in Czech)
BROHMER, E. & ULMER. Die Tierwelt Mitteleuropas. Ein Handbuch zu Ihrer Bestimmung .... Leipzig; 7 volumes. 1927 —.
IONÉSCU, V. Vertebratele din Romania. Bucuresti; 1-496. 1968.
MARCUZZI, G. Fauna della Dolomiti. Mem. Ist. Veneto, Sci. Mat. Nat., 31, 1-595. 1956.
RIEDL, R. Fauna und flora der Adria. Berlin; 1-640. 1963.
PESHEV, T. & BOEV, N. Vertebrates, In Fauna of Bulgaria. A short classification key. Pp. 7-83. 1962. (In Russian)
ZENKEVICH, L. A. Seas of the U. S. S. R., their fauna and flora, 2nd ed. Moskva; 1-424. 1956.
ZENKEVITCH, L. Biology of the seas of the U. S. S. R. London; 1-955. 1963.
GAYEVSKAYA, N. S. et al. Checklist of the fauna and flora of the northern seas of the U. S. S. R. Moskva; 1-737. 1948. (In Russian)
OLIFAN, V. I. Checklist of the fauna and flora of the Northern Seas of the U. S. S. R. (Reference not available; in Russian)
YASHNOV, B. A. Check list of the fauna and flora of the Northern Seas of the U. S. S. R. (Reference not available; in Russian)
VALKONOV, A. Catalogue of our Black Sea fauna. Trud. Morsk. Biol. Sta. G. Varna, 19, 1-62. 1957.
DAL, S. K. Animal world of the Armenian Republic. Vertebrates. Moskva; 1-415. 1954. (In Russian)
YANUSHEVICH, H. I. The animal world of the Kirghiz. Frunze; 1-108. 1957. (In Russian)
AFANASER, A. V. et al. The animals of Kazakhstan. Alma-Ata; 1-536. 1953. (In Russian)

Near East & Middle East

BODENHEIMER, F. S. Prodromus faunae Palestinae. Mem. Inst. Egypte, 32, 1-286. 1937. (Marine & freshwater)
MAHDI, N. & GEORG, P. V. A systematic list of the vertebrates of Iraq. Iraq Nat. Hist. Mus. Publ., 26, 1-104. 1969.

Africa

BOURGOIN, P. Animaux de chasse d'Afrique. Paris; 1-255. 1955.
BLANC, M. Faune Tunisienne. Tunis; 1-280. 1936.
DEKEYSER, P. L. & DERIVOT, J. La vie animale au Sahara. Paris; 1-220. 1959.
SARMENTO, A. A. Vertebrados da Madeira. 1. Mamiferos — Aves — Repteis — Batraquios, 2nd ed. Funchal; 1-317. 1948.
BASILIO, A. La vida animal en la Guinea Espanola. Madrid; 1-150. 1952.
BARNARD, K. H. South African shore life. Cape Town; 1-135. 1954.
BIGALKE, R. What animal is it? Pretoria; 1-187. 1954.
ASTLEY MABERLY, C. T. Animals of Rhodesia. Cape Town; 1-211. 1959.
ASTLEY MABERLY, C. T. Animals of East Africa. Cape Town; 1-211. 1960.

Madagascar

DECARY, R. La faune Malgache, son role dans les croyances et les usages indigenes. Paris; 1-236. 1950.

Indo-Malaya

BRANDER, A. A. D. Wild animals in central India. London; 1-296. 1923.
MENDIS, A. S. & FERNANDO, C. H. A guide to the freshwater fauna of Ceylon. Fish. Res. Stat. Dep. Fish. Ceylon, 12, 1-160. 1962.
SUVATTI, C. A check-list of aquatic fauna in Siam (excluding fishes). Bangkok; 1-116. 1938.
SUVATTI, C. Fauna of Thailand. Bangkok; 1-1100. 1950.
BOURRET, R. La faune de l'Indochine. Vertebres. Hanoi; 1-453. 1927.
TWEEDIE, M. W. F. & HARRISON, J. L. Malayan animal life. London; 1-237. 1954.
DAMMERMAN, K. W. The fauna of Krakatau, 1883-1933. Verh. K. Ned. Akad. Wet. Amsterdam, (2), 44, 1-594. 1948.
BRONGERSMA, L. D. The animal world of Netherlands New Guinea. Groningen; 1-71. 1958.

Australia & New Zealand

DAKIN, W. J. et al. Australian seashores. A guide .... London; 1-372. 1953.
LORD, C. E. & SCOTT, H. H. A synopsis of the vertebrate animals of Tasmania. Hobart; 1-340. 1924.
HUTTON, F. W. Index faunae Novae Zelandiae. London; 1-372. 1904.
POWELL, A. W. B. Native animals of New Zealand. Auckland; 1-96. 1947.

The Orient

RABOR, D. S. et al. Brief list of the land vertebrates of Negros Island. Silliman Journ., 5, 286-300. 1958.
CHEN, J. T. F. A synopsis of the vertebrates of Taiwan. Taipei. 1956.
OKADA, Y. An annotated list of animals of Okinawa Island. High School Naha, 34, (6), 1-384. Also as Naha City, Japan; 1-384 and Okinawa Biol. Educ. Res. Assoc. Shuri.
CHENG, T. H. List of common vertebrates of Fukien. Biol. Bull. Fukien Chr. Univ., 3, 111-142; 1947. (In Chinese)
MORI, T. A hand-list of the Manchurian and eastern Mongolian Vertebrata. (China); 1-209. 1927. (In Chinese)
LINDBERG, G. U. et al. A list of fauna of the sea waters of the South Sakhalin and South Kuril Islands. Issled. Dalnevost. Mar. SSSR, 6, (2), 173-253. 1959. (In Russian)
OKADA, Y. A catalogue of vertebrates of Japan. Tokyo; 1-412. 1938.
OKADA, Y. Annotated list of animals and plants of Mie Prefecture, Japan. Biol. Surv. Mie Prefecture, 1951, 1-352. 1951.

Oceania

EDMONDSON, C. E. Reef and shore fauna of Hawaii. Spec. Publ. Bishop Mus., 22, 1-295. 1933.
CHABOUIS, L. & CHABOUIS, F. Petite histoire naturelle des establishments francais de l'Oceanie, vol. 2, Zoology. Saint-Amand-Montroud, 2, 1-137. 1954.

EXPEDITIONS AND LOCALITIES

Expeditions

A bibliography of fishes, by B. Dean. New York; 3 volumes. (Voyages, 343-347) 1916-23.
A reference guide to the literature of travel, by E. G. Cox. Univ. Washington Publ. Langu. Lit., volumes 9, 10, 12. 1935, 1938, 1949. (Expeditions)
Scientific expeditions, by E. Terek. New York; 1-176. 1952.

Localities

(anonymous). Dictionnaire des bureaux de poste, 5th ed. Berne; 2 volumes. 1951. (Post offices of the world)
(anonymous). Preliminary N. I. S. gazetteer. Washington, D. C.; 167 parts & continuation. 1950 —. (National Intelligence Survey)
(anonymous). Webster's geographical dictionary, 2nd ed. Springfield, Massachusetts; 1-1293. 1962.
(anonymous). Directory of post offices (with ZIP codes). Washington, D. C.; 1-485. 1969. (Previous to 1962 as: U. S. Official Postal Guide, several editions)
BOUSFIELD, E. L. & McALLISTER, D. E. Station list ... South Eastern Alaska and Prince William Sound. Bull. Nat. Mus. Canada, 183, 76-103. 1962.
ORTH, D. J. Dictionary of Alaska place names. United States Geol. Surv. Prof. Pap., 567, 1-1084; 1967.
PEARCE, T. M. New Mexico place names. A geographical dictionary. Albuquerque; 1-187. 1965.
GRANGER, B. H. Will C. Barnes' Arizona Place Names. Tucson; 1-519. 1960.
WENZEL, R. L. & TIPTON, V. J. Ectoparasites of Panama. Chicago; 1-861. 1966. (Gazetteer)
LANA, G. et al. Glossary of geographical names / in six languages. Amsterdam; 1-184. 1967. (In English, French, Italian, Spanish, German, Dutch)
(anonymous). Concise survey of localities ... Turkey 1959. Zool. Meded., 38, 129-151. 1963.
BELL, B. D. List of New Zealand off-shore islands and their wildlife status. (New Zealand); 1-56. 1963.

SPECIAL SUBJECT LISTS

Cavernicolous

HAZELTON, M. & GLENNIE, E. A. A preliminary list of the cave fauna of Great Britain and Ireland. Publ. Cave Res. Group, 1, (2), 1-20. 1947.
FRANCISCOLO, M. E. Fauna cavernicola del Savonese. Ann. Mus. Stor. Nat. Genova, 67, 1-223. 1955.
MANSFIELD, R. W. et al. Mendip cave bibliography and survey catalogue. 1901-1963. Cave Res. Group Great Britain, 13, 1-164. 1965.

Dangerous

HALSTEAD, B. W. Dangerous marine animals. Cambridge, Maryland; 1-146. 1959.

Deep-sea

MARSHALL, N. B. Aspects of deep sea biology. London; 1-380. 1954.
GUNTHER, K. & DECKERT, K. Creatures of the deep sea. London; 1-222. 1956. (Translation of 1950)

Larvae

VANNUCCI, M. Catalogue of marine larvae. No. 1. Inst. Paulista Oceanogr., 1959, 1-44.

Trogloditic

NICHOLAS, B. G. Checklist of trogloditic organisms of Middle America. American Midl. Nat., 68, 165-188. 1962.

Venomous

MACHADO, O. Catalogo sistematico dos animais urticantes e peconhentos do Brasil. Bol. Inst. Vital Brasil, 25, 41-56. 1943.
KEEGAN, H. L. & MACFARLANE, W. V. Venomous and poisonous animals and noxious plants of the Pacific Region. Oxford; 1-456. 1963.
HALSTEAD, B. W. Poisonous and venomous marine animals of the world. Washington, D. C.; 3 volumes. 1965-70.
BUCHERL, W. et al. Venomous animals and their venoms. New York; 2 volumes. 1968-71.

METHODS

GREEN, T. L. Zoological technique, 2nd ed. London; 1-224. 1942.
JIROVEC, O. Zoologicha technika, 2nd ed. Praze; 1-303. 1947. (In Czech)
LEONE, C. A. The immunotaxonomic literature: The animal kingdom. Bull. Serol. Mus., 39, 1-23. 1968.
LOWEGREN, Y. Biologisk teknik. Malmö; 1-447. 1944.

von OSTERMANN, G. F. Manual of foreign languages for the use of librarians ..., 4th ed. New York; 1-414. 1952.
TOLMATCHEV, A. I. Principles of modern taxonomy and their application in palaeontology. Vert. Leningr. Gos. Univ., 1968, 3, 5-18. (In Russian)
ZIMMER, C. Zoologische Musealtechnik. In Methodik der wissenschaftlicher Biologie..., ed by T. Peterfi. Berlin; 2 volumes. 1928.

Collecting, preserving, curating

(anonymous). Directions for collecting and preserving specimens. United States Nat. Mus. Bull., 39, parts A - S. 1911.
(anonymous). A field collector's manual in natural history. Washington, D. C.; 1-118. 1944.
AIYAPPAN, A. & SATYAMURTI, S. T. Handbook of museum technique. Madras; 1-228. 1960.
ANDERSON, R. M. Methods of collecting and preserving vertebrate animals. Bull. Nat. Mus. Canada, 69, 1-141. 1932.
ANDERSON, R. M. Instructions for preserving animal specimens for scientific purposes. Spec. Contr. Nat. Mus. Canada, 43, (2), 1-34. 1943.
ANDERSON, R. M. Methods of collecting and preserving vertebrate animals. Bull. Nat. Mus. Canada, 69, 1-162. 1944.
ANDERSON, R. M. Methods of collecting and preserving vertebrate animals, 2nd ed. Bull. Nat. Mus. Canada, 69, Biol., 1-162. 1948.
ANDERSON, R. M. Methods of collecting and preserving vertebrate animals. Bull. Nat. Mus. Canada, Biol., 69. 1965.
BIANCO, S. L. The methods employed at the Naples Zoological Station for the preservation of marine animals. United States Nat. Mus. Bull., 39, (M), 1-42. 1899.
BOYDEN, A. A. Collecting serological samples. Atoll Res. Bull., 17, 96-99. 1953.
BURNS, N. J. Field manual for museums. Washington, D. C.; 1-426. 1943.
DAVIS, D. D. & GORE, U. R. Clearing and staining skeletons of small vertebrates. Field Mus. Nat. Hist., Tech. Ser., 4, 1-15. 1936.
FINLAYSON, H. H. A simple apparatus for degreasing bones for museum purposes. Trans. Roy. Soc. South Australia, 56, 172-174. 1932.
GENTILLE, J. Natural history specimens. A handbook for naturalists. Handb. Western Australian Nat. Club, 2, 1-40. 1951.
HORNADAY, W. T. Taxidermy and zoological collecting. New York; 1-362. 1891.
HOWER, R. O. The freeze-dry preservation of biological specimens. Proc. United States Nat. Mus., 119, (3549), 1-24. 1967.
KNUDSEN, J. W. Biological techniques. Collecting, preserving, and illustrating plants and animals. New York; 1-525. 1966.
LEGGE, M. M. Naturalist's handbook. A collector's guide. London; 1-64. 1947.
LUCAS, F. A. Notes on the preparation of rough skeletons. Circ. United States Nat. Mus., 33, 1-8. 1885.
MAYR, E. & GOODWIN, R. Biological materials. Part I. Preserved materials and museum collections. Washington, D. C. 1956. (NAS/NRC Publ. 399)
MENDES, F. P. Colheita e preparacao preliminar de animais para museu. Bull. Soc. Portug. Cienc. Nat., 13, Suppl. 2, 670-675. 1942.
MERYMAN, H. T. The preparation of biological museum specimens by freeze-drying. Curator, 3, 5-19. 1960.
MORTON, J. E. Collecting and preserving zoological specimens. Tuatara, 3, 104-114. 1950.
SCHRODER, G. Das Sammeln, Konservieren und Anstellen von Wirbeltieren .... Berlin; 1-93. 1936.
STOREY, M. & WILIMOVSKY, N. J. Curatorial practices in zoological research collections. 1. Preliminary report on containers and closures for storing specimens preserved in liquid. Circ. Nat. Hist. Mus. Stanford Univ., 3, 1-22. 1955.
TAYLOR, W. R. An enzyme method of clearing and staining small vertebrates. Proc. United States Nat. Mus., 122, (3596), 1-17. 1967.
WAGSTRAFFE, R. & FIDLER, J. H. The preservation of natural history specimens. Volume 2, Vertebrates, botany, geology. New York; 1-404. 1968.

Describing

BLACKWELDER, R. E. Taxonomy .... New York; 1-698. 1967.
FERRIS, G. F. The principles of systematic entomology. Stanford Univ. Publ. Biol. Sci., 5, 103-269. 1928.
MAYR, E., LINSLEY, E. G. & USINGER, R. L. Methods and principles of systematic zoology. New York; 1-328. 1953.
RENSCH, B. Kurze Anweizung fur zoologisch-systematische Studien. Leipzig; 1934.
SIMPSON, G. G. et al. Quantitative zoology, 2nd ed. New York; 1-440. 1960.
VOSS, E. G. The history of keys and phylogenetic trees in systematic biology. Journ. Sci. Lab. Denison Univ., 43, 1-25. 1952.

Citation

(anonymous). A manual of style ..., 12th ed. Chicago; 1-546. 1969. (University of Chicago)
(anonymous). Style manual for biological journals, 2nd ed. Washington, D. C.; 1-117. 1964. (American Institute of Biological Sciences)

ALRUTZ, R. W. A punched-card system for cataloguing a general bibliography in biology. Journ. Sci. Lab. Denison Univ., 45, 18-37. 1960.
BALL, I. R. Literature citations in taxonomic publications. Syst. Zool., 17, 217-218. 1968.
BLACKWELDER, R. E. Citing literature in the Coleopterists' Bulletin. Coleopt. Bull., 3, 55-59. 1949.
CROACH, W. G. & ZETLER, R. L. A guide to technical writing, 2nd ed. New York; 1-441. 1954.
ESDAILE, A. A student's manual of bibliography. London; 1-383. 1931.
McCRUM, B. & JONES, H. D. Bibliographical procedures and style. A manual for bibliographers in the Library of Congress. Washington, D. C.; 1-127. 1954.
MUNRO, I. S. R. An application of visible indexing to systematic zoology. Australian Journ. Marine Freshw. Res., 3, 92-100.

Illustration

CANNON, H. G. A method of illustration for zoological papers. (England); 1-37. 1936. (Association for British Zoology)
FERRIS, G. F. The principles of systematic zoology. Stanford Univ. Publ. Biol. Sci., 5, 103-269. 1928.
MAYR, E., LINSLEY, E. G., & USINGER, R. L. Methods and principles of systematic zoology. New York; 1-328. 1953. (Pp. 148-151, 295)
OLSEN, T. H. & MORROW, J. E. Guide for preparing figures. Bull. Bingham Oceanogr. Coll., 17, 147-153. 1959.
PAPP, C. S. An introduction to scientific illustration. Riverside, California; 1-212. 1963.
RIDGWAY, J. L. Scientific illustration. Stanford, California; 1-173. 1938.
ZWEIFEL, F. W. A handbook of biological illustration. Chicago; 1-137. 1961.

Experimental and numerical methods

BIANCHI, U. Utility significance and limitations of electrophoresis applied to particular taxonomic problems. Riv. Biol., 60, 227-245. 1967. (Title also in Italian; in Italian & English)
BOYDEN, A. A. Serology and animal systematics. American Nat., 77, 234-255. 1943.
HAWKES, J. G. Chemotaxonomy and serotaxonomy. London; 1-299. 1968.
MILLER, R. R. Utilization of X-rays as a tool in systematic zoology. Syst. Zool., 6, 29-40. 1957.
RYKE, P. A. J. Numerical taxonomy — principles, implications and applications. Tydskr. Natuurw., 4, (1), 82-944. 1964.
SOKAL, R. R. & SNEATH, P. H. A. Principles of numerical taxonomy. San Francisco; 1-359. 1963.
SOKAL, R. R. Statistical methods in systematics. Biol. Rev., 40, 337-391. 1965.
STEYSKAL, G. C. The number and kind of characters needed for significant numerical taxonomy. Syst. Zool., 17, 474-477. 1968.
WRIGHT, C. A. Experimental taxonomy. A review of some techniques and their applications. Int. Rev. Gen. Exp. Zool., 2, 1-42. 1966.

GLOSSARIES

[See also glossaries listed under the separate classes]

BAGUE, J. Glossary of animal biology. (English - Spanish). [Almanaque Agricola Puerto Rico, 1948-50] Pp. 143-310. 1950.
BAGUE, J. Glosario de biologie lanimal. (Español - Ingles) Almanaque Agricola Puerto Rico, 1951/1952, 195-373. 1952.
de BARROS e CUNHA, J. G. Vocabulario de termos gregos em zoologia e antropologia. Coimbra; 1-64. 1937.
BLACKWELDER, R. E. Taxonomy .... New York; 1-698. 1967. (Latin phrases, p. 266; abbreviations, p. 265)
BLAIR, W. F. et al. Vertebrates of the United States. New York; 1-819. 1957.
BOGORAD, V. B. Kratkii slovar' biologicheskikh terminov. Moskva; 1-236. 1963. (In Russian)
BORROR, D. J. Dictionary of word roots and combining forms. Palo Alto, California; 1-134. 1960.
BROWN, R. W. Composition of scientific words. Washington, D. C.; 1-882. 1954.
De SAINT-DENIS, E. Le vocabulaire des animaux marins en Latin classique. Paris; 1-120. 1947.
DITTLER, R. Handworterbuch der Naturwissenschafter, 2nd ed. Jena; Band I, 1-1078. 1931.
DUMBLETON, C. W. Russian - English biological dictionary. Edinburgh; 1-512. 1964.
HENDERSON, I. F. A dictionary of scientific terms. Pronunciation, derivation, and definition, 2nd ed. New York; 1-352. 1929.
HENDERSON, I. F. & HENDERSON, W. D. A dictionary of biological terms, 8th ed. Princeton, New Jersey; 1-656. 1963.
HUSSON, R. Glossaire de biologie animals. Paris; 1-280. 1964.
JAEGER, E. C. A dictionary of Greek and Latin combining forms used in zoological names, 2nd ed., London; 1-157. 1931.
JAEGER, E. C. A source-book of biological names and terms, 2nd ed. Springfield, Illinois; 1-287. 1950. (First edition 1944)
KEEN, A. M. & MULLER, S. W. Schenk & McMasters, Procedure in Taxonomy, 2nd ed. Stanford, California; 1-92. 1948. (Latin terms, 27-29)
KEEN, A. M. & MULLER, S. W. Schenk & McMasters, Procedure in Taxonomy, 3rd ed. Stanford, California; 1-149. 1956 (Abbreviations, 27-29)

LEFTWICH, A. W. A student's dictionary of zoology. London; 1-290. 1963.
LEFTWICH, A. W. A dictionary of zoology. London & Princeton, New Jersey; 1-319. 1967.
MALARET, A. Lexicon de fauna y flora. Bogota; 1-577. 1961.
MELANDER, A. L. Source book of biological terms. New York; 1-157. 1937.
MELANDER, A. L. & SPIETH, E. The source of biological terms. New York; 1-32. 1934.
NIEMANN, G. & HONIGMANN, H. L. Zoologisches Worterbuch .... Harz; 1-221. 1919.
NYBAKKEN, O. E. Greek and Latin in scientific terminology. Ames, Iowa; 1-321. 1959.
PENNAK, R. W. Collegiate dictionary of zoology. New York; 1-583. 1964.
PRATT, H. S. A manual of land and fresh water vertebrate animals of the United States (exclusive of birds). Philadelphia; 1-422. 1923. (Pp. 382-391)
SAVORY, T. H. Latin and Greek for biologists. London; 1-42. 1946.
SCHENK, E. T. & McMASTERS, J. H. Procedure in taxonomy. Stanford Univ., California; 1-72. 1936. (Latin terms, pp. 23-25)
TIERNO, J. C. Dicionario zoologico .... Lisboa; 1-772. 1954.
WOODS, R. S. The naturalist's lexicon. A list of classical Greek and Latin words suitable for use in biological nomenclature. Pasadena, California; 1-282. 1944. (Also Addenda to ....)

NOMENCLATORS

Archiv fur Naturgeschichte. Berlin. 1832-1922.
Biological abstracts. Philadelphia; volume 1 —. 1926 —.
Century dictionary. New York; 6 volumes (24 parts). 1889-91. (Names of taxa in higher categories)
BLACKWELDER, R. E. Classification of the animal kingdom. Carbondale, Illinois; 1-94. (Orders and above, pp. 80-94)
Concilium bibliographicum. Zurich; 3x5 cards. 1896-1940.
Encyclopaedia Britannica, 11th ed. & 12th ed. 1911-12, 1921-22.
Homonyms and nomenclators. American Midl. Nat., 16, 962-964. 1935. By P. H. Oehser.
Index-catalogue of medical and veterinary zoology. Supplement 15 .... Hosts. Washington, D. C.; 1-149. 1966. By J. M. Humphrey & D. B. Segal.
Index-catalogue of medical and veterinary zoology. Supplement 16, part 7 .... Hosts. Washington, D. C.; 1-276. 1968. By D. B. Segal et al.
Index-catalogue of medical and veterinary zoology. Trematoda.... Hosts. Washington, D. C.; 1-694. 1969. By M. A. Doss & M. M. Farr.
Index animalium ..., Sectio prima [1758-1800]. Cantabrigiae; 1-1195. 1902. (Species & genera)
Index animalium......, Sectio secunda [1800-1850]. London; 1-7056. 1922-23. (Species & genera) By C. D. Sherborn.
Index zoologicus [1864-1900]. London; 1-421. 1902. (Genera) By C. O. Waterhouse.
Index zoologicus No. 2. [1901-1910]. London; 1-324. 1912. By C. O. Waterhouse.
International catalogue of scientific literature. N. Zoology for 1904 to 1916. 1906-1919.
Key to the names of British fishes, mammals, amphibians, reptiles. London; 1-71. 1956. (Meaning & origin) By R. D. Macleod.
McGraw-Hill Encyclopedia of science and technology. New York; 15 volumes. 1960. (The commoner current names, in index)
Nomenclator animalium generum et subgenerum. Berlin; 25 Lief., 5 volumes. 1926-1954. (Genera) By F. E. Schulze, W. Kukenthal, & K. Heider.
Nomenclator zoologicus. Soloduri. 1842-46. (By classes) By L. Agassiz.
Nomenclator zoologicus .... Vindobonae; 1-482. 1873. (By groups) By A. de Marschall.
Nomenclator zoologicus [1758-1879]. United States Nat. Mus. Bull., 19, 1-376, 1-340. 1882. By S. H. Scudder. (Genera)
Nomenclator zoologicus: A list of the names of genera and subgenera from the Tenth Edition of Linnaeus 1758, to the end of 1935. London; 4 volumes, with Addenda. By S. A. Neave. 1939-40.
Nomenclator zoologicus ..., 1936-45. London; volume 5, with Addendum. 1950. By S. A. Neave.
Nomenclator zoologicus by S. A. Neave. London; volume 6 (Supplement) for the years 1946-1955. 1966. By M. Edwards & A. T. Hopwood.
Nomenclatoris zoologici. Index universalis .... Soloduri; 1-393. 1846. (Quarto) By L. Agassiz.
Nomenclatoris zoologici. Index universalis .... Soloduri; 1-1135. 1848. (Small octavo) By L. Agassiz.
Register zum Zoologischen Anzeiger. Leipzig; 8 volumes. 1889-1929. By J. V. Carus & E. Korschelt.
GALVAN, Y-J. Repertoire des noms de genres de Vertebres. Paris; 1-383. 1965.
ABEL, O. Die Stamme der Wirbeltiere. Berlin & Leipzig; 1-914. 1919. (Orders & families).
POCHE, F. Supplement zu C. D. Sherborn's Index Animalium. Festschr. 60. Geburtstage E. Strand, 5, 477-615. 1938. (Genera and species)
ROMER, A. S. Vertebrate paleontology. Chicago; 1-687. 1945. (Genera & families)
Zoological Record 1863 —. London; 104 volumes & continuation. 1864 —.

NOMENCLATURE

(serial). Bulletin of Zoological Nomenclature. London; 1-1943 —.
BLACKWELDER, R. E. Taxonomy .... New York; 1-698. 1967.
KEEN, A. M. & MULLER, S. W. Schenk & McMasters, Procedure in taxonomy, 2nd ed. Stanford, California; 1-92. 1948.
KEEN, A. M. & MULLER, S. W. Schenk & McMasters, Procedure in taxonomy, 3rd ed. Stanford, California; 1-149. 1956.

MAYR, E., LINSLEY, E. G. & USINGER, R. L. Methods and principles of systematic zoology. New York; 1-328. 1953.
RICHTER, R. Einfuhrung in die zoologische Nomenklatur durch Erlauterung der Internationalen Regeln, 2nd ed. Frankfurt a. M.; 1-252. 1948.
SAVORY, T. Naming the living world. An introduction to the principles of biological nomenclature. New York & London; 1-128. 1962.
SCHENK, E. T. & McMASTERS, J. H. Procedure in taxonomy. Stanford University, California; 1-72. 1931.
SIMPSON, G. G. The principles of classification and a classification of mammals. Bull. American Mus. Nat. Hist., 85, 1-350. 1945. (Pp. 24-33)

Common names

MARIA, H. A. Vocabulario de terminos vulgares en historia natural Columbiana. Rev. Acad. Columbiana Cienc., 4, 326-336, 5, 40-60, 149-170, 295-307. 1941-42.
von IHERING, R. Ensaio geografico sobre o vocabulario zoologico popular do Brasil. Rev. Brasileira Geogr., 1, 73-88. 1939.
DANJALSSON a RYGGI, M. D. List of Faroese animal names. In Jensen, Zoology of the Faroes, 3, (66), 1-10. 1942.
CADENAT, J. Noms vernaculaires des principales formes d'animaux marins des Cotes de l'Afrique Occidentale Francaise. Catal. Inst. Francaise Afrique Noire, 2, 1-56. 1947.
VILLIERS, A. Noms vernaculaires de quelques animaux de l'Air [French West Africa]. Not. Africaines, 40, 23-25. 1948.
MITCHELL, B. L. A naturalist in Nyasaland. Nyasaland Agric. Quart. Journ., 6, 1-23, 25-47. 1946.
SMITINAND, T. & SCHEIBLE, W. R. Edible and poisonouslplants and animals. Annex to survival manual for Thailand .... Bangkok; 1-250. 1966.
TWEEDIE, M. W. F. & HARRISON, J. L. Malayan animal life. London; 1-237. 1954.
JOHNSTON, T. H. Aboriginal names and utilization of the fauna in the Eyrean region. Trans. Roy. Soc. South Australia, 67, 244-311. 1943.

Rules of nomenclature

BANKS, N. & CAUDELL, A. N. The entomological code. Washington, D. C.; 1-31. 1912.
(anonymous). International Rules of Zoological Nomenclature. Proc. Biol. Soc. Washington, 39, 75-104. 1926.
FERRIS, G. F. The principles of systematic entomology. Stanford University Publ. Biol. Sci., 5, 103-269. 1928.
VAN CLEAVE, H. J. An index to the International Rules of Zoological Nomenclature. Trans. American Microscopical Soc., 52, 322-325. 1933.
SCHENK, E. T. & McMASTERS, J. H. Procedure in taxonomy. Stanford Univ., California; 1-72. 1936.
RICHTER, R. Einfuhrung in die zoologische Nomenklatur durch Erlauterung der Internationalen Regeln. Frankfurt a. M.; 1-154. 1943.
VAN CLEAVE, H. J. An index to the Opinions rendered by the International Commission on Zoological Nomenclature. American Midl. Nat., 30, 223-240. 1943.
RICHTER, R. Einfuhrung in die zoologische Nomenklatur durch Erlauterung der Internationalen Regeln, 2nd ed. Frankfurt a. M.; 1-252. 1948.
KEEN, A. M. & MULLER, S. W. Schenk & McMasters, Procedure in taxonomy, 2nd ed. Stanford, California; 1-93. 1948.
HEMMING, F. Copenhagen decisions on zoological nomenclature .... London; 1-135. 1953.
FOLLETT, W. I. An unofficial interpretation of the International Rules of Zoological Nomenclature as amended by the XIII International Congress of Zoology, Paris, 1948, and by the XIV International Congress of Zoology, Copenhagen, 1953. San Francisco; 1-99. 1956.
KEEN, A. M. & MULLER, S. W. Schenk & McMasters, Procedure in taxonomy, 3rd ed. Stanford, California; 1-149. 1956.
HEMMING, F. 1958. Official index of rejected and invalid family-group names in zoology. First instalment .... London; 1-38. 1959.
HEMMING, F. Official index of rejected and invalid generic names in zoology. First instalment.... London; 1-132. 1958.
HEMMING, F. Official index of rejected and invalid specific names in zoology. First instalment .... London; 1-73. 1958.
HEMMING, F. Official index of rejected and invalid works in zoological nomenclature. First instalment .... London; 1-14. 1959.
HEMMING, F. Official list of family-group names in zoology. First instalment .... London; 1-38. 1958.
HEMMING, F. Official list of generic names in zoology. First instalment .... London; 1-200. 1958.
HEMMING, F. Official list of specific names in zoology. First instalment .... London; 1-206. 1958.
HEMMING, F. Official list of works approved as available for zoological nomenclature. First instalment .... London; 1-12. 1958.
STOLL, N. R. et al. International Code of Zoological Nomenclature.... London; 1-176. 1961.
STOLL, N. R. et al. International Code of Zoological Nomenclature.... [Revised edition]. London; 1-176. 1964.
BLACKWELDER, R. E. Taxonomy .... New York; 1-698. 1967.

Types of genera & species

BLACKWELDER, R. E. The generic names of the beetle family Staphylinidae, with an essay on genotypy. United States Nat. Mus. Bull., 200, 1-483. 1952.
COOK, O. F. Terms relating to generic types. American Nat., 48, 308-314. 1914.
FRIZZELL, D. L. Terminology of types. American Midl. Nat., 14, 637-668. 1933.
MALAISE, R. Fabricius as the first designator and original inventor of genotypes. Ent. News, 48, 130-134. 1937.

Gender of names

BLACKWELDER, R. E. The gender of scientific names in zoology. Journ. Washington Acad. Sci., 31, 135-140. 1941.
GRENSTED, L. W. The formation and gender of generic names. Ent. Monthly Mag., 80, 229-233. 1944.

Pronunciation of names

(anonymous). Century dictionary. New York; 6 volumes. 1889-1891. (Of higher taxa)
BLACKWELDER, R. E. Taxonomy .... New York; 1-698. (Pp. 231-238)
HOPPER, H. P. The pronunciation and derivation of the names of genera and subgenera of the family Ichneumonidae ... in North America .... Proc. Ent. Soc. Washington, 61, 155-171. 1959.
JAEGER, E. C. The biologist's handbook of pronunciations. Springfield, Illinois; 1-317. 1960.
MELANDER, A. L. The pronunciation of insect names. Bull. Brooklyn Entom. Soc., 11, 93-101. 1916.
NYBAKKEN, O. E. Greek and Latin in scientific terminology. Ames, Iowa; 1-321. 1959. (Pp. 243-249)

CLASSIFICATIONS

COPE, E. D. Synopsis of families of Vertebrates [Chordata]. American Nat., 23, 849-877. 1889.
(various authors). Das Tierreich. Leipzig; 79 volumes 7 continuation. 1896 —. Ed. by F. E. Schulze.
GADOW, H. A classification of vertebrates, Recent and extinct. London; 1-82. 1898.
(various authors). Encyclopaedia Britannica, 11th ed. & 12th ed. 1911-12 & 1921-22.
ALLEE, W. C. Synoptic key to the phyla, classes and orders of animals ... North America. Chicago; 1-63. 1923.
HICKSON, S. J. Outline classification of the animal kingdom, 5th ed. Manchester; 1-28. 1924.
LAMEERE, A. Abrege de la classification zoologique. Ann. Soc. Zool. Belge, 57, 68-182. 1927.
LAMEERE, A. Abrege de la classification zoologique. Rec. Inst. Zool. Torley-Rousseau, 3, 165-283. 1931.
HYMAN, L. H. The invertebrates: Protozoa through Ctenophora. New York; 26-39. 1940. [Volume 1]
ROMER, A. S. Vertebrate paleontology. Chicago; 1-687. 1945.
von HUENE, F. Classification and phylogeny of tetrapods. Geol. Mag., 86, 189-195. 1949.
NAVARRO CANDIDO, A. Clasificacion de los animales. Madrid; 1-317. 1949.
PEARSE, A. S. Zoological names. A list of phyla, classes, and order, 4th ed. Durham, North Carolina; 1-24. 1949.
RINGUELET, R. A. Clasificacion moderna del reino animal incluyendo clases y subclases de Acordados vivientes y extinguidos. Ser. Tecn. Didact. La Plata, 3, 1-61. 1950.
BURT, C. E. Synopsis of the classification of the Animal Kingdom. Quivira Booklets, 3, 1-8. 1954.
ALVARADO, R. Los grandes grupos del reino animal y el sistema zoologica. Rev. Univ. Madrid, 6, 531-565. 1957.
PIVETEAU, J. Classification des vertebres. Traite de Zoologie, 13, (1), 1-10. 1958.
WHITTAKER, R. H. On the broad classification of organisms. Quart. Rev. Biol., 34, 210-226. 1959.
ROTHSCHILD, —. A classification of living animals. London & New York; 1-106. 1961.
BLACKWELDER, R. E. Classification of the animal kingdom. Carbondale, Illinois; 1-94. 1963.
MACKERRAS, I. M. The classification of animals. Proc. Linn. Soc. New South Wales, 88, (3), 324-325. 1964. WEICHERT, C. L. Anatomy of the chordates, ed. 3. New York; 1-758. 1965.
BARKLEY, F. A. Outline classification of organisms. Providence, Rhode Island; 1-25. 1967.
BARKLEY, F. A. Outline classification of organisms, 2nd ed. Providence, Rhode Island; 1-25. 1968.

CEPHALOCHORDATA

GENERAL BOOKS

DELAGE, Y. & HEROUARD, G. Traite de zoologie concrete, volume 8, Les Procordes. Paris; 1-379. 1898. (Pp. 68-131)
DRACH, P. Traite de Zoologie, volume XI. Paris; 931-1037. (Anatomy, physiology, development). 1948.
FRANZ, V. Haut, Sinnesorgane und Nervensystem der Akranier .... Jenaische Zeitschr. Naturwiss., 59, 401-526. 1923.
FRANZ, V. Morphologische und ontogenetische Akranier-studien ... Jenaische Zeitsch. Naturwiss., 61, 407-488. 1925.
HERDMAN, W. A. Leptocardi. In Cambridge Nat. Hist., vol. 7, 112-138. 1904.
JORDAN, D. S., Fishes, 2nd ed. New York; 1-773. 1925.
LAMEERE, A. Les Protochordes. Recueil Inst. Zool. Torley-Rousseau, 8, 1-116. 1940.
LONNBERG, E. Leptocardii. In Bronn's Klass. Ordn. Thierr., 6, (I), 1-240. 1901-1906.
MARSHALL, A. J. Parker & Haswell, A textbook of zoology, volume 2, 7th ed. London & New York; 1-952; 1962. (Pp. 47-68)
PERRIER, E. Traite de zoologie. Paris; 2137-2170. 1899.
PIETSCHMANN, V. Acrania - Cephalochordata. Handbuch der Zoologie, 6, (1), 1-124. 1929-1933.
YOUNG, J. Z. The life of vertebrates. Oxford; 1-767; 1950. (Pp. 24-46)

BIBLIOGRAPHIES

[See also many bibliographies on Vertebrata and Pisces]

BURCHARDT, E. Beitrage zur Kenntniss der Amphioxus lanceolatus, nebst einem ausfuhrlichen Verzeichniss der bisher uber Amphioxus veroffentlichten Arbeiten. Jenaische Zeitschr., 34, 719-832. 1900.

FAUNAL & FAMILY STUDIES

[Because there are so few references to be listed by family, all regional and revisionary works are cited here in geographical order. The families cited are indicated here as annotations]

GOLDSCHMIDT, R. Amphioxides Vertreter einer neuer Acranier families. Biol. Centralbl., 25, 235-240. (Amphioxididae) 1905.
PIETSCHMANN, V. Acrania - Cephalochordata. Handbuch der Zoologie, 6, (l), 1-112; 1929. (Monograph, as Branchiostomidae)
FRANZ, V. Acrania (Lanzett - Fische). Tabulae Biol., 6, 573-581. 1930. (Systematic revision)
SCHNAKENBECK, W. Acrania (Cephalochordata) - Cyclostoma - Pisces. Handbuch der Zoologie, VI, (1), 905-1000. 1960. (monograph)
HILDEBRAND, S. F. & SCHROEDER, W. C. Fishes of Chesapeake Bay. Bull. United States Bur. Fish. 43, (1), 1-366. 1928. (Branchiostomidae)
NICHOLS, J. T. The fishes of Porto Rico and the Virgin Islands. Branchiostomidae and Sciaenidae. Sci. Surv. Porto Rico Virgin Is., 10, 161-295.
MEEK, S. E. & HILDEBRAND, S. F. The marine fishes of Panama, Part I. Field Mus. Nat. Hist., Zool., 15, 1-330. 1923. (Branchiostomidae)
HILDEBRAND, S. F. A descriptive catalog of the shore fishes of Peru. United States Nat. Mus. Bull., 189, 1-530. 1946. (Branchiostomidae)
BIGELOW, H. B. & FARFANTE, I. P. Fishes of the western North Atlantic. Lancelets. Mem. Sears Found. Mar. Res., 1, (1), 1-28. 1948. (As Branchiostomidae & Epigonichthyidae)
SMITT, F. A. et al. A history of Scandinavian fishes, 2nd ed. Stockholm; 1-1240. 1895. (Amphioxidae)
FRANZ, V. Branchiostoma, in Grimpe & Wagler, Tierwelt Nord- und Ostsee, 12, (6), 1-46. 1927.
POLL, M. Faune de Belgique. Poissons marins. Bruxelles; 1-452. 1947. (Branchiostomidae)
TORTONESE, E. Fauna d'Italia. Leptocardia. Ciclostomata. Selachii. Bologna; 1-334. 1956. (Italia, as Branchiostomidae)
TORTONESE, E. Elenco dei Leptocardi del Mare Mediterraneo. Atti Soc. Italiana Sci. Nat., 97, 309-345. 1958.
TORTONESE, E. Elenco rveduto dei Leptocardi ... Mare Mediterraneo. Ann. Mus. Stor. Nat. Giacomo Doria, 74, 156-185. 1964.
RIEDL, R. Fauna und flora der Adria. Hamburg; 1-640. 1963.
FOWLER, H. W. The marine fishes of West Africa .... Bull. American Mus. Nat. Hist., 70, 1-1493. 1936. (Branchiostomidae)
WEBB, J. E. On the lancelets of West Africa. Proc. Zool. Soc. London, 125, 421-443. 1955.
WEBB, J. E. Cephalochordata of the coast of tropical West Africa. Atlantide Rep., 4, 167-182. 1956. (Branchiostomidae)
MONOD, T. Contribution a la faune du Cameroun. Pisces I. Faune Colon. Francaise, 1, 643-742. (Cephalochordata by Hubbs) 1927.

WEBB, J. E. On the lancelets of southeast Africa. Ann. South African Mus., 43, 249-270. 1958.

FOWLER, H. W. Fishes of the Red Sea and southern Arabia. I. Branchiostomida to Polynemida. Jerusalem; 1-240. 1956. (Branchiostomidae)

COOPER, C. F. Cephalochorda. Systematic and anatomical account. In J. S. Gardiner, Faun. Maldive Laccadive Archipelagoes, 1, 347-360. 1903.

PARKER, G. H. Maldive cephalochordates. Bull. Mus. Harvard, 46, 39-52. 1904.

DAY, F. The fauna of British India, including Ceylon and Burma, Fishes, vol. 1, London. 1889. (Amphioxidae)

PRASHAD, B. On a collection of Indian cephalochordates, with notes on the species from the Indian waters in the Indian Museum, Calcutta. Rec. Indian Mus., 36, 329-334. 1934.

FRANZ, V. Systematische Revision der Akranier. Jenaische Zeitschr. Naturw., 58, 369-452. 1922. (Ceylon)

SUVATTI, C. Fauna of Thailand. Bangkok; 1-1100. 1950.

PIYAKARNCHANA, T. Taxonomic studies on the lancelets found in the Gulf of Thailand. Nat. Hist. Bull. Siam Soc., 20, 95-107. 1962.

WHITLEY, G. P. The lancelets and lampreys of Australia. Australian Zool., 7, 256-264. 1932. (As Asymmetrontidae, Branchiostomidae, Epigonichthyidae)

OGILBY, J. D. Check-list of the cephalochordates ... of Queensland, Part I. Mem. Queensland Mus., 5, 70-98. 1916.

FRANZ, V. Acrania. In Michaelson & Hartmeyer, Fauna S. W. Australiens, 5, 219-222. 1927.

LORD, C. E. & SCOTT, H. H. A synopsis of the vertebrate animals of Tasmania. Hobart; 1-340. 1924.

FOWLER, H. W. A synopsis of the fishes of China. Part i. The sharks, rays and related fishes. Hong Kong Nat., 1, 24-34. 1930. (Branchiostomidae)

LINDBERGH, G. U. & LEGHEZA, M. I. Fishes of the Japanese Sea and the adjacent parts of the Okhotsk and Yellow Seas, part I. Tabl. Anal. Faune URSS., 68, 1-208. 1959. (In Russian, as Amphioxidae)

JORDAN, D. S. & SNYDER, J. O. A review of the lancelets, hag-fishes, and lampreys of Japan .... Proc. United States Nat. Mus., 23, 725-734. 1901. (Branchiostomidae)

(anonymous) Illustrated encyclopedia of the fauna of Japan ..., 2nd ed. Tokyo, 2 volumes. 1949.

SCHULTZ, L. P. et al. Fishes of the Marshall and Marianas Islands, volume 1. United States Nat. Mus. Bull., 202, 1-685. 1953. (As Asymmetrontidae)

NOMENCLATORS

PIETSCHMANN, V. Acrania - Cephalochordata. Handbuch der Zoologie, 6, (1), 1-124. 1929-1933. (Higher taxa only)

CATALOGS

HUBBS, C. L. A list of the lancelets of the world .... Occ. Pap. Univ. Michigan Mus. Zool., 105, 1-16. 1922.

FAMILIES

[Some students treat these families as one, under the name Amphioxidae or Branchiostomidae. Because there are so few references, they are combined above with the regional studies, and only the family names are cited here.]

Amphioxidae
Amphioxididae
Asymmetrontidae

Branchiostomatidae
Branchiostomidae

Epigonichthyidae

PISCES

GENERAL BOOKS

BARNARD, T. et al. La pesca en el mar. Enciclopedia manuel de la pesca. Barcelona, 1-353.
BERTIN, L. et al. Poissons. Pisces. Traité de Zool., 13, (1), 427-924. (Anatomy)
BIDEN, C. L. Sea-angling fishes of the Cape, 2nd ed. Cape Town; 1948.
BOULENGER, G. A. Fishes (systematic account of Teleostei). Cambridge Nat. Hist., VII, 537-727.
BREDER, C. M., Jr. & ROSEN, D. E. Modes of reproduction in fishes. Garden City, N. Y.; 1-941; 1966.
BRIDGE, T. W. Fishes (exclusive of the systematic account of Teleosti). Cambridge Nat. Hist., VII, 139-537; 1910.
CARAUSU, S. Tratat de ichtiologie. Bucuresti; 1-802; 1952. (In Rumanian)
COATES, C. W. & ATZ, J. W. Fishes of the world. In Drimmer, The animal kingdom. Garden City, New York; 1393-1640; 1954.
DANIEL, J. F. The elasmobranch fishes, 3rd ed. Berkeley, California; 1-332; 1934. (1st ed., 1922; 2nd ed., 1928)
GRAHAM, D. H. A treasury of New Zealand fishes. Wellington; 1-404; 1953.
GREGORY, W. K. Fish skulls: A study of the evolution of natural mechanisms. Trans. American Philos. Soc., 23, 75-481. (Recent & fossil; especially teleosts)
HALME, E. Pohjolen kalat Värikuvina. Helsinki; 1-136; 1954. (Fishes of the North in colour)
HARDER, W. Anatomie der Fische. Handb. Binnenfisch. Mitteleuropas, 6, (2A), 1-308.
HVASS, H. Alverdens fisk. København; 1-158; 1964.
HVASS, H. Fishes of the world. London; 1-156; 1965. (Also New York)
IUDKIN, I. I. Ikhtiologica. Moskva; 1-247; 1951. (In Russian)
JORDAN, D. S. A guide to the study of fishes. New York; vol. 1; 1905.
JORDAN, D. S. Fishes, 2nd ed. New York; 1-773; 1925. (1st ed., 1907)
LAGLER, K. F. et al. Ichthyology, the study of fishes. New York; 1-545; 1962.
LeDANOIS, E. et al. Fishes of the world. London; 1-190; 1957.
MANN F., G. Vida de los peces en aguas chilenas. Santiago, Chile; 1-342; 1954.
MARSHALL, N. B. The life of fishes. London; 1-402; 1965. (Cleveland, Ohio; 1966)
NIKOLSKI, G. V. Biology of fishes. Moskva; 1-231; 1944. (In Russian)
NIKOL'SKII, G. V. Systematic ichthyology. Moskva; 1-436; 1950. (In Russian)
NIKOL'SKII, G. V. Special ichthyology, Moskva; 1-538; 1954. (In Russian)
NIKOLSKI, G. W. Spezielle Fischkunde. Berlin; 1-632; 1957. (Translation of 1954)
NIKOL'SKII, G. V. Special ichthyology. Jerusalem; 1-538; 1961. (Translation of 1954)
NORMAN, J. R. A history of fishes. London; 1-463; 1931. (1st edition, 1931)
PIETSCHMANN, V. Cyclostoma. Marsipobranchii. Cyclostomata. Handb. der Zool., Band 6, Hälfte 1, I Teil, 125-547; 1933-1955. (In German)
ROTH, V. Fish life in British Guiana. Georgetown; 1-282; 1943.
ROUGHLEY, T. C. Fish and fisheries of Australia. Sydney; 1-328; 1966.
SANCHEZ ROIG, M. & GOMEZ de la MAZA, F. La pesca en Cuba. Habana; 1-272; 1952.
SCHINDLER, O. Freshwater fishes. New York & London; 1-243; 1957. (Translation)
SCHNAKENBECK, W. Pisces. Handb. der Zool., Band 6, Hälfte 1, Teil I, 551-1115; 1962. (In German)
SCHULTZ, L. P. & STERN, E. M. The ways of fishes. New York; 1-264; 1948.
STERBA, G. Süsswasserfische aus aller Welt. Berchtesgaden; 1-638; 1959.
STERBA, G. Freshwater fishes of the world. New York, 1-878; 1962. (Translation of 1959)
SUVOROV, E. K. Osnovy ikhtiologii. Leningrad; 1-580; 1948. (Principles of ichthyology; in Russian)
SUYEHIRO, Y. Text-book of ichthyology. Tokyo; 1-332; 1951. (In Japanese)
TARASOV, N. I. Outlines on general problems of ichthyology. Moskva; 1-319; 1953. (In Russian)
THOMAZI, A. Histoire de la pêche. Paris; 1-645; 1947.
TRET'YAKOV, D. K. Fishes and Cyclostomata, their life and significance. Moskva & Leningrad; 1-418; 1949; (In Russian)
TUGE, H. et al. An atlas of the brains of fishes of Japan. Tokyo; 1-240; 1968.
YUDKIN, I. I. Ichtiologia. Moskva; 1-323; 1955. (In Russian)

BIBLIOGRAPHIC WORKS

Bibliographies, separate works

[See also: Under ANIMALIA, above; in Zoological Record, in each volume under the heading Bibliographies]

(anonymous). Contribution to a bibliography on the osteology of fishes. Circ. Nat. Hist. Mus. Stanford Univ., 1, 1-87; 1956.
(anonymous). List of publications of the Fisheries Research Board of Canada, 1901-1949. Bull. Fish. Res. Board Canada, 87; 1950.
ALLOUSE, B. E. A bibliography of the vertebrate fauna of Iraq and neighbouring countries. IV. Fishes. Publ. Iraq Nat. Hist. Mus., 7, 1-32; 1955.

ALVAREZ, J. Ictiologia dulce acuicola Mexicana. II. Bibliografia taxonomica. Rev. Soc. Mexicana Hist. Nat., 11, 201-216; 1950.
ATZ, J. W. Dean bibliography of fishes, 1968. New York; 1-512; 1971.
BASSETT, D. A bibliography of the freshwater fishes of New Zealand. Nat. Libr. Serv., Biblio. Ser., 5, 1-46; 1961.
BLANCO, G. J. & MONTALBAN, H. R. A bibliography of Philippine fishes and fisheries. Philippine Journ. Fish., 1, 107-130.
BRIGGS, J. C. Illustrated works on fishes. Copeia, 1955, 243-245.
CABO, F. L. Bibliografia referente a los huevos y a las puestas de los peces. Comprendida entre los anos 1900 a 1924. Anal. Cienc. Nat. Madrid, 1941, 99-124; 1941.
DAWSON, C. E. A bibliography of anomalies of fishes. Gulf Res. Rep., 1, 308-399; 1964.
DEAN, B. A bibliography of fishes. New York; 3 volumes. (By author, by voyages, by expeditions, and by subject. Reprinted 1962)
DIAZ, E. L. Bibliographic material on the fishes of Colombia and northwestern South America. Fish. Tech. Pap. FAO, 53, 1-72; 1965.
DIEUZEIDE, R. Rapport sur les travaux récents d'ichthyologie mediterraneenne. Rapp. Comm. Int. Mer. Mediterraneenne, 14, 297-309; 1958; 15, 377-387; 1960.
EREMEEV, V. F. Bibliography of ganoid fishes (Chondrostei). Abstr. Works Inst. Moscou Univ., 3, 35-66; 1936.
GILL, T. Bibliography of the fishes of the Pacific Coast of the United States to the end of 1879. United States Nat. Mus. Bull., 11, 1-73; 1882.
GREGORY, H. E. Pacific fishes. Bull. Bernice P. Bishop Mus., 57, 24-25; 1928. (Oceanic)
JONES, S. & BENSAM, P. An annotated bibliography on the breeding habits and development of fishes of the Indian region. Fish. Res. Inst., 3, 1-154; 1968.
MacDONALD, R. M. E. An analytical subject bibliography of the publications of the Bureau of Fisheries 1871-1920. Washington, D. C.; 1-306; 1921.
MANSUETI, R. Partial bibliography of fish eggs, larvae, juveniles.... (Maryland); 1-55; 1954.
McDOWALL, R. M. A bibliography of the indigenous freshwater fishes of New Zealand. Trans. Roy. Soc. New Zealand, Zool., 5, 1-38; 1964.
McPHAIL, J. D. Annotated bibliography on arctic North American freshwater fishes. Inst. Fish. Univ. British Columbia, 6, 1-24; 1960.
MURTY, V. S. et al. Bibliography of marine fisheries and oceanography of the Indian Ocean. Bull. Cent. Mar. Fish. Res. Inst., 1, 1-1210; 1968.
NISSEN, C. Schöne Fischbücher. Kurze Geschichte der ichthyologischen Illustration. Bibliographie fischkundlicher Abbildungswerke. Stuttgart; 1-108; 1951.
OKADA, Y. & MATSUBARA, K. Bibliography of fishes in Japan (1612-1950). Mie; 1-228; 1953.
PHILLIPPS, W. J. Bibliography of New Zealand fishes. Bull. New Zealand Mar. Dep. Fish., 1, 1-68; 1927.
RASS, T. S. Biology of the Pacific Ocean. Book III. Fishes of the open waters. Moskva; 1-273; 1967. (Pp. 247-273; in Russian)
SOULIER, A. Preliminary bibliography of fish and fisheries for the Republic of Korea. Occ. Pap. Indo-Pacific Fish. Counc., 65, (6), 1-68; 1966.
WHITLEY, G. P. A survey of Australian ichthyology. Proc. Linn. Soc. New South Wales, 89, 11-127; 1964.
WHITLEY, G. P. & HALSTEAD, B. W. An annotated bibliography of the poisonous and venomous fishes of Australia. Rec. Australian Mus., 23, 211-217; 1955.

Bibliographies in other works

[See also: References in most large works]

BAUCHOT, M. La faune ichthyologique des eaux douces antillaires. Compt. Rent. Soc. Biogeogr., 36, 7-26; 1959.
BERG, L. S. Freshwater fishes of the U. S. S. R. and adjacent countries. Jerusalem; volume 3; 1965. (Pp. 351-442; translation of ed. 4, 1949)
BERTIN, L. & ARAMBOURG, C. Super-ordre des téléostéens (Teleosti). Traité de Zool., 13, (3), 2204-2500; 1958.
BLAIR, W. F. et al. Vertebrates of the United States, 2nd ed. New York; 1-616; 1968. (Pp. 158-165; 1st ed., 1957, pp. 206-210)
CERVIGON M., F. Las peces marinos de Venezuela. Caracas; 2 volumes; 1966. (Pp. 880-900)
CHU, Y. T. Bibliography of Chinese fishes. Bull. Peking Soc. Nat., 4, (4), 45-65; 1930. (Also published in Bull. Dep. Biol. Yenching Univ, 1, (4), 1-21; 1930)
CLEMENS, W. A. & WILBY, G. V. Fishes of the Pacific Coast of Canada, 2nd ed. Fish. Res. Board Canada, 68, 412-423; 1961. (1st ed., 1946)
EIGENMANN, C. H. Catalogue of the fresh-water fishes of the tropical and south temperate America. Rep. Princeton Univ. Exp. Patagonia, 1896-98, 3, Zool., 4, 375-511; 1910.
EIGENMANN, C. H. The freshwater fishes of British Guiana. Mem. Carnegie Mus., 5, 1-578; 1912. (Pp. 530-554) (South American fishes)
EIGENMANN, C. H. & ALLEN, W. R. Fishes of western South America. Lexington, Kentucky; 1-494; 1942. (Pp. 411-429)
GREENWOOD, P. H. et al. Phyletic studies of teleostean fishes. . . . Bull. American Mus. Nat. Hist., 131, (4), 436-444; 1966. (Families only)
HERRE, A. W. C. T. Addition to the fish fauna of Malaya. . . . Bull. Raffles Mus., 16, 27-61; 1940. (Pp. 56-61; Malayan fresh-water fishes)

HUBBS, C. A contribution to the classification of the blennioid fishes of the family Clinidae. . . . Stanford Ichth. Bull., 4, 42-165; 1952. (Pp. 156-165; superfamily Blenniicae)

LEIM, A. H. & SCOTT, W. B. Fishes of the Atlantic Coast of Canada. Fish. Res. Board Canada, 155, 436-459; 1966.

MENDIS, A. S. Fishes of Ceylon. Bull. Fish. Res. Sta. Ceylon, 2, 1-122; 1954.

OKADA, Y. Studies on the freshwater fishes of Japan. Tsu, Japan; 1-860; 1960. (Historical bibliography 3-230; general bibliography 763-846; also in: Journ. Fac. Fish. Univ. Mie, 4, 1-860)

RASS, T. S. Biology of the Pacific Ocean. Book III. Fishes of the open waters. Moskva; 1-273; 1967. (Pp. 247-273; in Russian; also in English translation)

SMITH, J. L. B. The sea fishes of southern Africa, 4th ed., Cape Town; 1-580; 1961. (Pp. 492-500a; previous editions 1949 and 1953)

WHITLEY, G. P. A survey of Australian ichthyology. Proc. Linn. Soc. New South Wales, 89, 11-127; 1964.

Personal bibliographies

[See under heading BIOGRAPHIES (PERSONS & INSTITUTIONS) below]

Institutional bibliographies

[See under ANIMALS, above]

Periodical lists

[See also under ANIMALIA]

(anonymous). World list of periodicals for aquatic sciences and fisheries. FAO Fish Biol. Tech. Pap., 19, vol. 1, 1-239; 1962. (Suppl. 1, 1963, 1-50; 2, 1964, 1-23)

ANDRIYASHEV, A. P. Fishes of the northern seas of the U. S. S. R. St. Petersburg; 1-566; 1954. (In Russian; Russian periodicals)

ANDRIYASHEV, A. P. Fishes of the northern seas of the U. S. S. R. Jerusalem; 1-617; 1964. (Translation of 1954; Russian serials pp. 612-617)

DEAN, B. A bibliography of fishes. New York; 3 vols. (Serials 348-353; reprinted 1962) 1916-1923

Indexes

LAGLER, K. F. & PUNPOKA, S. Index to the families and major groups in Weber and de Beaufort's Fishes of the Indo-Australian Archipelago, volumes i-xi, 1911-1962. Notes Fac. Fish. Kasetsart Univ., 1, 1-4; 1965.

REED, C. F. Index to Copeia. 1913-1954. Part I - Author index. Lancaster, Penna.; 1-106; 1955.

REED, C. F. Index to Copeia. 1913-1954. Part II - Subject index. Lancaster, Penna.; 1-332; 1956.

REED, C. F. Index to Copeia. Supplement. 1955-1964. Lancaster, Penna.; 1-329; 1965.

PERSONS AND INSTITUTIONS

Biographies & personal bibliographies

[See also: Zoological Record, in each volume, under heading: OBITUARIES)

BLEEKER, P. WEBER, M. & de Beaufort, L. F., The fishes of Indo-Australian Archipelago. I. Index of the ichthyological papers of P. Bleeker. Personal bibliography, 1-45; list of his species, 47-4410; 1911. (Reprinted 1964)

BOULENGER, G.-A. Liste des publications ichthyologiques et herpetologiques. 1877-1920 .... Ann. Soc. Roy. Zool. Malac. Belgique, 52, 11-88; 1921.

FOWLER, H. W. PHILLIPS, V. T. & Phillips, M. E., Writings of Henry Ward Fowler, published from 1897 to 1965. Proc. Acad. Nat. Sci. Philadelphia, 117, 173-212; 1965.

GRONOVIUS, L. T. Wheeler, A. C., The Gronovius fish collection. A catalogue and historical account. Bull. British Mus. (Nat. Hist.), Hist., 1, 187-249; 1958.

JORDAN, D. S. Hays, A. N., David Starr Jordan. A bibliography of his writings. 1871-1931. Stanford, California; 1-195; 1952.

MIRANDA RIBEIRO, A. Travassos, H., Bibliografia de Alipio de Miranda Ribeiro. Arqu. Mus. Nac. Rio de Janeiro, 41, xix-xxxvi; 1951.

MIRANDA RIBEIRO, P. Tipos das especies e subespecies de Prof. Alipio de Miranda Ribeiro depositados na Museu Nacional. (Com uma relacao dos genres, especies e subespecies descritos). Arqu. Mus. Nac. Rio de Janeiro, 42, 389-418; 1953.

MYERS, G. S. Annotated chronological bibliography of the publications of George Sprague Myers (to the end of 1969). Proc. California Acad. Sci., (4), 38, 19-52; 1970.

PELLEGRIN, J. 2e notice sur les titres et travaux scientifiques de M. Jacques Pellegrin. 1896-1936. Paris; 1-66; 1936.

REGAN, C. T. Burne, R. H. & Norman, J. R., Charles Tate Regan. Obit. Not. Fell. Roy. Soc., 4, 411-426; 1943.

RUSCONI. C. Lista de los generos y especies fundados por Carlos Rusconi. Rev. Mus. Hist. Nat. Mendoza, 9, 121-156; 1956.

SMITH, H. M. Schultz, L. P., Hugh McCormick Smith. Copeia, 1941, 194-209; 1941.
WHITLEY, G. P. Bibliography of published writings. . . . (Australia), 1-38; 1965.

Directories

[See also: Directories listed under ANIMALIA]

(anonymous). American Society of Ichthyologists and Herpetologists. List of members. San Francisco; 1-42; 1967.
(anonymous). Aquatic biologists of the world and their laboratories. Washington, D. C.; 2 parts; 1964.

Institutions

[See also under ANIMALIA, above]

(anonymous). Aquatic biologists of the world and their laboratories. Washington, D. C.; 2 parts; 1964.

Museum lists

North America

American Museum of Natural History

NICHOLS, J. T., Chinese freshwater fishes in the . . . collections. Bull. American Mus. Nat. Hist., 58, 1-62; 1928.

Carnegie Museum

FISHER, H. G., A list of the Hypophthalmidae, Diplomystidae and . . . Siluridae in the collection of the . . . Carnegie Museum. Ann. Carnegie Mus., 11, 405-427; 1917.

United States National Museum

BEAN, T. H., A list of European fishes in the collection . . . . Proc. United States. Nat. Mus., 2, 10-44; 1879.

South America

Museo del Instituto Oceanografico de la Universidad de Oriente

CARVAHAL, R. J., Catalogo de peces del Museo . . . . Primera entrega. Lagena, 6, 2-107; 1965.

Museu Nacional (Rio de Janeiro)

de MIRANDA RIBEIRO, P., Catalogo dos peixes do Museu . . ., parts I-XI. Publ. Avuls. Mus. Nac. Rio de Janeiro, 22, 1-10; 1954-1962.

Estacion de Biologia Marina de Montemar, Chile

de BUEN, F., Lampreas, tiburones, rayas y peces en la Estacion . . . . Rev. Biol. Mar. 9, 3-200; 1959.

Museo Nacional de Chile

QUIJADA, R., Catalogo de la collection de los peces Chilenos: Estrangeros del Museo . . . . Bol. Mus. Nac. Santiago, 4, 69-109; 1912.

Europe

British Museum (Natural History)

1850-1895, various lists and descriptive catalogue of fishes in the museum.
BOULENGER, G. A., Catalogue of the fresh-water fishes of Africa . . . . London, 4 columns; 1909-1916.
WHEELER, A. C., The Gronovius fish collection: A catalogue and historical account. Bull. British Mus. (Nat. Hist.), Hist., 1, 187-249; 1958.

Rijksmuseum van Natuurlijke Historie

POPTA, C. M. L., Lijst van de Nederlandsche visschen aanwesig in 's Rijks Museum van Natuurlijke Historie te Leiden. Zool. Meded., 8, 77-119; 1924.

[Museums in Leiden and Amsterdam]

Van der STIGCHEL, J. W. B., The South American Nematognathi of the museums . . . . Zool. Meded., 27, 1-204; 1947.

Muséum d'Histoire Naturelle, Paris

BERTIN, L., Revision des Stomiatiformes ... du Muséum. Bull. Mus. Hist. Nat. Paris, (2), 11, 378-382; 1939.

CHAUX, J. & FANG, R. W., Catalogue des Siluroidei d'Indochine de la collection du Laboratoire des Peches Coloniales au museum . . . . Bull. Mus. Nat. Hist. Nat. Paris, (2), 21, 194-201, 342-346; 1949.

Muséum d'Histoire Naturelle de Marseille

FABRE, F., Catalogue de la collection des poissons du Museum . . . . Bull. Mus. Hist. Nat. Marseille, 26, 85-115; 1966.

Musée Océanographique (de Monaco)

ROULE, L., Notice sur les Selaciens conserves dans des collections du Musée . . . . Res. Camp. Sci. Monaco, 103, 174-200; 1940.

Museu de Zoologia da Universidade de Coimbra

HELLING, H., Novo catalogo dos peixes de Portugal em coleciao no Museu . . . . Mem. Mus. Zool. Univ. Coimbra, (1), 115, 1-18; 149, 1-110; 1940-1943.

Museo ... de Historia Natural del Instituto de Valencia

PARDO, L., Las collecciones del peces del Museo . . . . Anal. Inst. Gen. Tec. Valencia, Trab. Lab. Hist. Nat., 9, 1-126; 1921.

Inst. Forest. Invest., Minist. Agric., Madrid

(anonymous), Las colecciones de peces de la section de biologia de las aguas continentales. Madrid; 1-136; 1952.

Museu Municipal do Funchal

MAUL, G. E., Monografia dos peixes do Museu . . . . Bol. Mus. Funchal, 2-14, 1946-1961.

Regia Museo Zoologico di Piza

BORRI, C., Catalogo delle collezioni di Vertebrati del R. Museo . . . , II. Squali. Mem. Soc. Toscana Sci. Nat., 44, 88-103; 1935.

Ungarischen Naturwissenschaftlichen Museums

MIHALYI, F., Revision der Süsswasserfische von Ungarn . . . . Anal. Hist. Nat. Mus. Hungarici, (n. s.), 5, 433-456; 1954.

Africa

Institut Francaise d'Afrique Noire

CADENAT, J., Notes d'ichthyologie ouest-africaine. XXXIV. Liste complementaire des especes de poissons de mer (provenant des côtes de l'Afrique occidentale) en collectionaa la section de biologie marine . . . . Bull. Inst. Francaise Afrique Noire, 23A, 231-245; 1961.

Centre d'Oceanographie de l'Institut d' Etudes Centralafricaines de Pointe Noire

COLLIGNON, J. et al., Mollusques, Crustaces, poissons marins des cotes d'A. E. F. en collection du Centre . . . . Paris; 1-369; 1957.

The Orient

Colombo Museum

de SILVA, P. H. D. H., The flying fishes . . . off Ceylon, with a list of specimens in the Colombo Museum, 4 parts. Spolia Zeylanica, 28, 27-54; 1956; and Ceylon Journ. Sci., 7C, 183-200; 1956.

Taiwan Fisheries Research Institute

LIANG, Y., A check-list of fish specimens in the . . . Institute. Rep. Taiwan Fish. Res. Inst., 3, 1-35; 1951.

Museum type lists

[See also this heading under ANIMALIA: see also: Zoological Record, each volume, under heading Institutions & Collections]

North America

National Museum of Canada

McALLISTER, D. E., Type specimens of fishes in the National Museum . . . . Nat. Hist. Pap. Nat. Mus. Canada, 31, 1-13; 1965.

Museum of Comparative Zoology, Harvard University

RIVERO, L. H. y., List of the fishes, types of Poey, in the Museum . . . . Bull. Mus. Comp. Zool., 82, 169-227; 1938.

New York Zoological Society

MEAD, G. W., A catalog of the type specimens of fishes formerly in the collections of the Department of Tropical Research . . . . Zoologica, 43, 131-134; 1958.

Carnegie Museum

HENN, A. W., List of types of Recent fishes in the collection of the Carnegie Museum on September 1, 1928. Ann. Carnegie Museum, 19, 47-99; 1928.

Chicago Natural History Museum [Field Museum of Natural History]

GREY, M., Catalogue of type specimens of fishes in Chicago Natural History Museum. Fieldiana, Zool., 32, 109-205; 1947.

Illinois State Laboratory of Natural History

SMITH, P. W. & BRIDGES, D. W., Ichthyological type specimens extant from the old Illinois State Laboratory of Natural History. Copeia, 1960, 253-254; 1960.

Stanford University Natural History Museum

BOHLKE, J., A catalogue of the type specimens of recent fishes in the . . . Museum. . . . Stanford Ichth. Bull., 5, 1-168; 1953.

South America

Museu Nacional (Rio de Janeiro)

de MIRANDA RIBEIRO, P., Tipos das especies e subespecies de Prof. Alipio de Miranda Ribeiro depositados no Museu Nacional. Arqu. Mus. Nac. Rio de Janeiro, 42, 389-418; 1953.

Europe

Hamburgischen Zoologischen Staatsinstituts und Zoologischen Museums

LADIGES, W. et al., Die Typen und Typoide der Fischsammlung des ... Museums. Mitt. Hamburgischen Zool. Mus., 56, 155-167; 1958.

Ubersee-Museums (Bremen)

von WAHLERT, G., Die Typen und Typoide des Ubersee-Museums Bremen, 2: Pisces. Veröff. Uberseemus. Bremen, 2A, 323-326; 1955.

Naturmuseums Senckenberg

KAUSEWITZ, W., Die Typen und Typoide des Naturmuseums Senckenberg. 23. . . . . Elasmobranchii. Senckenbergische Biol., 41, 189-196; 1960.

Staatlichen Museums für Tierkunde zu Dresden

SCHUZ, E., Verzeichnis der Typen des ... Museum .... I. Fische, Amphibien und Reptilien. Abh. Mus. Dresden, 17, (2), 1-16; 1929.

Zoological Museum, University of Amsterdam

HOEDEMAN, J. J., A list of type specimens of fishes in the ... Museum .... I. Order Mugiliformes. Beaufortia, 7, (87), 211-217; 1960.
MATTHES, H., A list of types of African freshwater fishes in the ... Museum .... Beaufortia, 10, 177-182; 1964.

Museum National d'Histoire Naturelle (Paris)

BAUCHOT, M. L. & BLANC, M., Catalogue critique des types de poissons du Museum. (Continuing series in: Publ. Mus. Hist. Nat., vol. 20 and Bull. Mus. Nat. Hist. Nat., vol. 23 etc.; 1961— (Various authors)
COLLETTE, B. B., Revue critique des types de Scombridae des collections . . . . Bull. Mus. Nat. Hist. Nat., (2), 38, 362-375; 1966.

Museum Oceanographique de Monaco

BELLOC, G., Catalogue des types de poissons du Museum . . . . Bull. Inst. Oceanogr. Monaco, 958, 1-23; 1949.

Museo Civico di Storia Naturale di Genova

TORTONESE, E., Catalogo dei tipi di pesci del Museo . . . , I-III. Ann. Mus. Civ. Stor. Nat. Genova, 72, 179-191, 1961; 73, 306-316, 333-350, 1963.

Museo di Zoologica ed Anatomia Comparata della Regia Universita degli Studi di Torino

TORTONESE, E., Elenco dei tipi esistenti nella collezione ittiologica del R. Museo . . . . Boll. Mus. Zool. Anat. Comp. Torino, 48, 133-144; 1946.

Australia

Australian Museum (Sydney)

WHITLEY, G. P., List of type-specimens of Recent fishes in the ... Museum, Sydney. Sydney; 1-40; 1957.

Western Australian Museum

(anonymous), Type specimens in the Western Australian Museum. Rep. Western Australian Museum, 1963-64, 35-37, 1964; 1964-65, 46-48, 1965; etc.

EXPLORATIONS AND LOCALITIES

[See also this heading under ANIMALIA]

DEAN, B. A bibliography of fishes. New York; 3 volumes. (Pp. 343-347)

METHODS

[See also: Zoological Record, each volume, under heading Techniques; also this heading under ANIMALIA, above]

(anonymous). Instructions for collectors. No. 3. Reptiles, amphibians, and fishes, 6th ed. London; 1-28; 1953. (5th edition, 1936)
BEAN, T. H. Directions for collecting and preserving fish. Proc. United States Nat. Mus., 4, 235-238; 1882.
GINSBURG, I. Contribution to a methodology in the caudal fin-ray count of fishes and its use in classification. Copeia, 1945, 133-142; 1945.
McCULLY, H. A new type of field key applied to the flatfishes of California. California Fish Game, 35, 11-13; 1949.
MYERS, G. S. Brief directions for preserving and shipping specimens of fishes, amphibians and reptiles, 2nd ed. Circ. Nat. Hist. Mus. Stanford Univ., 5; 1956.
MYERS, G. S. & STOREY, M. H. Curatorial practices in zoological research collections. 2. System followed in filing specimens of Recent fishes in the Natural History Museum of Stanford University. Circ. Nat. Hist. Mus. Stanford Univ., 6, 1-44; 1956.
NORRIS, K. S. et al. A survey of fish transportation methods and equipment. California Fish Game, 46, 5-33; 1960.
PARR, A. E. An approximate formula for stating taxonomically significant proportions of fishes with reference to growth changes. Copeia, 1949, 47-55; 1949.
RANDALL, J. E. Methods of collecting small fishes. Underw. Nat., 1, (2), 6-11, 32-36; 1963.
RICKER, W. E. & MERRIMAN, D. On the methods of measuring fish. Copeia, 1945, 184-191; 1945.
SCHULTZ, L. P. Directions for collecting, pressing, and shipping fishes. Atoll Res. Bull., 17, 90-95; 1953.
SMITH, J. L. B. The collecting and preservation of fishes. Bull. South African Mus. Assoc., 9, 202-206; 1968.
TCHERNAVIN, V. V. A revision of the subfamily Orestiinae. Proc. Zool. Soc. London, 114, 140-233; 1944. (Measuring and recording taxonomic characters)
WHITLEY, G. P. Ichthyological illustrations. Proc. Roy. Zool. Soc. New South Wales, 1955-1956, 56-711 1957.

GLOSSARIES

[See also many of the general works listed above, as well as Glossaries under ANIMALIA]

BOHLKE, J. E. & CHAPLIN, C. C. G. Fishes of the Bahamas . . . . Wynnewood, Pennsylvania; 1-771; 1968. (Pp. 731-738)
CLEMENS, W. A. & WILBY, G. V. Fishes of the Pacific Coast of Canada, 2nd ed. Bull. Fish. Res. Board Canada, 68, 1-443; 1961. (Pp. 404-411)
FREY, H. Illustrated dictionary of tropical fishes. Jersey City, New Jersey; 1-768; 1961.
HARRINGTON, R. W. The osteocranium of the American cyprinoid fish . . . , with annotated synonymy of teleost skull bones. Copeia, 1955, 267-290; 1955.
KOBAYASI, H. Revision of the technical terms of fish lepidology. Japan Journ. Ichth., 1, 175-181; 1950.
LEIM, A. H. & SCOTT, W. B. Fishes of the Atlantic Coast of Canada. Bull. Fish. Res. Board Canada, 155, 1-485; 1966. (Pp. 430-436)
SANCHEZ ROIG, M. & GOMEZ DE LA MAZA, F. La pesca en Cuba. Habana; 1-272; 1952.
THOMPSON, D. W. A glossary of Greek fishes. London; 1-302; 1947.

NOMENCLATURAL STUDIES

Common names

DEAN, B. A bibliography of fishes. New York; 3 volumes; 1916-1923. (Bibliography of sources)

North America

(anonymous). A list of common and scientific names of fishes from the United States and Canada, 2nd ed. American Fish. Soc. Spec. Publ. 2, 1-102; 1960.
HALKETT, A. Checklist of the fishes of the Dominion of Canada and Newfoundland. Ottawa; 1-138; 1913.
LEGENDRE, V. et al. French and English names of the Canadian Atlantic fishes. Rapp. Serv. Faune Quebec, 2, 1-178; 1964.
LAGLER, K. F. Freshwater fishery biology. Dubuque, Iowa; 1-360; 1952.
RIVAS, L. R. Checklist of Florida game and commercial fishes . . . , with approved common names. Board Conserv., Educ. Ser., 4, 1-39; 1949. (Gulf and West Indies)
(anonymous). Common and scientific names of fishes . . . . Fish Bull. Sacramento, 58, 45-47;1941; 89, 21-23; 1953. (California fishes)
ROEDEL, P. M. Official common names of certain marine fishes of California. California Fish Game, 39, 251-262; 1953.

West Indies

PALMA, F. Nombres de peces cubanos y floridanos. Habana; 1-78; 1941.
ZANEVELD, J. S. Index to vernacular names of fishes of Netherlands Indies. Caribbean Mar. Biol. Inst. Publ., 16, 1-24; 1964.

South America

MARIA, H. A. Vocabulario de terminos vulgares a historia natural Colombiano. Rev. Acad. Colombiana Cienc., 4, 326-336; 1941; 5, 40-60, 149-170, 1942; 295-307, 1943.
MILES, C. Los peces del Rio Magdalena. Bogota; 1-214; 1947. (Pp. xiv-xv)
ORCES, G. Nombres vulgares y su equivalenta cientifico de peces marinos de las costas del Ecuador. Cienc. y Med., 2, 15-19; 1959.
SANCHEZ R., J. F. Peces comunes de la costa Peruana. Serie Divulg. Cient. Peru, 6, 1-121; 1966.
PUYO, J. Poissons de la Guyane francaise. 1. Faune de l'Empire francaise, 12, 1-280; 1949.
de MENEZES, R. S. Lista dos nomes vulgares de peixes de aguas doces . . . Brasil. Arqu. Mus. Nac., 42, 343-388; 1953.
CARVALHO, J. P. C. Nota preliminar sobre a fauna ictiologica do litoral sul do Estado de Sao Paulo. Bol. Industr. Animal Sao Paulo, (n. s.), 4, (3-4), 27-81; 1941.
FOWLER, H. W. A list of the fishes known from the coast of Brazil. Arqu. Zool., 3, 115-184; 1942.
DEVINCENZI, G. J. & TEAGUE, G. W. Ictiofauna del Rio Uruguay medio. Anal. Mus. Montevideo, (2), 5, (4), 1-103; 1942.

Atlantic Ocean

METZELAAR, J. Over tropisch atlantische Visschen. Amsterdam; 1-314; 1919. (Pp. 174-179)

Europe

(anonymous). Names of fishes in the different languages together with their scientific names. Bull. Statist. Pech. Marit. Copenhagen, 37, 36-41; 1954. (Previous edition 1951/1952)
FARRAN, G. P. Local Irish names of fishes. Irish Nat. Journ., 8, 1-30, 344-347, 370-376, 420-433; 1946.
DANJALSSON á Ryggi, M. List of Faroese animal names. In Jenson, Zoology of the Faroes, 3, (66), 1-10; 1942.
NANCE, R. M. Celtic names of fish in Cornwall. Journ. Roy. Inst. Cornwall, (n. s.), 2, 74-87; 1954.
(anonymous). [Title not known]. Svensk. Fisk. Tidskr., 64- 12 0; 1955. (Swedish, Norwegian, Danish, Finnish)
(anonymous). Scandinavian fishing year-book. Copenhagen; 1-70; 1959.
BROFELDT, P. De finska fisknammen. Svensk Fisk. Tidskr., 64, 157-158; 1955.
DUNCKER, G. Die Fische der Nordmark. Abh. Naturw. Ver. Hamburg, (n. f.), 3, Suppl. 1-432; 1960.
LOZANO, F. Nomenclatura ictiologica. Nombres cientificos y vulgares de los peces espanoles. Trab. Inst. Esp. Oceanogr., 31, 1-270; 1963.
MAUL, G. E. Monografia dos peixes do Museu Municipal do Funchal. Ordem Isospondyli. Bol. Mus. Munic. Funchal, 3, (5), 5-41; 1948.
OSORIO de CASTRO, J. Glossario de nomes dos peixes. Lisboa; 1-249; 1954.
BRANDAO, J. M. Glossario de nomes dos pexes, portugues ingles, sistematico. Bol. Est. Pesca, 4, (4), 1-40; (5), 1-59; (6), 1-59; 1964
PALOMBI, A. & SANTARELLI, M. Gli animali commestibili dei mari d'Italia. Descrizione e nomi italiani, dilettali e straniere dei pesci .... Milano; 1-349; 1953.
BINI, G. Catalogue of names of fishes . . . . Rome; 1-407; 1965. (Mediterranean fishes)
(anonymous). Catalogue of the names of fishes of commercial importance in the Mediterranean. Rome; 1960. (In all Mediterranean languages)

STOYANOV, S. et al. The fishes of the Black Sea. Nauk Izst. Inst. Rib. Varna, 1-246; 1963. (In Russian)
ANDRIYASHEV. A. P. Fishes of the northern seas of the U. S. S. R. St. Petersbourg; 1-566; 1954. "
ANDRIYASHEV, A. P. Fishes of the northern seas of the U. S. S. R. Jerusalem; 1-617; 1964. (Translation of 1954; pp. 605-611)
BERG, L. S. Freshwater fishes of the U. S. S. R., vol. III, ed. 4. Jerusalem; 1-510; 1965. (Translation of 1949; pp. 485-510)

Africa

(anonymous). Tableaux statistiques des peches maritimes en Tunisie. Bull. Sta. Oceanogr. Salammbo, 31, 35-38; 1933-1937.
GIRGIS, S. A list of common fish of the upper Nile with their thilluk Dinka and Nuer names. Sudan Notes, 29, 120-125; 1948.
CADENAT, J. Noms vernaculaires des principales formes d'animaux marins des cotes de l'Afrique occidentale francaise. Cat. Inst. Francaise Afrique Noire, 2; 1947.
IRVINE, F. R. Fishes and fisheries of the Gold Coast. London; 1-352; 1947.
CADENAT, K. & PARAISO, F. Noms vernaculaires des principales formes d'animaux marins et de lagunes du Togo et du Dahomey. Notes Africaines, 49, 24-29; 1951.
DAGET, J. Noms vernaculaires de poissons du Moyen Niger. Cybium, 4, 68-72 .
WELMAN, J. B. Preliminary survey of the freshwater fisheries of Nigeria. Lagos; 1948.
BLACHE, J. Les poissons du bassin du Tchad et du bassin adjacent du Mayo Kebbi. Paris;1-483; 1964.
SMITH, J. L. B. The sea fishes of southern Africa. Cape Town; 1-580; 1961. (Pp. 560-564)
BARNARD, K. H. A pictorial guide to South African fishes marine and freshwater. Cape Town; 1-226; 1947. (Pp. 219-222)

Indo-Malayan

DERANIYAGALA, P. E. P. Names of some fishes from Ceylon. Bull. Ceylon Fish., 5, 79-111; 1933.
SETNA, S. B. & SARANGDHAR, P. N. Selachian fauna of the Bombay waters. Proc. Nat. Inst. Sci. India, 12, 243-259; 1946.
KULKARNI, C. V. Local and scientific names of commercial fishes of Bombay. Journ. Bombay Nat. Hist. Soc., 51, 917-925; 1953.
ROFEN, R. R. Handbook of the food fishes of the Gulf of Thailand. La Jolla, California; 1-235; 1963.
THIEMMEAH, J. Fishes of Thailand: Their English, scientific and Thai names. Fish. Res. Bull. Kasetsart Univ., 4, 1-212; 1966.
TWEEDIE, M. W. F. Malay names of fresh-water fishes. Journ. Malay British Asiatic Soc., 25, (1), 62-67; 1952.
TWEEDIE, M. W. F. Notes on Malayan fresh-water fisher. No. 3, The anabantoid fishes. No. 5, Malay names. Bull. Raffles Mus., 24, 63-95; 1952 .
BURDON, T. W. & GIBSON-HILL, C. A. Malay names of salt-water fish. Journ. Malay British Asiatic Soc., 27, (2), 175-180; 1954.
INGER, R. F. & KONG, C. P. The fresh-water fishes of North Borneo. Fieldiana, Zool., 45, 1-268; 1962.

Australasia

WHITLEY, G. The common names of fishes. Australian Mus. Mag., 10, 310-315; 1952.

The Orient

HERRE, A. W. & UMAL, A. F. English and local common names of Philippine fishes. Circ. United States Fish Wildl. Serv., 14, 1-128; 1948.
DASHDORZH, A. et al. Dictionary of names of economically important fishes in the western part of the Pacific Ocean. Peking; 1-509; 1964. (Latin, Russian, Chinese, Korean, Vietnamese, Mongolian, Japanese, English)
TAKU, J. & KOBAYASHI, K. Local names of fishes of the Ryukyu Islands. Bull. Fac. Fish. Hokkaido, 13, 21-35; 1962.
OKADA, Y. Fishes of Japan. Tokyo; 1-458, 1-16; 1966. (Pp. 1-16)
ASANO, N. Vernacular names of fishes in Ibaragi Prefecture. Japan Journ. Ichth., 5, 19-51; 1956.
HIKITA, T. The common names of salmonoid fishes and their related forms found in northern Japan and its adjacent waters. Sci. Rep. Hokkaido Fish Hatch., 9, 137-145; 1954.

Oceania

JUNE, F. C. & REINTJES, J. W. Common tuna-bait fishes of the central Pacific. Res. Rep. United States Fish. Serv., 34, 1-54; 1953. (In Hawaiian and Gilbertese)
TINKER, S. W. Hawaiian fishes. Honolulu; 1-404; 1944.
RANDALL. J. E. Fishes of the Gilbert Islands. Atoll Res. Bull., 47, 1-243; 1955.
DANIELSSON, B. Check list of the native names of fishes for Roraia Atoll. Atoll Res. Bull., 31, 102-109; 1954.
JORDAN, D. S. & SEALE, A. The fishes of Samoa. . . . Bull. United States Bur. Fish., 25, 173-455; 1906. (Pp. 446-455)

Nomenclators

[See also: Nomenclators listed under ANIMALIA]

BERG, L. S. Classification of fishes both Recent and fossil. Trav. Inst. Zool. Acad. Sci. U. R. S. S., 5, (2), 501-517; 1940. (In Russian; families and higher taxa; English edition, 1947)
BERG, L. S. Classification of fishes both Recent and fossil. Ann Arbor, Michigan;1-517; 1947. (Taxa above genus; in Russian and English; reprinted in Thailand, 1965)
GOLVAN, Y. -J. Catalogue systematique des noms de genres de poissons actuels de la Xe edition du "Systema Naturae" de Charles Linné jusqu'a la fin de l'annee 1959. Ann. Parasitol. Hum. Comp., 37, (6bis), 1-227; 1962. (Genera, alphabetically by families, and index)
GREENWOOD, P. H. et al. Phyletic studies of teleostean fishes, with a provisional classification of living forms. Bull. American Mus. Nat. Hist., 131, (4), 339-459; 1966. (Index to families, 445-455)
JORDAN, D. S. A classification of fishes including families and genera as far as known. Stanford Univ. Publ. Univ. Ser., Biol. Sci., 3, (2), 77-243; 1923. (Families and above, iv-x; reprinted 1963)
NORMAN, J. R. A draft synopsis of the orders, families and genera of Recent fishes. . . . London, 1-649; 1966. (Index to genera, pp. 607-649) (Also previously, 1957)
SMITH, H. M. & SCHULTZ, L. P. Index [to reprint of Jordan 1963, above]. (Genera, pp. 745-794; higher taxa, pp. 795-800)

Name studies

FOLLETT, W. I. & DEMPSTER, L. J. Ichthyological and herpetological names and works on which the International Commission on Zoological Nomenclature has ruled since the publication (in 1958) of the several Official Lists and Indexes. Copeia, 1965, 518-523; 1965.
GERD, A. S. Arrangement and unification of the Russian names of fishes. Vop. Ikhtiol., 8, 350-356; 1968. (In Russian; translated in Probl. Ichth., 8, 276-281)
MACLEOD, R. D. Key to the names of British fishes, mammals, amphibians and reptiles. London; 1-71; 1956. (Meaning and origin of names)
MYERS, G. S. Nomenclator of certain terms used for higher categories of fishes. Stanford Ichth. Bull., 7, 31-42; 1958. (Historical nomenclator of names above family level, for some fishes)
STENZEL, H. B. Proposed uniform endings for names of higher categories in zoological systematics. Science, 112, p. 94; 1950.
TRAVASSOS, H. Catalogo dos generos e subgenerso da suborden Characoidei . . . . Dusenia, 2, 205-224, 273-292, 341, 360, 419-434; 3, 141-180, 225-250, 313-328; 1951, 1952. (Generic)

Types of genera

JORDAN, D. S. The genera of fishes (1758-1920), 4 parts. Stanford, California; 1-576; 1917-1920. (Types of all genera cited; reprinted 1963)

FAUNAL STUDIES

Faunas, checklists, identification works

General

ANGEL, F. Petit atlas des poissons. Fasc. 4. Paris; 1-129; 1946. (Freshwater and exotique)
KENNEDY, M. The sea angler's fishes. London; 1-524; 1954. (Keys)
NORMAN, J. R. A draft synopsis of the orders, families and genera of Recent fishes. . . . London; 1-649; 1957. (Keys to genera)
NORMAN, J. R. & FRASER, F. C. Field book of giant fishes. New York;1-376; 1949. (Keys)
PERLMUTTER, A. Guide to marine fishes. New York; 1-431; 1961. (Keys)
STERBA, G. Süsswasserfische aus aller Welt. Berchtesgaden; 1-638; 1959. (Keys)
STERBA, G. Freshwater fishes of the world. New York; 1-877; 1963. (translation of 1959; also published in London, 1962; descriptive handbook, keys)

North America

JORDAN, D. S. & GILBERT, C. H. Synopsis of the fishes of North America. United States Nat. Mus. Bull., 16, 1-1018; 1882. (Faunal monograph)
GILL, T. Synopsis of the pediculate fishes of the east coast of extratropical North America. Proc. United States Nat. Mus., 1, 215-221; 1878. (Monograph)
BREDER, C. M., Jr. Field book of marine fishes of the Atlantic Coast. New York; 1-332; 1929. (Keys)
SCHRENKEISEN, R. Fieldbook of freshwater fishes of North America north of Mexico. New York; 1938. (Keys)
BEAN, T. H. Preliminary catalogue of the fishes of Alaskan and adjacent waters. Proc. United States Nat. Mus., 4, 129-272; 1882.
EVERMANN, B. W. & GOLDSBOROUGH, E. L. The fishes of Alaska. Bull. United States Bur. Fish., 26, 219-360; 1907.

WILIMOVSKY, N. J. List of the fishes of Alaska. Stanford Ichth. Bull., 4, 279-294; 1954.
WILIMOVSKY, N. J. Provisional keys to the fishes of Alaska. Juneau; 1-113; 1958.
McALLISTER, D. E. Keys to the marine fishes of Arctic Canada. Nat. Hist. Pap. Nat. Mus. Canada, 5, 1-21; 1960.
SCOTT, W. B. A checklist of the freshwater fishes of Canada and Alaska. Toronto;1-30; 1958.
HALKETT, A. Check list of the fishes of the Dominion of Canada and Newfoundland. Ottawa; 1-138; 1913.
McALLISTER, D. E. List of the marine fishes of Canada. Bull. Nat. Mus. Canada, 168, 1-76; 1960.
SLASTENENKO, E. P. Freshwater fishes of Canada. Toronto; 1-388; 1958.
DYMOND, J. R. A list of the freshwater fishes of Canada east of the Rocky Mountains, with keys. Roy. Ontario Mus. Zool., Misc. Publ., 1, 1-36; 1947. (List & keys)
SCOTT, W. B. Freshwater fishes of eastern Canada. Toronto; 1-137; 1967. (1st edition, 1954)
LEIM, A. H. & SCOTT, W. B. Fishes of the Atlantic Coast of Canada. Bull. Fish. Res. Board Canada, 155, 1-485; 1966. (Fauna with keys)
SCOTT, W. B. & SCOTT, M. G. Checklist for Canadian Atlantic fishes with keys for identification. Contr. Roy. Ontario Mus. Zool., 66, 1-106; 1965. (List & keys)
CLEMENS, W. A. & WILBY, G. V. Fishes of the Pacific Coast of Canada, 2nd ed. Bull. Fish. Res. Board Canada, 68, 1-443; 1961. (Keys; 1st edition 1946)
BACKUS, R. H. The fishes of Labrador. Bull. American Mus. Nat. Hist., 113, 273-332; 1957.
LIVINGSTONE, D. A. The fresh water fishes of Nova Scotia.. Proc. Nova Scotia Inst. Sci., 23, 1-90; 1953. (Keys)
SCOTT, W. B. & CROSSMAN, E. J. The freshwater fishes of New Brunswick: a checklist. Contr. Roy. Ontario Mus. Zool., 51, 3-37; 1959.
STENGER, J. Key to the eggs of the known species of Passamaquoddy Bay, New Brunswick. Ontario Field. Biol., 10, 22-25;1956.
BERGERON, J. Liste des poissons marins de l'estuarie et du Golfe St. Laurent. Contr. Dept. Pech. Quebec, 80, 1-27; 1960.
LEGENDRE, V. The freshwater fishes. Vol. 1. Key to game and commercial fishes of the Province of Quebec. Montreal; 1-180; 1954.
VLADYKOV, V. D. Preliminary list of marine fishes of Quebec. Nat. Canadienne, 88, 53-78; 1961.
CARL, G. C. & CLEMENS, W. A. The freshwater fishes of British Columbia. British Columbia Prov. Mus. Handb., 5, 1-132; 1948. (Keys)
CARL, G. C. et al. The fresh-water fishes of British Columbia. Handb. British Columbia Prov. Mus., 5, 7-192; 1959. (Keys)
CARL, G. C. Guide to marine life of British Columbia. Handb. British Columbia Prov. Mus., 21, 1-135; 1966.
BLAIR, W. F. et al. Vertebrates of the United States, 2nd ed. New York; 1-616; 1968. (Keys 21-165; 1st edition, 1957; to species)
EDDY, S. How to know the freshwater fishes, 2nd ed. Dubuque, Iowa;1-286; 1969.
BIGELOW, H. B. & SCHROEDER, W. C. Fishes of the Gulf of Maine. Fish. Bull. United States, 53, 1-577; 1953.
KENDALL, W. C. An annotated catalogue of the fishes of Maine. Portland Soc. Nat. Hist., 3, 1-198; 1914.
WHITWORTH, W. R. et al. Freshwater fishes of Connecticut. Bull. State Geol. Nat. Hist. Surv. Connecticut, 101, 1-134; 1968. (Keys)
RICHARDS, S. W. Pelagic fish eggs and larvae of Long Island Sound. Bull. Bingham Oceanogr. Coll., 17, (1), 95-124; 1959. (Keys)
FOWLER, H. W. A list of the fishes of New Jersey, with off-shore species. Proc. Acad. Nat. Sci. Philadelphia, 104, 89-151; 1952.
FOWLER, H. W. A study of the fishes of the southern piedmont and coastal plain. Mon. Acad. Nat. Sci. Philadelphia, 7, 1-408; 1945.
HILDEBRAND, S. F. & SCHROEDER, W. C. Fishes of Chesapeake Bay. Bull. United States Bur. Fish., 43, (1), 1-366; 1928.
MANSUETI, R. A partial bibliography of fish eggs, larvae and juveniles and a checklist of fishes of Maryland waters. Maryland Dep. Res. Educ., Chesapeake Biol. Lab., 1-55; 1954.
MANSUETI, A. J. & HARDY, J. D., Jr. Development of fishes of the Chesapeake Bay region; an atlas of egg, larval, and juvenile stages. College Park, Maryland; 1967.
SMITH-VANIZ, W. F. Freshwater fishes of Alabama. Auburn; 1-211; 1968. (Keys)
CARR, A. & Goin, C. J. Guide to the reptiles, amphibians and fresh-water fishes of Florida. Gainesville, Florida; 1-341; 1959.
TRAUTMANN, M. B. The fishes of Ohio with illustrated keys. Columbus, Ohio; 1-683; 1957.
GERKING, S. D. Key to the fishes of Indiana. Invest. Indiana Lakes, 4, 49-86; 1955.
NELSON, J. S. & GERKING, S. D. Annotated key to the fishes of Indiana. Bloomington, Indiana; 1-84; 1968.
FORBES, S. A. & RICHARDSON, R. E. The fishes of Illinois. Urbana, Illinois;1-357; 1908.
HUBBS, C. L. & LAGLER, K. List of fishes of the Great Lakes and tributary waters. Michigan Fish., 1, 1-6; 1957.
EDDY, S. & SURBER, T. Northern fishes. with special reference to the upper Mississippi Valley, 2nd ed. Minneapolis; 1-276; 1947. (Previous edition 1943)
BAILEY, R. M. A revised list of the fishes of Iowa with keys for identification. Des Moines, Iowa; 327-377; 1956. (Keys)
CROSS, F. B. Handbook of fishes of Kansas. Misc. Publ. Univ. Kansas Mus. Nat. Hist., 45, 1-357; 1967.
HOESE, H. D. A partially annotated checklist of the marine fishes of Texas. Publ. Inst. Mar. Sci. Univ. Texas, 5, 312-352; 1958.
HUBBS, C. A check-list of Texas fresh-water fishes. Texas Game Fish Comm., (1F), 3, 1-14; 1957.

JURGENS, K. C. & HUBBS, C. A checklist of Texas freshwater fishes. Texas Game Fish., 11, (4), 12-15; 1953.
BECKMAN, W. C. Guide to the fishes of Colorado. Leafl. Univ. Colorado Mus., 11, 1-110; 1968.
LaRIVERS, I. A key to Nevada fishes. Bull. Southern California Acad. Sci., 51, 86-102; 1952.
LaRIVERS, I. Fishes and fisheries of Nevada. Nevada State Fish Game Comm., 1-782; 1962. (Keys)
WINN, H. E. & MILLER, R. R. Native postlarval fishes of the lower Colorado River basin, with a key to their identification. California Fish Game, 40, 273-285; 1954.
SIGLER, W. F. & MILLER, R. R. Fishes of Utah. Salt Lake City; 1-203; 1963.
BOLIN, R. L. IN Light et al., Intertidal invertebrates of the central California coast. Berkeley; 1-446; 1954. (Pp. 313-322, keys)
JORDAN, D, S. & GILBERT, C. H. List of fishes of the Pacific Coast of the United States . . . . Proc. United States Nat. Mus., 3, 452-458; 1881.
STARKS, E. C. A key to the families of marine fishes of the West Coast. California Fish Game Comm. Fish. Bull., 5, 1-16; 1921.
WALFORD, L. A. Marine game fishes of the Pacific Coast, Alaska to the equator. Berkeley; 1-207; 1937. (Keys)
SCHULTZ, L. P. Keys to the fishes of Washington, Oregon. . . . Univ. Washington Publ. Zool., 2, 103-228; 1936.
BERRY, F. H. & PERKINS, H. C. Survey of pelagic fishes of the California current area. Fishery Bull. United States Fish. Wildl. Serv., 65, 625-682; 1966.
CLOTHIER, C. R. A key to some southern California fishes based on vertebral characters. Fish. Bull. Sacramento, 79, 1-83; 1951.
FITCH, J. E. & LAVENBERG, R. J. Deep-water fishes of California. Berkeley; 1-155; 1968. (List, key)
KIMSEY, J. B. & FISK, L. O. Keys to the freshwater and anadromous fishes of California. California Fish. Game, 46, 453-479; 1960.
ROEDEL, P. M. & RIPLEY, W. E. California sharks and rays. Fish. Bull. California, 75, 1-88; 1950.
SHAPOVALOV, L. et al. A revised check list of the freshwater and anadromous fishes of California. California Fish Game, 45, 159-180; 1959. (Previous edition 1950)
ULREY, A. B. & GREELEY, P. O. A list of the marine fishes (Teleostei) of southern California. Bull. Southern California Acad. Sci., 27, 1-53; 1928.

West Indies

BAUCHOT, M. La faune ichthyologique des eaux douces antillaises. Compt. Rend. Soc. Biogeogr., 36, No. 311, 7-26; 1959. (Faunal lists)
JORDAN, D. S. A preliminary list of the fishes of the West Indies. Proc. United States Nat. Mus., 9, 554-608; 1897.
METZELAAR, J. Over tropisch atlantische visschen. Part I. Amsterdam; 1-179; 1919.
RANDALL, J. E. Caribbean reef fishes. Jersey City, New Jersey; 1-318; 1968. (Fauna, keys)
BOHLKE, J. E. & CHAPLIN, C. C. G. Fishes of the Bahamas . . . . Wynnewood, Pennsylvania; 1-771; 1968. (Fauna, keys)
DUARTE-BELLO, P. P. Catalogo de pesces Cubanos. Lab. Biol. Mar., Monogr., 6, 7-208; 1959.
CALDWELL, D. K. Marine and freshwater fishes of Jamaica. Bull. Inst. Jamaica, Sci. Ser., 17, 1-120; 1966.
BEEBE, W. & TEE-VAN, J. The fishes of Port-au-Prince Bay, Haiti, with a summary of the known species of marine fish of the island of Haiti and Santo Domingo. Zoologica, 10, 1-279; 1928.
FOWLER, H. W. The fishes of Hispaniola. Mem. Soc. Cubana Hist. Nat., 21, 83-122; 1952. (Faunal list)
FOWLER, H. W. The fishes of Trinidad, Grenada and St. Lucia, British West Indies. Proc. Acad. Nat. Sci. Philadelphia, 67, 520-546; 1915. (Faunal list)
BOESEMAN, M. The fresh-water fishes of the island of Trinidad. Natuurw. Stud. Surinam, 21, 72-153; 1960. (Fauna, keys)

Central America & South America

EIGENMANN, C. H. Catalogue of the fresh-water fishes of tropical and south temperate America. Rep. Princeton Univ. Exp. Patagonia, 1896-99, 3, (Zool.), (4), 375-511; 1910.
GOSLINE, W. A. Catalogo das Nematognatos de agua doce da America do Sul e Central. Bol. Mus. Nac. Rio de Janeiro, (n. s.), 33, 1-138; 1945.

Central America

GOODE, G. B. & BEAN, T. H. A list of the species of fishes recorded as occurring in the Gulf of Mexico. Proc. United States Nat. Mus., 5, 234-240; 1882.
REGAN, C. T. Biologia Centrali-Americana. Pisces. London; 1-203; 1906-1908.
JORDAN, D. S. A list of the fishes known from the Pacific coast of tropical America, from the Tropic of Cancer to Panama. Proc. United States Nat. Mus., 8, 361-394; 1885.
EIGENMANN, C. H. Catalogue of the fresh-water fishes of Central America and southern Mexico. Proc. United States Nat. Mus., 16, 53-60; 1893.
ALVAREZ, J. Claves para determinacion de especies en los peces de las aguas continentales mexicanos. Mexico; 1-144; 1950.
de BUEN, F. Notas sobre ictiologia de aguas dulces de Mexico. III. La lista de peces y la moderna clasificacion. Invest. Estac. Limnol. Patscuaro, 11, 1-9; 1941. (Faunal list)

de BUEN, F. Investigaciones sobre ictiologia Mexicana. I. Catalogo de los peces de la region neartica en suelo Mexicano. Anal. Inst. Biol. Mexico, 18, 257-348; 1947. (Faunal list)
CANNON, R. The sea of Cortez. Menlo Park, California; 1-283; 1966. (Pp. 258-270; keys)
MEEK, S. E. The fresh-water fishes of Mexico north of the Isthmus of Tehuantepec. Field Mus., Zool., 51, 1-252; 1904.
CONTRERAS B., S. Lista de peces del estado de Nuevo Leon. Cuad. Invest. Inv. Cient. Univ. Nuevo Leon, 11, 1-12; 1967
HILDEBRAND, S. F. Fishes of the Republic of El Salvador, Central America. Bull. United States Bur. Fish., 41, 237-287; 1926.
MEEK, S. E. An annotated list of fishes known to occur in the fresh waters of Costa Rica. Publ. Field Mus. Nat. Hist., Zool., 10, 101-134; 1914.
BUSSING, W. A. New species . . . of Costa Rican freshwater fishes with a tentative list of species. Rev. Biol. Trop., 14, 205-249; 1967.
GILBERT, C. H. & STARKS, E. C. The fishes of Panama Bay. Mem. California Acad. Sci., 4, 1-304; 1904.
MEEK, S. E. & HILDEBRAND, S. F. The fishes of the fresh waters of Panama. Field Mus. Nat. Hist., Zool., 10, (15), 217-374; 1916.
MEEK, S. E. & HILDEBRAND, S. F. The marine fishes of Panama, 3 parts. Field Mus. Nat. Hist., Zool., 15, 1-1045; 1923-1928.
HILDEBRAND, S. F. A new catalogue of the fresh-water fishes of Panama. Field Mus. Publ., Zool., 22, 219-359; 1938. (Fauna and keys)

South America

EIGENMANN, C. H. & EIGENMANN, R. S. A catalogue of the fresh-water fishes of South America. Proc. United States Nat. Mus., 14, 1-81; 1891.
EIGENMANN, C. H. The fishes of western South America. Part I. The fresh-water fishes of northwestern South America . . . . Mem. Carnegie Mus., 9, 1-346; 1922.
EIGENMANN, C. H. & ALLEN, W. R. Fishes of western South America. Lexington, Kentucky; 1-494; 1942.
FOWLER, H. W. Lista de peces de Colombia. Rev. Acad. Colombiana Cienc., 5, 128-138; 1942.
FOWLER, H. W. The shore fishes of the Colombian Caribbean. Caldasia, 6, (27), 43-73; 1953. (List)
MILES, C. Los peces del Rio Magdalena. Bogota; 1-214; 1947. (Keys)
ORCES, G. Contribuciones al conocimiento de los peces marinos del Ecuador. Anal. Univ. Centro Ecuador, 88, 107-160; 1959.
OVCHYNNYK, M. M. Annotated list of the freshwater fishes of Ecuador. Zool. Anz., 181, 237-268; 1968.
HILDEBRAND, S. F. A descriptive catalogue of the shore fishes of Peru. United States National Mus. Bull., 189, 1-530; 1946.
EVERMANN, B. W. & RADCLIFFE, L. The fishes of the west coast of Peru and the Titicaca Basin. United States Nat. Mus. Bull., 95, 1-166; 1917.
FOWLER, H. W. Las peces del Peru. Catalogo sistematico de los peces que habitan en aguas peruanas. Bol. Mus. Hist. Nat. Javier Prado, 5, (7 parts); 1941-1944. (Fauna, list)
KOEPCKE, H. W. Lista de los peces marinos conocidos del Peru. Biota, 3-4, (6 parts); 1962-1964.
SCHULTZ, L. P. A further contribution to the ichthyology of Venezuela. Proc. United States Nat. Mus., 99, 1-211; 1949.
CERVIGON, F. Lista de los peces marinos de Venezuela. Lagena, 5, 8-71; 1965. (Fauna)
CERVIGON, F. Los peces marinos de Venezuela complemento. Mem. Soc. Cienc. Nat. La Salle, 80, 177-218; 1968.
LOWE, R. H. The fishes of the British Guiana continental shelf, Atlantic coast of South America. . . . Journ. Linn. Soc. London, Zool., 44, 669-700; 1962. (Faunal list)
BOESEMAN, M. A preliminary list of Surinam fishes not included in Eigenmann's enumeration of 1912. Zool. Meded., 31, 179-200; 1952.
PUYO, J. Poissons de la Guyane francaise. Faune Empire Franc., 12, 1-280; 1949.
FOWLER, H. W. A list of the fishes known from the coast of Brazil. Arqu. Zool., 3, 115-184; 1942.
de MIRANDA RIBEIRO, A. Fauna brasiliense. Peixes. (several parts) Mus. Nac. Rio de Janeiro, 1923; and Arch. Mus. Rio de Janeiro; 1912-1923.
FOWLER, H. W. Os peixes de agua doce do Brasil. 2 entrega. Arqu. Zool. Sao Paulo, 6, 205-404; 1950. Arch. Zool. Sao Paulo, 9, 1-400; 1954.
TRAVASSOS, H. Catalogo dos peixes do Vale do Rio Sao Francisco. Bol. Soc. Cear. Agron., 1, 1-66; 1960.
DEVINCENZI, G. J. Peces del Uruguay. Anal. Mus. Nac. Montevideo, (2), 1, 97-134, 139-290; 1920-1924.
DEVINCENZI, G. J. & TEAGUE, G. W. Ictiofauna del Rio Uruguay medio. Montevideo; 1-103; 1942.
POZZI, A. J. Sistematica y distribucion de los peces de agua dulce de la Republica Argentina. Gaea, 7, 239-292; 1945.
RINGUELET, R. & ARAMBURU, R. H. Peces marinos de la Republica Argentina. Clave para el reconocimiento de familias y generos. Catalogo critico abreviado. Agro, 2, (5), 1-141; 3, (7), 1-98; 1960.
RINGUELET, R. A. & ARAMBURU, R. H. Peces argentinas de agua dulce. Clave de reconocimiento y caracterizacion de familias y subfamilias, con glosario explicativo. Agro, 3, (7), 1-98; 1961.
RINGUELET, R. A. et al. Los peces argentinos de agua dulce. La Plata; 1-602; 1967. (Freshwater)
EIGENMANN, C. The fresh water fishes of Patagonia . . . . Rep. Princeton Univ. Exp. Patagonia, 1898-1899, 3, (3), 225-374; 1909.
NORMAN, J. R. Coast fishes. Part II. The Patagonian region. Disc. Rep., 16, 1-150; 1937.
EIGENMANN, C. H. The fresh-water fishes of Chile. Mem. Nat. Acad. Sci., 22, (2), 1-63; 1927.
FOWLER, H. W. Fishes of Chile. Systematic catalog. Rev. Chilena Hist. Nat., 45, 22-57; 1943. And 46-47, 15-116, 275-343; 1944.

MANN, G. Peces de Chile. Clave de determinacion de las especes importantes. Santiago de Chile; 1-48; 1950.
FOWLER, H. W. Analysis of the fishes of Chile. Rev. Chilena Hist. Nat., 51-53, 263-326; 1951. (Keys)
MANN F., G. Vida de los peces en aguas chilenas. Santiago de Chile; 1-342; 1954. (Fauna, keys)
de BUEN, F. Peces del suborden Scombroidei en aguas de Chile. Rev. Biol. Mar. Valparaiso, 7, 3-38; 1958. (Keys)
de BUEN, F. Peces de la superfamilia Clupeoidae en aguas de Chile. Rev. Biol. Mar. Valparaiso, 8, 83-110; 1958. (Keys)
de BUEN, F. Peces chilenos. Beloniformes, Syngnathiformes y Gobiidae. Bol. Soc. Biol. Concepcion, 35-36, 81-101; 1963.
SCHNEIDER, C. O. Catalogo de los peces marinos del litoral de Concepcion y Arauco. Bol. Soc. Biol. Concepcion, 17, 75-126; 1943.

Atlantic Ocean, tropical and south

POSTEL, E. Liste commentee des poissons signales dans l'Atlantique tropico-oriental nord . . . . Bull. Soc. Sci. Bretagne, 34, 129-170; 1958.
NIELSEN, J. Psettodoidea and Pleuronectoidea . . . . Atlantide Rep., 6, 101-127; 1961.
BAUCHOT, M. L. & BLANC, M. Poissons marins de l'Est Atlantique tropical. II. Percoidei. . . , 1st partie. Atlantide Rep., 6, 65-100; 1961.
METZELAAR, J. Over tropisch atlantische visschen. Amsterdam; 1-314; 1919.
BAUCHOT, M. L. Poissons marins de l'Est Atlantique tropical. III. Atlantide Rep., 9, 7-43, 63-91; 1966.
PENRITH, M. J. The fishes of Tristan da Cunha, Gough Island and the Vema seamount. Ann. South African Mus., 48, 523-548; 1967.
GOODE, G. B. Catalogue of the fishes of the Bermudas. United States Nat. Mus. Bull., 5, 1-82; 1876.
BEEBE, W. & TEE-VAN, J. Field book of the shore fishes of Bermuda. New York; 1-337; 1933.

North Atlantic Ocean

BIGELOW, H. B. et al. Fishes of the western North Atlantic. Mem. Sears Found. Mar. Res., 1, 7 parts; 1948-1966. (Faunal monograph)
AASEN, O. & MYKLEVOLL, S. De vanligste boreale or subtropiske haiarter i Nord-Atlanteren. Fisken Hav, 2, 1-40; 1965.
JOUBIN, — Faune ichthyologique de l'Atlantique Nord. Copenhagen; parts 1-18; 1929-1938.

Arctic

JENSEN, A. S. List of the fishes of Greenland. Rapp. Proc. -Verb. Expl. Mer, 39, 85-102; 1926. (Appendix)
SAEMUNDSSON, B. Synopsis of the fishes of Iceland. Rit Visindafelags Island, 2, 1-66; 1927.
SAEMUNDSSON, B. Zoology of Iceland. Marine Pisces, vol. IV, part 72, 1-150; 1949.
KNIPOVICH, N. M. Guide for determination of the fishes of Barents-Sea, White-Sea and Kara-Sea. Trans. Inst. Sci. Explor. North, 27, 1-182; 1926. (In Russian)
RENDALL, H. Fische aus dem östlichen Sibirischen Eismeer und dem Nordpazifik. Ark. Zool., 22, (A10), 1-81; 1931.
ANDRIYASHEV. A. P. Fishes of the northern seas of the U. S. S. R. St. Petersbourg; 1-566; 1954. (In Russian)
ANDRIYASHEV. A. P. Fishes of the northern seas of the U. S. S. R. Jerusalem; 1-617; 1964. (Faunal monograph; translation of 1954)
ANDRIASHEV, A. P. The fishes of the Bering Sea and neighbouring waters. . . . Leningrad; 1-187; 1939. (In Russian)
OKADA, S. & KOBAYASHI, K. Hokuyo - Gyorui - Zusetsu . . . . (Japan); 1-179; 1968. (In Japanese; Colored illustrations of pelagic and bottom fishes of Bering Sea)
WALTERS, V. Fishes of western Arctic American and eastern Arctic Siberia. . . . Bull. American Mus. Nat. Hist., 106, 255-368; 1955.

Europe

LADIGES, W. & VOGT, D. Die süsswasserfische Europas bis zum Ural und Kaspischen Meer. Hamburg; 1-250; 1965. (Keys)
SMITT, F. A. et al. A history of Scandinavian fishes, by B. Fries. Stockholm; 2 volumes; 1892-1895.
WOLLEBACK, A. Norges fisker. Kristiania; 1-239; 1924.
NYBELIN, O. Våra fiskar och hur man känner igen dem. Illustrerad fickbok . . . . Stockholm; Del I, III 1943-1945. (Keys)
CURRY-LINDAHL, K. Fiskarna i färg. Stockholm; 1-242; 1953/1966. (Fauna, keys)
BERG, L. S. A review of the fishes of the Finnish Gulf. Rep. All-Union Inst. Sci. Res. Lake Riv. Fish., 23, 3-45; 1940. (In Russian)
SCHNEIDER, G. Die Süsswasserfische des Ostbaltikums und ihre Verbreitung immerhalt des Gebietes. Arch. Hydrobiol. Stuttgart, 16, 133-155; 1925.
EHRENBAUM, E. Handbuch der Seefischerei Nordeuropas. Band II. Stuttgart; 1-337; 1936. (Keys)
DUNCKER, G. et al. Die Fische der Nord- und Ostsee. Leipzig; (many pages); 1929. (Fauna, keys)
DUNCKER, G. & LADIGES, W. Die Fische der Nordmark. Abh. Naturw. Ver. Hamburg, (n. f.), 3, Suppl., 1. 432; 1960.

OTTERSTRØM, C. V. Danmarks Fauna. 20. Fisk III. København; 1-168; 1917.
BRUUN, A. F. et al. List of Danish vertebrates. København; 1-180; 1950.
BONDESEN, P. et al. Bestemmelsennøgle over Dansk ferskvandsfisk. Nat. og Mus., 7, (1), 8-12; 1959. (Keys to Danish freshwater fishes)
DAY, F. The fishes of Great Britain and Ireland. London; 2 volumes; 1880-1884. (Faunal monograph)
REGAN, C. T. The freshwater fishes of the British Isles. London; 1-287; 1911.
SMART, J. Bibliography of keyworks for the identification of the British fauna and flora. Syst. Assoc. Publ., 1, 1-105; 1942.
JENKINS, J. T. The fishes of the British Isles both fresh water and salt. London; 1-408; 1950. (Previous editions 1925 and 1936)
TAYLOR, F. J. Fish of rivers, lakes and ponds. London; 1-88; 1961. (British)
BRACHEN, J. J. A key to the identification of the eggs and young stages of coarse fish in Irish waters. Scient. Proc. Roy. Dublin Soc., (B), 2, 97-108; 1967.
WHEELER, A. The fishes of the British Isles and north-west Europe. East Lansing, Michigan; 1-613; 1969. (List and keys)
DODERLEIN, L. Bestimmungsbuch für deutsche Land- und Süsswassertiere. Wirbeltiere . . ., 2nd ed. München; 1-304; 1955.
FERRANT, V. Faune du Grand-Duche de Luxembourg. Part 1. Poissons. Gesellsch. Luxembourg. Naturfreunde, Festschr. Feier der 25-Jahr Bestehens 1890-1915, 437-515; 1915. (List)
POPTA, C. M. L. Lijst van de Nederlandsche visschen aanwezig in 's Rijks Museum van natuurlijke historie te Leiden. Zool. Meded., 8, 77-119; 1924.
NIJSSEN, H. Zeewissen. Wet. Meded. K. nederl. natuurh. Veren., 65, 1-68; 1966.
POLL, M. Faune de Belgique. Poissons marins. Bruxelles; 1-452; 1947. (Faunal monograph)
MOREAU, E. Histoire naturelle des poissons de France. Paris; 3 volumes; 1881. (Faunal monograph)
PERRIER, R. La faune de France en tableaux synoptiques illustres. Vertebres. Poissons. Paris; 1-212; 1924. (Keys)
ROULE, L. Les poissons des eaux douces de la France. Paris; 1-228; 1925.
PARIS, P. Les captures de pecheur dans nos eaux douces. Tableaux de determination. Bull. Sci. Bourgogne, 5, 89-121; 1936.
BERTIN, L. Petit atlas des poissons. Paris; 3 fasc.; 1942. (Keys)
ANGEL, A. F. Petit atlas des poissons. III. Poissons des eaux douces. Especes francaise. Paris; 1-55; 1943.
GOSSOT, P. Ce qu'il faut savoir des poissons des eaux douces de France. Paris; 1-257; 1946. (Keys)
DOTTRENS, E. Poissons d'eau douce. Paris; 2 volumes; 1951-1952. (Keys)
BOUGIS, P. Atlas des poissons. Fasc. I. Poissons marins. Paris; 2 volumes; 1959. (Keys)
SPILLMANN, J. Poissons d'eau douce. Faune France, 65, 1-303; 1961.
LOZANO REY, L. Fauna Iberica. Peces. Tomo I. Madrid; 1-692; 1928.
NOBRE, A. Fauna marinha de Portugal. I. Vertebrados . . . . Porto; 1-579; 1935.
ALBUQUERQUE, R. M. Peixas de Portugal e ilhas adjacentas chave para a sua determinacao. Portug. Acta Biol., 5B, 1-1167; 1956. (Fauna, keys)
LOZANO REY, L. Los peces fluviales de Espana. Madrid; 1-387; 1935. (Faunal monograph)
LOZANO y REY, L. Los principales peces marinos y fluviales de Espana, 2nd ed. Madrid; 1-128; 1949.
PRIETO, A. R. Estudio bromatologico de los peces espanoles del orden Heterosomata. Arch. Zootec., 3, (11), 203-234; 1954.
LOZANO, F. Nomenclatura ictiologica. Nombres cientificos y vulgares de los peces espanoles. Trab. Inst. Espanola Oceanogr., 31, 1-271; 1963. (Faunal list)
de BUEN, F. Catalogo ictiologico Mediterranee espanol y de Marruecos . . . (Mar de Espana). Result. Campan. Real. Acuerd. Intern., 2, 1-221; 1926. (Also as Inst. Espanol Oceanogr., 2)
FAGE, L. Shore-fishes. In Rep. Danish Oceanogr. Exp. 1908-10 to Mediterranean, II, Biol., A3, 1-154; 1918.
TORTONESE, E. Elenco dei Leptocardii, Ciclostomi, Pesci cartilaginei ed ossi del Mare Mediterraneo. Atti Soc. Ital. Sci. Nat., 97, 309-345; 1958.
SUPINO, F. Determinazione sistematica pesci d'acqua dolce d'Italia. Natura, 24, 118-131; 1933. (Keys)
TORTONESE, E. Fauna d'Italia. Leptocardia, ciclostomata, Selachii. Bologna; 1-334; 1956.
BINI, G. Atlante dei pesci delle coste Italiane. I, 1967; VIII, 1968.
TORTONESE, E. & TROTTI, L. Catalogo dei pesci del Mare Ligure. Atti Accad. Ligure, 6, 1-118, 49-164; 1949-1950. (List)
TORTONESE, E. Pesci e cetacei del mar Ligure. Genova; 1-216; 1965.
LANFRANCO, G. G. A complete guide to the fishes of Malta. Malta; 1-74; 1958.
SOLJAN, T. Fauna et flora Adriatica. Fishes of the Adriatic. Inst. Oceanogr. Rib. Jugoslavia, 1, 1-437; 1948.
SOLJAN, T. Fishes of the Adriatic, 2nd ed. Zagreb; 1-428; 1961. (In Jugoslavian)
SOLJAN, T. Fishes of the Adriatic. Belgrade; 1-428; 1963. (Translation of 1948) (Keys)
KARAMAN, S. Pisces Macedoniae. Split; 1-90; 1924.
KONSULOFF, S. & DRENSKY, P. Die Fischfauna der Aegaeis. Ann. Univ. Sofia Fac. Phys. Math., 39, 293-308; 1943. (In Bulgarian)
HEIN, W. & WINTER, F. W. Susswasserfische Mittel-Europas. Leipzig; 1912.
KAHSBAUER, P. Cyclostomata, Teleostomi. Cat. Faun. Austria, Teil 21aa, 1-56; 1961.
VLADYKOV, V. Poissons de la Russie sous carpathique (Tchecoslovaquie). Mem. Soc. Zool. France, 29, 217-374; 1931.
STEPANEK, O. Klic nasich obratlovcu. Orbis-Praha; 1-250; 1950. (Key to Czech vertebrates)
HRABE, S. & OLIVA, O. Klic nasich ryb. Nakl. Cesk. Akad. Ved., Biol., 1953, 1-100; 1953. (Freshwater)
STROUHAL, H. Cyclostomata, Teleostomi (Pisces). Catal. Faun. Austriae, 21aa, 1-56; 1961.

STEINMANN, P. Die Fische der Schweiz. Aarau; 1-154; 1936. (Keys)
STEINMANN, P. Schweizerische Tischkunde. Aarau; 1-222; 1948. (Keys)
VUTSKITS, G. Fauna Regni Hungariae. Pisces. Budapest; 1-42; 1918.
MIHELYI, F. Revision der Süsswasserfische von Ungarn und den angrenzenden Gebieten in der Sammlung des Ungarischen Naturwissenschaftlichen Museums. Anal. Hist. Nat. Mus. Hungarici, (n. s.), 5, 433-456; 1954.
VUKOVIC, T. Ribe Bosni i Hercegovine. Sarajevo; 1-127; 1963. (Keys)
ANTIPA, G. Fauna ichtiologica a Romaniei. Bucuresti; 1-294; 1909.
VASILIU, G. D. Revizuire sistematica a faunei ichtyologice din Romania . . . . Notat. Biol. Bucuresti, 4, 204-300; 1946. (Faunal list)
BANARESCU. P. Fauna Republica Populare Romine. XIII. Pisces — Osteichthyes. Bucuresti; 1-959; 1964. (Faunal monograph; in Romanian)
BANARESCU, P. A revised list of the fishes of Romania. Bul. Inst. Cerc. Pisc., 27, (3), 5-18; 1968. (In Romanian)
DRENSKI, P. Synopsis and distribution of fishes in Bulgaria. Annu. Univ. Sofia, 44, (3), 11-71; 1948.
DRENSKI, P. Fauna of Bulgaria. No. 2. [Fishes]. Sofia; 1-270; 1951. (Faunal monograph)
PESHEV. T. & BOEV, N. Vertebrates in fauna of Bulgaria, a short classification key. Sofia; 1-520; 1962. (In Bulgarian; fishes pp. 15-70)
RHASIS ERAZI, R. A. Marine fishes found in the Sea of Marmora and in the Bosphorus. Rev. Fac. Sci. Univ. Istanbul, 7B, 103-115; 1942. (Faunal list)
DRENSKY, P. Bestimmungstabellen der Fische Schwarzen Meer. Trav. Soc. Bulg. Sci. Nat., 11, 50-63; 1924. (Keys)
SLASTENENKO, E. P. Catalogue of the fishes of the Black Sea and of the Azof Seas. Works V. M. Arnoldi Riol. Stat. Novorossisk, 2, 109-149; 1938.
SLASTENENKO, E. P. Les poissons de la mer Moire et de la mer d"Azof. Ann. Sci. Univ. Jassy, 25, (2), 1-196; 1939.
VODYANITZKI, V. A. & KAZANOVA, I. I. Key for identification of the pelagic eggs and fish larvae in the Black Sea. Trud. V. N. I. R. O., 28, 240-333; 1954. (In Russian)
ZHUKOV, P. I. Handbook of fishes of White Russian U. S. S. R. Minsk; 1-122; 1960. (In Russian)
SVETOVIDOV, A. N. Fishes of the Black Sea. Moskva; 1-550; 1964. (In Russian)
BERG, L. A catalogue of the freshwater fishes of Russia. Ann. Mus. Zool. Russ. Petrograd, 21, 222-242; 1916. (In Russian; list)
BERG, L. S. Faune de Russie. Poissons. I-III. St.-Petersbourg; 3 volumes; 1911-1933. (In Russian)
BERG, L. S. Les poissons des eaux douces de l'U. R. S. S. . . , 3rd ed. Leningrad; 2 volumes; 1932-33.
SVETOVIDOV, A. N. Fauna of U. S. S. R. Pisces. Gadiformes. Moskva; 1-221; 1948. (In Russian. Also as Zool. Inst. Acad. Sci., n. s., 34, (9), 1-221; 1948)
BERG, L. S. Freshwater fishes of Soviet Union and adjacent countries, 4th ed. Tabl. Anal. Faune URSS, vol. 27-30, 1-1382; 1948-1949. (3 parts)
BORISOV, P. E. & OVSYANNIKOV, N. S. A descriptive catalogue of industrial fish of S. S. S. R. Moskva, 1-177; 1951. (In Russian) (Keys)
BORISOV, P. G. & OVSYANNIKOV, N. S. Keys to fishes of economic importance in U. S. S. R. Moskva; 1-260; 1954. (In Russian; keys)
BERG, L. S. Freshwater fishes of the U. S. S. R., 4th ed. Jerusalem; 1-510; 1962-65. (Trans. of 1948-49)
NIKOLSKII, G. V. Ryby basseina Amura. Moskva; 1-552; 1956.

Near East & Middle East

PELLEGRIN, J. Les poissons des eaux douces d'Asia-mineure. Paris;1-134; 1928.
AKSIRAY, F. Turkiye deniz Baliklari. Tayin Anahtari. Istanbul; 1-277; 1954.
BECKMAN, W. C. The freshwater fishes of Syria . . . . FAO Fish. Biol. Techn. Pap., 8, 1-297; 1962.
BERG, L. S. Freshwater fish of Iran. . . . Trav. Inst. Zool. Acad. Sci. U. R. S. S., 8, 783-858; 1949. (In Russian)
MAHDI, N. Fishes of Iraq. Baghdad; 1-82; 1961.
KHALAF, K. T. The marine and fresh water fishes of Iraq. Baghdad; 1-164; 1961.
BLEGVAD, H. & LØPPENTHIN, B. Fishes of Iranian Gulf. Danish Sci. Invest. Iran, 3, 1-247; 1944.
BEN-TUVIA, A. Mediterranean fishes of Israel. Bull. Sea Fish. Res. Sta., 8, 1-40; 1953. (Faunal list)
STEINITZ, H. The freshwater fishes of Palestine. An annotated list. Bull. Res. Counc. Israel, 3, 207-227; 1953.
KLUNZIGER, C. B. Synopsis der Fische des Rothen Meeres. I. Teil. Percoiden — Mugiloiden. Verh. k.-K. Zool. Bot. Ges. Wien, 20, 669-1368; 1870. (Reprinted 1964)
FOWLER, H. W. The fishes of the Red Sea. Sudan Notes, 26, 113-137; 1945.
FOWLER, H. W. Fishes of the Red Sea and southern Arabia. I. Branchiostomida to Polynemida. Jerusalem; 1-240; 1956. (Keys)
WHITEHEAD, P. J. P. A review of the elopoid and clupeoid fishes of the Red Sea and adjacent regions. Bull. British Museum (Nat. Hist.), Zool., 12, 225-281; 1965.

North Africa & Atlantic Islands

JUBB, R. A. A revised list of the freshwater fishes of northern Africa. Ann. Cape Prov. Mus., 3, 5-39; 1963.
BOULENGER, G. A. Zoology of Egypt: The fishes of the Nile. London; 1-578; 1909.
TORTONESE, E. Appunti di ittiologica libica: Pesci di Tripoli. Ann. Mus. Libico Stor. Nat., 1, 359-379; 1939. (Faunal list)

DIEUZEIDE, R. et al. Catalogue des poissons des cotes algeriennes. Bull. Sta. Aquic. Peche Castiglione, (n. s.), 4, 5, 6; 1953-1955.
de NORONHA, A. C. & SARMENTO, A. A. Vertebrados da Madeira, vol. 2, Peixes, 2nd ed. Funchal; 1-181; 1948.
MAUL, G. E. Lista sistematica dos peixes assinalados nos mares da Madeira e indice alfabetico. Funchal; 1949.
ABREU NUNES, A. Peixes da Madeira. Funchal; 1-274; 1953.
MAUL, G. E. Monografia dos peixes do Museu Municipal do Funchal. Ordem Isospondyli. Bol. Mus. Munic. Funchal, 3, (5), 5-41; 1948; and 4, (9), 5-20; 1949.
COLLINS, B. L. Lista de peixes dos mares das Acores. Acoreana, 5, (2), 1-40; 1954. (Faunal list)

Africa

POLL, M. Les genres des poissons d'eau douce de l'Afrique. Ann. Mus. Roy. Congo Belge, 54, 1-191; 1957. (Keys to genera)
METZELAAR, J. Over tropisch atlantische visschen. Amsterdam; 1-314; 1919.
FOWLER, H. W. The marine fishes of West Africa. Bull. American Mus. Nat. Hist., 70, 1-1493; 1936.
BLACHE, J. et al. Cles de determination des familles, genres, especes de poissons selaciens signales dans le Golfe de Guinee. Paris; 1-479; 1970.
PELLEGRIN, J. Les poissons des eaux douces de l'Afrique occidentale (du Senegal au Niger). Paris; 1-373; 1923.
CADENAT, J. Poissons de Mer du Senegal. Dakar; 1-345; 1950. (Keys)
SCHULTZ, L. P. The fresh-water fishes of Liberia. Proc. United States Nat. Mus., 92, 301-348; 1942.
DAGET, J. & ILTIS, A. Poissons de cote d'Ivoire (eaux douces et saumatres). Dakar; 1-385; 1965. (Also as: Mem. Inst. Francaise d'Afrique Noire, 74)
IRVINE, F. R. The fishes and fisheries of the Gold Coast. London; 1-352; 1947. (Keys)
WELMAN, J. B. Preliminary survey of the freshwater fisheries of Nigeria. Lagos; 1948. (Keys to families)
BOESEMAN, M. An annotated list of fishes from the Niger delta. Zool. Verh. (Leiden), 61, 1-48; 1963.
DAGET, J. Les poissons du Niger Superieur. Mem. Inst. Francaise Afrique Noire, 36, 1-391; 1954.
GOSSE, J.-P. Les poissons du Bassin de l' Ubangi. Doc. Zool. Mus. r. Afr. Centr., 13, 1-56; 1968.
BLACHE, J. Les poissons du bassin du Tchad et du bassin adjacent du Mayo Kebbi. Paris; 1-483; 1964.
THYS van der AUDENAERDE, D. F. E. List of the freshwater fishes presently known from the island of Fernando Po. Bonn Zool. Beitr., 16, 316-317; 1965.
THYS van der AUDENAERDE, D. F. E. The freshwater fishes of Fernando Po. Bruxelles; 1-167; 1967.
DAGET, J. Les poissons du Fouta Dialon et de la Basse Guinee. Dakar; 1-210; 1962.
BOULENGER, G. Les poissons du basin du Congo. Bruxelles; 1-537; 1901.
POLL, M. Poissons. IV. Teleosteens. Acanthopterygiens. I part. Res. Sci. Exp. Oceanogr. Belg. Cot. Afr. Atl. Sud., 1948-1949, 4, (3A), 1-390; 1954.
MONTEIRO, R. Heterosomata de Angola. 1 . . . . Anal. Junta Invest. Ultramar Lisbon, 12, Tom2, 87-130; 1958.
SMITH, J. L. B. The sea fishes of southern Africa, 4th ed. Cape Town; 1-580; 1961. (Fauna & keys; previous editions 1949 & 1953)
JUBB, R. A. Freshwater fishes of southern Africa. Cape Town; 1-248; 1967.
BARNARD, K. H. A pictorial guide to South African fishes marine and freshwater. Cape Town; 1-226; 1947. (Keys)
GILCHRIST, J. D. F. & THOMPSON, W. W. A catalogue of the sea fishes recorded from Natal, I, II. Ann. Natal Mus., 1, 1255-290, 1916; 291-431, 1917.
CRASS, R. S. Freshwater fishes of Natal. Pietermarizburg; 1-167; 1964. (Keys)
HEWITT, J. A guide to the vertebrate fauna of the eastern Cape Province, South Africa. II. Reptiles and amphibians and freshwater fishes. Grahamstown; 1-141; 1937.
BARNARD, K. H. Revision of the indigenous freshwater fishes of the S. W. Cape region. Ann. South African Mus., 36, 101-262; 1943.
JUBB, R. A. Freshwater fishes of Cape Province. Ann. Cape Prov. Mus., 4, 1-72; 1965. (Keys)
COPLEY, H. Common freshwater fishes of East Africa. London; 1-172; 1958. (Keys)
LOSSE, G. F. The elopoids and clupeoid fishes of East African coastal waters. Journ. East African Nat. Hist Soc. Nat. Mus., 27, 77-115; 1968.
JACKSON, P. B. N. The fishes of Northern Rhodesia. A check-list of indigenous species. Lusaka; 1-140; 1961. (Faunal list)
MAAR, A. Introductory checklist of fish of the Zambezi sub-region of the Ethiopian region. Proc. 1st. Fed. Sci. Congr. Salisbury, 1960, 1-8; 1960. (Faunal list)
JUBB, R. A. An illustrated guide to the freshwater fishes of the Zambezi River, Lake Kariba, Pungwe, Sabi, Lunchi, and Limpopo Rivers. Bulawayo; 1-171; 1971. (Keys)
JACKSON, P. B. N. Check list of the fishes of Nyasalandd Occ. Pap. Nat. Mus. S. Rhodesia, 3, 535-621; 1961.
LOCKLEY, G. J. The families of freshwater fishes of Tanganyika Territory, with a key to their identification. East African Agric. Journ., 14, 212-218; 1949.
POLL, M. Revision de la faune ichthylogique du lac Tanganyika. Ann. Mus. Congo., Zool., (1), 4, 141-364; 1946.
POLL, M. Poissons non Cichlidae. Res. Sci. Expl. Hydrob. Lac Tanganyika, 1946-47, 3, (5A), 1-251; 1953.
GREENWOOD, P. H. The fishes of Uganda, 2nd ed. Kampala; 4 parts; 1966; (1st edition, 1955-57)
COPLEY, H. A short account of the freshwater fishes of Kenya. Nairobi; 1-24; 1946.

CLARK, E. & GOHAR, A. F. The fishes of the Red Sea: Order Plectognathi. Publl. Mar. Biol. Sta. Ghardagu, 8, 1-80; 1953.

Madagascar

PELLEGRIN, J. Les poissons des eaux douces de Madagascar et des iles voisines (Comores, Seychelles, Mascareignes). Mem. Acad. Malgache, 14, 1-222; 1933.
PELLEGRIN, J. La faune ichtyologique des eaux douces de Madagascar. Ann. Sci. Nat., Zool., (10), 17, 425-432; 1934. (Faunal list)
FOURMANOIR, P. Poissons telesteens des eaux malgaches du Canal de Mozambique. Mem. Inst. Sci. Madagascar, (F Oceanogr.), 1, 1-316; 1957. (Annotated faunal list)
ARNOULT, J. Poissons des eaux douces. Faune Madagascar, 10, 1-163; 1959.
FOURMANOIR, P. Liste complementaire des poissons du Canal de Mozambique. Mem. Inst. Sci. Madagascar, 4F, 83-107; 1961.

Indian Ocean

BAISSAC, J. de B. Contribution a l'etude des poissons de l'Ile Maurice. VII. Proc. Roy. Soc. Art Sci. Mauritius, 2, 1-37; 1959. (And other parts)
FOURMANOIR, P. Ichthyologie et peche aux Comores. Mem. Inst. Sci. Madagascar, 9A, 187-239; 1955. (Faunal list)
HERRE, A. W. C. T. A list of the fishes known from the Andaman Islands. Mem. Indian Mus., 13, 331-403; 1941.
SMITH, J. L. B. The fishes of Aldabra. Parts I - IX. Ann. Mag. Nat. Hist., (12), 8, 304-312, 689-697, 886-896, 928-937; continued to (13), 1; 1955-1958.
SMITH, J. L. B. & SMITH, M. M. The fishes of the Seychelles. Grahamstown; 1-215; 1963. (List)

India

DAY, F. The fauna of British India. Fishes. London; 2 volumes; 1889. (Faunal monograph)
MISRA, K. S. A check-list of the fishes of India, Burma and Ceylon, 2 parts. Rec. Indian Mus., 45, 1-46, 377-431; 1947-1949.
MISRA, K. S. An aid to the identification of the fishes of India, Burma and Ceylon, I, II. Rec. Indian Mus., 49, 89-137, 1952; 50, 367-422, 1953.
DAY, F. The fishes of India. London; 2 volumes; 1958. (Reprint of 1889)
QURESHI, M. R. A field-key for the identification of fishes . . . . Scientist, 3, 4, 7; 1959-1965.
JONES, S. & SILAS, E. G. A systematic review of the scombroid fishes of India. Symp. Scombroid Fishes, (ser. 1) part 1, 1-105; 1964. (Faunal monograph)
(anonymous). Marine fishes of Karachi and the coasts of Sind and Mahran. Karachi; 1-80; 1955.
BAL, D. V. & MOHMED, K. H. A systematic account of the eels of Bombay. Journ. Bombay Nat. Hist. Soc., 54, 732-740; 1957. (Key to genera and species)
SHAW, G. E. & SHEBBEARE, E. O. The fishes of northern Bengal. Journ. Roy. Asiatic Soc. Bengal, Sci., 3, 1-137; 1937.
SRIVASTAVA, G. J. Fishes of eastern Uttar Pradesh. Varanasi, India; 1-163; 1968. (Fauna & keys)
MENON, A. G. K. A distributional list of fishes of the Himalayas. Journ. Zool. Soc. India, 14, 23-32; 1963.
DUNCKER, G. Die Süsswasserfische Ceylons. Jahrb. Wiss. Anst., 29, 241-272; 1912. (Faunal list)
MENDIS, A. S. Fishes of Ceylon. Bull. Fish. Res. Sta. Ceylon, 2, 1-122; 1954. (Fauna, list, keys)
MUNRO, I. S. R. The marine and fresh-water fishes of Ceylon. Canberra; 1-351; 1955. (Fauna, keys)
DERANIYAGALA, P. E. P. A colored atlas of some vertebrates from Ceylon. Vol. 1 — Fishes. Colombo; 1-149; 1952. (Keys)
MENDIS, A. S. & FERNANDO, C. H. A guide to the freshwater fauna of Ceylon. Bull. Fish. Res. Sta. Ceylon, 12, 1-160; 1962.
AHMAD, N. Fish fauna of East Pakistan. Pakistan Journ. Sci., 5, 18-24; 1953. (Faunal list)

Indo-Malaya

TINT HLAING, U. A classified list of fishes of Burma. Occ. Pap. Indo-Pacific Fish Counc., 67, (10), 1-24; 1967.
STAUCH, A. & d'AUBENTON, F. Poissons Pleuronectiformes du Cambodge. Cah. ORSTOM, Oceanogr., 4, 137-158; 1966.
PUNPOKA, S. A review of the flatfishes of the Gulf of Thailand. . . . Kasetsart Univ. Fish. Res. Bull., 1, 1-86; 1964. (Keys)
FOWLER, H. W. A list of the fishes known from Malaya. Fishery Bull., Singapore, 1, 1-268; 1938. (List)
SCOTT, J. S. An introduction to the sea-fishes of Malaya. Kuala Limpur; 1-180; 1959. (Keys)
LINDSEY, C. C. Guide to families of Malaysian fishes. Singapore; 1-60; 1963.
CHEVEY, P. Inventoire de la faune ichtyologique de l'Indochine. Notes Inst. Oceanogr. Indochine, 19, 1-31; 1932. (Faunal list)
SUVATTI, C. Index to fishes of Siam. Bangkok; 1-226; 1936. (Faunal list)
SMITH, H. M. The fresh-water fishes of Siam, or Thailand. United States Nat. Mus. Bull., 188, 1-622; 1945. (Faunal monograph)
ROFEN, R. R. Handbook of the food fishes of the Gulf of Thailand. La Jolla, California; 1-235; 1963. (Also as: Spec. Rep. Aqu. Res. Inst. Stockton, 1, 1-236) (Fauna and keys)
ALFRED, E. R. Singapore fresh-water fishes. Malay Nat. Journ., 15, 1-19; 1961. (Keys)

ALFRED, E. R. The fresh-water fishes of Singapore. Zool. Verh., 78, 1-68; 1966.

Malay Archipelago

WEBER, M. & de BEAUFORT, L. F. The fishes of the Indo-Australian Archipelago. Leiden, 11 volumes; 1911-1964. (Faunal monograph) (Some additional authors)
SCHUSTER, W. H. et al. Handleiding voor het determineren van vissen. Buitenzorg; 1-69; 1949. (Keys)
HERRE, A. W. A checklist of fishes from Sandakan, British North Borneo. Journ. Pan-Pacific Res. Inst., 8, (4), 2-5; 1933. (Faunal list)
INGER, R. F. & KONG, C. P. The fresh-water fishes of North Borneo. Fieldiana, Zool., 45, 1-268; 1962.
MUNRO, I. S. R. The fishes of the New Guinea region. A check-list . . . . Fish. Bull. Papua, 1, 97-369; 1958. (Faunal list and synonymic catalog)
MUNRO, I. S. R. The fishes of New Guinea. Port Moresby; 1-650; 1967. (Fauna and keys)

Australia

ROUGHLEY, T. C. Fishes of Australia and their technology. Sydney; 1-296; 1916. (Keys)
McCULLOCH, A. R. A check-list of the fishes recorded from Australia, I, II, III. Mem. Australian Mus., 5, 1-144, 145-329, 329-436; 1929. (Faunal list)
WHITLEY, G. P. The fishes of Australia. Part 1. The sharks. . . . Sydney; 1-280; 1940. (Keys)
MUNRO, I. S. R. Handbook of Australian fishes. Nos. 1-7. Fish. Newsl., vol. 14-20; 1956-1957. (Keys)
WHITLEY, G. P. List of the native freshwater fishes of Australia. Proc. Roy. Zool. Soc. New South Wales, 1954-55, 39-47; 1956. (Faunal list)
WHITLEY, G. P. The freshwater fishes of Australia. Monogr. Biol., 8, 136-149; 1959. (Faunal list)
PARROTT, A. W. Sea angler's fishes of Australia. Melbourne; 1-208; 1959. (Keys)
WHITLEY, G. Marine fishes. Brisbane; 2 volumes; 1962. (Keys)
WHITLEY, G. P. A survey of Australian ichthyology. Proc. Linn. Soc. New South Wales, 89, 11-127; 1964. (Faunal list)
OGILBY, J. D. Check-list of the cephalochordates, selachians and fishes of Queensland. Part I. Mem. Queensland Mus., 5, 70-98; 1916. (Faunal list)
McCULLOCH, A. R. & WHITLEY, G. P. A list of the fishes recorded from Queensland waters. Mem. Queensland Mus., 8, 125-182; 1925. (Faunal list)
BLEAKLEY, M. C. & GRANT, E. M. Key to the common freshwater fishes of southern Queensland. Queensland Nat., 15, (1-2), 21-26; 1954. (Keys)
MARSHALL, T. C. et al. Know your fishes. An illustrated guide to principal commercial fishes . . . of Queensland. Brisbane; 1-113; 1959. (Keys)
GRANT, E. M. Guide to fishes. Queensland; 1-280; 1965. (Keys)
MARSHALL, T. C. Tropical fishes of the Great Barrier Reef. Sydney; 1-239; 1966. (Keys)
McCULLOCH, A. R. Checklist of the fishes of New South Wales, 3 parts. Australian Zool., 1-2; 1920-22.
McCULLOCH, A. R. The fishes and fish-like animals of New South Wales. Sydney; 1-104; 1934. (Also previous editions, 1927 and ?)
LORD, C. A list of the fishes of Tasmania. Pap. Proc. Roy. Soc. Tasmania, 1922, 60-73; 1923. (List)
LORD, C. List of the fishes of Tasmania. Journ. Pan-Pacific Res. Inst., 2, (4), 11-16; 1927. (Faunal list)
SCOTT, E. O. G. Observations on some Tasmanian fishes. Part X. Pap. Roy. Soc. Tasmania, 95, 49-65; 1961. (Keys to genera and species)
WAITE, E. R. Catalogue of the fishes of South Australia. Rec. South Australian Mus., 2, 1-199; 1921.
WAITE, E. R. The fishes of South Australia. Handbooks of the flora and fauna of South Australia. Adelaide; 1-243; 1923. (Keys)
SCOTT, T. D. The sharks and rays of South Australia with keys to the species. South Australian Nat., 29, 55-66; 1955.
SCOTT, T. D. The marine and fresh water fishes of South Australia. Adelaide; 1-338; 1962. (Keys)
WHITLEY, G. P. A list of the fishes of Western Australia. Fish. Bull. Western Australia, 2, 1-35; 1948.

New Zealand

PHILLIPPS, W. J. A review of the elasmobranch fishes of New Zealand. No. 1. New Zealand Journ. Sci. Tech., 6, 257-269; 1924.
PHILLIPPS, W. J. Sharks of New Zealand. No. 2. New Zealand Journ. Sci. Tech., 10, 221-226; 1928.
PHILLIPPS, W. J. The fishes of New Zealand, volume 1. New Plymouth, New Zealand; 1-87; 1940.
PHILLIPPS, W. J. Sharks of New Zealand. Dom. Mus. Rec., Zool., 1, 5-20; 1946. (Keys)
RICHARDSON, L. R. & GARRICK, J. A. F. A guide to the lesser chordates and the cartilaginous fishes. Tuatara, 5, 22-37; 1953.
STOKELL, G. Fresh water fishes of New Zealand. Christchurch; 1-145; 1955. (Keys)
McDOWALL, R. M. A guide to the identification of New Zealand fresh-water fishes. Tuatara, 14, 89-104; 1966. (Keys)
WHITLEY, G. P. A check-list of the fishes recorded from the New Zealand region. Australian Zool., 15, 1-102; 1968.

The Orient

FOWLER, H. W. & BEAN, B. A. The fishes of the ... Philippine seas and adjacent waters. United States Nat. Mus. Bull., 100, vol. 7, 8, 10, 11, 12, 13; 1928-1941.
FOWLER, H. W. A list of Philippine fishes. Copeia, 58, 62-65; 1918. (Faunal list)

ROXAS, H. A. & MARTIN, C. A check list of Philippine fishes. Philippine Dep. Agric. Comm., Tech. Bull., 6, 1-314; 1937. (Faunal list)
UMALI, A. F. Key to the families of commercial fishes in the Philippines. Res. Rep. United States Fish. Serv., 21, 1-47; 1951. (Keys)
HERRE, A. W. Check list of Philippine fishes. Res. Rep. United States Fish Wildl. Serv., 20, 1-977; 1953.
JORDAN, D. S. & SEALE, A. Fishes of the islands of Luzon and Panay. Bull. Bur. Fish., 1906, 26, 1-48; 1907.
OSHIMA, M. Contributions to the study of the fresh water fishes of the island of Formosa. Ann. Carnegie Mus., 12, 169-328; 1919.
CHEN, J. T. F. Check-list of the species of fishes known from Taiwan (Formosa). Quart. Journ. Taiwan Mus., 4, 181-210, 1951; 5, 305-341, 1952; 6, 102-140, 1953. (Faunal list)
CHEN, J. T. A review of the sharks of Taiwan. Biol. Bull. Dep. Biol. Coll. Sci. Tunghai Univ., Ichthyol. Ser., 1, 1-102; 1963.
CHEN. J. T. F. & WING, H. T. C. A review of the apodal fishes of Taiwan. Biol. Bull. Tunghai Univ., 32, 1-86; 1967.
CHEN, J. T. F. et al. A review of the pediculate fishes of Taiwan. Biol. Bull. Tunghai Univ., 33, 1-23; 1967.
TANG, D. S. The elasmobranchiate fishes of Amoy. Nat. Sci. Bull. Univ. Amoy, 1, 29-111; 1934.
SHEN, S. C. [Shih-Chien, s.] A list of the fishes from Hong Kong. Part I. Quart. Journ. Taiwan Mus., 17, 187-208; 1964.
POPE, C. H. & NICHOLS, J. T. The fishes of Hainan. Bull. American Mus. Nat. Hist., 54, 321-394; 1927.
LIN, S. Y. Freshwater fishes of Hong Kong. Journ. Hong Kong Fish. Res. Sta., 2, (1), 75-101; 1949.
CHAN, W. L. Marine fishes of Hong Kong. Part I. Hong Kong; 1-129; 1968. (Fauna, keys)
RENDALL, H. Beiträge zur Kenntnis der chinesischen Süsswasserfische. I. Systematischer Teil. Ark. Zool., 20A, 1-194; 1928. (Faunal list)
FOWLER, H. W. A synopsis of the fishes of China. Parts I - X, in many issues. Hong Kong Nat., 1-10; Journ. Hong Kong Fish. Res. Sta., 2; Quart. Journ. Taiwan Mus., 6-11; 1930-1962.
REEVES, C. D. Manual of the vertebrate animals of northeastern and central China exclusive of birds. Shanghai; 1-806; 1933. (Keys)
CHEVEY, P. & LEMASSON, J. Contribution a l'etude poissons des eaux douces Tonkinoises. Hanoi; 1-183; 1937.
NICHOLS, J. T. The fresh-water fishes of China. In Natural History of Central China, vol. IX. New York; 1-322; 1943.
JORDAN, D. S. & STARKS, E. C. List of fishes recorded from Okinawa or the Riu Kiu Islands of Japan. Proc. United States Nat. Mus., 32, 491-504; 1907. (Faunal list)
SNYDER, J. O. The fishes of Okinawa, one of the Riu Kiu Islands. Proc. United States Nat. Mus., 42, 487-519; 1912.
SCHMIDT, P. Fishes of the Riu-Kiu Islands. Trans. Pacific Comm. Acad. Sci. U. S. S. R., 1, 19-156; 1930.
AOYAGI, H. Notes on the fishes of the Ryukyu Islands. V. The marine fishes of Miyako Island. Biogeographica, 3, 287-301; 1941. (Faunal List)
OKADA, Y. Annotated list of animals of Okinawa Islands. Naka City, Japan; 1-384; 1959.
MORI, T. The fresh water fishes of Jehol. Rep. First Sci. Exped. Manchoukuo, (V), 1, 1-61; 1934.
JORDAN, D. S. & METZ, A. A catalog of the fishes known from the waters of Korea. Mem. Carnegie Mus., 6, 1-65; 1913. (Faunal list)
MORI, T. Check list of the fishes of Korea. Mem. Hyogo Univ. Agric., 1, (3), 1-228; 1952. (Faunal list)
SOLDATOV, V. K. & LINDBERG, G. J. A review of the fishes of the seas of the Far East. Vladivostok; 1-576; 1930. (In Russian)
TARANETZ, A. J. Handbook for identification of Soviet Far East and adjacent waters. Vladivostok; 1-200; 1937. (In Russian)
RASS, T. S. Deep-sea fish of the far-east seas of U. S. S. R. Zool. Zh., 33, 1312-1324; 1954. (In Russian)
SCHMIDT, P. Y. Fishes of the Sea of Okhotsk. Trans. Pacific Comm. Leningrad, 6, 1-320; 1950. (In Russian)
SHMIDT, P. Y. Fishes of the Sea of Okhotsk. Jerusalem; 1-392; 1965. (Translation of 1950)
LINDBERG, G. U. & LEGHEZA, M. I. Fishes of the Japanese Sea and the adjacent parts of the Okhotsk and Yellow Seas. 3 parts. Moskva. (Part 1 as Tabl. Anal. Faune U. R. S. S., 68, (1), 1-208; 1959-1969. (In Russian)
LINDBERG, G. U. & LEGEZA, M. I. Fishes of the Sea of Japan and the adjacent areas of the Sea of Okhotsk and the Yellow Sea. I. Jerusalem; 1-198; 1967. (Translation of 1959) (Keys)
YOSHIDA, H. & ITO, T. Fish fauna of the Japan Sea. Journ. Shimonoseki Coll. Fish., 6, 261-270; 1957. (Faunal list)
TANAKA, S. Figures and descriptions of the fishes of Japan. Tokyo; 48 parts. (In Japanese & English) (2nd edition of 30 parts, 1935)
JORDAN, D. S. et al. A catalogue of the fishes of Japan. Journ. Coll. Sci. Tokyo Imp. Univ., 33, (1), 1-497; 1913. (Synonymic catalogue)
TOMIYAMA, I. & ABE, T. Figures and descriptions of fishes of Japan. Tokyo, vols. 49-56; 1953-1954.
MATSUBARA, K. Fish morphology and hierarchy. Tokyo; 3 volumes; 1955. (In Japanese)
KAMOHARA, T. Coloured illustrations of the fishes of Japan. Osaka; 1-135; 1955.
KAMOHARA, T. A catalogue of fishes of Kochi Prefecture (Province Tosa), Japan. Rep. USA Mar. Biol. Sta., 5, (1), 1-76; 1958.
OKADA, Y. Studies on the freshwater fishes of Japan. Journ. Fac. Fish. Univ. Mie, 4, 1-860; 1960. (Fauna and list; also as separate book)
OCHIAI, A. Fauna Japonica. Soleina. Tokyo; 1-114; 1963. (Faunal monograph)

KAMOHARA, T. On the fishes of the family Scaridae of Japan, including the Riukiu Islands. Rep. USA Biol. Sta., 10, (1), 1-24; 1963.
NAKAMURA, M. Keys to the freshwater fishes of Japan fully illustrated in colors. Tokyo; 1-258; 1963.
KAMOHARA, T. Revised catalogue of fishes of Kochi Prefecture, Japan. Rep. USA Mar. Biol. Sta. Kochi Univ., 11, (1), 1-99; 1964.
OKADA, Y. Fishes of Japan. Illustrations and descriptions of fishes of Japan., 2nd ed. Tokyo; 1-474; 1966. (Keys; 1st edition, 1955)
KAMOHARA, T. Fishes of Japan in color. Osaka; 1-135; 1967. (Keys)

Oceania

FOWLER, H. W. A list of the sharks and rays of the Pacific Ocean. Proc. 4th Pacific Sci. Congr., 481-508; 1929. (Faunal list)
RASS, T. S. Biology of the Pacific Ocean. Book III. Fishes of the open waters. Moskva; 1-273; 1967. (In Russian)
RASS, T. S. Fishes of the open waters of the Pacific Ocean. Jerusalem; 1-266; 1967. (Translation of previous item)
HIKITA, T. A preliminary list of fishes from the northern Pacific Ocean. Journ. Fac. Sci. Hokkaido Univ., Zool., 13, 46-48; 1957. (Faunal list)
FOWLER, H. W. The fishes of Oceania. Mem. Bernice P. Bishop Mus., 10, 1-540; 1928. Three supplements, 1931, 1934, 1949; faunal list)
KAMI, H. T. et al. Check-list of Guam fishes. Micronesica, 4, 95-131; 1968. (Faunal list)
JORDAN, D. S. & EVERMANN, B. W. The shore fishes of the Hawaiian Islands. Bull. United States Fish Comm., 23, (I), 1-574; 1905.
JORDAN, D. S. & JORDAN, E. K. A list of the fishes of Hawaii. . . . Mem. Carnegie Mus., 10, 1-92; 1922.
FOWLER, H. W. A list of Hawaiian fishes. Copeia, 112, 82-84; 1922. (Faunal list)
FOWLER, H. W. et al. Fishes of Hawaii, Johnston Island and Wake Island. Bull. Bernice P. Bishop Mus., 26, 1-31; 1925. (Faunal list)
JORDAN, D. S. & EVERMANN, B. W. A check-list of the fishes of Hawaii. Journ. Pan-Pacific Res. Inst., 1, 3-15; 1926. (Faunal list)
TINKER, S. W. Hawaiian fishes. A handbook of the fishes found among the islands of the Central Pacific Ocean. Honolulu; 1-404; 1944. (Keys)
CLARK, E. Notes on some Hawaiian plectognath fishes. . . . American Mus. Novit., 1397, 1-22; 1949. (Keys to families and species)
JUNE, F. C. & REINTJES, J. W. Common tuna-bait fishes of the central Pacific. Res. Rep. United States Fish. Serv., 34, 1-54; 1953. (Hawaiian & Gilbert Islands)
GOSLINE, W. A. The inshore fish fauna of Johnston Island, a central Pacific atoll. Pacific Sci., 9, 442-480; 1955.
GOSLINE, W. A. & BROCK, V. E. Handbook of Hawaiian fishes. Honolulu; 1-372; 1960. (Fauna and faunal list and keys)
SCHULTZ, L. P. et al. Fishes of the Marshall and Marianas Islands. United States Nat. Mus. Bull. 202, 3 volumes; 1953-1966. (Faunal monograph)
RANDALL, J. E. Fishes of the Gilbert Islands. Atoll Res. Bull., 47, 1-243; 1955.
WHITLEY, G. P. A check list of the fishes of the Santa Cruz Archipelago, Melanesia. Journ. Pan-Pacific Res. Inst. Honolulu, 3, 11-13; 1928.
FOWLER, H. W. Fishes of Fiji. Suva; 1-670; 1959. (Faunal monograph)
JORDAN, D. S. & SEALE, A. The fishes of Samoa. . . . Bull. United States Bur. Fish., 25, 173-455; 1906.
HERRE, A. W. Check list of fishes recorded from Tahiti. Journ. Pan-Pacific Res. Inst., 7, 2-11; 1932
WILHELM, O. E. & HULOT, A. L. Pesca y peces di la Isla de Pascua. Bol. Soc. Biol. Concepcion, 32, 139-152; 1957. (Faunal list of Easter Island)

Antarctica

NORMAN, J. R. Coast fishes. Part III. The Antarctic zone. Discovery Rep., 18, 1-104; 1938.
BLANC, M. Les poissons des terres australes et antarctiques francaises. Mem. Inst. Sci. Madagascar, 4F, 109-159; 1961.
TAYLOR, W. R. Fishes of Arnhem Land. Rec. American-Australian Sci. Exped. Arnhem Land, 4, 45-307; 1964.
ANDRIASHEV, A. P. A general review of the Antarctic fish fauna. Biol. Res. Soviet Antarctic Exp. (1955-1958) Issled. Faun. Morei, 2, 335-386; 1964. (Also in: Monographia Biol., 15, 491-550)
ANDRIYASHEV, A. P. Field key to coastal species of fishes of east Antarctica. Inf. Byull. Soviet Antarkt. Eksped., 5, 392-395; 1966. (Keys)

SPECIAL LISTS

[See also, under Monographs, below]

Abyssal. GREY, M. The distribution of fishes found below a depth of 2000 meters. Fieldiana, Zool., 36, 75-337; 1956.
Abyssal. NYBELIN, O. Deep-sea bottom fishes. Rep. Sweish Deep-sea Exped. 1947-48, 2, Zool. 3, 247-345; 1957.
Asymmetrical. CHABANAUD, P. Catalogue systematique et chorologique des Teleosteens dyssymetriques du Globe. Bull. Inst. Oceanogr., 763, 1-31; 1939.

Blind. THINES, G. Les poissons aveugles. Ann. Soc. Roy. Zool. Belgique, 86, 1-128; 1955.
Circumtropical. BRIGGS, J. C. Fishes of worldwide (circumtropical) distribution. Copeia, 1960, 171-180; 1960.
Euryhaline. GUNTER, G. A list of the fishes of the mainland of North and Middle America recorded from both freshwater and sea water. American Midl. Nat., 28, 305-326; 1942.
Euryhaline. GUNTER, G. A revised list of euryhalin fishes of North and Middle America. American Midl. Nat., 56, 345-354; 1956.
Hybrids. SLASTENENKO, E. P. Una lista de los hibridos naturales de peces del mundo. Rev. Soc. Mexicana Hist. Nat., 17, 63-84; 1956.
Oceanic. NORMAN, J. R. Oceanic fishes . . . collected in 1925-27. Parts I, II. Discovery Rep., 2, 263-357; 1930.
Pelagic. WILLIAMS, F. Preliminary survey of the pelagic fishes of East Africa. Fish Publ., London, 8, 1-68; 1956.

MONOGRAPHIC, DESCRIPTIVE, & REVISIONARY

Monographs

GOODE, G. B. & BEAN, T. H. Oceanic ichthyology. A treatise on the deep-sea and pelagic fishes of the world. United States Nat. Mus. Spec. Bull, 2, 2 volumes; 1895.
BRIDGE, T. W. Fishes (except of the systematic account of Teleosti). Cambridge Nat. Hist., volume VII, 139-537; 1910.

Monographs by groups

[These comprehensive works cover groups of families of various size. They are arranged here in approximately the order of Berg's 1947 "Classification of Fishes." Most are not listed also under the included families, because they include no revisions of the species or genera of the families but only studies of the families themselves in relation to others.]

Cyclostomata. KAHSBAUER, P. Cat. Faun. Austriae, Teil 21aa, 1-56; 1961. FONTAINE, M. et al. Superordre des Petromyzonoidea et des Myxinoidea. Anatomie. Traite de Zool., 13, fasc. 1, 13-144; 1958.
Placodermi. ARAMBOURG, C. Traite de Zool., 13, fasc. 3, 1990-2009; 1958. (Bibliography and classification to family)
Acanthodii. ARAMBOURG, C. Traite de Zool., 13, fasc. 3, 1984-1989; 1958. (Classification to family)
Elasmobranchii. WHITE, E. G. A classification and phylogeny of the elasmobranch fishes. American Mus. Novit., 837, 1-16; 1936. WHITE, E. G. Interrelationships of the elasmobranchs with a key to the order Galea. Bull. American Mus. Nat. Hist., 74, 25-138; 1937. (Classification to family)
Chondrichthyes. ARAMBOURG, C. & BERTIN, L. Traite de Zool., 13, fasc. 3, 2010-2015; 1958. (Classification to family)
Selachii. REGAN, C. T. A classification of the selachian fishes. Proc. Zool. Soc. London, 1906, 722-758. BERTIN, L. Essai de classification et de nomenclature des poissons de la sous-classe des Selaciens. Bull. Inst. Oceanogr. Monaco, 775, 1-24; 1939. (Classification, key to orders, suborders, families) SETNA, S. B. & SARANGDHAR, P. N. Selachian fauna of the Bombay waters. Proc. Nat. Inst. Sci. India, 12, 243-259; 1946. (Key to families and genera) GLIKMAN, L. S. On the classification of sharks. Leningrad; 1-20; 1958. (In Russian)
Plagiostomia. GARMAN, S. The Plagiostomia (sharks, skates, and rays). Mem. Mus. Comp. Zool., 36, 1-515; 1913. (Monograph)
Galea. WHITE, E. G. Interrelationships of the elasmobranchs with a key to the order Galea. Bull. American Mus. Nat. Hist., 74, 25-138; 1937. (Key to families)
Bradyodonti. ARAMBOURG, C. & BERTIN, L. Traite de Zool., 3, fasc. 3, 2057-2067; 1958. (Bibliography and classification to family)
Dipneusti. ARAMBOURG, C. & GUIBE, J. Traite de Zool., 13, fasc. 3, 2522-2540; 1958. (Classification to family)
Crossopterygii. ARAMBOURG, C. Traite de Zool., 13, fasc. 3, 2541-2552; 1958. (Classification to family)
Chondrostei. LEHMAN, J. P. Traite de Zool., 13, fasc. 3, 2130-2172; 1958. (Classification to family) GARDINER, B. G. Further notes on palaeoniscoid families with a classification of the Chondrostei. Bull. British Mus. (Nat. Hist.), Geol., 14, 143-206; 1967. (Classification)
Holostei. ARAMBOURG, C. & BERTIN, L. Traite de Zool., 13, fasc. 3, 2173-2203; 1958. (Classification to family)
Teleosti. REGAN, C. T. The classification of teleostean fishes. Ann. Mag. Nat. Hist., (18), 3, 75-86; 1909. BOULENGER, G. A. Teleosti (systematic part). Cambridge Nat. Hist., volume VII. London; 539-727; 1910. (Monograph) KYLE, H. M. The classification and phylogeny of the Teleosti anteriores. Wiss. Meeresuntersuch. Kiel, (n. f.), 14, Abt. Helgoland, 201-239; 1923. TRETIAKOV, D. K. The classification of primitive Teleostei. Bull. Acad. Sci. U. R. S. S., Biol., 1945, 49-55; 1945. BERTIN, L. & ARAMBOURG, C. Traite de Zool., 13, fasc. 3, 2204-2500; 1958. (Classification to family) GREENWOOD, P. H. et al. Phyletic studies of teleostian fishes, with a provisional classification of living forms. Bull. American Mus. Nat. Hist., 131, 339-456; 1966. GREENWOOD, P. H. et al. Named main divisions of teleostean fishes. Proc. Biol. Soc. Washington, 80, 227-228; 1967.

Halecostomi. ARAMBOURG, C. & BERTIN, L. [See above under Holostei]
Isospondyli. GOSLINE, W. A. Contribution toward a classification of modern isospondylous fishes. Bull. British Mus. (Nat. Hist), Zool., 6, 327-365; 1960. (Classification to family)
Clupeoidea. WHITEHEAD, P. J. P. A contribution to the classification of clupeoid fishes. Ann. Mag. Nat. Hist., (13), 5, 737-750; 1963.
Esocoidei. BERG, L. S. The suborder Esocoidei.... Bull. Inst. Rech. Biol. Perm., 10, 385-391; 1936.
Stomiatoidei. REGAN, C. T. The classification of the stomiatoid fishes. Ann. Mag. Nat. Hist., (9), 11, 612-614; 1923.
Iniomi. REGAN, C. T. The anatomy and classification of the teleostean fishes of the order Iniomi. Ann. Mag. Nat. Hist., (8), 7, 120-133; 1911. PARR, A. E. A contribution to the osteology and classification of the orders Iniomi . . . . Occ. Pap. Bingham Oceanogr. Coll., 2, 1-45; 1929. MARSHALL, N. B. Studies of alepisaud fishes. Discovery Rep., 27, 303-336; 1955.
Saccopharyngiformes. GILL, T. & RYDER, J. A. On the literature and systematic relations of the saccopharyngoid fishes. Proc. United States Nat. Mus., 7, 48-65; 1884.
Lyomeri. The anatomy and classification of the teleostean fishes of the order Lyomeri. Ann. Mag. Nat. Hist., (8), 10, 347-349; 1912.
Ostariophysi. REGAN, C. T. The classification of the teleostean fishes of the order Ostariophysi. Ann. Mag. Nat. Hist., (8), 8, 553-577; 1911.
Nematognathi. EIGENMANN, C. H. & EIGENMANN, R. S. A revision of the South American Nematognathi or cat-fishes. Occ. Pap. California Acad. Sci., 1, 1-508;1890.
Apodes. REGAN, C. T. The osteology and classification of the teleostean fishes of the order Apodes. Ann. Mag. Nat. Hist., (8), 10, 377-387; 1912. TREWAVAS, E. A contribution to the classification of the fishes of the order Apodes. Proc. Zool. Soc. London, 1932, 639-659; 1932.
Synentognathi. The classification of the teleostean fishes of the order Synentognathi. Ann. Mag. Nat. Hist., (8), 7, 327-335; 1911.
Anacanthini. REGAN, C. T. On the systematic position and classification of the gadoid or anacanthine fishes. Ann. Mag. Nat. Hist., (7), 11, 459-466; 1903.
Gadiformes. SVETOVIDOV, A. N. Über die Klassification der Gadiformes oder Anacanthini. Bull. Acad. Sci. U. R. S. S., Biol., 1937, 1281-1288; 1937. SVETOVIDOV, A. N. Fauna of U. S. S. R., Fishes, vol. IX, No. 4. Washington, D. C.; 1-304; 1962. (Translation of 1948)
Allotriognathi. REGAN, C. T. On the anatomy, classification, and systematic position of the teleostean fishes of the suborder Allotriognathi. Proc. Zool. Soc. London, 1907, 634-643; 1907.
Veliferoidei. Synopsis of the lampridiform suborder Veliferoidei. Copeia, 1960, 245-247; 1960. WALTERS, V. Synopsis of the lampridiform suborder Veliferoidei. Copeia, 1960, 245-247; 1960.
Trachipteroidei. WALTERS, V. & FITCH, J. E. The families and genera of the lampridiform (Allotriognath) suborder Trachipteroidei. California Fish Game, 46, 441-451; 1960.
Microcyprini. REGAN, C. T. The osteology and classification of the teleostean fishes of the order Microcyprini. Ann. Mag. Nat. Hist., (8), 7, 320-327; 1911.
Phallostethiformes. ARBOCCO, G. Morfologia, biologia e systematica dei Phallostethiformes. Natura, 47, 51-60; 1956.
Xenoberyces. REGAN, C. T. The anatomy and classification of the teleostean fishes of the orders ... Xenoberyces. Ann. Mag. Nat. Hist., (8), 7, 1-9; 1911. PARR, A. E. A contribution to the osteology and classification of the orders Iniomi and Xenoberyces. Occ. Pap. Bingham Oceanogr. Coll., 2, 1-45; 1929.
Berycomorphi. REGAN, C. T. The anatomy and classification of the teleostean fishes of the orders Berycomorphi . . . . Ann. Mag. Nat. Hist., (8), 7, 1-9; 1911.
Zeomorphi. REGAN, C. T. The anatomy and classification of the teleostean fishes of the order Zeomorphi. Ann. Mag. Nat. Hist., (8), 6, 481-484; 1910.
Mugiliformes. GOSLINE, W. A. Systematic position and relationships of the percesocine fishes. Pacific Sci., 16, 207-217; 1962.
Atheriniformes. ROSEN, D. E. The relationships and taxonomic position of the halfbeaks, killifishes, silverside, and their relatives. Bull. American Mus. Nat. Hist., 127, 217-267; 1964.
Perciformes. GOSLINE, W. A. The suborders of perciform fishes. Proc. United States Nat. Mus., 124, (3647), 1-78; 1968.
Percoidei. REGAN, C. T. The classification of the percoid fishes. Ann. Mag. Nat. Hist., (8), 12, 111-145; 1913.
Blennioidea. REGAN, C. T. The classification of the blennioid fishes. Ann. Mag. Nat. Hist., (8), 10, 265-280; 1912.
Scombroidei. REGAN, C. T. On the anatomy and classification of the scombroid fishes. Ann. Mag. Nat. Hist., (8), 3, 66-75; 1909. LeDANOIS, E. & LeDANOIS, Y. L'ordre des Scombres. Mem. Inst. Francaise Afrique Noire, 68, 153-192; 1963. (Classification)
Gobioidei. REGAN, C. T. The osteology and classification of the gobioid fishes. Ann. Mag. Nat. Hist., (8), 8, 729-733; 1911. KOUMANS, F. P. A preliminary revision of the genera of the gobioid fishes with united ventral fins. Lisse; 1-174; 1932.
Scleroparei. REGAN, C. T. The osteology and classification of the teleostean fishes of the order Scleroparei. Ann. Mag. Nat. Hist., (8), 11, 169-184; 1913.
Heterosomi. WU, H. -W. Contribution a l'etude morphologique, biologique et systematique des poissons Heterosomes de la Chine. Paris; 1-179; 1932.
Discocephali. REGAN, C. T. The anatomy and classification of the teleostean fishes of the order Discocephali. Ann. Mag. Nat. Hist., (8), 10, 634-637; 1912.
Plectognathi. GILL, T. Synopsis of the plectognath fishes. Proc. United States Nat. Mus., 7, 411-427; 1884. REGAN, C. T. On the classification of the fishes of the suborder Plectognathi. Proc. Zool. Soc. London, 1902, II, 284-303; 1902.

Triacanthoidea. TYLER, J. C. Monograph on plectognath fishes of the superfamily Triacanthoidea. Acad. Nat. Sci. Philadelphia Mon., 16, 1-364; 1968.
Pediculati. REGAN, C. T. The classification of teleostean fishes of the order Pediculati. Ann. Mag. Nat. Hist., (8), 9, 277-289; 1912.
Orbiculati. LeDANOIS, Y. Sur le remaniement des sous-ordre des plectognathes et la les orbiculates. Compt. Rend. Acad. Sci. Paris, 240, 1933-1934; 1955.

Classifications

[See also works listed under Animalia; also works cited under GROUPS, above]

JORDAN, D. S. A classification of fishes including families and genera so far as known. Stanford, California; 77-243; 1923. (Reprinted 1963)
SCHMIDT, H. Einführung in die Palaeontologie. Stuttgart; 1-253; 1935. (Pp. 130-148)
TRETIAKOV, D. K. Outlines of the phylogeny of fishes. Moskva; 1-176; 1944. (In Russian)
LeDANOIS, E. Oceanographie, biologie marine et peches. Remarques ichthyologiques. Rev. Trav. Peches Marit., 13, 55-175; 1944.
ROMER, A. S. Vertebrate paleontology. Chicago; 1-687; 1945. (Previous edition 1933)
BERG, L. S. Classification of fishes, both Recent and fossil. Ann Arbor, Michigan; 1-517; 1947. (In Russian and English)
BERG, L. S. The systematics of fishes, Recent and past, 2nd ed. Trav. Inst. Zool. Acad. Sci. U. R. S. S., 20, 1-286; 1955. (In Russian)
MATSUBARA, K. Fish morphology and hierarchy. Tokyo; 3 volumes; 1955. (In Japanese)
BLAIR, W. F. et al. Vertebrates of the United States. New York; 1-819; 1957. (To orders)
NORMAN, J. R. A draft synopsis of orders, families, and genera of Recent fishes. London; 1-649; 1957.
BERG, L. S. System der rezenten und fossilen Fischartigen und Fische. Berlin; 1-310; 1958. (Translated from 1955)
BERTIN, L. & ARAMBOURG, C. Systematique des poissons. Traite de Zool., 13, fasc. 3, 1967-1983; 1958. (To family)
BERG, L. S. Classification of fishes both Recent and fossil. Bangkok; 1965.
KLEE, A.-J. A new classification of fishes. Parts 1, 2, 3. Aquarium, 1, (12), 14-15; 2, (1), 14-16; 2, (2), 22-23, 72-76; 1968.
BLAIR, W. F. et al. Vertebrates of the United States, 2nd ed. New York; 1-616; 1968.
McALLISTER, D. E. Evolution of branchiostegals and classification of teleostome fishes. Bull. Nat. Mus. Canada, 221, 1-239; 1968.

Descriptive catalogs

GÜNTHER, A. C. L. G. Catalogue of the . . . fishes in the collection of the British Museum. London; 7 volumes; 1859-1870
GÜNTHER, A. Catalogue of the fishes in the British Museum (Natural History). 8 volumes; 4268 pp.; 1964. (Reprint of 1859-1870)
JORDAN, D. S. & EVERMANN, B. W. The fishes of North and Middle America. A descriptive catalogue . . . , 4 parts. United States Nat. Mus. Bull., 47, 1-3313; 1896-1900.
BOULENGER, G. A. Catalogue of the fresh-water fishes of Africa in the British Museum (Natural History). London; 4 volumes; 1909-1915. (Reprinted 1964)

Synonymic catalogs

JORDAN, D. S. & EVERMANN, B. W. A check-list of the fishes . . . of North and Middle America. Rep. United States Comm. Fish Fish., 1895, App. 5, 207-584; 1896.
EIGENMANN, C. H. Catalogue of fresh-water fishes of tropical and south temperate America, volume 3, part 4, 375-777; 1909.
JORDAN, D. S. The genera of fishes. Stanford, California; 1-576; 1917-1920. (Reprinted 1963)
JORDAN, D. S. et al. Checklist of the fishes . . . of North and Middle America. Rep. United States Comm. Fish., 1928, part II, 1-670; 1930.
HERRE, A. W. Check list of Philippine fishes. United States Fish Wildl. Serv. Res. Rep., 20, 1-977; 1953.
GOSLINE, W. A. & BROCK, V. E. Handbook of Hawaiian fishes. Honolulu; 308-346; 1960.
FOWLER, H. W. A catalog of world fishes, 5 parts. Quart. Journ. Taiwan Mus., 17-19; 1964-1966.

Revisionary works by family

These lists of works were made up primarily from two sources: (1) listings in the Zoological Record for the years 1910 to 1968, which were not checked further if the title appeared to be explicit; and (2) library shelves and the office libraries of various specialists in each field; these were examined for appropriateness and, if revisionary, checked for included families. The works listed by families are therefore of two sorts, so far as examination by the compiler is concerned, those cited in the title as referring to one or a few named families (not usually examined or verified) and those directly examined and checked and listed as considered to be appropriate.

Works cited by author

In order to save space in the list of works on individual families, the following studies, which are each referred to more than a few times, are cited by author and date and number only. Citations are otherwise given by title, if a book, or by the serial reference, if not a separate work. The titles of journal articles are not usually given, but the relevant limitations are shown by annotations.

ANDRIYASHEV, A. P., No. 1, 1954. Fishes of the northern seas of the U. S. S. R. St. Petersbourg; 1-566. (In Russian; translated 1964)

ARNOULT, J., No. 1, 1959. Poissons des eaux douces. Faune de Madagascar, 10, 1-163.

BANARESCU, P., No. 1, 1964. Fauna Republici Populare Romine. Pisces — Osteichthyes, vol. XIII. Bucuresti; 1-659.

BAUCHOT, M. L. & BLANC, M., No. 1, 1961. Poissons marins de l'Est Atlantique tropical . . . . 1 ére partie. Atlantide Rep., 6, 43-100.

de BEAUFORT, L. F., No. 8, 1940. The fishes of the Indo-Australian Archipelago. VIII. Leiden; 1-508.

de BEAUFORT, L. F. & BRIGGS, J. C., No. 11, 1962. The fishes of the Indo-Australian Archipelago. XI. Leiden; 1-481.

de BEAUFORT, L. F. & CHAPMAN, W. M., No. 9, 1951. The fishes of the Indo-Australian Archipelago. IX. Leiden; 1-484.

BEEBE, W. & TEE-VAN, J., No. 1, 1928. The fishes of Port-au-Prince Bay, Haiti, with a summary of the known species of marine fish of the island of Haiti and Santo Domingo. Zoologica, 10, 1-279.

BERG, L. S., No. 7, 1911. Poissons (Marsipobranchii et Pisces). Volume I. Faune de la Russie; 1-337. (In Russian)

BERG, L. S., No. 1, 1932. Les poissons des eaux douces de l'U. R. S. S., 3rd ed. Leningrad; I, 1-543.

BERG, L. S., No. 2, 1933. Les poissons des eaux douces de l'U. R. S. S., 3rd ed. Leningrad; II, 547-899.

BERG, L. S., No. 3, 1948. Fresh-water fishes of Soviet Union.... Tabl. Anal. Faune U. R. S. S., I, vol. 27, 1-466. (In Russian)

BERG, L. S., No. 4, 1949. Same, II, vol. 29, 467-925.

BERG, L. S., No. 5, 1949. Same, III, vol. 30, 927-1382.

BERG. L. S., No. 6, 1949. Fresh-water fish of Iran and neighbouring countries. Trav. Inst. Zool. Acad. Sci. U. R. S. S., 8, 783-858. (In Russian)

BERG, L. S., No. 8, 1962. Freshwater fishes of the U. S. S. R. and adjacent countries. Volume I. Washington, D. C.; 1-504. (Translation of 1948)

BERG, L. S., No. 9, 1964. Same; volume II, 1-496. (Translation of 1949)

BERG, L. S., No. 10, 1965. Same; volume III, 1-510. (Translation of 1949)

BERTELSEN, E., No. 1, 1951. The ceratioid fishes. Ontogeny, taxonomy, distribution and biology. Dana Rep., 39, 1-281.

BIGELOW, H. B. & SCHROEDER, W. C., No. 1, 1948. Fishes of the western North Atlantic. Sharks. Mem. Sears Found. Oceanogr. Res., 1, (1), 59-546.

BIGELOW, H. B. & SCHROEDER, W. C., No. 2, 1953. Same; part 2. Mem., 1, (2), 1-588.

BIGELOW, H. B. et al., No. 3, 1964. Same; Mem., 1, (4), 1-599.

BIGELOW, H. B. et al., No. 4, 1966. Same; Mem., 1, (5), 1-647.

BIGELOW, H. B. et al., No. 5, 1963. Same; Mem., 1, (3), 1-630.

BOULENGER, G., No. 1, 1901. Les poissons du bassin du Congo. Bruxelles; 1-537.

BOULENGER, G. A., No. 2, 1907. Zoology of Egypt: The fishes of the Nile. London; 1-578.

BOULENGER, G. A., No. 3, 1909. Catalogue of the fresh-water fishes of Africa in the British Museum (Natural History). London; I, 1-373.

BOULENGER, G. A., No. 4, 1911. Same; II, 1-373.

BOULENGER, G. A,, No. 5, 1915. Same; III, 1-526.

BOULENGER, G. A., No. 6, 1919. Same; IV, 1-392.

CHAN, W. L., No. 1, 1968. Marine fishes of Hong Kong. Part I. Hong Kong; 1-129.

CHEN, J. T. F., No. 1, 1948. A synopsis of the Platosomeae of China. Quart. Journ. Taiwan Mus., 1, 23-50.

DAGET, J., No. 1, 1954. Les poissons du Niger Superieur. Mem. Inst. Francaise Afrique Noire, 36, 1-391.

DAGET, J., No. 2, 1962. Les poissons du Fouta Dialon et de la Basse Guinee. Dakar; 1-210.

DAGET, J. & ILTIS, A., No. 1, 1965. Poissons de Cote d'Ivoire (eaux douces et saumatres). Dakar; 1-385.

DAY, F., No. 1, 1889. The fauna of British India, including Ceylon and Burma. Fishes, vol. 1, 1-548.

DAY, F., No. 2, 1889. Same; vol. 2, 1-509.

DEVINCENZI, G. J., No. 1, 1920. Peces del Uruguay. Anal. Mus. Nac. Montevideo, (2), 1, 97-134.

DEVINCENZI, G. J., No. 2, 1924. Same; Anal., (2), 2, 139-290.

DUNCKER, G. & LADIGES, W., No. 1, 1960. Die Fische der Nordmark. Abhandl. Naturw. Ver. Hamburg, (n. f.), 3, Suppl., 1-432.

EIGENMANN, C. H., No. 1, 1912. The freshwater fishes of British Guiana. Mem. Carnegie Mus., 5, 1-578.
EIGENMANN, C. H., No. 2, 1927. The fresh-water fishes of Chile. Mem. Nat. Acad. Sci., 22, (2), 1-63.
EIGENMANN, C. H. & EIGENMANN, R. S., No. 1, 1890. A revision of the South American Nematognathi or cat-fishes. Occ. Pap. California Acad. Sci., 1, 1-508.
FAGE, L., No. 1, 1918. Shore-fishes. . . . In Rep. Danish Oceanogr. Exp. 1908-10 to Mediterranean and adjacent seas, vol. II, Biology, 1-154.
FOWLER, H. W., No. 1, 1945. A study of the fishes of the southern piedmont and coastal plain. Acad. Nat. Sci. Philadelphia, Mon. 7, 1-408.
FOWLER, H. W., No. 2, 1930. A synopsis of the fishes of China, part I. Hong Kong Nat., 1, 24-34, 79-88, 129-138, 177-189.
FOWLER, H. W., No. 3, 1931. Same, part II. Hong Kong Nat., 2, 69-79, 111-123, 198-108.
FOWLER, H. W., No. 4, 1931. Same, part III. Hong Kong Nat., 3, 46-63, 126-144.
FOWLER, H. W., No. 5, 1932. Same, part IV. Hong Kong Nat., 3, 247-279.
FOWLER, H. W., No. 6, 1933-35. Same, part V. Hong Kong Nat., 4, 156-175; 5, 54-67, 146-155, 210-225, 304-319; 6, 62-77. 132-147, 276-284.
FOWLER, H. W., No. 7, 1936. Same, part VI. Hong Kong Nat., 7, 61-80, 186-202, 271-315.
FOWLER, H. W., No. 8, 1936. The marine fishes of West Africa. . . . Bull. American Mus. Nat. Hist., 70, 1-1493.
FOWLER, H. W., No. 9, 1937-41. A synopsis of the fishes of China. Part VII. Hong Kong Nat., 8, 124-145, 249-289; 9, 58-86, 141-164, 203-220; 10, 39-62, 109-121, 205-222.
FOWLER, H. W., No. 10, 1948-51. Os peixes de agua doce do Brasil (2 entrega). Arqu. Zool. Sao Paulo, 6, 1-625.
FOWLER, H. W., No. 11, 1949. A synopsis of the fishes of China. Part VII continued. Journ. Hong Kong Fish. Res. Sta., 2, 3-65.
FOWLER, H. W., No. 12, 1953. Same. Quart. Journ. Taiwan Mus., 6, 1-77.
FOWLER, H. W., No. 13, 1954. Os peixes de agua doce do Brasil. Vol. 2. Arch. Zool. Sao Paulo, 9, 1-400.
FOWLER, H. W., No. 14, 1954. A synopsis of the fishes of China. Part VII continued. Quart. Journ. Taiwan Mus., 7, 1-110.
FOWLER, H. W., No. 15, 1956. Same, completed. Quart. Journ. Taiwan Mus., 9, 161-354.
FOWLER, H. W., No. 16, 1958. Same. Part VIII. Quart. Journ. Taiwan Mus., 11, 147-339.
FOWLER, H. W., No. 17, 1959. Fishes of Fiji. Suva; 1-670.
FOWLER, H. W., No. 18, 1956. Fishes of the Red Sea and southern Arabia. I. Jerusalem; 1-240. (Volumes II and III not seen)
FOWLER, H. W., No. 19, 1931. The fishes . . . collected by the United States Bureau of Fisheries steamer "Albatross", chiefly in the Philippine Seas and adjacent waters. United States Nat. Mus. Bull., 100, (11), 1-388.
FOWLER, H. W., No. 20, 1933. Same. Bull. 100, (12), 1-465.
FRASER-BRUNNER, A., No. 1, 1943. Notes on the plectognath fishes. VIII. — The classification of the suborder Tetraodontoidea, with a synopsis of the genera. Ann. Mag. Nat. Hist., (11), 10, 1-18.
GARMAN, S., No. 1, 1913. The Plagiostomia (sharks, skates, and rays). Mem. Mus. Comp. Zool., 36, 1-515.
GILL, T., No. 1, 1884. Synopsis of the plectognath fishes. Proc. United States Nat. Mus., 7, 411-427.
GOHAR, H. A. F. & MAZHAR, F. M., No. 1, 1964. The elasmobranchs of the north-western Red Sea. Publ. Mar. Biol. Sta. Al-Ghardaqa, 13, 1-144.
HERRE, A. W., No. 1, 1927. Gobies of the Philippines and the China Sea. Manila; 1-352.
HILDEBRAND, W. F., No. 1, 1946. A descriptive catalog of the shore fishes of Peru. United States Nat. Mus. Bull., 189, 1-530.
HILDEBRAND, S. F. & SCHROEDER, W. C., No. 1, 1923. Fishes of Chesapeake Bay. Bull. United States Bur. Fish., 43, (1), 1-366.
INGER, R. F. & KONG, C. P., No. 1, 1962. The fresh-water fishes of North Borneo. Fieldiana, Zool., 45, 1-268.
JORDAN, D. S. & HERRE, A. C., No. 1, 1906. A review of the herring-like fishes of Japan. Proc. United States Nat. Mus., 31, 613-645.
JORDAN, D. S. & SNYDER, J. O., No. 1, 1901. A review of the apodal fishes or eels of Japan. . . . Proc. United States Nat. Mus., 23, 837-890.
KOUMANS, F. P., No. 1, 1932. A preliminary revision of the genera of the gobioid fishes with united ventral fins. Lisse; 1-174.
KOUMANS, F. P., No. 10, 1953. Gobioidea. In Weber & de Beaufort, Fishes of the Indo-Australian Archipelago. X Leiden; 1-423.
LeDANOIS, Y., NO. 1, 1959. Etude osteologique, myologique et systematique des poissons du sous-ordre des Orbiculata. Ann. Inst. Oceanogr. Paris, (n. s.), 36, (1), 1-273.
LINDBERG, G. U. & KRASJUKOVA, Z. V., No. 1, 1969. Fishes of the Sea of Japan. . . . Part III. Moskva; 1-479. (In Russian)
LINDBERG, G. U. & LEGHEZA, M. I., No. 1, 1959. Fishes of the Japanese Sea and the adjacent parts of the Okhotsk and Yellow Seas. Part I. Tabl. Anal. Faune U. R. S. S., 68, 1-208. (In Russian)
LINDBERG, G. U. & LEGHEZA, M. I., No. 2, 1965. Fishes of the Japanese Sea . . . . Part 2. Moskva; 1-391. (In Russian)
LINDBERG, G. U. & LEGEZA, M. I., No. 3, 1969. Fishes of the Sea of Japan. . . . Part 2. Jerusalem; 1-389. (Translation of No. 2)
LOZANO REY. L., No. 1, 1935. Los peces fluviales de Espana. Madrid; 1-387.
LOZANO y REY, D. L., No. 2, 1952. Peces fisoclistos, subserie toracicos. Primera partie. Mem. Real Acad. Madrid Cienc. Nat., 14, 1-378.

LOZANO y REY, D. L., No. 3, 1952. Same. Segunda partie. Pp. 379-705.

LOZANO REY, D. L., No. 4, 1947. Peces ganoideos y fisostomos. Mem. Roy. Acad. Madrid, 11, 1-839.

LOZANO y REY, L., No. 5, 1960. Same. Tercera parte. 14, 1-613.

MAKUSHOK, V. M., No. 1, 1958. Morphology and systematics of Stichaeoidae, Blennioidei, Pisces. Trud. Inst. Zool., 25, 3-129. (In Russian & English)

MARSHALL, T. C., No. 1, 1964. Fishes of the Great Barrier Reef and coastal waters of Queensland. Narberth, Pennsylvania; 1-566.

MEEK, S. E., No. 1, 1904. The fresh-water fishes of Mexico north of the Isthmus of Tehuantepec. Field Columbian Mus., Publ. 93, 1-252.

MEEK, S. E. & HILDEBRAND, S. R., No. 1, 1916. The fishes of the fresh waters of Panama. Field Mus. Nat. Hist., Zool., 10, 217-374.

MEEK, S. E. & HILDEBRAND, S. R., No. 2, 1923. The marine fishes of Panama. Part I. Field Mus. Nat. Hist., Zool., 15, 1-330.

MEEK, S. E. & HILDEBRAND, S. F., No. 3, 1925. The marine fishes of Panama. Part II. Field Mus. Nat. Hist., Zool., 15, 331-707.

MEEK, S. E. & HILDEBRAND, S. F., No. 4, 1928. The marine fishes of Panama. Part III. Field Mus. Nat. Hist., Zool., 15, 709-1045.

MEEK, S. E. & HILDEBRAND, S. F., No. 6, 1925. The marine fishes of Panama. Part II. Field Mus. Nat. Hist., Zool., 15, 331-707.

NICHOLS, J. T., No. 1, 1943. The fresh-water fishes of China. (Volume IX of Natural History of Central Asia). New York; 1-322.

NOBRE, A., No. 1, 1935. Fauna marinha de Portugal. I. Vertebrados (Mamiferos, reptis e peixes). Porto; 1-579.

NORMAN, J. R., No. 1, 1930. Oceanic fishes and flatfishes collected in 1925-27. Part I. Oceanic fishes. Discovery Rep., 2, 263-357.

NORMAN, J. R., No. 2, 1937. Coast fishes. Part II. The Patagonian region. Discovery Rep., 16, 1-150.

NORMAN, J. R., No. 3, 1938. Coast fishes. Part III. The Antarctic Zone. Discovery Rep., 18, 1-104.

OKADA, Y., No. 1, 1960. Studies on the freshwater fishes of Japan. Tsu; 1-880.

OKADA, Y., No. 2, 1960. Studies on the freshwater fishes of Japan. Journ. Fac. Fish. Univ. Mie, 4, 1-860.

OSHIMA, M., No. 1, 1919. Contributions to the study of the fresh water fishes of the island of Formosa. Ann. Carnegie Mus., 12, 169-328.

PELLEGRIN, J., No. 1, 1933. Les poissons des eaux douces de Madagascar et des iles voisines (Comores Seychelles, Mascareignes). Mem. Acad. Malgache, 14, 1-222.

PELLEGRIN, J., No. 2, 1923. Les poissons des eaux douces de l'Afrique occidentale (du Senegal au Niger). Paris; 1-373.

POLL, M., No. 1, 1946. Revision de la faune ichthyologique du lac Tanganyika. Ann. Mus. Congo, Zool., (1), 4, 141-364.

POLL, M., No. 2, 1947. Faune de Belgique. Poissons marins. Bruxelles; 1-452.

POLL, M., No. 3, 1953. Poissons non Cichlidae. Res. Sci. Expl. Hydrobiol. Lac Tanganyika (1946-47), 3, (5A), 1-251.

POLL, M., No. 4, 1954. Poissons IV. Teleosteens Acanthopterygiens (Premiere partie). Res. Sci. Exp. Oceanogr. Belgique eaux cotes Afrique Atlantique sud 1948-49, 4, (3A), 1-390.

POPE, C. H. & NICHOLS, J. T., No. 1, 1927. The fishes of Hainan. Bull. American Mus. Nat. Hist., 54, 321-394.

PUYO, J., No. 1, 1949. Poissons de la Guyane francaise. Faune de l'Empire francaise, 12, 1-280.

REGAN, C. T., No. 1, 1906-1908. Pisces. Biologia Centrali-Americana. London; 1-203.

SCHULTZ, L. P., No. 1, 1942. The fresh-water fishes of Liberia. Proc. United States Nat. Mus., 92, 301-348.

SCHULTZ, L. P., No. 2, 1944. The catfishes of Venezuela . . . . Proc. United States Nat. Mus., 94, 173-388.

SCHULTZ, L. P. et al., No. 1, 1966. Fishes of the Marshall and Marianas Islands. United States Nat. Mus. Bull., 202, vol. 3, 1-176.

SCHULTZ, L. P. et al., No. 2, 1953. Same, volume 1; 1-685.

SCHULTZ, L. P. et al., No. 3, 1960. Same, volume 2; 1-438.

SCHMIDT, P. Y., No. 1, 1950. Fishes of the Sea of Okhotsk. Trans. Pacific Comm. Leningrad, 6, 1-370. (In Russian)

SHMIDT, P. Y., No. 2, 1965. Fishes of the Sea of Okhotsk. Jerusalem; 1-392. (Translation of 1950)

de SILVA, P. H. D. H., No. 1, 1956. The beak-mouthed and tube-mouthed fishes (orders Synentognathi and Aulostomi) off Ceylon, with a list of the specimens in the Colombo Museum. Spol. Zeylanica, 28, 47-54.

de SILVA, P. H. D. H., No. 2, 1956. The flatfishes (Heterosomata) of Ceylon with a list of the specimens in the Colombo Museum. Ceylon Journ. Sci., 7C, 183-200.

SLASTENENKO, E. P., No. 1, 1958. The freshwater fishes of Canada. Toronto; 1-388.

SMITH, H. M., No. 1, 1945. The fresh-water fishes of Siam, or Thailand. United States Nat. Mus. Bull., 188, 1-622.

SMITH, J. L. B., No. 1, 1961. The sea fishes of southern Africa, 4th edition. Cape Town; 1-590. (Previous editions in 1949 and 1953)

SMITT, F. A. et al., No. 1, 1892. A history of Scandinavian fishes, by B. Fries et al.; 2nd edition. Stockholm; 2 volumes, 1-566.

SMITT, F. A. et al., No. 2, 1895. Same, volume 2, 567-1240.

SOLDATOV, V. K. & LINDBERG, G. J., No. 1, 1930. A review of the fishes of the seas of the Far East. Bull. Pacific Inst. Fish., 5, 1-576. (In Russian)

SPILLMANN, J., No. 1, 1961. Poissons d'eau douce. Faune France, 65, 1-303.

SVETOVIDOV, A. N., No. 1, 1948. Fauna of U. S. S. R. Pisces. Gadiformes. Zool. Inst. Acad. Sci., (n. s.), 34, (9), 1-221. (In Russian; translated into English in 1962)
SVETOVIDOV, A. N., No. 2, 1964. Fish of the Black Sea. Moskva; 1-550. (In Russian)
TORTONESE, E., No. 3, 1956. Fauna d'Italia. Leptocardia Ciclostomata Selachii. Bologne; 1-334.
VAN DER STIGCHEL, J. W. B., No. 1, 1946. The South American Nematognathi of the museums at Leiden and Amsterdam. Zool. Meded., 27, 1-204.
WALFORD, L. A., No. 1, 1937. Marine game fishes of the Pacific coast, Alaska to the equator. Berkeley; 1-207.
WEBER, M. & de BEAUFORT, L., No. 2, 1913. Fishes of the Indo-Australian Archipelago. II. Leiden; 1-404.
WEBER, M. & de BEAUFORT, L., No. 3, 1916. Same; III, 1-456.
WEBER, M. & de BEAUFORT, L., No. 4, 1922. Same; IV, 1-410.
WEBER, M. & de BEAUFORT, L., No. 5, 1929. Same; V, 1-458.
WEBER, M. & de BEAUFORT, L., No. 6, 1931. Same; VI, 1-448.
WEBER, M. & de BEAUFORT, L., No. 7, 1936. Same; VII, 1-607.

Families alphabetically

Abudefdufidae, as Pomacentridae
Abyssocottidae, as Cottocomephoridae
Acanthaspidae, as Macropetalichthyidae (only as fossils)
Acanthoclinidae. CHEN, J. T. F. & LIANG, Y. -S., Quart. Journ. Taiwan Mus., 1, (3), 31-34 (key to genera); SMITH, J. L. B., No. 1, 1961 (marine, South Africa)
Acanthocybiidae, as Scombridae
Acanthodidae (incl. Ácanthoessidae) (only as fossils)
Acanthoessidae, as Acanthodidae (only as fossils)
Acanthopsidae, as Cobitidae
Acanthorhinidae (only as fossils)
Acanthuridae (incl. Acronuridae, Harpurldae, Hepatidae, Nasidae, Teuthidae, Teuthididae in part, Teuthridae, Zanclidae) GILL, T., 1884, Proc. United States Nat. Mus., 7, 275-281 (as Teuthididae) synopsis of genera); RANDALL, J. E., 1955, Pacific Sci., 9, 359-367 (analysis of genera); BEEBE & TEE-VAN, NO. 1, 1928 (Hispaniola); MEEK & HILDEBRAND, No. 7, 1928 (marine, Panama); FOWLER, H. W., No. 8, 1936 (marine, West Africa, as Hepatidae); DAGET & ILTIS, No. 1, 1965 (Cote d'Ivoire); SMITH, J. L. B., 1966, Ichth. Bull. Rhodes Univ., 32, 635-682 (subf. Nasinae & Prionurinae); SMITH, J. L. B., No. 1, 1961 (marine, South Africa, also as Zanclidae); SMITH, J. L. B., 1955, South African Journ. Sci., 51, 169-174 (key to genera & species, as Nasidae); DAY, F., No. 2, 1889 (India, also as Teuthididae); de BEAUFORT & CHAPMANS, No. 9, 1951 (Indo-Australian Arch.); SMITH, J. L. B., 1951,, Ann. Mag. Nat. Hist., (12), 4, 1126-1132 (key to Indo-Pacific genera & species); MARSHALL, T. C., No. 1, 1964 (Great Barrier Reef; also as Zanclidae); HERRE, A. W., 1917, Philippine Journ. Sci., 34, 403-478 (Philippines; as Teuthididae & ZANCLIDAE); SCHULTZ, L. P. et al., 1953, (Marshal & Marianas Is.; also as Zanclidae); FOWLER, H. W., No. 17, 1958, (Fiji; as Hepatidae)
Acentrophoridae (only as fossils)
Aceratiidae, as Linophrynidae
Acestrorhynchidae, as Characidae
Achiridae, as Soleidae
Acinaceidae, as Gempylidae
Acipenseridae. HOLLY, M., 1936, Das Tierreich, 67, 1-65; BAILEY, R. M. & CROSS, F. B., 1954, Pap. Michigan Acad. Sci., 30, 169-208(key to American); SLASTENENKO, E. P., No. 1, 1958 (Canada); HILDEBRAND & SCHROEDER, No. 1, 1928 (Chesapeake Bay); FOWLER, H. W., No. 1, 1945 (southeastern U. S.); BIGELOW, H. B., No. 5, 1963 (western North Atlantic); SMITT, F. A. et al., No. 2, 1895 (Scandinavia); POLL, M., No. 2, 1947 (marine, Belgique); SPILLMANN, J., No. 1, 1961 (France); NOBRE, A., No. 1, 1935 (marine, Portugal); LOZANO REY, L., No. 1, 1935 (fresh-water, Espana); DUNCKER & LADIGES, No. 1, 1960 (Nordmark); BANARESCU, P., No. 1, 1964 (Romania); BERG, L. S., No. 7, 1911 (Russia); BERG, L. S., No. 1, 1932 (fresh-water, U. R. S. S.); BERG, L. S., No. 3, 1948 (fresh-water, Soviet Union); BERG, L. S., No. 8, 1962 (fresh-water, U. S. S. R.); SVETOVIDOV, A. N., No. 2, 1964 (Black Sea); ANDRIYASHEV, A. P., No. 1, 1954 (marine, northern Russia); NICHOLS, J. T., No. 1, 1943 (fresh-water, China); SOLDATOV & LINDBERG, No. 1, 1930 (Far East seas); SCHMIDT, P. Y., No. 1, 1950 (Sea of Okhotsk); SHMIDT, P. Y., No. 2, 1965 (Sea of Okhotsk); LINDBERG & LEGHEZA, No. 2, 1965 (Sea of Japan); LINDBERG & LEGEZA, M. I., No. 3, 1969 (Sea of Japan); JORDAN, D. S. & SNYDER, J. O., 1906, Proc. United States Nat. Mus., 30, 397-398 (Japan); OKADA, Y., No. 1 & 2, 1960 (fresh-water, Japan)
Acrolepidae (only as fossils)
Acrolepididae, incl. Acrolepidae (as Palaeoniscidae) (only as fossils)
Acronuridae, as AS Acanthuridae
Acropomatidae (including Acropomidae). SMITH, J. L. B., No. 1, 1961 (marine, South Africa); LINDBERG & KRASJUKOVA, No. 1, 1969 (Sea of Japan)
Acropomidae, as Acropomatidae
Acrotidae, as Icosteidae

Actobatidae, as Aetobatidae
Adiposiidae, as Cobitidae
Adrianichthyidae. WEBER & de BEAUFORT, No. 4, 1922 (Indo-Australian Arch.)
Aeduellidae (no revisionary references noted)
Aeschynichthyidae, as Diceratiidae
Aetheodontidae (only as fossils)
Aëtobatidae, as Aetobatidae
Aetobatidae (incl. Actobatidae, Aëtobatidae) (Family not identified). HILDEBRAND, S. F., No. 1, 1946 (shore, Peru); SMITH, J. L. B., No. 1, 1961 (marine, South Africa); GOHAR & MAZHAR, No. 1, 1964 (Red Sea)
Agamidae. SMITT, F. A. et al., No. 1, 1892 (Scandinavia)
Ageneiosidae (incl. Ageniosidae). VAN DER STIGCHEL, J. W. B., No. 1, 1946 (South America); SCHULTZ, L. P., No. 2, 1944 (Venezuela); FOWLER, H. W., No. 10, 1948-51 (fresh-water, Brasil)
Ageniosidae, as Ageneiosidae
Agonidae (incl. Aspidophoroididae, Aspidophoridae). NORMAN, J. R., No. 2, 1937 (marine, Patagonia); DUNCKER & LADIGES, No. 1, 1960 (Nordmark) ; POLL, M., No. 2, 1947 (marine, Belgique); NOBRE, A., No. 1, 1935 (marine, Portugal); ANDRIYASHEV, A. P., No. 1, 1954 (northern seas of U. S. S. R.); SOLDATOV & LINDBERG, No. 1, 1930 (Far East Seas); SCHMIDT, P. Y. No. 1, 1950 (Sea of Okhotsk); SHMIDT, P. Y., No. 2, 1965 (Sea of Okhotsk); JORDAN, D. S. & STARKS, E. C., 1904, Proc. United States Nat. Mus., 27, 575-599 (Japan)
Agriopidae, as Congiopodidae
Agrostichthyidae, as Regalecidae
Akysidae. SMITH, H. M., No. 1, 1945 (fresh-water, Siam); WEBER & HILDEBRAND, No. 2, 1913 (Indo-Australian Arch.); INGER & KONG, No. 1, 1962 (fresh-water, North Borneo)
Alabetidae (incl. Alabidae, Cheilobranchidae, Chilobranchidae). MARSHALL, T. C., No. 1, 1964 (Great Barrier Reef)
Alabidae, as Alabetidae
Albulidae (incl. Bathythrissidae, Pterothrissidae). BEEBE & TEE-VAN, No. 1, 1928 (Hispaniola); MEEK & HILDEBRAND, No. 2, 1923 (marine, Panama); BIGELOW et al., No. 5, 1963 (western North Atlantic); FOWLER, H. W., No. 8, 1936 (marine, West Afrika); SMITH, J. L. B., No. 1, 1961 (marine, South Africa); FOWLER, H. W., No. 18, 1956 (Red Sea); WEBER & de BEAUFORT, No. 2, 1913 (Indo-Australian Arch.); MARSHALL, T. C., No. 1, 1964 (Great Barrier Reef); LINDBERG & LEGHEZA, No. 2, 1965 (Sea of Japan; also as Pterothrissidae); LINDBERG & LEGEZA, No. 3, 1969 (Sea of Japan); JORDAN & HERRE, No. 1, 1906 (Japan; also as Pterothrissidae)
Alepidosauridae, as Alepisauridae
Alepisauridae (incl. Alepidosauridae, Plagyodontidae). BIGELOW, H. B. et al., No. 4, 1966 (western North Atlantic); NOBRE, A., No. 1, 1935 (marine, Portugal); LOZANO REY, D. L., No. 4, 1947 (Espana); FOWLER, H. W., No. 8, 1936 (marine, West Africa); SMITH, J. L. B., No. 1, 1961 (marine, South Africa); SOLDATOV & LINDBERG, No. 1, 1930 (Far East Seas)
Alepocephalidae (incl. Platyproctidae, Platytroctidae, Searsidae, Searsiidae). PARR, A. E., 1937, Bull. Bingham Oceanogr. Coll., 3, (1), 1-79 (synopsis of genera); PARR, A. E, 1951, American Mus. Novit., 1531, 1-21 (revision; as Searsidae also); PARR, A. E., 1960, Dana Rep., 51, 1-108 (classification; key to genera & species; as Searsidae); NORMAN, J. R., No. 1, 1930 (oceanic); de BUEN, F., 1961, Montemar, 1, 1-90 (Chile); LOZANO REY, D. L., No. 4, 1947 (Espana); FOWLER, H. W., No. 8, 1936 (marine, West Africa); SMITH, J. L. B., No. 1, 1961 (marine, South Africa); FOWLER, H. W:, No. 18, 1956 (Red Sea); WEBER & de BEAUFORT, No. 2, 1913 (Indo-Australian Arch.); LINDBERG & LEGHEZA, No. 2, 1965 (Sea of Japan); LINDBERG & LEGEZA, No. 3, 1969 (Sea of Japan); JORDAN & HERRE, No. 1, 1906 (Japan)
Aleuteridae, as Balistidae
Alopidae, as Alopiidae
Alopiidae (incl. Alopidae). BIGELOW & SCHROEDER, No. 1, 1948 (western North Atlantic); TORTONESE, E., No. 3, 1956 (Italy); FOWLER, H. W., No. 8, 1936 (Marine, West Africa); SMITH, J. L. B., No. 1, 1961 (marine, South Africa); FOWLER, H. W., No. 18, 1956 (Red Sea); GOHAR & MAZHAR, No. 1, 1964 (Red Sea); MARSHALL, T. C., No. 1, 1964 (Great Barrier Reef); FOWLER, H. W., No. 2, 1930 (China); LINDBERG & LEGHEZA, No. 1, 1959 (Japan Sea); JORDAN, D. S. & FOWLER, H. W., 1903, Proc. United States Nat. Mus., 26, 593-674 (Japan)
Aluteridae, as Balistidae
Ambassidae, as Centropomidae
Ambassiidae, as Centropomidae
Ambiotocidae, as Embiotocidae
Amblycepidae, as Amblycipitidae
Amblycipitidae (incl. Amblycepidae). SMITH, H. M., No. 1, 1945 (fresh-water, Siam)
Amblyopidae, as Gobioididae
Amblyopsidae (incl. Hypsaeidae, Hypsocidae). FOWLER, H. W., No. 1, 1945 (southeastern U. S.); WOODS, L. P. & INGER, R. F., 1957, American Midl. Nat., 58, 232-256 (central & eastern U. S.)
Amblypteridae (only as fossils)
Ameiuridae, as Ictaluridae
Amiatidae, as Amiidae
Amiidae (incl. Amiatidae, Liodesmidae). HOLLY, M., 1936, Das Tierreich, 67, 1-65; SLASTENENKO, E. P., No. 1, 1958 (Canada); FOWLER, H. W., No. 1, 1945 (southeastern U. S.); BEEBE & TEE-VAN, No. 1, 1928 (Hispaniola); LOZANO y REY, D. L., No. 2, 1952 (Espana); FOWLER, H. W., No. 8, 1936 (marine, West Africa); FOWLER, H. W., No. 9, 1937-41 (China); FOWLER,

H. W., 1930, United States Nat. Mus., Bull., 100, (10), 1-334 (Philippine seas); FOWLER, H. W., No. 17, 1959 (Fiji)

Amiuridae, as Ictaluridae

Ammodytidae (incl. Bleekeridae, Bleekeriidae). DUNCKER, G. & MOHR, E., 1939, Mitt. Zool. Mus. Berlin, 24, 8-31 (revision); SMITT, F. A. et al., No. 2, 1895 (Scandinavia); DUNCKER & LADIGES, No. 1, 1960 (Nordmark); POLL, M., No. 2, 1947 (marine, Belgique); NOBRE, A., No. 1, 1935 (marine, Portugal); LOZANO y REY, L., No. 5, 1960 (España); FAGE, L., No. 1, 1918 (Mediterranean); BANARESCU, P., No. 1, 1964 (Romania); SVETOVIDOV, A. N., No. 2, 1964 (Black Sea); ANDRIYASHEV, A. P., No. 1, 1954 (northern seas of U. S. S. R.); FOWLER, H. W., No. 8, 1936 (marine, West Africa); SMITH, J. L. B., No. 1, 1961 (marine, South Africa); MARSHALL, T. C., No. 1, 1964 (Great Barrier Reef); FOWLER, H. W., 1958, Quart. Journ. Taiwan Mus., 12, 67-97 (China); SOLDATOV & LINDBERG, No. 1, 1930 (Far East Seas); SCHMIDT, P. Y., No. 1, 1950 (Sea of Okhotsk); SHMIDT, P. Y., No. 2, 1965 (Sea of Okhotsk); JORDAN, D. S., 1906, Proc. United States Nat. Mus., 30, 715-719 (Japan)

Amphacanthidae, as Siganidae

Amphiaspidae (only as fossils)

Amphicentridae (only as fossils)

Amphiliidae. HARRY, R. R., 1953, Rev. Zool. Bot. Africaine, 47, 177-232 (Africa); DAGET & ILTIS, No. 1, 1965 (Côte d'Ivoire); DAGET, J., No. 1, 1954 (Upper Niger); DAGET, J., No. 2, 1962 (Guinea); POLL, M., No. 1, 1946 (Lake Tanganyika); POLL, M., No. 3, 1953 (L. Tanganyika)

Amphipnoidae. (No revisionary references noted)

Amphiprionidae, as Pomacentridae

Amphisilidae, as Centriscidae

Amphistiidae, as Monodactylidae

Anabantidae (incl. Anabatidae). FORSELIUS, S., 1957, Zool. Bidr. Uppsala, 32, 93-587 (studies I - III); BLANC, M., 1963, Bull. Mus. Nat. Hist. Nat., (2), 35, 70-77 (types in Museum); REGAN, C. T., 1909, Proc. Zool. Soc. London, 1909, 767-787 (Asiatic); BOULENGER, G. A., No. 6, 1916 (fresh-water, Africa); BOULENGER, G. A., No. 2, 1907 (Egypt); PELLEGRIN, J., No. 2, 1923 (western Africa); DAGET, J., No. 2, 1962 (Guinea); SCHULTZ, L. P., No. 1, 1942 (Liberia); DAGET & ILTIS, No. 1, 1965 (Côte d'Ivoire); DAGET, J., No. 1, 1954 (Upper Niger); BOULENGER, G. A., No. 1, 1901 (Congo basin); POLL, M., No. 1, 1946 (Lake Tanganyika); POLL, M., No. 3, 1953 (Lake Tanganyika); BARNARD, K. H., 1943, Ann. South African Mus., 36, 101-262 (southwest Cape region); ARNOULT, J., No. 1, 1959 (fresh-water, Madagascar); DERANIYAGALA, P. E. P., 1929, Ceylon Journ. Sci., 15, 79-111 (Labyrinthici of Ceylon); SMITH, H. M., No. 1, 1945 (fresh-water, Thailand); TWEEDIE, M. W. F., 1952, Bull. Raffles Mus., 24, 63-95 (Malay Peninsula); WEBER & de BEAUFORT, No. 4, 1922 (Indo-Australian Arch.); INGER & KONG, No. 1, 1962 (fresh-water, North Borneo); POPE & NICHOLS, No. 1, 1927 (Hainan I.); NICHOLS, J. T., No. 1, 1943 (fresh-water, China); OKADA, Y., No. 2, 1960 (Japan)

Anabatidae, as Anabantidae

Anablepidae (incl. Anablepsidae). (No revisionary references noted)

Anablepsidae, as Anablepidae

Anacanthidae, as Balistidae

Anacanthobatidae. BIGELOW & SCHROEDER, No. 2, 1953 (western North Atlantic); BIGELOW, H. B. & SCHROEDER, W., 1962, Bull. Mus. Comp. Zool., 128, 161-244 (keys, western Atlantic)

Anaethalionidae (only as fossils)

Anarhichadidae (incl. Anarichodidae, Anarrhichadidae, Anarrhichthyidae). MAKUSHOK, V. M., No. 1, 1958 (morphology & systematics); SMITH, F. A. et al., No. 1, 1892 (Scandinavia): DUNCKER & LADIGES, No. 1, 1960 (Nordmark); POLL, M., No. 2, 1947 (marine, Belgique); BARSUKOV, V. V., 1959, Fauna U. S. S. R., (n. s.), 73, 1-171; ANDRIYASHEV, A. P., No. 1, 1954 (northern seas of U. S. S. R.); SOLDATOV & LINDBERG, No. 1, 1930 (Far East seas); SCHMIDT, P. Y., No. 1, 1950 (Sea of Okhotsk); SHMIDT, P. Y., No. 2, 1965 (Sea of Okhotsk); FOWLER, H. W., No. 16, 1958 (China)

Anarichodidae, as Anarhichadidae

Anarrhichadidae, as Anarhichadidae

Anarrhichthyidae, as Anarhichadidae

Ancylostylidae (only as fossils)

Anglaspidae (only as fossils)

Angillavidae (only as fossils)

Angillichthyidae, as Moringuidae

Anguillidae. SLASTENENKO, E. P., No. 1, 1958 (Canada); HILDEBRAND & SCHROEDER, No. 1, 1928 (Chesapeake Bay); BEEBE & TEE-VAN, No. 1, 1928 (Hispaniola); REGAN, C. T., No. 1, 1906-08 (Central America); MEEK, S. E., No. 1, 1904 (fresh-water, Mexico); MEEK & HILDEBRAND, No. 2, 1923 (marine, Panama); SMITT, F. A. et al., No. 2, 1895 (Scandinavia); DUNCKER & LADIGES, No. 1, 1960 (Nordmark); POLL, M., No. 2, 1947 (marine, Belgique); SPILLMANN, J., No. 1, 1961 (France); LOZANO REY, L., No. 1, 1935 (fresh-water, España); BANARESCU, P., No. 1, 1964 (Romania); SVETOVIDIV, A. N., No. 2, 1964 (Black Sea); BERG, L. S., No. 2, 1933 (fresh-water, U. R. S. S.); BERG, L. S., No. 5, 1949 (fresh-water, Soviet Union); BERG, L. S, No. 10, 1965 (fresh-water, U. S. S. R.); ANDRIYASHEV, A. P., No. 1, 1954 (northern seas of U. S. S. R.); BOULENGER, G. A., No. 5, 1915 (fresh-water, Africa); BOULENGER, G, A., No. 2, 1907 (Egypt); PELLEGRIN, J. No. 2, 1923 (western Africa); SCHULTZ, L. P., No. 1, 1942 (Liberia); BARNARD, K. H., 1943, Ann. South African Mus., 36, 101-262 (southwest Cape region); ARNOULT, J., No. 1, 1958 (fresh-water, Madagascar); FOWLER, H. W., No. 18, 1956 (Red Sea); SMITH, H. M., No. 1, 1945 (fresh-water, Thailand); WEBER & de BEAUFORT, No. 3, 1916 (Indo-Australian Arch.); INGER & KONG, No. 1, 1962 (fresh-water, North Borneo); MARSHALL, T. C., No. 1, 1964 (Great Barrier Reef); OSHIMA, M., No. 1, 1919 (fresh-water, Formosa); POPE & NICHOLS, No. 1, 1927 (Hainan I.); NICHOLS, J. T., No. 1, 1943 (fresh-water, China); LINDBERG & LEGHEZA, No. 2, 1965 (Sea of Japan); LINDBERG & LEGEZA, No. 3, 1969 (Sea of Japan); JORDAN & SNYDER, No. 1, 1901 (Japan); OKADA, Y., No. 1 & 2, 1960 (fresh-water, Japan); SCHULTZ, L. P. et al., No. 2, 1953 (Marshall & Marianas Is.)

Anisochromidae. SMITH, J. L. B., 1954, Ann. Mag. Nat. Hist., (12), 7, 298-302 (East Africa)

Anodontidae, as Curimatidae

Anogmiidae, as Plethodidae (only as fossils)

Anomalopidae. SILVESTER, C. F. & FOWLER, H. W., 1926, Proc. Acad. Nat. Sci. Philadelphia, 78, 245-247 (key to genera); WEBER & de BEAUFORT, No. 5, 1929 (Indo-Australian Arch.); FOWLER, H. W., No. 17, 1959 (Fiji)

Anoplogasteridae (incl. Caulolepidae). (no revisionary references noted)

Anoplopomatidae (incl. Anoplopomidae, Erilepidae) (no revisionary references noted)

Anoplopomidae, as Anoplopomatidae

Anostomidae. FERNANDEZ-YEPEZ, A., 1949, Mem. Soc. Cienc. Nat. La Salle, 9, (25), 353-355 (key to genera)

Anotopteridae. MARSHALL, N. B., 1955, Discovery Rep., 27, 303-336 (deep-sea); BIGELOW, H. B. et al., No. 4, 1966 (western North Atlantic); FOWLER, H. W., No. 8, 1936 (marine, West Africa)

Antennariidae. SCHULTZ, L. P., 1957, Proc. United States Nat. Mus., 107, 47-105; HILDEBRAND & SCHROEDER, No. 1, 1928 (Chesapeake Bay); BEEBE & TEE-VAN, No. 1, 1929 (Hispaniola); MEEK & HILDEBRAND, No. 4, 1928 (marine, Panama); HILDEBRAND, S. F., No. 1, 1946 (shore, Peru); BARBOUR, T., 1942, Proc. New England Zool. Club, 19, 21-40 (revision); ANDRIYASHEV, A. P., No. 1, 1954 (northern seas of U. S. S. R.); FOWLER, H. W., No. 8, 1936 (marine, West Africa); DAGET & ILTIS, No. 1, 1965 (Côte d'Ivoire); SMITH, J. L. B., No. 1, 1961 (marine, South Africa); de BEAUFORT & BRIGGS, No. 11, 1962 (Indo-Australian Arch.); MARSHALL, T. C., No. 1, 1964 (Great Barrier Reef); SOLDATOV & LINDBERG, No. 1, 1930 (Far East seas); JORDAN, D. S., 1902, Proc. United States Nat. Mus., 24, 361-381 (Japan); SCHULTZ, L. P. et al., No. 1, 1966 (Marshall & Marianas Is.); FOWLER, H. W., No. 17, 1959 (Fiji)

Anthiidae, as Serranidae

Antigoniidae, as Caproidae

Aoteidae (incl. Aoteridae) (no revisionary references noted)

Aoteridae, as Aoteidae

Aphareidae, as Lutjanidae

Aphredoderidae (incl. Erismatopteridae). FOWLER, H. W., No. 1, 1945 (southeastern U. S.)

Aphyonidae, as Ophidiidae

Aploactidae, as Aploactinidae

Aploactinidae (incl. Aploactidae, Bathyaploactidae). de BEAUFORT & BRIGGS, No. 11, 1962 (Indo-Australian Arch., also as Bathyaploactidae); FOWLER, H. W., No. 17, 1959 (Fiji)

Aplochitonidae (incl. Haplochitonidae, Prototroctidae). EIGENMANN, C. H., No. 2, 1927 (fresh-water, Chile); NORMAN, J. R., No. 2, 1937 (marine, Patagonia)

Aplodactylidae (incl. Haplodactylidae). HILDEBRAND, S. F., No. 1, 1946 (shore, Peru); FOWLER, H. W., No. 15, 1956 (China); LINDBERG & KRASJUKOVA, No. 1, 1969 (Sea of Japan)

Apocrypteidae, as Gobiidae

Apogonichthyidae, as Apogonidae

Apogonidae (incl. Apogonichthyidae, Dinolestidae, Cheilodipteridae, Epigonidae, Gymnapogonidae, Henichthyidae, Henicichthyidae, Ostorhinchidae). SCHULTZ, L. P., 1940, Proc. United States Nat. Mus., 88, 403-423 (synopsis of genera, as Chilodipteridae); NORMAN, J. R., No. 1, 1930 (oceanic, as Chilodipteridae); MEEK & HILDEBRAND, No. 3, 1925 (marine, Panama); HILDEBRAND, S. F., No. 1, 1946 (shore, Peru); DEVINCENZI, G. L., No. 2, 1924 (Uruguay, as Cheilodipteridae); BOHLKE & RANDALL, 1968, Proc. Acad. Nat. Sci. Philadelphia, 120, 175-206 (western Atlantic); NOBRE, A., No. 1, 1935 (marine, Portugal, as Cheilodipteridae); FAGE, L., No. 1, 1918 (Mediterranean, as Chilodipteridae); POLL, M., No. 4, 1954 (coastal South Africa, as Chilodipteridae); SMITH, J. L. B., No. 1, 1961 (marine, South Africa); KLAUSEWITZ, W., 1959, Senckenbergische Biol., 40, 251-262 (Red Sea); SMITH, J. L. B., 1961, Ichth. Bull. Rhodes Univ., 22, 373-418 (western Indian Ocean & Red Sea); WEBER & de BEAUFORT, No. 5, 1929 (Indo-Australian Arch.); MARSHALL, T. C., No. 1, 1964 (Great Barrier Reef); LINDBERG & KRASJUKOVA, No. 1, 1969 (Sea of Japan); JORDAN, D. S. & SNYDER, J. O., 1901, Proc. United States Nat. Mus., 23, 891-913 (Japan); SCHULTZ, L. P. et al., No. 2, 1953 (Marshall & Marianas Is.)

Apolectidae, as Formionidae

Apteronotidae (incl. Sternarchidae, Sternopygidae). FOWLER, H. W., No. 10, 1948-51 (fresh-water, Brasil)

Aracanidae, as Ostraciontidae

Arapaimidae, as Osteoglossidae

Archaeomaenidae (only as fossils)

Arctolepidae (incl. Acanthaspidae in part, Jaekelaspidae, Monaspidae) (only as fossils)

Argentinidae (incl. Xenophthalmichthyidae). CHAPMAN, W. M., 1942, Journ. Washington Acad. Sci., 32, 104-117 (osteology); COHEN, D. M., 1958, Bull. Florida State Mus., Biol. Sci., 3, 93-172 (revision of Argentininae); NORMAN, J. R., No. 1, 1930 (oceanic); BIGELOW, H. B. et al., No. 3, 1964 (western North Atlantic); LOZANO REY, D. L., No. 4, 1947 (España);

ANDRIYASHEV, A. P., No. 1, 1954 (northern seas of U. S. S. R.); FOWLER, H. W., No. 8, 1936 (marine, West Africa); SMITH, J. L. B., No. 1, 1961 (marine, South Africa); SCHMIDT, P. Y., No. 1, 1950 (Sea of Okhotsk); SHMIDT, P. Y., No. 2, 1965 (Sea of Okhotsk); LINDBERG & LEGHEZA, No. 2, 1965 (Sea of Japan); LINDBERG & LEGEZA, No. 3, 1969 (Sea of Japan)

Argidae, as Astroblepidae

Argiidae, as Astroblepidae

Ariidae (incl. Bagreidae, Doiichthyidae, Tachysuridae). HILDEBRAND & SCHROEDER, No. 1, 1928 (Chesapeake Bay); VAN DER STIGCHEL, J. W. B., No. 1, 1946 (South America); HILDEBRAND, S. F., No. 1, 1946 (shore, Peru); SCHULTZ, L. P., No. 2, 1944 (Venezuela, as Bagreidae); FOWLER, H. W., No. 10, 1948-51 (fresh-water, Brasil, as Tachysuridae); FOWLER, H. W., No. 8, 1936 (marine, West Africa, as Tachysuridae); DAGET & ILTIS, No. 1, 1965 (Côte d'Ivoire); DAGET, J., No. 1, 1954 (Upper Niger); SMITH, J. L. B., No. 1, 1961 (marine, South Africa, as Tachysuridae); FOWLER, H. W., No. 18, 1956 (Red Sea, as Tachysuridae); SMITH, H. M., No. 1, 1945 (fresh-water, Thailand, as Tachysuridae); WEBER & de BEAUFORT, No. 2, 1913 (Indo-Australian Arch.); INGER & KONG, No. 1, 1962 (fresh-water, North Borneo); MARSHALL, T. C., No. 1, 1964 (Great Barrier Reef, as Tachysuridae); HERRE, A. W. C. T., 1926 Philippine Journ. Sci., 31, 385-411 (Philippines); FOWLER, H. W., No. 5, 1932 (China, as Tachysuridae); LINDBERG & LEGHEZA, No. 2, 1965 (Sea of Japan); LINDBERG & LEGEZA, No. 3, 1969 (Sea of Japan)

Ariommidae (no revisionary references noted)

Arripidae, as Arripididae

Arripididae (incl. Arripidae). FOWLER, H. W., No. 20, 1933 (Philippine seas)

Asarotidae (only as fossils)

Ascelichthyidae, as Cottidae

Asineopidae, as Aphredoderidae

Aspidophoridae, as Agonidae

Aspidophoroididae, as Agonidae

Aspidorhynchidae (incl. Rhynchodontidae) (only as fossils)

Aspredinidae (incl. Bunocephalidae). EIGENMANN & EIGENMANN, No. 1, 1890 (South America, as Bunocephalidae); MYERS, G. S., 1942, Stanford Ichth. Bull., 2, 89-114 (key to South American genera, as Bunocephalidae); VAN DER STIGCHEL, J. W. B., No. 1, 1946 (South America, as Bunocephalidae also); FERNANDEZ-YEPES, A., 1953, Noved. Cient. Mus. Hist. Nat. Caracas, 11, 1-6 (key to genera, as Bunocephalidae); MYERS, G. S., 1960, Stanford Ichth. Bull., 7, 132-139 (South America); SCHULTZ, L. P., No. 2, 1944 (Venezuela, as Bunocephalidae); EIGENMANN, C. H., No. 1, 1912 (British Guiana); PUYO, J., No. 1, 1949 (French Guiana); FOWLER, H. W., No. 13, 1954 (fresh-water, Brasil)

Asterodermidae, as Rhinobatidae

Asterolepidae (only as fossils)

Asterosteidae, as Gemuendinidae (only as fossils)

Astraspidae (only as fossils)

Astroblepidae (incl. Argidae, Argiidae, Cyclopidae, Cyclopiidae). MEEK & HILDEBRAND, No. 1, 1916 (fresh-water, Panama, as Cyclopidae); EIGENMANN & EIGENMANN, No. 1, 1890 (South America); EIGENMANN, C. H., 1922, Mem. Carnegie Mus., 9, 1-346 (fresh-water, northwestern South America); SCHULTZ, L. P., No. 2, 1944 (Venezuela); FOWLER, H. W., No. 13, 1954 (fresh-water, Brasil)

Astronesthidae. PARR, A. E., 1927, Bull. Bingham Oceanogr. Coll., 3, (2), 1-123 (revision of species); REGAN, C. T. & TREWAVAS, E., 1929, Oceanogr. Rep. Dana Exp. 1920-22, 5, 1-39 (monograph); NORMAN, J. R., No. 1, 1930 (oceanic); BIGELOW, H. B. et al., No. 3, 1964 (western North Atlantic); LOZANO REY, D. L., No. 4, 1947 (España); FOWLER, H. W., No. 8, 1936 (marine, West Africa); SMITH, J. L. B., No. 1, 1961 (marine, South Africa); FOWLER, H. W., No. 18, 1956 (Red Sea)

Astroscopidae, as Uranoscopidae

Ateleaspidae, as Cephalaspidae (only as fossils)

Ateleopidae, as Ateleopodidae

Ateleopodidae (incl. Ateleopidae, Podatelidae). RIVERO, L. H., 1935, Mem. Soc. Cubana Hist. Nat., 9, 91-106 (West Indies); BARNARD, K. H., 1948, Ann. South African Mus., 36, 341-406 (Key to South African genera); SMITH, J. L. B., No. 1, 1961 (marine, South Africa); WEBER & de BEAUFORT, No. 5, 1929 (Indo-Australian Arch.); LINDBERG & LEGHEZA, No. 2, 1965 (Sea of Japan); LINDBERG & LEGEZA, No. 3, 1969 (Sea of Japan)

Atelomycteridae, as Scyliorhinidae

Atherinidae (incl. Bedotiidae, Pseudomugilidae). JORDAN, D. S. & HUBBS, C. L., 1919, Stanford Univ. Publ., Univ. Ser., Biol. Sci., 1919, 1-87 (monograph); SCHULTZ, L. P., 1948, Proc. United States Nat. Mus., 98, 1-48; SLASTENENKO, E. P., No. 1, 1958 (Canada); HILDEBRAND & SCHROEDER, No. 1, 1928 (Chesapeake Bay); FOWLER, H. W., No. 1, 1945 (southeastern U. S.); BEEBE & TEE-VAN, No. 1, 1928 (Hispaniola); REGAN, C. T., No. 1, 1906-08 (Central America); MEEK, S. E., No. 1, 1904 (fresh-water, Mexico); de BUEN, F., 1945, Anal. Inst. Biol. Mexicana, 16, 475-532 (Mexico); MEEK & HILDEBRAND, No. 2, 1923 (marine, Panama); HILDEBRAND, S. F., No. 1, 1946 (shore, Peru); FOWLER, H. W., No. 13, 1954 (fresh-water, Brasil); DEVINCENZI, G. L., No. 2, 1924 (Uruguay); de BUEN, F., 1953, Bol. Inst. Oceanogr. Sao Paulo, 4, (1-2), 3-80 (Uruguay); LAHILLE, F., 1929, Bol. Minist. Agric. Nac. Argentina, 28, 261-395 (Argentina, Chile, Peru, "etc."); EIGENMANN, C. H., No. 2, 1927 (fresh-water, Chile); NORMAN, J. R., No. 2, 1937 (marine, Patagonia); DUNCKER & LADIGES, No. 1, 1960 (Nordmark); POLL, M., No. 2, 1947 (marine, Belgique);

SPILLMANN, J., No. 1, 1961 (France); NOBRE, A., No. 1, 1935 (marine, Portugal); LOZANO REY, L., No. 1, 1935 (fresh-water, España); LOZANO REY, D. L., No. 4, 1947 (España); FAGE, L., No. 1, 1918 (Mediterranean); BANARESCU, P., No. 1, 1964 (Romania); SVETOVIDOV, A. N., No. 2, 1964 (Black Sea); BERG, L. S., No. 2, 1933 (fresh-water, U. R. S. S.); BERG. L. S., No. 5, 1949 (fresh-water, Soviet Union); BERG, L. S., No. 10, 1965 (fresh-water, U. S. S. R.); BOULENGER, G. A., No. 6, 1916 (fresh-water, Africa); BOULENGER, G. A., No. 2, 1907 (Egypt); FOWLER, H. W., No. 8, 1936 (marine, West Africa); SMITH, J. L. B., No. 1, 1961 (marine, South Africa); ARNOULT, J., No. 1, 1959 (fresh-water, Madagascar); PELLEGRIN, J., No. 2, 1960 (fresh-water, Madagascar); SMITH, J. L. B., 1965, Ichth. Bull. Rhodes Univ., 31, 601-632 (Red Sea & western Indian Ocean); DAY, F., No. 2, 1889 (India); WEBER & de BEAUFORT, No. 4, 1922 (Indo-Australian Arch.); MARSHALL, T. C., No. 1, 1964 (Great Barrier Reef); WHITLEY, G. P., 1943, Proc. Linn. Soc. New South Wales, 68, 114-144 (key to Australian genera); WHITLEY, G. P., 1955, Western Australian Nat., 5, 25-31 (fresh-water, Western Australia); FOWLER, H. W., No. 6, 1933-35 (China); LINDBERG & LEGHEZA, No. 2, 1965 (Sea of Japan); LINDBERG & LEGEZA, No. 3, 1969 (Sea of Japan); JORDAN, D. S. & STARKS, E. C., 1901, Proc. United States Nat. Mus., 24, 199-206 (Japan); SCHULTZ, L. P., No. 2, 1953 (Marshall & Marianas Is.); FOWLER, H. W., No. 17, 1959 (Fiji)

Atherstoniidae (only as fossils)

Atopoclinidae, as Blenniidae

Auchenipteridae (incl. Trachycorystidae). VAN DER STIGCHEL, J. W. B., No. 1, 1946 (South America); SCHULTZ, L. P., No. 2, 1944 (Venezuela); FOWLER, H. W., No. 10, 1948-51 (fresh-water, Brasil)

Aulopidae, as Aulopodidae

Aulopodidae (incl. Aulopidae). BIGELOW, H. B. et al., No. 4, 1966 (western North Atlantic); LOZANO REY, D. L. No. 4, 1947 (España); FOWLER, H. W., No. 8, 1936 (marine, West Africa); MARSHALL, T. C., No. 1, 1964 (Great Barrier Reef); LINDBERG & LEGHEZA, No. 2, 1965 (Sea of Japan); LINDBERG & LEGEZA, No. 3, 1969 (Sea of Japan)

Aulorhynchidae. LINDBERG & LEGHEZA, No. 2, 1965 (Sea of Japan); LINDBERG & LEGEZA, No. 3, 1969 (Sea of Japan); JORDAN, D. S. & STARKS, E. C., 1902, Proc. United States Nat. Mus., 26, 57-73 (Japan)

Aulostomatidae, as Aulostomidae

Aulostomidae (incl. Aulostomatidae). BEEBE & TEE-VAN, No. 1, 1928 (Hispaniola); FOWLER, H. W., No. 8, 1936 (marine, West Africa); SMITH, J. L. B., No. 1, 1961 (marine, South Africa); DAY, F., No. 2, 1889 (India); de SILVA, P. H. D. H., No. 1, 1956 (marine, Ceylon); WEBER & de BEAUFORT, No. 4, 1922 (Indo-Australian Arch.); MARSHALL, T. C., No. 1, 1964 (Great Barrier Reef); FOWLER, H. W., No. 6, 1933-35 (China); JORDAN, D. S. & STARKS, E. C., 1902, Proc. United States Nat. Mus., 26, 57-73 (Japan); SCHULTZ, L. P., et al., No. 2, 1953 (Marshall & Marianas Is.)

Avocettinidae, as Nemichthyidae

Avocettinopsidae, as Nemichthyidae

Badidae. BARLOW, G. W. et al., 1968, Journ. Zool., 156, 415-447 (new family)

Bagaridae, as Sisoridae

Bagariidae, as Sisoridae

Bagreidae, as Ariidae

Bagridae (incl. Mystidae, Porcidae). JAYARAM, K. C., 1955, Proc. Nat. Inst. Sci. India, 21B, 120-128 (generic review); SPILLMANN, J., No. 1, 1961 (France); BERG, L. S., No. 2, 1933 (fresh-water, U. R. S. S.); BERG, L. S., No. 4, 1949 (fresh-water, Soviet Union); BERG, L. S., No. 9, 1964 (fresh-water, U. S. S. R.); BERG, L. S., No. 6, 1949 (fresh-water, Iran); JAYARAM, K. C., 1966, Bull. Inst. Francaise Afrique Noire, 28A, 1064-1139 (Africa); DAGET, J., No. 2, 1962 (Guinea); DAGET & ILTIS, No. 1, 1965 (Côte d'Ivoire); DAGET, J., No. 1, 1954 (Upper Niger); BARNARD, K. H., 1943, Ann. South African Mus., 36, 101-262 (S. W. Cape region); POLL, M., No. 1, 1946 (Lake Tanganyika); POLL, M., No. 3, 1953 (Lake Tanganyika); ARNOULT, J., No. 1, 1959 (fresh-water, Madagascar); SMITH, H. M., No. 1, 1945 (fresh-water, Thailand); WEBER & de BEAUFORT, No. 2, 1913 (Indo-Australian Arch.); INGER & KONG, No. 1, 1962 (fresh-water, North Borneo); SOLDATOV & LINDBERG, No. 1, 1930 (Far East Seas); OKADA, Y., No. 1 & 2, 1960 (fresh-water, Japan)

Balistidae (incl. Aleuteridae, Aluteridae, Anacanthidae, Ballistidae, Monacanthidae, Psilocephalidae). GILL, T., No. 1, 1884 (synopsis); JORDAN, D. S. & FOWLER, H. W., 1902, Proc. United States Nat. Mus., 25, 251-286 (review, also as Monacanthidae); FRASER-BRUNNER, A., 1935, Ann. Mag. Nat. Hist., (10), 15, 658-663 (synopsis of genera); FRASER-BRUNNER, A., 1941, Ann. Mag. Nat. Hist., (11)8, 176-199 (synopsis, as Aluteridae); HILDEBRAND & SCHROEDER, No. 1, 1928 (Chesapeake Bay, also as Monacanthidae); BEEBE & TEE-VAN, No. 1, 1928 (Hispaniola, also as Monacanthidae); MEEK & HILDEBRAND, No. 4, 1928 (marine, Panama, also as Monacanthidae); FERNANDEZ-YEPES, A., 1955, Mem. Soc. Cienc. Nat. La Salle, 15, 137-139 (Venezuela, as Monacanthidae); HILDEBRAND, S. F., No. 1, 1946 (shore, Peru); DEVINCENZI, G. F., No. 2, 1924 (Uruguay); MOORE, D., 1967, Bull. Mar. Sci., 17, 689-722 (western Atlantic); BERRY, F. H. & VOGELLE, L. E., 1961, Bull. United States Fish Comm., 61, 61-109 (western North Atlantic, as Monacanthidae); SMITT. F. A. et al., No. 2, 1895 (Scandinavia); POLL, M., No. 2, 1947 (marine, Belgique); NOBRE, A., No. 1, 1935 (marine, Portugal); LOZANO REY, D. L., No. 2, 1952 (España); SVETOVIDOV, A. N., No. 2, 1964 (Black Sea); FOWLER, H. W., No. 8, 1936 (marine, West Africa, also as

Monacanthidae); SMITH, J. L. B., No. 1, 1961 (marine, South Africa, also as Aluteridae, & Monacanthidae); de BEAUFORT & BRIGGS, No. 11, 1962 (Indo-Australian Arch., also as Monacanthidae); MARSHALL, T. C., No. 1, 1964 (Great Barrier Reef, also as Aluteridae, Anacanthidae, & Monacanthidae); SOLDATOV & LINDBERG, No. 1, 1930 (Far East Seas, as Monacanthidae); BERRY, F. H. & BALDWIN, W. J., 1966, Proc. California Acad. Sci., (4), 34, 429-474 (eastern Pacific, as Balistidae); SCHULTZ, L. P. et al., No. 1, 1966 (Marshall & Marianas Is., as Aluteridae & Balistidae); FOWLER, H. W., No. 17, 1959 (Fiji, as Balistidae & Monacanthidae)

Ballistidae, as Balistidae

Banjosidae. FOWLER, H. W., No. 20, 1933 (Philippine seas); FOWLER, H. W., No. 9, 1937-41 (China); LINDBERG & KRASJUKOVA, No. 1, 1969 (Sea of Japan)

Barbourisidae, as Barbourisiidae

Barbourisiidae (incl. Barbourisidae). (No revisionary references noted)

Bathyaploactidae, as Aploactinidae

Bathyclupeidae. WEBER & de BEAUFORT, No. 6, 1931 (Indo-Australian Arch.)

Bathdraconidae. NORMAN, J. R., No. 3, 1938 (revision, Antarctic)

Bathylaconidae. NIELSON, J. G. & LARSEN, V., 1968, Galathea Rep., 9, 221-238 (synopsis); BIGELOW H. B., No. 3, 1964 (western North Atlantic)

Bathylagidae (incl. Microstomatidae, Microstomidae). CHAPMAN, W. M., 1943, Journ. Washington Acad. Sci., 33, 147-160 (osteology & relationships); CHAPMAN, W. M., 1948, Proc. California Acad. Sci., (4), 26, 1-22 (osteology & relationships, as Microstomidae); BIGELOW, H. B. et al., No. 3, 1964 (western North Atlantic); LOZANO REY, D. L., No. 4, 1947 (España); also as Microstomidae); FOWLER, H. W., No. 8, 1936 (marine, West Africa, as Microstomidae); SMITH, J. L. B., No. 1, 1961 (marine, South Africa, as Microstomidae); SCHMIDT, P. Y., No. 1, 1950 (Sea of Okhotsk); SHMIDT, P. Y., No. 2, 1965 (Sea of Okhotsk)

Bathymasteridae. FOWLER, H. W., No. 16, 1958 (China); SOLDATOV & LINDBERG, No. 1, 1930 (Far East seas); SCHMIDT, P. Y., No. 1, (Sea of Okhotsk); SHMIDT, P. Y., No. 2, 1965 (Sea of Okhotsk); LINDBERG & KRASJUKOVA, No. 1, 1969 (Sea of Japan)

Bathypteridae, as Bathypteroidae

Bathypteroidae (incl. Bathypteridae, Benthosauridae). MEAD, G. W., 1966, inBigelow, H. B. et al., No. 4 (western North Atlantic); LOZANO REY, D. L., No. 4, 1947 (España); FOWLER, H. W., No. 8, 1936 (marine, West Africa, also as Benthosauridae); SMITH, J. L. B., No. 1, 1961 (marine, South Africa); FOWLER, H. W., No. 18, 1956 (Red Sea)

Bathysauridae, as Synodontidae

Bathythrissidae, as Albulidae

Batrachidae, as Batrachoididae

Batrachoididae (incl. Batrachidae). COLLETTE, B. B., 1966, Copeia, 1966, 846-864 (review of Thalassophryninae); HILDEBRAND & SCHROEDER, No. 1, 1928 (Chesapeake Bay); MEEK & HILDEBRAND, No. 4, 1928 (marine, Panama); HILDEBRAND, S. F., No. 1, 1946 (shore, Peru); PUYO, J., No. 1, 1949 (French Guiana); FOWLER, H. W., No. 13, 1954 (fresh-water, Brasil); DEVINCENZI, G. F., No. 2, 1924 (Uruguay); SEÑORANS, J. S., 1958, Rev. Fac. Hum. Cienco, 16, 275-285 (Uruguay); SMITT, F. A. et al., No. 1, 1892 (Scandinavia); NOBRE, A., No. 1, 1935 (marine, Portugal); LOZANO y REY, L., No. 5, 1960 (España); FOWLER, H. W., No. 8, 1936 (marine, West Africa); ROUX, C., 1971, Bull. Mus. Nat. Hist. Nat., (2), 42, 626-643 Côte occidentale Africaine); SMITH, J. L. B., 1952, Ann. Mag. Nat. Hist., (12), 5, 313-339 (South & East Africa); SMITH, J. L. B., No. 1, 1961 (marine, South Africa); DAY, F., No. 2, 1889 (India); de BEAUFORT & BRIGGS, No. 11, 1962 (Indo-Australian Arch.); MARSHALL, T. C., No. 1, 1964 (Great Barrier Reef)

Bdellostomatidae (incl. Heptatretidae). (No revisionary references noted)

Bedotiidae, as Atherinidae

Beerichthyidae (only as fossils)

Belonidae (incl. Esocesidae, Esocidae, Petalichthyidae, Tylosuridae). FERNANDEZ-YEPES, A., 1948, Mem. Soc. Cienc. Nat. La Salle, 8, 142-144; MARTIN, F., 1954, Noved. Cient. Mus. Hist. Nat. Caracas, Zool., 4, 1-8, 1954 (key to genera); MEES, G. F., 1962, Zool. Verh., 54, 1-96 (revision); COLLETTE, B. B. & BERRY, F. H., 1965, Copeia, 1965, 386-392; COLLETTE, B. B., 1966, American Mus. Novit., 2274, 1-22 (generic characters); CRESSEY, R. F. & COLLETTE, B. B., 1970, Fish. Bull., United States Fish Wildl. Serv., 68, 347-432 (list of speces, parasitic copepods of each); JORDAN, D. S. & FORDICE, M. W., 1887 (American); WALFORD, L. A., No. 1, 1937 (Alaska to Equator); HILDEBRAND & SCHROEDER, No. 1, 1928 (Chesapeake Bay); FOWLER, H. W., No. 1, 1945 (southeastern U. S.); BEEBE & TEE-VAN, No. 1, 1928 (Hispaniola); MEEK, S. F., No. 1, 1904 (fresh-water, Mexico); MEEK & HILDEBRAND, No. 2, 1923 (marine, Panama); HILDEBRAND, S. F., No. 1, 1946 (shore, Peru); EIGENMANN, C. H., No. 1, 1912 (British Guiana); FOWLER, H. W., No. 13, 1954 (fresh-water, Brasil); BERRY, F. H. & RIVAS, L. R., 1962, Copeia, 1962, 152-160 (western Atlantic); COLLETTE, B. B., 1970, Atlantide Rep., 11, 1-60 (eastern Atlantic); DUNCKER & LADIGES, No. 1, 1960 (Nordmark); POLL, M., No. 2, 1947 (marine, Belgique); LOZANO REY, D. L., No. 4, 1947 (España); BANARESCU, P., No. 1, 1964 (Romania); SVETOVIDOV, A. N., No. 2, 1964 (Black Sea); ANDRIYASHEV, A. P., No. 1, 1954 (northern seas of U. S. S. R.); FOWLER, H. W., No. 8, 1936 (marine, West Africa); DAGET & ILTIS, No. 1, 1965 (Côte d'Ivoire); SMITH, J. L. B., No. 1, 1961 (marine, South Africa, as Petalichthyidae & Tylosuridae); FOWLER, H. W., No. 18, 1956 (Red Sea); de SILVA, P. H. D. H., No. 1, 1956 (marine, Ceylon); SMITH, H. M., No. 1, 1945 (fresh-water, Thailand); WEBER & de BEAUFORT, No. 4, 1922 (Indo-Australian Arch.); MARSHALL, T. C., No. 1, 1964 (Great Barrier Reef); FOWLER,

H. W., No. 5, 1932 (China); SOLDATOV & LINDBERG, No. 1, 1930 (Far East seas); LINDBERG & LEGHEZA, No. 2, 1965 (Sea of Japan); LINDBERG & LEGEZA, No. 3, 1969 (Sea of Japan); JORDAN, D. S. & STARKS, E. C., 1903, Proc. United States Nat. Mus., 26, 525-544 (Japan); PARIN, N. V., 1967, Trudy Okeanologii Inst. 84, 3-83 (review of marine, western Pacific & Indian Oceans); SCHULTZ, L. P., No. 2, 1953 (Marshall & Marianas Is.); FOWLER, H. W., No. 17, 1959 (Fiji)

Belonorhynchidae, as Saurichthyidae (only as fossils)

Belontidae, as Belontiidae

Belontiidae (incl. Belontidae, Polyacanthidae). (No revisionary references noted)

Bembradidae, as Platycephalidae

Bembridae, as Platycephalidae

Bembropidae, as Percophididae

Bembropsidae, as Percophididae

Benthophilidae, as Gobiidae

Benthosauridae, as Bathypteroidae

Berycidae. SMITT, F. A. et al., No. 1, 1892 (Scandinavia); NOBRE, A., No. 1, 1935 (marine, Portugal); LOZANO REY, D. L., No. 2, 1952 (España); FOWLER, H. W., No. 8, 1936 (marine, West Africa); SMITH. J. L. B., No. 1, 1961 (marine, South Africa); DAY, F., No. 2, 1889 (India); MARSHALL, T. C., No. 1, 1964 (Great Barrier Reef); LINDBERG & LEGHEZA, No. 2, 1965 (Sea of Japan); LINDBERG & LEGEZA, No. 3, 1969 (Sea of Japan); JORDAN, D. S. & FOWLER, H. W., 1902, Proc. United States Nat. Mus., 26, 1-21 (Japan)

Berycopsidae (only as fossils)

Birgeriidae (incl. Xenesthidae) (only as fossils)

Birkeniidae (only as fossils)

Bleekeridae, as Ammodytidae

Bleekeriidae, as Ammodytidae

Blennidae, as Blenniidae

Blenniidae (incl. Atopoclinidae, Blennidae, Nemophididae, Salariidae, Xiphasiidae). NORMAN, J. R., 1943, Ann. Mag. Nat. Hist., (11), 10, 793-812 (genera); STEINITZ, H., 1949, Rev. Fac. Sci. Univ. Istanbul, B14, 129-152, 170-197; SPRINGER, V. G., 1959, Texas Journ. Sci., 11, 321-334 (key to genera); SPRINGER, V. G., 1968, United States Nat. Mus. Bull., 284, 1-85 (osteology & classification); HUBBS, C., 1953, Copeia, 1953, 11-23 (Chaenopsinae, key to tribes); HILDEBRAND & SCHROEDER, No. 1, 1928 (Chesapeake Bay); BEEBE & TEE-VAN, No. 1, 1928 (Hispaniola); MEEK & HILDEBRAND, No. 4, 1928 (marine, Panama); HILDEBRAND, S. F., No. 1, 1946 (shore, Peru); DEVINCENZI, G. F., No. 2, 1924 (Uruguay); SMITT, F. A. et al., No. 1, 1892 (Scandinavia); POLL, M., No. 2, 1947 (marine, Belgique); SPILLMANN, J., No. 1, 1961 (France); NOBRE, A., No. 1, 1935 (marine, Portugal); LOZANO REY, L., No. 1, 1935 (fresh-water, España); LOZANO y REY, L., No. 5, 1960 (España); FAGE, L., No. 1, 1918 (Mediterranean); BANARESCU, P., No. 1, 1964 (Romania); RHASIS ERAZI, R. A., 1941, Rev. Fac. Sci. Univ. Istanbul, 6B, 118-127 (Bosphorus & Mer de Marmora); SVETOVIDOV, A. N., No. 2, 1964 (Black Sea); BOULENGER, G. A., No. 6, 1916 (fresh-water, Africa); FOWLER, H. W., No. 8, 1936 (marine, West Africa); BARNARD, K. H., 1948, Ann. South African Mus., 36, 341-406 (marine, South Africa); SMITH, J. L. B., No. 1, 1961 (marine, South Africa, also as Salaridae); PELLEGRIN, J., No. 2, 1960 (fresh-water, Madagascar); SMITH, J. L. B., 1959, Ichth. Bull. Rhodes Univ., 14, 229-252 (western Indian Ocean, also as Salariidae); DAY, F., No. 2, 1889 (India); de BEAUFORT & CHAPMANS, No. 9, 1951 (Indo-Australian Arch.); MARSHALL, T. C., No. 1, 1964 (Great Barrier Reef); HERRE, A. W. C. T., 1939, Philippine Journ. Sci., 70, 315-372 (Philippines, also as Xiphasiidae); FOWLER, H. W., No. 16, 1958 (China); SOLDATOV & LINDBERG, No. 1, 1930 (Far East seas); JORDAN, D. S. & SNYDER, J. O., 1902, Proc. United States Nat. Mus., 25, 441-504 (Japan); STRASBURG, D. W., 1956, Pacific Sci., 10, 241-267 (key to Hawaiian species); SCHULTZ, L. P., No. 3, 1960 Marshall & Marianas Is.); FOWLER, H. W., No. 17, 1959 (Fiji); KREJSA, R. J., 1960, Copeia, 1960, 322-336 (key to tropical Pacific Blenniinae)

Blepisiidae, as Cottidae

Blepsiidae, as Cottidae

Blochiidae (only as fossils)

Bobasatraniidae, (only as fossils)

Bodianidae, as Labridae

Boreolepidae as Palaeoniscidae (only as fossils)

Boreol epididae, as Palaeoniscidae

Boreosomidae (only as fossils)

Bostockiidae, as Serranidae

Bothidae (incl. Paralichthyidae). NORMAN, J. R., 1930, Discovery Rep., 2, 358-369 (oceanic); WU, H.-W., 1932, Contribution a l'etude. . . Heterosomes. Paris;1-179 (monograph); NORMAN, J. R., 1934, A systematic monograph of the flatfishes. . . . I. London;1-459; BEEBE & TEE-VAN, No. 1, 1928 (Hispaniola); HILDEBRAND, S. F., No. 1, 1946 (shore, Peru); de BUEN, F., Montemar, 1, 1-90 (Chile); NORMAN, J. R., No. 2, 1937 (marine, Patagonia); DUNCKER & LADIGES, No. 1, 1960 (Nordmark); POLL, M., No. 2, 1947 (marine, Belgique); POLL, M., No. 2, 1947 (marine, Belgique); BANARESCU, P., No. 1, 1964 (Romania); SVETOVIDOV, A. N., No. 2, 1964 (Black Sea); ANDRIYASHEV, A. P., No. 1, 1954 (northern seas of U. S. S. R.); FOWLER, H. W., No. 8, 1936 (marine, West Africa); DAGET & ILTIS, No. 1, 1965 (Côte d'Ivoire); SMITH, J. L. B., No. 1, 1961 (marine, South America); FOWLER, H. W., No. 18, 1956 (Red Sea); NORMAN, J. R., 1927, Rec. Indian Mus., 29, 7-48(fauna &

museum list); de SILVA, P. H. D. H., No. 2, 1956 (Ceylon); WEBER & de BEAUFORT, No. 5, 1929 (Indo-Australian Arch.); MARSHALL, T. C., No. 1, 1964 (Great Barrier Reef); CHAN, J. T. F. & WENG, H. T. C., 1965, Biol. Bull. Tunghai Univ., 25 & 27 (Taiwan); FOWLER, H. W., No. 6, 1933-35 (China); SCHULTZ, L. P. et al., No. 1, 1966 (Marshall & Marianas Isl.) FOWLER, H. W., No. 17, 1959 (Fiji); NORMAN, J. R., No. 3, 1938 (Antarctic)

Bothriolepidae (only as fossils)

Bovichtidae, as Bovichthyidae

Bovichthyidae (incl. Bovichtidae, Bovictidae, Pseudaphritidae). NORMAN, J. R., No. 2, 1937 (marine, Patagonia)

Bovictidae, as Bovichthyidae

Brachionichthyidae. (No revisionary references noted)

Brachydeglmidae (only as fossils)

Brachydiridae (only as fossils)

Bramidae (incl. Lepidotidae, Lepodidae, Pteraclidae, Pteraclididae, Steinegeriidae, Trachyberycidae). ABE, T., 1961, Japan Journ. Ichth., 8, 91-99 (key to genera of Braminae); MEAD, G. W., 1957, Zoologica, 42, 51-61 (Gulf of Mexico); de BUEN, F., 1958, Invest. Zool. Chile, 4, 132-134 (Chile); SMITT, F. A. et al., No. 1, 1892 (Scandinavia); POLL, M., No. 2, 1947 (marine, Belgique); LOZANO REY, D. L., No. 3, 1952 (España); SCHMIDT, J. & STRUBBERG, A., 1918, Rep. Danish Oceanogr. Exp. 1908-10, 2, (Biol. A6), 1-15 (Mediterranean); ANDRIYASHEV, A. P., No. 1, 1954 (northern seas of U. S. S. R.); FOWLER, H. W., No. 8, 1936 (marine, West Africa, also as Pteraclidae); SMITH, J. L. B., No. 1, 1961 (marine, South Africa, also as Pteraclidae); FOWLER, H. W., No. 7, 1936 (China); LINDBERG & KRASJUKOVA, No. 1, 1969 (Sea of Japan, as Pteraclididae also)

Branchiostegidae (incl. Latilidae, Malacanthidae). GILL, T., No. 1, 1884 (synopsis, as Molacanthidae); HILDEBRAND & SCHROEDER, No. 1, 1928 (Chesapeake Bay); BEEBE & TEE-VAN, No. 1, 1928 (Hispaniola, as Malacanthidae); HILDEBRAND, S. F., No. 1, 1946 (shore, Peru, as Malacanthidae); DEVINCENZI, G. F., No. 2, 1924 (Uruguay, as Malacanthidae); SMITH, J. L. B., No. 1, 1961 (marine, South Africa, also as Malacanthidae); POLL, M., No. 4, 1954 (coast of South America, as Latilidae); DAY, F., No. 2, 1889 (India, as Malacanthidae); WEBER & de BEAUFORT, No. 7, 1936 (Indo-Australian Arch.) as Malacanthidae); MARSHALL, T. C., No. 1, 1964 (Great Barrier Reef); FOWLER, H. W., No. 11, 1949 (China, also as Malacanthidae); LINDBERG & KRASJUKOVA, No. 1, 1969 (Sea of Japan)

Bregmacerotidae. D'ANCONA, U. & CAVINATO, G., 1965, Dana Rep., 64, 1-91; BEEBE & TEE-VAN, No. 1, 1928 (Hispaniola); MEEK & HILDEBRAND, No. 4, 1928 (marine, Panama); SVETOVIDOV, A. N., No. 1, 1948 (U. S. S. R.); SVETOVIDOV, A. N., 1962, Fauna U. S. S. R., Fishes IX, 4. Washington, D. C., 1-304 (U. S. S. R.); MAUL, G. E., 1952, Bol. Mus. Munic. Funchal, 6, 1-62(museum list); SMITH, J. L. B., No. 1, 1961 (marine, South Africa); FOWLER, H. W., No. 18, 1956 (Red Sea); LINDBERG & LEGHEZA, No. 2, 1965 (Sea of Japan); LINDBERG & LEGEZA, No. 3, 1969 (Sea of Japan); SCHULTZ, L. P. et al., No. 2, 1953 (Marshall & Marianas Is.); FOWLER, H. W., No. 17, 1959 (Fiji)

Brookvaliidae (only as fossils)

Brotulidae, as Ophidiidae

Brotulophidae, as Ophidiidae

Bunocephalidae, as Aspredinidae

Caesiodidae, as Lutjanidae

Caesionidae, as Lutjanidae

Calichthyidae, as Callichthidae

Callichthiidae, as Callichthyidae

Callichthyidae (incl. Callichthiidae, Calichthyidae, Callichtyidae). ELLIS, M. D., 1913, Ann. Carnegie Mus., 8, 384-413 (checklist, museum list of Carnegie Museum & Indiana University); GOSLINE, Stanford Ichth. Bull., 2, 1-29 (neotropical, key to genera, bibliography); GOSLINE, W. A., 1940, Stanford Ichth. Bull., 2, 1-29 (neotropical); MEEK & HILDEBRAND, No. 1, 1916 (fresh-water, Panama); EIGENMANN, & EIGENMANN, No. 1, 1890 (South America); VAN DER STIGCHEL, J. W. B., No. 1, 1946 (South America); SCHULTZ, L. P., No. 2, 1944 (Venezuela); ORCES, V. G., 1960, Cienc. y Nat., Quito, 3, 2-6 (Ecuador); EIGENMANN, C. H., No. 1, 1912 (British Guiana); HOEDEMANN, J. J., 1952, Beaufortia, 12, 1-11 (keys to genera & species); PUYO, J., No. 1, 1949 (French Guiana); FOWLER, H. W., No. 13, 1954 (fresh-water, Brasil); DEVINCENZI, G. J., No. 2, 1924 (Uruguay)

Callichtyidae, as Callichthyidae

Calliodontidae, as Scaridae

Callionymidae (incl. Draconettidae). SCHULTZ, L. P. & WOODS, L. P., 1948; Journ. Washington Acad. Sci., 38, 419-420 (key to genera); BRIGGS, J. C. & BERRY, F. H., 1959, COPEIA, 1959, 123-133 (review of Draconettidae); DAVIS, W. P., 1966, Bull. Mar. Sci., 16, 834-862 (western Atlantic); SMITT. F. A. et al., No. 1, 1892 (Scandinavia); DUNCKER & LADIGES, No. 1, 1960 (Nordmark); POLL, M., No. 2, 1947 (marine, Belgique); NOBRE, A., No. 1, 1935 (marine, Portugal); LOZANO y REY, L., No. 5, 1960 (España); FAGE, L., No. 1, 1918 (Mediterranean); BANARESCU, P., No. 1, 1964 (Romania); SVETOVIDOV, A. N., No. 2, 1964 (Black Sea); SMITH, J. L. B., No. 1, 1961 (marine, South Africa); SMITH, J. L. B., 1963, Ichth. Bull. Rhodes Univ., 28, 547-564 (Red Sea, also as Draconettidae); SMITH, J. L. B., 1963, Ichth. Bull. Rhodes Univ., 28, 1-17 (Red Sea & western Indian Ocean, also as Draconettidae); DAY, F., No. 2, 1889 (India); FOWLER, H. W., 1941, Proc. United States Nat. Mus., 90, 1-31 (key to Indo-Pacific genera); de BEAUFORT & CHAPMANS, No. 9, 1951 (Indo-Australian Arch.);

MARSHALL, T. C., No. 1, 1964 (Great Barrier Reef); MEES, G. F., 1963, Journ. Roy. Soc. New South Wales, 46, 93-99 (Western Australia); FOWLER, H. W., 1958, Quart. Journ. Taiwan Mus., 12, 67-97 (China); JORDAN, D. S. & FOWLER, H. W., 1903, Proc. United States Nat. Mus., 25, 939-959 (Japan); SCHULTZ, L. P. et al., No. 3, 1960 (Marshall & Marianas Is.); FOWLER, H. W., No. 17, 1959 (Fiji)

Callipterygidae, as Trachinidae

Callophysidae, as Pimelodidae

Callorhinchidae, as Callorhynchidae

Callorhynchidae (incl. Callorhinchidae). BIGELOW & SCHROEDER, No. 2, 1953 (western North Atlantic); HILDEBRAND, S. F., No. 1, 1946 (shore, Peru); SMITH, J. L. B., No. 1, 1961 (marine, South Africa)

Callyodontidae, as Scaridae

Calophysidae, as Pimelodidae

Canobiidae, as Palaeoniscidae (only as fossils)

Canthigasteridae, as Tetraodontidae

Caproidae (incl. Antigoniidae, Caprophonidae). NOBRE, A., No. 1, 1935 (marine, Portugal); LOZANO REY, D. L., No. 2, 1952 (España); FAGE, L., No. 1, 1918 (Mediterranean); FOWLER, H. W., No. 8, 1936 (marine, West Africa); POLL, M., No. 4, 1954 (coastal, South Africa); SMITH, J. L. B., No. 1, 1961 (marine, South Africa); WEBER & de BEAUFORT, No. 5, 1929 (Indo-Australian Arch.); MARSHALL, T. C., No. 1, 1964 (Great Barrier Reef)

Caprophonidae, as Caproidae

Caracanthidae. SMITH, J. L. B., 1958, Ichth. Bull. Rhodes Univ., 12, 167-181 (western Indian Ocean); de BEAUFORT & BRIGGS, No. 11, 1962 (Indo-Australian Arch.); SCHULTZ, L. P. et al., No. 1, 1966 (Marshall & Marianas Is.); FOWLER, H. W., No. 17, 1959 (Fiji)

Carangidae (incl. Juvenellidae, Nematistiidae, Seriolidae). JORDAN, D. S. & GILBERT, C. H., 1883, Proc. United States Nat. Mus., 6, 188-207 (America); WALFORD, L. A., No. 1, 1937 (Alaska to Equator); HILDEBRAND & SCHROEDER, No. 1, 1928 (Chesapeake Bay); BEEBE & TEE-VAN, No. 1, 1928 (hispaniola); GINSBURG, I., 1952, Publ. Inst. Mar. Sci. Univ. Texas, 2, (2), 43-117 (Gulf of Mexico); REGAN, C. T., No. 1, 1906-08 (Central America); MEEK & HILDEBRAND, No. 2, 1923 (marine, Panama, as Nematistiidae); MEEK & HILDEBRAND, No. 3, 1925 (marine, Panama); HILDEBRAND, S. F., No. 1, 1946 (shore, Peru, as Nematistiidae); PUYO, J., No. 1, 1948 (Guyane); FOWLER, H. W., No. 13, 1954 (fresh-water, Brasil); DEVINCENZI, G. L. No. 2, 1924 (Uruguay); NORMAN, J. R., No. 2, 1937 (marine, Patagonia); BAUCHOT, M. L. & BLANC, M., 1963, Atlantide Rep., 7, 37-61 (eastern tropical Atlantic); WHEELER, A. C., 1963, Ann. Mag. Nat. Hist., (13), 5, 529-540 (Europe, Trachinotinae); SMITT, F. A. et al., No. 1, 1892 (Scandinavia); DUNCKER & LADIGES, No. 1, 1960 (Nordmark); POLL, M., No. 2, 1947 (marine, Belgique); NOBRE, A., No. 1, 1935 (marine, Portugal); LOZANO REY, D. L., No. 3, 1952 (España, also as Seriolidae); TORTONESE, G., 1951-52, Ann. Mus. Civ. Stor. Nat. Genova, 65, 259-324 (Mediterranean Carangini); BANARESCU, P., No. 1, 1964 (Romania); SVETOVIDOV, A. N., No. 2, 1964 (Black Sea); BOULENGER, G. A., No. 6, 1916 (fresh-water, Africa); PELLEGRIN, J., No. 2, 1923 (western Africa); PELLEGRIN, J., 1923 (western Africa); FOWLER, H. W., No. 8, 1936 (marine, West Africa); BAUCHOT, M. L. & BLANC, M., 1963, Atlantide Rep., 7, 37-61 (eastern tropical Atlantic); DAGET & ILTIS, No. 1, 1965 (Côte d'Ivoire); MONOD, T., 1927, Faune Colon. Francaise, 1, 643-742 (Cameroun); POLL, M., No. 4, 1954 (coastal South Africa); SMITH, J. L. B., No. 1, 1961 (marine, South Africa); WILLIAMS, F., 1958, Ann. Mag. Nat. Hist., (13), 1, 369-430 (British East African waters); PELLEGRIN, J., No. 2, 1960 (fresh-water, Madagascar); SMITH, J. L. B., 1959, Ichth. Bull. Rhodes Univ., 15, 255-261 (western Indian Ocean, as Seriolidae); DAY, F., No. 2, 1889 (INdia); WEBER & de BEAUFORT, No. 6, 1931 (Indo-Australian Arch); MARSHALL, T. C., No. 1, 1964 (Great Barrier Reef); GRIFFIN, L. T., 1932, Rec. Auckland Mus., 1, 123-134 (New Zealand, also as Seriolidae); ROXAS, H. A. & AGCO, A. G., 1941, Philippine Journ. Sci., 74, 1-82 (Philippines); OSHIMA, M., 1924, Proc. Pan-Pacific Sci. Congr., 2, 1571-1577 (Taiwan); OSHIMA, M., 1925, Philippine Journ. Sci., 26, 345-412 (Taiwan); CHAN, W. L., No. 1, 1968 (Hong Kong); FOWLER, H. W., No. 7, 1936 (China); NICHOLS, J. T., 1945, Lingnan Sci. Journ., 21, 81-86 (Caranginae of China); SOLDATOV & LINDBERG, No. 1, 1930 (Far East seas); LINDBERG & KRASJUKOVA, No. 1, 1969 (Sea of Japan); WAKIYA, Y., 1924, Ann. Carnegie Mus., 15, 139-244 (Japan); SUZUKI, K., 1963, Rep. Fac. Fish. Univ. Mie, 4, 43-232 (Japan); SCHULTZ, L. P. et al., No. 2, 1953 (Marshall & Marianas Is.); FOWLER, H. W., No. 17, 1959 (Fiji)

Carapidae (incl. Disparichthyidae, Fierasferidae). ARNOLD, D. C., 1956, Bull. British Mus. (Nat. Hist.), Zool., 4, 245-307 (revision); ARNOLD, D. C., 1956, Bull. Inst. Oceanogr. Monaco, 1084, 1-5 (museum list); MEEK & HILDEBRAND, No. 4, 1928 (marine, Panama); LOZANO y REY, L., No. 5, 1960 (España); FOWLER, H. W., No. 8, 1936 (marine, West Africa); SMITH, J. L. B., No. 1, 1961 (marine, South Africa); SMITH, J. L. B., 1955, Ann. Mag. Nat. Hist., (12), 8, 401-416 (western Indian Ocean); de BEAUFORT & CHAPMANS, No. 9, 1951 (Indo-Australian Archipelago); MARSHALL, T. C., No. 1, 1964 (Great Barrier Reef); SCHULTZ, L. P. et al., No. 3, 1960 (Marshall & Marianas Is.); FOWLER, H. W., No. 17, 1959 (Fiji)

Carbovelidae. (No revisionary references noted)

Carcharhinidae (incl. Carcharidae, Carchariidae, Carcharinidae, Eulamiidae, Galeorhinidae, Galeidae, Mustelidae, Triakidae). GARMAN, S., No. 1, 1913 (monograph, as Carcharinidae & Galeorhinidae also); HILDEBRAND & SCHROEDER, No. 1, 1928 (Chesapeake Bay, as Galeidae); BEEBE & TEE-VAN, No. 1, 1928 (Hispaniola, as Carcharhinidae & Galeorhinidae);

REGAN, C. T., No. 1, 1906-08 (Central America); MEEK & HILDEBRAND, No. 2, 1923 (marine, Panama, also as Galeorhinidae); HILDEBRAND, S. F., No. 1, 1946 (shore, Peru) as Galeorhinidae); PUYO, J., No. 1, 1949 (Guyane, as Galeidae); FOWLER, H. W., 1948-51, Arqu. Zool. Sao Paulo, 6, 1-625 (fresh-water, Brasil, as Galeorhinidae); DEVINCENZI, G. J., No. 1, 1920 (Uruguay, as Galeidae & Galeorhinidae); BIGELOW & SCHROEDER, No. 1, 1948 (western North Atlantic, also as Carchariidae & Triakidae); SMITT, F. A. et al., No. 2, 1895 (Scandinavia, as Carchariidae); DUNCKER & LADIGES, No. 1, 1960 (Nordmark, as Triakidae also); POLL, M., No. 2, 1947 (marine, Belgique); NOBRE, A., No. 1, 1935 (marine, Portugal); TORTONESE, E., No. 3, 1956 (Italia, also as Carchariidae & Triakidae); BERG, L. S., No. 7, 1911 (Russia, as Carchariidae); ANDRIYASHEV, A. P., No. 1, 1954 (northern seas of U. S. S. R.); BERG, L. S., No. 6, 1949 (fresh-water, Iran); BOULENGER, G. A., No. 3, 1909 (fresh-water, Africa, as Carchariidae); FOWLER, H. W., No. 8, 1936 (marine, West Africa, as Carchariidae, Eulamiidae, & Mustelidae); SMITH, J. L. B., No. 1, 1961 (marine, South Africa, as Carchariidae & Galeorhinidae); FOWLER, H. W., No. 18, 1956 (Red Sea, as Carchariidae, Galeorhinidae, & Triakidae); GOHAR & MAZHAR, No. 1, 1964 (Red Sea, also as Triakidae); DAY, F., No. 1, 1889 (India, as Carchariidae); SMITH, H. M., No. 1, 1945 (fresh-water, Thailand); MARSHALL, T. C., No. 1, 1964 (Great Barrier Reef, as Carchariidae, Galeidae, & Mustelidae); FOWLER, H. W., No. 2, 1930 (China, as Carchariidae & Galeorhinidae); SOLDATOV & LINDBERG, No. 1, 1930 (Far East seas); LINDBERGH & LEGHEZA, No. 1, 1959, (Japan Sea, also as Carchariidae); JORDAN, D. S., & FOWLER, H. W., 1903, Proc. United States Nat. Mus., 26, 593-674 (Japan, as Carchariidae); SCHULTZ, L. P. et al., No. 2, 1953 (Marshall & Marianas Is., also as Triakidae); FOWLER, H. W., No. 17, 1959 (Fiji, as Eulamiidae)

Carcharidae, as Carcharhinidae

Carchariidae, as Odontaspidae (but see also Carcharhinidae)

Carcharinidae, as Carcharhinidae

Cardipeltidae. (No revisionary references noted)

Caristiidae (incl. Elephenoridae). REGAN, C. T., 1912, Ann. Mag. Nat. Hist., (8, 10, 637-638; NORMAN, J. R., No. 1, 1930 (oceanic); LOZANO REY, D. L., No. 2, 1952 (España)

Catopteridae, as Redfieldiidae (only as fossils)

Catostomidae. HUBBS, C. L., 1930, Misc. Publ. Mus. Zool., Univ. Michigan, 20, 1-47 (eastern North America); SLASTENENKO, E. P., No. 1, 1958 (Canada); HILDEBRAND & SCHROEDER, No. 1, 1928 (Chesapeake Bay); FOWLER, H. W., No. 1, 1945 (southeastern U. S.); MEEK, S. E., No. 1, 1904 (fresh-water, Mexico); BERG, L. S., 1912, Faune de la Russie; Poissons, III. 1, 1-336 (Russie); BERG, L. S., No. 1, 1932 (fresh-water, U. R. S. S.); BERG, L. S., No. 4, 1949 (fresh-water, Soviet Union); BERG, L. S., No. 9, 1964 (fresh-water, U. S. S. R.); NICHOLS, J. T., No. 1, 1943 (fresh-water, China)

Catulidae, as Scyliorhinidae

Caulolepidae, as Anoplogasteridae

Caulophrynidae. BERTELSEN, E., 1951, No. 1 (revision); FOWLER, H. W., No. 8, 1936 (marine, West Africa)

Cebedichthyidae, as Stichaeidae

Centracanthidae, as Emmelichthyidae

Centracantidae, as Emmelichthyidae

Centraciontidae, as Heterodontidae

Centrarchidae (incl. Elassomatidae, Elassomidae, Eucentrarchidae, Grystidae, Micropteridae). McKAY, C. L., 1881, Proc. United States Nat. Mus., 4, 87-93 (review of genera & species); SLASTENENKO, E. P., No. 1, 1958 (fresh-water, Canada); HILDEBRAND & SCHROEDER, No. 1, 1928 (Chesapeake Bay); FOWLER, H. W., No. 1, 1945 (fresh-water, southeastern U. S.); FOWLER, H. W., No. 1, 1945 (as Elassomidae also); REGAN, C. T., No. 1, 1906-08 (Central America); MEEK, S. E., No. 1, 1904 (fresh-water, Mexico); DEVINCENZI, G. L., No. 2, 1924 (Uruguay); DUNCKER & LADIGES, No. 1, 1960 (Nordmark); SPILLMANN, J., No. 1, 1961 (France); BANARESCU, P., No. 1, (Romania); SVETOVIDOV, A. N., No. 2, 1964 (Black Sea); BERG, L. S., No. 5, 1949 (fresh-water, Soviet Union)); BERG, L. S., No. 10, 1965 (fresh-water, U. S. S. R.); BOULENGER, G. A., No. 5, 1915 (fresh-water, Africa); PELLEGRIN, J., No. 1, 1933 (fresh-water, Madagascar); ARNOULT, J., No. 1, 1959 (fresh-water, Madagascar); OKADA, Y., No. 2, 1960 (fresh-water, Japan)

Centriscidae (incl. Amphisilidae). MOHR, E., 1937, Dana Rep., 13, 1-70 (revision); KAHSBAUER, P., 1951, Ann. Naturh. Mus. Wien, 58, 118-121 (Systematik); SMITH, J. L. B., No. 1, 1961 (marine, South Africa); FOWLER, N. W., No. 18, 1956 (Red Sea); DAY, F., No. 2, 1889 (India); de SILVA, P. H. D. H., No. 1, 1956 (marine, Ceylon); WEBER & de BEAUFORT, No. 4, 1922 (Indo-Australian Arch.); MARSHALL, T. C., No. 1, 1964 (Great Barrier Reef); OSHIMA, M., 1921, Ann. Carnegie Mus., 13, 260-264 (Taiwan); FOWLER, H. W., No. 6, 1933-35 (China); JORDAN, D. S. & STARKS, E. C., 1902, Proc. United States Nat. Mus., 26, 57-73 (Japan)

Centrolepidae, as Centrolepididae

Centrolepididae (incl. Centrolepidae) (No revisionary references noted)

Centrolophidae (incl. Icichthyidae). NORMAN, J. R., No. 2, 1937 (marine, Patagonia); TORTONESE, E., 1960, Ann. Mus. Civ. Stor. Nat. Genova, 71, 57-82 (Mare Ligure)

Centrophrynidae. BERTELSEN, E., No. 1, 1951 (as Centrophrynidae, new family)

Centropomidae (incl. Ambassidae, Ambassiidae, Chandidae, Latidae, Oxylabracidae). FRASER-BRUNNER, A., 1955, Bull. Raffles Mus., 25, 185-213 (synopsis of Chandinae); BEEBE & TEE-VAN, No. 1, 1928 (Hispaniola); REGAN, C. T., No. 1, 1906-08 (Central America); MEEK, S. E., No. 1, 1904 (fresh-water, Mexico); MEEK & HILDEBRAND, No. 3, 1925 (marine, Panama);

HILDEBRAND, S. F., No. 1, 1946 (shore, Peru); EIGENMANN, C. H., No. 1, 1912 (British Guiana); PUYO, J., No. 1, 1949 (Guyane); FOWLER, H. W., No. 13, 1954 (fresh-water, Brasil) DAGET, J., No. 2, 1962 (Guinea); DAGET & ILTIS, No. 1, 1965 (côte d'Ivoire); DAGET, J., No. 1, 1954 (Upper Niger); SMITH, J. L. B., No. 1, 1961 (marine, South Africa, as Ambassidae); POLL, M., No. 1, 1946 (Lake Tanganyika); POLL, M., No. 3, 1953 (Lake Tanganyika); SMITH, H. M., No. 1, 1945 (fresh-water, Thailand); WEBER & de BEAUFORT, No. 5, 1929 (Indo-Australian Arch.); INGER & KONG, No. 1, 1962 (fresh-water, North Borneo, as Ambassidae); SCHULTZ, L. P., 1945, Proc. United States Nat. Mus., 96, 115-121 key to New Guinea genera & species); MARSHALL, T. C., No. 1, 1964 (Great Barrier Reef); WHITLEY, G. P., 1956, Australian Zool., 12, 251-261 (key to fresh-water chandids); FOWLER, H. W., 1930, United States Nat. Mus. Bull., 100, (10), 1-334 (Philippine seas, as Chandidae); CHAN, W. L., No. 1, 1968 (Hong Kong, as Ambassidae); POPE & NICHOLS, No. 1, 1927 (Hainan I., as Ambassidae); FOWLER, H. W., No. 9, 1937-41 (China, as Chandidae); Nichols, j. t., No. 1, 1943 (fresh-water, China, as Ambassidae); FOWLER, H. W., No. 17, 1959 (Fiji, as Chandidae)

Cephalacanthidae, as Dactylopteridae

Cephalaspidae (incl. Ateleaspidae) (only as fossils)

Cephalopholidae, as Serranidae

Cephalopteridae, as Mobulidae

Cephaloxenidae (only as fossils)

Cepolidae. NOBRE, A., No. 1, 1935 (marine, Portugal); LOZANO REY, D. L., No. 2, 1952 (España); FAGE, L., No. 1, 1918 (Mediterranean); SMITH, J. L. B., No. 1, 1961 (marine, South Africa); DAY, F., No. 2, 1889 (India); WEBER & de BEAUFORT, No. 7, 1936 (Indo-Australian Arch); HERRE, A. W., 1927, Philippine Journ. Sci., 34, 281-284 (Philippines); FOWLER, H. W., No. 11, 1949 (China); LINDBERG & KRASJUKOVA, No. 1, 1969 (Sea of Japan); JORDAN, D. S. & FOWLER, H. W., 1903, Proc. United States Nat. Mus., 26, 699-702 (Japan)

Ceraspidae (incl. Ceratolepidae) (only as fossils)

Ceratidae, as Ceratiidae

Ceratiidae (incl. Ceratidae). REGAN, C. T., 1926, Oceanogr. Rep. Danish "Dana" Exp. 1920-22, 2, 1-45; REGAN, C. T. & TREWAVAS, E., 1932, Rep. Carlsberg Oceanogr. Exp. 1928-30, 2, 1-113 (deep-sea); BERTELSEN, E., No. 1, 1951 (revision); NORMAN, J. R., No. 1, 1930 (oceanic); NOBRE, A., No. 1, 1935 (marine, Portugal); FOWLER, H. W., No. 8, 1936 (marine, West Africa); SMITH, J. L. B., No. 1, 1961 (marine, South Africa); de BEAUFORT & BRIGGS, No. 11, 1962 (Indo-Australian Arch.)

Ceratodidae. HOLLY, M., 1933, Das Tierreich, 61, 1-20.

Ceratolepidae, as Ceraspidae (only as fossils)

Cerdalidae, as Microdesmidae

Cestracionidae, As Heterodontidae

Cestraciontidae, as Heterodontidae

Cetomimidae. PARR, A. E., 1934, Bull. Bingham Oceanogr. Coll., 4, (6), 1-59 (revison); FOWLER, H. W., No. 8, 1936 (marine, West Africa); SMITH, J. L. B., No. 1, 1961 (marine, South Africa)

Cetopsidae. VAN DER STIGCHEL, J. W. B., No. 1, 1946 (South America); SCHULTZ, L. P., No. 2, 1944 (Venezuela); FOWLER, H. W., No. 13, 1954 (fresh-water, Brasil)

Cetorhinidae, as Lamnidae

Chacidae. WEBER & de BEAUFORT, No. 2, 1913 (Indo-Australian Arch.)

Chaenichthyidae, as Channicthyidae

Chaenopsidae (incl. Emblemariidae). STEPHENS, J. S., Jr., 1963, Univ. California Publ. Zool., 68, 1-133 (monograph); BOHLKE, J., 1957, Proc. Acad. Nat. Sci. Philadelphia, (Bahama Is.); BEEBE & TEE-VAN, No. 1, 1928 (Hispaniola); STEPHENS, J., 1961, Notul. Nat. Acad. Philadelphia, 349, 1-8 (western Atlantic)

Chaetodipteridae, as Ephippidae

Chaetodontidae (incl. Pomacanthidae). AHL, E., 1923, Arch. Naturg., 89, (A5), 1-205 (revision, synopsis of genera); FRASER-BRUNNER, A., 1933, Proc. Zool. Soc. London, 1933, 543-599 (revision); HILDEBRAND & SCHROEDER, No. 1, 1928 (Chesapeake Bay); BEEBE & TEE-VAN, No. 1, 1928 (Hispaniola); MEEK & HILDEBRAND, No. 4, 1928 (marine, Panama); MARTIN, S. F., 1956, in Mendez, A. et al., El Archipelago de los Roques y la Orchila; Caracas, 87-144 (key to genera); PUYO, J., No. 1, 1949 (Guyane); HILDEBRAND, S. F., No. 1, 1946 (shore, Peru); BAUCHOT, M. L., & BLANC, M., No. 1, 1961 (eastern tropical Atlantic); FOWLER, H. W., No. 8, 1936 (marine, West Africa); POLL, M., No. 4, 1954 (coastal, South Africa); SMITH, J. L. B., No. 1, 1961 (marine, South Africa, also as Pomacanthidae); ARNOULT, J., No. 1, 1959 (fresh-water, Madagascar); PELLEGRIN, J., No. 2, 1960 (fresh-water, Madagascar); WEBER & de BEAUFORT, No. 7, 1936 (Indo-Australian Arch.); MARSHALL, T. C., No. 1, 1964 (Great Barrier Reef); HERRE, A. W. & MONTALBAN, H. R, 1927, Philippine Journ. Sci., 34, 1-113 (Philippines); FOWLER, H. W., No. 12, 1953 (China, also as Pomacanthidae); LINDBERG & KRASJUKOVA, No. 1, 1969 (Sea of Japan); JORDAN, D. S. & FOWLER, H. W., 1902, Proc. United States Nat. Mus., 25, 513-563 (Japan); SMITH, J. L. B., 1955, Ann. Mag. Nat. Hist., (12, 8, 377-384 (western Pacific Ocean); SCHULTZ, L. P. et al., 1953, No. 2 (Marshall & Marianas Is.); FOWLER, H. W., No. 17, 1959 (Fiji)

Champsodontidae. SMITH, J. L. B., No. 1, 1961 (marine, South Africa); de BEAUFORT. L. F. &CHAPMAN, W. M., No. 9, 1951 (Indo-Australian Arch.); FOWLER, H. W., No. 15, 1956 (China); LINDBERG & KRASJUKOVA, No. 1, 1969 (Sea of Japan)

Chandidae, as Centropomidae

Chanidae (incl. Chanoidae). REGAN, C. T., No. 1, 1906-08 (Central America); SMITH, J. L. B., No. 1, 1961 (marine, South Africa); PELLEGRIN, J., No. 1, 1933 (fresh-water, Madagascar); ARNOULT, J., No. 1, 1959 (fresh-water, Madagascar); FOWLER, H. W., No. 18, 1956 (Red Sea); WEBER & de BEAUFORT, No. 2, 1913 (Indo-Australian Arch.); MARSHALL, T. C., No. 1, 1964 (Great Barrier Reef); FOWLER, H. W., No. 3, 1931 (China); JORDAN & HERRE, No. 1, 1906 (Japan); FOWLER, H. W., No. 17, 1959 (Fiji)

Channicthyidae (incl. Chaenichthyidae, Channichthyidae). NORMAN, J. R., No. 2, 1937 (marine, Patagonia); NORMAN, J. R., No. 3, 1938 (Antarctic); NYBELIN, O., 1947, Sci. Res. Norwegian Antarct. Exp. 1927-28, 2, (26), 1-76 (key to Antarctic genera)

Channichthyidae, as Channicthyidae

Channidae (incl. Ophicephalidae, Ophiocephalidae, Parophiocephalidae). (All these references are under the name Ophicephalidae) BLANC, M., 1963, Bull. Mus. Nat. Hist. Nat., (2), 35, 70-77 (museum type list); BERG, L. S., No. 2, 1933 (fresh-water, U. R. S. S.); BERG, L. S., No. 5, 1949 (fresh-water, Soviet Union); BERG, L. S., No. 10, 1965 (fresh-water, U. S. S. R.); BERG, L. S., No. 6, 1949 (fresh-water, Iran); BOULENGER, G. A., No. 6, 1916 (fresh-water, Africa); BOULENGER, G. A., No. 2, 1907 (Egypt); PELLEGRIN, J., No. 2, 1923 (western Africa); SCHULTZ, L. P., No. 1, 1942 (Liberia); DAGET & ILTIS, No. 1, 1965 (côte d'Ivoire); DAGET, J., No. 1, 1954 (Upper Niger); BOULENGER, G. A., No. 1, 1901 (Congo Basin); DAY, F., No. 2, 1889 (India); SMITH, H. M., No. 1, 1945 (fresh-water, Thailand); WEBER & de BEAUFORT, No. 4, 1922 (Indo-Australian Arch.); INGER & KONG, No. 1, 1962 (fresh-water, North Borneo); OSHIMA, M., No. 1, 1919 (fresh-water, Taiwan); POPE & NICHOLS, No. 1, 1927 (Hainan I.); NICHOLS, No. 1, 1943 (fresh-water, China); OKADA, Y., No. 2, 1960 (fresh-water, Japan)

Chanoidae, as Chanidae

Characidae (incl. Acestrorhynchidae, Characinidae, Creagrutidae, Glandulocaudidae, Serrasalmidae, Tetragonopteridae). EIGENMANN, C. H. & OGLE, F., 1907, Proc. United States Nat. Mus., 33, 1-36 (museum lists, U. S. N. M. & Indiana Univ., as Characinidae); GREGORY, W. K., & CONRAD, G. M., 1938, Zoologica, 23, 319-360 (phylogeny, as Characinidae); GERY, J., 1959, Bull. Mens. Soc. Linn. Lyon, 28, 248-260 (world list); GOSLINE, W. A., 1951, Proc. California Acad. Sci., (4), 27, 17-58 (key to genera of Serrasalminae); BOHLKE, J., 1954, Ann. Mag. Nat. Hist., (12), 7, 97-104 (Iguanodectinae); BOHLKE, JEE., 1958, Proc. Acad. Nat. Sci. Philadelphia,, 110, 1-121(key to genera of Glandulocaudini); WEITZMAN, S. H., 1960, Stanford Ichth. Bull., 7, 217-239 (key to Gasteropelecinae); EIGENMANN, C. H., 1917-29, Mem. Mus. Comp. Zool., 43, 1-558 (American); FOWLER, H. W., No. 1, 1945 (fresh-water, southeastern U. S.); FERNANDEZ-YEPEZ, A., 1955, Rev. Fac. Agr. Maracay, 1, (4), 1-11 neotropical, as Acestrorhynchidae); REGAN, C. T., No. 1, 1906-08 (Central America, as Characinidae); MEEK, S. E., No. 1, 1904 (fresh-water, Mexico, as Characinidae); MEEK & HILDEBRAND, No. 1, 1916 (fresh-water, Panama); EIGENMANN, C. H., 1915, Mem. Carnegie Mus., 7, 1-99 (South American Cheirodontinae); EIGENMANN, C. H., 1915, Ann. Carnegie Mus., 9, 226-272 (South American Serrasalminae & Mylinae); MYERS, G. S., 1950 (generic revision of Anostominae); SCHULTZ, L. P., 1950, in Hatch, M. H., Studies in Honor of Trevor Kincaid, 44-73 (keys, South America); WEITZMAN, S. H., 1954, Stanford Ichth. Bull., 4, 213-263 (key to genera of Gasteopelecinae); MYERS, G. S. & BOHLKE, J., 1956, Stanford Ichth. Bull., 7, 6-12 (South American Xenurobryconini); BOHLKE, J. E., 1958, Copeia, 1958, 318-325 (key to Amazon Xenurobryconini); SCHULTZ, L. P., 1944, Proc. United States Nat. Mus., 95, 235-367 (Venezuela) as Characinidae); EIGENMANN, No. 1, 1912 (British Guiana); PUYO, J., No. 1, 1949 (Guyane) as Characinidae); FOWLER, H. W., No. 10, 1948-51 (fresh-water, Brasil, as Tetragonopteridae); CAMPOS, A. A., 1944, Pap. Avuls. Dep. Zool. Sao Paulo, 4, 197-212 (Mylinae, museum list); GOMES, A. L., 1947, Misc. Publ. Mus. Zool. Univ. Michigan, 67, 1-39 (Rio Grande do Sul); DEVINCENZI, G. J., No. 2, 1924 (Uruguay); ALONSO de ARAMBURU, A. S., 1953, Not. Mus. Univ. Nac. Eva Peron, 16, Zool., 297-320 (llamados cientudos); EIGENMANN, C. H., No. 2, 1927 (fresh-water, Chile); BOULENGER, G. A., No. 3, 1909 (fresh-water, Africa, as Characinidae); POLL, M., 1967, Ann. Mus. Rep. Afrique Central, Zool., 162-1-158 (revision of African); POLL, N., 1945, Rev. Zool. Bot. Africaines, 39, 36-77 (African genera); MONOD, T., 1950, Bull. Inst. Francaise Afrique Noire, 12, 71 (African genera); BOULENGER, G. A., No. 2, 1907 (Egypt, as Characinidae); PELLEGRIN, J., No. 2, 1923 (West Africa, as Characinidae); DAGET, J., No. 2, 1962 (Guinea); SCHULTZ, L. P., No. 1, 1942 (Liberia, as Characinidae); DAGET & ILTIS, No. 1, 1965 (Côte d'Ivoire); DAGET, J., No. 1, 1954 (Upper Niger, as Characinidae); BOULENGER, G. A., No. 1, 1901 (Congo basin, as Characinidae); POLL, M., No. 1, 1946 (Lake Tanganyika); POLL, M., No. 3, 1953 (Lake Tanganyika)

Characinidae, as Characidae

Characodontidae, as Goodeidae

Chaudhuriidae. (No revisionary references noted)

Chauliodidae, as Chauliodontidae

Chauliodontidae (incl. Chauliodidae). REGAN, C. T. & TREWAVAS, E., 1929, Oceanogr. Rep. Dana Exp. 1920-22, 5, 1-39 (monograph); EGE, V., 1948, Dana Rep., 31, 1-148 (systematics, phylogeny, & geography); NORMAN, J. R., No. 1, 1930 (oceanic); BIGELOW, H. B. et al., No. 3, 1964 (western North Atlantic); LOZANO REY, D. L., No. 4, 1947 (España); FOWLER, H. W., No. 8, 1936 (marine, West Africa); SMITH, J. L. B., No. 1, 1961 (marine, South Africa); FOWLER, H. W., No. 18, 1956 (Red Sea); SOLDATOV & LINDBERG, No. 1, 1930 (Far East seas)

Chaunacidae. LOZANO y REY, L., No. 5, 1960 (España); SMITH, J. L. B., No. 1, 1961 (marine, South Africa) de BEAUFORT & BRIGGS, No. 11, 1962 (Indo-Australian Arch.)

Cheilobranchidae, as Alabetidae
Cheilodactylidae (incl. Chilodactylidae). HILDEBRAND, S. F., No. 1, 1946 (shore, Peru); DEVINCENZI, G. E., No. 2, 1924 (Uruguay); NORMAN, J. R., No. 2, 1937 (marine, Patagonia); SMITH, J. L. B., No. 1, 1961 (marine, South Africa); MARSHALL, T. C., No. 1, 1964 (Great Barrier Reef)
Cheilodipteridae, as Apogonidae
Cheimarrhichthyidae, as Cheimarrichthyidae
Cheimarrichthyidae (incl. Chemarrhichthyidae, Chimarrhichthyidae, but not Chimarrichthyidae); (No revisionary references noted)
Cheiracanthidae (only as fossils)
Cheirolepidae, as Cheirolepididae (only as fossils)
Cheirolepididae (incl. Cheirolepidae) (only as fossils)
Chiasmodontidae. NORMAN, J. R., 1929, Ann. Mag. Nat. Hist. 10, 3, 529-544; NORMAN, J. R., No. 1, 1930 (oceanic); NOBRE, A., No. 1, 1935 (marine, Portugal); LOZANO y REY, L., No. 5, 1960 (España); FOWLER, H. W., No. 8, 1936 (marine, West Africa); SMITH, J. L. B., No. 1, 1961 (marine, South Africa); de BEAUFORT & CHAPMANS, No. 9, 1951 (Indo-Australian Archipelago)
Chilobranchidae, as Alabetidae
Chilodactylidae, as Cheilodactylidae
Chilodipteridae, as Apogonidae
Chilodontidae. GERY, J., 1964, Trop. Fish Hobby, 12, (9), 5-10, 63-64 (key to Chilodinae)
Chilorhinidae, as Xenocongridae
Chimaeridae (incl. Chimeridae). DEVINCENZI, G. J., No. 2, 1924 (Uruguay); NORMAN, J. R., no. 2, 1937 (marine, Patagonia); BIGELOW, H. B. & SCHROEDER, W. C., 1950, Bull. Mus. Comp. Zool., 103, 385-408 (Atlantic genera); BIGELOW & SCHROEDER, No. 2, 1953 (Western North Atlantic); SMITT, F. A. et al., NO. 2, 1895 (Scandinavia); POLL, M., No. 2, 1947 (marine, Belgique); NOBRE, A., No. 1, 1935 (marine, Portugal); TORTONESE, E., No. 3, 1956 (Italia); BERG, L. S., No. 7, 1911 (Russia); ANDRIYASHEV, A. P., No. 1, 1954 (northern seas of U. S. S. R.); FOWLER, H. W., No. 8, 1936 (marine, West Africa); SMITH, J. L. B., No. 1, 1961 (Marine, South Africa); SOLDATOV & LINDBERG, No. 1, 1930 (Far East seas); LINDBERGH & LEGHEZA, No. 1, 1959 (Sea of Japan); JORDAN, D. S. & FOWLER, H. W., 1903, Proc. United States Nat. Mus., 26, 593-674 (Japan)
Chimaeropsidae (only as fossils)
Chimarrhichthyidae, as Cheimarrichthyidae
Chimarrichthyidae. (Not same as Cheimarrichthyidae; family not identified)
Chimeridae, as Chimaeridae
Chiridae, as Hexagrammidae
Chirocentridae. SMITH, J. L. B., No. 1, 1961 (marine, South Africa); FOWLER, H. W., No. 18, 1956 (Red Sea); DAY, F., No. 1, 1889 (India); FOWLER, H. W., No. 3, 1931 (China); SOLDATOV & LINDBERG, No. 1, 1930 (Far East seas); LINDBERG & LEGHEZA, No. 2, 1965 (Sea of Japan); LINDBERG & LEGEZA, No. 3, 1969 (Sea of Japan); JORDAN & HERRE, No. 1, 1906 (Japan); FOWLER, H. W., No. 17, 1959 (Fiji)
Chirolophidae, as Stichaeidae
Chironemidae. MARSHALL, T. C., No. 1, 1944 (Great Barrier Reef)
Chirothrichidae (only as fossils)
Chlamydoselachidae. GARMAN, S., No. 1, 1913 (monograph); NOBRE, A., No. 1, 1935 (marine, Portugal); BERG, L. S., No. 7, 1911 (Russia); ANDRIYASHEV, A. P., No. 1, 1954 (northern seas of U. S. S. R.); FOWLER, H. W., No. 8, 1936 (marine, West Africa); SMITH, J. L. B., No. 1, 1961 (marine, South Africa); SOLDATOV & LINDBERG, No. 1, 1930 (Far East seas); JORDAN, D. S. & FOWLER, H. W., 1903, Proc. United States Nat. Mus., 26, 593-674 (Japan)
Chlopsidae, as Xenocongridae
Chlorophthalmidae. KAMOHARA, T., 1956, Res. Rep. Kochi Univ., 5, (15), 1-8 (Chlorophthalmidae, as subf. of Sudidae); MEAD, G. W., 1966, Mem. Sears Found. Mar. Res., 1, (5), 19-29, 103-189 (western North Atlantic); FOWLER, H. W., No. 8, 1936 (marine, West Africa); SMITH, J. L. B. No. 1, 1961 (marine, South Africa); FOWLER, H. W., No. 18, 1956 (Red Sea); LINDBERG & LEGHEZA, No. 2, 1965 (Sea of Japan); LINDBERG & LEGEZA, No. 3, 1969 (Sea of Japan); KAMOHARA, T., 1953, Japanese Journ. Ichth., 3, 1-6 (Japan); FOWLER, H. W., No. 17, 1959 (Fiji)
Chonarhinidae, as Tetraodontidae
Chondrenchelyidae (only as fossils)
Chondrosteidae (only as fossils)
Chonerhinidae, as Tetraodontidae
Chromidae, as Cichlidae (also as Pomacentridae)
Chromileptidae, as Serranidae
Cibiidae, as Scombridae
Cichlidae (incl. Chromidae, Priscacaridae). BLANC, M., 1962, Bull. Mus. Nat. Hist. Nat., (2), 34, 202-227, 1962 (museum type list); WICKLER, W., 1963, Senckenbergische Biol., 44, 83-96; THYS van den AUDENAERDE, D. F. E., 1968, Docum. Zool., 14, 1-406 (bibliography of Tilapia); EIGENMANN, C. H. & BRAY, W. L., 1894, Ann. New York Acad. Sci., 7, 607-624 (American); TEE-VAN, J., 1935, Zoologica, 10, 281-300 (West Indies); REGAN, C. T., No. 1, 1906-08 Central America); MEEK, S. E., No. 1, 1904 (fresh-water, Mexico); MEEK & HILDEBRAND, No. 1, 1916 (fresh-water, Panama); EIGENMANN, C. H., No. 1, 1912 (British Guiana); PUYO, J., No. 1, 1949 (Guyane); FOWLER, H. W., No. 13, 1954 (fresh-water, Brasil); DEVINCENZI,

G. F., No. 2, 1924 (Uruguay); BAUCHOT & BLANC, No. 1, 1961 (eastern tropical Atlantic); TREWAVAS, E., 1942, Ann. Mag. Nat. Hist., (11), 9, 526-536 (Syria & Palestine); REGAN, C. T., 1922, Ann. Mag. Nat. Hist., (9), 10, 249-264 (synopsis of genera of Africa & Syria); BOULENGER, G. A., No. 5, 1915 (fresh-water, Africa); BOULENGER, G. A., No. 2, 1907 (Egypt); PELLEGRIN, J., No. 2, 1923 (western Africa); DAGET, J., No. 2, 1962 (Guinea); SCHULTZ, L. P., No. 1, 1942 (Liberia); DAGET & ILTIS, No. 1, 1965 (Côte d'Ivoire); DAGET, J., No. 1, 1954 (Upper Niger); TREWAVAS, E., 1962, Bonn Zool. Beitr., 13, 146-192 (key, Cameroons); BOULENGER, G. A., No. 1, 1901 (fresh-water, Congo basin); BARNARD, K. H., 1948, Ann. South African Mus., 36, 407-458 (key to South African genera); REGAN, C. T., 1920, Ann. Mag. Nat. Hist., (9), 5, 33-53 (Tanganyika genera); POLL, M., No. 1, 1946 (Lake Tanganyika); POLL, M., 1956, Explor. Hydrobiol. Lac Tanganyika (1946-47), 3, 5B, 1-619; TREWAVAS, E., 1935, Ann. Mag. Nat. Hist., (10), 16, 65-118 (Lake Nyasa); REGAN, C. T., 1922, Proc. Zool. Soc. London, 1922, 157-191 (Lake Victoria); GREENWOOD, P. H., 1956, Bull. British Museum (Nat. Hist.), Zool., 3, 195-333 (Lake Victoria); ARNOULT, J., No. 1, 1959 (fresh-water, Madagascar); PELLEGRIN, J., No. 2, 1960 (fresh-water, Madagascar); MARSHALL, T. C., No. 1, 1964 (Great Barrier Reef, as Chromidae); FOWLER, H. W., No. 14, 1954 (China, as Chromidae)

Ciprinidae, as Cyprinidae

Cirrhitidae. SMITH, J. L. B., 1951, Ann. Mag. Nat. Hist., (12), 4, 625-652; RANDALL, J. E., 1963, Proc. United States Nat. Mus., 114, 389-451 (review); TEE-VAN, J., 1940, Zoologica, 25, 53-64 (American); SMITH, J. L. B., No. 1, 1961 (marine, South Africa); DAY, F., No. 2, 1889 (India); de BEAUFORT, L. F., No. 8, 1940 (Indo-Australian Arch.); MARSHALL, T. C., No. 1, 1964 (Great Barrier Reef); FOWLER, H. W., No. 15, 1956 (China); LINDBERG & KRASJUKOVA, No. 1, 1969 (Sea of Japan); JORDAN, D. S. & HERRE, A. C., 1907, Proc. United States Nat. Mus., 33, 157-167 (Japan); SCHULTZ, L. P. et al., No. 3, 1960 (Marshall & Marianas Is.); FOWLER, H. W., No. 17, 1959 (Fiji)

Citharidae. (No revisionary references noted)

Citharinidae. MONOD, T., 1950, Bull. Inst. Francaise Afrique Noire, 12, 1-71 (African genera); DAGET, J., No. 2, 1962 (Guinea); DAGET & ILTIS, No. 1, 1965 (Côte d'Ivoire); DAGET, J., No. 1, 1954 (Upper Niger); POLL, M., No. 1, 1946 (Lake Tanganyika); POLL, M., No. 3, 1953 (Lake Tanganyika)

Cladodontidae (only as fossils)

Cladoselachidae (only as fossils)

Clariidae. DAGET, J., No. 2, 1962 (Guinea); DAGET & ILTIS, No. 1, 1965 (Côte d' Ivoire); DAGET, J., No. 1, 1954 (Upper Niger); JACKSON, P. B. N., 1959, Proc. Zool. Soc. London, 132, 109-128 (Nyasaland); POLL, M., No. 1, 1946 (Lake Tanganyika); POLL, M., No. 3, 1953 (Lake Tanganyika); MENON, A. G. K., 1950, Rec. Indian Mus., 48, 59-66 (key to Indian genera); SMITH, H. M., No. 1, 1945 (fresh-water, Thailand); WEBER & de BEAUFORT, No. 2, 1913 (Indo-Australian Arch.); INGER & KONG, No. 1, 1962 (fresh-water, North Borneo)

Cleithrolepidae, as Cleithrolepididae (only as fossils)

Cleithrolepididae (incl. Cleithrolepidae) (only as fossils)

Climatiidae (only as fossils)

Clinidae (incl. Paraclinidae, Pterygocephalidae, Xenopoclinidae). HUBBS, C., 1952, Stanford Ichth. Bull., 4, 41-165 (classification, also as Pterygocephalidae); SPRINGER, V. G., 1958, Publ. Inst. Mar. Sci. Univ. Texas, 5, 417-492 (Labrisomini); SMITH, J. L. B., 1961, Ichth. Bull. Rhodes Bull., 20, 351-356 (as Xenopoclinidae); AL-UTHMAN, H. S., 1960, Texas Journ. Sci., 12, 163-175 (Starksiidi of Pacific); BEEBE & TEE-VAN, No. 1, 1928 (Hispaniola); SCHULTZ, L. P., 1950, Journ. Washington Acad. Sci., 40, 266-268 (key to genera of American tropical Pacific); HILDEBRAND, S. F., No. 1, 1946 (shore, Peru); LOZANO y REY, L., No. 5, 1960 (España); de BUEN, F., 1962, Montemar, No. 2, 53-96 (keys, Chile); BANARESCU, P., No. 1, 1964 (Romania); SMITH, J. L. B., 1946, Ann. Mag. Nat. Hist., (11)12, 535-546 (South Africa); SMITH, J. L. B., No. 1, 1961 (marine, South Africa); de BEAUFORT & CHAPMAN, No. 9, 1951 (Indo-Australian Arch.); MILWARD, N. E., 1967, Journ. Roy. Soc. Western Australia, 50, 1-9 (Western Australia); SCOTT, E. O. G., 1955, Pap. Roy. Soc. Tasmania, 89, 131-146 (key to Tasmanian genera & species); HERRE, A. W. C. T., 1939, Philippine Journ. Sci., 70, 315-372 (Philippines); FOWLER, H. W., No. 16, 1958 (China); SCHULTZ, L. P. et al., No. 3, 1960 (Marshall & Marianas Is.)

Clupanodontidae, as Clupeidae

Clupeidae (incl. Clupanodontidae, Clupidae, Congothrissidae, Dorosomatidae, Dorosomidae, Dussumeriidae, Dussumieridae, Dussumieriidae, Pristigasteridae). BERTIN, L., 1943, Bull. Inst. Oceanogr., Monaco, 853, 1-32 (revue, as Dussumieriidae); WHITEHEAD, P. J. P., 1963, Bull. British Mus. (Nat. Hist.), Zool., 10, 305-380 (revision, as Dussumieriidae); MILLER, R. R., 1960, United States Fish. Bull., 60, 371-392 (key to Orosomatinae); SLASTENENKO, E. P., No. 1, 1958 (fresh-water, Canada); HILDEBRAND & SCHROEDER, No. 1, 1928 (Chesapeake Bay, also as Dorosomidae); FOWLER, H. W., No. 1, 1945 (fresh-water, southeastern U. S., also as Dorosomidae); McHUGH, J. L. & FITCH, J. E., 1951, California Fish Game, 37, 491-495 (list, Pacific Coast); BEEBE & TEE-VAN, No. 1, 1928 (Hispaniola, also as Dussumieriidae); REGAN, C. T., No. 1, 1906-08 (Central America); MEEK, S. E., No. 1, 1904 (fresh-water, Mexico, as Dorosomatidae); MEEK & HILDEBRAND, No. 2, 1923 (marine, Panama); HILDEBRAND, S. F., No. 1, 1946 (shore, Peru); WEIBEZAHN, F. H., 1952, Bol. Pesc. Caracas, 3, 1-21 (Venezuela); MARTIN, S. F., 1955, Mem. Soc. Cienc. Nat. La Salle, 15, 184-188 (Venezuela, as Dussumieridae); EIGENMANN, C. H., No. 1, 1912 (British Guiana); PUYO, J., No. 1, 1949 (Guyane); FOWLER, H. W., No. 10, 1948-51 (Brasil); DEVINCENZI, No. 2, 1924

(Uruguay); NORMAN, J. R., No. 2, 1937 (marine, Patagonia); STOREY, M., 1938, Stanford Ichth. Bull., 1, 3-56 (key to western Atlantic genera); BIGELOW, H. B. et al., No. 5, 1963 (western North Atlantic); SMITT, F. A. et al., No. 2, 1895 (Scandinavia); DUNCKER & LADIGES, No. 1, 1960 (North Sea); POLL, M., No. 2, 1947 (marine, Belgique); SPILLMANN, J., No. 1, 1961 (fresh-water, France); NOBRE, A., No. 1, 1935 (marine, Portugal); LOZANO REY, L., No. 1, 1935 (fresh-water, España); LOZANO REY, D. L., No. 4, 1947 (España); LOZANO-REY, L., 1950, Rapp. Cons. Explor. Mer Copenhagen, 126, 7-20 (España & North Africa); LOZANO REY, L., 1929, Mem. Soc. Esp. Hist. Nat., 15, 647-666 ("Peninsula Iberica y del Rif"); FAGE, L., 1920, Rep. DanishOceanogr. Exp. 1908-10 to Mediterranean, II, Biol., A9, 1-140 (Mediterranean); BANARESCU, P., No. 1, 1964 (Romania); SVETOVIDOV, A. N., No. 2, 1964 (Black Sea); SVETOVIDOV, A. N., 1943, Zool. Zhurn., 22, 222-233 (Caspian & Black Seas); SMIRNOV, A. N., 1954, Trud. Zool. Inst. Azerbaidzh. Akad. Nauk S. S. S. R., 17, 113-195 (Caspian Sea); BERG, L. S., No. 1, 1932 (fresh-water, U. R. S. S.); BERG, L. S., No. 3, 1948 (fresh-water, Soviet Union); SVETOVIDOV, A. N., 1952, Tabl. Anal. Faune URSS, (n. s.), 48, 2, (1), 1-331 (U. S. S. R., in Russian); BERG, L. S., 1962 (fresh-water, U. S. S. R.); SVETOVIDOV, A. N., 1963, Fauna of U. S. S. R., II, 1, Clupeidae (faunal monograph); ANDRIYASHEV, A. P., No. 1, 1954 (northern seas of U. S. S. R.); SCHMIDT, P. Y., No. 1, 1950 (Siberia); BOULENGER, G. A., No. 3, 1909 (fresh-water, Africa); BOULENGER, G. A., No. 2, 1907 (Egypt); PELLEGRIN, J., No. 2, 1923 (western Africa); FOWLER, H. W., No. 8, 1936 (West Africa); DAGET, J., No. 2, 1962 (Guinea); SCHULTZ, L. P., No. 1, 1942 (Liberia); DAGET & ILTIS, No. 1, 1965 (Côte d'Ivoire); DAGET, J., No. 1, 1954 (Upper Niger); BOULENGER, G. A., No. 1, 1901 (Congo); POLL, M., 1948, Bull. Mus. Hist. Nat. Belgique, 24, (21), 1-24 (key to Congo genera and species); GILCHRIST, J. D. F., 1913, South African Mar. Biol. Dep. Rep., 1, 46-66 (review South African); SMITH, J. L. B., No. 1, 1961 (marine, southern Africa, also as Dorosomidae); BARNARD, K. H., 1943, Ann. South African Mus., 36, 101-262 (s. w. Cape region); POLL, M., No. 1, 1946 (fresh-water, Lake Tanganyika); POLL, M., No. 3, 1953 (Lake Tanganyika); LOSSE, G. F., 1966, Journ. East African Nat. Hist. Soc. Nat. Mus., 25, 166-178 (list, East African coastal); FOWLER, H. W., No. 18, 1956 (Red Sea, also as Dorosomidae); WHITEHEAD, P. J. P., 1965, Bull. British Mus. (Nat. Hist.), Zool., 12, 227-281 (Red Sea, also as Dussumieridae); DAY, F., No. 1, 1889 (India); NAIR, R. V., 1953, Journ. Zool. Soc. India, 5, 108-138 (India); SMITH, H. M., No. 1, 1945 (fresh-water, Thailand); WHITEHEAD, P. J. P., 1962, Bull. British Mus. (Nat. Hist.), Zool., 9, 87-102 (key to Indo-Pacific, as Dorosomatidae); WHITEHEAD, P. J. P., 1965, Bull. British Mus. (Nat. Hist.), Zool. 12, 115-156 (Indo-Pacific Alosinae); WEBER & de BEAUFORT, No. 2, 1913 (Indo-Australian Arch.); INGER & KONG, No. 1, 1962 (North Borneo); MARSHALL, T. C., No. 1, 1964 (Great Barrier Reef); CHU, K. & TSAI, C., 1958, Quart. Journ. Taiwan Mus. 11, 103-125 (Taiwan); FOWLER, H. W., No. 3, 1931 (China, as Dorosomidae); NICHOLS, J. T. No. 1, 1943 (fresh-water, China); SOLDATOV & LINDBERG, No. 1, 1930 (Far East seas); SHMIDT, P. Y., No. 2, 1965 (Sea of Okhotsk); LINDBERG & LEGHEZA, No. 2, 1965 (Sea of Japan); LINDBERG & LEGEZA, No. 3, 1969 (Sea of Japan); JORDAN, D. S. & HERRE, A. C., 1906, Proc. United States Nat. Mus., 31, 613-645 (Japan, also as Dorosomatidae); OKADA, Y., No. 1, 1960 (fresh-water, Japan); OKADA, Y., No. 2, 1960 (fresh-water, Japan); SCHULTZ, L. P. et al., No. 2, 1953 (Marshall & Marianas Is., as Dussumieridae); FOWLER, H. W., No. 7, 1959 (Fiji), also as Dussumieriidae)

Clupidae, as Clupeidae

Clupisudidae, as Osteoglossidae

Cobitidae (incl. Acanthopsidae, Adiposiidae). NALBANT, T., 1963, Trav. Mus. Hist. Nat. "G. Antipa", 4, 343-379 (genera of Botiinae, & Cobitinae); SMITT, F. A., et al., No. 2, 1895 (Scandinavia); DUNCKER & LADIGES, No. 1, 1960 (Nordmark); SPILLMANN, J., No. 1, 1961 (France); LOZANO REY, L., No. 1, 1935 (fresh-water, España); LOZANO REY, D. L., No. 4, 1947 (España); ZANANDREA, G. et al., 1966, Arch. Zool. Ital., 50, 233-259(Italia); BANARESCU, P., No. 1, 1964 (Romania); BERG, L. S., No. 2, 1933 (fresh-water, U. R. S. S.); BERG, L. S., No. 4, 1949 (fresh-water, Soviet Union); BERG, L. S., No. 9, 1964 (fresh-water, U. S. S. R.); BERG, L. S., 1907, Proc. United States Nat. Mus., 32, 435-438 (Amur); TORTONESE, E., 1951-52, Boll. Ist. Mus. Zool. Univ. Torino, 3, (8), 1-14 (Asia Minor); BERG, L. S., No. 6, 1949 (fresh-water, Iran); DERANIYAGALA, P. E. P., 1930, Ceylon Journ. Sci., (B), 16, 1-41 (Ceylon); SMITH, H. M., No. 1, 1945 (fresh-water, Thailand); TWEEDIE, M. W. F., 1956, Bull. Raffles Mus., 27, 56-64(Malayan, fresh-water); WEBER & de BEAUFORT, No. 3, 1916 (Indo-Australian Arch.); INGER & KONG, No. 1, 1962 (fresh-water, North Borneo); OSHIMA, M., No. 1, 1919 (fresh-water, Formosa); POPE & NICHOLS, No. 1, 1927 (Hainan I.); NICHOLS, J. T., No. 1, 1943 (fresh-water, China); JORDAN, D. S. & FOWLER, H. W., 1903, Proc. United States Nat. Mus., 26, 765-774 (Japan); KOBAYASI, H., 1936, Bot. Zool., 4, 855-862 (key, in Japanese)

Coccocephalichthyidae. (No revisionary references noted)

Coccodontidae (only as fossils)

Coccolepidae, as Coccolepididae (only as fossils)

Coccolepididae (incl. Coccolepidae) (only as fossils)

Coccosteidae (incl. Pholidosteidae) (only as fossils)

Cochliodontidae (only as fossils)

Coelacanthidae (only as fossils)

Coelolepidae (incl. Thelodontidae) (only as fossils)

Colobodontidae, as Perleididae (only as fossils)

Colocongridae, as Congridae

Colomesidae, as Tetraodontidae
Comephoridae. BERG, L. S., No. 2, 1933 (fresh-water, U. R. S. S.); BERG, L. S., No. 5, 1949 (fresh-water, (Soviet Union); BERG, L. S., No. 10, 1965 (fresh-water, U. S. S. R.); TALIEV, D. N., 1955, ACad. Sci. U. S. S. R., 1955, 1-603 (Lake Baical)
Comentryidae (only as fossils)
Conchopomidae (only as fossils)
Congeridae, as Congridae
Congiopodidae (incl. Agriopidae). HILDEBRAND, S. F., No. 1, 1946 (shore, Peru); NORMAN, J. R., No. 2, 1937 (marine, Patagonia); SMITH, J. L. B., No. 1, 1961 (marine, South Africa); NORMAN, J. R., No. 3, 1938 (Antarctic)
Congrothrissidae, as Clupeidae
Congridae (incl. Colocongridae, Congeridae, Heterocongridae, Leptocephalidae). KLAUSEWITZ, W. & EIBL-EIBESFELDT, I., 1959, Senckenbergische Biol., 40, 135-153 (key, as Heterocongridae) BOHLKE, J., 1957, Proc. Acad. Nat. Sci. Philadelphia, 109, 59-79 (western Atlantic Heterocongrinae); HILDEBRAND & SCHROEDER, No. 1, 1928 (Chesapeake Bay); MEEK & HILDEBRAND, No. 2, 1923 (marine, Panama, as Leptocephalidae); DEVINCENZI, G. J., No. 2, 1924 (Uruguay, as Leptocephalidae); DUNCKER & LADIGES, No. 1, 1-60 (Nordmark); POLL, M., No. 2, 1947 (marine, Belgique); LOZANO Rey, D. L., No14, 1947 (España); SVETOVIDOV, A. N. No. 2, 1964 (Black Sea); FOWLER, H. W., No. 8, 1936 (marine, West Africa, also as Heterocongridae); SMITH, J. L. B., No. 1, 1961 (marine, South Africa); CASTLE, P. H. J., 1968, Ichth. Bull. Rhodes Univ., 33, 685-726 (western Indian Ocean & Red Sea); FOWLER, H. W., No. 18, 1956 (Red Sea); WEBER & de BEAUFORT, No. 3, 1916 (Indo-Australian Arch., also as Heterocongridae); MARSHALL, T. C., No. 1, 1964 (Great Barrier Reef); GRIFFIN, L. T., 1936, Trans. Roy. Soc. New Zealand, 66, 12-26 (New Zealand); FOWLER, H. W., No. 4, 1931 (China); LINDBERG & LEGHEZA, No. 2, 1965 (Sea of Japan); LINDBERG & LEGEZA, No. 3, 1969 (Sea of Japan); JORDAN & SNYDER, No. 1, 1901 (Japan, as Leptocephalidae); SCHULTZ, L. P. et al., No. 2, 1953 (Marshall & Marianas Is.); CASTLE, P. H. J., 1967, Cah. Off. Rech. Sci. Techn. Outre-Mer. (Oceanogr.), 4, (4), 51-71 (southwest Pacific, as Leptocephalidae); FOWLER, H. W., No. 17, 1959 (Fiji)
Congrogadidae (incl. Halidesmidae, Haliophidae). SMITH, J. L. B., 1952, Ann. Mag. Nat. Hist., (12), 5, 85-101 (as Haliophidae); SMITH, J. L. B., No. 1, 1961 (marine, South Africa, as Halidesmidae); de BEAUFORT & CHAPMANS, No. 9, 1951 (Indo-Australian Arch.); MARSHALL, T. C., No. 1, 1964 (Great Barrier Reef); HERRE, A. W. C. T., 1939, Philippine Journ. Sci., 70, 315-372 (Philippines); FOWLER, H. W., No. 16, 1958 (China)
Copodontidae (only as fossils)
Coracinidae (incl. Dichistiidae). SMITH, J. L. B., No. 1, 1961 (marine, South Africa)
Coregonidae, as Salmonidae
Coridae, as Labridae
Cornuboniscidae. (No revisionary references noted)
Coryphaenidae. GIBBS, R. H., Jr. & COLLETTE, B. B., 1959, Bull. Mar. Sci. Gulf Caribbean, 9, 117-152 (keys); WALFORD, L. A., No. 1, 1937 (Alaska to Equator); BEEBE & TEE-VAN, No. 1, 1928 (Hispaniola); MEEK & HILDEBRAND, No. 3, 1925 (marine, Panama); HILDEBRAND, S. F., No. 1, 1946 (shore, Peru); NOBRE, A., No. 1, 1935 (marine, Portugal); LOZANO REY, D. L., No. 3, 1952 (España); FOWLER, H. W., No. 8, 1936 (marine, West Africa); POLL, M., No. 4, 1954 (coastal South Africa); SMITH, J. L. B., No. 1, 1961 (marine, South Africa); DAY, F., No. 2, 1889 (India); WEBER & de BEAUFORT, No. 6, 1931 (Indo-Australian Arch.); MARSHALL, T. C., No. 1, 1964 (Great Barrier Reef); AGCO, A. G., 1940, Philippine Journ. Sci., 72, 313-315 (Philippines); LINDBERG & KRASJUKOVA, No. 1, 1961 (Sea of Japan); SCHULTZ, L. P. et al., No. 2, 1953 (Marshall & Marianas Is.)
Coryphaenoididae, as Macrouridae
Cosmolepididae (only as fossils)
Cosmoptychidae (only as fossils)
Cottidae (incl. Ascelichthyidae, Blepisiidae, Blepsiidae, Hemitripteridae, Jordaniidae, Neophrynichthyidae, Rhamphocottidae, Sclerogenidae part, Scorpaenichthyidae, Synchiridae). TARANETZ, A. J., 1941, Bull. Acad. Sci. U. R. S. S., Biol., 1941, 427-447 (origin & taxonomy); TARANETZ, A. Y., 1959, Inst. Fish. Univ. British Columbia Mus. Contr., 5, 1-28 (classification); HUBBS, C. L., 1926, Occ. Pap. Mus. Zool. Univ. Michigan, 171, 1-18 (Oligocottinae); SLASTENENKO, E. P., No. 1, 1958 (Canada); HILDEBRAND & SCHROEDER, No. 1, 1928 (Chesapeake Bay, as Hemitripteridae); FOWLER, No. 1, 1945 (southeastern U. S.); BOLIN, R. L., 1944, Stanford Ichth. Bull., 3, 1-135 (fauna, keys to genera, bibliography, California); HILDEBRAND, S. F., No. 1, 1946 (shore, Peru); FOWLER, H. W., No. 10, 1948-51 (fresh-water, Brasil, as Serrasalmidae); DEVINCENZI, G. F., No. 2, 1924 (Uruguay); SMITT, F. A. et al., No. 1, 1892 (Scandinavia); DUNCKER & LADIGES, No. 1, 1960 (Nordmark); POLL, M., No. 2, 1947 (marine, Belgique); SPILLMANN, J., No. 1, 1961 (France); NOBRE, A., No. 1, 1935 (marine, Portugal); LOZANO REY, L., No. 1, 1935 (fresh-water, España); LOZANO REY, D. L., No. 2, 1952 (España); BANARESCU, P., No. 1, 1964 (Romania); BERG, L. S., No. 2, 1933 (fresh-water, U. R. S. S.); BERG, L. S., No. 5, 1949 (fresh-water, Soviet Union); BERG, L. S., No. 10, 1965 (fresh-water, U. S. S. R.); TALIEV, D. N., 1955, The Cottoidei of Lake Baical, Moskva, 1-603; ANDRIYASHEV, A. P., No. 1, 1954 (northern of U. S. S. R.); DAY, F., No. 2, 1889 (India); NICHOLS, J. T., No. 1, 1943 (fresh-water, China); BERG, L. S., 1932, Copeia, 1932, 17-20 (Pacific slope of Asia); SOLDATOV & LINDBERG, No. 1, 1930 (Far East seas); SCHMIDT, P. Y., No. 1, 1950 (Sea of Okhotsk, also as Blepsiidae) & Hemitripteridae); SHMIDT, P. Y., No. 2, 1965 (Sea of Okhotsk, also as Blepsiidae & Hemitripteridae);

JORDAN, D. S. & STARKS, E. C., 1904, Proc. United States Nat. Mus., 27, 231-335 (Japan); OKADA, Y., No. 1, 1960 (fresh-water, Japan); OKADA, Y., No. 2, 1960 (Japan); WATANABE, M., 1960, Fauna Japonica, Cottidae, Tokyo, 1-218.

Cottocomephoridae (incl. Abyssocottidae). BERG, L. S., No. 2, 1933 (fresh-water, U. R. S. S.); BERG, L. S., No. 5, 1949 (fresh-water, Soviet Union); BERG, L. S., No. 10, 1965 (fresh-water, U. S. S. R.)

Cottunculidae. ANDRIYASHEV, A. P., No. 1, 1954 (northern seas of U. S. S. R.); SMITH, J. L. B., No. 1, 1961 (marine, South Africa); de BEAUFORT & BRIGGS, No. 11, 1962 (Indo-Australian Arch.)

Cotulidae. GARMAN, S., No. 1, 1913 (monograph)

Cranoglanidae, as Cranoglanididae

Cranoglanididae (incl. Cranoglanidae). JAYARAM, K. C., 1955, Proc. Nat. Inst. Sci. India, 21B, 256-363 (China)

Cratoselachidae (only as fossils)

Creagrutidae, as Characidae

Creediidae. SMITH, J. L. B., No. 1, 1961 (marine, South Africa)

Crenuchidae, as Characidae)

Cristidae, as Centrarchidae

Cromeriidae, as Kneriidae

Crossognathidae (only as fossils)

Cryphiolepididae (only as fossils)

Cryptacanthidae, as Stichaeidae

Cryptacanthodidae, as Stichaeidae

Ctenacanthidae, (only as fossils)

Ctenaspidae (only as fossils)

Ctenodontidae (only as fossils)

Ctenolabridae, as Pomacentridae

Ctenoluciidae (incl. Xiphostomatidae, Xiphostomidae). (No revisionary references noted)

Ctenothrissidae (only as fossils)

Curimatidae (incl. Anodontidae). FERNANDEZ-YEPEZ, A., 1948, Bol. Taxon. Lab. Pesqu. Caiguire Caracas, 1, 1-79 (South America)

Cyathaspidae (only as fossils)

Cybiidae, as Scombridae

Cyclopasteridae, as Cyclopteridae

Cyclolabridae, as Labridae

Cyclopidae, as Astroblepidae

Cyclopiidae, as Astroblepidae

Cyclopteridae (incl. Cyclopasteridae, Eutelichthyidae, Liparidae, Lipariidae, Liparopidae, Rhodichthyidae). POPOV, A. M., 1930, Ann. Mag. Nat. Hist., (10)6, 69-76 (short review); LINDBERG, G. U. & LEGEZA, M. I., 1955, Trav. Inst. Zool. Acad. Sci. U. R. S. S., 18, 389-458 (review of Cyclopterinae); BURKE, V., 1930, United States Nat. Mus. Bull., 150, 1-204 (revision, as Liparidae); NORMAN, J. R., No. 1, 1930 (oceanic, as Liparidae); HILDEBRAND & SCHROEDER, No. 1, 1928 (Chesapeake Bay); NORMAN, J. R., No. 2, 1937 (marine, Patagonia, as Liparidae); de BUEN, F., 1961, Montemar, 1. 1-90 (Chile, as Liparidae); SMITT, F. A. et al., No. 1, 1892 (Scandinavia); DUNCKER & LADIGES, No. 1, 1960 (Nordmark, also as Liparidae); POLL, M., No. 2, 1947 (marine, Belgique, also as Liparidae); NOBRE, A., No. 1, 1935 (marine, Portugal); LOZANO REY, D. L., No. 2, 1952 (España); FAGE, L., No. 1, 1918 (Mediterranean); ANDRIYASHEV, A. P., No. 1, 1954 (northern seas of U. S. S. R., also as Liparidae); BERG, L. S., No. 10, 1965 (fresh-water, U. S. S. R.); FOWLER, H. W., No. 8, 1936 (marine, West Africa); SMITH, J. L. B., No. 1, 1961 (marine, South Africa, as Liparidae); SOLDATOV, & LINDBERG, No. 1, 1930 (Far East seas, as Liparidae); POPOV, A., 1931, Bull. Acad. Leningrad, 1931, 85-99 (Pacific Ocean, synopsis of genera); SCHMIDT, P. Y., No. 1, 1950 (Sea of Okhotsk); SHMIDT, P. Y., No. 2, 1965 (Sea of Okhotsk); JORDAN, D. S. & SNYDER, J. O., 1902, Proc. United States Nat. Mus., 24, 343-351 (Japan, also as Lipariidae); UENO, T., 1970 (Japan); NORMAN, J. R., No. 3, 1938 (Antarctic, as Liparidae)

Cyemidae. BERTIN, L., 1942, Bull. Soc. Zool. France, 67, 101-111 (osteology & revision); LOZANO REY, D. L., No. 4, 1947 (España); FOWLER, H. W., No. 8, 1936 (marine, West Africa)

Cynodontidae. (No revisionary references noted)

Cynoglossidae. NORMAN, J. R., 1930, Discovery Rep., 2, 358-369 (oceanic); NIELSEN, J. G., 1963, Atlantide Rep., 7, 7-37; HILDEBRAND & SCHROEDER, No. 1, 1928 (Chesapeake Bay); BEEBE & TEE-VAN, No. 1, 1928 (Hispaniola); HILDEBRAND, S. F., No. 1, 1946 (shore, Peru) FOWLER, H. W., No. 13, 1954 (fresh-water, Brasil); CHABANAUD, P., 1949, Bull. Mus. Hist. Nat., (2), 21, 60-66, 347-353 (Atlantique oriental); LOZANO y REY, L., No. 5, 1960 (España); DAGET & ILTIS, No. 1, 1965 (Côte d'Ivoire); SMITH, J. L. B., No. 1, 1961 (marine, South Africa); FOWLER, H. W., No. 18, 1956 (Red Sea); de SILVA, P. H. D. H., No. 2, 1956 (Ceylon); SMITH, H. M., No. 1, 1945 (fresh-water, Thailand); MARSHALL, T. C., No. 1, 1964 (Great Barrier Reef); CHEN, J. T. F. & WENG, H. T. C., 1965, Biol. Bull. Tunghai Univ., 25 & 27 (Taiwan); WU, H. -W., 1932, Contribution a l'etude morphologiques . . . de la Chine, Paris, 1-179, 1932 (monograph); FOWLER, H. W., No. 6, 1933-35 (China); OCHIAI, A., 1963, Fauna Japonica, Soleina, Tokyo, 1-114 (Japan)

Cyphosidae, as Kyphosidae

Cyprinidae (incl. Ciprinidae, Gobiobotidae, Medidae). TRETIAKOV, D. K., 1946, Zool. Zhurn., 25, 149-156 (systematic groups, in Russian); KRISHANOVSKI, S. G., 1947 (classification, in Russian; also as Kryzanovsky, S.); BALINSKY, B. I., 1948, Proc. Zool. Soc. London, 118, 335-344 (key to larvae); SLASTENENKO, E. P., No. 1, 1958 (Canada); LEGENDRE, V., 1960, Jeune Nat., 10, 178-210 (key, Quebec); HILDEBRAND & SCHROEDER, No. 1, 1928 (Chesapeake Bay); FOWLER, H. W., No. 1, 1945 (southeastern U. S.); REGAN, C. T., No. 1, 1906-08 (Central America); MEEK, S. E., No. 1, 1904 (fresh-water, Mexico); SMITT, F. A. et al., No. 2, 1895 (Scandinavia); DUNCKER & LADIGES, No. 1, 1960 (Nordmark); SPILLMANN, J., No. 1, 1961 (France); NOBRE, A., No. 1, 1935 (marine, Portugal); LOZANO REY, L., No. 1, 1935 (fresh-water, España); LOZANO REY, D. L., No. 4, 1947 (España); OLIVA, O., 1952, Bull. Int. Acad. Prague, 53, 1-61 (Ceskoslovenska); BANARESCU, P., No. 1, 1964 (Romania); BERG, L. S., 1912, Faune de la Russie, Poissons, III, 1, 1-336 (in Russian); BERG, L. S., 1914, Faune de la Russie, Poissons, III, 2, 337-704 (in Russian); BERG, L. S., No. 1, 1932 (fresh-water, U. R. S. S.); BERG, L. S., 1933, Faune de l'URSS, Poissons, III, 3, Leningrad; BERG, L. S., No. 4, 1949 (fresh-water, Soviet Union); BERG, L. S., No. 9, 1964 (fresh-water, U. S. S. R.); BATTALGIL, F., 1941, Rev. Fac. Sci. Univ. Istanbul, B6, 170-186 (Asia Minor); BERG, L. S., No. 6, 1949 (fresh-water, Iran); BOULENGER, G. A., No. 3, 1909 (fresh-water, Africa); BOULENGER, G. A., No. 2, 1907 (Egypt); PELLEGRIN, J., No. 2, 1923 (Western Africa); DAGET, J., No. 2, 1962 (Guinea); SCHULTZ, L. P., No. 1, 1942 (Liberia); DAGET & ILTIS, No. 1, 1965 (Côte d'Ivoire); DAGET, J., No. 1, 1954 (Upper Niger); BOULENGER, G. A., No. 1, 1901 (Congo Basin); BARNARD, K. H., 1943, Ann. South African Mus., 36, 101-262 (s. w. Cape region); POLL, M., No. 1, 1946 (Lake Tanganyika); POLL, M., No. 3, 1953 (Lake Tanganyika); PELLEGRIN, J., No. 1, 1933 (fresh-water, Madagascar); ARNOULT, J., No. 1, 1959 (fresh-water, Madagascar); DAY, F., No. 1, 1889 (India); DERANIYAGALA, P. E. P., 1930, Ceylon Journ. Sci., (B), 16, 1-41 (Ceylon); SMITH, H. M., No. 1, 1945 (fresh-water, Thailand); WEBER & de BEAUFORT, No. 3, 1916 (Indo-Australian Arch.); INGER & KONG, No. 1, 1962 (fresh-water, North Borneo); OSHIMA, M., No. 1, 1919 (fresh-water, Taiwan); POPE & NICHOLS, No. 1, 1927 (Hainan I.); TCHANG, T. -L., 1933, Zool. Sinica, B. Chordata of China, II, (1), 1-247; FANG, P. W., 1936, Sinensia, 7, 686-712 (fresh-water, China); NICHOLS, J. T., No. 1, 1943 (fresh-water, China); WU, H. W. et al., 1964, Cyprinid fishes of China, Shanghai, 1-228; SOLDATOV & LINDBERG, No. 1, 1930 (Far East seas); SCHMIDT, P. Y., No. 1, 1950 (Sea of Okhotsk); SHMIDT, P. Y., No. 2, 1965 (Sea of Okhotsk); LINDBERG & LEGHEZA, No. 2, 1965 (Sea of Japan); LINDBERG & LEGEZA, No. 3, 1969 (Sea of Japan); JORDAN, D. S. & FOWLER, H. W., 1903, Proc. United States Nat. Mus., 26, 811-862 (Japan); OKADA, Y., No. 1, 1960 (fresh-water, Japan); OKADA, Y., No. 2, 1960 (fresh-water, Japan)

Cyprinodontidae (incl. Empetrichthyidae, Fundulidae, Orestiidae). POLL, M., 1951, Rev. Zool. Bot. Africaines, 45, 157-171 (Congo Belge, Rivulini, museum list); MYERS, G. S., 1955, Trop. Fish Mag., 1955, p. 7 (list of subfamilies & genera); REGAN, C. T., 1913, Proc. Zool. Soc. London, 1913, 977-1018 (revision of Poeciliinae); TCHERNAVIN, V. V., 1944, Proc. Zool. Soc. London, 114, 140-233 (revision Orestiinae); SCHEEL, J., 1968, Rivulins of the Old World, Jersey City, New Jersey, 1-473 (Rivulinae); SLASTENENKO, E. P., No. 1, 1958 (Canada); HILDEBRAND & SCHROEDER, No. 1, 1928 (Chesapeake Bay); FOWLER, H. W., No. 1, 1945 (fresh-water, southeastern U. S.); MILLER, R. R., 1948, Misc. Publ. Mus. Zool. Univ. Michigan, 68, 1-155 (Death Valley system); MYERS, G. S., 1935, Zoologica, 10, 301-316 (faunal list, Hispaniola); RIVERO, L. H. & RIVAS, L. R., 1944, Torreia, 12, 1-19 (key to Cuban); TREWAVAS, E., 1948, Proc. Zool. Soc. London, 118, 408-415 (San Domingo); REGAN, C. T., No. 1, 1906-08 (Central America); de BUEN, F., 1941, Trab. Estac. Limnol. Patzcuaro, 4, 1-31 (Mexico); HOEDEMAN, J. J., 1959, Natuurw. Stud. Suriname Ned-Antillen, 20, 44-98 (Suriname &Guyanas); PUYO, J., No. 1, 1949 (British Guiana); von IHERING, R., 1930, Rev. Biol. Hyg. Sao Paulo, 2, 125-135 (genera, Brasil); von IHERING, R., 1931, Arch. Inst. Biol. Sao Paulo, 4, 243-280 (Brasil); FOWLER, H. W., No. 13, 1954 (fresh-water, Brasil); SPILLMANN, J., No. 1, 1961 (France); LOZANO REY, L., No. 1, 1935 (fresh-water, España); LOZANO REY, D. L., No. 4, 1947 (España); BERG, L. S., No. 10, 1965 (fresh-water, U. S. S. R.) SOZER, F., 1942, Rev. Fac. Sci. Univ. Istanbul, 7B, 307-316 (Turkiye); AKSIRAY, F., 1948, Rec. Fac. Sci. Univ. Istanbul, 13B, 97-138 (Turkiye); BERG, L. S., No. 6, 1949 (fresh-water, Iran); BOULENGER, G. A., No. 5, 1915 (fresh-water, Africa); BOULENGER, G. A., No. 2, 1907 (Egypt); PELLEGRIN, J., No. 2, 1923 (western Africa); FOWLER, H. W., No. 8, 1936 (marine, West Africa); DAGET, J., No. 2, 1962 (Guinea); SCHULTZ, L. P., No. 1, 1942 (Liberia); DAGET & ILTIS, No. 1, 1965 (Côte d'Ivoire); DAGET, J., No. 1, 1954 (Upper Niger); BOULENGER, G. A., No. 1, 1901 (Congo basin); POLL, M., 1952, Zooleo, (n. s.), 14, 227-232 (Congo); POLL, M., No. 1, 1946 (Lake Tanganyika); POLL, M., No. 3, 1953 (Lake Tanganyika); WHITEHEAD, P. J. P., 1962, Bull. British Mus. (Nat. Hist.), Zool., 9, 103-137 (East Africa, Pantanodontinae); PELLEGRIN, J., No. 1, 1933 (fresh-water, Madagascar); ARNOULT, J., No. 1, 1959 (fresh-water, Madagascar); FOWLER, H. W., No. 18, 1956 (Red Sea); DAY, F., No. 1, 1889 (India); SMITH, H. M., No. 1, 1945 (fresh-water, Thailand); WEBER & de BEAUFORT, No. 4, 1922 (Indo-Australian Arch.); NICHOLS, J. T., No. 1, 1943 (fresh-water, China); LINDBERG & LEGHEZA, No. 2, 1965 (Sea of Japan); LINDBERG & LEGEZA, No. 3, 1969 (Sea of Japan); OKADA, Y., No. 1, 1960 (fresh-water, Japan); OKADA, Y., No. 2, 1960 (fresh-water, Japan)

Cyttidae, as Zeidae

Cyttopsidae, as Zeidae

Dactylopteridae (incl. Cephalacanthidae). HILDEBRAND & SCHROEDER, No. 1, 1928 (Chesapeake Bay as Cephalacanthidae); BEEBE & TEE-VAN, No. 1, 1928 (Hispaniola, as Cephalacanthidae); MEEK & HILDEBRAND, No. 4, 1928 (marine, Panama, as Cephalacanthidae); DEVINCENZI, G. F., No. 2, 1924 (Uruguay, as Cephalacanthidae); LOZANO REY, D. L., No. 2, 1952 (España, as Cephalacanthidae); FOWLER, H. W., No. 8, 1936 (marine, West Africa, as Cephalacanthidae); SMITH, J. L. B., No. 1, 1961 (marine, South Africa, as Cephalacanthidae); de SILVA, P. H. D. H., 1956, Spolia Zeylanica, 28, 27-34 (marine, Ceylon, & museum list); de BEAUFORT & BRIGGS, No. 11, 1962 (Indo-Australian Arch.); MARSHALL, T. C., No. 1, 1964 (Great Barrier Reef, as Cephalacanthidae); JORDAN, D. S, & RICHARDSON, R. E., 1908, Proc. United States Nat. Mus., 33, 619-670 (Japan, as Cephalacanthidae)

Dactyloscopidae. MEEK & HILDEBRAND, No. 4, 1928 (marine, Panama); HILDEBRAND, S. F., No. 1, 1946 (shore, Peru)

Dalatiidae. BIGELOW & SCHROEDER, No. 1, 1948 (western North Atlantic); FOWLER, H. W., No. 8, 1936 (marine, West Africa); SMITH, J. L. B., No. 1, 1961 (marine, South Africa); JORDAN, D. S. & FOWLER, H. W., 1903, Proc. United States Nat. Mus., 26, 593-674 (Japan)

Dalliidae, as Umbridae

Dapediidae, as Semionotidae

Dartmuthiidae (only as fossils)

Dasyatidae, as Trygonidae

Dasybatidae, as Trygonidae

Deltoptychiidae (only as fossils)

Denaeidae (only as fossils)

Denticidae, as Sparidae

Denticipitidae (incl. Igborichthyidae). CLAUSEN, H. S., 1959, Vidensk. Medd. Dansk Naturh. Foren. København, 121, 141-151 (new family, fresh-water, West Africa)

Dercetidae (incl. Stratodontidae). (Only as fossils)

Derepodichthyidae, as Zoarcidae

Derichthyidae. MEEK & HILDEBRAND, No. 2, 1923 (marine, Panama)

Derrhiidae (only as fossils)

Diademichthyidae, as Gobiesocidae

Dianidae, as Luvaridae

Dicellopygidae. (No revisionary references noted)

Diceratiidae (incl. Aeschynichthyidae, Laevoceratiidae). BERTELSEN, E., No. 1, 1951 (revision); de BEAUFORT & BRIGGS, No. 11, 1962 (Indo-Australian Arch.)

Dichistiidae, as Coracinidae

Dictyonaspidae (only as fossils)

Dictyopygidae, as Redfieldiidae (only as fossils)

Didymaspidae (only as fossils)

Dinaspidae (incl. Irregularaspidae) (only as fossils)

Dinolestidae, as Apogonidae

Dinopterygidae (only as fossils)

Diodontidae. GILL, T., No. 1, 1884 (synopsis); FRASER-BRUNNER, A., No. 1, 1943 (classification, synopsis of genera); LeDANOIS, Y., No. 1, 1959 (osteology, myology, systematics); HILDEBRAND & SCHROEDER, No. 1, 1928 (Chesapeake Bay); BEEBE & TEE-VAN, No. 1, 1929 (Hispaniola) MEEK & HILDEBRAND, No. 4, 1928 (marine, Panama); DEVINCENZI, G. F., No. 2, 1924 (Uruguay); LOZANO REY, D. L., No. 2, 1952 (España); FOWLER, H. W., No. 8, 1936 (marine, West Africa); SMITH, J. L. B., No. 1, 1961 (marine, South Africa); de SILVA, P. H. D. H., 1958, Spol. Zeylanica, 28, 145-150 (keys, Ceylon); de BEAUFORT & BRIGGS, No. 11, 1962 (Indo-Australian Arch.); MARSHALL, T. C., No. 1, 1964 (Great Barrier Reef); JORDAN, D. S. & SNYDER, J. O., 1901, Proc. United States Nat. Mus., 24, 229-264 (Japan); SCHULTZ, L. P. et al., No. 1, 1966 (Marshall & Marianas Is.); FOWLER, H. W., No. 17, 1959 (Fiji)

Diplacanthidae (only as fossils)

Diplaspidae (only as fossils)

Diplocercidae (only as fossils)

Diplomystidae. EIGENMANN & EIGENMANN, No. 1, 1890 (South America); EIGENMANN, C. H., No. 2, 1927 (fresh-water, Chile)

Diploprionidae, as Serranidae

Dipnorhynchidae (only as fossils)

Dipterichthyidae (only as fossils)

Dipteridae (only as fossils)

Dipterygonotidae, as Emmelichthyidae

Diretmidae. NORMAN, J. R., No. 1, 1930 (oceanic); LOZANO REY, D. L., No. 2, 1952 (España); FOWLER, H. W., No. 8, 1936 (marine, West Africa); SMITH, J. L. B., No. 1, 1961 (marine, South Africa); FOWLER, H. W., No. 18, 1956 (Red Sea)

Discobatidae (incl. Platyrhinidae, Zanobatidae). FOWLER, H. W., No. 8, 1936 (marine, West Africa); CHEN, J. T. F., No. 1, 1948 (China, as Platyrhinidae)

Disparichthyidae, as Carapidae

Distichodontidae. (No revisonary references noted)

Ditremidae, as Embiotocidae

Doiichthyidae, as Ariidae

Dolichopterygidae, as Opisthoproctidae

Doliichthyidae, as Gobiidae

Doradidae. EIGENMANN, C. H., 1925, Trans. American Philos. Soc., 22, 280-365 (South America); VAN DER STIGCHEL, J. W. B., No. 1, 1946 (South America); FOWLER, H. W., No. 10, 1948-51 (fresh-water, Brasil); DEVINCENZI, G. J., No. 2, 1924 (Uruguay)
Dorosomatidae, as Clupeidae
Dorosomidae, as Clupeidae
Dorypteridae (only as fossils)
Draconettidae, as Callionymidae
Drepanaspidae (only as fossils)
Drepanichthyidae, as Ephippidae
Drepanidae, as Ephippidae
Duleidae, as Kuhliidae
Dussumeriidae, as Clupeidae
Dussumieridae, as Clupeidae
Dussumieriidae, as Clupeidae
Dysommidae. SMITH, J. L. B., No. 1, 1961 (marine, South Africa)
Dysominidae, as Dysomminidae
Dysomminidae (incl. Dysominidae). (No revisionary references noted)

Echelidae (probably part is Ophichthidae, part is Xenocongridae). LOZANO REY, D. L., No. 4, 1947 (España); FOWLER, H. W., No. 18, 1956 (Red Sea); LINDBERG & LEGHEZA, No. 2, 1965 (Sea of Japan); LINDBERG & LEGEZA, No. 3, 1969 (Sea of Japan); SCHULTZ, L. P. et al., No. 2, 1953 (Marshall & Marianas Is.)
Echeneidae (incl. Echeneididae). HILDEBRAND & SCHROEDER, No. 1, 1928 (Chesapeake Bay); BEEBE & TEE-VAN, No. 1, 1928 (Hispaniola); MEEK & HILDEBRAND, No. 4, 1928 (marine, Panama); HILDEBRAND, S. F., No. 1, 1946 (shore, Peru); DEVINCENZI, G. F., (Uruguay); SZIDAT, L. & NANI, A., 1951, Rev. Inst. Invest. Mus. Argentina Cienc. Nat. Zool., 2, (6), 385-417 (key to genera, Atlantico Austral); NOBRE, A., No. 1, 1935 (marine, Portugal); LOZANO y REY, L., No. 5, 1960 (España); MAUL, G. E., 1956, Bol. Mus. Munic. Funchal, 9, (23-24), 5-71, 75-96, etc. (revision, keys); FOWLER, H. W., No. 8, 1936 (marine, West Africa); DAGET & ILTIS, No. 1, 1965 (Côte d'Ivoire); SMITH, J. L. B., No. 1, 1961 (marine, South Africa); de BEAUFORT & BRIGGS, No. 11, 1962 (Indo-Australian Arch.); MARSHALL, T. C., No. 1, 1964 (Great Barrier Reef); CHU, K. Y., 1957, Quart. Journ. Taiwan Mus., 10, (1), 47-53 (Taiwan); SOLDATOV & LINDBERG, No. 1, 1930 (Far East seas); SCHULTZ, L. P. et al., No. 1, 1966 (Marshall & Marianas Is.); FOWLER, H. W., No. 17, 1959 (Fiji)
Echeneididae, as Echeneidae
Echeneiidae, as Echeneidae
Echidnidae, as Muraenidae
Echinorhinidae. GARMAN, S., No. 1, 1913 (monograph); BIGELOW & SCHROEDER, No. 1, 1948 (western North Atlantic); FOWLER, W. W., No. 8, 1936 (marine, West Africa); FOWLER, H. W., No. 18, 1956 (Red Sea)
Ectosteorhachidae (only as fossils)
Edestidae (only as fossils)
Elacatidae, as Rachycentridae
Elasssomatidae, as Centrarchidae
Elassomidae, as Centrarchidae
Electrophoridae. FOWLER, H. W., No. 10, 1948-51 (fresh-water, Brasil)
Eleotridae, as Gobiidae
Elephenoridae, as Caristiidae
Elonichthyidae, as Palaeoniscidae
Elopidae (incl. Elopsidae part, Raphiosauridae, Spaniodontidae). HILDEBRAND & SCHROEDER, No. 1, 1928 (Chesapeake Bay); BEEBE & TEE-VAN, No. 1, 1928 (Hispaniola); REGAN, C. T., No. 1, 1906-08 (Central America); MEEK & HILDEBRAND, No. 2, 1923 (marine, Panama); HILDEBRAND, S. F, No. 1, 1946 (shore, Peru); EIGENMANN, C. H., No. 1, 1912 (British Guiana); PUYO, J., No. 1, 1949 (Guyane); BIGELOW, H. B., et al., No. 5, 1963 (western North Atlantic); BOULENGER, G. A., No. 3, 1909 (fresh-water, Africa); FOWLER, H. W., No. 8, 1936 (marine, West Africa); PELLEGRIN, J., No. 2, 1923 (western Africa); DAGET & ILTIS, No. 1, 1965 (Côte d'Ivoire); BOULENGER, G. A., No. 1, 1901 (Congo basin); SMITH, J. L. B., No. 1, 1961 (marine, South Africa); LOSSE, G. F., 1966, Journ. East African Nat. Hist. Soc. Nat. Mus., 25, 166-178 (list, coast of East Africa); PELLEGRIN, J., No. 1, 1933 (fresh-water, Madagascar); ARNOULT, J., No. 1, 1959 (fresh-water, Madagascar); FOWLER, H. W., No. 18, 1956 (Red Sea); SMITH, H. M., No. 1, 1945 (fresh-water, Thailand); WEBER & de BEAUFORT, No. 2, 1913 (Indo-Australian Arch.); MARSHALL, T. C., No. 1, 1964 (Great Barrier Reef); POPE & NICHOLS, No. 1, 1927 (Hainan I.); FOWLER, H. W., No. 3, 1931 (China); NICHOLS, J. T., No. 1, 1943 (fresh-water, China); LINDBERG & LEGHEZA, No. 2, 1965 (Sea of Japan); LINDBERG & LEGEZA, No. 3, 1969 (Sea of Japan); JORDAN & HERRE, No. 1, 1906 (Japan); OKADA, Y., No. 1, 1960 (fresh-water, Japan); OKADA, Y., No. 2, 1960 (fresh-water, Japan); FOWLER, H. W., No. 17, 1959 (Fiji)
Elopsidae, part as Elopidae, part as Megalopidae
Embiotocidae (incl. Ambiotocidae, Ditremidae, Holconotidae, Hysterocarpidae). HUBBS, C. L., 1918, Proc. Biol. Soc. Washington, 31, 9-14 (revision); TARP, F. H., 1952, Fish. Bull. Sacramento, 88, 1-99 (revision); FOWLER, H. W., No. 11, 1949 (China); SOLDATOV & LINDBERG, No. 1, 1930 (Far East seas); SCHMIDT, P. Y., No. 1, 1950 (Sea of Okhotsk); SHMIDT. P, Y., No. 2, 1965 (Sea of Okhotsk); LINDBERG & KRASJUKOVA, No. 1, 1969 (Sea of Japan); JORDAN,

D. S., 1911, Proc. United States Nat. Mus., 24, 353-359 (Japan)
Emblemariidae, as Chaenopsidae
Emmelichthyidae (incl. Centracanthidae, Centracantidae, Erythrichthyidae, Erythroclidae, Dipterygonotidae, Inermiidae, Maenidae, Merolepidae, Spicaridae) SCHULTZ, L. P., 1945, Journ. Washington Acad. Sci., 35, 132-136 (key, Bahama Is.); NOBRE, A., No. 1, 1935 (marine, Portugal, as Maenidae); LOZANO REY, D. L., No. 2, 1952 (España, as Centracanthidae); LOZANO CABO, F., 1953, Mem. R. Acad. Madrid Cienc. Nat., 17, (2), 1-128 (Mediterranean, as Centracanthidae); LOZANO CABO, F., 1953, Bol. Inst. Esp. Oceanogr., 59, 1-126 (Mediterranean, as Centracanthidae); ZEI, M., 1941, Acta Adriatica, 2, 135-190 (Adriatic, as Maenidae); BANARESCU, P., No. 1, 1964 (Romania, as Maenidae); BERG, L. S., No. 5, 1949 (fresh-water, Soviet Union, as Maenidae); BERG, L. S., No. 10, 1965 (fresh-water, U. S. S. R., as Maenidae); FOWLER, H. W., No. 8, 1936 (marine, West Africa, as Centracantidae); NORMAN, J. R., 1931, Ann. Mag. Nat. Hist., (10), 7, 352-357 (key to genera, as Maenidae); SMITH, J. L. B, No. 1, 1961 (marine, South Africa, also as Centracantidae); POLL, M., No. 4, 1954 (Coastal South Africa, also as Maenidae); WEBER & de BEAUFORT, No. 6, 1931 (Indo-Australian Arch., also as Maenidae); FOWLER, H. W., No. 20, 1933 (Philippine seas); LINDBERG & KRASJUKOVA, No. 1, 1969 (Sea of Japan)
Empetrichthyidae, as Cyprinodontidae
Encheliidae (only as fossils)
Enchodontidae (only as fossils)
Endeiolepidae (only as fossils)
Engraulidae (incl. Engraulididae, Stolephoridae). JORDAN, D. S. & SEALE, A., 1926, Bull. Mus. Comp. Zool., 67, 355-418 (review); HILDEBRAND, S. F., 1943, Bull. Bingham Oceanogr. Coll., 8, (2), 1-165 (American); HILDEBRAND & SCHROEDER, No. 1, 1928 (Chesapeake Bay); FOWLER, H. W., No. 1, 1945 (southeastern U. S.); BEEBE & TEE-VAN, No. 1, 1928 (Hispaniola); MEEK & HILDEBRAND, No. 2, 1923 (marine, Panama); HILDEBRAND, S. F., No. 1, 1946 (shore, Peru); EIGENMANN, C. H., No. 1, 1912 (British Guiana); PUYO, J., No. 1, 1949 (Guyane); FOWLER, H. W., No. 10, 1948-51 (fresh-water, Brasil); DEVINCENZI, G. J., No. 2, 1924 (Uruguay, as Stolephoridae); BIGELOW, H. B., et al., No. 5, 1963 (western North Atlantic); DUNCKER & LADIGES, No. 1, 1960 (Nordmark); LOZANO REY, D. L., No. 4, 1947 (España); LOZANO-REY, L., 1950, Rapp. Cons. Explor. Mer. Copenhagen, 126, 7-20 (Espagne, Maroc, Sahara espagnols); FAGE, L., 1920, Rep. Danish Oceanogr. Exp. 1908-10 to Medit., II, Biol., A9, 1-40 (Mediterranean); BANARESCU, P., No. 1, 1964 (Romania); SVETOVIDOV, A. N., No. 2, 1964 (Black Sea); FOWLER, H. W., No. 8, 1936 (marine, West Africa); DAGET & ILTIS, No. 1, 1965 (Côte d' Ivoire); SMITH, J. L. B., No. 1, 1964 (marine, South Africa, also as Stolephoridae); FOWLER, H. W., No. 18, 1956 (Red Sea, also as Stolephoridae); WHITEHEAD, P. J. P., 1965, Bull. British Mus. (Nat. Hist), Zool., 12, 227-281 (Red Sea); SMITH, H. M., No. 1, 1945 (fresh-water, Thailand); INGER & KONG, No. 1, 1962 (fresh-water, North Borneo); FOWLER, H. W., No. 3, 1931 (China); NICHOLS, J. T., No. 1, 1943 (fresh-water, China); SOLDATOV & LINDBERG, No. 1, 1930 (Far East seas); LINDBERG & LEGHEZA, No. 2, 1965 (Sea of Japan); LINDBERG & LEGEZA, No. 3, 1969 (Sea of Japan); JORDAN & HERRE, No. 1, 1906 (Japan); OKADA, Y., No. 1, 1960 (fresh-water, Japan); OKADA, Y., No. 2, 1960 (fresh-water, Japan); STRASBURG, D. W., 1960 (keys, Hawaii); FOWLER, H. W., No. 17, 1959 (Fiji)
Engraulididae, as Engraulidae
Enoplosidae (incl. Enoposidae). MARSHALL, T. C., No. 1, 1964 (Great Barrier Reef); FOWLER, H. W., No. 20, 1933 (Philippine seas)
Enoposidae, as Enoplosidae
Eobuglossidae (only as fossils)
Ephippidae (incl. Chaetodipteridae, Drepanichthyidae, Drepanidae, Ilarchidae, Platacidae). HILDEBRAND & SCHROEDER, No. 1, 1928 (Chesapeake Bay); BEEBE & TEE-VAN, No. 1, 1928 (Hispaniola, as Chaetodipteridae); MEEK & HILDEBRAND, No. 4, 1928 (marine, Peru); HILDEBRAND, S. F., No. 1, 1946 (shore, Peru); BAUCHOT & BLANC, No. 1, 1961 (eastern tropical Atlantic); FOWLER, H. W., No. 8, 1936 (marine, West Africa); DAGET & ILTIS, NO. 1, 1965 (Côte d'Ivoire); SMITH, J. L. B., No. 1, 1961 (marine, South Africa, as Drepanidae & Platacidae); POLL, M., No. 4, 1954 (coastal South Africa); HERRE, A. W. & MONTALBAN, H. R., 1927, Philippine Journ. Sci., 34, 1-113 (Philippines, also as Drepanidae & Platacidae); FOWLER, H. W., No. 12, 1953 (China, also as Drepanidae & Platacidae); LINDBERG & KRASJUKOVA, No. 1, 1969 (Sea of Japan, as Platacidae); SCHULTZ, L. P. et al., No. 2, 1953 (Marshall & Marianas Is., as Platacidae); FOWLER, H. W., No. 17, 1959 (Fiji, as Platacidae)
Epigonidae, as Apogonidae
Epinephelidae, as Serranidae
Eptatretidae. JORDAN, D. S. & SNYDER, J. O., 1901, Proc. United States Nat. Mus., 23, 725-734 (Japan)
Equulidae, as Leiognathidae
Eretmophoridae, as Muridae
Ereuniidae, as Icelidae
Erilepidae, as Anoplopomatidae
Erismatopteridae, as Aphredoderidae (only as fossils)
Errolichthyidae. (No revisionary references noted)
Erromenosteidae (only as fossils)
Erythrichthyidae, as Emmelichthyidae
Erythrinidae. GERY, J., 1961, Bonn Zool. Beitr., 12, 332-342 (key to Hemiodontinae); FOWLER, H. W., No. 10, 1948-51 (fresh-water, Brasil)

Erythrinolepidae (only as fossils)
Erythroclidae, as Emmelichthyidae
Esocesidae, as Belonidae
Esocidae (part as Belonidae) (incl. Luciidae). SLASTENENKO, E. P., No. 1, 1958 (Canada); HILDEBRAND & SCHROEDER, No. 1, 1928 (Chesapeake Bay); FOWLER, H. W., No. 1, 1945 (southeastern U. S.); BIGELOW, H. B., et al., No. 3, 1964 (western North Atlantic); SMITT, F. A. et al., No. 2, 1895 (Scandinavia); DUNCKER & LADIGES, No. 1, 1960 (Nordmark); SPILLMANN, J., No. 1, 1961 (France); BANARESCU, P., No. 1, 1964 (Romania); BERG, L. S., No. 2, 1933 (fresh-water, U. R. S. S.); BERG, L. S., No. 3, 1948 (fresh-water, Soviet Union); BERG, L. S., No. 8, 1962 (fresh-water, U. S. S. R.); ANDRIYASHEV, A. P., No. 1, 1954 Northern seas of U. S. S. R.); SOLDATOV & LINDBERG, No. 1, 1930 (Far East seas)
Etelidae, as Lutjanidae
Etheostomatidae, as Percidae
Etheostomidae, as Percidae
Eucentrarchidae, as Centrarchidae
Eucinostomidae, as Gerridae
Eugnathidae (incl. Furidae) (only as fossils)
Eulamiidae, as Carcharhinidae
Euleptaspididae (only as fossils)
Euphaneropidae (only as fossils)
Eurypharyngidae. NORMAN, J. R., No. 1, 1930 (oceanic); BIGELOW, H. B., No. 4, 1966 (western North Atlantic); LOZANO REY, D. L., No. 4, 1947 (España); FOWLER, H. W., No. 8, 1936 (marine, West Africa)
Eutaeniophoridae (incl. Taeniophoridae). (No revisionary references noted)
Eutelichthyidae, as Cyclopteridae
Euthacanthidae (only as fossils)
Evermannellidae (incl. Odontostomidae). BIGELOW, H. B., No. 4, 1966 (western North Atlantic); LOZANO REY, D. L., No. 4, 1947 (España); FOWLER, H. W., No. 8, 1936 (marine, West Africa)
Evolantiidae, as Exocoetidae
Exocetidae, as Exocoetidae
Exocoetidae (incl. Evolantiidae, Exocetidae, Hemiramphidae, Hemirhamphidae, Oxyporhamphidae); PARIN, N. V., 1961, Trud. Inst. Okeanol. Moskva, 43, 92-183 (classification, also as Oxyporhamphidae); COLLETTE, B. B., 1966, American Mus. Novit., 2274, 1-22 (generic characters, also as Hemiramphidae); HUBBS, C. L. & KAMPA, E. M., 1946, Copeia, 1946, 188-218 (keys to Cypselurinae); HILDEBRAND & SCHROEDER, No. 1, 1928 (Chesapeake Bay, also as Hemiramphidae); BEEBE & TEE-VAN, No. 1, 1928 (Hispaniola, also as Hemiramphidae); MEEK & HILDEBRAND, No. 2, 1923 (marine, Panama, also as Hemirhamphidae); HILDEBRAND, S. F., No. 1, 1946 (shore, Peru, also as Hemiramphidae); DEVINCENZI, G. J., No. 2, 1924 (Uruguay, as Hemirhamphidae); BRUUN, A., 1935, Dana Rep., 6, 1-108 (Atlantic); LOZANO REY, D. L., No. 4, 1947 (España, also as Hemirhamphidae); TORTONESE, E., 1951-52, Boll. Ist. Mus. Zool. Univ. Torino, 3, (3), 1-8 (Mediterranean, as Hemirhamphidae); BERG, L. S., No. 5, 1949 (fresh-water, Soviet Union, as Hemirhamphidae); BERG, L. S., No. 10, 1965 (fresh-water, U. S. S. R., as Hemirhamphidae); FOWLER, H. W., No. 8, 1936 (marine, West Africa, also as Hemiramphidae); COLLETTE, B. B., 1965, Atlantide Rep., 8, 217-235 (keys, West Africa, as Hemiramphidae); SMITH, J. L. B., No. 1, 1961 (marine, South Africa, also as Hemirhamphidae); ARNOULT, J., No. 1, 1959 (fresh-water, Madagascar, as Hemiramphidae); FOWLER, H. W., No. 18, 1956 (Red Sea, as Hemiramphidae); PARIN, N. V, 1961, Trud. Inst. Okeanol. Moskva, 43, 40-91 (Pacific & Indian Oceans); de SILVA, P. H. D. H., 1956, Spolia Zeylanica, 28, 27-34 (off Ceylon & museum list); de SILVA, P. H. D. H., No. 1, 1956 (marine, Ceylon, as Hemiramphidae); SMITH, H. M., No. 1, 1945 (fresh-water, Thailand, as Hemiramphidae); WEBER & de BEAUFORT, No. 4, 1922 (Indo-Australian Arch., also as Hemirhamphidae); INGER & KONG, No. 1, 1962 (fresh-water, North Borneo, as Hemiramphidae); MARSHALL, T. C., No. 1, 964 (Great Barrier Reef, also as Hemiramphidae); HERRE, A. W. C. T., 1944, Stanford Univ. Publ. Biol., 9, 41-86 (Philippines, as Hemiramphidae); FOWLER, H. W., No. 5, 1932 (China, as Hemiramphidae also); NICHOLS, J. T., No. 1, 1943 (fresh-water, China, as Hemiramphidae); SOLDATOV & LINDBERG, No. 1, 1930 (Far East seas, also as Hemirhamphidae); LINDBERG & LEGHEZA, No. 2, 1965 (Sea of Japan, also as Hemirhamphidae & Oxyporhamphidae); LINDBERG & LEGEZA, No. 3, 1969 (Sea of Japan, also as Hemirhamphidae & Oxyporhamphidae); JORDAN, D. S. & STARKS, E. C., 1903, Proc. United States Nat. Mus., 26, 525-544 (Japan, also as Hemiramphidae); IMAI, S., 1956, Mem. Fac. Fish. Kagoshima Univ., 5, 91-102 (Japan); OKADA, Y., No. 1, 1960 (fresh-water, Japan, as Hemiramphidae); OKADA, Y., No. 2, 1960 (fresh-water, Japan, as Hemiramphidae); PARIN, N. V., 1960, Trud. Inst. Okeanol., 31, 205-285 (northwest Pacific Ocean); SCHULTZ, L. P. et al., No. 2, 1953 (Marshall & Marianas Is., also as Hemiramphidae); FOWLER, H. W., No. 17, 1959 (Fiji, also as Hemiramphidae)

Fierasferidae, as Carapidae
Fistularidae, as Fistulariidae
Fistulariidae (incl. Fistularidae). HILDEBRAND & SCHROEDER, No. 1, 1928 (Chesapeake Bay); BEEBE & TEE-VAN, No. 1, 1928 (Hispaniola); MEEK & HILDEBRAND, No. 2, 1923 (marine, Panama); HILDEBRAND, S. F., No. 1, 1946 (shore, Peru); FOWLER, H. W. No. 8, 1936 (marine, West Africa); SMITH, J. L. B., No. 1, 1961 (marine, South Africa); FOWLER, H. W., No. 18, 1956 (Red Sea); de SILVA, P. H. D. H., No. 1, 1956 (marine, Ceylon); MARSHALL, T. C., No. 1,

1964 (Great Barrier Reef); FOWLER, H. W., No. 6, 1933-35 (China); LINDBERG & LEGHEZA No. 2, 1965 (Sea of Japan); LINDBERG & LEGEZA, No. 3, 1969 (Sea of Japan); JORDAN, D. S., & STARKS, E. C., 1902, Proc. United States Nat. Mus., 26, 57-73 (Japan); SCHULTZ, L. P. et al., No. 2, 1953 (Marshall & Marianas Is.); FOWLER, H. W., No. 17, 1959 (Fiji)

Fitzroyidae, as Jenynsiidae

Fitzroyiidae, as Jenynsiidae

Fleurantiidae (only as fossils)

Flutidae, as Synbranchidae

Forficidae (only as fossils)

Formiidae, as Formionidae

Formionidae (incl. Apolectidae, Formiidae). SMITH, J. L. B., No. 1, 1961 (marine, South Africa, as Apolectidae); LINDBERG & KRASJUKOVA, No. 1, 1969 (Sea of Japan)

Fundulidae, as Cyprinodontidae

Furidae, as Eugnathidae (only as fossils)

Gadidae (incl. Gaidropsaridae, Ranicipitidae). SVETOVIDOV, A. N., 1946, Bull. Acad. Sci. U. R. S. S., Biol., 1946, 2-3, 183-198 (classification); MAUL, G. E., 1952, Bol. Mus. Munic. Funchal, 6, 1-62 (museum list); NORMAN, J. R., No. 1, 1930 (oceanic); SLASTENENKO, E. P., No. 1, 1958 (Canada); HILDEBRAND & SCHROEDER, No. 1, 1928 (Chesapeake Bay); NORMAN, J. R., No. 2, 1937 (key to South American genera); DEVINCENZI, G. F., No. 2, 1924 (Uruguay); NORMAN, J. R., No. 2, 1937 (marine, Patagonia); SMITT, F. A. et al., No. 1, 1892 (Scandinavia); DUNCKER & LADIGES, No. 1, 1960 (Nordmark); POLL, M., No. 2, 1947 (marine, Belgique); SPILLMANN, J., No. 1, 1961 (France); NOBRE, A., No. 1, 1935 (marine, Portugal); LOZANO y REY, L., No. 5, 1960 (España); D'ANCONA, U., 1931, Arch. Zool. Torino, 15, 291-303 ("forme larvali e giovanili," Mediterranean); BANARESCU, P., No. 1, 1964 (Romania); SVETOVIDOV, A. N., No. 2, 1964 (Black Sea); BERG, L. S., No. 2, 1933 (fresh-water, U. R. S. S.); SVETOVIDOV, A. N., 1948, Fauna S. S. S. R. Ryby (Gadiformes) IX. 4. Moskva; 1-221 (S. S. S. R.); BERG, L. S., No. 5, 1949 (fresh-water, Soviet Union); SVETOVIDOV, A. N. 1962, Fauna U. S. S. R. Fishes. IX. 4. Gadiformes. Washington, D. C., 1-304 (U. S. S. R.); BERG, L. S., No. 10, 1965 (fresh-water, U. S. S. R.); ANDRIYASHEV, A. P., No. 1, 1954 (northern seas of U. S. S. R.); FOWLER, H. W., No. 8, 1936 (marine, West Africa); SMITH, J. L. B., No. 1, 1961 (marine, South Africa); FOWLER, H. W., No. 18, 1956 (Red Sea); DAY, F., No. 2, 1889 (India); WEBER & de BEAUFORT, No. 5, 1929 (Indo-Australian Arch.); FOWLER, H. W., No. 6, 1933-35 (China); SOLDATOV & LINDBERG, No. 1, 1930 (Far East seas); SCHMIDT, P. Y., No. 1, 1950 (Sea of Okhotsk); SHMIDT, P. Y., No. 2, 1965 (Sea of Okhotsk); LINDBERG & LEGHEZA, No. 2, 1965 (Sea of Japan); LINDBERG & LEGEZA, No. 3, 1969 (Sea of Japan)

Gadopsidae. (No revisionary references noted)

Gaidropsaridae, as Gadidae

Galaxidae, as Galaxiidae

Galaxiidae (incl. Galaxidae, Paragalaxiidae). SCOTT, E. O. G., 1966, Australian Zool., 13, 244-258 (genera); EIGENMANN, C. H., No. 2, 1927 (fresh-water, Chile); BOULENGER, G. A., No. 5, 1915 (fresh-water, Africa); BARNARD, K. H., 1943, Ann. South African Mus., 36, 101-262 (southwest Cape region); DAY, F., No. 1, 1889 (India); FRANKENBERG, R., 1966, Australian Nat. Hist., 15, 161-164; also Fisherman, 2, (5), 21-23; STOKELL, G., 1945, Trans. Proc. Roy. Soc. New Zealand, 75, 124-137 (generic classification, New Zealand); STOKELL, G., 1949, Trans. Proc. Roy. Soc. New Zealand, 77, 472-496 (new Zealand); McDOWALL, R. M., 1970, Bull. Mus. Comp. Zool., 139, 341-431 (New Zealand)

Galeaspidae (only as fossils)

Galeidae, as Carcharhinidae

Galeorhinidae, as Carcharhinidae

Ganolytidae (only as fossils)

Gasterochismidae, as Scombridae

Gasteropelecidae, as Characidae

Gasterosteidae (incl. Gastrosteidae, Sclerogenidae part). BERTIN, L., 1925, Ann. Inst. Oceanogr. Monaco, 2, 1-204; EIGENMANN, C. H., 1886, Proc. Acad. Nat. Sci. Philadelphia, 1886, 233-252 (American); SLASTENENKO, E. P., No. 1, 1958 (Canada); HILDEBRAND & SCHROEDER, No. 1, 1928 (Chesapeake Bay); SMITT, F. A. et al., No. 2, 1895 (Scandinavia); DUNCKER & LADIGES, No. 1, 1960 (Nordmark); POLL, M., No. 2, 1947 (marine, Belgique); SPILLMAN, J., No. 1, 1961 (France); NOBRE, A., No. 1, 1935 (marine, Portugal); LOZANO REY, L., No. 1, 1935 (fresh-water, España); LOZANO REY, D. L., No. 4, 1947 (España); BANARESCU, P., No. 1, 1964 (Romania); SVETOVIDOV, A. N., No. 2, 1964 (Black Sea); BERG, L. S., No. 2, 1933 (fresh-water, U. R. S. S.); BERG, L. S., No. 5, 1949 (fresh-water, Soviet Union); BERG, L. S., No. 10, 1965 (fresh-water, U. S. S. R.); ANDRIYASHEV, A. P., No. 1, 1954 (northern seas of U. S. S. R.); BOULENGER, G. A, No. 6, 1916 (fresh-water, Africa); NICHOLS, J. T., No. 1, 1943 (fresh-water, China); SOLDATOV & LINDBERG, No. 1, 1930 (Far East seas); SCHMIDT, P. Y., No. 1, 1950 (Sea of Okhotsk); SHMIDT, P. Y., No. 2, 1965 (Sea of Okhotsk); LINDBERG & LEGHEZA, No. 2, 1965 (Sea of Japan); LINDBERG & LEGEZA, No. 3, 1969 (Sea of Japan); JORDAN, D. S. & STARKS, E. C., 1902, Proc. United States Nat. Mus., 26, 57-73 (Japan); OKADA, Y., No. 1, 1960 (fresh-water, Japan); OKADA, Y., No. 2, 1960 (fresh-water, Japan)

Gastromyzonidae, as Homalopteridae

Gastromyzontidae, as Homalopteridae

Gastropelecidae, as Characidae
Gastrosteidae, as Gasterosteidae
Gaterinidae, as Pomadasyidae
Gavialicipitidae, as Serrivomeridae
Gelididae, as Nototheniidae
Gempylidae (incl. Acinaceidae, Lemnisomidae, Ruvettidae). GREY, M., 1953, Copeia, 1953, 135-141; BOESEMAN, M., 1962, Journ. Mar. Biol. Assoc. India, 4, 214-216 (types in Leiden Museum); NORMAN, J. R., No. 1, 1930 (oceanic); HILDEBRAND, S. F., No. 1, 1946 (shore, Peru); NORMAN, J. R., No. 2, 1937 (marine, Patagonia); NOBRE, A., No. 1, 1935 (marine, Portugal); LOZANO REY, D. L., No. 3, 1952 (España); FOWLER, H. W., No. 8, 1936 (marine, West Africa); SMITH, J. L. B., No. 1, 1961 (marine, South Africa); de BEAUFORT & CHAPMANS, No. 9, 1951 (Indo-Australian Arch.); FOWLER, H. W., No. 7, 1936 (China); KAMOHARA, T., 1938, Annot. Zool. Japan, 17, 45-50 (Japan); MATSUBARA, K. & IWAI, T., 1952, Pacific Sci., 6, 193-212 (Japan); MATSUBARA, K. & IWAI, T., 1958, Mem. Coll. Agric. Kyoto Univ., Fish. Ser., Spec. No. 23-54 (Japan)
Gemuendinidae (incl. Asterosteidae). (Only as fossils)
Geotriidae, as Petromyzonidae
Gerridae (incl. Eucinostomidae, Xystaemidae). HILDEBRAND & SCHROEDER, No. 1, 1928 (Chesapeake Bay); BEEBE & TEE-VAN, No. 1, 1928 (Hispaniola); MEEK & HILDEBRAND, No. 3, 1925 (marine, Panama); HILDEBRAND, S. F., No. 1, 1946 (shore, Peru); FOWLER, H. W., No. 13, 1954 (fresh-water, Brasil); DEVINCENZI, G. F., No. 2, 1924 (Uruguay); FOWLER, H. W., No. 8, 1936 (marine, South Africa); DAGET & ILTIS, No. 1, 1965 (Côte d'Ivoire); POLL, M., No. 4, 1954 (coastal South Africa); SMITH, J. L. B., No. 1, 1961 (marine, South Africa); PELLEGRIN, J., No. 2, 1960 (fresh-water, Madagascar); MARSHALL, T. C., No. 1, 1964 (Great Barrier Reef); FOWLER, H. W., No. 20, 1933 (Philippine seas); FOWLER, H. W., No. 9, 1937-41 (China); LINDBERG & KRASJUKOVA, No. 1, 1969 (Sea of Japan); JORDAN, D. S., 1907, Proc. United States Nat. Mus., 32, 245-248 (Japan); FOWLER, H. W., No. 17, 1959 (Fiji)
Gibberichthyidae. PARR, A. E., 1933, Bull. Bingham Oceanogr. Coll., 3, (6), 1-51 (new family, Bahamas & Bermuda); PARR, A. E., 1934, Bull. Bingham Oceanogr. Coll., 4, (6), 1-59.
Gigantactidae. (No revisionary references noted)
Gigantactinidae. BERTELSEN, E., No. 1, 1951 (revision)
Gigantodontidae. (Family not identified)
Giganturidae. BIGELOW, H. B. et al., No. 3, 1964 (western North Atlantic); FOWLER, H. W., No. 8, 1936 (marine, West Africa)
Ginglystomidae, as Orectolobidae
Girellidae, as Kyphosidae
Glandulocaudidae, as Characidae
Glaucosomidae. MARSHALL, T. C., No. 1, 1964 (Great Barrier Reef); LINDBERG & KRASJUKOVA, No. 1, 1969 (Sea of Japan)
Glyphidodontidae, as Pomacentridae
Glyphiodontidae, as Pomacentridae
Glyptopomidae (only as fossils)
Gnathacanthidae, as Pataecidae
Gnathanacanthidae, as Pataecidae
Gobiesocidae. (incl. Diademichthyidae). BRIGGS, J. C., 1955 (Stanford Ichth. Bull., 6, 1-224 (monograph); SCHULTZ, L. P. 1944, Proc. United States Nat. Mus., 96, 47-77 (American); HILDEBRAND & SCHROEDER, No. 1, 1928 (Chesapeake Bay); BEEBE & TEE-VAN, No. 1, 1928 (Hispaniola); REGAN, C. T, No. 1, 1906-08 (Central America); MEEK & HILDEBRAND, No. 4, 1928 (marine, Panama); HILDEBRAND, S. F., No. 1, 1946 (shore, Peru); BRIGGS, J. C., 1957, Copeia, 1957, 204-208 (key to Lepadogastrinae genera of eastern Atlantic); SMITT, F. A. et al., No. 1, 1892 (Scandinavia); NOBRE, A., No. 1, 1935 (marine, Portugal); LOZANO y REY, L., No. 5, 1960 (España); TORTONESE, E., 1960, Rapp. Comm. Int. Mer Mediterranean, 15, (2), 161-163 (Golfe de Gênes); BANARESCU, P., No. 1, 1964 (Romania); SVETOVIDOV, A. N., No. 2, 1964 (Black Sea); FOWLER, H. W., No. 8, 1936 (marine, West Africa); SMITH, J. L. B., No. 1, 1961 (marine, South Africa); SMITH, J. L. B., 1964 (Ich. Bull. Rhodes Univ., 30, 581-597 (western Indian Ocean & Red Sea); de BEAUFORT & BRIGGS, No. 11, 1962 (Indo-Australian Arch.); SCOTT, T. D., 1954, Rec. South Australian Mus., 11, 105-112 (keys, South Australia); MARSHALL, T. C., No. 1, 1964 (Great Barrier Reef); JORDAN, D. S. & FOWLER, H. W., 1902, Proc. United States Nat. Mus., 25, 413-426 (Japan); SCHULTZ, L. P. et al., No. 1, 1966 (Marshall & Marianas Is.); FOWLER, H. W., No. 17, 1959 (Fiji); BRIGGS, J. C., 1951, Proc. California Zool. Club, 1, 57-108)
Gobiidae (incl. Apocrypteidae, Benthophilidae, Doliichthyidae, Eleotridae, Gobiomoridae, Milyeringidae, Periophthalmidae, Sicydiaphiidae). ILJIN, B. S., 1930, Trab. Inst. Español Oceanogr., 2, 1-63 (Le systeme); KOUMANS, F. P., No. 1, 1932 (revision, also as Benthophilidae, Eleotridae); ILYIN, B., 1927, Trav. Soc. Nat. Leningrad. C. R., 57, 73-76 (museum list); EGGERT, B., 1935, Zool. Jahrb., Syst. 67, 29-116 (Periophthalminae); JORDAN, D. S, & EIGENMANN, C. H., 1887, Proc. United States Nat. Mus., 9, 477-518 (North America); HILDEBRAND & SCHROEDER, No. 1, 1928 (Chesapeake Bay); BEEBE & TEE-VAN, No. 1, 1928 (Hispaniola; also as Eleotridae); REGAN, C. T., No. 1, 1906-08 (Central America); MEEK, S. E., No. 1, 1904 (fresh-water, Mexico); MEEK & HILDEBRAND, No. 1, 1916 (fresh-water, Panama); MEEK & HILDEBRAND, No. 4, 1928 (marine, Panama); HILDEBRAND, S. F., No. 1, 1946 (shore, Peru); EIGENMANN, C. H., No. 1, 1912 (British Guiana); PUYO, J., No. 1, 1949

(Guyane); FOWLER, H. W., no. 13, 1954 (fresh-water, Brasil, also as Eleotridae); de BUEN, F., 1963, Bol. Soc. Biol, 35-36, 81-101 (Chile); BOHLKE, J. E. & ROBINS, C. R., 1968, Proc. Acad. Nat. Sci. Philadelphia, 120, 45-174 (keys, western Atlantic); de BUEN, F., 1931, Bull. Soc. Sci. Nat. Maroc, 10, 120-147 (synopsis European); de BUEN, F., 1940, Bull. Inst. Oceanogr. Monaco, 790, 1-16 (pelagic, western Europe); SMITT, F. A. et al., No. 1, 1892 (Scandinavia); DUNCKER & LADIGES, No. 1, 1960 (Nordmark); POLL, M., No. 2, 1947 (marine, Belgique); SPILLMANN, J., No. 1, 1961 (France); de BUEN, F., 1918, Bol. Pescas Madrid, 1918, 291-337 (Pen. Iberica & Baleares); de BUEN, F., 1931, Not. Res. Minist. Marina Madrid, (II), 54, 1-76 (Iberian Pen.); NOBRE, A., No. 1, 1935 (marine, Portugal); LOZANO REY, L., No. 1, 1935 (fresh-water, España); LOZANO y REY, L., No. 5, 1960 (España); CAVINATO, P. G., 1950, Arch. Oceanogr. Limnol. Roma, 7, 157-212 (revision, Italia); FAGE, L., No. 1, 1918 (Mediterranean); BANARESCU, P., No. 1, 1964 (Romania); SVETOVIDOV, A. N., No. 2, 1964 (Black Sea); BERG, L. S., No. 5, 1949 (fresh-water, Soviet Union, also as Eleotridae); BERG, L. S., No. 2, 1933 (fresh-water, U. S. S. R., also as Eleotridae); BERG, L. S., No. 10, 1965 (fresh-water, U. S. S. R., also as Eleotridae); SPANOVSKAYA, V. D., 1953, Zool. Zhurn, 32, 259-271 (Amur); SOZER, F., 1941, Rev. Fac. Sci. Univ. Istanbul, 6, 128-169 (Turkiye); BERG, L. S., No. 6, 1949 (fresh-water, Iran); BOULENGER, G. A., No. 6, 1916 (fresh-water, Africa); BOULENGER, G. S., No. 2, 1907 (Egypt); PELLEGRIN, J., No. 2, 1923 (western Africa); FOWLER, H. W., No. 8, 1936 (marine, West Africa, also as Eleotridae); DELAIS, M., 1951, Bull. Inst. Francaise Afrique Noire, 13, 343-370 (museum list); DAGET, J., No. 2, 1962 (Guinea, as Eleotridae & Periophthalmidae); SCHULTZ, L. P., No. 1, 1942 (Liberia, also as Eleotridae); DAGET & ILTIS, No. 1, 1965 (Côte d'Ivoire, also as Eleotridae & Periophthalmidae); DAGET, J., No. 1, 1954 (Upper Niger, as Eleotridae); SMITH, J. L. B., 1960, Ichth. Bull. Rhodes Univ., 18, 299-314 (South Africa); SMITH, J. L. B., No. 1, 1961 (marine, South Africa, also as Eleotridae & Periophthalmidae); BARNARD, K. H., 1943, Ann. South African Mus., 36, 101-262 (southwest Cape region); ARNOULT, J., No. 1, 1959 (fresh-water, Madagascar, also as Eleotridae & Periophthalmidae); PELLEGRIN, J., No. 2, 1960 (fresh-water, Madagascar); SMITH, J. L. B., 1958 Ichthy. Bull. Rhodes Univ., 11, 137-163 (western Indian Ocean, as Eleotridae); SMITH, J. L. B., 1959, Ichth. Bull. Rhodes Univ., 13, 185-225 (western Indian Ocean, also as Periophthalmidae); DAY, F., No. 2, 1889 (India); SMITH, H. M., No. 1, 1945 (fresh-water, Thailand, also as Apocrypteidae, Eleotridae, Periophthalmidae); KOUMANS, F. P., No. 10, 1953 (Indo-Australian Arch., also as Eleotridae); WEBER & de BEAUFORT, No. 2, 1913 (Indo-Australian Arch., as Doiichthyidae); INGER & KONG, No. 1, 1962 (fresh-water, North Borneo, also as Eleotridae); McCULLOCH, A. R. & OGILBY J. D., 1919, Rec. Australian Mus., 12, 193-291 (Australian); also as Eleotridae); MARSHALL, T. C., No. 1, 1964 (Great Barrier Reef); WHITLEY, G. P., 1954, Australian Mus. Mag., 11, 150-155 (keys, Australian, as Eleotridae); HEREE, A. W. C. T., 1954, Philippine Journ. Sci., 82, 345-373 (Philippines, as Eleotridae); AURICH, H. J., 1938, Int. Rev. Hydrobiol., 38, 125-183 (East Indies, Philippines, also as Eleotridae); HERRE, A. W., No. 1, 1927 (Philippines & China Sea, also as Eleotridae, Periophthalmidae); OSHIMA, M., No. 1, 1919 (fresh-water, Taiwan); POPE & NICHOLS, No. 1, 1927 (Hainan I., as Eleotridae); TCHANG, T. L., 1939, Bull. Fan Mem. Inst. Biol. Zool., 9, 263-288 (China); NICHOLS, J. T., No. 1, 1943 (fresh-water, China); FOWLER, H. W., 1962, Quart. Journ. Taiwan Mus., 13-15. 1960-62 (China, also as Eleotridae); SOLDATOV & LINDBERG, No. 1, 1930 (Far East seas, also as Eleotridae); SCHMIDT, P. Y., No. 1, 1950 (Sea of Okhotsk); SHMIDT, P. Y., No. 2, 1965 (Sea of Okhotsk); TARANETZ, A. J., 1936, Trav. Inst. Zool. Acad. Sci. U. R. S. S., 4, 483-540 (Japan Sea); JORDAN, D. S. & SNYDER, J. O., 1901, Proc. United States Nat. Mus., 24, 33-132 (Japan); TOMIYAMA, I., 1936, Japan Journ. Zool., 7, 37-112 (Japan); TAKAGI, K., 1953, Journ. Tokyo Univ. Fish., 39, 231-253 (keys, Japan); OKADA, Y., No. 1, 1960 (fresh-water, Japan, also as Eleotridae); OKADA, Y., No. 2, 1960 (fresh-water, Japan, also as Eleotridae); PIETSCHMANN, V., 1938, Bull. Bishop Mus., 156, 3-55 (Hawaii, as Eleotridae); FOWLER, H. W., No. 17, 1959 (Fiji, also as Eleotridae)

Gobiobotidae, as Cyprinidae

Gobioididae (incl. Amblyopidae, Taenioidae, Taenioididae). KOUMANS, F. P., No. 1, 1931 (revision, as Taenioididae); HILDEBRAND, S. F., No. 1, 1946 (shore, Peru); FOWLER, H. W., No. 13, 1954 (fresh-water, Brasil); SMITH, J. L. B., 1959, Ichth. Bull. Rhodes Univ., 13, 185-225 (western Indian Ocean, as Taenioididae); SMITH, H. W., No. 1, 1945 (fresh-water, Thailand, as Gobioididae); KOUMANS, F. P., No. 10, 1953 (Indo-Australian Arch., as Taenioididae); INGER & KONG, No. 1, 1962 (fresh-water, North Borneo, as Taenioididae); HERRE, A. W., No. 1, 1927 (Philippines & China Sea); FOWLER, H. W., 1962, Quart. Journ. Taiwan Mus., 15, 1-77 (China, as TAenioididae)

Gonatodidae (only as fossils)

Gonorynchidae (incl. Gonorhynchidae) FOWLER, H. W., No. 8, 1936 (marine, West Africa); SMITH, J. L. B. No. 1, 1961 (marine, South Africa); MARSHALL, T. C., No. 1, 1964 (Great Barrier Reef); JORDAN & HERRE, No. 1, 1906 (Japan)

Gonorhynchidae, as Gonorynchidae

Gonostomatidae (incl. Gonostomidae, Maurolicidae). GREY, M., 1960, Bull. Mus. Comp. Zool., 122, (2), 57-125 (review); NORMAN. J. R., No. 1, 1930 (oceanic); HILDEBRAND, S. F., No. 1, 1946 (shore, Peru, as Maurolicidae); BIGELOW, H. B. et al., No. 3, 1964 (western North Atlantic) POLL, M., No. 2, 1947 (marine, Belgique, as Maurolicidae); LEGENDRE, R., 1934, Ann. Inst. Oceanogr. Paris, 14, 249-418 (Golfe de Gascogne, genera); LOZANO REY, D. L., No. 4 1947 (España, also as Maurolicidae); ANDRIYASHEV, A. P., No. 1, 1954 (northern seas of U. S. S. R.); FOWLER, H. W., No. 8, 1936 (marine, West Africa); SMITH, J. L. B., No. 1, 1961 (marine, South Africa, also as Maurolicidae); FOWLER, H. W., No. 18, 1956 (Red Sea, also as Maurolicidae); SCHMIDT, P. Y., No. 1, 1950 (Sea of Okhotsk); SHMIDT, P. Y., No. 2, (Sea of Okhotsk); LINDBERG & LEGHEZA, No. 2, 1965 (Sea of Japan); LINDBERG & LEGEZA, No. 3, 1969 (Sea of Okhotsk)

Gonostomidae, as Gonostomatidae

Goodeidae (incl. Characodontidae) (No revisionary references noted)

Grammicolepidae, as Grammicolepididae

Grammicolepididae (incl. Grammicolepidae). MYERS, G. S., 1937, Proc. United States Nat. Mus., 84, 145-156 (deep-sea); SMITH, J. L. B., 1960, Ann. Mag. Nat. Hist., (13, 3, 231-235 (South Africa, key to genera); SMITH, J. L. B., No. 1, 1961 (marine, South Africa)

Grammidae (incl. Stigmatonotidae). (No revisionary references noted)

Grammistidae (incl. Rypticidae). (No revisionary references noted)

Grasseichthyidae, as Kneriidae

Gregoryinidae. (No revisionary references noted)

Groenlandaspididae (only as fossils)

Grystidae, as Centrarchidae

Guerichosteidae (only as fossils)

Gunnellichthyidae, as Microdesmidae

Gymnopogonidae, as Apogonidae

Gymnarchidae. PELLEGRIN, J., No. 2, 1923 (western Africa); DAGET & ILTIS, No. 1, 1965 (Côte d'Ivoire) DAGET, J., No. 1, 1954 (Upper Niger)

Gymnodontidae, as Tetraodontidae

Gymnoniscidae (only as fossils)

Gymnotidae. EIGENMANN. C. H. & WARD, D. P., 1905, Proc. Washington Acad. Sci., 7, 159-188; ELLIS, M. M., 1913, Mem. Carnegie Mus., 6, (3), 109-195 (tropical America); REGAN, C. T., No. 1, 1906-08 (Central America); MEEK & HILDEBRAND, No. 1, 1916 (fresh-water, Mexico); MEEK & HILDEBRAND, No. 8, 1916 (fresh-water, Panama); EIGENMANN, C. H. & ALLEN, W. R., 1942, Fishes of western South America, Lexington, Kentucky, 1-494 (revision); EIGENMANN, C. H. & FISHER, A. G., 1914, Contr. Zool. Lab. Indiana Univ., 141, 235-237 (trans-Andean Colombia & Ecuador); EIGENMANN, C. H., No. 1, 1912 (British Guiana); PUYO, J., No. 1, 1949 (Guyane); FOWLER, H. W., No. 10, 1948-51 (fresh-water, Brasil); DEVINCENZI, G. J., No. 2, 1924 (Uruguay)

Gymnuridae. BIGELOW & SCHROEDER, No. 2, 1953 (western North Atlantic); LINDBERGH & LEGHEZA, No. 1, 1959 (Japan Sea)

Gyracanthidae (only as fossils)

Gyrinocheilidae. SMITH, H. M., No. 1, 1945 (fresh-water, Thailand)

Gyrodontidae (only as fossils)

Gyrolepidotidae (only as fossils)

Habroichthyidae (only as fossils)

Hadrosteidae (only as fossils)

Haemulidae, as Pomadasyidae

Haemulonidae, as Pomadasyidae

Halaeluridae, as Scyliorhinidae

Halidesmidae, as Congrogadidae

Haliophidae, as Congrogadidae

Hapalocarcinidae. FIZE, A & SERENE, R., 1957, Les Hapalocarcinides du Viet-Nam, 1-202.

Halosauridae. LOZANO REY, D. L., No. 4, 1947 (España); FOWLER, H. W., No. 8, 1936 (marine, West Africa); SMITH, J. L. B., No. 1, 1961 (marine, South Africa); FOWLER, H. W., No. 18, 1956 (Red Sea); WEBER & de BEAUFORT, No. 4, 1922 (Indo-Australian Arch.)

Halsydridae. SMITH, J. L. B., No. 1,1961 (marine, South Africa)

Haplochitonidae, as Aplochitonidae

Haplodactylidae, as Aplodactylidae

Harpadontidae (incl. Harpodontidae). FOWLER, H. W., No. 18, 1956 (Red Sea); MARSHALL, T. C., No. 1, 1964 (Great Barrier Reef)

Harpagiferidae, as Nototheniidae

Harpidae, as Labridae

Harpodontidae, as Harpadontidae

Harpuridae, as Acanthuridae

Harriottidae. FOWLER, H. W., No. 18, 1956 (Red Sea)

Helodontidae (only as fossils)

Helogeneidae (incl. Hologenidae). VAN DER STIGCHEL, J. W. B., No. 1, 1946 (South America); EIGENMANN, C. H., No. 1, 1912 (British Guiana); FOWLER, H. W., No. 13, 1954 (fresh-water, Brasil)

Helogenidae. (No revisionary references noted)

Helostomatidae (incl. Helostomidae). (No revisionary references noted)

Helostomidae, as Helostomatidae

Hemerocoetidae, as Percophilididae

Hemiodidae, as Hemiodontidae

Hemiodontidae (incl. Bivibranchiidae, Hemiodidae). (No revisionary references noted)

Hemiramphidae, as Exocoetidae

Hemirhamphidae, as Exocoetidae

Hemiscylliidae, as Orectolobidae

Hemitripteridae, as Cottidae
Henichthyidae, as Apogonidae
Henicichthyidae, as Apogonidae
Hepatidae, as Acanthuridae
Hepsetidae (incl. Ctenoleucidae part). (No revisionary references noted)
Heptatretidae, as Bdellostomatidae
Heptranchidae, as Hexanchidae
Heptranchiidae, as Hexanchidae
Heterenchelidae, as Heterenchelyidae
Heterenchelyidae (incl. Heterenchelidae). BEN-TUVIA, A., 1956, Ann. Mag. Nat. Hist., (12), 9, 401-4081 FOWLER, H. W., No. 8, 1936 (marine, West Africa)
Heterocongridae, as Congridae
Heterodontidae (incl. Centraciontidae, Cestraciontidae). REGAN, C. T., 1908, Ann. Mag. Nat. Hist., (8), 1, 493-497 (synopsis, as Cestraciontidae); GARMAN, S., No. 1, 1913 (monograph, as Centraciontidae &Cestracionidae); MEEK & HILDEBRAND, No. 2, 1923 (marine, Panama, as Cestraciontidae); HILDEBRAND, S. F., No. 1, 1946 (shore, Peru); SMITH, J. L. B., No. 1, 1961 (marine, South Africa); MARSHALL, T. C., No. 1, 1964 (Great Barrier Reef); FOWLER, H. W., No. 2, 1930 (China); LINDBERGH & LEGHEZA, No. 1, 1959 (Japan Sea); JORDAN, D. S. &FOWLER, H. W., 1903, Proc. United States Nat. Mus., 26, 593-674 (Japan)
Heteromyridae, as Muraenidae
Heteropneustidae (incl. Saccobranchidae). SMITH, H. M., No. 1, 1945 (fresh-water, Thailand)
Heterostiidae (only as fossils)
Heterotidae, as Osteoglossidae
Hexagrammidae (incl. Chiridae, Ophiodontidae, Oxylebiidae). RUTENBERG, E. P., 1962, Trud. Inst. Okeanol., 58, 1-100 (classification); ANDRIYASHEV, A. P., No. 1, 1954 (northern seas of U. S. S. R.); SOLDATOV & LINDBERG, No. 1, 1930 (Far East seas); SCHMIDT, P. Y., No. 1, 1950 (Sea of Okhotsk); SHMIDT, P. Y., No. 2, 1965 (Sea of Okhotsk); JORDAN, D. S. & STARKS, E. C., 1903, Proc. United States Nat. Mus., 26, 1003-1013 (Japan)
Hexanchidae (incl. Heptranchidae, Heptranchiidae, Hexeptranchidae, Notidanidae). GARMAN, S., No. 1, 1913 (monograph, as Hexeptranchidae); DEVINCENZI, G. J., No. 1, 1920 (Uruguay, as Hexeptranchidae & Notidanidae); BIGELOW & SCHROEDER, No. 1, 1948 (western North Atlantic); POLL, M., No. 2, 1947 (marine, Belgique); NOBRE, A., No. 1, 1935 (marine, Portugal, as Notidanidae); TORTONESE, E., No. 3, 1956 (Italia); FOWLER, H. W. No. 8, 1936 (marine, West Africa, as Heptranchiidae); SMITH, J. L. B., No. 1, 1961 (marine, South Africa, as Heptranchidae); DAY, F., No. 1, 1889 (India, as Notidanidae); FOWLER, H. W., No. 2, 1930 (China); LINDBERGH & LEGHEZA, No. 1, 1959 (Japan Sea); JORDAN, D. S. & FOWLER, H. W., 1903, Proc. United States Nat. Mus., 26, 593-674 (Japan)
Hexeptranchidae, as Hexanchidae
Himantolophidae. BERTELSEN, E., No. 1, 1951 (revision); NORMAN, J. R., No. 1, 1930 (oceanic); LOZANO y REY, L., NO. 5, 1960 (España); FOWLER, H. W., No. 8, 1936 (marine, West Africa) SMITH, J. L. B., No. 1, 1961 (marine, South Africa)
Hiodontidae (incl. Hyodontidae). SLASTENENKO, E. P., No. 1, 1958 (Canada); FOWLER, H. W., No. 1, 1945 (fresh-water, southeastern U. S.)
Hippocampidae, as Syngnathidae
Hippoglossidae, as Pleuronectidae
Histiophoridae, as Istiophoridae
Histiopteridae, as Pentacerotidae
Holconotidae, as Embiotocidae
Holocentridae (incl. Holocenthridae). RIVERO, L. H., 1941, Torreia, 6, 1-7 (synopsis of genera); BEEBE & TEE-VAN, No. 1, 1928 (Hispaniola); MEEK & HILDEBRAND, No. 2, 1923 (marine, Panama); FOWLER, H. W., No. 8, 1936 (marine, West Africa); POLL, M., No. 4, 1954 (coastal South Africa); SMITH, J. L. B., No. 1, 1961 (marine, South Africa); FOWLER, H. W., No. 18, 1956 (Red Sea); WEBER & de BEAUFORT, No. 5, 1929 (Indo-Australian Arch.); MARSHALL, T. C., No. 1, 1964 (Great Barrier Reef); MONTILLA, J. R., 1938, Philippine Journ. Sci., 67, 207-229 (Philippines); YU, M. J., 1963, Biol. Bull. Dep. Biol. Coll. Sci. Tunghai Univ., 19, Ichthyol., Ser. 2, 1-20 (Taiwan); FOWLER, H. W., No. 6, 1933-35 (China); AOYAGI, H., 1941, Zool. Mag. Tokyo, 53, 468-470 (Riu Kiu Is.); LINDBERG & LEGHEZA, No. 2, 1965 (Sea of Japan); LINDBERG & LEGEZA, No. 3, 1969 (Sea of Japan); SCHULTZ, L. P. et al., No. 2, 1953 (Marshall & Marianas Is.); FOWLER, H. W., No. 17, 1959 (Fiji)
Hologenidae, as Helogeneidae
Holonemidae (only as fossils)
Holoptychiidae (only as fossils)
Holosauridae, as Halosauridae
Holuridae (only as fossils)
Homalopteridae (incl. Gastromyzonidae, Gastromyzontidae, Lepidoglanidae). HORA, S. L., 1931, Rec. Indian Mus., 33, 67-69 (classification); HORA, S. L., 1932, Mem. Indian Mus., 12, 263-330 classification); SILAS, E. G., 1953, Rec. Indian Mus., 50, 173-263 (classification, also as Gastromyzonidae); HORA, S. L., 1920, Rec. Indian Mus., 19, 195-215 (Indian); HORA, S. L., 1941, Rec. Indian Mus., 43, 221-232 (peninsular India, key to genera); SMITH, H. W., No. 1, 1945 (fresh-water, Thailand); WEBER & de BEAUFORT, No. 3, 1916 (Indo-Australian Arch.); INGER & KONG, No. 1, 1962 (fresh-water, North Borneo, also as Gastromyzontidae); OSHIMA, M., No. 1, 1919 (fresh-water, Taiwan); FANG, P. W., 1935, Sinensia, 6, 44-97 (China); CHEN, J. T. F. & LIANG, Y. -S., 1949, Quart. Journ. Taiwan Mus., 2, 157-169 (China)

Homostiidae (only as fossils)
Hoplegnathidae, as Oplegnathidae
Hoplichthyidae (incl. Oplichthyidae). SMITH, J. L. B., No. 1, 1961 (marine, South Africa); de BEAUFORT & BRIGGS, No. 11, 1962 (Indo-Australian Arch.); MARSHALL, T. C., No. 1, 1964 (Great Barrier Reef); JORDAN, D. S. & RICHARDSON, R. E., 1908, Proc. United States Nat. Mus., 33, 629-670 (Japan
Hoplopagridae, as Lutjanidae
Hoplopterygidae, as Trachichthyidae
Horaichthyidae. (No revisionary references noted)
Hussakofiidae (only as fossils)
Hybodontidae (only as fossils)
Hydrocynidae. AMARAL CAMPOS, A., 1945, Arqu. Zool. Est. Sao Paulo, 4, 467-484 (Brasil)
Hyodontidae, as Hiodontidae
Hypophthalmidae. EIGENMANN & EIGENMANN, No. 1, 1890 (South America); VAN DER STIGCHEL, J. W. B., No. 1, 1946 (South America); EIGENMANN, C. H., No. 1, 1912 (British Guiana); FOWLER, H. W., No. 13, 1954 (fresh-water, Brasil)
Hypoplectrodidae, as Serranidae
Hypoptychidae. (No revisionary references noted)
Hypostomidae, as Loricariidae
Hypsaeidae, as Amblyopsidae
Hypsocidae, as Amblyopsidae
Hysterocarpidae, as Embiotocidae

Icelidae (incl. Ereuniidae, Marukawichthyidae). (No revisionary references noted)
Ichthyboridae (incl. Icthyoboridae). (No revisionary references noted)
Ichthodectidae (only as fossils)
Icichthyidae, as Centrolophidae
Icosteidae (incl. Acrotidae). REGAN, C. T., 1923, Ann. Mag. Nat. Hist., (9), 11, 610-612..
Ictaluridae (incl. Ameiuridae, Amiuridae). TAYLOR, W. R., 1969, United States Nat. Mus. Bull., 282, 1-315 (analysis of higher groups); SLASTENENKO, E. P., No. 1, 1958 (Canada); HILDEBRAND & SCHROEDER, No. 1, 1928 (Chesapeake Bay, as Ameiuridae); FOWLER, H. W., No. 1, 1945 (southeastern U. S., as Ameiuridae); de BUEN, F., 1958, Invest. Zool. Chile, 4, 146-148 (keys); DUNCKER & LADIGES, No. 1, 1960 (Nordmark, as Ameiuridae); BANARESCU, P., No. 1, 1964 (Romania); BER, L. S., No. 4, 1949 (fresh-water, Soviet Union); BERG, L. S., No. 9, 1964 (fresh-water, U. S. S. R., as Amiuridae)
Icthoboridae, as Ichthyboridae
Idiacanthidae (incl. Stylophthalmidae, Stylophthalmoidae). PARR, A. E., 1927, Bull. Bingham Oceanogr. Coll., 3, (2), 1-123 BIGELOW, H. W., et al., No. 3, 1964 (western North Atlantic); LOZANO REY, D. L., No. 4, 1947 (España); FOWLER, H. W., No. 8, 1936 (marine, West Africa); SMITH, J. L. B., No. 1, 1961 (marine, South Africa, also as Stylophthalmidae)
Igborichthyidae, as Denticipitidae
Ilarchidae, as Ephippidae
Ilyophidae, as Synaphobranchidae
Indostomidae. (No revisionary references noted)
Inermiidae, as Emmelichthyidae
Ipnopidae. ANDERSEN, K. P., 1966, Galathea Rep., 8, 77-90 (numerical classification); NIELSEN, J. G., 1966, Galathea Rep., 8, 49-75 (synopsis); MEAD, G. W., 1966, Mem. Sears Found. Mar. Res., 1, (5), 19-29, 103-189 (western North Atlantic); SMITH, J. L. B., No. 1, 1961 (marine, South Africa)
Irregularaspidae, as Dinaspidae (only as fossils)
Ischnacanthidae (only as fossils)
Isonidae. (No revisionary references noted)
Istiophoridae (incl. Histiophoridae, Makairidae, Tetrapturidae). JORDAN, D. S. & EVERMANN, B. W., 1926, Occ. Pap. California Acad. Sci., 12, 1-113 (review); LAMONTE, F. R., 1955, Bull. American Mus. Nat. Hist., 107, 319-358 (key to genera); LOZANO REY, D. L., No. 3, 1952 (España); FOWLER, H. W., no. 8, 1936 (marine, West Africa); SMITH, J. L. B., No. 1, 1961 (marine, South Africa); SMITH, J. L. B., 1956, Ichth. Bull. Rhodes Univ., 2, 25-34 (key to African genera); DERANIYAGALA, P. E. P., 1951, Spolia Zeylanica, 26, 137-142 (Ceylon); VINELL, G. D., 1956, Journ. Malay Angl. Assoc., 3, 86-92 (key to species); de BEAUFORT & CHAPMAN, No. 9, 1951 (Indo-Australian Arch.); MARSHALL, T. C., No. 1, 1964 (Great Barrier Reef); WHITLEY, G. P., 1955, Australian Mus. Mag., 11, 292-297 (key, Australian); HIRASAKA, K. & NAKAMURA, H., 1947, Bull. Oceanogr. Inst. Taipeh, 3, 11-24 (keys, Taiwan); FOWLER, H. W., No. 7, 1936 (China); SOLDATOV & LINDBERG, No. 1, 1930 (Far East seas); HOWARD, J. K. & UEYANAGI, S., 1963, Occ. Pap. Indo-Pacific Fish. Counc., 63, (12), 1-3 (Japan, common names); NAKAMURA, I. et al., 1968, Kyoto Univ. Spec. Rep., 4, 1-95 (keys, classification, synonymic catalog)
Isuridae, as Lamnidae

Jaekelaspidae, as Arctolepidae (only as fossils)
Jagorinidae (only as fossils)
Janassidae (only as fossils)
Jenynsiidae (incl. Fitzroyidae, Fitzroyiidae). FOWLER, H. W., No. 13, 1954 (fresh-water, Brasil, as Fitzroyididae)

Joleaudichthyidae (only as fossils)
Jordaniidae, as Cottidae
Juvenellidae, as Carangidae

Kasidoroidae, as Kasidoridae
Kasidoridae (incl. Kasidoroidae). (No revisionary references noted)
Katsuwonidae, as Scombridae
Kneriidae (incl. Cromeriidae, Grasseichthyidae). BOULENGER, G. A., No. 3, 1909 (fresh-water, Africa, also as Cromeriidae); PELLEGRIN, J., 1935, Arch. Mus. Hist. Nat. Paris, (6), 12, 461-463 (Africa, as Cromeriidae); BOULENGER, G. A, No. 2, 1907 (Egypt, as Cromeriidae); DAGET, J., No. 1, 1954 (Upper Niger, as Cromeriidae); POLL, M., No. 1, 1946 (Lake Tanganyika); POLL, M., No. 3, 1953 (Lake Tanganyika)
Korsogasteridae. PARR, A. E., 1933, Bull. Bingham Oceanogr. Coll., 3, (6), 1-51 (new family, deep-sea)
Kraemeriidae (incl. Psammichthyidae). KOUMANS, F. P., No. 1, 1932 (revision, as Psammichthyidae); SMITH, J. L. B., 1959, Ichth. Bull. Rhodes Univ., 13, 185-225 (western Indian Ocean); SCHULTZ, L. P. et al., No. 1, 1966 (Marshall & Marianas Is.); ROFEN, R. R., 1958, Nat. Hist. Rennel Islands, 1, 149-218 (keys, Rennell I.)
Kuhlidae, as Kuhliidae
Kuhliidae (incl. Duleidae, Kuhlidae, Nannopercidae, Nannatherinidae). FOWLER, H. W., No. 8, 1936 (marine, West Africa, as Duleidae); SMITH, J. L. B., No. 1, 1961 (marine, South Africa, as Duleidae) WEBER & de BEAUFORT, No. 5, 1929 (Indo-Australian Arch.); MARSHALL, T. C., No. 1, 1964 (Great Barrier Reef); FOWLER, H. W., 1930, United States Nat. Mus. Bull., 100, (10), 1-334 (Philippine seas, as Duleidae); OSHIMA, M., No. 1, 1919 (fresh-water, Taiwan); FOWLER, H. W., No. 9, 1937-41 (China, as Duleidae); LINDBERG & KRASJUKOVA, No. 1, 1969 (Sea of Japan); OKADA, Y., No. 1, 1960 (fresh-water, Japan); OKADA, Y., No. 2, 1960 (fresh-water, Japan); SCHYLTZ, L. P. et al., No. 2, 1953 (Marshall & Marianas Is.); FOWLER, H. W., No. 17, 1959 (Fiji, as Duleidae)
Kurtidae. DAY, F., No. 2, 1889 (India); de BEAUFORT & CHAPMAN, No. 9, 1951 (Indo-Australian Arch.); FOWLER, H. W., No. 7, 1936 (China).
Kyphosidae (incl. Cyphosidae, Girellidae, Parascorpidae, Scorpidae, Scorpididae). HILDEBRAND & SCHROEDER, No. 1, 1928 (Chesapeake Bay); BEEBE & TEE-VAN, No. 1, 1928 (Hispaniola); MEEK & HILDEBRAND, No. 3, 1925 (marine, Panama); HILDEBRAND, S. F., No. 1, 1946 (shore, Peru); BAUCHOT, M. L. & BLANC, M., No. 1, 1961 (eastern tropical Atlantic); BOULENGER, G. A, No. 5, 1915 (fresh-water, Africa, as Scorpididae); PELLEGRIN, J., No. 2, 1923 (western Africa, as Scorpididae); FOWLER, H. W., No. 8, 1936 (marine, West Africa, also as Girellidae); SMITH, J. L. B., No. 1, 1961 (marine, South Africa, also as Parascsorpidae & Scorpidae); PELLEGRIN, J., No. 2, 1960 (fresh-water, Madagascar, as Scorpididae); WEBER & de BEAUFORT, No. 7, 1936 (Indo-Australian Arch.); MARSHALL, T. C., No. 1, 1964 (Great Barrier Reef); HERRE, A. W. & MONTALBAN, H. R., 1927, Philippines); FOWLER, H. W., No. 20, 1933 (Philippine seas, also as Girellidae); FOWLER, H. W., No. 9, 1937-41 (China, as Girellidae); LINDBERG & KRASJUKOVA, No. 1, 1969 (Sea of Japan, also as Girellidae & Scorpididae); SCHULTZ, L. P. et al., No. 2, 1953 (Marshall & Marianas Is.); FOWLER, H. W., No. 17 (Fiji)

Labracoglossidae. LINDBERG & KRASJUKOVA, No. 1, 1969 (Sea of Japan)
Labridae (incl. Bodianidae, coridae, Cyclolabridae, Harpidae, Neolabridae, Pharyngodopilidae, Phyllodontidae). JORDAN, D. S, & HUGHES, E. G., 1886, Proc. United States Nat. Mus., 9, 56-70 (American Julidinae); HILDEBRAND & SCHROEDER, No. 1, 1928 (Chesapeake Bay); BEEBE & TEE-VAN, No. 1, 1928 (Hispaniola, also as Coridae); MEEK & HILDEBRAND, No. 4, 1928 (marine, Panama); HILDEBRAND, S. F., No. 1, 1946 (shore, Peru); DEVINCENZI, G. F., No. 2, 1924 (Uruguay); BAUCHOT, M. L. & BLANC, M., 1961, Atlantide Rep., 6, 43-64 (eastern tropical Atlantic); BAUCHOT & BLANC, No. 1, 1961 (eastern tropical Atlantic); SMITT, F. A. et al., No. 1, 1892 (Scandinavia); DUNCKER & LADIGES, No. 1, 1960 (Nordmark); POLL, M., No. 2, 1947 (marine, Belgique); NOBRE, A., No. 1, 1935 (marine, Portugal); LOZANO y REY, D. L., No. 3, 1952 (España); FAGE, L., No. 1, 1918 (Mediterranean); de CAPORIACCO, L., 1921, Atti Soc. Italiana Milan, 60, 49-101 (Mediterranean); BANARESCU, P., No. 1, 1964 (Romania); RHASIS ERAZI, R. A., 1943, Rev. Fac. Sci. Univ. Istanbul, 8B, 141-160 (Bosphorus & Sea of Marmora); SVETOVIDOV, A. M., No. 2, 1964 (Black Sea); FOWLER, H. W., No. 8, 1936 (Marine, West Africa); SMITH, J. L. B., No. 1, 1961 (marine, South Africa); SMITH, J. L. B., 1957, Ichth. Bull. Rhodes Univ., 7, 99-114 (western Indian Ocean); DAY, F., No. 2, 1889 (India); de BEAUFORT, L. F., No. 8, 1940 (Indo-Australian Arch.); MARSHALL, T. C., No. 1, 1964 (Great Barrier Reef); FOWLER, H. W. & BEAN, B. A., 1928, United States Nat. Mus. Bull., 100, (7), 1-525 (Philippine seas); YU, M. -J., 1968, Biol. Bull. Dep. Biol. Tunghai Univ., 30, 1-236 (Taiwan); FOWLER, H. W., No. 15, 1956 (China); LINDBERG & KRASJUKOVA, No. 1, 1969 (Sea of Japan); JORDAN, D. S. & SNYDER, J. O., 1902, Proc. United States Nat. Mus., 24, 595-662 (Japan); MASUDA, T. & TANAKA, K., 1962, Journ. Tokyo Univ. Fish., 48.1-98 (key to larvae of Pacific Coast of Japan); SCHULTZ, L. P. et al., No. 3, 1960 (Marshall & Marianas Is.); FOWLER, H. W., No. 17, 1959 (Fiji)
Labyrinthicidae, as Osphronemidae
Lactariidae. WEBER & de BEAUFORT, No. 6, 1931 (Indo-Australian Arch.); CHAN, W. L., No. 1, 1968 (Hong Kong); FOWLER, H. W., No. 7, 1936 (China)
Laevoceratiidae, as Diceratiidae

Lagocephalidae, as Tetraodontidae
Lamiostomatidae (only as fossils)
Lamnidae (incl. Cetorhinidae, Isuridae, Vulpeculidae). GARMAN, S., No. 1, 1913 (monograph, as Isuridae & Vulpeculidae); HILDEBRAND & SCHROEDER, No. 1, 1928 (Chesapeake Bay); MEEK & HILDEBRAND, No. 2, 1923 (marine, Panama, as Isuridae & Vulpeculidae); HILDEBRAND, S. F., No. 1, 1946 (shore, Peru, as Isuridae & Cetorhinidae); NORMAN, J. R., No. 2, 1937 (marine, Patagonia); de BUEN, F., 1958, Invest. Zool. Chile, 4, 201-207 (marine, Chile); BIGELOW & S, No. 1, 1948 (western North Atlantic, as Isuridae & Cetorhinidae); SMITT, F. A. et al., No. 2, 1895 (Scandinavia); DUNCKER & LADIGES, No. 1, 1960 (Nordmark, as Isuridae); POLL, M., No. 2, 1947 (marine, Belgique, as Isuridae); NOBRE, A., No. 1, 1935 (marine, Portugal); TORTONESE, E., No. 3, 1956 (Italia, as Cetorhinidae & Isuridae); BERG, L. S., No. 7, 1911 (Russia); ANDRIYASHEV, A. P., No. 1, 1954 (northern seas of U. S. S. R.); FOWLER, H. W., No. 8, 1936 (marine, West Africa, as Isuridae); SMITH, J. L. B., No. 1, 1961 (marine, South Africa, as Isuridae); FOWLER, H. W., No. 18, 1956 (Red Sea, as Isuridae); GOHAR, H. A. F. & MAZHAR, F. M., No. 1, 1964 (Red Sea, as Isuridae); DAY, F., No. 1, 1889 (India); MARSHALL, T. C., No. 1, 1964 (Great Barrier Reef, as Isuridae); SOLDATOV & LINDBERG, No. 1, 1930 (Far East seas, also as Cetorhinidae); JORDAN, D. S. & FOWLER, H. W., 1903, Proc. United States Nat. Mus., 26, 593-674 (Japan, also as Cetorhinidae); LINDBERGH & LEGHEZA, No. 1, 1959 (Japan Sea, also as Cetorhinidae); SCHULTZ, L. P. et al., No. 2, 1953 (Marshall & Marianas Is., as Isuridae)
Lampridae (incl. Lamprididae); SMITT, F. A. et al., No. 1, 1892 (Scandinavia); POLL, M., No. 2, 1947 (marine, Belgique); NOBRE, A., No. 2, 1935 (marine, Portugal); LOZANO REY, D. L., No. 4, 1947 (España); ANDRIYASHEV, A. P., No. 1, 1954 (northern seas of U. S. S. R.); FOWLER, H. W. No. 8, 1936 (marine, West Africa); SMITH, J. L. B., No. 1, 1961 (marine, South Africa); FOWLER, H. W., No. 6, 1033-35 (China); LINDBERG & LEGHEZA, No. 2, 1965 (Sea of Japan); LINDBERG & LEGEZA, No. 3, 1969 (Sea of Japan)
Lamprididae, as Lampridae
Lasaniidae (only as fossils)
Latidae, as Centropomidae
Latilidae, as Branchiostegidae
Latimeriidae. SMITH, J. L. B., No. 1, 1961 (marine, South Africa)
Latridae (incl. Latrididae). (No revisionary references noted)
Latrididae, as Latridae
Laugiidae (only as fossils)
Lawniidae. (No revisionary references noted)
Lebiasinidae (incl. Nannostomidae). HOEDEMAN, J. J., 1954, Beaufortia, 4, 81-89 (keys, Suriname)
Leiognathidae (incl. Equulidae, Liognathidae). REGAN, C. T., No. 1, 1906-08 (Central America); BAUCHOT & BLANC, No. 1, 1961 (eastern tropical Atlantic); SMITH, J. L. B., No. 1, 1961 (marine, South Africa); WEBER & de BEAUFORT, No. 6, 1931 (Indo-Australian Arch.); WHITLEY, G. P., 1932, Mem. Queensland Mus., 10, 99-116; MARSHALL, T. C., No. 1, 1964 (Great Barrier Reef); FOWLER, H. W., No. 7, 1936 (China); LINDBERG & KRASJUKOVA, No. 1, 1969 (Sea of Japan); SCHULTZ, L. P. et al.; No. 2, 1953 (Marshall & Marianas Is.); FOWLER, H. W., No. 17, 1959 (Fiji)
Lemnisomidae, as Gempylidae
Lepidoglanidae, as Homalopteridae
Lepidopidae, as Trichiuridae
Lepidosirenidae. HOLLY, M., 1933, Das Tierreich, 61, 1-20; FOWLER, H. W., No. 10, 1948-51 (fresh-water, Brasil); BOULENGER, G. A., No. 3, 1909 (fresh-water, Africa); BOULENGER, G. A., No. 2, 1907 (Egypt); PELLEGRIN, J., No. 2, 1923 (western Africa); DAGET, J., No. 2, 1962 (Guinea); DAGET & ILTIS, No. 1, 1965 (Côte d'Ivoire); BOULENGER, G. A., No. 1, 1901 (Congo basin); POLL, M., No. 1, 1946 (Lake Tanganyika); POLL, M., No. 3, 1953 (Lake Tanganyika)
Lepidosteidae. HOLLY, M., 1936, Das Tierreich, 67, 1-65; SLASTENENKO, E. P., No. 1, 1958 (Canada); HILDEBRAND & SCHROEDER, No. 1, 1928 (Chesapeake Bay); FOWLER, H. W., No. 1, 1945 (southeastern U. S.); REGAN, C. T., No. 1, 1906-08 (Central America); MEEK, S. E., No. 1, 1904 (fresh-water, Mexico); BIGELOW, H. B., et al., No. 5, 1963 (western North Atlantic); HERRE, A. W., 1928, Philippine Journ. Sci., 36, 215-232 (Philippines)
Lepidotidae, as Bramidae
Lepisosteidae, as Lepidosteidae
Lepodidae, as Bramidae
Leptobramidae, as Pempheridae
Leptobranchidae. (No revisionary references noted)
Leptocephalidae, as Congridae
Leptolepidae (only as fossils)
Leptoscopidae. TAKI, I., 1953, Journ. Sci. Hiroshima Univ., Zool., 14, 201-212 (key to Japanese genera)
Leptosteidae (only as fossils)
Lethrinidae (incl. Monotaxidae, Neolethrinidae). AKAZAKI, M., 1962, Studies on spariform fishes, Osaka, 1-368 (keys, in Japanese); SMITH, J. L. B., No. 1, 1961 (marine, South Africa); SMITH, J. L. B, Ichthyl. Bull. Rhodes Univ., 17, 285-295 (western Indian Ocean); WHEELER, J. F. G., 1961, Fish. Publ., 15, 1-51 (keys to East African, & western Indian Ocean); FOWLER, H. W., No. 20, 1933 (Philippine seas); FOWLER, H. W., No. 9, 1937-41 (China); LINDBERG & KRASJUKOVA, No. 1, 1969 (Sea of Japan); FOWLER, H. W., No. 17, 1959, (Fiji)
Limnichthidae, as Limnichthyidae

Limnichthyidae (incl. Limnichthidae). (No revisionary references noted)
Linophrynidae (incl. Aceratiidae, Photocorynidae). BERTELSEN, E., No. 1, 1951 (revision); NORMAN, J. R., No. 1, 1930 (oceanic, also as Aceratiidae); FOWLER, H. W., No. 8, 1936 (marine, West Africa, as Aceratiidae); de BEAUFORT & BRIGGS, No. 11, 1962 (Indo-Australian Arch.)
Liodesmidae, as Amiidae
Liognathidae, as Leiognathidae
Liparidae, as Cyclopteridae
Lipariidae, as Cyclopteridae
Liparopidae, as Cyclopteridae
Lipogenidae, as Lipogenyidae
Lipogenyidae (incl. Lipogenidae). (No revisionary references noted)
Lobotidae. HILDEBRAND & SCHROEDER, No. 1, 1928 (Chesapeake Bay); BEEBE & TEE-VAN, No. 1, 1928 (Hispaniola); MEEK & HILDEBRAND, No. 3, 1925 (marine, Panama); DEVINCENZI, G. J., No. 2, 1924 (Uruguay); BAUCHOT & BLANC, No. 1, 1961 (eastern tropical Atlantic); FOWLER, H. W., No. 8, 1936 (marine, West Africa); DAGET & ILTIS, No. 1, 1965 (Côte d'Ivoire); SMITH, J. L. B., No. 1, 1961 (marine, South Africa); SMITH, H. M., No. 1, 1945 (fresh-water, Thailand); WEBER & de BEAUFORT, No. 7, 1936 (Indo-Australian Arch.); MARSHALL, T. C., No. 1, 1964 (Great Barrier Reef); FOWLER, H. W., No. 19, 1931 (Philippine seas); CHAN, W. L., No. 1, 1968 (Hong Kong); FOWLER, H. W., No. 9, 1937-41 (China); LINDBERG & KRASJUKOVA, No. 1, 1969 (Sea of Japan); JORDAN, D. S. & THOMPSON, W. F., 1911, Proc. United States Nat. Mus., 39, 435-471 (Japan)
Lophiidae. HILDEBRAND & SCHROEDER, No. 1, 1928 (Chesapeake Bay); HILDEBRAND, S. F., No. 1, 1946 (shore, Peru); PUYO, J., No. 1, 1949 (Guyane); DEVINCENZI, G. F., No. 2, 1924 (Uruguay); SMITT, F. A. et al., No. 1, 1892 (Scandinavia); DUNCKER & LADIGES, No. 1, 1960 (Nordmark); POLL, M., No. 2, 1947 (marine, Belgique); NOBRE, A., No. 1, 1935 (marine, Portugal); LOZANO y REY, L., No. 5, 1960 (España); SVETOVIDOV, A. N., No. 2, 1964 (Black Sea); ANDRIYASHEV, A. P., No. 1, 1954 (northern seas of U. S. S. R.); FOWLER, H. W. No. 8, 1936 (marine, West Africa); SMITH, J. L. B., No. 1, 1961 (marine, South Africa); de BEAUFORT & BRIGGS, No. 11, 1962 (Indo-Australian Arch); MARSHALL, T. C., No. 1, 1964 (Great Barrier Reef); SOLDATOV & LINDBERG, No. 1, 1930 (Far East seas); JORDAN, D. S., 1902, Proc. United States Nat. Mus., 24, 361-381 (Japan)
Lophotidae. LOZANO REY, D. L., No. 4, 1947 (España); SMITH, J. L. B., No. 1, 1961 (marine, South Africa) LINDBERG & LEGHEZA, No. 2, 1965 (Sea of Japan); LINDBERG & LEGEZA, No. 3, 1969 (sea of Japan)
Loricariidae (incl. Hypostomidae). REGAN, C. T., 1904, Trans. Zool. Soc. London, 17, 191-350 (monograph) GOSLINE, W. A., 1948, Arqu. Mus. Nac. Rio de Janeiro, 41, 79-134 (classification); REGAN, C. T., No. 1, 1906-08 (Central America); MEEK & HILDEBRAND, No. 1, 916 (fresh-water, Panama); EIGENMANN & EIGENMANN, No. 1, 1890 (South America); VAN DER STIGCHEL, J. W. B., No. 1, 1946 (South America); SCHULTZ, L. P., 1944, Proc. United States Nat. Mus., 94, 173-338 (Venezuela); EIGENMANN, C. H., No. 1, 1912 (fresh-water, British Guiana); PUYO, J., No. 1, 1949 (Guyane); FOWLER, H. W., No. 13, 1954 (fresh-water, Brasil); DEVINCENZI, G. J., No. 2, 1924 (Uruguay)
Luciidae, as Esocidae
Luciocephalidae. WEBER & de BEAUFORT, No. 4, 1922 (Indo-Australian Archipelago)
Luganoiidae (only as fossils)
Lumpenidae, as Stichaeidae
Luthianidae, as Lutjanidae
Lutianidae, as Lutjanidae
Lutjanidae (incl. Aphareidae, Caesiodidae, Caesionidae, Etelidae, Hoplopagridae, Luthianidae, Lutianidae, Verilidae). JORDAN, D. S. & SWAIN, J., 1884, Proc. United States Nat. Mus., 7, 427-474 (American waters); HILDEBRAND & SCHROEDER, No. 1, 1928 (Chesapeake Bay); WALFORD, L. A., No. 1, 1937 (Alaska to Equator); BEEBE & TEE-VAN, No. 1, 1928 (Hispaniola); MEEK & HILDEBRAND, No. 3, 1925 (marine, Panama); HILDEBRAND, No. 1, 1946 (shore, Peru); PUYO, J., No. 1, 1949 (Guyane); ANDERSON, W. D., 1967, Circ. United States Fish.Wildl. Serv., 252, 1-14 (field guide, western Atlantic); BAUCHOT & BLANC, No. 1, 1961 (eastern tropical Atlantic); FOWLER, H. W., No. 8, 1936 (marine, West Africa); DELAIS, M., 1952, Bull. Inst.Francaise l'Afrique Noire, 14, 1214-1227 (Ouest-Africaine, keys); DAGET & ILTIS, No. 1, 1965 (Côte d'Ivoire); SMITH, J. L. B., No. 1, 1961 (marine, South Africa, also as Caesiodidae); POLL, M., No. 4, 1954 (coastal South Africa); WEBER & de BEAUFORT, No. 7, 1936 (Indo-Australian Arch.); INGER & KONG, No. 1, 1962 (fresh-water, North Borneo); MARSHALL, T. C., No. 1, 1964 (Great Barrier Reef); FOWLER, H. W. No. 19, 1931 (Philippine seas); FOWLER, H. W., No. 9, 1937-41 (China); LINDBERG & KRASJUKOVA, No. 1, 1969 (Sea of Japan, also as Caesionidae); JORDAN, D. S., & THOMPSON, W. F., 1911, Proc. United States Nat. Mus., 39, 435-471 (Japan); JORDAN, D. S. & JORDAN, E. K., 1922, Mem. Carnegie Mus., 10, 1902 (revision of Etelinae); SCHULTZ, L. P. et al., No. 2, 1953 (Marshall & Marianas Is.); FOWLER, H. W., No. 17, 1959 (Fiji)
Luvaridae (incl. Dianidae). de BUEN, F., 1957, Invest. Zool. Chilenos, 4, 83-88 (Chile); LOZANO REY, D. L., No. 3, 1952 (España); FOWLER, H. W., No. 8, 1936 (marine, West Africa)
Lycoclupeidae (only as fossils)
Lycodapodidae, as Zoarcidae
Lycodidae, as Zoarcidae

Lyconidae, as Macrouridae
Lycopteridae (only as fossils)

Maccullochellidae, as Serranidae
Macquariidae, as Serranidae
Macristiidae. FOWLER, H. W., No. 8, 1936 (marine, West Africa)
Macrocephenchelidae, as Macrocephenchelyidae
Macrocephenchelyidae (incl. Macrocephenchelidae). (No revisionary references noted)
Macropetalichthyidae (only as fossils)
Macropinnidae, as Opisthoproctidae
Macroramphosidae, as Macrorhamphosidae
Macrorhampasidae, as Macrorhamphosidae
Macrorhamphosidae (incl. Macroramphosidae, Macrorhampasidae, Rhamphosidae). DEVINCENZI, G. L., No. 2, 1924 (Uruguay); NOBRE, A., No. 1, 1935 (marine, Portugal); LOZANO REY, D. L., No. 4, 1947 (España); FAGE, L., No. 1, 1918 (Mediterranean); FOWLER, H. W., No. 8, 1936 (marine, West Africa); SMITH, J. L. B., No. 1, 1961 (marine, South Africa); WEBER & de BEAUFORT, No. 4, 1922 (Indo-Australian Arch.); MARSHALL, T. C., No. 1, 1964 (Great Barrier Reef); FOWLER, H. W., No. 6, 1933-35 (China); LINDBERG & LEGHEZA, No. 2, 1965 (Sea of Japan); LINDBERG & LEGEZA, No. 3, 1969 (Sea of Japan); JORDAN, D. S. & STARKS, E. C., 1902, Proc. United States Nat. Mus., 26, 57-73 (Japan)
Macrosemiidae (only as fossils)
Macrouridae (incl. Coryphaenoididae, Lyconidae, Macrouroididae, Macruridae). NORMAN, J. R., No. 1, 1930 (oceanic); PARR, A. E., 1946, Bull. Bingham Oceanogr. Coll., 10, 1-99 (western North Atlantic & Caribbean); DEVINCENZI, G. F., No. 2, 1924 (Uruguay, as Coryphaenoididae); NORMAN. J. R., No. 2, 1937 (marine, Patagonia); SMITT, F. A. et al., No. 2, 1895 (Scandinavia); NOBRE, A., No. 1, 1935 (marine, Portugal); LOZANO y REY, L., No. 5, 1960 (España, as Coryphaenoididae); MAUL, G. E., 1951, Bol. Mus. Nac. Funchal, 5, (12), 5-55 (museum list); ANDRIYASHEV, A. P., No. 1, 1954 (northern seas of U. S. S. R.); FOWLER, H. W., No. 8, 1936 (marine, West Africa, as Coryphaenidae); IWAMOTO, T., 1970, Stud. Trop. Oceanogra., 4, 316-431 (Gulf of Guinea); SMITH, J. L. B., No. 1, 1961 (marine, South Africa) as Coryphaenoididae); FOWLER, H. W., No. 18, 1956 (Red Sea, as Coryphaenoididae); WEBER & de BEAUFORT, No. 5, 1929 (Indo-Australian Arch., as Coryphaenoididae); GILBERT, C. H. & HUBBS, C. L., 1920, United States Nat. Mus. Bull., 100, (1, 7), 369-588 (Philippines & East Indies, as Coryphaenoididae); SOLDATOV & LINDBERG, No. 1, 1930 (Far East seas); SCHMIDT, P. Y., No. 1, 1950 (Sea of Okhotsk); SHMIDT, P. Y., No. 2, 1965 (Sea of Okhotsk); LINDBERG & LEGHEZA, No. 2, 1965 (Sea of Japan); LINDBERG & LEGEZA, No. 3, 1969 (Sea of Japan); GILBERT, C. H. & HUBBS, C. L., 1916, Proc. United States Nat. Mus., 51, 135-214 (Japan, synopsis of genera); OKAMURA, O., 1970, Fauna Japonica, Macrourina, Tokyo, 1-216 (Japan, also as Macrouroididae); FOWLER, H. W., No. 17, 1959 (Fiji)
Macrouroididae, as Macrouridae
Macruridae, as Macrouridae
Macrurocyttidae (incl. Zeniontidae). (No revisionary references noted)
Maenidae, as Emmelichthyidae
Makairidae, as Istiophoridae
Malacanthidae, as Branchiostegidae
Malacosteidae. NORMAN. J. R., No. 1, 1930 (oceanic); REGAN, C. T. & TREWAVAS, E., 1930, Dana Oceanogr. Rep., 6, 1-143; BIGELOW, H. B. et al., No. 3, 1964 (western North Atlantic); NOBRE, A., No. 1, 1935 (marine, Portugal); LOZANO REY, D. L., No. 4, 1947 (España); SMITH, J. L. B., No. 1, 1961 (marine, South Africa); FOWLER, H. W. No. 18, 1956 (Red Sea)
Malapteruridae (incl. Malopteruridae, Torpedinidae part). DAGET, J., No. 2, 1962 (Guinea); DAGET & ILTIS, No. 1, 1965 (Côte d'Ivoire); DAGET, J., No. 1, 1954 (Upper Niger); POLL, M., No. 1, 1946 (Lake Tanganyika); POLL, M., No. 3, 1953 (Lake Tanganyika)
Malopteruridae, as Malapteruridae
Maltheidae, as Ogcocephalidae
Malthidae, as Ogcocephalidae
Mantidae, as Mobulidae
Marukawichthyidae, as Icelidae
Mastacembelidae (incl. Mastocembelidae, Rhynchobdellidae). BERG, L. S., No. 6, 1949 (fresh-water, Iran); BOULENGER, G. A., No. 6, 1916 (fresh-water, Africa); BOULENGER, G. A., No. 2, 1907 (Egypt); PELLEGRIN, J., No. 2, 1923 (western Africa); DAGET, J., No. 2, 1962 (Guinea); SCHULTZ, L. P., No. 1, 1942 (Liberia); DAGET & ILTIS, No. 1, 1965 (Côte d'Ivoire); DAGET, J., No. 1, 1954 (Upper Niger); BOULENGER, G. A., No. 1, 1901 (Congo basin); POLL, M., No. 1, 1946 (Lake Tanganyika); POLL, M., No. 3, 1953 (Lake Tanganyika); DAY, F., No. 2, 1889 (India, as Rhynchobdellidae); SUFI, S. M. K., 1956, Bull. Raffles Mus., 27, 93-146 (Oriental); SMITH, H. M., No. 1, 1945 (fresh-water, Thailand); de BEAUFORT & BRIGGS, No. 11, 1962 (Indo-Australian Arch.); INGER & KONG, No. 1, 1962 (fresh-water, North Borneo); POPE & NICHOLS, No. 1, 1927 (Hainan I.); NICHOLS, J. T., No. 1, 1943 (fresh-water, China)
Mastocembelidae, as Mastacembelidae
Maurobiidae. FOWLER, H. W., No. 8, 1936 (marine, West Africa)
Maurolicidae, as Gonostomatidae
Mediaspidae, as Phlyctaenaspidae (only as fossils)
Medidae, as Cyprinidae
Megalichthyidae, As Rhizodontidae

Megalomycteridae. MYERS, G. S. & FREIHOFER, W. C., 1966, Stanford Ichth. Bull., 8, 193-207 (new family, North Atlantic)
Megalopidae (incl. Elopsidae part). BEEBE & TEE-VAN, NO. 1, 1928 (Hispaniola); FOWLER, H. W., No. 10, 1948-51 (fresh-water, Brasil); LINDBERG & LEGHEZA, No. 2, 1965 (Sea of Japan); LINDBERG & LEGEZA, No. 3, 1969 (Sea of Japan)
Melamphaeidae (incl. Melamphaidae, Melamphasidae). EBELING, W. W., 1962, Dan Rep., 58, 1-164 (key); NORMAN, J. R., No. 1, 1930 (oceanic); LOZANO REY, D. L., No. 2, 1952 (España); FOWLER, H. W. No. 8, 1936 (marine, West Africa); SMITH, J. L. B., No. 1, 1961 (marine, South Africa); POLL, M., No. 4, 1954 (coast of South Africa); FOWLER, H. W., No. 18, 1956 (Red Sea); WEBER & de BEAUFORT, No. 5, 1929 (Indo-Australian Archipelago)
Melamphaidae, as Melamphaeidae
Melamphasidae, as Melamphaeidae
Melanocetidae. BERTELSEN, E., No. 1, 1951 (revision); NORMAN, J. R., No. 1, 1930 (oceanic); LOZANO y REY, L., No. 5, 1960 (España); FOWLER, H. W., No. 8, 1936 (marine, West Africa); SMITH, J. L. B., No. 1, 1961 (marine, South Africa); de BEAUFORT & BRIGGS, No. 11, 1962 (Indo-Australian Archipelago)
Melanocoetidae. (Family not identified)
Melanostomiatidae. PARR, A. E., 1927, Bull. Bingham Oceanogr. Coll., 3, (2), 1-123; BIGELOW, H. B. et al., No. 3, 1964 (western North Atlantic); LOZANO REY, D. L., No. 4, 1947 (España)
Melanotaeniidae (incl. Neoatherinidae, Zanteclidae). (No revisionary references noted)
Menaspidae (only as fossils)
Menidae. SMITH, J. L. B., No. 1, 1961 (marine, South Africa); MARSHALL, T. C., No. 1, 1964 (Great Barrier Reef); CHAN, W. L., No. 1, 1968 (Hong Kong); FOWLER, H. W., No. 7, 1936 (China); LINDBERG & KRASJUKOVA, No. 1, 1969 (Sea of Japan)
Merlucciidae. HILDEBRAND & SCHROEDER, No. 1, 1928 (Chesapeake Bay); HILDEBRAND, S. F., No. 1, 1946 (shore, Peru); DEVINCENZI, G. F., No. 2, 1924 (Uruguay); NORMAN, J. R., No. 2, (marine, Patagonia); DUNCKER & LADIGES, No. 1, 1960 (Nordmark); POLL, M., No. 2, 1947 (marine, Belgique); MAUL, G. E., 1951, Bol. Mus. Nac. Funchal, 5, (12), 5-55 (museum list); FOWLER, H. W., No. 8, 1936 (marine, West Africa); SMITH, J. L. B., No. 1, 1961 (marine, South Africa); FOWLER, H. W., No. 18, 1936 (Red Sea)
Merolepidae, as Emmelichthyidae
Mesacanthidae (only as fossils)
Microbrachiidae (only as fossils)
Microdesmidae (incl. Cerdalidae, Gunnellichthyidae, Paragobioididae, Pholidichthyidae). REID, E. D., 1936, Proc. United States Nat. Mus., 84, 55-72 (revision); DAWSON, C. E., 1968, Yearb. American Philos. Soc., 1967, 256-257; MEEK & HILDEBRAND, No. 4, 1928 (marine, Panama, as Cerdalidae); de BEAUFORT & CHAPMANS, No. 9, 1951 (Indo-Australian Arch., as Pholidichthyidae); SCHULTZ, L. P. et al., No. 1, 1966 (Marshall & Marianas Is.)
Micropteridae, as Centrarchidae
Microstomatidae, as Bathylagidae
Microstomidae, as Bathylagidae
Milyeringidae, as Gobiidae
Mirapinnatidae, as Mirapinnidae
Mirapinnidae (incl. Mirapinnatidae). (No revisionary references noted)
Mitsukurinidae, as Odontaspidae
Mjollnirulidae (only as fossils)
Mobulidae (incl. Cephalopteridae, Mantidae). CADENAT, J., 1958, Notes Africaines, 80, 116-120 (marine); HILDEBRAND & SCHROEDER, No. 1, 1928 (Chesapeake Bay); BEEBE & TEE-VAN, No. 1, 1928 (Hispaniola); MEEK & HILDEBRAND, No. 2, 1923 (marine, Panama); BIGELOW & SCHROEDER, No. 2, 1953 (western North Atlantic); TORTONESE, E., No. 3, 1956 (Italia); CADENAT, J., 1960, Bull. Inst. Francaise Afrique Noire, 22A, 1053-1084 (keys, Côte occidentale d'Afrique); FOWLER, H. W., No. 8, 1936 (marine, West Africa); SMITH, J. L. B., No. 1, 1961 (marine, South Africa); FOWLER, H. W., No. 18, 1956 (Red Sea); GOHAR & MAZHAR, No. 1, 1964 (Red Sea); MARSHALL, T. C., No. 1, 1964 (Great Barrier Reef); CHEN, J. T. F., No. 1, 1948 (China); LINDBERGH & LEGHEZA, No. 1, 1959 (Japan Sea); JORDAN, D. S. & FOWLER, H. W., 1903, Proc. United States Nat. Mus., 26, 593-674 (Japan); SCHULTZ, L. P. et al., No. 2, 1953 (Marshall & Marianas Is.)
Mochochidae, as Mochokidae
Mochocidae, as Mochokidae
Mochockidae, as Mochokidae
Mochokidae (incl. Mochochidae, Mochocidae, Mochockidae, Synodidae). DAGET, J., No. 2, 1962 (Guinea); DAGET & ILTIS, No. 1, 1965 (Côte d'Ivoire); DAGET, J., No. 1, 1954 (Upper Niger); POLL, M., No. 1, 1946 (Lake Tanganyika); POLL, M., No. 3, 1953 (Lake Tanganyika)
Molacanthidae, as Malacanthidae
Molidae (incl. Orthagoriscidae, Triuridae). GILL, T., No. 1, 1884 (synopsis); FRASER-BRUNNER, A., No. 1, 1943 (classification); FRASER-BRUNNER, A., 1951, Bull. British Mus. (Nat. Hist.), Zool., 1, 89-121 (oceanic); DEVINCENZI, G. F., No. 2, 1924 (Uruguay); de BUEN, F., 1957, Invest. Zool. Chile, 4, 66-76 (Chile); SMITT, F. A. et al., No. 2, 1895 (Scandinavia, as Orthagoriscidae); DUNCKER & LADIGES, No. 1, 1960 (Nordmark); POLL, M., No. 2, 1947 (marine, Belgique); NOBRE, A., No. 1, 1935 (marine, Portugal); LOZANO REY, D. L., No. 2, 1952 (España); ANDRIYASHEV, A. P., No. 1, 1954 (northern seas of U. S. S. R.); CADENAT, J., 1959, Bull. Inst. Francaise Afrique Noire, 21A, 1112-1122 (ouest-africaines); FOWLER, H. W., No. 8, 1936 (marine, West Africa); SMITH, J. L. B., No. 1, 1961 (marine, South Africa);

de BEAUFORT & BRIGGS, No. 11, 1962 (Indo-Australian Arch.); MARSHALL, T. C., No. 1, 1964 (Great Barrier Reef); JORDAN, D. S. & SNYDER, J. O., 1901, Proc. United States Nat. Mus., 24, 229-264 (Japan)

Monacanthidae, as Balistidae

Monaspidae (only as fossils)

Monocentridae. SMITH, J. L. B., No. 1, 1961 (marine, South Africa); WEBER & de BEAUFORT, No. 5, 1929 (Indo-Australian Arch.); MARSHALL, T. C., No. 1, 1964 (Great Barrier Reef); FOWLER, H. W., No. 6, 1933-35 (China); LINDBERG & LEGHEZA, No. 2, 1965 (Sea of Japan); LINDBERG & LEGEZA, No. 3, 1969 (Sea of Japan)

Monodactylidae (incl. Psettidae). BAUCHOT & BLANC, No. 1, 1961 (eastern tropical Atlantic); FOWLER, H. W., No. 8, 1936 (marine, West Africa); DAGET & ILTIS, No. 1, 1965 (Côte d'Ivoire); SMITH, J. L. B., No. 1, 1961 (marine, South Africa); POLL, N., No. 4, 1954 (coastal South Africa); ARNOULT, J., No. 1, 1959 (fresh-water, Madagascar); WEBER & de BEAUFORT, No. 7, 1936 (Indo-Australian Arch.); MARSHALL, T. C., No. 1, 1964 (Great Barrier Reef); FOWLER, H. W., No. 12, 1953 (China); FOWLER, H. W., No. 17, 1959 (Fiji)

Monognathidae. (No revisionary references noted)

Monopteridae, as Synbranchidae

Monotaxidae, as Lethrinidae

Moridae (incl. Eretmophoridae, Tripterophycidae). SVETOVIDOV, A. N., No. 1, 1948 (U. S. S. R.); SVETOVIDOV, A., 1962, Fauna U. S. S. R. Fishes IX, 4, Gadiformes, Washington, D. C., 1-304 (U. S. S. R.); WHITLEY, G. P., 1948, Rec. Australian Mus., 22, 70-94 (new family); SCHMIDT, P. Y., No. 1, 1950 (Sea of Okhotsk); SHMIDT, P. Y., No. 2, 1965 (Sea of Okhotsk); LINDBERG & LEGHEZA, No. 2, 1965 (Sea of Japan); LINDBERG & LEGEZA, No. 3, 1969 (Sea of Japan)

Moringuidae (incl. Anguillichthyidae, Ratabouridae, Stilbiscidae). CASTLE, P. H. J., 1968, Spec. Publ. Dep. Ichth. Rhodes Univ., 3, 1-29 (revision); WEBER & de BEAUFORT, No. 3, 1916 (Indo-Australian Arch.); CASTLE, P. H. J., 1965, Trans. Roy. Soc. New Zealand, Zool., 7, 125-133 (Australia); FOWLER, H. W., No. 4, 1931 (China, as Ratabouridae); JORDAN & SNYDER, No. 1, 1901 (Japan); GOSLINE, W. A., 1956, Copeia, 1956, 9-18(Hawaii); SCHULTZ, L. P. et al., No. 2 1953 (Marshall & Marianas Is.); FOWLER, H. W., No. 17, 1959 (Fiji)

Mormyridae. BOULENGER, G. A., No. 3, 1909 (fresh-water, Africa); BOULENGER, G. A., No. 2, 1907 (Egypt); PELLEGRIN, J., No. 2, 1923 (western Africa); DAGET, J., No. 2, 1962 (Guinea); SCHULTZ, L. P., No. 1, 1942 (Liberia); DAGET & ILTIS, No. 1, 1965 (Côte d'Ivoire); DAGET, J., No. 1, 1954 (Upper Niger); BOULENGER, G. A., No. 1, 1901 (Congo basin); POLL, M., No. 1, 1946 (Lake Tanganyika); POLL, M., No. 3, 1953 (Lake Tanganyika)

Moronidae, as Serranidae

Mugilidae. SCHULTZ, L. P., 1946, Proc. United States Nat. Mus., 96, 377-395 (generic revision); THOMSON, J. M., 1964, A bibliography of . . . grey mullets, Melbourne, 1-127; JORDAN, D. S. & SWAIN, J., 1884, Proc. United States Nat. Mus., 7, 261-275 (marine, America); HILDEBRAND & SCHROEDER, No. 1, 1928 (Chesapeake Bay); BEEBE & TEE-VAN, No. 1, 1928 (Hispaniola); REGAN, C. T., No. 1, 1906-08 (Central America); MEEK, S. E., No. 1, 1904 (fresh-water, Mexico); MEEK & HILDEBRAND, No. 1, 1916 (fresh-water, Panama); MEEK & HILDEBRAND, No. 2, 1923 (marine, Panama); HILDEBRAND, S. F., No. 1, 1946 (shore, Peru); EIGENMANN, C. H., No. 1, 1912 (British Guiana); PUYO, J., No. 1, 1949 (Guyane); FOWLER, H. W., No. 13, 1954 (fresh-water, Brasil); DEVINCENZI, G. L., No. 2, 1924 (Uruguay); EIGENMANN, C. H., No. 2, 1927 (fresh-water, Chile); POPOV, A., 1930, Trav. Sta. Biol. Sebastopol, 2, 47-125 (Europe, in Russian); SMITT, F. A. et al., No. 1, 1892, (Scandinavia); DUNCKER & LADIGES, No. 1, 1960 (Nordmark); POLL, M., No. 2, 1947 (marine, Belgique); SPILLMANN, J., No. 1, 1961 (France); NOBRE, A., No. 1, 1935 (marine, Portugal); LOZANO REY., D. L., No. 1, 1935 (España); LOZANO REY, D. L., No. 4, 1947 (España); MOROVIC, D., 1957, Jadranski Mugilidi . . . l'Adriatique, Zagreb, 1-22 (Key, bibliography, Adriatic); BANARESCU, P., No. 1, 1964 (Romania); BORCEA, I., 1934, Ann. Sci. Univ. Jassy, 19, 1-231 (Romanian Black Sea); SVETOVIDOV, A. N., No. 2, 1964 (Black Sea); POPOV, A., 1929, Compt. Rend. Acad. Sci. Leningrad, 1929, 243-247 (Russian); BERG, L. S., No. 2, 1933 (fresh-water, U. R. S. S.); BERG, L. S., No. 5, 1949 (fresh-water, Soviet Union); BERG, L. S., No. 10, 1965 (fresh-water, U. S. S. R.); BERG, L. S., No. 6, 1949 (fresh-water, Iran); BOULENGER, G. A., No. 6, 1916 (fresh-water, Africa); BOULENGER, G. A., No. 2, 1907 (Egypt); PELLEGRIN, J., No. 2, 1923 (western Africa); FOWLER, H. W., No. 8, 1936 (marine, West Africa); CADENAT, J., 1955, Rapp. Cons. Explor. Mer, 137, 59-62 (keys, West Africa); DAGET & ILTIS, No. 1, 1965 (Côte d'Ivoire); BOULENGER, G. A., No. 1, 1901 (Congo basin); SMITH, J. L. B., 1948, Ann. Mag. Nat. Hist., (11), 14, 833-843 (South Africa); SMITH, J. L. B., No. 1, 1961 (marine, South Africa); BARMARD, K. H., 1943, Ann. South Africa Mus., 36, 101-262 (southwestern Cape region); ARNOULT, J., No. 1, 1959 (fresh-water, Madagascar); PELLEGRIN, J., No. 2, 1960 (fresh-water, Madagascar); DAY, F., No. 2, 1889 (India); PILLAY, S. R., 1962, Journ. Bombay Nat. Hist. Soc., 59, 254-270, 547-576 (India); WEBER & de BEAUFORT, No. 4, 1922 (Indo-Australian Arch.); THOMSON, J. M., 1954, Australian Journ. Mar. Freshw. Res., 5, 70-131 (Australia & adjacent seas); THOMSON, J. M., 1964, Tech. Pap. Div. Fish. Australia, 16, 1-127 (Australia); MARSHALL, T. C., No. 1, 1964 (Great Barrier Reef); ROXAS, H. A., 1934, Philippine Journ. Sci., 54, 393-431 (Philippines); OSHIMA, M., No. 1, 1919 (fresh-water, Taiwan); OSHIMA, M., 1921, Ann. Carnegie Mus., 13, 240-259 (Taiwan, also as 1922); FOWLER, H. W., No. 6, 1933-35 (China); SOLDATOV & LINDBERG, No. 1, 1930 (Far East seas); LINDBERG & LEGHEZA, No. 2, 1965 (Sea of Japan); LINDBERG & LEGEZA, No. 3, 1969 (Sea of Japan); OKADA, Y.,

No. 1, 1960 (fresh-water, Japan); OKADA, Y., No. 2, 1960 (fresh-water, Japan); SCHULTZ, L. P. et al., No. 2, 1953 (Marshall & Marianas Is.); FOWLER, H. W., No. 17, 1959 (Fiji); EBELING, A. W., 1961 (keys, eastern Pacific)

Mugiloididae (incl. Parapercichthyidae, Parapercidae, Pinguipedidae). NORMAN, J. R., No. 2, 1937 (marine, Patagonia, as Pinguipedidae); HILDEBRAND, S. F., No. 1, 1946 (shore, Peru); SMITH, J. L. B., No. 1, 1961 (marine, South Africa, as Parapercidae); de BEAUFORT & CHAPMANS, No. 9, 1951 (Indo-Australian Arch., as Pinguipedidae); MARSHALL, T. C., No. 1, 1964 (Great Barrier Reef, as Parapercidae); FOWLER, H. W., No. 15, 1956 (China, as Parapercidae); LINDBERG & KRASJUKOVA, No. 1, 1969 (Sea of Japan); KAMOHARA, T., 1960 Rep. USA Mar. Biol. Sta., 7, (2), 1-14 (Japan, as Parapercidae); SCHULTZ, L. P. et al., No. 3, 1960 (Marshall & Marianas Is.); FOWLER, H. W., No. 17, 1959 (Fiji, as Parapercidae)

Mullidae. BEEBE & TEE-VAN, No. 1, 1928 (Hispaniola); MEEK & HILDEBRAND, No. 2, 1923 (marine, Panama); HILDEBRAND, S. F., No. 1, 1946 (shore, Peru); SMITT, F. A. et al., No. 1, 1892 (Scandinavia); DUNCKER & LADIGES, No. 1, 1960 (Nordmark); POLL, M., No. 2, 1947 (marine, Belgique); NOBRE, A., No. 1, 1935 (marine, Portugal); LOZANO REY, D. L., No. 2, 1952 (España); FAGE, L., No. 1, 1918 (Mediterranean); BANARESCU, P., No. 1, 1964 (Romania); SVETOVIDOV, A. N., No. 2, 1964 (Black Sea); FOWLER, H. W., No. 8, 1936 (marine, West Africa); POLL, M., No. 4, 1854 (Coastal South Africa); SMITH, J. L. B., No. 1, 1961 (marine, South Africa); DAY, F., No. 2, 1889 (India); WEBER & de BEAUFORT, No. 6, 1931 (Indo-Australian Arch.); MARSHALL, T. C., No. 1, 1964 (Great Barrier Reef); MEES, G. F., 1964, Fish. Bull. Western Australia, 9, 31-55 (keys, Western Australia); HERRE, A. W. & MONTALBAN, H. R., 1928, Philippine Journ. Sci., 36, 95-136 (Philippines); FOWLER, H. W. No. 20, 1933 (Philippine seas); FOWLER, H. W., No. 9, 1937-41 (China); LINDBERG & KRASJUKOVA, No. 1, 1969 (Sea of Japan); SNYDER, J. O., 1907, Proc. United States Nat. Mus., 32, 87-102 (Japan); SCHULTZ, L. P. et al., No. 3, 1960 (Marshall & Marianas Is.); FOWLER, H. W., No. 17, 1959 (Fiji)

Muraenesocidae (incl. Muroenesocidae, Sauromuraenesocidae). BEEBE & TEE-VAN, No. 1, 1928 (Hispaniola); MEEK & HILDEBRAND, No. 2, 1923 (marine, Panama); FOWLER, H. W., No. 8, 1936 (marine, West Africa); SMITH, J. L. B., No. 1, 1961 (marine, South Africa); FOWLER, H. W., No. 18, 1956 (Red Sea); SMITH, H. M., No. 1, 1945 (fresh-water, Thailand); MARSHALL, T. C., No. 1, 1964 (Great Barrier Reef); FOWLER, H. W., No. 4, 1931 (China); LINDBERG & LEGHEZA, No. 2, 1965 (Sea of Japan); LINDBERG & LEGEZA, No. 3, 1969 (Sea of Japan); JORDAN & SNYDER, No. 1, 1901 (Japan); FOWLER, H. W., No. 17, 1959 (Fiji)

Muraenichthyidae, as Xenocongridae

Muraenidae (incl. Echidnidae, Heteromyridae). FOWLER, H. W., No. 1, 1945 (fresh-water, southeastern U. S.); BEEBE & TEE-VAN, No. 1, 1928 (Hispaniola); REGAN, C. T., No. 1, 1906-08 (Central America); MEEK & HILDEBRAND, No. 2, 1923 (marine, Panama); HILDEBRAND, S. F., No. 1 1946 (shore, Peru); DEVINCENZI, G. J., No. 2, 1924 (Uruguay); de BUEN, F., 1961, Montemar, 1, 1-90 (Chile); POLL, M., No. 2, 1947 (marine, Belgique); NOBRE, A., No. 1, 1935 (marine, Portugal); LOZANO REY, D. L., No. 4, 1947 (España); LOZANO REY, D. L., No. 4, 1947 (España); FOWLER, H. W., No. 8, 1936 (marine, West Africa, also as Echidnidae); SMITH, J. L. B., No. 1, 1961 (marine, South Africa, also as Echidnidae); FOWLER, H. W, No. 18, 1956 (Red Sea); DAY, F., No. 1, 1889 (India); WEBER & de BEAUFORT, No. 3, 1916 (Indo-Australian Arch.); CASTLE, P. H. J., 1965, Trans. Roy. Soc. New Zealand, Zool., 7, 57-84 (Australia); MARSHALL, T. C., No. 1, 1964 (Great Barrier Reef); FOWLER, H. W., No. 4, 1931 (China, also as Echidnidae); LINDBERG & LEGHEZA, No. 2, 1965 (Sea of Japan); LINDBERG & LEGEZA, No. 3, 1969 (Sea of Japan); JORDAN & SNYDER, No. 1, 1901 (Japan); SCHULTZ, L. P. et al., No. 2, 1953 (Marshall & Marianas Is); FOWLER, H. W., No. 17, 1959 (Fiji, also as Echidnidae)

Muraenolepidae, as Muraenolepididae

Muraenolepididae (incl. Muraenolepidae). NORMAN, J. R., N9. 2, 1937 (marine, Patagonia); SVETOVIDOV, A. N., No. 1, 1948 (U. S. S. R.); SVETOVIDOV, A. N., 1962, Fauna of U. S. S. R., Fishes, IX, 4, Gadiformes, Washington D. C., 1-304 (U. S. S. R.); NORMAN, J. R., No. 3, 1938 (Antarctic)

Muroenesocidae, as Muraenesocidae

Mustelidae, as Carcharhinidae

Myctophidae (incl. Scopelidae). ESTEVE, R., 1947, Bull. Mus. Nat. Hist. Nat., (2), 19, 67-69 (museum type list); FRASER-BRUNNER, A., 1949, Proc. Zool. Soc. London, 118, 1019-1106 (classification); NORMAN, J. R., No. 1, 1930 (oceanic); BOLIN, R. L., 1939, Stanford Ichth. Bull., 1, 89-156 Pacific Coast of U. S. & Baja California); HILDEBRAND, S. F., No. 1, 1946 (shore, Peru); SMITT, F. A. et al., No. 2, 1895 (Scandinavia, as Scopelidae); LEGENDRE, R., 1934, Ann. Inst. Oceanogr. Paris, 14, 249-418 (keys, Atlantic); NOBRE, A., No. 1, 1935 (marine, Portugal, also as Scopelidae); LOZANO REY, D. L., No. 4, 1947 (España); TANING, A. V., 1918, Rep. Danish Oceanogr. Exp. 1908-10, 2, (Biol. A7), 1-154 (genera, Mediterranean, also as Scopelidae); KARLOVAC, J., 1953, Rep. Inst. Oceanogr. Split, 5, (2B), 1-46 (Adriatic Sea, as Scopelidae); ANDRIYASHEV, A. P., No. 1, 1954 (northern seas of U. S. S. R.); FOWLER, H. W., No. 8, 1936 (marine, West Africa); SMITH, J. L. B., No. 1, 1961 (marine, South Africa); FOWLER, HW.. No. 18, 1956 (Red Sea); DAY, F., No. 1, 1889 (India, as Scopelidae); Sarenas, A. M., 1954, Philippine Journ. Sci., 82, 375-427 (Philippines); SCHMIDT, P. Y., No. 1, 1950 (Sea of Okhotsk); SHMIDT, P. Y., No. 2, 1965 (Sea of Okhotsk); LINDBERG & LEGHEZA, No. 2, 1965 (Sea of Japan); LINDBERG & LEGEZA, No. 3, 1969 (Sea of Japan); SCHULTZ, L. P. et al., No. 2, 1953 (Marshall & Marianas Is.); FOWLER, H. W., No. 17, 1959 (Fiji); ANDRIYASHEV, A. P., 1962, Acad. Sci. USSR. Explor. Fauna Seas, 1, 216-294 (keys, Antarctic)

Myliobatidae (incl. Rhinopteridae). HILDEBRAND & SCHROEDER, No. 1, 1928 (Chesapeake Bay, also as Rhinopteridae); BEEBE & TEE-VAN, No. 1, 1928 (Hispaniola); MEEK & HILDEBRAND, No. 2, 1923 (marine, Panama); PUYO, J., No. 1, 1949 (Guyane); DEVINCENZI, G. J., No. 1, 1920 (Uruguay); BIGELOW & SCHROEDER, No. 2, 1953 (western North Atlantic, also as Rhinopteridae); SMITT, F. A. et al., No. 2, 1895 (Scandinavia); POLL, M., No. 2, 1947 (marine, Belgique); NOBRE, A., No. 1, 1935 (marine, Portugal); TORTONESE, E., No. 3, 1956 (Italia); FOWLER, H. W., No. 8, 1936 (marine, West Africa); FOWLER, H. W., No. 18, 1956 (Red Sea); FOWLER, H. W., No. 18, 1956 (Red Sea, as Rhinopteridae); DAY, F., No. 1, 1889 (India); MARSHALL, T. C., No. 1, 1964 (Great Barrier Reef); FOWLER, H. W., No. 2, 1930 (China); CHEN, J. T. F., No. 1, 1948 (China); LINGBERGH & LEGHEZA, No. 1, 1959 (Japan Sea); JORDAN, D. S. & FOWLER, H. W., 1903, Proc. United States Nat. Mus., 26, 593-674 (Japan)

Mylomyridae (only as fossils)

Mylostomidae (only as fossils

Myriacanthidae (only as fossils)

Myridae, part as Xenocongridae, part as Ophichthidae

Myrocongridae. FOWLER, H. W., No. 8, 1936 (marine, West Africa)

Myrophidae, as Ophichthidae

Mystidae, as Bagridae

Myxinidae. HOLLY, M., 1933, Das Tierreich, 59, 1-62; NORMAN, J. R., no. 2, 1937 (marine, Patagonia); de BUEN, F., 1961, Invest. Zool. Chile, 7, 101-124 (Chile); BIGELOW & SCHROEDER, 1953, Mem. Sears Found. Mar. Res., 1, (2), 1-588 (western North Atlantic); SMITT, F. A. et al., No. 2, 1895 (Scandinavia); NOBRE, A., No. 1, 1935 (marine, Portugal); BERG, L. S., No. 7, 1911 (Russia); ANDRIYASHEV, A. P., No. 1, 1954 (northern seas of U. S. S. R.); LINDBERGH & LEGHEZA, No. 1, 1959 (Japan Sea); JORDAN, D. S. & SNYDER, J. O., 1901, Proc. United States Nat. Mus., 23, 725-734 (Japan); NORMAN, J. R., No. 3, 1938 (Antarctic)

Nandidae (incl. Polycentridae, Pristolepidae). EIGENMANN, C. H., No. 1, 1912 (British Guiana, as Polycentridae); FOWLER, H. W., No. 13, 1954 (fresh-water, Brasil, as Polycentridae); BOULENGER, G. A., No. 5, 1915 (fresh-water, Africa); PELLEGRIN, J., No. 2, 1923 (western Africa); DAGET & ILTIS, No. 1, 1965 (Côte d'Ivoire); DAY, F., No. 2, 1889 (India); SMITH, H. M., No. 1, 1945 (fresh-water, Thailand); WEBER & de BEAUFORT, No. 7, 1936 (Indo-Australian Arch.); INGER & KOCH, No. 1, 1962 (fresh-water, North Borneo)

Nannatherinidae, as Kuhliidae

Nannopercidae, as Kuhliidae

Nannostomidae, as Lebiasinidae

Narcaciontidae (incl. Narcationtidae). MEEK & HILDEBRAND, No. 2, 1923 (marine, Panama)

Narcationtidae, as Narcaciontidae

Narcobatidae. JORDAN, D. S. & FOWLER, H. W., 1903, Proc. United States Nat. Mus., 26, 593-674

Narkidae. LINDBERGH & LEGHEZA, No. 1, 1959 (Japan Sea)

Nasidae, as Acanthuridae

Neenchelyidae. FOWLER, H. W., No. 18, 1956 (Red Sea); WEBER & de BEAUFORT, No. 3, 1916 (Indo-Australian Arch.)

Nematistiidae, as Carnagidae

Nematogenyidae, as Trichomycteridae

Nemichthyidae (incl. Avocettinidae, Avocettinopsidae). ROULE, L. & BERTIN, L., 1929, Oceanogr. Rep. Danish Dana Exp. 1920-22, 4, 1-113 (monograph); BERTIN, L., 1942, Bull. Soc. Zool. France, 67, 101-111 (revision, also as Avocettinidae & Avocettinopsidae); NORMAN, J. R., no. 1, 1930 (oceanic); LOZANO REY, D. L., No. 4, 1947 (España); FOWLER, H. W., No. 8, 1936 (marine, West Africa); SMITH, J. L. B., No. 1, 1961 (marine, South Africa); WEBER & de BEAUFORT, No. 3, 1916 (Indo-Australian Arch.); CASTLE, P. H. J., 1965, Trans. Roy. Soc. New Zealand, Zool., 5, 131-146 (Australia)

Nemipteridae (incl. Scolopsidae). SMITH, J. L. B., No. 1, 1961 (marine, South Africa, also as Scolopsidae); LINDBERG & KRASJUKOVA, No. 1, 1969 (Sea of Japan); AKAZAKI, M., 1962, Studies on Spariform fishes . . ., Osaka, 1-368 (key, Japan)

Nemophididae, as Blenniidae

Neoatherinidae, as Melanotaeniidae

Neoceratiidae. BERTELSEN, E., No. 1, 1951 (revision)

Neodacidae, as Neodaciidae

Neodaciidae(incl. Neodacidae, Nesodacidae). (No revisionary references noted)

Neolabridae, as Labridae

Neolethrinidae, as Lethrinidae

Neoodacidae, as Odacidae

Neophrynichthyidae, as Cottidae

Neoscopelidae. SMITH, J. L. B., No. 1, 1961 (marine, South Africa)

Neostethidae. (No revisionary references noted)

Nessorhamphidae. (No revisionary references noted)

Nettastomatidae (incl. Nettastomidae). LOZANO REY, D. L., No. 4, 1947 (España); FOWLER, H. W., No. 8, 1936 (marine, West Africa); SMITH, J. L. B., No. 1, 1961 (marine, South Africa); WEBER & de BEAUFORT, No. 3, 1916 (Indo-Australian Arch.); CASTLE, P. H. J., 1965, Trans. Roy. Soc. New Zealand, Zool., 5, 131-146 (Australia); JORDAN & SNYDER, No. 1, 1901 (Japan)

Nettastomidae, as Nettastomatidae

Niobraridae (only as fossils)

Niphonidae, as Serranidae

Nomeidae (incl. Psenidae). BEEBE & TEE-VAN, No. 1, 1928 (Hispaniola)
Normanichthyidae. (No revisionary references noted)
Notacanthidae. LOZANO REY, D. L., No. 4, 1947 (España); FOWLER, H. W., No. 8, 1936 (marine, West Africa); SMITH, J. L. B., No. 1, 1961 (marine, South Africa); FOWLER, H. W., No. 18, 1956 (Red Sea)
Notidanidae, as Hexanchidae
Notogeneidae. (No revisionary references noted)
Notograptidae (incl. Stichariidae). de BEAUFORT & CHAPMANS, No. 9, 1951 (Indo-Australian Arch.)
Notopteridae. BOULENGER, G. A., No. 3, 1909 (fresh-water, Africa); BOULENGER, G. A, No. 2, 1907 (Egypt); PELLEGRIN, J., No. 2, 1923 (western Africa); DAGET, J., No. 2, 1962 (Guinea); SCHULTZ, L. P., No. 1, 1942 (Liberia); DAGET & ILTIS, No. 1, 1965 (Côte d'Ivoire); DAGET, J., No. 1, 1954 (Upper Niger); BOULENGER, G. A., No. 1, 1901 (Congo basin); DAY, F., No. 1, 1889 (India); SMITH, H. M., No. 1, 1945 (fresh-water, Thailand); WEBER & de BEAUFORT, No. 2, 1913 (Indo-Australian Arch.)
Notosudidae, as Scopelosauridae
Nototheniidae (incl. Gelididae, Harpagiferidae). BLANC, M., & HUREAU, J. C., 1962, Bull. Mus. Nat. Hist. Nat., (2), 34, 339-342 (museum list, also as 1963); NORMAN, J. R., No. 2, 1937 (marine, Patagonia); NORMAN, J. R., No. 3, 1938 (Antarctic, also as Harpagiferidae); DeWITT, H. H., 1962 Copeia, 1962, 826-833 (key to genera)
Novumbridae, as Umbridae
Nybelinidae (only as fossils)

Obrucheviidae (only as fossils)
Odacidae (incl. Neodacidae, Neodaciidae, Neoodacidae). MARSHALL, T. C., No. 1, 1964 (Great Barrier Reef)
Odontaspidae (incl. Carchariidae Garman, Mitsukurinidae). DEVINCENZI, G. J., No. 1, 1920 (Uruguay); BIGELOW & SCHROEDER, No. 1, 1948 (western North Atlantic, as Scapanorhynchidae); JORDAN, D. S. & FOWLER, H. W., 1903, Proc. United States Nat. Mus., 26, 593-674 (Japan)
Odontostomidae, as Evermannillidae
Oeselaspidae (only as fossils)
Ogcocephalidae (incl. Onchocephalidae, Oncocephalidae, Maltheidae, Malthidae). BRADBURY, M. G., 1967 Copeia, 1967, 399-422 (genera); HILDEBRAND & SCHROEDER, No. 1, 1928 (Chesapeake Bay); BEEBE & TEE-VAN, No. 1, 1929 (Hispaniola); MEEK & HILDEBRAND, No. 4, 1928 (marine, Panama); FOWLER, H. W., No. 8, 1936 (marine, West Africa); SMITH, J. L. B., No. 1, 1961 (marine, South Africa); de BEAUFORT & BRIGGS, No. 11, 1962 (Indo-Australian Arch.); MARSHALL, T. C., No. 1, 1964 (Great Barrier Reef); SOLDATOV & LINDBERG, No. 1, 1930 (Far East seas); JORDAN, D. S., 1902, Proc. United States Nat. Mus., 24, 361-381 (Japan)
Olbiaspididae (only as fossils)
Oligopleuridae (only as fossils)
Oligoridae, as Serranidae
Olyridae. (No revisionary references noted)
Omosudidae. BIGELOW, H. B. et al., No. 4, 1966 (western North Atlantic); LOZANO REY, D. L., No. 4, 1947 (España); FOWLER, H. W., No. 8, 1936 (marine, West Africa); SMITH, J. L. B., No. 1, 1961 (marine, South Africa)
Onchocephalidae, as Ogcocephalidae
Oncocephalidae, as Ogcocephalidae
Oneirodidae. BERTELSEN, E., No. 1, 1951 (revision); NORMAN, J. R., No. 1, 1930 (oceanic); SMITH, J. L. B., No. 1, 1961 (marine, South Africa); de BEAUFORT & BRIGGS, No. 11, 1962 (Indo-Australian Archipelago)
Onychodontiae. (No revisionary references noted)
Ophicephalidae, as Channidae
Ophichthidae (incl. Echelidae part, Cephichthyidae, Myridae part, Myrophidae). GINSBURG, I., 1951, Texas Journ. Sci., 3, 431-484 (Gulf Coast of U. S.); BEEBE & TEE-VAN, NO. 1, 1928 (Hispaniola); MEEK & HILDEBRAND, No. 2, 1923 (marine, Panama); HILDEBRAND, S. F., No. 1, 1946 (shore, Peru); NORMAN, J. R., No. 2, 1937 (marine, Patagonia); LOZANO REY, D. L., No. 4, 1947 (España); TORTONESE, E., 1960, Ann. Mus. Civ. Stor. Nat. Genova, 71, 233-247 (Mediterranean); FOWLER, H. W., No. 8, 1936 (marine, West Africa); DAGET & ILTIS, No. 1, 1965 (Côte d'Ivoire); SMITH, J. L. B., No. 1, 1961 (marine, South Africa); FOWLER, H. W., No. 18, 1956 (Red Sea); MENON, A. G. K., 1961, Journ. Zool. Soc. India, 13, 13-15 (keys, India); SMITH, H. M., No. 1, 1945 (fresh-water, Thailand); WEBER & de BEAUFORT, No. 3, 1916 (Indo-Australian Arch.); CASTLE, P. H. J., 1965 (Australia); MARSHALL, T. C., No. 1, 1964 (Great Barrier Reef, also as Myridae); FOWLER, H. W., No. 4, 1931 (China); LINDBERG & LEGHEZA, No. 2, 1965 (Sea of Japan); LINDBERG & LEGEZA, No. 3, 1969 (Sea of Japan); JORDAN & SNYDER, No. 1, 1901 (Japan); GOSLINE, W. A., 1951, Pacific Sci., 5, 298-320 (Hawaii); SCHULTZ, L. P. et al., No. 2, 1953 (Marshall & Marianas Is.); FOWLER, H. W., No. 17, 1959 (Fiji)
Ophichthyidae, as Ophichthidae
Ophiclinidae (incl. Ophioclinidae). (No revisionary references noted)
Ophidiidae (incl. Brotulidae, Brotulophidae, Aphyonidae). NORMAN, J. R., 1939, Sci. Rep. John Murray Exp., 7, 1-116 (synopsis of oceanic genera, as Brotulidae); HILDEBRAND & SCHROEDER, No. 1, 1928 (Chesapeake Bay); BEEBE & TEE-VAN, No. 1, 1928 (Hispaniola); MEEK & HILDEBRAND, No. 4, 1928 (marine, Panama, also as Brotulidae); HILDEBRAND, S. F., No. 1, 1946 (shore, Peru, also as Brotulidae); DEVINCENZI, G. F., No. 2, 1924 (Uruguay); NORMAN J. R., No. 2, 1937, (marine, Patagonia, also as Brotulidae); SMITT, F. A. et al., 1895 (Scandi-

navia); LOZANO y REY, L., No. 5, 1960 (España, also as Brotulidae); FAGE, L., No. 1, 1918 Mediterranean); BANARESCU, P., No. 1, 1964 (Romania); SVETOVIDOV, A. N., No. 2, 1964 (Black Sea); NIELSEN, J. G. & NYBELIN, O., 1963, Atlantide Rep., 7, 195-213, as Brotulidae); FOWLER, H. W., No. 8, 1936 (marine, West Africa, also as Brotulidae); SMITH, J. L. B., No. 1, 1961 (marine, South Africa, also as Brotulidae); DAY, F., No. 2, 1889 (India); de BEAUFORT & CHAPMANS, No. 9, 1951 (Indo-Australian Arch., also as Brotulidae); MARSHALL, T. C., No. 1, 1964 (Great Barrier Reef, as Brotulidae); FOWLER, H. W., No. 16, 1958 (China, as Brotulidae); JORDAN, D. S. & FOWLER, H. W., 1902, Proc. United States Nat. Mus., 25, 743-766 (Japan); KAMOHARA, T., 1954, Rep. USA Mar. Biol. Sta., 1, (2), 1-14 (Japan, as Brotulidae); GOSLINE, W. A., 1953, Copeia, 1953, 215-225 (Hawaii, shallow water, as Brotulidae); SCHULTZ, L. P., et al., No. 3, 1960 (Marshall & Marianas Is., as Brotulidae)

Ophidiontidae. FOWLER, H. W., No. 16, 1958 (China)

Ophiocephalidae, as Channidae

Ophioclinidae, as Ophiclinidae

Ophiodontidae, as Hexagrammidae

Opisthocentridae, as Pholididae

Opisthognathidae. MYERS, G. S., 1935, Smithsonian Misc. Coll., 91, (23), 1-5 (synopsis of genera); BOHLKE, J., 1955, Notul. Nat. Acad. Philadelphia, 281, 1-6 (Bahama Is.); MEEK & HILDEBRAND, No. 4, 1928 (marine, Panama); SMITH, J. L. B., No. 1, 1961 (marine, South Africa); de BEAUFORT & CHAPMANS, No. 9, 1951 (Indo-Australian Arch.); MARSHALL, T. C., No. 1, 1964 (Great Barrier Reef); LINDBERG & KRASJUKOVA, No. 1, 1969 (Sea of Japan);

Opisthomyzonidae (only as fossils)

Opisthoproctidae (incl. Dolichopterygidae, Macropinnidae, Winteridae, Winteriidae). BIGELOW, H. B. et al., No. 3, 1964 (western North Atlantic); LOZANO REY, D. L., No. 4, 1947 (España); FOWLER, H. W., No. 8, 1936 (marine, West Africa)

Oplegnathidae (incl. Hoplegnathidae). HILDEBRAND, S. F., No. 1, 1946 (shore, Peru); SMITH, J. L. B., No. 1, 1961 (marine, South Africa); FOWLER, H. W., No. 20, 1933 (Philippine seas); FOWLER, H. W., No. 9, 1937-41 (China); SOLDATOV & LINDBERG, No. 1, 1930 (Far East seas); LINDBERG & KRASJUKOVA, No. 1, 1969 (Sea of Japan); JORDAN, D. S. & FOWLER, H. W., 1902, Proc. United States Nat. Mus., 25, 75-78 (Japan)

Oplichthyidae, as Hoplichthyidae

Orectolobidae (incl. Ginglystomidae, Hemiscyllidae, Rhineodontidae). REGAN, C. T., 1908, Proc. Zool. Soc. London, 1908, 347-364 (revision); GARMAN. S, No. 1, 1913 (monograph); HILDEBRAND & SCHROEDER, No. 1, 1928 (Chesapeake Bay); BEEBE & TEE-VAN, No. 1, 1928 (Hispaniola); MEEK & HILDEBRAND, No. 2, 1923 (marine, Panama); BIGELOW & SCHROEDER, No. 1, 1948 (western North Atlantic); FOWLER, H. W., No. 8, 1936 (marine, West Africa); SMITH, J. L. B., No. 1, 1961 (marine, South Africa); LOWLER, H. W., No. 18, 1956 (Red Sea); GOHAR & MAZHAR, No. 1, 1964 (Red Sea); MARSHALL, T. C., No. 1, 1964 (Great Barrier Reef); FOWLER, H. W., No. 2, 1930 (China); LINDBERGH & LEGHEZA, No. 1, 1959 (Japan Sea); JORDAN, D. S. & FOWLER, H. W., 1903, Proc. United States Nat. Mus., 26, 593-674) as Hemiscyllidae andl Rhineodontidae); SCHULTZ, L. P. et al., No. 2, 1953 (Marshall and Marianas Is.)

Oreosomatidae. MYERS, G. S., 1940, Stanford Ichth. Bull., 7, 89-98 (key to genera)

Orestidae, as Cyprinodontidae

Orthagoriscidae, as Molidae

Oryziatidae. (No revisionary references noted)

Osmeridae. McALLISTER, D. E., 1963, Bull. Nat. Mus. Canada, 191, 1-53 (revision); SLASTENENKO, E. P., No. 1, 1958 (Canada); HART, J. L. & McHUGH, J. L., 1944, Bull. Fish. Res. Board Canada, 64, 1-27 (British Columbia); BIGELOW, H. B. et al., No. 5, 1963 (western North Atlantic); DUNCKER & LADIGES, No. 1, 1960 (Nordmark); SPILLMANN, J., No. 1, 1961 (France); POLL, M., No. 2, 1947 (marine, Belgique); BERG, L. S., No. 1, 1932 (fresh-water, U. R. S. S.); BERG, L. S., No. 3, 1948 (fresh-water, Soviet Union); BERG, L. S., No. 8, 1962 (fresh-water, U. S. S. R.); ANDRIYASHEV, A. P., No. 1, 1954 (northern seas of U. S. S. R.); SOLDATOV & LINDBERG, No. 1, 1930 (Far East Seas); SCHMIDT, P. Y., No. 1, 1950 (Sea of Okhotsk); SHMIDT, P. Y., No. 2, 1965 (Sea of Okhotsk); TARANETZ, A. J., 1936, Trav. Inst. Zool. Acad. Sci. U. R. S. S., 4, 483-540 (Japan Sea); LINDBERG, & LEGHEZA, No. 2, 1965 (Sea of Japan); LINDBERG & LEGEZA, No. 3, 1969 (Sea of Japan); OKADA, Y., No. 1, 1960 (fresh-water, Japan); OKADA, Y., No. 2, 1960 (fresh-water, Japan); HUBBS, C. L., 1925, Proc. Biol. Soc. Washington, 38, 49-56 (northern Pacific)

Osorioichthyidae (only as fossils)

Osphromenidae, as Osphronemidae

Osphronemidae (incl. Labyrinthicidae, Osphromenidae). BOULENGER, G. A., No. 6, 1916 (fresh-water, Africa); PELLEGRIN, J., No. 2, 1960 (fresh-water, Madagascar); NICHOLS, J. T., No. 1, 1943 (fresh-water, China)

Ospiidae (only as fossils)

Osteoglossidae (incl. Arapaimidae, Clupisudidae, Heterotidae, Phareodidae). EIGENMANN, C. H., No. 1, 1912 (British Guiana, also as Arapaimidae); PUYO, J., No. 1, 1949 (Guyane); FOWLER, H. W., No. 10, 1948-51 (fresh-water, Brasil, also as Arapaimidae); BOULENGER, G. A., No. 3, 1909 (fresh-water, Africa); BOULENGER, G. A., No. 2, 1907 (Egypt); PELLEGRIN, No. 2, 1923 (western Africa); DAGET & ILTIS, No. 1, 1965 (Côte d'Ivoire); DAGET, J., No. 1, 1954 (Upper Niger); SMITH, H. M., No. 1, 1945 (fresh-water, Thailand); WEBER & de BEAUFORT, No. 2, 1913 (Indo-Australian Arch.)

Osteolepidae (only as fossils)

Ostorhinchidae, as Apogonidae
Ostraciidae, as Ostraciontidae
Ostracionidae, as Ostraciontidae
Ostraciontidae (incl. Aracanidae, Ostraciidae, Ostracionidae). GILL, T., No. 1, 1884 (synopsis); FRASER-BRUNNER, A., 1935, Ann. Mag. Nat. Hist., (10), 16, 313-320 (synopsis of genera); FRASER-BRUNNER, A., 1914, Ann. Mag. Nat. Hist., (11), 8, 306-313 (as Aracanidae); LeDANOIS, Y., 1961, Bull. Mus. Nat. Hist. Nat., (2), 33, 176-281 (museum lists, as Aracanidae); LeDANOIS, Y., 1961, BullMus. Nat. Hist. Nat., (2), 33, 276-281 (museum lists); HILDEBRAND & SCHROEDER, No. 1, 1928 (Cheaapeake Bay); BEEBE & TEE-VAN, No. 1, 1928 (Hispaniola); MEEK & HILDEBRAND, No. 4, 1928 (marine, Panama); FOWLER, H. W., No. 8, 1936 (marine, West Africa); SMITH, J. L. B., No. 1, 1961 (marine, South Africa, also as Aracanidae); de BEAUFORT & BRIGGS, No. 11, 1962 (Indo-Australian Arch.); MARSHALL, T. C., No. 1, 1964 (Great Barrier Reef, also as Aracanidae); JORDAN, D. S. & FOWLER, H. W., 1902, Proc. United States Nat. Mus., 25, 251-286 (Japan); SCHULTZ, L. P. et al., NO. 1, 1966 (Marshall & Marianas Is.); FOWLER, H. W., No. 17, 1959 (Fiji)
Ostracoberycidae, as Serranidae
Otodontidae (only as fossils)
Otolithidae, as Sciaenidae
Ovoididae, as Tetraodontidae
Owstoniidae. SMITH, J. L. B., No. 1, 1961 (marine, South Africa)
Oxudercidae. FOWLER, H. W., No. 15, 1956 (China)
Oxygnathidae (only as fossils)
Oxylabracidae, as Centropomidae
Oxylebiidae, as Hexagrammidae
Oxynotidae. BIGELOW, . H. B. & SCHROEDER, W. C., 1957, Bull. Mus. Comp. Zool., 117, 1-150.
Oxyosteidae (only as fossils)
Oxyporhamphidae, as Exocoetidae

Pachycormidae (incl. Microlepidoti). (Only as fossils)
Pachyosteidae, as Selenosteidae (only as fossils)
Pachyrhizodontidae, as Elopidae
Palaeaspidae (only as fossils)
Palaeodontidae (only as fossils)
Palaeoesocidae (only as fossils)
Palaeoniscidae (incl. Acrolepidae, Boreolepidae, Canobiidae, Elonichthyidae, Pygopteridae, Scanilepidae, Styracopteridae, Trissolepidae). (Only as fossils)
Palaeorhynchidae (only as fossils)
Palaeospinacidae (only as fossils)
Palaeospondylidae (only as fossils)
Pampidae, as Stromateidae
Pangasiidae. WEBER & de BEAUFORT, No. 2, 1913 (Indo-Australian Archipelago)
Pantodontiae. BOULENGER, G. A., No. 3, 1909 (fresh-water, Africa); PELLEGRIN, J., No. 2, 1923 (western Africa); BOULENGER, G. A., No. 1, (Congo basin); WHITEHEAD, P. J. P., 1962, Bull. British Mus. (Nat. Hist.), Zool., 9, 103-137 (East Africa)
Paramblypteridae (only as fossils)
Paracentropristidae, as Serranidae
Paraclinidae, as Clinidae
Paradichthyidae, as Sparidae
Paradicichthyidae, as Sparidae
Paragalaxiidae, as Galaxiidae
Paragobioididae, as Microdesmidae
Paralepidae, as Paralepididae
Paralepididae (incl. Paralepidae, Sudidae). NORMAN. J. R., No. 1, 1930 (oceanic, as Sudidae); HARRY, R. R., 1953, Pacific Sci., 7, 219-249 (bathypelagic); EGE, V., 1953, Dana Rep., 40, 1-184; MARSHALL, N. B., 1955, Discovery Rep., 27, 303-336 (deep-sea); BIGELOW, H. B. et al., No. 4, 1966 (western North Atlantic); LOZANO REY, D. L., No. 4, 1947 (España, as Sudidae); TORTONESE, E., 1951, Boll. Pesca Piscic. Idrobiol., 27, 177-186 (subfamily Sudinae, Mediterranean); ANDRIYASHEV, A. P., No. 1, 1954 (northern seas of U. S. S. R., as Sudidae); MAUL, G. E., 1945, Bol. Mus. Funchal, 1, 1-37 (museum list, as Sudidae); FOWLER, H. W., No. 8, 1936 (marine, West Africa, as Sudidae); FOWLER, H. W., No. 18, 1956 (Red Sea, as Sudidae); HARRY, R. R., 1953, Proc. Acad. Nat. Sci. Philadelphia, 105, 169-230 (bathypelagic, northern Pacific)
Paralichthodidae, as Pleuronectidae
Paralichthyidae, as Bothidae
Paramyxinidae. (no revisionary references noted)
Parapercichthyidae, as Mugiloididae
Parapercidae, as Mugiloididae
Paraplesiobatidae (only as fossils)
Paraplesiolepidae (only as fossils)
Parascorpidae, as Kyphosidae
Parasemionotidae (only as fossils)
Paratrygonidae. FOWLER, H. W., No. 10, 1948-51 (fresh-water, Brasil)
Parazenidae. (No revisionary references noted)

Parechelidae (only as fossils)
Parexidae (only as fossils)
Parodontidae. (no revisionary references noted)
Parophiocephalidae, as Channidae
Pataecidae (incl. Gnathacanthidae, Gnathanacanthidae). MARSHALL, T. C., No. 1, 1964 (Great Barrier Reef)
Pegasidae. SMITH, . J. L. B., No. 1, 1961 (marine, South Africa); FOWLER, N. W., No. 18, 1956 (Red Sea); de SILVA, P. H. D. H., 1956, Spol. Zeylanica, 28, 27-34 (museum list & marine of Ceylon); de BEAUFORT, & BRIGGS, No. 11, 1962 (Indo-Australian Arch.); MARSHALL, T. C., No. 1, 1964 (Great Barrier Reef); FOWLER, H. W., No. 6, 1933-35 (China); JORDAN, D. S. & SNYDER, J. O., 1901, Proc. United States Nat. Mus., 24, 1-20 (Japan)
Peipiaosteidae (only as fossils)
Pelamidae. NUMANN, W., 1955, Publ. Hydrobiol. Res. Inst. Istanbul, 3B, 75-127 (Black Sea)
Peltopleuridae. (No revisionary references noted)
Pelycorapidae. (No revisionary references noted)
Pempheridae (incl. Leptobramidae). SMITH, J. L. B., No. 1, 1961 (marine, South Africa); WEBER & de BEAUFORT, No. 7, 1936 (Indo-Australian Arch.);MARSHALL, T. C., No. 18, 1964 (Breat Barrier Reef); FOWLER, H. W., No. 19, 1931 (Philippine seas); FOWLER, H. W., No. 9, 1937-41 (China); LINDBERG & KRASJUKOVA, No. 1, 1969 (Sea of Japan); TOMINAGA, Y., 1965, Japan Journ. Ichth., 12, 33-56 (new family, as Leptobramidae); SCHULTZ, L. P. et al., No. 2, 1953 (Marshall & Marianas Is.)
Pentacerotidae (incl. Histiopteridae). SMITH, J. L. B., 1964, Ichth. Bull. Rhodes Univ., 29, 567-578); SMITH, J. L. B., No. 1, 1961 (marine, South Africa, as Histiopteridae); SMITH, J. L. B., 1964, Ichth. Bull. Rhodes Univ., 29, 567-578 (South Africa); SMITH, J. L. B., 1951, Ann. Mag. Nat. Hist., (12), 4, 873-882 (key to genera of western Indian Ocean, as Histiopteridae); LINDBERG & KRASJUKOVA, No. 1, 1969 (Sea of Japan); JORDAN, D. S., 1907, Proc. United States Nat. Mus., 32, 235-239 (Japan, as Histiopteridae)
Pentanchidae,, as Scyliorhinidae
Pentapodidae. SMITH, J. L. B., No. 1, 1961 (marine, South Africa)
Percichthyidae, as Serranidae
Percidae (incl. Etheostomatidae, Etheostomidae). COLLETTE, B. B., 1963, Copeia, 1963, 615-623 (subfamilies, tribes, genera); BOULENGER, G. A., 1895, Catalogue of the perciform fishes in the British Museum, vol. 1, 2nd ed. (descriptive catalog); COLLETTE, B. B. & KNAPP, L. W., 1966, Proc. United States Nat. Mus., 119, 3550, 1-88 (museum type list); SLASTENENKO, E. P. No. 1, 1958 (fresh-water, Canada); HILDEBRAND & SCHROEDER, No. 1, 1928 (Chesapeake Bay, also as Etheostomidae); FOWLER, H. W., No. 1, 1945 (southeastern U. S.); REGAN, C. T., No. 1, 1906-08 (Central America); MEEK, S. E., No. 1, 1904 (fresh-water, Mexico); SMITT, F. A. et al., NO. 1, 1892 (Scandinavia); DUNCKER & LADIGES, No. 1, 1960 (Nordmark); SPILLMANN, J., No. 1, 1961 (France); Oliva, A., 1953, Vest. Kraloski Ceske Spol. Nauk, 1952, (8), 1-13 (Ceskoslovenska); BANARESCU, P., No. 1, 1964 (Romania); BERG, L. S., No. 2, 1933 (fresh-water, U. R. S. S.); BERG, L. S., No. 5, 1949 (fresh-water, Soviet Union); BERG, L. S., No. 10, 1965 (fresh-water, U. S. S. R.); DAY, F., No. 1, 1889 (India)
Perciliidae, as Serranidae
Percophidae, as Percophididae
Percophididae (incl. Bembropidae, Bembropsidae, Hemerocoetidae, Percophidae, Pteropsaridae). GINSBERG, I., 1955, Proc. United Statees Nat. Mus., 104, 623-639 (coasts of eastern U. S. & West Indies); DEVINCENZI., G. F., No. 2, 1924 (Uruguay); SMITH, J. L. B., No. 1, 1961 (marine, South Africa, as Bembropsidae); de BEAUFORT & CHAPMAN, No. 9, 1951 (Indo-Australian Arch., as Hemerocoetidae)
Percopsidae. REGAN, C. T., 1911, Ann. Mag. Nat. Hist., (8), 8, 294-296 (anatomy & classification); SLASTENENKO, E. P., No. 1, 1958 (Canada); FOWLER, H. W., No. 1, 1945 (southeastern U. S.)
Periophthalmidae, as Gobiidae
Peripristidae, as Petalodontidae
Peristedidae, as Peristediidae
Peristediidae, as Triglidae
Peristediontidae, as Triglidae
Perleididae (incl. Colodontidae). (No revisionary references noted)
Peronedyidae, as Peronedysidae
Peronedysidae (incl. Peronedyidae). (No revisionary references noted)
Petalodontidae (incl. Peripristidae, Pristodontidae). (Only as fossils)
Petromyzonidae (incl. Geotriidae, Petromyzontidae, Ptetromyzantidae). REGAN, C. T., 1911, Ann. Mag. Nat. Hist., (8), 7, 193-204 (synopsis); HOLLY, M., 1933, Das Tierreich, 59, 1-62; BERG, L. S., 1931, Ann. Mus. Zool. Acad. Leningrad, 32, 87-116 (Northern Hemisphere); CREASER, C. W. & HUBBS, C. L., 1922, Occ. Pap. Mus. Zool. Univ. Michigan, 120, 1-14 (Holarctic); VLADYKOV, W. D. & FOLLETT, W. I., 1967, Journ. Fish. Res. Board Canada, 24, 1067-1075 (key to Holarctic genera); SLASTENENKO, E. P., No. 1, 1958 (Canada); HUNTSMAN, A. G., 1917, Ottawa, Nat., 31, 12-17 (eastern Canada); HILDEBRAND & SCHROEDER, No. 1, 1928 (Chesapeake Bay); FOWLER, H. W., No. 1, 1945 (southeastern U. S.); REGAN, C. T., No. 1, 1906-08 (Central America); MEEK, S. E., No. 1, 1904 (fresh-water, Mexico); EIGENMANN, C. H., No. 2, 1927 (fresh-water, Chile); as Geotriidae); NORMAN, J. R., No. 2, 1937 (marine, Patagonia); BIGELOW & SCHROEDER, 1953, Mem. Sears Found. Mar. Res., 1, (2), 1-588 (western North Atlantic); VLADYKOV, V. D., 1959, Proc. Inst. Congr. Zool., 15, 207-208 (ammocoetes larvae of Europe); SMITT, F. A. et al., No. 2, 1895 (Scandinavia); MacDONALD,

T. H., 1959, Glasgow Nat., 18, 91-95 (British); DUNCKER & LADIGES, No. 1, 1960 (Nordmark); POLL, M., No. 2, 1947 (marine, Belgique); NOBRE, A., No. 1, 1935 (marine, Portugal) LOZANO REY, D. L., No. 1, 1935 (España); TORTONESE, E., No. 3, 1956 (Italia); ZANANDREA, G., 1957, Arch. Zool. Napoli, 42, 249-305 (Italia); OLIVA, O., 1953, Vest. Kraloskí Ceske Spol. Naul, 1952, (9), 1-19 (Ceskoslovenska); ZANANDREA, G., 1958, Biol. Glasn., 11, 45-54 (Zagabria e Lubiana); BERG, L. S., No. 7, 1911 (Russia); BERG, L. S., No. 1, 1932 (fresh-water, U. R. S. S.); BERG, L. S., No. 2, 1948 (fresh-water, Soviet Union); BERG, L. S. No. 8, 1962 (fresh-water, U. S. S. R.); ANDRIYASHEV, A. P., No. 1, 1954 (northern seas of U. S. S. R.); FOWLER, H. W., No. 8, 1936 (marine, West Africa); WHITLEY, G. P., 1932, Australian Zool., 7, 256-264 (Australia, as Geotriidae); SOLDATOV & LINDBERG, No. 1, 1930 (Far East seas); SCHMIDT, P. Y., No. 1, 1950 (Sea of Okhotsk); SHMIDT, P. Y., No. 2, 1965 (Sea of Okhotsk); LINDBERGH & LEGHEZA, No. 1, 1959 (Japan Sea); JORDAN, D. S. & SNYDER, J. O., 1901, Proc. United States Nat. Mus., 23, 725-734 (Japan); OKADA, Y., No. 1, 1960 (fresh-water, Japan); OKADA, Y., No. 2, 1960 (fresh-water, Japan)

Petromyzontidae (as Petromyzonidae)

Phallostethidae. MYERS, G. S., 1937, Proc. United States Nat. Mus., 84, 137-143 (key to genera); SMITH, H. M., No. 1, 1945 (fresh-water, Thailand); WEBER & de BEAUFORT, No. 4, 1922 (Indo-Australian Arch.); VILLADOLID, D. & MANACOP, P., 1934, Philippine Journ. Sci., 55, 193-220 (Philippines); HERRE, A. W. C. T., 1942, Stanford Ichth. Bull., 2, 137-156 (key to genera, Philippines)

Phaneropleuridae (only as fossils)

Phanerorhynchidae (only as fossils)

Phareodidae, as Osteoglossidae

Pharopterygidae, as Plesiopidae

Pharyngodopilidae, as Labridae (only as fossils)

Pharyngolepidae (only as fossils)

Phlebolepidae (only as fossils)

Phlyctaenaspidae (incl. Acanthaspidae part, Mediaspidae). (Only as fossils)

Pholidae, as Pholididae

Pholidichthyidae, as Microdesmidae

Pholididae (incl. Opisthocentridae, Pholidae). MAKUSHOK, V. M., No. 1, 1958 (morphology & systematics); DUNCKER & LADIGES, No. 1, 1960 (Nordmark); POLL, M., No. 2, 1947 (marine, Belgique); ANDRIYASHEV, A. P., No. 1, 1954 (northern seas of U. S. S. R.); FOWLER, H. W., No. 16, 1958 (China, also as Opisthocentridae); SCHMIDT, P. Y., No. 1, 1950 (Sea of Okhotsk); SHMIDT, P. Y., No. 2, 1965 (Sea of Okhotsk)

Pholidophoridae (only as fossils)

Pholidopleuridae (only as fossils)

Pholidosteidae, as Coccosteidae (only as fossils)

Photocorynidae, as Linophrynidae

Phractolaemidae. BOULENGER, G. A., No. 3, 1909 (fresh-water, Africa); PELLEGRIN, J., No. 2, 1923 (West Africa)

Phyllodontidae, as Labridae (only as fossils)

Phyllolepidae (only as fossils)

Pimelepteridae, as Sparidae

Pimelodidae (inc. Callophysidae, Pseudopimelodidae). GOSLINE, W. A., 1941, Stanford Ichth. Bull., 2, 83-88 (synopsis); VAN DER STIGCHEL, J. W. B., No. 1, 1946 (South America, also as Callophysidae); SCHULTZ, L. P., No. 2, 1944 (Venezuela, also as Callophysidae); FOWLER, H. W. No. 10, 1948-51 (fresh-water, Brasil); FOWLER, H. W., No. 13, 1954 (fresh-water, Brasil, as Callophysidae)

Pinguipedidae, as Mugiloididae

Plagiodontidae, as Alepisauridae

Planidae, as Pleuronectidae

Platacidae, as Ephippidae

Platycephalidae (incl. Bembradidae, Bembridae, Parabembridae). WHITELEY, G. P., 1931, Rec. Australian Mus., 18, 138-160 (key to subfamilies); FOWLER, H. W., No. 8, 1936 (marine, West Africa); SMITH, J. L. B., No. 1, 1961 (marine, South Africa, also as Bembridae); de BEAUFORT & BRIGGS, No. 11, 1962 (Indo-Australian Arch.); MARSHALL, T. C., No. 1, 1964 (Great Barrier Reef); JORDAN, D. S. & RICHARDSON, R. E., 1908, Proc. United States Nat. Mus., 33, 629-670 (Japan, also as Bembridae); MATSUBARA, K. & OCHIAE, A., 1955, Mem. Coll. Agric. Kyoto Univ., 68, 1-109 (Japan); SCHULTZ, L. P. et al., No. 1, 1966 (Marshall & Marianas Is.)

Platyproctidae, as Alepocephalidae

Platypteridae, as Rhyacichthyidae

Platyrhinidae, as Discobatidae

Platyrhinoididae. FOWLER, H. W., No. 2, 1930 (China)

Platysiagidae (only as fossils)

Platysomidae (only as fossils)

Platytroctidae, as Alepocephalidae

Plecoglossidae. LINDBERG & LEGHEZA, No. 2, 1965 (Sea of Japan); LINDBERG & LEGEZA, No. 3, 1969 (Sea of Japan); OKADA, Y., No. 1, 1960 (fresh-water, Japan); OKADA, Y., No. 2, 1960 (fresh-water, Japan)

Plectorhynchidae, as Pomadasyidae

Plectroplitidae, as Serranidae

Plectropomidae, as Serranidae
Plesiopidae (incl. Pharopterycidae). SMITH, J. L. B., No. 1, 1961 (marine, South Africa); SMITH, J. L. B., 1952, Ann. Mag. Nat. Hist., (12), 5, 139-151 (keys, Africa); WEBER & de BEAUFORT, No. 5, 1929 (Indo-Australian Arch.); MARSHALL, T. C., No. 1, 1964 (Great Barrier Reef); FOWLER, H. W., No. 9, 1937-41 (China)
Plethodidae (incl. Anogmiidae, Plethodontidae, Thryptodontidae). (Only as fossils)
Plethodontidae, as Plethodidae (only as fossils)
Pleuracanthidae, as Xenacanthidae
Pleuronectidae (incl. Hippoglossidae, Paralichthodidae, Planidae, Rhombosoleidae). NORMAN, J. R., 1934, A systematic monograph of the flatfishes. . ., vol. 1, London, 1-459 (monograph); NORMAN, J. R., 1933, Ann. Mag. Nat. Hist., (10), 11, 214-222 (genera of Pleuronectinae); HILDEBRAND & SCHROEDER, No. 1, 1928 (Chesapeake Bay); LOCKINGTON, W. N., 1879, Proc. United States Nat. Mus., 2, 69-108 (San Francisco); REGAN, C. T., No. 1, 1906-08 (Central America); MEEK & HILDEBRAND, No. 4, 1928 (marine, Panama); PUYO, J., No. 1, 1949 (Guyane); FOWLER, H. W., No. 13, 1954 (fresh-water, Brasil); DEVINCENZI, G. F., No. 2, 1924 (Uruguay); SMITT, F. A. et al., No. 1, 1892 (Scandinavia); DUNCKER & LADIGES, No. 1, 1960 (Nordmark); POLL, M., No. 2, 1947 (marine, Belgique); NOBRE, A., No. 1, 1935 (marine, Portugal); LOZANO REY, D. L., No. 1, 1935 (España); LOZANO y REY, L., No. 5, 1960 (España); BANARESCU, P., No. 1, 1964 (Romania); SVETOVIDOV, A. N., No. 2, 1964 (Black Sea); BERG, L. S., No. 2, 1933 (fresh-water, U. R. S. S.); BERG, L. S., No. 5, 1949 (fresh-water, Soviet Union); BERG, L. S., No. 10, 1965 (fresh-water, U. S. S. R.); ANDRIYASHEV, A. P., No. 1, 1954 (northern seas of U. S. S. R.); PELLEGRIN, J., No. 2, 1923 (western Africa); SMITH, J. L. B., No. 1, 1961 (marine, South Africa); FOWLER, H. W., No. 18, 1956 (Red Sea); DAY, F., No. 2, 1889 (India); NORMAN, J. R., 1927, Rec. Indian Mus., 29, 7-48 (India & museum list); de SILVA, P. H. D. H., No. 2, 1956 (Ceylon, also as Samaridae) WEBER & de BEAUFORT, No. 5, 1929 (Indo-Australian Arch.); MARSHALL, T. C., No. 1, 1964 (Great Barrier Reef); CHEN, J. T. F. & WENG, H. T. C., 1965, Biol. Bull. Tunghai Univ., 25 & 27 (Taiwan); WU, H. -W., 1932, Contributions a . . . Heterosomes . . . de la Chine, Paris, 1-179 (China); FOWLER, H. W., No. 6, 1933-35 (China, also as Hippoglossidae); SOLDATOV & LINDBERG, No. 1, 1930 (Far East seas); SCHMIDT, P. Y., No. 1, 1950 (Sea of Okhotsk); SHMIDT, P. Y., No. 2, 1965 (Sea of Okhotsk); JORDAN, D. S. & STARKS, E. C., 1906, Proc. United States Nat. Mus., 31, 161-246 (Japan); OKADA, Y., No. 1, 1960 (fresh-water, Japan); OKADA, Y., No. 2, 1960 (fresh-water, Japan); SCHULTZ, L. P. et al., No. 1, 1966 (Marshall & Marianas Is.)
Pleuroscopidae, as Uranoscopidae
Plotosidae. SMITH, J. L. B., No. 1, 1961 (marine, South Africa); FOWLER, H. W., No. 18, 1956 (Red Sea); SMITH, H. M., No. 1, 1945 (fresh-water, Thailand); WEBER & de BEAUFORT, No. 2, 1913 (Indo-Australian ARch.); MARSHALL, T. C., No. 1, 1964 (Great Barrier Reef); HERRE, A. W. C. T., 1926, Philippine Journ. Sci., 31, 385-411 (Philippines); FOWLER, H. W., No. 5, 1932 (China); LINDBERG & LEGHEZA, No. 2, 1965 (Sea of Japan); LINDBERG & LEGEZA, No. 3, 1969 (Sea of Japan); FOWLER, H. W., No. 17, 1959 (Fiji)
Podatelidae, as Ateleopodidae
Poecilidae, as Poeciliidae
Poeciliidae (incl. Poecilidae, Poecilliidae, Tomeuridae). ROSEN, D. E. & BAILEY, R. M., 1963, Bull. American Mus. Nat. Hist., 126, (1), 1-176; ROSEN, D. E., 1952 (revision of Alfarinae); HILDEBRAND & SCHROEDER, No. 1, 1928 (Chesapeake Bay); FOWLER, H. W., No. 1, 1945 (fresh-water, southeastern U. S.); RIVAS, L. R., 1958, Proc. American Philos. Soc., 102, 281-320 (Girardinini of Cuba); MEEK, S. E., No. 1, 1904 (fresh-water, Mexico); ROSEN, D. E. 1960, Bull. Florida State Mus., Biol. Sci., 5, 57-242 (to genus, Central America); MEEK & HILDEBRAND, No. 1, 1916 (fresh-water, Panama); EIGENMANN, C. H., No. 1, 1912 (British Guiana); FOWLER, H. W., No. 13, 1954 (fresh-water, Brasil); EIGENMANN, C. H., 1907, Proc. United States Nat. Mus., 32, 415-433 (Rio Grande do Sul & La Plata); EIGENMANN, C. H., No. 2, 1927 (fresh-water, Chile); SPILLMANN, J., No. 1, 1961 (France); LOZANO REY, D. L., No. 1, 1935 (España); LOZANO REY, D. L., No. 4, 1947 (España); BANARESCU, P., No. 1, 1964 (Romania); SVETOVIDOV, A. N., No. 2, 1964 (Black Sea); BERG, L. S., No. 5, 1949 (fresh-water, Soviet Union); BERG, L. S., No. 10, 1965 (fresh-water, U. S. S. R.); ARNOULT, J., No. 1, 1959 (fresh-water, Madagascar); SMITH, H. M., No. 1, 1945 (fresh-water, Thailand); INGER & KONG, No. 1, 1962 (fresh-water, North Borneo); OSHIMA, M., (fresh-water, Taiwan); POPE & NICHOLS, No. 1, 1927 (Hainan I.); JORDAN, D. S. & SNYDER, J. O., 1906, (Proc. United States Nat. Mus., 31, 287-290 (Japan); OKADA, Y., No. 2, 1960 (fresh-water, Japan)
Poecilliidae, as Poeciliidae
Polyacanthidae, as Belontiidae
Polyacrodontidae .(only as fossils)
Polyaspidae (only as fossils)
Polybranchiaspidae (only as fossils)
Polycentridae, as Nandidae
Polydontidae, as Polyodontidae
Polymerichthyidae (only as fossils)
Polymixiidae. FOWLER, H. W., No. 8, 1936 (marine, West Africa); SMITH, J. L. B., No. 1, 1961 (marine, South Africa); LINDBERG & LEGHEZA, No. 2, 1965 (Sea of Japan); LINDBERG & LEGEZA, No. 3, 1969 (Sea of Japan)

Polynemidae. HILDEBRAND & SCHROEDER, No. 1, 1928 (Chesapeake Bay); BEEBE & TEE-VAN, No. 1, 1928 (Hispaniola); REGAN, C. T., No. 1, 1906-08 (Central America); MEEK & HILDEBRAND No. 2, 1923 (marine, Panama); HILDEBRAND, S. F., No. 1, 1946 (shore, Peru); PUYO, J., No. 1, 1949 (Guyane); BOULENGER, G. A., No. 6, 1916 (fresh-water, Africa); PELLEGRIN, J., No. 2, 1923 (western Africa); FOWLER, H. W., No. 8, 1936 (marine, West Africa); DAGET & ILTIS, No. 1, 1965 (Côte d'Ivoire); BOULENGER, G. A., No. 1, 1901 (Congo basin); SMITH, J. L. B., No. 1, 1961 (marine, South Africa); FOWLER, H. W., No. 18, 1956 (Red Sea); DAY, F. No. 2, 1889 (India); SMITH, H. W., No. 1, 1945 (fresh-water, Thailand); WEBER & de BEAUFORT, No. 4, 1922 (Indo-Australian Arch.); MARSHALL, T. C., No. 1, 1964 (Great Barrier Reef); FOWLER, H. W., No. 6, 1933-35 (China); LINDBERG & LEGHEZA, No. 2, 1965 (Sea of Japan); LINDBERG & LEGEZA, No. 3, 1969 (Sea of Japan); SCHULTZ, L. P. et al., No. 2, 1953 (Marshall & Marianas Is.); FOWLER, H. W., No. 17, 1959 (Fiji)

Polyodontidae. HOLLY, M., 1936, Das Tierreich, 67, 1-65; SLASTENENKO, E. P., No. 1, 1958 (Canada); FOWLER, H. W., No. 1, 1945 (fresh-water, southeastern U. S.); NICHOLS, J. T., No. 1, 1943 (fresh-water, China)

Polypteridae. HOLLY, M., 1933, Das Tierreich, 63, 1-22; POLL, M., 1942, Rev. Zool. Bot. Africaines, 35, 141-179, 269-317; BOULENGER, G. A., No. 3, 1909 (fresh-water, Africa); BOULENGER, G. A., No. 2, 1907 (Egypt); PELLEGRIN, J., No. 2, 1923 (western Africa); DAGET, J., No. 2, 1962 (Guinea); DAGET & ILTIS, No. 1, 1965 (Côte d'Ivoire); SCHULTZ, L. P., No. 1, 1942 (Liberia); DAGET, J., No. 1, 1954 (Upper Niger); BOULENGER, G. A., No. 1, 1901 (Congo basin); POLL, M., No. 1, 1946 (Lake Tanganyika); POLL, M., No. 3, 1953 (Lake Tanganyika)

Pomacanthidae, as Chaetodontidae

Pomacentridae (incl. Abudefdufidae, Amphiprionidae, Chromidae, Ctenolabridae, Glyphidodontidae, Glyphiodontidae, Premnidae). BEEBE & TEE-VAN, No. 1, 1928 (Hispaniola); MEEK & HILDEBRAND, No. 3, 1925 (marine, Panama); HILDEBRAND, S. F., No. 1, 1946 (shore, Peru); LOZANO REY, D. L., No. 3, 1952 (España); FAGE, L., No. 1, 1918 (Mediterranean); BANARESCU, P., No. 1, 1964 (Romania); SVETOVIDOV, A. N., No. 2, 1964 (Black Sea); BAUCHOT, M. L. & BLANC, M., 1961, Atlantide Rep., 6, 43-64 (eastern tropical Atlantic); FOWLER, H. W., No. 8, 1936 (marine, West Africa); SMITH, J. L. B., No. 1, 1961 (marine, South Africa, as Amphiprionidae & Abudefdufidae); SMITH, J. L. B., 1960, Ichth. Bull. Rhodes Univ., 19, 317-349 (western Indian Ocean & Red Sea); DAY, F., No. 2, 1889 (India, as Glyphidodontidae) de BEAUFORT, L. F., No. 8, 1940 (Indo-Australian Archipelago); MARSHALL, T. C., No. 1, 1964 (Great Barrier Reef, also as Amphiprionidae); WHITLEY, G. P., 1929, Mem. Queensland Mus., 9, 207-246; MONTALBAN, H. R., 1927, Pomacentridae of the Philippine Islands, Manila, 1-117 (Philippines); FOWLER, H. W. & BEAN, B. A., 1928, United States Nat. Mus. Bull., 100, (7), 1-525 (Philippine seas); FOWLER, H. W., No. 14, 1954 (China, as Amphiprionidae); LINDBERG & KRASJUKOVA, No. 1, 1969 (Sea of Japan); AOYAGI, H., 1941, Biogeographica, 4, 157-279 (Japan); SCHULTZ, L. P. et al., No. 3, 1960 (Marshall & Marianas Is.); FOWLER, H. W., No. 17, 1959 (Fiji)

Pomadasidae, as Pomadasyidae

Pomadasydae, as Pomadasyidae

Pomadasyidae (incl. Gaterinidae, Haemulidae, Haemulonidae, Plectorhynchidae, Pomadasidae, Pomadasydae, Pristipomatidae, Pristipomidae, Xenichthyidae). HILDEBRAND & SCHROEDER, No. 1, 1928 (Chesapeake Bay); BEEBE & TEE-VAN, No. 1, 1928 (Hispaniola); REGAN, C. T., No. 1, 1906-08 (Central America); MEEK, S. F., No. 1, 1904 (fresh-water, Mexico, as Haemulidae); MEEK & HILDEBRAND, No. 3, 1925 (marine, Panama, as Haemulidae); HILDEBRAND, S. F., No. 1, 1946 (shore, Peru); PUYO, J., No. 1, 1949 (Guyane, as Pristipomatidae); FOWLER, H. W., No. 13, 1954 (fresh-water, Brasil); DEVINCENZI, G. F., No. 2, 1924 (Uruguay, as Haemulidae); NOBRE, A., No. 1, 1935 (marine, Portugal, as Haemulidae); LOZANO REY, D. L., No. 2, 1952 (España); BOULENGER, G. A., No. 5, 1915 (fresh-water, Africa, as Pristipomatidae); PELLEGRIN, J., No. 2, 1923 (western Africa, as Pristipomatidae); FOWLER, H. W., No. 8, 1936 (marine, West Africa); DAGET & ILTIS, No. 1, 1965 (Côte d'Ivoire); BOULENGER, G. A., No. 1, 1901 (Congo basin, as Pristipomatidae); SMITH, J. L. B., No. 1, 1961 (marine, South Africa, also as Plectorhynchidae); POLL, M., No. 4, 1954 (coastal South Africa); PELLEGRIN, J., No. 2, 1960 (fresh-water, Madagascar, as Pristipomatidae); SMITH, J. L. B., 1962, Ichth. Bull. Rhodes Univ., 25, 469-502, as Gaterinidae); FOWLER, H. W., No. 19, 1931 (Philippine seas); FOWLER, H. W., No. 9, 1937-41 (China); LINDBERG & KRASJUKOVA, No. 1, 1969 (Sea of Japan); FOWLER, H. W., No. 17, 1959 (Fiji)

Pomatomidae (incl. Scombropidae, Scombropsidae). HILDEBRAND & SCHROEDER, No. 1, 1928 (Chesapeake Bay); MEEK & HILDEBRAND, No. 3, 1925 (marine, Panama); LOZANO REY, D. L., No. 3, 1952 (España); BANARESCU, P., No. 1, 1964 (Romania); SVETOVIDOV, A. N., No. 2, 1964 (Black Sea); BAUCHOT, M. L. & BLANC, M., 1963, Atlantide Rep., 7, 37-61 (eastern tropical Atlantic); FOWLER, H. W., No. 8, 1936 (marine, West Africa); POLL, M., No. 4, 1954 (coastal South Africa); SMITH, J. L. B., No. 1, 1961 (marine, South Africa); WEBER & de BEAUFORT, No. 6, 1931 (Indo-Australian Arch.); MARSHALL, T. C., No. 1, 1964 (Great Barrier Reef); LINDBERG & KRASJUKOVA, No. 1, 1969 (Sea of Japan)

Poraspidae (only as fossils)

Porcidae, as Bagridae

Porolepidae (only as fossils)

Potamotrygonidae. (No revisionary references noted)

Premnidae, as Pomacentridae

Prenidae, as Scatophagidae

Priacanthidae. HILDEBRAND & SCHROEDER, No. 1, 1928 (Chesapeake Bay); BEEBE & TEE-VAN, No. 1, (Hispaniola); MEEK & HILDEBRAND, No. 3, 1925 (marine, Panama); DEVINCENZI, G. F., No. 2, 1924 (Uruguay); CALDWELL, D. K., 1962, Copeia, 1962, 417-424 (key to western Atlantic); FOWLER, H. W., No. 8, 1936 (marine, West Africa); POLL, M., No. 4, 1954 (coastal South Africa); SMITH, J. L. B., No. 1, 1961 (marine, South Africa); WEBER & de BEAUFORT No. 5, 1929 (Indo-Australian Arch.); MARSHALL, T. C., No. 1, 1964 (Great Barrier Reef); FOWLER, H. W., No. 19, 1931 (Philippine seas); CHAN. W. L., No. 1, 1968 (Hong Kong); FOWLER, H. W., No. 9, 1937-41 (China); SOLDATOV & LINDBERG, No. 1, 1930 (Far East seas); LINDBERG & KRASJUKOVA, No. 1, 1969 (Sea of Japan); SCHULTZ, L. P. et al., No. 2 1953 (Marshall & Marianas Is.); FOWLER, H. W., No. 17, 1959 (Fiji)

Priscacaridae, as Cichlidae

Pristidae. HILDEBRAND & SCHROEDER, No. 1, 1928 (Chesapeake Bay); REGAN, C. T., No. 1, 1906-08 (Central America); MEEK & HILDEBRAND, No. 2, 1923 (marine, Panama); PUYO, J., No. 1, 1949 (Guyane); BIGELOW & SCHROEDER, No. 2, 1953 (western North Atlantic); NOBRE, A., No. 1, 1935 (marine, Portugal); TORTONESE, E., No. 3, 1956 (Italia); BOULENGER, G. A., No. 3, 1909 (fresh-water, Africa); PELLEGRIN, J., No. 2, 1923 (western Africa); FOWLER, H. W., No. 8, 1936 (marine, West Africa); SMITH, J. L. B., No. 1, 1961 (marine, South Africa); FOWLER, H. W., No. 18, 1956 (Red Sea); GOHAR & MAZHAR, No. 1, 1964 (Red Sea); DAY, F. No. 1, 1889 (India); SMITH, H. M., No. 1, 1945 (fresh-water, Thailand); MARSHALL, T. C., No. 1, 1964 (Great Barrier Reef); FOWLER, H. W., No. 2, 1930 (China); CHEN, J. T. F., No. 1, 1948 (China); LINDBERGH & LEGHEZA, No. 1, 1959 (Japan Sea)

Pristigasteridae, as Clupeidae

Pristiophoridae. GARMAN, S., No. 1, 1913 (monograph); SMITH, J. L. B., No. 1, 1961 (marine, South Africa); FOWLER, H. W., No. 2, 1930 (China); SOLDATOV & LINDBERG, No. 1, 1930 (Far East seas); LINDBERGH & LEGHEZA, No. 1, 1959 (Japan Sea); JORDAN, D. S. & FOWLER, H. W. 1903, Proc. United States Nat. Mus., 26, 593-674 (Japan)

Pristipomatidae, as Pomadasyidae

Pristipomidae, as Pomadasyidae

Pristodontidae, as Petalodontidae (only as fossils)

Pristolepidae, as Nandidae

Prochilodontidae. (No revisionary references noted)

Protodontidae (only as fossils)

Protopteridae. DAGET, J., No. 1, 1954 (Upper Niger)

Protosphyraenidae (only as fossils)

Protospinacidae (only as fossils)

Protosyngnathidae (only as fossils)

Prototroctidae, as Aplochitonidae

Psammichthyidae, as Kraemeriidae

Psammodontidae (only as fossils)

Psammosteidae (only as fossils)

Psenidae, as Nomeidae

Psettidae, as Monodactylidae

Psettodidae. WU, H. -W., 1932, Contribution a l'etude . . . Heterosomes de la Chine, Paris, 1-179 (China); NORMAN, J. R., 1934, Systematic monograph of the flatfishes, I, London, 1-459; FOWLER, H. W., No. 8, 1936 (marine, West Africa); SMITH, J. L. B., No. 1, 1961 (marine, South Africa); FOWLER, H. W., No. 18, 1956 (Red Sea); de SILVA, P. H. D. H., No. 2, 1956 (Ceylon); WEBER & de BEAUFORT, No. 5, 1929 (Indo-Australian Arch.); MARSHALL, T. C., No. 1, 1964 (Great Barrier Reef); FOWLER, H. W., No. 6, 1933-35 (China)

Pseudaphritidae, as Bovichthyidae

Pseudoberycidae (only as fossils)

Pseudochromidae. BEEBE & TEE-VAN, No. 1, 1928 (Hispaniola); SMITH, J. L. B., No. 1, 1961(marine, South Africa); SMITH, J. L. B., 1954, Ann. Mag. Nat. Hist., (12), 7, 195-208 (keys to genera, South & East Africa); DAY, F., No. 2, 1889 (India); SCOTT, T. D., 1959, Trans. Roy. Soc. South Australia, 82, 73-91 (Western Australian); FOWLER, H. W., No. 19, 1931 (Philippine seas); FOWLER, H. W., No. 9, 1937-41 (China); LINDBERG & KRASJUKOVA, No. 1, 1969 (Sea of Japan); AOYAGI, H., 1941, Annot. Zool. Japan., 20, 41-54 (Riu-kiu & Palau); SCHULTZ, L. P. et al., No. 2, 1953 (Marshall & Marianas Is.); FOWLER, H. W., No. 17, 1959 (Fiji)

Pseudodontichthyidae (only as fossils)

Pseudogrammidae (incl. Rhegmatidae). SMITH, J. L. B., 1953, Ann. Mag. Nat. Hist., (12), 6, 548-560 (East Africa)

Pseudomugilidae, as Atherinidae

Pseudopimelodidae, as Pimelodidae

Pseudoplesiopidae. SMITH, J. L. B., No. 1, 1961 (marine, South Africa); WEBER & de BEAUFORT, No. 5, 1929 (Indo-Australian Arch.)

Pseudotriakidae, as Scyliorhinidae

Psilocephalidae, as Balistidae

Psilonotidae. GILL, T., No. 1, 1884 (synopsis)

Psilorhynchidae. (No revisionary references noted)

Psuchrolutidae, as Psychrolutidae

Psychrolutidae (inc. Psuchrolutidae). NORMAN, J. R., No. 2, 1937 (marine, Patagonia); SCHMIDT, P. Y., No. 1, 1950 (Sea of Okhotsk); SHMIDT, P. Y., No. 2, 1965 (Sea of Okhotsk)

Pteraclidae, as Bramidae

Pteraclididae, as Bramidae
Pteraspidae (only as fossils)
Pterolepidae (only as fossils)
Pteropsauridae, as Percophididae
Pterothrissidae, as Albulidae
Pterygocephalidae, as Clinidae (only as fossils)
Ptetromyzantidae, as Petromyzonidae
Ptilichthyidae. MAKUSHOK, V. M., No. 1, 1958 (morphology & systematics)
Ptychodontidae (only as fossils)
Ptycholepididae. (No revisionary references noted)
Ptyctodontidae (only as fossils)
Pycnodontidae (only as fossils)
Pygaeidae (only as fossils)
Pygidiidae, as Trichomycteridae
Pygopteridae, as Palaeoniscidae
Pyramodontidae. (No revisionary references noted)

Rachycentridae (incl. Elacatidae, Rhachycentridae). HILDEBRAND & SCHROEDER, No. 1, 1928 (Chesapeake Bay); MEEK & HILDEBRAND, No. 3, 1925 (marine, Panama); FOWLER, H. W., No. 8, 1936 (marine, West Africa); SMITH, J. L. B., No. 1, 1961 (marine, South Africa); WEBER & de BEAUFORT, No. 6, 1931 (Indo-Australian Arch.); MARSHALL, T. C., No. 1, 1964 (Great Barrier Reef); CHAN, W. L., No. 1, 1968 (Hong Kong); FOWLER, H. W., No. 7, 1936 (China); LINDBERG & KRASJUKOVA, No. 1, 1969 (Sea of Japan)
Radamantidae (incl. Rhadamantidae).
Radotinidae (only as fossils)
Raiidae, as Rajidae
Rainfordiidae, as Serranidae
Rajidae (incl. Raiidae). HUBBS, C. L. & ISHIYAMA, R., 1968, Copeia, 1968, 483-491 (methods of study); HILDEBRAND & SCHROEDER, No. 1, 1928 (Chesapeake Bay); MEEK & HILDEBRAND, No. 2, 1923 (marine, Panama); HILDEBRAND, S. F., No. 1, 1946 (shore, Peru); DEVINCENZI, G. J., No. 1, 1920 (Uruguay); DEVINCENZI, G. J., No. 1, 1924 (Uruguay); NORMAN, J. R., No. 2, 1937 (Marine, Patagonia); BIGELOW & SCHROEDER, No. 2, 1953 (western North Atlantic); SMITT, F. A. et al., No. 2, 1895 (Scandinavia); STEVEN, G. A., 1947, Sci. Progr., 35, 220-236 DUNCKER & LADIGES, No. 1, 1960 (Nordmark); POLL, M., No. 2, 1947 (marine, Belgique); NOBRE, A., No. 1, 1935 (marine, Portugal); TORTONESE, E., No. 3, 1956 (Italia); SVETOVIDOV, A. N., No. 2, 1964 (Black Sea); BERG, L. S., No. 7, 1911 (Russia); ANDRIYASHEV, A. P., No. 1, 1954 (northern seas of U. S. S. R.); FOWLER, H. W., No. 8, 1936 (Marine, West Africa); SMITH, J. L. B., No. 1, 1961 (marine, South Africa); FOWLER, H. W., No. 18, 1956 (Red Sea); DAY, F., No. 1, 1889 (India); MARSHALL, T. C., No. 1, 1964 (Great Barrier Reef); FOWLER, H. W., No. 2, 1930 (China); CHEN, J. T. F., No. 1, 1948 (China); SOLDATOV & LINDBERG, No. 1, 1930 (Far East seas); SCHMIDT, P. Y., No. 1, 1950 (Sea of Okhotsk); SHMIDT, P. Y., No. 2, 1965 (Sea of Okhotsk); LINDBERGH & LEGHEZA, No. 1, 1959 (Japan Sea); JORDAN, D. S. & FOWLER, H. W., 1903, Proc. United States Nat. Mus., 26, 593-674 (Japan); ISHIYAMA, R., 1958, Bull. Mus. Comp. Zool., 118, 1-24 (Japan); ISHIYAMA, R., 1958, Journ. Shimonoseki Coll. Fish., 7, 193-394 (Japan); ISHIYAMA, R., 1967, Fauna Japo-Nica, Rajidae, Tokyo, 1-84; NORMAN, J. R., No. 3, 1938 (Antarctic)
Ramphosidae, as Macrorhamphosidae (only as fossils)
Ranicipitidae, as Gadidae
Raphiosauridae, as Elopidae (only as fossils)
Ratabouridae, as Moringuidae
Redfieldiidae (incl. Catopteridae, Dictyopygidae). (Only as fossils)
Regalecidae (incl. Agrostichthyidae). ANDRIYASHEV, A. P., No. 1, 1954 (northern seas of U. S. S. R.); MARSHALL, T. C., No. 1, 1964 (Great Barrier Reef); LINDBERG & LEGHEZA, No. 2, 1965 (Sea of Japan); LINDBERG & LEGEZA, No. 3, 1969 (Sea of Japan)
Remigolepidae (only as fossils)
Retropinnidae. (No revisionary references noted)
Rhabdolepididae. (No revisionary references noted)
Rhachycentridae, as Rachycentridae
Rhadamantidae, as Radamantidae
Rhadinichthyidae (only as fossils)
Rhamphichthyidae. CURRA, R. A. & de MIRANDA RIBEIRO, P., 1961, Bol. Inst. Oceanogr. Cumana, 1, 474-482 (key to genera); FOWLER, H. W., No. 10, 1948-51 (fresh-water, Brasil)
Rhamphobatidae, as Rhinobatidae
Rhamphocottidae, as Cottidae
Rhamphosidae, as Macrorhamphosidae (only as fossils)
Rhegmatidae, as Pseudogrammidae
Rhiacichthyidae, as Rhyacichthyidae
Rhincodontidae (incl. Rhinodontidae). GARMAN, S., No. 1, 1913 (monograph); HILDEBRAND, S. F., 1946 No. 1, 1946 (shore, Peru); BIGELOW & SCHROEDER, No. 1, 1948 (western North Atlantic); SMITH, J. L. B., No. 1, 1961 (marine, South Africa); FOWLER, H. W., No. 18, 1956 (Red Sea); DAY, F., No. 1, 1889 (India); LINDBERGH & LEGHEZA, No. 1, 1959 (Japan Sea)
Rhinellidae (only as fossils)
Rhineodontidae, as Orectolobidae

Rhinidae, as Rhinobatidae, part as Squatinidae
Rhinobatidae (incl. Asterodermidae, Rhamphobatidae, Rhinidae part, Rhinobatidae, Rhynchobatidae) GARMAN, S., No. 1, 1913 (monograph, as Rhinidae); NORMAN, J. R., 1926, Proc. Zool. Soc. London, 1926, 941-982 (synopsis); GARMAN, S., 1881, Proc. United States Nat. Mus., 3, 516-523 (American); MEEK & HILDEBRAND, No. 2, 1923 (marine, Panama); HILDEBRAND, S. F., No. 1, 1946 (shore, Peru); DEVINCENZI, G. J., No. 1, 1920 (Uruguay); BIGELOW & SCHROEDER, No. 2, 1953 (western North Atlantic, also as Rhynchobatidae); NOBRE, A., No. 1, 1935 (marine, Portugal); TORTONESE, E., No. 3, 1956 (Italia); FOWLER, H. W., No. 8, 1936 (marine, West Africa); SMITH, J. L. B., No. 1, 1961 (marine, South Africa); FOWLER, H. W., No. 18, 1956 (Red Sea); GOHAR & MAZHAR, No. 1, 1964 (Red Sea, also as Rhynchobatidae); DAY, F., No. 1, 1889 (India); MARSHALL, T. C., No. 1, 1964 (Great Barrier Reef); HERRE, A. W. C. T., 1955, Philippine Journ. Sci., 83, 381-400 (Philippines); HERRE, A. W. C. T., 1957, Proc. 8th Pacific Sci. Congr., 3A, 1595-1620 (Philippines); TENG, H. T., 1959, Rep. Lab. Fish. Biol. Taiwan, 10, 1-15 (Taiwan); FOWLER, H. W., No. 2, 1930 (China); CHEN, J. T. F., No. 1, 1948 (China); LINDBERGH & LEGHEZA, No. 1, 1959 (Japan Sea, also as Rhynchobatidae); JORDAN, D. S. & FOWLER, H. W., 1903, Proc. United States Nat. Mus., 26, 593-674 (Japan)
Rhinochimaeridae. BIGELOW & SCHROEDER, No. 2, 1953, (western North Atlantic); FOWLER, H. W., No. 8, 1936 (marine, West Africa); SMITH, J. L. B., No. 1, 1961 (marine, South Africa); JORDAN, D. S. & FOWLER, H. W., 1903, Proc. United States Nat. Mus., 26, 593-674 (Japan)
Rhinodontidae, as Rhincodontidae
Rhinoprenidae. (No revisionary references noted)
Rhinopteridae, as Myliobatidae
Rhizodontidae (incl. Megalichthyidae). (Only as fossils)
Rhizodopsidae (only as fossils)
Rhodichthyidae, as Cyclopteridae
Rhombosoleidae, as Pleuronectidae
Rhyacichthyidae (incl. Platypteridae). KOUMANS, F. P., No. 1, 1932 (revision); HERRE, A. W., No. 1, 1927 (Philippines, China Sea); FOWLER, H. W., 1962, Quart Journ. Taiwan Mus., 15, 1-77 (China)
Rhynchobatidae, as Rhinobatidae
Rhynchobdellidae, as Mastacembelidae
Rhynchodipteridae (only as fossils)
Rhynchodontidae, as Aspidorhynchidae (Only as fossils)
Rhyncholepidae (only as fossils)
Rogeniidae (only as fossils)
Rondeletiidae. (No revisionary references noted)
Rosauridae. (No revisionary references noted)
Runulidae, as Blenniidae
Ruvettidae, as Gempylidae
Rypticidae, as Grammistidae

Saccobranchidae, as Heteropneustidae
Saccopharyngidae. BIGELOW, H. B. et al., No. 4, 1966 (western North Atlantic); NOBRE, A., No. 1, 1935 (marine, Portugal); FOWLER, H. W., No. 8, 1936 (marine, West Africa)
Salangidae. WAKIYA, Y. & TAKAHASI, N., 1937, Journ. Coll. Agric. Tokyo, 14, 265-296 (revision); REGAN, C. T., 1908, Ann. Mag. Nat. Hist., (8), 2, 444-446 (synopsis of Salanginae); BERG, L. S., No. 1, 1932 (fresh-water, U. R. S. S.); BERG, L. S., No. 3, 1948 (fresh-water, Soviet Union); BERG, L. S., No. 8, 1962 (fresh-water, U. S. S. R.); OSHIMA, M., No. 1, 1919 (fresh-water, Taiwan); FANG, P. W., 1934, Sinensia, 4, 231-268 (China); NICHOLS, J. T., No. 1, 1943 (fresh-water, China); SOLDATOV & LINDBERG, No. 1, 1930 (Far East seas); SCHMIDT, P. Y., No. 1, 1950 (Sea of Okhotsk); SHMIDT, P. Y., No. 2, 1965 (Sea of Okhotsk); LINDBERG & LEGHEZA, No. 2, 1965 (Sea of Japan); LINDBERG & LEGEZA, No. 3, 1969 (Sea of Japan); WAKIYA, Y., 1936, Dobuts. Zasshi Tokyo, 48, 835-845 (Japan); OKADA, Y., No. 1, 1960 (fresh-water, Japan); OKADA, Y., No. 2, 1060 (fresh-water, Japan)
Salariidae, as Blenniidae
Salmonidae (incl. Coregonidae, Thymallidae). NORDEN, C. R., 1961, Journ. Fish. Res. Board Canada, 18, (5), 679-791 (comparative osteology to subfamilies); VLADYKOV, V. D., 1963, Trans. Roy. Soc. Canada, (4), 1, 459-504 (review of genera); SLASTENENKO, E. P., No. 1, 1958 (Canada, also as Coregonidae & Thymallidae); LEGENDRE, V., 1949, Bibl. Journ. Nat. Soc. Canad. Hist. Nat., 91, 1-4 (keys, Quebec); SCHULTZ, L. P., 1934, Proc. Fifth Pan-Pacific Sci. Congr., 5, 3777-3782 (northwestern U. S.); FOWLER, H. W., No. 1, 1945 (fresh-water, southeastern U. S.); REGAN, C. T., No. 1, 1906-08 (Central America); MEEK, S. E., No. 1, 1904 (fresh-water, Mexico); BIGELOW, H. B. et al., No. 5, 1963 (western North Atlantic, also as Coregonidae); DOTTRENS, E., 1959, Rev. Suisse Zool., 66, 1-66 (western Europe, as Coregonidae); SMITT, F. A. et al., No. 2, 1895 (Scandinavia); OTTERSTRØM, C. V., 1948, Journ. Conseil Copenhagen, 15, 223-231 (keys, Baltic); DUNCKER & LADIGES, No. 1, 1960 (Nordmark, also as Coregonidae & Thymallidae); POLL, M., No. 2, 1947 (marine, Belgique, also as Coregonidae); SPILLMANN, J., No. 1, 1961 (France); NOBRE, A., No. 1, 1935 (marine, Portugal); LOZANO REY, D. L., No. 1, 1935 (España); LOZANO REY, D. L., No. 4, 1947 (España); NERESHEIMER, E., 1937, in Demoll & Maier, Handbuch Binnfisch Mitteleurop., 3, 219-370 (Central Europe); STEINMANN, P., 1950-51, Schweiz. Zeitschr. Hydrol., 12, 1090189, 340-491, 13, 54-155, 162-191 (Suisse, as Coregonidae); KARAMAN,

S., 1927, Bull. Soc. Sci. Skopge, 2, 253-268 (Balkans); BANARESCU, P., No. 1, 1964 (Romania, also as Thymallidae); SVETOVIDOV, A. N., No. 2, 1964 (Black Sea); BERG, L. S., No. 1, 1932 (fresh-water, U. R. S. S., also as Thymallidae); BERG, L. S., No. 3, 1948 (fresh-water, Soviet Union, also as Thymallidae); BERG, L. S., No. 8, 1962 (fresh-water, U. S. S. R., also as Thymallidae); ANDRIYASHEV, A. P., No. 1, 1954 (northern seas of U. S. S. R., also as Thymallidae); BERG, L. S., No. 6, 1949 (fresh-water, Iran); BOULENGER, G. A., No. 3, 1909 (fresh-water, Africa); PELLEGRIN, J., 1924, Comp. Rend. Acad. Sci. Paris, 178, 970-972 (Maroc); PELLEGRIN, N., No. 1, 1933 (fresh-water, Madagascar); ARNOULT, J., No. 1, 1959 (fresh-water, Madagascar); OSHIMA, M., No. 1, 1919 (fresh-water, Taiwan); NICHOLS, J. T., No. 1, 1943 (fresh-water, China); SOLDATOV & LINDBERG, No. 1, 1930 (Far East seas); SCHMIDT, P. Y., No. 1, 1950 (Sea of Okhotsk, also as Coregonidae); SHMIDT, P. Y., No. 2, 1965 (Sea of Okhotsk, also as Coregonidae); TARANETZ, A. J., 1936, Trav. Inst. Zool. Acad. Sci. U. R. S. S., 4, 483-540 (Japan Sea); LINDBERG & LEGHEZA, No. 2, 1965 (Sea of Japan); LINDBERG & LEGEZA, No. 3, 1969 (Sea of Japan); JORDAN, D. S. & SNYDER, J. O., 1902, Proc. United States Nat. Mus., 24, 567-593 (Japan); JORDAN, D. S. & McGREGOR, E. A., 1925, Mem. Carnegie Mus., 10, 93-346 (Japan); OKADA, Y., No. 1, 1960 (fresh-water, Japan); OKADA, Y., No. 2, 1960 (fresh-water, Japan)

Samaridae, as Pleuronectidae

Sardidae, as Scombridae

Saurichthyidae (incl. Belonorhynchidae). (Only as fossils)

Sauridae, as Synodontidae

Saurocephalidae, as Saurodontidae

Saurodontidae (incl. Saurocephalidae). (Only as fossils)

Sauromuraenesocidae, as Muraenesocidae

Scanilepidae (only as fossils)

Scanilepididae, as Palaeoniscidae. (Only as fossils)

Scapanorhynchidae, as Odontaspidae

Scarichthyidae, as Scaridae

Scaridae (incl. Calliodontidae, Callyodontidae, Scarichthyidae, Sparisomidae). SCHULTZ, L. P., 1958, United States Nat. Mus. Bull., 214, 1-143 (review); BAUCHOT, M. L. & GUIBE, J., 1960, Bull. Mus. Nat. Hist. Nat., (2)32, 290-300 (museum type list); HILDEBRAND & SCHROEDER, No. 1, 1928 (Chesapeake Bay); BEEBE & TEE-VAN, No. 1, 1928 (Hispaniola); MEEK & HILDEBRAND, No. 4, 1928 (marine, Panama); HILDEBRAND, S. F., No. 1, 1946 (shore, Peru); LOZANO REY, D. L., No. 3, 1952 (España, as Calliodontidae); BAUCHOT, M. L. & BLANC, M., 1961, Atlantide Rep., 6, 43-64 (eastern tropical Atlantic); FOWLER, H. W., No. 8, 1936 (marine, West Africa, as Callyodontidae); SMITH, J. L. B., No. 1, 1961 (marine, South Africa, as Callyodontidae); SMITH, J. L. B., 1956, Ichth. Bull. Rhodes Univ., 1, 1-23 (western Indian Ocean, as Callyodontidae); de BEAUFORT, L. F., No. 8, 1940 (Indo-Australian Arch.,); MARSHALL, T. C., No. 1, 1964 (Great Barrier Reef, as Callyodontidae); FOWLER, H. W. & BEAN, B. A., 1928, United States Nat. Mus. Bull., 100, (7), 1-525 (Philippine seas, as Callyodontidae); FOWLER, H. W., No. 15, 1956 (China); LINDBERG & KRASJUKOVA, No. 1, 1969 (Sea of Japan); KAMOHARA, T., 1963, Rep. USA Biol. Sta., 10, (1), 1-24 (Riu-kiu Is. & Japan); MASUDA, T. & TANAKA, K., 1962, Journ. Tokyo Univ. Fish., 48, 1-98 (key to larvae, Pacific coasts of Japan); SCHULTZ, L. P. et al., No. 3, 1960 (Marshall & Marianas Is.); FOWLER, H. W., No. 17, 1959 (Fiji, as Callyodontidae)

Scatophagidae (incl. Prenidae). SMITH, J. L. B., No. 1, 1961 (marine, South Africa); HERRE, A. W. & MONTALBAN, H. R., 1927, Philippine Journ. Sci., 34, 1-113 (Philippines); FOWLER, H. W., No. 12, 1953 (China)

Scaumenacidae (only as fossils)

Schilbeidae. DAGET, J., No. 2, 1962 (Guinea); DAGET & ILTIS, No. 1, 1965 (Côte d'Ivoire); TREWAVAS, E., 1943, Proc. Zool. Soc. London, 113B, 164-171(Gold Coast, synopsis of African genera); DAGET, J., No. 1, 1954 (Upper Niger); POLL, M., No. 1, 1946 (Lake Tanganyika); SMITH, H. M., No. 1, 1945 (fresh-water, Thailand); INGER & KONG, No. 1, 1962 (fresh-water, North Borneo)

Schindleriidae. SCHULTZ, L. P. et al., No. 3, 1960 (Marshall & Marianas Is.)

Sciaenidae (incl. Otolithidae). WALFORD, L. A., No. 1, 1937 (Alaska to Equator); SLASTENENKO, E. P., No. 1, 1958 (Canada); HILDEBRAND & SCHROEDER, No. 1, 1928 (Chesapeake Bay, also as Otolithidae); FOWLER, H. W., No. 1, 1945 (fresh-water, southeastern U. S.); BEEBE & TEE-VAN, No. 1, 1928 (Hispaniola, also as Otolithidae); NICHOLS, J. T., 1929 (Puerto Rico & Virgin Is.); REGAN, C. T., No. 1, 1906-08 (Central America); MEEK, S. E., No. 1, 1904 (fresh-water, Mexico); MEEK & HILDEBRAND, No. 3, 1925 (marine, Panama); HILDEBRAND, S. F., No. 1, 1946 (shore, Peru); EIGENMANN, C. H., No. 1, 1912 (British Guiana); LOWE (McConnell), R. H., 1966, Bull. Mar. Sci. Gulf Caribb., 16, 20-57 (British Guiana); PUYO, J., No. 1, 1949 (Guyane); CAMPOS, A. A., 1942, Arqu. Mus. Paranense, 2, 9-22 (Brasil); FOWLER, H. W., No. 13, 1954 (fresh-water, Brasil); TRAVASSOS, H & PAIVA, M. P., 1957, Bol. Inst. Oceanogr. Sao Paulo, 8, 139-164 (marine, Brasil); DEVINCENZI, G. F., No. 2, 1924 (Uruguay); de BUEN, F., 1961, Montemar, 1, 1-90 (Chile); COLLIGNON, J., 1959, Bull. Inst. Oceanogr. Monaco, 1155, 1-11 (eastern Atlantic); SMITT, F. A. et al., No. 1, 1892 (Scandinavia); DUNCKER & LADIGES, No. 1, 1960 (Nordmark); POLL, M., No. 2, 1947 (marine, Belgique); NOBRE, A., No. 1, 1935 (marine, Portugal); LOZANO REY, D. L., No. 2, 1952 (España); BANARESCU, P., No. 1, 1964 (Romania); SVETOVIDOV, A. N., No. 2, 1964 (Black Sea); BOULENGER, G. A., No. 5, 1915 (fresh-water, Africa); PELLEGRIN, J., No. 2, 1923 (western Africa); TREWAVAS, E., 1961, CCTA/CSA West Coast, (60), 8, 2-5 (keys to genera of West Africa); FOWLER, H. W., No. 8, 1936 (marine, West Africa); DAGET & ILTIS, No. 1, 1965 (Côte d'Ivoire); BOULENGER, G. A., No. 1, 1901 (Congo basin); SMITH, J. L. B., No. 1, 1961 (marine, South Africa); POLL, M., No. 4, 1954 (coastal South Africa); DAY, F., No. 2, 1889 (India); WEBER & de BEAUFORT, No. 7, 1936 (Indo-Australian Arch.); INGER & KONG, No. 1, 1962 (fresh-water, North Borneo); MARSHALL, T. C., No. 1, 1964 (Great Barrier Reef); FOWLER, H. W., No. 20, 1933 (Philippine seas); FOWLER, H. W., No. 11, 1949 (China); CHU, Y. T. et al., 1963, Study of classification of sciaenoid fishes of China, Shanghai, 1-100; LINDBERG & KRASJIKOVA, No. 1, 1969 (Sea of Japan); JORDAN, D. S., 1911, Proc. United States Nat. Mus., 39, 241-261 (Japan); McPHAIL, J. D., 1958, Inst. Fish. Univ. British Columbia Mus. Contr., 2, 1-20 (keys, eastern Pacific)

Scienidae, as Sciaenidae

Sclerodidae (only as fossils)

Sclerogenidae (family not identified)

Scolopsidae, as Nemipteridae

Scomberesocidae (incl. Scombresocidae). COLLETTE, B. B., 1966, American Mus. Novit., 2274, 1-22 (generic characters); HILDEBRAND & SCHROEDER, No. 1, 1928 (Chesapeake Bay); REGAN, C. T., No. 1, 1906-08 (Central America); Puyo, J., No. 1, 1949 (Guyane); DEVINCENZI, G. J., No. 2, 1924 (Uruguay); PARIN, N. V., 1968, Atlantide Rep., 10, 275-290 (eastern Atlantic); SMITT, F. A. et al., No. 1, 1892 (Scandinavia); DUNCKER & LADIGES, No. 1, 1960 (Nordmark) POLL, M., No. 2, 1947 (marine, Belgique); NOBRE, A., No. 1, 1935 (marine, Portugal); LOZANO REY, D. L., No. 4, 1947 (España); ANDRIYASHEV, A. P., No. 1, 1954 (northern seas of U. S. S. R.); BOULENGER, G. A., No. 5, 1915 (fresh-water, Africa); PELLEGRIN, J., No. 2, 1923 (western Africa); FOWLER, H. W., No. 8, 1936 (marine, West Africa); SMITH, J. L. B., No. 1, 1961 (marine, South Africa); PELLEGRIN, J., No. 1, 1933 (fresh-water, Madagascar); FOWLER, H. W., No. 18, 1956 (Red Sea); DAY, F., No. 1, 1889 (India); SOLDATOV & LINDBERG, No. 1, 1930 (Far East seas); LINDBERG & LEGHEZA, No. 2, 1965 (Sea of Japan); LINDBERG & LEGEZA, No. 3, 1969 (Sea of Japan); JORDAN, D. S. & STARKS, E. C., 1903, Proc. United States Nat. Mus., 26, 525-544 (Japan)

Scomberidae, as Scombridae

Scomberomoridae, as Scombridae

Scombresocidae, as Scomberesocidae

Scombridae (incl. Acanthocybiidae, Cibiidae, Cybiidae, Gasterochismidae, Katsuwonidae, Sardidae, Scomberidae, Scomberomoridae, Thunnidae). PARONA, C., 1919, Mem. R. Com. Talass. Ital., 68, 1-265; KISHINOUYE, K., 1923, Journ. Coll. Agric. Imp. Univ. Tokyo, 8, 293-475 (comparative study); JORDAN, D. S. & EVERMANN, B. W., 1926, Occ. Pap. California Acad. Sci., 12, 1-113 (review); FRASER-BRUNNER, A., 1950, Ann. Mag. Nat. Hist., (12), 3, 131-163; ROSA, H., 1950, Scientific and common names applied to tunas . . . , Washington, D. C., 1-235; MORICE, J., 1953, Rev. Trav. Off. Pêches Marit., 18, (1), 65-74 (Essai systematique); as Cybiidae, Katsuwanidae, & Thunnidae); Van CAMPEN, W. & HOVEN,, E. E., 1956, Fish. Bull. U. S. F. W. S., 57, 173-249 (annotated bibliography); IWAI, T. et al., 1965, Spec. Rep. Misaki Mar. Biol. Inst., 2, 1-51 (taxonomic studies); COLLETTE, B. B., 1966, Bull. Mus. Nat. Hist. Nat., (2), 38, 362-375 (museum type list); GIBBS, R. H., Jr. & COLLETTE, B. B., 1967, Fish. Bull. U. S. F. W. S., 12, 65-130 (comparative anatomy & systematics); FOREST, J. & LEGENDRE, R., 1950, Bull. Inst. Oceanogr. Monaco, 966, 1-12 (key to subfamilies); HILDEBRAND & SCHROEDER, No. 1, 1928 (Chesapeake Bay); WALFORD, L. A., No. 1, 1937 (Alaska to Equator, also as Acanthocybiidae, Cybiidae, Katsuwonidae, Thunnidae). GODSIL, H. C., 1945, California Fish Game, 31, 185-194 (key, California); RIVERO, L. H., 1953, Los escombridos en Cuba. . . , Habana, 1-105 (Cuba); BEEBE & TEE-VAN, No. 1, 1928 (Hispaniola, also as Thunnidae); MEAD, G. W., 1951, Fish. Bull. U. S. F. W. S., 63, 121-127 (key to postlarvae of Pacific Coast of Central America); MEEK & HILDEBRAND, No. 2, 1923 (marine, Panama); FERNANDEZ-YEPEZ, A & SANTUELLA, E. J., 1956, Occ. Pap. Soc. Venezolana Act. Sub., 1, 1-29 (key, Venezuela); HILDEBRAND, S. F., No. 1, 1946 (shore, Peru); PUYO, J., No. 1, 1949 (Guyane); DEVINCENZI, G. L., No. 2, 1924 (Uruguay); NORMAN, J. R., No. 2, 1937 (marine, Patagonia); RIVAS, L. R., 1951, Bull. Mar. Sci. Gulf Caribbean, 1. 209-230 (western North Atlantic); SMITT, F. A. et al., No. 1, 1892 (Scandinavia); DUNCKER & LADIGES, No. 1, 1960 (Nordmark); POLL, M., No. 2, 1947 (marine, Belgique); NOBRE, A., No. 1, 1935 (marine, Portugal); LOZANO REY, D. L., No. 3, 1952 (España) LOZANO CABO, F., 1958, Trab. Inst. Esp. Oceanogr., 25, 6-254 (aguas españolas y marroques); COLLETTE, B. B., 1970, Bull. Sea Fish Sta., 54, 3-6 (Key, Mediterranean); BANARESCU, P., No. 1, 1964 (Romania, also as Cybiidae & Thunnidae); SVETOVIDOV, A. N., No. 2, 1964 (Black Sea); ANDRIYASHEV, A. P., No. 1, 1954 (northern seas of U. S. S. R., also as Thunnidae); FOWLER, H. W., No. 8, 1936 (marine, West Africa); POSTEL, E., 1949, Bull. Serv. Elev. Industr. Anim. Afrique Occidentale Francaise, 2, (4), 39-46 (keys, West Africa); DAGET & ILTIS, No. 1, 1965 (Côte d'Ivoire); SMITH, J. L. B., No. 1, 1961 (marine, South Africa, also as Scomberomoridae); MERRETT, N. R. & THORP, C. H., 1966, Ann. Mag. Nat. Hist., (13), 8, 367-384 (keys, East AFrica); COLLETTE, B. B. & GIBBS, R. H., Jr., Preliminary field guide to the (Scombridae) of the Indian Ocean, Washington, D. C., 1963, 1-48; DAY, F., No. 2, 1889 (India); JONES, S. & SILAS, E. G., 1962, Indian Journ. Fish., 8, 189-206 (keys, Indian waters); de BEAUFORT & CHAPMAN, No. 9, 1951 (Indo-Australian Arch.); MARSHALL, T. C., No. 1, 1964 (Great Barrier Reef); FOWLER, H. W., No. 7, 1936 (China, also as Thunnidae); SOLDATOV & LINDBERG, No. 1, 1930 (Far East seas); SHIMADA, B. M., 1951, Fish Bull. U. S. F. W. S., 58, 1-58 (annotated bibliography of Pacific tunas); BROCK, V. E., 1949, Pacific Sci., 3, 271-277 (keys, Hawaii); SCHULTZ, L. P. et al., No. 3,

(Marshall & Marianas Is.); FOWLER, H. W., No. 17, 1959 (Fiji, also as Thunnidae)
Scombropidae, as Pomatomidae
Scombropsidae, as Pomatomidae
Scopelarchidae MARSHALL, N. B., 1955, Discovery Rep., 27, 303-336 (deep-sea); BIGELOW, H. B. et al., No. 4, 1966 (western North Atlantic); LOZANO REY, D. L., No. 4, 1947 (España); SMITH, J. L. B., No. 1, 1961 (marine, South Africa)
Scopelidae, as Myctophidae
Scopelosauridae (incl. Notosudidae). BIGELOW, H. B. et al., No. 4, 1966 (western North Atlantic)
Scophthalmidae. LOZANO y REY, L., No. 5, 1960 (España)
Scorpaenichthyidae, as Cottidae
Scorpaenidae (incl. Sclerogenidae part, Tetrarogidae). EIGENMANN, C. H. & BEESON, C. H., 1894, Proc. United States Nat. Mus., 17, 375-407 (Pacific Coast of America); ALVERSON, D. L. & WELANDER, A. D., 1952, Copeia, 1952, 138-142 (key, Washington); PHILLIPS, J. B., 1957, Fish. Bull. Sacramento, 104, 1-158 (California); BEEBE & TEE-VAN, No. 1, 1928 (Hispaniola); MEEK & HILDEBRAND, No. 4, 1928 (marine, Panama); HILDEBRAND, S. F., No. 1, 1946 (shore, Peru); DEVINCENZI, G. F., No. 2, 1924 (Uruguay); de BUEN, F., 1961, Montemar, 1, 1-90 (Chile); NORMAN, J. R., No. 2, 1937 (marine, Patagonia); GINSBURG, I., 1953, Smithsonian Misc. Coll., 121, (8), 1-103 (western Atlantic); LEGENDRE, R., 1934, Ann. Inst. Oceanogr. Paris, 14, 249-418 (eastern Atlantic); SMITT, F. A. et al., No. 1, 1892 (Scandinavia); DUNCKER & LADIGES, No. 1, 1960 (Nordmark); POLL, M., No. 2, 1947 (marine, Belgique); NOBRE, A., No. 1, 1935 (marine, Portugal); LOZANO REY, D. L., No. 2, 1952 (España); FAGE, L., No. 1, 1918 (Mediterranean); BANARESCU, P., No. 1, 1964 (Romania); SVETOVIDOV, A. N., No. 2, 1964 (Black Sea); ANDRIYASHEV, A. P., No. 1, 1954 (northern seas of U. S. S. R.); BOUTIERE, H., 1958, Trav. Inst. Sci. Cherifien, 15, 1-84 (Maroc); FOWLER, H. W., No. 8, 1936 (marine, West Africa); SMITH, J. L. B., No. 1, 1961 (marine, South Africa, also as Tetrarogidae); SMITH, J. L. B., 1957, Ichth. Bull. Rhodes Univ., 4, 49-69 (western Indian Ocean); SMITH, J. L. B., 1958, Ichth. Bull., 12, 167-181 (western Indian Ocean); DAY, F., No. 2, 1889 (India); de BEAUFORT & BRIGGS, No. 11, 1962 (Indo-Australian Archip.); MARSHALL, T. C., No. 1, 1964 (Great Barrier Reef); SCOTT, E. O. G., 1960, Pap. Roy. Soc. Tasmania, 94, 87-100 (Tasmania, IX, keys); HERRE, A. W., 1952 (Philippine Journ. Sci., 80, 381-482 (Philippines); SOLDATOV & LINDBERG, No. 1, 1930 (Far East seas); SCHMIDT, P. Y., No. 1, 1950 (Sea of Okhotsk); SHMIDT, P. Y., No. 2, 1965 (Sea of Okhotsk); JORDAN, D. S. & STARKS, E. C., 1904, Proc. United States Nat. Mus., 27, 91-175 (Japan); MATSUBARA, K., 1943, Trans. Sigenkaguku Kenkyusyo, 1 & 2, 1-486 (keys, Japan); SCHULTZ, L. P. et al., No. 1, 1966 (Marshall & Marianas Is.); FOWLER, H. W., No. 17, 1959 (Fiji)
Scorpidae, as Kyphosidae
Scorpididae, as Kyphosidae
Scyliorhinidae (incl. Atelomycteridae, Catulidae, Pentanchidae, Pseudotriakidae, Halaeluridae, Scyllidae, Scylliidae, Scylliorhinidae). REGAN, C. T., 1908, Ann. Mag. Nat. Hist., (8), 1, 453-465 (synopsis); GARMAN, S., No. 1, 1913 (monograph, as Pseudotriakidae); HILDEBRAND, S. F., No. 1, 1946 (shore, Peru); NORMAN, J. R., No. 2, 1937 (marine, Patagonia); BIGELOW & SCHROEDER, No. 1, 1948 (Western North Atlantic, also as Pseudotriakidae); SPRINGER, S., 1966, Fish. Bull. U. S. F. W. S., 65, 581-624 (western Atlantic); SMITT, F. A. et al., No. 2, 1895 (Scandinavia, as Scylliidae); DUNCKER & LADIGES, No. 1, 1960 (Nordmark); POLL, M., No. 2, 1947 (marine, Belgique); NOBRE, A., No. 1, 1935 (marine, Portugal, as Scylliidae); TORTONESE, E., No. 3, 1956 (Italia); SVETOVIDOV, A. N., No. 2, 1964 (Black Sea); BERG, L. S., No. 7, 1911 (Russia); ANDRIYASHEV, A. P., No. 1, 1954 (northern seas of U. S. S. R.); FOWLER, H. W., No. 8, 1936 (marine, West Africa, also as Pseudotriakidae); SMITH, J. L. B. No. 1, 1961 (marine, South Africa); FOWLER, H. W., No. 18, 1956 (Red Sea); DAY, F., No. 1, 1889 (India, as Scyllidae); FOWLER, H. W., No. 2, 1930 (China); SOLDATOV & LINDBERG, No. 1, 1930 (Far East seas); LINDBERGH & LEGHEZA, No. 1, 1959 (Japan Sea); JORDAN, D. S. & FOWLER, H. W., 1903, Proc. United States Nat. Mus., 26, 593-674 (Japan)
Scyllidae, as Scyliorhinidae
Scylliidae, as Scyliorhinidae
Scylliogaliidae. SMITH, J. L. B., 1957, South African Journ. Sci., 53, 353-359 (South Africa)
Scylliorhinidae, as Scyliorhinidae
Scymnidae. NOBRE, A., No. 1, 1935 (marine, Portugal)
Scymnorhinidae. GARMAN, S., No. 1, 1913 (monograph)
Scytalinidae (incl. Scytaliscidae). SOLDATOV & LINDBERG, No. 1, 1930 (Far East seas)
Scytaliscidae, as Scytalinidae
Searsidae, as Alepocephalidae
Searsiidae, as Alepocephalidae
Selenosteidae (inc. Pachosteidae) (only as fossils)
Semionotidae (incl. Dapediidae, Sphaerodontidae, Stylodontidae) (only as fossils)
Semiophoridae. (Family not identified)
Seriolidae, as Carangidae
Serranidae (incl. Anthiidae, Bostockiidae, Cephalophoridae, Chromileptidae, Diploprionidae, Epinephelidae, Hypoplectrodidae, Maccullochellidae, Macquariidae, Moronidae, Niphonidae, Oligoridae, Ostracoberycidae, Paracentropristidae, Percichthyidae, Perciliidae, Plectroplitidae, Plectropomidae, Rainfordiidae). BOHLKE, J. E., 1960, Notul. Ent., Philadelphia, 330, 1-11 (key to families of serranoids); SMITH, J. L. B., 1961, Ichth. Bull. Rhodes Univ., 21, 359-369; GOSLINE, W. A., 1966, Proc. California Acad. Sci., (4), 33, 91-112 (limits of family); SLASTENENKO. E. P., No. 1, 1958 (fresh-water, Canada); HILDEBRAND & SCHROEDER,

No. 1, 1928 (Chesapeake Bay, also as Epinephelidae & Moronidae); FOWLER, H. W., No. 1, 1945 (southeastern U. S.); WALFORD, LA., No. 1, 1937 (Alaska to Equator); BEEBE & TEE-VAN, No. 1, 1928 (Hispaniola, also as Epinephelidae); MEEK & HILDEBRAND, No. 3, 1925 (marine, Panama); HILDEBRAND, S. F., No. 1, 946 (shore, Peru); PUYO, J., No. 1, 1949 (Guyane); FOWLER, H. W., No. 13, 1954 (fresh-water, Brasil); DEVINCENZI, G. L., No. 2, 1924 (Uruguay); EIGENMANN, C. H., No. 2, 1927 (fresh-water, Chile); BAUCHOT & BLANC, No. 1, 1961 (eastern tropical Atlantic); ROBINS, C. R. & STARCK, W. A., 1961 (western North Atlantic); DUNCKER & LADIGES, No. 1, 1960 (Nordmark); POLL, M., No. 2, 1947 (marine, Belgique); SPILLMANN, J., No. 1, 1961 (France); NOBRE, A., No. 1, 1935 (marine, Portugal); LOZANO REY, D. L., No. 1, 1935 (España); LOZANO REY, D. L., No. 2, 1952 (España, also as Moronidae); FAGE, L., No. 1, 1918 (Mediterranean); Banarescu, P., No. 1, 1964 (Romania); SVETOVIDOV, A. N., No. 2, 1964 (Black Sea); BERG, L. S., No. 2, 1933 (fresh-water, U. R. S. S.); BERG, L. S., No. 5, 1949 (fresh-water, Soviet Union); BERG, L. S., No. 10, 1965 (fresh-water, U. S. S. R.); ANDRIYASHEV, A. P., No. 1, 1954 (northern seas of U. S. S. R.); BOULENGER, G. A., No. 5, 1915 (fresh-water, Africa); BOULENGER, G. A., No. 2, 1907 (Egypt); PELLEGRIN, J., No. 2, 1923 (western Africa); CADENAT, J., 1935, Rev. Trav. Pêches Maritimes, 8, 377-422 (Côte occidentale d'Afrique); FOWLER, H. W., No. 8, 1936 (marine, West Africa); DAGET & ILTIS, No. 1, 1965 (Côte d'Ivoire); BOULENGER, G. A, No. 1, 1901 (Congo basin); POLL, M., No. 4, 1954 (coastal South Africa); SMITH, J. L. B., No. 1, 1961 (Marine, South Africa, also Anthiidae, Plectropomidae); ARNOULT, J., No. 1, 1959 (fresh-water, Madagascar); PELLEGRIN, J., No. 1, 1960 (fresh-water, Madagascar); SMITH, J. L. B., 1955, Ann. Mag. Nat. Hist., (12), 8, 337-350 (western Indian Ocean, as Anthiidae); WEBER & de BEAUFORT, No. 6, 1931 (Indo-Australian Arch.); MARSHALL, T. C., No. 1, 1964 (Great Barrier Reef); FOWLER, H. W., 1930, United States Nat. Mus. Bull., 100, (10), 1-334 (Philippine seas); POPE & NICHOLS, No. 1, 1927 (Hainan I.); CHAN, W. L., No. 1, 1968 (Hong Kong); FOWLER, H. W., No. 9, 1937-41 (China); NICHOLS, J. T., No. 1, 1943 (fresh-water, China); SOLDATOV & LINDBERG, No. 1, 1930 (Far East seas) LINDBERG & KRASJUKOVA, No. 1, 1969 (Sea of Japan); JORDAN, D. S. & RICHARDSON, R. E., 1910, Proc. United States Nat. Mus., 37, 421-474 (Japan); OKADA, Y., No. 1, 1960 (fresh-water, Japan); OKADA, Y, , No. 2, 1960 (fresh-water, Japan); KATAYAMA, M., 1960 Fauna Japonica, Serranidae, Tokyo, 1-189 (Japan); SCHULTZ, L. P. et al., No. 2, 1953 (Marshall & Marianas Is.); FOWLER, H. W., No. 17, 1959 (Fiji); SMITH, J. L. B., 1965, Ann. Mag. Nat. Hist., (13), 7, 533-537 (Cook I., key to genera)

Serrasalmidae, as Characidae

Serrivomeridae (incl. Gavialicipitidae). BERTIN, L., 1944, Bull. Mus. Nat. Hist. Nat., (2), 16, 101-108 (osteology & synonymy); CASTLE, P. H. J., 1965, Trans. Roy. Soc. Zew Zealand, Zool., 5, 131-146 (Australian)

Sicydiaphiidae, as Gobiidae

Siganidae (incl. Theutyidae, Teuthididae, Amphacanthidae). GILL, T., 1884, Proc. United States Nat. Mus., 7, 275-281 (Synopsis); SMITH, J. L. B., No. 1, 1961 (marine, South Africa); de BEAUFORT & CHAPMAN, No. 9, 1951 (Indo-Australian Arch.); MARSHALL, T. C., No. 1, 1964 (Great Barrier Reef); HERRE, A. W. & MONTALBAN, H. R., 1928, Philippine Journ. Sci., 35, 151-184 (Philippines); SCHULTZ, L. P. et al., No. 2, 1953 (Marshall & Marianas Is.); FOWLER, H. W., No. 17, 1959 (Fiji)

Sillaginidae. SMITH, J. L. B., No. 1, 1961 (marine, South Africa); WEBER & de BEAUFORT, No. 6, 1931 (Indo-Australian Arch.); MARSHALL, T. C., No. 1, 1964 (Great Barrier Reef); FOWLER, H. W., No. 20, 1933 (Philippine seas); FOWLER, H. W., No. 11, 1949 (China); LINDBERG & KRASJUKOVA, No. 1, 1969 (Sea of Japan)

Siluridae. HAIG, J., 1952, Rec. Indian Mus., 48, 59-116 (Orient, Palearctic); REGAN, C. T., No. 1, 1906-1908 (Central America); MEEK, S. E., No. 1, 1904 (fresh-water, Mexico); MEEK & HILDEBRAND, No. 1, 1916 (fresh-water, Panama); MEEK & HILDEBRAND, No. 2, 1923 (marine, (Panama); EIGENMANN & EIGENMANN, No. 1, 1890 (South America); EIGENMANN, C. H., No. 1, 1912 (British Guiana); PUYO, J., No. 1, 1949 (Guyane); DEVINCENZI, G. J., No. 2, 1924 (Uruguay); SMITT, F. A. et al., No. 2, 1895 (Scandinavia); DUNCKER & LADIGES, No. 1, 1960 (Nordmark); SPILLMANN, J., No. 1, 1961 (France); BANARESCU, P., No. 1, 1964 (Romania); BERG, L. S., No. 2, 1933 (fresh-water, U. S. S. R.); BERG, L. S., No. 4, 1949 (fresh-water, Soviet Union); BERG, L. S., No. 9, 1964 (fresh-water, U. S. S. R.); BERG, L. S., No. 6, 1949 (fresh-water, Iran); BOULENGER, G. A., No. 4, 1911 (fresh-water, Africa); BOULENGER, G. A., No. 2, 1907 (Egypt); PELLEGRIN, J., No. 2, 1923 (western Africa); SCHULTZ, L. P., No. 1, 1942 (Liberia); BOULENGER, G. A., No. 1, 1901 (Congo basin); PELLEGRIN, J., No. 2, 1960 (fresh-water, Madagascar); DAY, F., No. 1, 1889 (India); SMITH, H. M., No. 1, 1945 (fresh-water, Thailand); HORA, S. L. & GUPTA, J. C., 1941, Bull. Raffles Mus., 17, 12-43 (key to Malay genera); WEBER & de BEAUFORT, No. 2, 1913 (Indo-Australian Arch.); INGER & KONG, No. 1, 1962 (fresh-water, North Borneo); OSHIMA, M., No. 1, 1919 (fresh-water, Taiwan); POPE & NICHOLS, No. 1, 1927 (Hainan I.); NICHOLS, J. T., No. 1, 1943 (fresh-water, China); SOLDATOV & LINDBERG, No. 1, 1930 (Far East seas); JORDAN, D. S. & FOWLER, H. W., 1903, Proc. United States Nat. Mus., 26, 897-911 (Japan); OKADA, Y., No. 1, 1960 (fresh-water, Japan); OKADA, Y., No. 2, 1960 (fresh-water, Japan)

Simenchelidae, as Simenchelyidae

Simenchelydae, as Simenchelyidae

Simenchelyidae (incl. Simenchelidae, Simenchelydae). LOZANO REY, D. L., No. 4, 1947 (España); FOWLER, H. W., No. 8, 1936 (marine, West Africa); SMITH, J. L. B., No. 1, 1961 (marine, South Africa)

Sinamiidae (only as fossils)
Singididae (only as Fossils)
Siphonognathidae, as Odacidaeacidae
Siphostomidae, as Syngnathidae
Sisoridae (incl. Bagaridae, Bagariidae). HORA, S. L. & SILAS, E. G., 1952, Rec. Indian Mus., 49, 5-29 (revision); BERG, L. S., No. 2, 1933 (fresh-water, U. R. S. S.); BERG, L. S., No. 4, 1949 (fresh-water, Soviet Union); BERG, L. S., No. 9, 1964 (fresh-water, U. S. S. R.); BERG, L. S., No. 6, 1949 (fresh-water, Iran); SMITH, H. M., No. 1, 1945 (fresh-water, Thailand); WEBER & de BEAUFORT, No. 2, 1913 (Indo-Australian Arch., as Bagaridae); INGER & KONG, No. 1, 1962 (fresh-water, North Borneo)
Soleidae (incl. Achiridae, Synapturidae, Trinectidae). NORMAN, J. R., 1930, Discovery Rep., 2, 358-369 (oceanic); WU, H. -W., 1932, Contribution a l'etude . . . . Heterosomes . . ., Paris, 1-179 (monograph); NIELSEN, J. G., 1963, Atlantidae Rep., 7, 7-37; JORDAN, D. S., 1923, Univ. California Publ. Zool., 26, 1-14 (revision of Achirinae); HILDEBRAND & SCHROEDER, No. 1, 1928 (Chesapeake Bay, as Achiridae); FOWLER, H. W., No. 1, 1945 (southeastern U.S., as Achiridae); BEEBE & TEE-VAN, No. 1, 1928 (Hispaniola, as Achiridae); MEEK, S. E., No. 1, 1904 (fresh-water, Mexico); MEEK & HILDEBRAND, No. 4, 1928 (marine, Panama); HILDEBRAND, S. F., No. 1, 1946 (shore, Peru); EIGENMANN, C. H., No. 1, 1912 (British Guiana); FOWLER, H. W., No. 13, 1954 (fresh-water, Brasil, as Achiridae); DEVINCENZI, G. F., no. 2, 1924 (Uruguay); CHABANAUD, P., 1927, Bull. Inst. Oceanogr. Monaco, 488, 1-67 (Atlantique oriental Nord); DUNCKER & LADIGES, No. 1, 1960 (Nordmark); POLL, M., No. 2, 1947 (marine, Belgique); LOZANO y REY, L., No. 5, 1960 (España); BANARESCU, P., No. 1, 1964 (Romania); SVETOVIDOV, A. N., No. 2, 1964 (Black Sea); FOWLER, H. W., No. 8, 1936 (marine, West Africa); DAGET & ILTIS, No. 1, 1965 (Côte d'Ivoire); SMITH, J. L. B., No. 1, 1961 (marine, South Africa); FOWLER, H. W., No. 18, 1956 (Red Sea); de SILVA, P. H. D. H., No. 2, 1956 (Ceylon); SMITH, H. M., No. 1, 1945 (fresh-water, Thailand, as Synapturidae); WEBER & de BEAUFORT, No. 5, 1929 (Indo-Australian Arch.); MARSHALL, T. C., No. 1, 1964 (Great Barrier Reef); CHEN, J. T. F. & WENG, H. T. C., 1965, Biol. Bull. Tunghai Univ., 25, & 27 (Taiwan); FOWLER, H. W., No. 6, 1933-35 (China); SOLDATOV & LINDBERG, No. 1, 1930 (Far East seas); JORDAN, D. S. & STARKS, E. C., 1906, Proc. United States Nat. Mus., 31, 161-246 (Japan); OCHIAE, A., 1963, Fauna Japonica, Soleina, Tokyo, 1-114 (Japan); SCHULTZ, L. P. et al., No. 1, 1966 (Marshall & Marianas Is.)
Solenichthyidae, as Solenostomidae
Solenostomatichthyidae, as Solenostomidae
Solenostomatidae, as Solenostomidae
Solenostomidae (incl. Solenichthyidae, Solenostomatichthyidae, Solenostomatidae). SMITH, J. L. B., No. 1, 1961 (marine, South Africa); FOWLER, H. W., No. 18, 1956 (Red Sea, as Solenostomatichthyidae); de SILVA, P. H. D. H., 1956, Spol. Zeylanica, 28, 35-45 (Ceylon & museum list); WEBER & de BEAUFORT, No. 4, 1922 (Indo-Australian Arch.); FOWLER, H. W., No. 6, 1933-35 (China, as Solenostomatichthyidae); JORDAN, D. S. & SNYDER, J. O., 1901, Proc. United States Nat. Mus., 24, 1-20 (Japan); SCHULTZ, L. P. et al., No. 2, 1953 (Marshall & Marianas Is.)
Somniosidae. SOLDATOV & LINDBERG, No. 1, 1930 (Far East seas)
Sorosichthyidae, as Trachichthyidae
Spaniodontidae, as Elopidae
Sparidae (incl. Denticidae, Paradichthyidae, Paradicichthyidae, Pimelepteridae). AKAZAKI, M., 1962, Studies on spariform fishes . . . , Osaka, 1-368 (keys, Japan, in Japanese); HILDEBRAND & SCHROEDER, No. 1, 1928 (Chesapeake Bay); BEEBE & TEE-VAN, No. 1, 1928 (Hispaniola); MEEK & HILDEBRAND, No. 3, 1925 (marine, Panama); HILDEBRAND, S. F., No. 1, 1946 (shore, Peru); FOWLER, H. W., No. 13, 1954 (fresh-water, Brasil); DEVINCENZI, G. F., No. 2, 1924 (Uruguay); SMITT, F. A. et al., No. 1, 1892 (Scandinavia); POLL, M., No. 2, 1947 (marine, Belgique); NOBRE, A., No. 1, 1935 (marine, Portugal); LOZANO REY, D. L., No. 2, 1952 (España); FAGE, L., No. 1, 1918 (Mediterranean); BANARESCU, P., No. 1, 1964 (Romania); SVETOVIDOV, A. N., No. 2, 1964 (Black Sea); BERG, L. S., No. 5, 1949 (fresh-water, Soviet Union); BERG, L. S., No. 10, 1965 (fresh-water, U. S. S. R.); BOULENGER, G. A., No. 5, 1915 (fresh-water, Africa); FOWLER, H. W., No. 8, 1936 (marine, West Africa); SMITH, J. L. B., No. 1, 1961 (marine, South Africa, also as Denticidae); POLL, M., No. 4, 1954 (Coastal South Africa); DAY, F., No. 2, 1889 (India); WEBER & de BEAUFORT, No. 7, 1936 (Indo-Australian Arch.); MUNRO, I. S. R., 1949, Mem. Queensland Mus., 12, 182-223 (Australian); HERRE, A. W. & MONTALBAN, H. R., 1927, Philippine Journ. Sci., 33, 397-439 (Philippines); FOWLER, H. W., No. 20, 1933 (Philippine seas); OSHIMA, M., 1927, Japan Journ. Zool., 1, 127-155 (Taiwan); FOWLER, H. W., No. 9, 1937-41 (China); SOLDATOV & LINDBERG, No. 1, 1930 (Far East seas); LINDBERG & KRASJUKOVA, No. 1, 1969 (Sea of Japan); JORDAN, D. S. & THOMPSON, W. F., 1912, Proc. United States Nat. Mus., 41, 521-601 (Japan); FOWLER, H. W., No. 17, 1959 (Fiji)
Sparisomidae, as Scaridae
Sphaerodontidae, as Semionotidae (only as fossils)
Sphoeroididae, as Tetraodontidae
Sphrynidae, as Sphyrnidae
Sphyraenidae (incl. Sphyrenidae, Sphyrinidae). HILDEBRAND & SCHROEDER, No. 1, 1928 (Chesapeake Bay); BEEBE & TEE-VAN, No. 1, 1928 (Hispaniola); REGAN, C. T., No. 1, 1906-08 (Central America); MEEK & HILDEBRAND, No. 2, 1923 (marine, Panama); HILDEBRAND, S. F., No. 1, 1946 (shore, Peru); NOBRE, A., No. 1, 1935 (marine, Portugal); LOZANO REY, D. L.,

No 4, 1947 (España); BANARESCU, P., No. 1, 1964 (Romania); SVETOVIDOV, A. N., No. 2, 1964 (Black Sea); BOULENGER, G. A., No. 6, 1916 (fresh-water, Africa); PELLEGRIN, J., No. 2, 1923 (western Africa); FOWLER, H. W., No. 8, 1936 (marine, West Africa); DAGET & ILTIS, No. 1, 1965 (Côte d'Ivoire); BOULENGER, G. A., No. 1, 1901 (Congo basin); SMITH, J. L. B., No. 1, 1961 (marine, South Africa); SMITH, J. L., 1956, Ichth. Bull. Rhodes Univ., 3, 37-45 western Indian Ocean); DAY, F., No. 2, 1889 (India); WEBER & de BEAUFORT, No. 4, 1922 (Indo-Australian Arch.); MARSHALL, T. C., No. 1, 1964 (Great Barrier Reef); FOWLER, H. W., No. 6, 1933-35 (China); LINDBERG & LEGHEZA, No. 2, 1965 (Sea of Japan); LINDBERG & LEGEZA, No. 3, 1969 (Sea of Japan); SCHULTZ, L. P. et al., No. 2, 1953 (Marshall & Marianas Is.); FOWLER, H. W., No. 17, 1959 (Fiji)

Sphyrenidae, as Sphyraenidae

Sphyrinidae, as Sphyraenidae

Sphyrnidae (incl. Sphrynidae). GILBERT, C. R., 1967, Proc. United States Nat. Mus., 119, 3539, 1-88 (revision); GILBERT, C. R. et al., 1967, Sharks, skates, and rays, Baltimore, 1-624 (Synopsis); BEEBE & TEE-VAN, No. 1, 1928 (Hispaniola); HILDEBRAND, S. F., No. 1, 1946 (shore, Peru); PUYO, J., No. 1, 1949 (Guyane); DEVINCENZI, G. J., No. 1, 1920 (Uruguay); BIGELOW & SCHROEDER, No. 1, 1948 (western North Atlantic); TORTONESE, E., 1949-50, Boll. Mus. Zool. Torino, 2, (2), 1-39 (Mediterranean); TORTONESE, E., No. 3, 1956 (Italia); FOWLER, H. W., No. 8, 1936 (marine, West Africa); SMITH, J. L. B., No. 1, 1961 (marine, South Africa); FOWLER, H. W., No. 18, 1956 (Red Sea); GOHAR & MAZHAR, No. 1, 1964 (Red Sea); MARSHALL, T. C., No. 1, 1964 (Great Barrier Reef); FOWLER, H. W., No. 2, 1930 (China); SOLDATOV & LINDBERG, No. 1, 1930 (Far East seas); LINDBERGH & LEGHEZA, No. 1, 1959 (Japan Sea); JORDAN, D. S. & FOWLER, H. W., 1903, Proc. United States Nat. Mus., 26, 593-674 (Japan); FOWLER, H. W., No. 17, 1959 (Fiji)

Spicaridae, as Emmelichthyidae

Spinacanthidae (only as fossils)

Spinacidae, as Squalidae

Squalidae (incl. Spinacidae). REGAN, C. T., 1908, Ann. Mag. Nat. Hist., (8)2, 39-56 (synopsis); GARMAN, S., No. 1, 1913 (monograph); BIGELOW, H. B. & SCHROEDER, W. C., 1957, Bull. Mus. Comp. Zool., 117, 1-150; HUBBS, C. L. & McHUGH, J. L., 1951, Proc. California Acad. Sci., (4), 27, 159-176 (key to genera of Dalatinae); HILDEBRAND & SCHROEDER, No. 1, 1928 (Chesapeake Bay); MEEK & HILDEBRAND, No. 2, 1923 (marine, Panama); DEVINCENZI, G. J., No. 1, 1920 (Uruguay); NORMAN, J. R., No. 2, 1937 (marine, Patagonia); BIGELOW & SCHROEDER, No. 1, 1948 (western North Atlantic); SMITT, F. A. et al., No. 2, 1895 (Scandinavia, as Spinacidae); DUNCKER & LADIGES, No. 1, 1960 (Nordmark); POLL, M., No. 2, 1947 (marine, Belgique); NOBRE, A., No. 1, 1935 (marine, Portugal, as Spinacidae); TORTONESE, E., No. 3, 1956 (Italia); SVETOVIDOV, A. N., No. 2, 1964 (Black Sea); BERG, L. S., No. 7, 1911 (Russia); ANDRIYASHEV, A. P., No. 1, 1954 (northern seas of U. S. S. R.); FOWLER, H. W., No. 8, 1936 (marine, West Africa); SMITH, J. L. B., No. 1, 1961 (marine, South Africa); FOWLER, H. W., No. 2, 1930 (China); SOLDATOV & LINDBERG, No. 1, 1930 (Far East seas); SCHMIDT, P. Y., No. 1, 1950 (Sea of Okhotsk); SHMIDT, P. Y., No. 2, 1965 (Sea of Okhotsk); LINDBERGH & LEGHEZA, No. 1, 1959 (Japan Sea); JORDAN. D. S. & FOWLER, H. W., 1903, Proc. United States Nat. Mus., 26, 593-674 (Japan)

Squalorajidae (only as fossils)

Squatinidae (incl. Rhinidae). HILDEBRAND & SCHROEDER, No. 1, 1928 (Chesapeake Bay); HILDEBRAND S. F., No. 1, 1946 (shore, Peru); DEVINCENZI, G. J., No. 1, 1920 (Uruguay); MARINI, T. L., 1936, Physis, 12, 19-30 (Argentina); NORMAN, J. R., No. 2, 1937 (marine, Patagonia); BIGELOW & SCHROEDER, No. 1, 1948 (western North Atlantic); POLL, M., No. 2, 1947 (marine, Belgique); TORTONESE, E., No. 3, 1956 (Italia); FOWLER, H. W., No. 8, 1936 (marine, West Africa); SMITH, J. L. B., No. 1, 1961 (marine, South Africa); FOWLER, H. W., No. 18, 1956 (Red Sea); FOWLER, H. W., No. 2, 1930 (China); LINDBERGH & LEGHEZA, No. 1, 1959 (Japan Sea); JORDAN, D. S. & FOWLER, H. W., 1903, Proc. United States Nat. Mus., 26, 593-674 (Japan)

Stegotrachelidae. (No revisionary references noted)

Steinegeriidae, as Bramidae

Stensiöellidae (only as fossils)

Stephaloberycidae. (No revisionary references noted)

Sternarchidae, as Apteronotidae

Sternoptychidae (incl. Sternoptychiidae). SCHULTZ, L. P., 1961, Proc. United States Nat. Mus., 112, 587-649 (marine); NORMAN, J. R., No. 1, 1930 (oceanic); BIGELOW, H. B. et al., No. 3, 1964 (western North Atlantic); NOBRE, A., No. 1, 1935 (marine, Portugal); LOZANO REY, D. L., No. 4, 1947 (España); KARLOVAC, J., 1953, Rep. Inst. Oceanogr. Split, 5, (2B), 1-46 (Adriatic Sea); ANDRIYASHEV, A. P., No. 1, 1954 (northern seas of U. S. S. R.); FOWLER, H. W., No. 8, 1936 (marine, West Africa); SMITH, J. L. B., No. 1, 1961 (marine, South Africa); FOWLER, H. W., No. 18, 1956 (Red Sea); HAIG, J., 1955, Pacific Sci., 9, 318-323 (key to Hawaiian species)

Sternoptychiidae, as Sternoptychidae

Sternopygidae, as Apteronotidae

Stevardiidae. FOWLER, H. W., No. 10, 1948-51 (fresh-water, Brasil)

Stichaeidae (incl. Cebedichthyidae, Chirolophidae, Cryptacanthidae, Cryptacanthodidae, Lumpenidae, Sticheidae, Xiphidiontidae, Xiphisteridae). MAKUSHOK, V. M., 1958, Trud. Inst. Zool., 25, 3-129 (morphology & systematics, in Russian); DUNCKER & LADIGES, No. 1, 1960 (Nordmark, also as Lumpenidae); ANDRIYASHEV, A. P. No. 1, 1954 (northern seas of U. S. S. R., also as Lumpenidae); FOWLER, H. W., No. 16, 1958 (China, also as Cryptacanthodidae & Lumpenidae); SOLDATOV & LINDBERG, No. 1, 1930 (Far East seas, as Cryptacanthodidae); SCHMIDT, P. Y., No. 1, 1950 (Sea of Okhotsk, also as Chirolophidae & Lumpenidae); SHMIDT, P. Y., No. 2, 1965 (Sea of Okhotsk, also as Chirolophidae & Lumpenidae)

Stichariidae, as Notograptidae

Sticheidae, as Stichaeidae

Stigmatonotidae, as Grammidae

Stilbiscidae, as Moringuidae

Stolephoridae, as Engraulidae

Stomiatidae. REGAN, C. T. & TREWAVAS, E., 1930, Dana Oceanogr. Rep., 6, 1-143; NORMAN, J. R., No. 1, 1930 (oceanic); ZUGMAYER, E., 1940, Res. Camp. Sci. Monaco, 103, 201-205 (key to genera); BIGELOW, H. B. et al., No. 3, 1964 (western North America); LOZANO REY, D. L., No. 4, 1947 (España); KARLOVAC, J., 1953, Rep. Inst. Oceanogr. Split, 5, (2B), 1-46 (Adriatic Sea); FOWLER, H. W., No. 8, 1936 (marine, West Africa); SMITH, J. L. B., No. 1, 1961 (marine, South Africa); FOWLER, H. W., No. 18, 1956 (Red Sea); WEBER & de BEAUFORT, No. 2, 1913 (Indo-Australian Arch.)

Stratodontidae, as Dercetidae (Only as fossils)

Stromateidae (incl. Pampidae). REGAN, C. T., 1902, Ann. Mag. Nat. Hist., (7), 10, 115-131, 194-207 (revision); HAEDRICH, R. L., Bull. Mus. Comp. Zool., 135, 31-139 (systematics & classification); NORMAN, J. R., No. 1, 1930 (oceanic); HILDEBRAND & SCHROEDER, No. 1, 1928 (Chesapeake Bay); BEEBE & TEE-VAN, No. 1, 1928 (Hispaniola); MEEK & HILDEBRAND, No. 3, 1925 (marine, Panama); HILDEBRAND, S. F., No. 1, 1946 (shore, Peru); PUYO, J., No. 1, 1949 (Guyane); DEVINCENZI, G. L., No. 2, 1924 (Uruguay); NORMAN, J. R., No. 2, 1937 (marine, Patagonia); POLL, M., No. 2, 1947 (marine, Belgique); NOBRE, A., No. 1, 1935 (marine, Portugal); LOZANO REY, D. L., No. 3, 1952 (España); FOWLER, H. W., No. 8, 1936 (marine, West Africa); SMITH, J. L. B., 1949, Ann. Mag. Nat. Hist., (12), 2, 839-851 (South Africa); SMITH, J. L. B., No. 1, 1961 (marine, South Africa); DAY, F., No. 2, 1889 (India); de BEAUFORT & CHAPMAN, No. 9, 1951 (Indo-Australian Arch.); FOWLER, H. W., No. 7, 1936 (China); SOLDATOV & LINDBERG, No. 1, 1930 (Far East seas); SCHULTZ, L. P. et al., No. 3, 1960 (Marshall & Marianas Is)

Stylephoridae (incl. Stylophroidae). NORMAN, J. R., No. 1, 1930 (oceanic)

Stylodontidae, as Semionotidae (only as fossils)

Stylophoridae, as Stylephoridae

Stylophthalmidae, as Idiacanthidae

Stylophthalmoidae, as Idiacanthidae

Styracopteridae, as Palaeoniscidae (only as fossils)

Sudidae, as Paralepididae

Syllaemidae (incl. Pelycorapidae). (Only as fossils)

Symbranchidae, as Synbranchidae

Sananceidae, as Synancejidae

Synancejidae (incl. Synanceidae, Synanciidae). SMITH, J. L. B., No. 1, 1961 (marine, South Africa); SMITH, J. L. B., 1958, Ichth. Bull. Rhodes Univ., 12, 167-181 (western Indian Ocean); de BEAUFORT & BRIGGS, No. 11, 1962 (Indo-Australian Arch.); FOWLER, H. W., No. 17, 1959 (Fiji)

Synanciidae, as Synancejidae

Synaphobranchidae (incl. Ilyophidae, Synapobranchidae).). BRUUN, A. F., 1937, Dana Rep., 9, 1-31 (life histories); CASTLE, P. H. J., 1964, Galathea Rep., 7, 29-42 (monograph); NOBRE, A., No. 1, 1935 (marine, Portugal); LOZANO REY, D. L., No. 4, 1947 (España); FOWLER, H. W., No. 8, 1936 (marine, West Africa); SMITH, J. L. B., No. 1, 1961 (marine, South Africa, also as Ilyophidae); WEBER & de BEAUFORT, No. 3, 1916 (Indo-Australian Arch.); CASTLE, P. H. J., 1965, Trans. Roy. Soc. New Zealand, Zool., 5, 131-146 (Australia); JORDAN & SNYDER, No. 1, 1901 (Japan)

Synapobranchidae, as Synaphobranchidae

Synapturidae, as Soleidae

Synaucheniidae (only as fossils)

Synbranchidae (incl. Flutidae, Monopteridae, Symbranchidae). REGAN, C. T., 1912, Ann. Mag. Nat. Hist., (8), 9, 387-390 (anatomy & classification); REGAN, C. T., No. 1, 1906-08 (Central America); MEEK, S. E., No. 1, 1904 (fresh-water, Mexico); MEEK & HILDEBRAND, No. 2, 1923 (marine Panama); EIGENMANN, C. H., No. 1, 1912 (British Guiana); FOWLER, H. W., No. 13, 1954 (fresh-water, Brasil); DEVINCENZI, G. J., No. 2, 1924 (Uruguay); BERG, L. S., No. 2, 1933 (fresh-water, U. R. S. S.); BERG, L. S., No. 5, 1949 (fresh-water, Soviet Union); BERG, L. S., No. 10, 1965 (fresh-water, U. S. S. R.); BOULENGER, G. A., No. 5, 1915 (fresh-water, Africa); PELLEGRIN, J., No. 2, 1923 (western Africa); DAY, F., No. 1, (India); SMITH, H. M., No. 1, 1945 (fresh-water, Thailand, also as Flutidae); WEBER & de BEAUFORT, No. 3, 1916 (Indo-Australian Arch.); INGER & KONG, No. 1, 1962 (fresh-water, North Borneo); HERRE, A. W. C. T., 1923, Philippine Journ. Sci., 23, 123-236 (Philippines); OSHIMA, M., No. 1, 1919 (fresh-water, Taiwan, as Monopteridae); POPE & NICHOLS, No. 1, 1927 (Hainan I.); NICHOLS, J. T., No. 1, 1943 (fresh-water, China); as Monopteridae); JORDAN, & SNYDER, No. 1, 1901 (Japan, as Monopteridae); OKADA, Y., No. 1, 1960 (fresh-water, Japan, as Flutidae); OKADA, Y., No. 2, 1960 (fresh-water, Japan, as Flutidae)

Synchiridae, as Cottidae

Syngnathidae (incl. Hippocampidae, Siphostomidae, Syngnatidae). DUNCKER, G., 1912, Mitt. Naturh. Mus. Hamburg, 29, 219-240 (genera; also in Jahrb. Wiss. Anst., 29, 219-240); DUNCKER, G., 1915, Mitt. Naturh. Mus. Hamburg, 32, 9-120 (revision); KAHSBAUER, P., 1950, Ann. Naturh. Mus.

Wien, 57, 263-272; HERALD, E. S., 1942, Stanford Ichth. Bull., 2, 125-134 (keys to Atlantic America); HERALD, E. S., 1940, Rep. Allen Hancock Pac. Exp. 1932-38, 9, 51-64 (keys to Pacific America); SWAIN, J., 1882, Proc. United States Nat. Mus., 5, 307-315 (United States); HILDEBRAND & SCHROEDER, No. 1, 1928 (Chesapeake Bay); FOWLER, H. W., No. 1, 1945 (fresh-water, southeastern U. S.); BEEBE & TEE-VAN, No. 1, 1928 (Hispaniola); REGAN, C. T., No. 1, 1906-08 (Central America); MEEK, S. E., No. 1, 1904 (fresh-water, Mexico); MEEK & HILDEBRAND, No. 2, 1923 (marine, Panama); HILDEBRAND, S. F., No. 1, 1946 (shore, Peru); EIGENMANN, C. H., No. 1, 1912 (British Guiana); DEVINCENZI, G. J., No. 2, 1924 (Uruguay, also as Hippocampidae); NORMAN, J. R., No. 2, 1937 (marine, Patagonia); SMITT, F. A. et al., No. 2, 1895 (Scandinavia); DUNCKER & LADIGES, No. 1, 1960 (Nordmark); POLL, M., No. 2, 1947 (marine, Belgique); NOBRE, A., No. 1, 1935 (marine, Portugal); LOZANO REY, D. L., No. 1, 1935 (España); LOZANO REY, D. L., No. 4, 1947 (España); di CAPORIACCO, L., 1948, Boll. Pesca Piscicolt. Idrobiol. Roma, 24, 73-95 (Italia); RAUTHER, M., 1925, Pubbl. Staz. Zool. Napoli, Mon, 36, 1-365 (Golfes von Neapel); BANARESCU, P., No. 1, 1964 (Romania); SVETOVIDOV, A. N., No. 2, 1964 (Black Sea); BERG, L. S., No. 2, 1933 (fresh-water, U. R. S. S.); BERG, L. S., No. 5, 1949 (fresh-water, Soviet Union); BERG, L. S., No. 10, 1965 (fresh-water, U. S. S. R.); ANDRIYASHEV, A. P., No. 1, 1954 (northern seas of U. S. S. R.); BOULENGER, G. A., No. 5, 1915 (fresh-water, Africa); PELLEGRIN, J., No. 2, 1923 (western Africa); FOWLER, H. W., No. 8, 1936); (Marine, West Africa); DAGET, J., No. 2, 1962 (Guinea); DAGET & ILTIS, No. 1, 1965 (Côte d'Ivoire); SMITH, J. L. B., No. 1, 1961 (marine, South Africa); ARNOULT, J., No. 1, 1959 (fresh-water, Madagascar); PELLEGRIN, J., No. 2, 1960 (fresh-water, Madagascar); FOWLER, N. W., No. 18, 1956 (Red Sea); SMITH, J. L. B., 1963, Ichth. Bull. Rhodes Univ., 27, 515-543 (Red Sea); SMITH, J. L. B., 1963, Icht. Bull. Rhodes Univ., 27, 515-543 (Red Sea & western Indian Ocean); DAY, F., No. 2, 1889 (India); de SILVA, P. H. D. H., 1956, Spol. Zeylanica, 28, 35-45 (Ceylon & museum list); SMITH, H. M., No. 1, 1945 (fresh-water, Thailand); WEBER & de BEAUFORT, No. 4, 1922 (Indo-Australian Arch.); INGER & KONG, No. 1, 1962 (fresh-water, North Borneo); MARSHALL, T. C., No. 1, 1964 (Great Barrier Reef); FOWLER, H. W., No. 6, 1933-35 (China); CHEN, J. T. F., 1935, Biol. Bull. Norm. Coll. Shiang Chyn Univ., 1, 1-22 (China); SOLDATOV & LINDBERG, No. 1, 1930 (Far East seas); LINDBERG & LEGHEZA, No. 2, 1965 (Sea of Japan); LINDBERG & LEGEZA, No. 3, 1969 (Sea of Japan); JORDAN, D. S. & SNYDER, J. O., 1901, Proc. United States Nat. Mus., 24, 1-20 (Japan); SCHULTZ, L. P. et al., No. 2, 1953 (Marshall & Marianas Is.); FOWLER, H. W., No. 17, 1959 (Fiji)

Syngnatidae, as Syngnathidae

Synodidae (lizard fishes, marine), as Synodontidae

Synodidae (catfishes, fresh-water, Africa), as Mochokidae

Synodontidae (incl. Bathysauridae, Sauridae, Synodidae). HILDEBRAND & SCHROEDER, No. 1, 1928 (Chesapeake Bay); BEEBE & TEE-VAN, No. 1, 1928 (Hispaniola); MEEK & HILDEBRAND, No. 2, 1923 (marine, Panama); HILDEBRAND, S. F., No. 1, 1946 (shore, Peru); BIGELOW, H. B. et al., No. 4, 1966 (western North Atlantic, also as Bathysauridae); MEAD, G. W., 1966, Mem. Sears Found Mar. Res., 1, (5), 19-29, 103-189 (western North Atlantic, as Bathysauridae); GIBB, R. H., 1959, Copeia, 1959, 232-236 (post-larvae, western Atlantic); ANDERSON, W. W. et al., 1966, Circ. United States Fish. Wildl. Serv., 245, 1-12 (field guide, western Atlantic); LOZANO REY, D. L., No. 4, 1947 (España); FOWLER, H. W., No. 8, 1936 (marine, West Africa); SMITH, J. L. B., No. 1, 1961 (marine, South Africa); FOWLER, H. W., No. 18, 1956 (Red Sea); MARSHALL, T. C., No. 1, 1964 (Great Barrier Reef); FOWLER, H. W., No. 5, 1932 (China); LINDBERG & LEGHEZA, No. 2, 1965 (Sea of Japan); LINDBERG & LEGEZA, No. 3, 1969 (Sea of Japan); JORDAN, D. S. & HERRE, A. C., 1907, Proc. United States Nat. Mus., 32, 513-524 (Japan); SCHULTZ, L. P. et al., No. 2, 1953 (Marshall & Marianas Is.); FOWLER, H. W., No. 17, 1959 (Fiji)

Tachysuridae, as Ariidae

Taenioidae, as Gobioididae

Taenioididae, as Gobioididae

Taeniophoridae, as Eutaeniophoridae

Tamiobatidae, (only as fossils)

Tannuaspididae (only as fossils)

Tarrasiidae (only as fossils)

Tegiolepididae (only as fossils)

Teleopterinidae (only as fossils)

Teraponidae, as Theraponidae

Terapontidae, as Theraponidae

Tesapontidae, as Theraponidae

Tetragonopteridae, as Characidae

Tetragonuridae. LOZANO REY, D. L., No. 3, 1952 (España); FOWLER, H. W., No. 8, 1936 (marine, West Africa)

Tetraodontidae (incl. Canthigasteridae, Chonarhinidae, Colomesidae, Gymnodontidae, Lagocephalidae, Ovoididae, Sphoeroididae, Tetrodontidae, Tropidichthyidae, Xenopteridae). GILL, T., No. 1, 1884 (synopsis, also as Chonerhinidae); FRASER-BRUNNER, A., No. 1, 1943 (classification, also as Canthigasteridae, Chonerhinidae, Colomesidae, Lagocephalidae); BENL, G., 1956, Aquar. Terrar. Z., 9, 141-147, 172-175, 202-205 (fresh-water, keys); LeDANOIS, Y., No. 1, 1959 (osteology, myology, systematics, also as Colomesidae, Lagocephalidae, Xenopteridae);

LeDANOIS, Y., 1961, Bull. Mus. Nat. Hist. Nat., (2), 33, 276-281 (musuem type list, as Canthigasteridae & Xenopteridae); JORDAN, D. S. & EDWARDS, C. L., 1886, Proc. United State Nat. Mus., 9, 230-247 (America); HILDEBRAND & SCHROEDER, No. 1, 1928 (Chesapeake Bay); BEEBE & TEE-VAN, No. 1, 1928 (Hispaniola, also as Canthigasteridae); MEEK & HILDEBRAND, No. 4, 1928 (marine, Panama); HILDEBRAND, S. F., No. 1, 1946 (shore, Peru); EIGENMANN, C. H., No. 1, 1912 (British Guiana); PUYO, J., No. 1, 1949 (Guyane); FOWLER, H. W., No. 13, 1954 (fresh-water, Brasil); DEVINCENZI, G. F., No. 2, 1924 (Uruguay); LOZANO REY, D. L., No. 2, 1952 (España); BERG, L. S., No. 5, 1949 (fresh-water, Soviet Union); BERG, L. S., No. 10, 1965 (fresh-water, U. S. S. R.); BOULENGER, G. A., No. 6, 1916 (fresh-water, Africa); BOULENGER, G. A., No. 2, 1907 (Egypt); PELLEGRIN, J., No. 2, 1923 (western Africa); FOWLER, H. W., No. 8, 1936 (marine, West Africa, also as Canthigasteridae); DAGET, J., No. 2, 1962 (Guinea); DAGET & ILTIS, No. 1, 1965 (Côte d'Ivoire, also as Lagocephalidae); DAGET, J., No. 1, 1954 (Upper Niger); BOULENGER, G. A., No. 1, 1901 (Congo basin); SMITH, J. L. B., No. 1, 1961 (marine, South Africa, also as Canthigasteridae & Lagocephalidae); SMITH, J. L. B., 1958, Ann. Mag. Nat. Hist., (13), 1, 156-160 (South & East Africa); POLL, M., No. 3, 1953 (Lake Tanganyika); SMITH, H. W., No. 1, 1945 (fresh-water, Thailand); de BEAUFORT & BRIGGS, No. 11, 1962 (Indo-Australian Arch.); INGER & KONG, No. 1, 1962 (fresh-water, North Borneo); MARSHALL, T. C., No. 1, 1964 (Great Barrier Reef, also as Canthigasteridae); NICHOLS, J. T., No. 1, 1943 (fresh-water, China); SOLDATOV & LINDBERG, No. 1, 1930 (Far East seas); JORDAN, D. S. & SNYDER, J. O., 1921, Proc. United States Nat. Mus., 24, 229-264 (Japan, also as Tropidichthyidae); ABE, T., 1949, Bull. Biogeogr. Soc. Japan, 14, (1), 1-15, (13), 89-140s(Japan); SCHULTZ, L. P. et al., No. 1, 1966, (Marshall & Marianas Is., also as Canthigasteridae); FOWLER, H. W., No. 17, 1959 (Fiji, Canthigasteridae)

Tetrapturidae, as Istiophoridae

Tetrarogidae, as Scorpaenidae

Tetrodontidae, as Tetraodontidae

Teuthidae, as Acantharidae

Teuthididae, part as Acanthuridae, part as Siganidae

Teuthrididae, as Acanthuridae

Thaumaturidae (only as fossils)

Thelodontidae, as Coelolepidae (Only as fossils)

Theraponidae (incl. Teraponidae, Terpontidae, Tesapontidae). NOBRE, A., No. 1, 1935 (marine, Portugal); SMITH, J. L. B., No. 1, 1961 (marine, South Africa); WEBER & de BEAUFORT, No. 6, 1931 (Indo-Australian Arch.); MARSHALL, T. C., No. 1, 1964 (Great Barrier Reef); FOWLER, H. W., No. 19, 1931 (Philippine seas); CHAN, W. L., No. 1, 1968 (Hong Kong); FOWLER, H. W., No. 9, 1937-41 (China); LINDBERG & KRASJUKOVA, No. 1, 1969 (Sea of Japan); OKADA, Y., No. 1, 1960 (fresh-water, Japan); OKADA, Y., No. 2, 1960 (fresh-water, Japan); FOWLER, H. W., No. 17, 1959 (Fiji)

Theutyidae, as Siganidae

Thryptodontidae, as Plethodidae (only as fossils)

Thunnidae, as Scombridae

Thyestidae (only as fossils)

Thymallidae, as Salmonidae

Titanichthyidae (only as fossils)

Tjalvidae (only as fossils)

Todaridae. (No revisionary references noted)

Tolypelepidae (only as fossils)

Tomeuridae, as Poeciliidae

Tomognathidae (only as fossils)

Torpedinidae (not an electric ray), as Malapteruridae

Torpedinidae (incl. Narcationtidae). HILDEBRAND & SCHROEDER, No. 1, 1928 (Chesapeake Bay); HILDEBRAND, S. F., No. 1, 1946 (shore, Peru); FOWLER, H. W., No. 10, 1948-51 (fresh-water, Brasil); DEVINCENZI, G. J., No. 2, 1924 (Uruguay); NORMAN, J. R., No. 2, 1937 (marine, Patagonia); BIGELOW & SCHROEDER, No. 2, 1953 (western North Atlantic); POLL, M., No. 2, 1947 (marine, Belgique); NOBRE, A., No. 1, 1935 (marine, Portugal); TORTONESE, E., No. 3, 1956 (Italia); FOWLER, H. W., No. 8, 1936 (marine, West Africa); SMITH, J. L. B., No. 1, 1961 (marine, South Africa); FOWLER, H. W., No. 18, 1956 (Red Sea); GOHAR & MAZHAR, No. 1, 1964 (Red Sea); DAY, F., No. 1, 1889 (India); MARSHALL, T. C., No. 1, 1964 (Great Barrier Reef); FOWLER, H. W., No. 2, 1930 (China); CHEN, J. T. F., No. 1, 1948 (China)

Toxotidae. SMITH, H. M., No. 1, 1945 (fresh-water, Thailand); WEBER & de BEAUFORT, No. 7, 1936 (Indo-Australian Arch.); INGER & KONG, No. 1, 1962 (fresh-water, North Borneo); MARSHALL, T. C., No. 1, 1964 (Great Barrier Reef); WHITLEY, G. P., 1950, Rec. Australian Mus., 22, 234-245 (keys, Australia);

Trachichthyidae (incl. Hoplopterygidae, Sorosichthyidae, Trachychthyidae). LOZANO REY, D. L., No. 2, 1952 (España); FOWLER, H. W., No. 8, 1936 (marine, West Africa); POLL, M., No. 4, 1954 (coastal South Africa); SMITH, J. L. B., No. 1, 1961 (marine, South Africa); FOWLER, H. W., No. 18, 1956 (Red Sea); WEBER & de BEAUFORT, No. 5, 1929 (Indo-Australian Arch.); MARSHALL, T. C., No. 1, 1964 (Great Barrier Reef)

Trachinidae (incl. Callipterygidae). DEVINCENZI, G. F., No. 2, 1924 (Uruguay); SMITT, F. A. et al., No. 1, 1892 (Scandinavia); DUNCKER & LADIGES, No. 1, 1960 (Nordmark); POLL, M., No. 2, 1947 (marine, Belgique); NOBRE, A., No. 1, 1935 (marine, Portugal); LOZANO y REY, L., No. 5,

No. 5, 1960 (España); FAGE, L., No. 1, 1918 (Mediterranean); BANARESCU, P., No. 1, 1964 (Romania); SVETOVIDOV, A. N., No. 2, 1964 (Black Sea); FOWLER, H. W., No. 8, 1936 (marine, West Africa); DAY, F., No. 2, 1889 (India); JORDAN, D. S. & SNYDER, J. O., 1902, Proc. United States Nat. Mus., 24, 461-497 (Japan)

Trachipteridae, (incl. Trachypteridae). SMITT, F. A. et al., No. 1, 1892 (Scandinavia); NOBRE, A., No. 1, 1935 (marine, Portugal); LOZANO REY, D. L., No. 4, 1947 (España); PALMER, G., 1961, Bull. British Mus. (Nat. Hist.), Zool., 7, 335-351 (northeastern Atlantic and Mediterranean); ANDRIYASHEV, A. P., No. 1, 1954 (northern seas of U. S. S. R.); FOWLER, H. W., No. 8, 1936 (marine, West Africa); SMITH, J. L. B., No. 1, 1961 (marine, South Africa); DAY, F., No. 2, 1889 (India); WEBER & de BEAUFORT, No. 5, 1929 (Indo-Australian Arch.); MARSHALL, T. C., No. 1, 1964 (Great Barrier Reef); LINDBERG,& LEGHEZA, No. 2, 1965 (Sea of Japan); LINDBERG & LEGEZA, No. 3, 1969 (Sea of Japan)

Trachyberycidae, as Bramidae

Trachychthyidae, as Trachichthyidae

Trachycorystidae, as Auchenipteridae

Trachypteridae, as Trachipteridae

Traquairaspidae (only as fossils)

Tremataspidae (only as fossils)

Trematosteidae (only as fossils)

Triacanthidae (incl. Triacanthodidae). GILL, T., No. 1, 1884 (synopsis); FRASER-BRUNNER, A., 1941 Ann. Mag. Nat. Hist., (11), 7, 42-430 (families, also as Triaconthodidae); TYLER, J. C., 1968, Acad. Nat. Sci. Philadelphia Monogr., 16, 1-364 (monograph, also as Triacanthodidae); SMITH, J. L. B., No. 1, 1961 (marine, South Africa, as Triacanthodidae); de BEAUFORT & BRIGGS, No. 11, 1962 (Indo-Australian Arch., also as Triacanthodidae); MARSHALL, T. C., No. 1, 1964 (Great Barrier Reef); JORDAN, D. S.,& FOWLER, H. W., 1902, Proc. United States Nat. Mus., 25, 251-286 (Japan); KAMOHARA, T., 1937, Annot. Zool. Japan, 16, 5-8 (Japan)

Triacanthodidae, as Triacanthidae

Triakidae, as Carcharhinidae

Trichiuridae (incl. Lepidopidae). TUCKER, D. W., 1956, Bull. British Mus. (Nat. Hist.), Zool., 4, 73-130 (revision); BOESEMAN, M., 1962, Journ. Mar. Biol. Assoc. India, 4, 214-216 (Leiden museum type list); HILDEBRAND & SCHROEDER, No. 1, 1928 (Chesapeake Bay); BEEBE & TEE-VAN, No. 1, 1928 (Hispaniola); MEEK & HILDEBRAND, No. 2, 1923 (marine, Panama); HILDEBRAND, S. F., No. 1, 1946 (shore, Peru); PUYO, J., No. 1, 1949 (Guyane); DEVINCENZI, G. L., No. 2, 1924 (Uruguay); NOBRE, A., No. 1, 1935 (marine, Portugal, also as Lepidopidae); LOZANO REY, D. L., No. 3, 1952 (España, also as Lepidopidae); SCHMIDT, J., & STRUBBERG, A., 1918, Rep. Danish Oceanogr. Exp. 1908-10, 2, (Biol. A6), 1-15 (Mediterranean); FOWLER, H. W., No. 8, 1936 (marine, West Africa, also as Lepidopidae); DAGET & ILTIS, No. 1, 1965 (Côte d'Ivoire); SMITH, J. L. B., No. 1, 1961 (marine, South Africa, also as Lepidopidae); DAY, F., No. 2, 1889 (India); JAMES, P. S. B. R., 1960, Journ. Mar. Biol. Assoc. India, 1, 139-142 (key, Indian waters); JAMES, P. S. B. R., 1962, Journ. Mar. Biol. Assoc. India, 3, 215-248 (comparative osteology, Indian waters); de BEAUFORT & CHAPMAN, No. 9, 1951 (Indo-Australian Arch.); MARSHALL, T. C., No. 1, 1964 (Great Barrier Reef); FOWLER, H. W., No. 7, 1936 (China); SOLDATOV & LINDBERG, No. 1, 1930 (Far East seas)

Trichodontidae. SOLDATOV & LINDBERG, No. 1, 1930 (Far East seas); SCHMIDT, P. Y., No. 1, 1950 (Sea of Okhotsk); SHMIDT, P. Y., No. 2, 1965 (Sea of Okhotsk); LINDBERG & KRASJUKOVA, No. 1, 1969 (Sea of Japan)

Trichomycteridae (incl. Nematogenyidae, Pygidiidae). de MIRANDO RIBEIRO, P., 1954, Publ. Avuls. Mus. Nac. Brasil, 15, 1-17 (museum list); MEEK & HILDEBRAND, No. 1, 1916 (fresh-water, Panama, as Pygidiidae); EIGENMANN & EIGENMANN, No. 1, 1890 (South America, as Pygidiidae); EIGENMANN, C. H., 1918, Mem. Carnegie Mus., 7, 259-373 (South America, as Pygidiidae); VAN DER STIGCHEL, J. W. B., No. 1, 1946 (South America, as Pygidiidae); SCHULTZ, L. P., No. 2, 1944 (Venezuela, as Pygidiidae); EIGENMANN, C. H., No. 1, 1912 (British Guiana); FOWLER, H. W., No. 13, 1954 (fresh-water, Brasil); MYERS, G. S., 1944, Proc. California Acad. Sci., (4), 23, 591-602 (Rio Negro, Brasil, as Pygidiidae); EIGENMANN, C. H., No. 2, 1927 (fresh-water, Chile, as Nematogenyidae & Pygidiidae)

Trichonotidae. SMITH, J. L. B., No. 1, 1961 (marine, South Africa); DAY, F., No. 2, 1889 (India); de BEAUFORT & CHAPMAN, No. 9, 1951 (Indo-Australian Arch.); FOWLER, H. W., No. 15, 1956 (China, as SCHULTZ, L. P. et al., No. 3, 1960 (Marshall & Marianas Is.)

Triglidae (incl. Peristediidae, Peristediontidae, Sclerogenidae part). TEAGUE, G. W., 1961, Anal. Mus. Hist. Nat. Montevideo, (2), 7, (2), 1-27 (America, as Peristediidae); HILDEBRAND & SCHROEDER, No. 1, 1928 (Chesapeake Bay); BEEBE & TEE-VAN, No. 1, 1928 (Hispaniola); MEEK & HILDEBRAND, No. 4, 1928 (marine, Panama); HILDEBRAND, S. F., No. 1, 1946 (shore, Peru); DENVINCENZI, G. F., No. 2, 1924 (Uruguay); GINSBURG, I., 1950, Texas Journ. Sci., 2, 489-527 (western Atlantic); RICHARDS, W. J., 1968, Atlantide Rep., 10, 77-114 (eastern Atlantic); DUNCKER & LADIGES, No. 1, 1960 (Nordmark); POLL, M., No. 2, 1947 (marine, Belgique); NOBRE, A., No. 1, 1935 (marine, Portugal); LOZANO REY, D. L., No. 2, 1952 (España, as Peristediidae); FAGE, L., No. 1, 1918 (Mediterranean); BANARESCU, P., No. 1, 1964 (Romania); SVETOVIDOV, A. N., No. 2, 1964 (Black Sea); SVETOVIDOV, A. N., 1936, Fauna S. S. S. R., VI, 9, Triglidae, Moskva, 1-24 (S. S. S. R.); ANDRIYASHEV, A. P., No. 1, 1954 (northern seas of U. S. S. R.); FOWLER, H. W., No. 8, 1936 (marine, West Africa, also as Peristediidae); SMITH, J. L. B., No. 1, 1961 (marine, South Africa, also as Peristediidae);

de BEAUFORT & BRIGGS, No. 11, 1962 (Indo-Australian Arch.); WHITLEY, G. P., 1953, Australian Mus. Mag., 11, 24-29 (key to Australian genera); MARSHALL, T. C., No. 1, 1964 (Great Barrier Reef); SOLDATOV & LINDBERG, No. 1, 1930 (Far East seas); JORDAN, D. S. & RICHARDSON, R. E., 1908, Proc. United States Nat. Mus., 33, 629-670 (Japan, also as Peristediidae); MATSUBARA, K. & HIYAMA, J., 1932, Journ. Imp. Fish. Inst., 28, 3-67 (Japan); KURONUMA, K., 1929, Bull. Biogeogr. Soc. Tokyo, 9, 223-260 (Japan)

Trigonodontidae (only as fossils)

Trinectidae, as Soleidae

Triodontidae. GILL, T., No. 1, 1884 (synopsis); de BEAUFORT & BRIGGS, No. 11, 1962 (Indo-Australian Arch.); JORDAN, D. S. & SNYDER, J. O., 1901, Proc. United States Nat. Mus., 24, 229-264 (Japan)

Tripterophycidae, as Moridae

Tripterygiidae (incl. Tripterygiontidae). SVETOVIDOV, A. N., No. 2, 1964 (Black Sea); SMITH, J. L. B., No. 1, 1961 (marine, South Africa)

Tripterygiontidae, as Tripterygiidae

Trissolepidae, as Palaeoniscidae (only as fossils)

Trissolepididae, as Palaeoniscidae (only as fossils)

Tristychiidae (only as fossils)

Triuridae, as Molidae

Tropidichthyidae, as Tetraodontiae

Trygonidae (incl. Dasyatidae, Dasybatidae). HILDEBRAND, & SCHROEDER, No. 1, 1928 (Chesapeake Bay, as Dasyatidae); BEEBE & TEE-VAN, No. 1, 1928 (Hispaniola, as Dasyatidae); MEEK & HILDEBRAND, No. 2, 1923 (marine, Panama, as Dasybatidae); HILDEBRAND, S. F, No. 1, 1946 (shore, Peru, as Dasyatidae); EIGENMANN, C. H., No. 1, 1912 (British Guiana); BOESEMAN, M., 1948, Zool. Meded., 30, 31-47 (key, Suriname, as Dasyatidae); PUYO, J., No. 1, 1949 (Guyane); DEVINCENZI, G. F., No. 1, 1920 (Uruguay, as Dasybatidae); BIGELOW & SCHROEDER, No. 2, 1953 (western North Atlantic, as Dasyatidae); SMITT, F. A. et al., No. 2, 1895 (Scandinavia); DUNCKER & LADIGES, No. 1, 1960 (Nordmark, as Dasyatidae); POLL, M., No. 2, 1947 (marine, Belgique, as Dasyatidae); NOBRE, A., No. 1, 1935 (marine, Portugal); TORTONESE, E., No. 3, 1956 (Italia, as Dasyatidae); SVETOVIDOV, A. N., No. 2, 1964 (Black Sea, as Dasyatidae); BERG, L. S., No. 7, 1911 (Russia); FOWLER, H. W., No. 8, 1936 (marine, West Africa, as Dasyatidae); DAGET & ILTIS, No. 1, 1965 (Côte d'Ivoire, as Dasyatidae); SMITH, J. L. B., No. 1, 1961 (marine, South Africa, as Dasyatidae); FOWLER, H. W., No. 18, 1956 (Red Sea, as Dasyatidae); GOHAR & MAZHAR, No. 1, 1964 (Red Sea, as Dasyatidae); DAY, F., No. 1, 1889 (India); SMITH, H. M., No. 1, 1945 (freshwater, Thailand, as Dasyatidae); MARSHALL, T. C., No. 1, 1964 (Great Barrier Reef, as Dasyatidae); SCOTT, E. O. G., 1957, Pap. Roy. Soc. Tasmania, 91, 145-156 (key, Tasmania, as Dasyatidae); FOWLER, H. W., No. 2, 1930 (China, as Dasyatidae); CHEN, J. T. F., No. 1, 1948 (China, as Dasyatidae); SOLDATOV & LINDBERG, No. 1, 1930 (Far East seas); LINDBERGH & LEGHEZA, No. 1, 1959 (Japan Sea, as Dasyatidae); JORDAN, D. S. & FOWLER, H. W., 1903, Proc. United States Nat. Mus., 26, 593-674 (Japan, as Dasyatidae); SCHULTZ, L. P. et al., No. 2, 1953 (Marshall & Marianas Is., as Dasyatidae); FOWLER, H. W., No. 17, 1959 (Fiji, as Dasyatidae)

Trypauchenidae. SMITH, J. L. B., No. 1, 1961 (marine, South Africa); SMITH, J. L. B., 1959 (western Indian Ocean); HERRE, A. W., No. 1, 1927 (Philippines & China Sea)

Tungusichthyidae (only as fossils)

Turiniidae (only as fossils)

Turseoidae (only as fossils)

Tylosuridae, as Belonidae

Umbridae (incl. Dalliidae, Novumbridae). SLASTENENKO, E. P., No. 1, 1958 (Canada); FOWLER, H. W., No. 1, 1945 (southeastern U. S.); SPILLMANN, J., No. 1, 1961 (France); BANARESCU, P., No. 1, 1964 (Romania); BERG, L. S., No. 2, 1933 (fresh-water, U. R. S. S.); BERG, L. S., No. 3, 1948 (fresh-water, Soviet Union, also as Dalliidae); BERG, L. S., No. 8, 1962 (fresh-water, U. S. S. R., also as Dalliidae)

Uranoscopidae (incl. Astroscopidae, Pleuroscopidae). HILDEBRAND & SCHROEDER, No. 1, 1928 (Chesapeake Bay); MEEK & HILDEBRAND, No. 4, 1928 (marine, Panama); HILDEBRAND, S. F., No. 1, 1946 (shore, Peru); DEVINCENZI, G. F., No. 2, 1924 (Uruguay); BERRY, F. H. & ANDERSON, W. W., 1961, Proc. United States Nat. Mus., 112, 563-586 (keys, western North Atlantic); LOZANO y REY, L., No. 5, 1960 (España); FAGE, L., No. 1, 1918 (Mediterranean); BANARESCU, P., No. 1, 1964 (Romania); SVETOVIDOV, A. N., No. 2, 1964 (Black Sea); FOWLER, H. W., No. 8, 1936 (marine, West Africa); SMITH, J. L. B., No. 1, 1961 (marine, South Africa, also as Pleuroscopidae); DAY, F., No. 2, 1889 (India); de BEAUFORT & CHAPMAN, No. 9, 1951 (Indo-Australian Arch.); MARSHALL, T. C., No. 1, 1964 (Great Barrier Reef); MEES, G. F., 1960, Journ. roy. Soc. Western Australia, 43, 46-58 (Western Australia); LIANG, Y., 1955, Quart. Journ. Taiwan Mus., 8, 169-176 (keys, Taiwan); FOWLER, H. W., No. 15, 1956 (China); LINDBERG & KRASJUKOVA, No. 1, 1969 (Sea of Japan); FOWLER, H. W., No. 17, 1959 (Fiji)

Urenchelyidae (only as fossils)

Urolophidae. BIGELOW & SCHROEDER, No. 2, 1953 (western North Atlantic); LINDBERGH & LEGHEZA, No. 1, 1959 (Japan Sea)

Uronemidae (only as fossils)

Urosphenidae. (No revisionary references noted)

Urosthenidae (only as fossils)

Veliferidae. SMITH, J. L. B., No. 1, 1961 (marine, South Africa); SMITH, J. L. B., 1951, Ann. Mag. Nat. Hist., (12), 4, 497-510 (south-east Africa); LINDBERG, & LEGHEZA, No. 2, 1965 (Sea of Japan); LINDBERG & LEGEZA, No. 3, 1969 (Sea of Japan)
Verilidae, as Lutjanidae
Vorhisiidae (only as fossils)
Vulpeculidae, as Lamnidae

Weigeltaspidae (only as fossils)
Winteridae, as Opisthoproctidae
Winteriidae, as Opisthoproctidae
Wudinolepidae (Only as fossils)

Xenacanthidae (incl. Pleuracanthidae). (Only as fossils)
Xenesthidae, as Birgeriidae (Only as fossils)
Xenichthyidae, as Pomadasyidae
Xenocephalidae. (No revisionary references noted)
Xenocongridae (incl. Chilorhinidae, Chlopsidae, Echelidae part, Muraenichthyidae, Myridae part): BOHLKE, J., 1956, Proc. Acad. Nat. Sci. Philadelphia, 108, 61-95 (synopsis, also as Chilorhinidae & Chlopsidae); BOHLKE, J. E. & SMITH, D. G., 1968, Proc. Acad. Nat. Sci. Philadelphia, 120, 25-43 (key to genera); BEEBE & TEE-VAN, No. 1, 1928 (Hispaniola, as Echelidae); MEEK & HILDEBRAND, No. 2, 1923 (marine, Panama, as Myridae); FOWLER, H. W., No. 8, 1936 (marine, West Africa, as Echelidae); SMITH, J. L. B., No. 1, 1961, as Muraenichthyidae); SMITH, J. L. B., 1965, Occ. Pap. Dep. Ichth. Rhodes Univ., 5, 45-54 (western Indian Ocean); WEBER & de BEAUFORT, No. 3, 1916 (Indo-Australian Arch., as Myridae); FOWLER, H. W., No. 4, 1931, (China, as Echelidae); JORDAN & SNYDER, No. 1, 1901 (Japan, as Myridae); SCHULTZ, L. P. & WOODS, L. P., 1949 (key to genera of Pacific, as Echelidae); FOWLER, H. W., No. 17, 1959 (Fiji, as Echelidae)
Xenophthalmichthyidae, as Argentinidae
Xenopoclinidae, as Clinidae
Xenopteridae, as Tetraodontidae
Xiphasiidae, as Blenniidae
Xiphidae, as Xiphiidae
Xiphidiontidae, as Stichaeidae
Xiphiidae. GOODE, G. B., 1882, Proc. United States Nat. Mus., 4, 315-433 (taxonomy & distribution); JORDAN, D. S. & EVERMANN, B. W., 1926, Occ. Pap. California Acad. Sci., 12, 1-113 (review); HILDEBRAND & SCHROEDER, No. 1, 1928 (Chesapeake Bay); HILDEBRAND, S. F., No. 1, 1946 (shore, Peru); SMITT, F. A. et al., No. 1, 1892 (Scandinavia); DUNCKER & LADIGES, No. 1, 1960 (Nordmark); POLL, N., No. 2, 1947 (marine, Belgique); NOBRE, A., No. 1, 1935 (marine, Portugal); LOZANO REY, D. L., No. 3, 1952 (España); BANARESCU, P., No. 1, 1964 (Romania); SVETOVIDOV, A. N., No. 2, 1964 (Black Sea); ANDRIYASHEV, A. P., (northern seas of U. S. S. R.); FOWLER, H. W., No. 8, 1936 (marine, West Africa); SMITH, J. L. B., No. 1, 1961 (marine, South Africa); DAY, F., No. 2, 1889 (India); DERANIYAGALA, P. E. P., 1951, Spol. Zeylanica, 26, 137-142 (Ceylon)
Xiphiodontidae, as Stichaeidae
Xiphiorhynchidae (only as fossils)
Xiphisteridae, as Stichaeidae
Xiphostomatidae, as Ctenoluciidae
Xiphostomidae, as Ctenoluciidae
Xystaemidae, as Gerridae

Zanclidae, as Acanthuridae
Zaniolepidae, as Zaniolepididae
Zaniolepididae (incl. Zaniolepidae). (No revisionary references noted)
Zanobatidae, as Discobatidae
Zanteclidae, as Melanotaeniidae
Zaproridae. SCHMIDT, P. Y., No. 1, 1950 (Sea of Okhotsk); SHMIDT, P. Y., No. 2, 1965 (Sea of Okhotsk)
Zeidae (incl. Cyttidae, Cyttopsidae, Zenidae). DEVINCENZI, G. L., No. 2, 1924 (Uruguay); SMITT, F. A. et al., No. 1, 1892 (Scandinavia) as Cyttidae); DUNCKER & LADIGES, No. 1, 1960 (Nordmark) POLL, M., No. 2, 1947 (marine, Belgique); LOZANO REY, D. L., No. 2, 1952 (España); BANARESCU, P., No. 1, 1964 (Romania); SVETOVIDOV, A. N., No. 2, 1964 (Black Sea); FOWLER, H. W., No. 8, 1936s(marine, West Africa); POLL, M., No. 4, 1954 (coastal South Africa); SMITH, J. L. B., No. 1, 1961 (marine, South Africa); WEBER & de BEAUFORT, No. 5, 1929 (Indo-Australian Archipelago); MARSHALL, T. C., No. 1, 1964 (Great Barrier Reef); FOWLER, H. W., No. 6, 1933-35 (China); LINDBERG & LEGHEZA, No. 2, 1962 (Sea of Japan); LINDBERG & LEGEZA, No. 3, 1969 (Sea of Japan)
Zenidae, as Zeidae
Zeniontidae, as Macrurocyttidae
Zeudae, as Zeusidae
Zeusidae (incl. Zeudae). NOBRE, A., No. 1, 1935 (marine, Portugal)

Zoarcidae (incl. Derepodichthyidae, Lycodapodidae, Lycodidae). DEVINCENZI, G. F., No. 2, 1924 (Uruguay, as Lycodidae); NORMAN, J. R., No. 2, 1937 (marine, Patagonia, also as Lycodapodidae); SMITT, F. A. et al., No. 2, 1895 (Scandinavia, as Lycodidae); DUNCKER & LADIGES, No. 1, 1960 (Nordmark); POLL, M., No. 2, 1947 (marine, Belgique); LOZANO y REY, L., No. 5, 1960 (España); BERG, L. S., No. 5, 1949 (fresh-water, Soviet Union); BERG, L. S., No. 10, 1965 (fresh-water, U. S. S. R.); ANDRIYASHEV, A. P., 1955, Trav. Inst. Zool. Acad. Sci. U. R. S. S., 18, 349-384 (keys, U. S. S. R. seas); ANDRIYASHEV, A. P., 1955, Trud. Zool. Inst. Leningrad, 18, 349-384 (keys, seas of U. S. S. R.); ANDRIYASHEV, A. P., 1960, Trans. Ser. Fish. Res. Board Canada, 278 (keys, seas of U. S. S. R.); ANDRIYASHEV, A. P., No. 1, 1954 (northern seas of U. S. S. R.); FOWLER, H. W., No. 8, 1936 (marine, West Africa); SMITH, J. L. B., No. 1, 1961 (marine, South Africa); FOWLER, H. W., No. 16, 1958 (China, also as Lycodapodidae); SOLDATOV & LINDBERG, No. 1, 1930 (Far East seas); SCHMIDT, P. Y., No. 1, 1950 (Sea of Okhotsk); SHMIDT, P. Y., No. 2, 1965 (Sea of Okhotsk); NORMAN, J. R., No. 3, 1938 (Antarctica); DeWITT, H. H., 1962, Copeia, 1962, 819-826 (key to Patagonian genera)

AMPHIBIA

GENERAL BOOKS

BOGERT, C. M. Amphibians and reptiles of the world. In Drimmer, the Animal Kingdom. Garden City, New York; 11-89-1390; 1954.
BUHLER, W. Amphibien und reptilian. Aarau; 1-127; 1966.
COCHRAN, D. M. Living amphibians of the world. London; 1-199; 1961.
DICKERSON, M. C. The frog book . . . New York; 1-253; 1906.
DICKERSON, M. C. The frog book. Garden City, New York; 1-253; 1920.
DOTTRENS, E. Batraciens et reptiles d'Europe. Neuchatel;1-263; 1963.
FREIBERG, M. A, Vida de batracios y reptiles sudamericanos. Buenos Aires; 1-192; 1954.
GOIN, C. J. & GOIN, O. B. Introduction to herpetology, 2nd ed. San Francisco; 1-353; 1971. (1st ed., 1-341; 1962)
HVASS, H. Reptiles and amphibians of the world. London; 1-125; 1964.
INGER, R. F. Amphibia. In South African Animal Life, by B. Hanstrom et al. Stockholm; 510-553;1959.
MATTHEWS, L. H. The British Reptilia and Amphibia. London; 1-54; 1952.
MERTENS, R. The world of amphibians and reptiles. London & New York; 1-207; 1960. (Translation)
MOORE, J. A. Physiology of Amphibia. New York;1-654; 1964.
NOBLE, G. K. The biology of the Amphibia, 2nd ed. New York; 10577; 1944. (1st ed., 1931)
OLIVER, J. A. The natural history of North American amphibians and reptiles. Princeton, New Jersey; 1-359; 1955.
PERRIER, E. Les Batraciens. Traite de Zool., 8, 2725-2882; 1928.
REICHENBACH-KLINKE, H. H. Krankheiten der Amphibien. Stuttgart; 1-100; 1961.
ROSE, W. The reptiles and amphibians of southern Africa. Cape Town; 1-494; 1962. (1st ed., 1950)
SMITH, M. A. The British amphibians and reptiles, 3rd ed. London; 1-322; 1964. (Prev. eds., 1951, 1954)
TERENT'EV, P. V. Herpetology. A study of amphibians and reptiles. Moskva; 1-336; 1961. (In Russian)
TERENT'EV, P.V. Herpetology.Amanual of amphibians and reptiles. Jerusalem; 1-313; 1965. (Translation of 1961)
WERNER, F. Die Lurche und Kriechtiere von Alfred Brehm (Neubearbeiter). Brehms Tierleben. Leipzig; 4, 1-572; 1925; 5, 1-598; 1925. (Previous edition 1912)
WERNER, F. Allgemeine Einleitung in die Naturgeschichte der Amphiba. Handbuch der Zoologie, 6, 1-92; 1931.

BIBLIOGRAPHIC WORKS

Bibliographies, separate works

[See also: Under ANIMALIA, above; see also Zoological Record, under heading Bibliographies, in each volume]

(anonymous). Bibliography of amphibians and reptiles. New York; American Museum of Natural History 1-11; 1958. (Supplement of 4 pages, 1958)
ALLOUSE, B. G. . A bibliography on the vertebrate fauna of Iraq and neighbouring countries. III. Reptiles and amphibians. Publ. Iraq Nat. Hist. Mus., 6, 1-23; 1955.
BANTA, B. H. An annotated chronological bibliography of the herpetology of the state of Nevada. Wasmann Journ. Biol., 23, 1-224; 1965.
GANS, C. A bibliography of the herpetology of Japan. Bull. American Mus. Nat. Hist., 93, 391-496; 1949.
PETERS, J. A. Supplemental list of titles of papers concerning the herpetology of Ecuador. In Biblio. Cient. del Ecuador, by Carlos Larrea, 5, 1067-1076; 1953.
SCHMIDT, K. P. Annotated bibliography of marine ecological relations of living amphibians and reptiles (except turtles). Mar. Life, 1, 43-54; 1951.
TAYLOR, E. H. A bibliography of Mexican amphibiology. Univ. Kansas Sci. Bull., 31, 543-589; 1947.
VALENTINE, B. D. Index and references to the taxonomy and external morphology of larval (gilled) salamanders. Columbus, Ohio; 1967.

Bibliographies in other works

[See also: References in most large works]

BERGER, L. et al. Catalogus faunae Poloniae. Amphibia et Reptilia. Warszawa; 1-73; 1969.
BESKOW, V. & BERON, P. Catalogue et bibliographie des amphibiens et des reptiles en Bulgarie. Sofia; 1-40; 1964.
BISHOP, S. C. Handbook of salamanders. Ithaca, New York; 1-555; 1943. (Pp. 479-545) (Reprinted in 1962)
BLAIR, W. F. et al. Vertebrates of the United States. New York; 1-616; 1968. (1st edition, 1957)
BORING, A. M & LEROY, P. Chinese amphibians living and fossil forms. Publ. Inst. Geo-Biol. Pekin, 13, 1-151; 1945.
BOURRET, R. Les batraceens de l'Indochine. Mem. Inst. Oceanogr. Indochine, 6, 1-547; 1942.
COCHRAN, D. M. Frogs of southeastern Brazil. United States Nat. Mus. Bull., 206, 1-423; 1955.
COCHRAN, D. M & GOIN, C. J. Frogs of Colombia. United States Nat. Mus. Bull., 288, 1-655; 1970.

COPE, E. D. The Batrachia of North America. United States Nat. Mus. Bull., 34, 1-515; 1889.
NEILL, W. T. The occurrence of amphibians and reptiles in saltwater areas, and a bibliography. Bull. Mar. Sci. Gulf Caribbean, 8, 1-97; 1958.
NOBLE, G. K. Contributions to the herpetology of the Belgian Congo based on the collection of the American Museum Congo Expedition, 1909-1915. Bull. American Mus. Nat. Hist., 49, 147-347; 1924. (Pp. 284-303)
PENNEY, J. T. Distribution and bibliography of the amphibians and reptiles of South Carolina. Univ. South Carolina Publ., Ser. III, Biol., 1, 3-28; 1952.
SMITH, H. M. Handbook of amphibians and reptiles of Kansas. Misc. Publ. Univ. Kansas Mus. Nat. Hist., 9, 1-356; 1956. (1st edition, 1950)
STEBBINS, R. C. Amphibians of western North America. Berkeley; 1-539; 1951. (Pp. 515-530)
TAYLOR, E. H. The caecilians of the world. Lawrence, Kansas; 1-848; 1968.
TAYLOR, E. H. The turtles and crocodiles of Thailand . . . , with a synoptic herpetological bibliography. Univ. Kansas Sci. Bull., 49, 87-179; 1970
WRIGHT, A. H. & WRIGHT, A. A. Handbook of frogs and toads of the United States and Canada, 3rd ed., Ithaca, New York; 1-640; 1949. (Pp. 584-621) (Previous editions 1933 & 1942)

Indexes

CLARK, H. Synopsis of herpetological literature in the Proceedings of the Iowa Academy of Science, 1891-1944. Herpetologica, 3, 131-135; 1946.
MINTON, S. A. An annotated list of herpetological papers from the Proceedings of the Indiana Academy of Sciences. Herpetologica, 3, 195-201; 1947.
REED, C. F. Index to Copeia, 1913-1954. Part I. — Author index. Lancaster, Pennsylvania; 1-106; 1955.
REED, C. F. Index to Copeia, 1913-1954. Part II. — Subject index. Lancaster, Penn.; 1-332; 1956.
REED, C. F. Index to Herpetologica, 1936-1955. Lancaster, Pennsylvania; 1-132; 1956.
REED, C. F. Index to Copeia. Supplemeht, 1955-1964. Lancaster, Pennsylvania; 1-329; 1965.

Personal bibliographies

[See under heading BIOGRAPHIES (Persons and Institutions), below]

Institutional bibliographies

[See also under Animalia]

PETERS, J. A. A list of the herpetological publications of the United States National Museum. 1853-1965. Washington, D. C.; 1-12; 1965.
PETERS, J. A. A list of the herpetological publications of the United States National Museum. 1853-1968. Washington, D. C.; 1-14; 1968.

Periodical lists

[See under Animalia and Vertebrata]

PERSONS AND INSTITUTIONS

Biographies & personal bibliographies

[See also: Zoological Record under heading Obituaries, in each volume]

BOULENGER, G.-A. Liste des publications ichthyologiques et herpetologiques (1877-1920) de G.-A. Boulenger. Ann. Soc. Roy. Zool. Malac. Belgique, 52, 11-88; 1921.
COCHRAN, D. M. Bibliography of Doris M. Cochran; 1922-1952; list of new forms described. Herpetologica, 8, 115-120; 1953.
DUGES, A. Alfredo Duges' types of Mexican reptiles and amphibians, by H. M. Smith & W. L. Necker. Anal. Esc. Nac. Cienc. Biol. Mexico, 3, 179-189; 1943. Early foundations of Mexican herpetology. An annotated and indexed bibliography of the herpetological publications of Alfredo Duges, 1826-1910, by H. M. Smith & R. B. Smith. Urbana, Illinois; 1-85; 1969.
DUNN, E. R. Herpetological publications of Emmett Reid Dunn, by A. M. Dunn. (Priv. publ.) 1-296; 1957.
MEHELY, L. Die wissenschaftliche und literarische Tatigkeit von Ludwig Mehely aus dem Gebiete der Zoologie, by O. G. Dely. Vert. Hung., 9, 21-64; 1967.
MYERS, G. S. Annotated bibliography of the publications of George Sprague Myers (to the end of 1969). Proc. California Acad. Sci., (4), 38, 19-52; 1970.
NOBLE, G. K. Gladwyn Kingsley Noble, 1894-1940: A herpetological bibliography, by W. L. Necker. Herpetologica, 2, 47-55; 1940.
POPE, C. H. Bibliography of Clifford H. Pope. Escondido, California; 1-8; 1967.
SMITH, M. A. Bibliography of the works of Malcolm Arthur Smith, by R. H. Ahrenfeldt. British Journ. Herpet., 2, 146-189; 1959.
STEJNEGER, L. A herpetological bibliography of Leonard Stejneger (1851-1943), by W. L. Necker. Herpetologica, 2, 86-92; 1943.
WALL, F. Bibliography of the herpetological papers of Frank Wall (1868-1950) 1898-1928, by S. M. Campdon-Main. Washington, D. C.; 1-7; 1969.

Directories

[See also: Directories listed under Animalia]

(anonymous). American Society of Ichthyologists and Herpetologists. List of members. San Francisco; 1-42; 1967.

Institutions

[See also under Animalia]

DOWLING, H. G. & GILBOA, I. Herpetological societies. Amateur & professional. Herpetological Rev., 2, (3), 6-9; 1970.

Museum lists

North America

Museum of Comparative Zoology, Harvard University

DUNN, E. R. The collection of Amphibia Caudata of the Museum of Comparative Zoology. Bull. Mus. Comp. Zool., 62, 445-471; 1948.

LOVERIDGE, A. New Guinean reptiles and amphibians in the Museum of Comparative Zoology and the United States National Museum. Bull. Mus. Comp. Zool., 101, 305-430; 1948.

United States National Museum

YARROW, H. C. Check list of North American Reptilia and Batrachia, with catalogue of specimens in U. S. National Museum. United States Nat. Mus. Bull., 24, 1-249; 1882.

LOVERIDGE, A. East African reptiles and amphibians in the United States National Museum. United States Nat. Mus. Bull., 151, 1-135; 1929.

KELLOGG, R. Mexican tailless amphibians in the United States National Museum. United States Nat. Mus. Bull., 160, 1-224; 1932.

LOVERIDGE, A. New Guinean reptiles and amphibians in the Museum of Comparative Zoology and the United States National Museum. Bull. Mus. Comp. Zool., 101, 305-430; 1948.

West Virginia Biological Survey

GREEN, N. B. The herpetological collections of the West Virginia Biological Survey. Proc. West Virginia Acad. Sci., 20, 57-64; 1949.

Field Museum of Natural History (Chicago Natural History Museum)

LOVERIDGE, A. African reptiles and amphibians in Field Museum of Natural History. Field Mus. Publ., Zool., 22, (1), 1-111; 1936.

Texas Cooperative Wildlife Collection

SMITH, H. M. & LAUFE, L. E. Mexican amphibians and reptiles in the Texas Cooperative Wildlife collections. Trans. Kansas Acad. Sci., 48, 325-354; 1946.

Central America

Museo Alfredo Duges.

MALDONADO-KOERDELL, M. Las colecciones de anfibios del Museo Alfredo Duges en la Universidad de Guanajuato. I - Urodeles. Mem. Acad. Nac. Cienc. Mexico, 56, 185-226; 1948.

South America

Sociedad de Ciencias La Salle de Venezuela

RIVERO, J. A. Salientos (Amphibia) en la coleccion de la Sociedad de Ciencias La Salle de Venezuela. Caribbean Journ. Sci., 4, 297-305; 1964.

Instituto Oswaldo Cruz

LUTZ, B. Anfibios anuros da colecao Adolpho Lutz do Instituto Oswaldo Cruz. Mem. Inst. Oswaldo Cruz, 46, 295-313; 1948.

Europe

British Museum (Natural History)

BOULENGER, G. A. Catalogue of the Batrachia Gradientia s. Caudata and Batrachia Apoda in the collection of the British Museum. London; 1-128; 1882. (Reprinted 1965)

Musee Zoologique Strasbourg

ANGEL, F. Reptiles et amphibiens de Madagascar . . . du Musee Zoologique de Strasbourg. Bull. Mus. Nat. Hist. Nat., (2), 22, 553-558; 1950.

The Orient

Colombo Museum

de SILVA, P. H. D. H. A list of Amphibia recorded from Ceylon with a report on the Amphibia in the Colombo Museum. Spol. Zeylanica, 27, 243-250; 1955.

Museum type lists

TAYLOR, E. H. Present location of certain herpetological and other type specimens. Univ. Kansas Sci. Bull., 30, 117-187; 1944.

DOWLING, H. G. et al. A list of herpetological type lists. Herpet. Rev., 2, 53-54; 1970.

North America

Museum of Comparative Zoology, Harvard University

BARBOUR, T. & LOVERIDGE, A. Typical reptiles and amphibians. Bull. Mus. Comp. Zool. 69, 205-360; 1929.

BARBOUR, T. & LOVERIDGE, A. First supplement to Typical Reptiles and Amphibians. Bull. Mus. Comp. Zool., 96, 59-214; 1946.

United States National Museum

COCHRAN, D. M. Type specimens of reptiles and amphibians in the United States National Museum. United States Nat. Mus. Bull., 220, 1-291; 1961.

Carnegie Museum

McCOY, C. J. & RICHMOND, N. D. Herpetological type specimens in Carnegie Museum. Ann. Carnegie Mus., 38, 233-264; 1966.

University of Michigan Museum of Zoology

PETERS, J. A. Catalogue of type specimens in the herpetological collections of the University of Michigan Museum of Zoology. Occ. Pap. Mus. Zool. Univ. Michigan, 539, 1-55; 1952.

Field Museum of Natural History (Chicago Museum of Natural History)

MARX, H. Catalogue of type specimens of reptiles and amphibians in Chicago Natural History Museum. Fieldiana, Zool., 36, 409-496; 1958.

University of Illinois Museum of Natural History

SMITH, H. M. et al. Herpetological type-specimens in University of Illinois Museum of Natural History. Illinois Biol. Monogr., 32, 1-80; 1964.

University of Kansas Museum of Natural History

DUELLMAN, W. E. & BERG, B. Type specimens of amphibians and reptiles in the Museum of Natural History, the University of Kansas. Publ. Mus. Nat. Hist. Univ. Kansas, 15, 183-204; 1962.

California Academy of Sciences

SLEVIN, J. R. & LEVITON, A. E. Holotype specimens of reptiles and amphibians in the collection of the California Academy of Sciences. Proc. California Acad. Sci., (4), 28, 529-560; 1956.

Museum of Vertebrate Zoology, University of California

CRIPPEN, R. G. Holotype specimens of amphibians and reptiles in the Museum of Vertebrate Zoology, University of California, Berkeley. Herpetologica, 18, 187-194; 1962.

Stanford University Natural History Museum

LEVITON, A. E. Catalogue of the amphibians and reptile types in the Natural History Museum of Stanford University. Herpetologica, 8, 121-132; 1953.

LEVITON, A. E. & BANTA, B. H. Catalogue of the amphibian and reptile types . . . . Supplement number 1. Herpetologica, 12, 213-219; 1956.

Europe

British Museum (Natural History)

CONDIT, J. M. A list of the types of hylid frogs in the collection of the British Museum (Natural History). Journ. Ohio Herp. Soc., 4, 85-98; 1964.

Musee d'Histoire Naturelle (Musee National d'Histoire . . .)

GUIBE, J. Catalogue des types d'amphibiens du Museum National d'Histoire Naturelle. Paris; 1-71; 1950.

Museo Civico di Storia Naturale

CAPOCACCIA, L. Catalogo dei tipi di anfibi del Museo Civico di Storia Naturale Genova. Ann. Mus. Civ. Stor. Nat. Genova, 69, 208-222; 1958.

Naturhistorisches Museum Basel

FORCART, L. Katalog des Typusexemplare in der Amphibien-sammlung des Naturhistorichen Museums zu Basel. Verh. Nat. Gesell. Basel, 57, 118-142; 1946.

Senckenberg Natur-Museums und Forschungs-Institutes

MERTENS, R. Verzeichnis der Typen in der herpetologischen Sammlung des Senckenbergischen Museums. Senckenbergiana, 4, 162-183; 1922.

MERTENS, R. Die herpetologische Sektion des Natur-Museums und Forschungs-Institutes Senckenberg in Frankfurt a. M. nebst einem Verzeichnis ihrer Typen. Senckenbergisches Biol., 48, 1-106; 1967.

Staatliches Museum Dresden

SCHUZ, E. Verzeichnis der Typen des Staatlichen Museums fur Tierkunde zu Dresden. I. Teil. Fische, Amphibien und Reptilien. . . . Abh. Mus. Dresden, 17, (2), 1-16; 1929.

Zoological Museum, Amsterdam

DAAN, S. & HILLENIUS, D. Catalogue of the type specimens of amphibians and reptiles in the Zoological Museum, Amsterdam. Beaufortia, 13, 117-144; 1966.

Australia

Macleay Museum, University of Sydney

GOLDMAN, J. et al. Type specimens in the Macleay Museum, University of Syndey. II. Amphibians and reptiles. Proc. Linn. Soc. New South Wales, 93, 427-438; 1969.

National Museum of Victoria

COVENTRY, A. L. Reptile and amphibian type specimens housed in the National Museum of Victoria. Mem. Nat. Mus. Victoria, 31, 115-124; 1970.

EXPLORATION AND LOCALITIES

[See also this heading under Animalia]

SMITH, H. M. & TAYLOR, E. H. Type localities of Mexican reptiles and amphibians. Univ. Kansas Sci. Bull., 33, 313-380; 1950.

COCHRAN, D. M. & GOIN, C. J. Frogs of Colombia. United States Nat. Mus. Bull., 288, 1-655; 1970

PETERS, J. A. Herpetological type localities in Ecuador. Rev. Ecuatoriana Ent. Par., 2 - 3/4, 335-352; 1955.

BOKERMANN, W. C. Lista anotada das localidades tipo de anfibios Brasileiros. Sao Paulo; 11-183; 1966.

STEJNEGER, L. The herpetology of Japan and adjacent territory. United States Nat. Mus. Bull., 58, 1-577; 1907.

METHODS

[See also: Zoological Record under heading Techniques, in each volume; see also this heading under Animalia]

(anonymous). Instructions for collectors. No. 3, Reptiles, amphibians and fishes, 6th edition. London; 1-28; 1953. [British Museum (Natural History)]

COOK, F. R. Collecting and preserving amphibians and reptiles. Bull. Nat. Mus. Canada, 69, 128-151; 1965.

DUELLMAN, W. E. Directions for preserving amphibians and reptiles. Misc. Publ. Univ. Kansas Mus. Nat. Hist., 30, 37-40; 1962.

GLOYD, H. K. Methods of preserving and labeling amphibians and reptiles for scientific study. Turtox News, 16, 49-53, 66-67; 1938.

MYERS, G. S. Manual of tropical herpetological collecting. Stanford, California; 1-12; 1951.

MYERS, G. S. Manual of tropical herpetological collecting. Stanford Univ. Nat. Hist. Mus. Circ., 4, 2nd edition; 1956.

MYERS, G. S. Brief directions for preserving and shipping specimens of fishes, amphibians and reptiles. 2nd edition. Circ. Stanford Univ. Nat. Hist. Mus., 5; 1956.

NEILL, W. T. How to preserve reptiles and amphibians for scientific study. Ross Allan Rept. Inst., Spec. Publ., 2, 1-15; 1950.

SMITH, A. G. Making a collection of amphibians and reptiles. Contr. Dep. Biol. Sci Loyola Univ., 1, 1-5; 1951.

SMITH, H. M. Handbook of amphibians and reptiles of Kansas, 2nd edition. Misc. Publ. Univ. Kansas Mus. Nat. Hist., 9, 1-356; 1956.

STEBBINS, R. C. Amphibians and reptiles of western North America. New York; 1-528; 1954.

STEJNEGER, L. Directions for collecting reptiles and batrachians. United States Nat. Mus. Bull., 39, (E), 1-13; 1891.
VOGEL, Z. Reptiles and amphibians; their care and behaviour. London; 1-228; 1964.
ZWEIFEL, R. G. Guidelines for the care of a herpetological collection. Curator, 9, 24-35; 1966.

GLOSSARIES

[See also: Many of the general books listed above, as well as Glossaries listed under Animalia]

HIRTZ, M. Dictionary of national [Croatian] zoological names. Part I. (Amphibia and Reptilia). Zagreb; 1-197; 1928. (In Croatian)
PETERS, J. A. Dictionary of herpetology. New York; unpaged; 1964.
STEBBINS, R. C. Amphibians of western North America. Berkeley; 1-539; 1961. (Pp. 509-512)

NOMENCLATURAL STUDIES

Name studies

FOLLETT, W. I. & DEMPSTER, L. J. Ichthyological and herpetological names and works on which the International Commission on Zoological Nomenclature has ruled since the publication (in 1958) of the several Official Lists and Indexes. Copeia, 1965, 518-523; 1965.
KUHN, O. Amphibien und Reptilien. Katalog der ... Taxa .... Stuttgart; 1-124; 1967. (Families and more inclusive taxa)
MACLEOD, R. D. Key to the names of British fishes, mammals, amphibians, and reptiles. London; 1-71; 1956. (Origin and meaning)

Common names

(anonymous). Common names for North American amphibians and reptiles. Copeia, 1956, 172-185; 1956.
WRIGHT, A. H. & WRIGHT, A. A. Handbook of frogs and toads of the United States and Canada, 3rd ed. Ithaca, New York; 1-640; 1949.
BARRAN, E. F. & FREIBURG, M. A. Nombres vulgares de reptiles y batracios del Argentina. Physis, 20, (58), 303-319; 1951.
STEWARD, J. W. The tailed amphibians of Europe. Newton Abbot, England; 1-180; 1969. (In English, French, German, Dutch, Italian, Polish, Czech, Russian)
MACLEOD, R. D. Key to the names of British fishes, mammals, amphibians, and reptiles. London; 1-71; 1956.
BRODMANN, P. et al. Die amphibien der Basler region. Veroff. Naturh. Mus. Basel, 4, 3-32; 1966.
FROMMHOLD, E. Wir bestimmen Lurche und Kreichtiere mitteleuropas. Radebeul; 1-219; 1959.
PASTEUR, G. & BONS, J. Les batraciens du Maroc. Trav. Inst. Sci. Cherifien, Zool., 17, 1-241; 1959.
MINTON, S. A. A contribution to the herpetology of West Pakistan. Bull. American Mus. Nat. Hist., 134, 27-184; 1966.
MOORE, J. A. The frogs of eastern New South Wales. Bull. American Mus. Nat. Hist., 121, 149-386; 1961. (Amphibians of Australia)

Nomenclators

[See also: Nomenclators listed under Animalia]

KUHN, O. Amphibien und reptilien. Katalog der Subfamilien und hoheren Taxa mit Nachweis des ersten Auftretens. Stuttgart; 1-124; 1967.
ROMER, A. S. Vertebrate paleontology. Chicago; 1-468; 1966; 3rd edition. (Genera alphabetically) (1st edition, 1933; 2nd edition, 1945)

REGIONAL STUDIES

Faunas, checklists, identification works

North America

COPE, E. D. Check-list of North American Batrachia and Reptilia, with a systematic list of the higher groups . . . . United States Nat. Mus. Bull., 1, 1-104; 1875.
COPE, E. D. The Batrachia of North America. United States Nat. Mus. Bull., 34, 1-525; 1889.
YARROW, H. C. Check list of North American Reptilia and Batrachia, with catalogue of specimens in U. S. National Museum. United States National Museum Bull., 24, 1-249; 1882.
WRIGHT, A. H. Synopsis and description of North American tadpoles. Proc. United States Nat. Mus., 74, Art. 11), 1-70; 1929.
STEJNEGER, L. H. & BARBOUR, T. A check list of North American amphibians and reptiles, 5th ed. Bull. Mus. Comp. Zool., 93, 1-260; 1943. (Previous editions, 1917, 1923, 1933, 1939,

1943; Cambridge, Massachusetts; Harvard University Press)
BISHOP, S. C. Handbook of salamanders. The salamanders of the United States, of Canada, and of Lower California. Ithaca, New York; 1-555; 1943. (Reprinted 1962)
WRIGHT, A. H. & WRIGHT, A. A. Handbook of frogs and toads of the United States and Canada; 3rd ed. Ithaca, New York; 1-652; 1949.
ORTON, G. L. Key to the genera of tadpoles in the United States and Canada. American Midl. Nat., 47, 382-395; 1952.
ZIM, H. S. & SMITH, H. M. Reptiles and amphibians. A guide to familiar American species. New York; 1-157; 1953.
SCHMIDT, K. P. A check list of North American amphibians and reptiles, 6th ed. Chicago;1-280; 1953.
CONANT, R. A field guide to the reptiles and amphibians of the United States and Canada east of the 100th meridian. Boston; 1-366; 1958.
KENNEDY, J. Field guide to Reptilia and Amphibia of the United States and Canada east of the 100th meridian. Southwest. Nat., 3, 238-246; 1959.
VALENTINE, B. D. A preliminary key to the families of salamanders and sirenids with gills or gill slits. Copeia, 1964, 582-583; 1964.
BOULENGER, G. A. Contributions to American herpetology, collected papers. Reprints by Ohio Herpetological Society; since 1966, Society for the Study of Amphibians and Reptiles. Part 1, 1877-1881, 1965; 2, 1882-1883, 1966; 3, 1884-1885, 1966; 4, 1886-1887, 1967; 5, 1888, 1967; 6, 1889-1890, 1968; 7, 1891-1893, 1968; 8, 1894, 1969; 9, 1895-1896, 1969; 10, 1897, 1970; 11, 1898, 1970; [projected: 12 & 13, 1971; 14,15, & 16, 1972]
ALTIG, R. A key to the tadpoles of the continental United States and Canada. Herpetologica, 26, 180-207; 1970.
COCHRAN, D. M. & GOIN, C. J. The new field book of reptiles and amphibians. New York; 1-359; 1970. (United States)
LOGIER, E. B. S. & TONER, G. E. Checklist of the amphibians and reptiles of Canada and Alaska. Contr. Roy. Ontario Mus. Zool., 41, 1-88; 1955.
LOGIER, E. B. S. & TONER, G. C. Checklist of the amphibians and reptiles of Canada and Alaska. Contr. Life Sci. Div. Roy. Ontario Mus., 53, 1-92; 1961. (Revision of Contribution 41)
MILLS, R. C. A check list of the reptiles and amphibians of Canada. Herpetologica, 4, (suppl. 2, 1-15;1948.
BLEAKNEY, S. S. A zoogeographical study of the amphibians and reptiles of eastern Canada. Bull. Nat. Mus. Canada, 155, 1-119; 1958.
LOGIER, E. B. S. The frogs, toads and salamanders of eastern Canada. Toronto; 1-127; 1952.
BLEAKNEY, S. The amphibians and reptiles of Nova Scotia. Canadian Field Nat., 66, 125-129; 1952.
LOGIER, E. B. S. The amphibians of Ontario. Handb. Roy. Ontario Mus. Zool., 3, 1-16; 1937.
COWAN, I. M. A review of the reptiles and amphibians of British Columbia. Rep. Prov. Mus. Nat. Hist. Victoria, 1936, 16-25; 1937.
CARL, G. C. The amphibians of British Columbia. Hand. British Columbia Prov. Mus., 2, 1-62; 1943.
COOK, F. R. An analysis of the herpetofauna of Prince Edward Island. Bull. Nat. Mus. Canada, 212, 1-60; 1967.
LIVEZEY, R. L. & WRIGHT, A. H. A synoptic key to the salientian eggs of the United States. American Midl. Nat., 37, 179-222; 1947.
BLAIR, W. F. et al. Vertebrates of the United States, 2nd ed. New York; 1-616; 1966. (1st edition, New York, 1-819, 1957) (Amphibia, pp. 167-212)
WHITAKER, J. O., Jr. Keys to the vertebrates of the eastern United States, exclusive of birds. Minneapolis, Minnesota; 1-256; 1969.
WRIGHT, A. H. & WRIGHT, A. A. A key to the eggs of the Salientia east of the Mississippi River. American Midl. Nat., 58, 375-381; 1924.
CONANT, R. Salamanders of the north-eastern states. Fauna, 4, 8-11; 1942.
CONANT, R. Reptiles and amphibiansof the northeastern states. A non-technical resume ..., 3rd ed. Philadelphia; 1-40; 1958. (Previous edition, 1947)
BURT, C. E. The frogs and toads of the south-eastern United States. Trans. Kansas Acad. Sci., 41, 331-367; 1938.
BABBITT, L. H. The Amphibia of Connecticut. Bull. Connecticut Geol. Nat. Hist. Surv., 57, 5-50; 1937.
NETTING, M. G. The amphibians and reptiles of Pennsylvania. (Philadelphia), 1-29; 1946.
MANSUETI, R. A descriptive catalogue of the amphibians and reptiles found in and around Baltimore City, Maryland, within a radius of twenty miles. Proc. Nat. Hist. Soc. Maryland, 7, 1-53; 1941. (Keys)
GREEN, N. B. A key to the eggs of West Virginia Salientia. Proc. West Virginia Acad. Sci., 24, 36-38; 1953.
DePOE, C. E. et al. The reptiles and amphibians of North Carolina. A preliminary check list and bibliography. Journ. Elisha Mitchel Sci. Soc., 77, 125-136; 1961.
CHERMOCK, R. L. A key to the amphibians and reptiles of Alabama. Mus. Rep. Geol. Surv. Alabama, 33, 1088; 1952.
CARR, A & GOIN, C. J. Guide to the reptiles, amphibians and fresh-water fishes of Florida. Gainesville, Florida; 1-341; 1959.
GEHLBACH, F. R. Comments on the study of Ohio salamanders with key to their identification. Journ. Ohio Herpet. Soc., 2, 10-15; 1960. (Keys)
GOODNIGHT, C. J. A key to the adult salamanders of Illinois. Trans. Illinois State Acad. Sci., 30, 300-302; 1937.
SMITH, P. W. The amphibians and reptiles of Illinois. Bull. Illinois Nat. Hist. Surv., 28, 1-298; 1961.
POPE, C. H. Amphibians and reptiles of the Chicago area. Chicago; 1-275; 1944.

SMITH, A. G. Key to the amphibians and reptiles of the Chicago area. Contr. Dep. Biol. Sci. Loyola Univ., 7, 1-4; 1951.
KEISER, E. D., Jr. & WILSON, L. D. Checklist and key to the herpetofauna of Louisiana. Lafayette Nat. Hist. Mus. Tech. Bull., 1, 1-51; 1969.
SMITH, H. M. The amphibians of Kansas. American Midl. Nat., 15, 377-516; 1934.
SMITH, H. M. Handbook of amphibians and reptiles of Kansas, 2nd ed. Misc. Publ. Univ. Kansas Mus. Nat. Hist., 9, 1-356; 1956. (1st edition, Misc. Publ., 2, 1-366; 1950)
CLARKE, R. F. Salamanders in Kansas and vicinity. Kansas School Nat., 16, (4), 1-16; 1970. (Keys)
HUDSON, G. E. The amphibians and reptiles of Nebraska. Nebraska Cons. Bull., 23, (4), 1-146; 1942.
SLEVIN, J. R. The amphibians of western North America. Occ. Pap. California Acad. Sci., 16, 1-152; 1928.
STEBBINS, R. C. Amphibians of western North America. Berkeley; 1-539; 1951. (Keys, eggs, larvae)
STEBBINS, R. C. Amphibians and reptiles of western North America. New York; 1-528; 1954.
STEBBINS, R. C. A field guide to western reptiles and amphibians. . . . Boston; 1-279; 1966.
WRIGHT, A. H. & WRIGHT, A. A. Amphibians of Texas. Trans. Texas Acad. Sci., 21, 5-35; 1938.
BURT, C. E. Contributions to Texan herpetology. VII. The salamanders. American Midl. Nat., 20, 374-380; 1938.
BROWN, B. C. An annotated check list of the reptiles and amphibians of Texas. Waco, Texas; 1-250; 1950.
RODECK, H. G. Guide to the Amphibia of Colorado. Univ. Colorado Mus. Leaflet, 2, 108; 1943.
MASLIN, T. P. An annotated check list of the amphibians and reptiles of Colorado. Univ. Colorado Stud., Biol., 6, 1-78; 1959.
TANNER, V. M. A synoptical study of Utah Amphibia. Proc. Utah Acad. Sci., 8, 159-198; 1931.
SLEVIN, J. R. A handbook of reptiles and amphibians of the Pacific states including certain eastern species. San Francisco; 1-71; 1934.
PICKWELL, G. Amphibians and reptiles of the Pacific states. Stanford, California; 1-236; 1947.
SLATER, J. R. A key to the adult amphibians of Washington State. Occ. Pap. Dep. Biol. Coll. Puget Sound, 25, 235-242; 1964.
GORDON, K. The Amphibia and Reptilia of Oregon. Oregon State Monogr., Zool., 1, 1-82; 1939.
STORER, T. I. A synopsis of the Amphibia of California. Univ. California Publ. Zool., 27, 1-342; 1925.
BASEY, H. Sierra Nevada amphibians. Three Rivers, California; 1-27; 1969.

West Indies

BARBOUR, T. A contribution to the zoogeography of the West Indies, with especial reference to amphibians and reptiles. Mem. Mus. Comp. Zool., 44, 205-359; 1914. (List)
BARBOUR, T. A list of Antillean reptiles and amphibians. Zoologica, 11, 61-116; 1920.
BARBOUR, T. A second list of Antillean reptiles and amphibians. Zoologica, 19, 77-141; 1935.
BARBOUR, T. A third list of Antillean reptiles and amphibians. Bull. Mus. Comp. Zool., 82, 77-166; 1937.
COPE, E. D. List of the Batrachia and Reptilia of the Bahama Islands. Proc. United States Nat. Mus., 10, 436-439; 1887.
BARBOUR, T. & RAMSDEN, C. T. The herpetology of Cuba. Mem. Mus. Comp. Zool., 47, 69-213; 1919.
BUIDE, M. S. Lista de los anfibios y reptiles de Cuba. Torreia, (n. s.), 1, 7-60; 1967.
GRANT, C. The herpetology of the Cayman Islands . . . . Bull. Inst. Jamaica, Sci., 2, 1-65; 1940.
BARBOUR, T. An annotated list of the Amphibia and Reptilia of Jamaica. Handb. of Jamaica for 1922.
DUNN, E. R. The frogs of Jamaica. Proc. Boston Soc. Nat. Hist., 38, 111-130; 1926. (Adults & tadpoles)
LYNN, W. G. The herpetology of Jamaica. I. Amphibians. Bull. Inst. Jamaica, Sci., 1, 1-60; 1940.
COCHRAN, D. M. The herpetology of Hispaniola. United States Nat. Mus. Bull., 177, 1-398; 1941.
STEJNEGER, L. The herpetology of Porto Rico. Ann. Rep. United States Nat. Mus., 1902, 549-727; 1904.
SCHMIDT, K. P. Amphibians and land reptiles of Porto Rico. Sci. Surv. Porto Rico Virgin Is., 10, (1), 21-160; 1928.
GRANT, C. The herpetology of Barbados. Herpetologica, 15, 97-101; 1959. (List) t)
PARKER, H. W. List of the frogs and toads of Trinidad. Trop. Agric., 10, 8-12; 1933.

Central America

GUNTHER, A. C. L. G. Biologia Centrali-Americana. Reptilia and Batrachia. London; 1-326; 1885-1902.
COPE, E. D. Catalogue of batrachians and reptiles of Central America and Mexico. United States Nat. Mus. Bull., 32, 1-98; 1887.
KELLOGG, R. Mexican tailless amphibians in the United States National Museum. United States Nat. Mus. Bull., 160, 1-224; 1932.
TAYLOR, E. H. Tadpoles of Mexican Anura. Univ. Kansas Sci. Bull., 28, 37-55; 1942. (Keys)
SMITH, H. M. & TAYLOR, E. H. An annotated check list and key to the Amphibia of Mexico. United States Nat. Mus. Bull., 194, 1-118; 1948.
BOGERT, C. M. & OLIVER, J. A. A preliminary analysis of the herpetofauna of Sonora. Bull. American Mus. Nat. Hist., 83, 303-425; 1945.
HARDY, L. M. & McDIARMID, R. W. The amphibians and reptiles of Sinaloa, Mexico. Publ. Univ. Kansas Mus. Nat. Hist., 18, 39-252; 1969.
PETERS, J. A. The amphibians and reptiles of the coast and coastal sierra of Michoacan, Mexico. Occ. Pap. Mus. Zool. Univ. Michigan, 554, 1-37; 1954.
DUELLMAN, W. E. A distributional study of the amphibians of the Isthmus of Tehuantepec, Mexico. Publ. Univ. Kansas Mus. Nat. Hist., 13, 19-72; 1960.
DUELLMAN, W. E. A preliminary analysis of the herpetofauna of Colima, Mexico. Occ. Pap. Mus.

Zool. Univ. Michigan, 589, 1-22; 1958.
ZWEIFEL, R. G. Results of the Puritan-American Museum of Natural History Expedition to Western Mexico. 9. Herpetology of the Tres Marias Islands. Bull. American Mus. Nat. Hist., 119, 77-128; 1960.
STUART, L. C. A checklist of the herpetofauna of Guatemala. Misc. Publ. Mus. Zool. Univ. Michigan, 122, 1-150; 1963.
STUART, L. C. The amphibians and reptiles of Alta Verapaz, Guatemala. Misc. Publ. Mus. Zool. Univ. Michigan, 69, 1-109; 1948.
STUART, L. C. The herpetofauna of the Guatemalan plateau. . . . Contr. Lab. Vert. Biol. Univ. Michigan, 49, 1-71; 1951.
STUART, L. C. Herpetofauna of the southeastern highlands of Guatemala. Contr. Lab. Vert. Biol. Univ. Michigan, 68, 1-65; 1954.
SCHMIDT, K. P. The amphibians and reptiles of British Honduras. Publ. Field Mus. Nat. Hist., 512, 475-510; 1941.
MERTENS, R. Die Amphibien und Reptilien von El Salvador, auf Grund der Reisen von R. Mertens und A. Zilch. Abh. Senckenbergische Naturf. Ges., 487, 1-83; 1952.
TAYLOR, E. H. A review of the frogs and toads of Costa Rica. Univ. Kansas Sci. Bull., 35, 577-942; 1952.
TAYLOR, E. H. The salamanders and caecilians of Costa Rica. Univ. Kansas, Sci. Bull., 34, 695-771; 1952.
DUNN, E. R. The amphibians of Barro Colorado Island. Occ. Pap. Boston Soc. Nat. Hist., 5, 403-421; 1931.
MYERS, C. W. & RAND, A. S. Checklist of amphibians and reptiles of Barro Colorado Island, Panama, with comments on faunal change and sampling. Smithsonian Contrib. Zool., 10, 1-21; 1969.

South America

BRAME, A. H. & WAKE, D. B. The salamanders of South America. Los Angeles; 1-72; 1963.
DONOSO-BARROS, R. Les salamandras Sudamericanos. · Mus. Nat. Hist. Nat. Santiago, 14; 1966.
MEDEM, F. Rev. Acad. Colombiana Cienc. Ex. Fis. Nat., 13, 149-199. (Colombia)
VELLARD, J. Distribucion de los batracios en las altas regiones Andinas del Peru y Bolivia. Actas Trab. I-Congr. Sudamericana Zool., 1959, 1, 279-292; 1961.
GINES, H. Familias y generos de anfibios — Amphibia — de Venezuela. Mem. Soc. Cienc. Nat. La Salle, 19, (53), 85-146; 1959.
RIVERO, J. A. Salientia of Venezuela. Bull. Mus. Comp. Zool., 126, 3-207; 1961.
KENNY, J. S. The Amphibia of Trinidad. Stud. Faun. Curacao Caribbean Isl., 29, 1-78; 1969.
PARKER, H. W. The frogs, lizards, and snakes of British Guiana. Proc. Zool. Soc. London, 1935, 505-530; 1935.
COCHRAN, D. M. Frogs of southeastern Brazil. United States Nat. Mus. Bull., 206, 1-423; 1955.
SANTOS, E. Anfibios e repteis do Brasil (vida e costumes). Rio de Janeira; 1-280; 1942.
MYERS, G. S. Lista provisoria dos anfibios do Distrito Federal, Brasil. Bol. Mus. Nac. Rio Janeiro, 55, 1-36; 1946.
LUTZ, B. Fauna anura Argentina-Brasilia. Acta Zool. Lilloana, 23, 147-152; 1967.
VELLARD, J. Batracios del Chaco Argentina. Acta Zool. Lillo, Tucuman, 5, 137-174; 1948.
KLAPPENBACH, N. A. & OREJA-MIRANDA, B. Anfibios y reptiles. Nuestra Tiera, 11, 1-68; 1969.
SCHOUTEN, G. B. Contribuciones al conocimiento de la fauna herpetologica del Paraguay y de los paises limitrofes. Rev. Soc. Cienc. Paraguay, 3, 5-32; 1931.
SCHOUTEN, G. B. Fauna herpetologica del Paraguay. Nov. Reun. Soc. Argentina Patol. Reg., 2, 1218-1232; 1937. (List)
FREIBERG, M. A. Enumeracion sistematica y distribucion geografica de los batracos argentinas. Physis, 19, 219-240; 1942.
CEI, J. M. Nueva lista sistematica de los batrachios de Argentina . . . . Invest. Zool. Chilenos, 3, 35-38; 1956.
VELLARD, J. Repartition des batraciens dans les Andes au sud de l'Equateur. Trav. Inst. Francaise Etude Antilles, 5, 141-162; 1956.
CAPURRO, L. Lista preliminar de los anfibios de Chile y breves apuntes sobre su distribucion y biologia. Invest. Zool. Chilenos, 4, 289-299; 1958.
CEI, J. M. Batracios de Chile. Santiago de Chile; 1-228, i-cviii; 1962.
DONOSO-BARROS, R. Catalogo herpetologico chileno.. Bol. Mus. Nac. Hist. Nat. Chile, 31, 42-124; 1970.

Europe

BOULENGER, G. A. The tailless batrachians of Europe, 2 parts. London; 1-376; 1897.
BOULENGER, G. A. Les batraciens et principalement ceux d'Europe. In Encyclopedie bibliotheque de zoologie. Paris; 1-305; 1910.
SCHREIBER, E. Herpetologia europaea. Eine systematische Bearbeitung der Amphibien und Reptilien . . . , 2. Auflage. Jena; 1-960; 1912.
MERTENS, R. & MULLER, L. Liste der Amphibien und Reptilien Europas. Abh. Senckenbergische Naturf. Ges., 41, 1-62; 1928.
MERTENS, R. & MULLER, L. Die Amphibien und Reptilien Europas. Zweite Liste. Frankfurt-am-Main; 1940.
HELLMICH, W. Die Lurche und Kriechtiere Europas. Winters Naturwiss. Taschenb., 26, 1-166; 1956.
MERTENS, R. & WERMUTH, H. Die Amphibien und Reptilien Europas. (Dritte Liste, nach dem Stand vom I. January 1960). Frankfurt-am-Main; 1-264; 1960.

HELLMICH. W. Reptiles and amphibians of Europe. London; 1-160; 1962. (Transl. of Die Lurche, 1956)
THORN, R. Les salamandres d'Europe, d'Asie, et d'Afrique du Nord. Paris; 1-376; 1969.
STEWARD, J. W. The tailed amphibians of Europe. Newton Abbott, England; 1-180; 1969.
BOULENGER, G. A. British batrachians. Proc. South London Ent. Nat. Hist. Soc., 1919, 28-31; 1919.
TAYLOR, R. H. R. The distribution of reptiles and amphibia in the British Isles. . . . British Journ. Herpet., 1, 1-38; 1948.
TAYLOR, R. H. R. The distribution of amphibians and reptiles in England and Wales, Scotland and Ireland and the Channel Isles; a revised survey. British Journ. Herpet., 3, 95-116; 1963.
FITTER, R. S. R. A check-list of the mammals, reptiles and amphibia of the London area 1900-1949. London Nat., 28, 98-115; 1949.
BAAL, H. J. The indigenous mammals, reptiles and amphibians of the Channel Islands. Bull. Soc. Jersiaise, 15, 101-110; 1949.
MERTENS, R. Amphibia, Reptilia. In Grimpe & Wagler, Die Tierwelt der Nord- und Ostsee, 3. Teil, 12, 1-20; 1926. MERTENS, R. Amphibia, Reptilia (Nachtrage). In Tierwelt der Nord- und Ostsee, Lief. 32, 1-2; 1938.
BRUUN, A. F. et al. List of Danish vertebrates. København; 1-180; 1950.
SCHAEFER, H. Die Artbestimmung der deutschen Anuren nach dem Skelet. Zeit. Anat. Entw. Ges. Berlin, 97, 767-779; 1932.
MERTENS, R. Neue Liste der Amphibien und Reptilien Deutschlands. Bl. Aquar. Terrar., 12, 177-178; 1938.
WERNER, F. & HERTER, K. Fauna von Deutschland. Amphibia & Reptilia. Leipzig; 492-501; 1944.
WERMUTH, H. Taschenbuch der heimischen Amphibien und Reptilien. Leipzig; 1-106; 1957.
FREYTAG, G. E. Lurche—Amphibia. In Exkursionfauna von Deutschland. Band III. Wirbeltiere. Berlin; 79-114; 1961.
MERTENS, R. Welches Tier is das? Kriechtiere und Lurche, 3rd ed. Stuttgart; 1-98; 1964.
van KAMPEN, P. N. Fauna van Nederland. Amphibia en Reptilia. Leiden; 1-64; 1927.
VAN de BUND, C. F. De Nederlandse Amfibien. Wet. Meded. K. Ned. Natuurh. Veren., 73, 1-32; 1968.
de WITTE, G. F. Amphibiens et reptiles. Faune de Belgique, 2nd ed., Bruxelles; 1-321; 1948.
ROSE, L. Clef dichotomique pour la determination des batraciens adultes de la region vervietoise. Rev. Vervietoise Hist. Nat., 18, 96-98; 1961.
FERRANT, V. Faune du Grand-Duche de Luxembourg. Deuxieme partie: Amphibies et reptiles. Luxembourg; 1055; 1922.
ANGEL, F. Reptiles et amphibiens. Faune de France, 45, 1-204; 1946.
BETHENCOURT FERREIRA, J. & de SEABRA, A. F. Catalogue systematique des vertebres du Portugal. Reptiles et amphibiens. Bull. Soc. Port. Sci. Nat., 5, 97-128; 1913.
ARMANDO THEMIDO. A, Repteis e batraquios das Colonios Portuguesas. Mem. Mus. Zool. Univ. Coimbra, (1), 119, 1-31; 1941.
ARMANDO THEMIDO, A. Anfibios e repteis de Portugal. Mem. Mus. Zool. Univ. Coimbra, (1), 133, 1-49; 1942.
BETHENCOURT FERREIRA, J. Revisao sistematico dos anfibios da fauna Portuguesa. Mem. Mus. Zool. Univ. Coimbra, 144, 1-32; 1943.
LADEIRO, J. M. Anfibios de Portugal. (notas para a sua classificacao). Mem. Mus. Zool. Univ. Coimbra, 243, 1-36; 1956.
SERRA, J. A. & ALBUQUERQUE, R. M. Anfibios de Portugal. Rev. Portug. Zool., 4, 75-227; 1963.
MERTENS, R. Die Amphibien und Reptilien Korsikas. Senckenbergische Biol., 38, 175-192; 1958.
STERNFELD, R. Die Reptilien und Amphibien Mitteleuropas. Heidelberg; 1-94; 1952.
FROMMHOLD, E. Wir bestimmen Lurche und Kriechtiere Mitteleuropas. Radebeul; 1-209; 1959. (In eight languages)
BERGER, L. et al. Catalogue faunae Poloniae. Amphibia et Reptilia. Warszawa; 1-73; 1969.
STEPANEK, O. Klic nasich obratlovcu. Praha; 1-250; 1950.
EISELT, J. Amphibia, Reptilia. In Strouhal, Cat. Faun. Austriaca, 21b, 1-21; 1961.
MEHELY, A. L. Reptilia et Amphibia. Fauna Regni Hungariae, 1918, 1-12; 1918.
RADOVANOVIC, M. Vodozemci i gmizavci nase zemije. Belgrade; 1-249; 1951. (Amphibien und Reptilien Jugoslaviens)
POZZI, A. Geonemia e catalogo ragionato degli Anfibi e dei Rettili della Jugoslavia. Natura, 57, 5-55; 1966.
KOPSTEIN, F. & WETTSTEIN, O. Reptilien und Amphibien aus Albanien. Verh. Zool. Bot. Ges. Wien, 70, 387-457; 1921.
SOCHUREK. E. Liste der Lurche und Kriechtiere Karntens. Carinthia II, 67, 150-152; 1958.
MERTENS, R. Die Amphibien und Reptilien der Insel Korfu. Senckenbergische Biol., 42, 1-29; 1961.
WETTSTEIN, O. Herpetologia aegaea. Sitz. Ber. Ost. Akad. Wiss., Abt. I, 162, 651-833; 1953.
KIRITESCU, C. Cercetari asupra faunei herpetologice a Romaniei. Bucuresti; 1-117; 1930.
CALINESCU, R. I. Contributiuni sistematice si zoogeografice la studiul amphibulor si reptilelor din Romania. Mem. Sec. Stun. Acad. Romana, (3), 7, 119-172; 1931.
FUHN, I. E. Amphibia. Fauna Rep. Pop. Romane, 14, (1), 1-288; 1960.
IONESCU, V. Vertebratele din Romania. Bucuresti; 1-496; 1968.
BURESCH, I. & ZONKOW, J. Untersuchungen uber die Verbreitung der Reptilien und Amphibien in Bulgarien und der Balkanhalbinsel. II. Schlangen. Bull. Inst. Roy. Hist. Nat. Sofia, 7, 106-188; 1934. (In Bulgarian)
BESKOV, V. & BERON, P. Catalogue et bibliographie des amphibiens et des reptiles en Bulgarie. Sofia; 1-40; 1964.
von WETTSTEIN, O. Herpetologie der Insel Korfu. Ann. Naturhist. Mus. Wien, 45, 159-172; 1931.

BODENHEIMER, F. S. Introduction into the knowledge of the Amphibia and Reptilia of Turkey. Rev. Fac. Sci. Univ. Istanbul, 9B, 1-83; 1944.
MERTENS, R. Amphibien und Reptilien aus der Turkei. Rev. Fac. Sci. Univ. Istanbul, 17B, 41-75; 1952.
NIKOLSKI, A. M. Amphibiens (Amphibia). Faune de la Russie; 1-309; 1918. (In Russian)
TERENTIEV, P. V. & CERNOV, S. A. Key to the Amphibia and reptiles of the U. S. S. R. Moskva; 1-95; 1936. (In Russian)
TERENTIEV, P. V. & CHERNOV, S. A. A brief key to the reptiles and amphibia of the U. S. S. R. Leningrad; 1-184; 1940. (In Russian)
TERENT'EV, P. V. Herpetology: a study of amphibians and reptiles. Moskva; 1-336; 1961. (In Russian)
NIKOL'SKII, A. M. Amphibians. Fauna of U. S. S. R. Jerusalem; 1-220; 1962. (Translation of 1918)
TERENT'EV, P. V. & CHERNOV, S. A. Key to amphibians and reptiles, 3rd ed. Jerusalem; 1-315; 1965. (Translation of 1949)
TARASHCHUK, V. I. Amphibians and reptiles. Fauna Ukraini, 7, 1-246; 1959. (In Ukrainian)
CHERNOV, S. A. Systematic review of the animals of Crimea. III. Reptilia and Amphibia. Zhivat. Mir., 5, 73-78; 1958. (In Russian)
NIKOLSKI, A. M. Herpetologia caucasica. Tiflis; 1-272; 1913. (In Russian)
CERNOV, S. A. Herpetological fauna of the Armenian S. S. R. and of the Mehichevansk A. S. S. R. Trud. Biol. Inst., Armenian Sect. Acad. Sci. USSR, III, 1-194; 1939. (In Russian)
GUMILEVSKY, B. A. The batrachians of Armenia and Nachichevan A. R. S. S. Zool. Pap. Biol. Inst. Erevan, 1, 1-24; 1939. (In Russian)
CERNOV, S. A. On the herpetological fauna of Armenia and Nachichevan A. R. S. S. Zool. Pap. Biol. Inst. Erevan, 1, 77-194; 1939. (In Russian)

Near East

AHARONI, I. Reptilia and Amphibia. In Luke & Keith-Roach, The Handbook of Palestine and Trans-Jordan. London; 418-422; 1934.
SCHMIDT, K. P. Amphibians and reptiles of Yemen. Fieldiana, Zool., 34, 253-261; 1953.
SCHMIDT, K. P. & MARX, H. The herpetology of Sinae. Fieldiana, Zool., 39, (4), 21-40; 1956.

North Africa & Atlantic Islands

PELLEGRIN, J. Les batraciens urodeles de l'Afrique du Nord. Rev. Hist. Nat. Appliqu. Paris, 7, 145-148; 1926. (List)
ANDERSON, J. Reptilia and Batrachia. Zoology of Egypt; 1, 1-371; 1898.
FLOWER, S. S. Notes on the recent reptiles and amphibians of Egypt, with a list of the species recorded from that Kingdom. Proc. Zool. Soc. London, 1933, 735-851; 1933.
ZAVATTARI, E. Erpetologia della Cirenaica. Arch. Zool. Italiana Torino, 14, 253-287; 1930.
PELLEGRIN, J. Les Reptiles et les Batraciens de l'Afrique du Nord Francaise. Compt. Rend. Assoc. Francaise Avanc. Sci., 51, 260-264; 1927. (List)
DOMERGUE, C. A. Liste des batraciens, cheloniens et sauriens. Bull. Soc. Sci. Nat. Tunis, 9-10, 75-79; 1967.
AELLEN, V. Contribution a l'herpetologie du Maroc. Bull. Soc. Sci. Nat. Maroc, 31, 153-199; 1952.
PASTEUR, G. & BONS, J. Les batraciens du Maroc. Trav. Inst. Sci. Cherifien, Zool., 17, 1-241; 1959.
SARMENTO, A. A. Repteis e batraquios do Arquipelago da Madeira. Funchal; 1-45; 1940.
SARMENTO, A. A. Vertebrados da Madeira. 1. Mamiferos — Aves — Repteis — Batraquios, 2nd ed. Funchal; 1-317; 1948.

Africa

MANACAS, S. Batraquios aglosses da Guine Portuguesa. Ann. Junta Miss. Geogr. Lisboa, 2, 63-69; 1947.
MANACAS, S. Batraquios agrossus da Guine Portuguesa. Trav. 2a Conf. Int. Afrique Occidentale Bissau, 3, 287-290; 1951.
SCHIØTZ, A. A preliminary list of amphibians collected in Sierra Leone. Vidensk. Medd. Dansk. Naturn. Foren., 127, 19-33; 1964.
BARBOUR, T. & LOVERIDGE, A. Reptiles and amphibians from Liberia. In Strong, African Republic of Liberia & the Belgian Congo. Cambridge, Massachusetts; 769-785; 1930. (List)
SCHIØTZ, A. A preliminary list of the amphibians collected in Ghana. Vidensk. Medd. Dansk Naturh. Foren., 127, 1-17; 1964.
SCHIØTZ, A. The amphibians of Nigeria. Vidensk. Medd. Dansk Naturh. Foren., 125, 1-92; 1963.
PERRET, J.-L. Les amphibiens du Cameroun. Zool. Jahr., Syst., 93, 289-464; 1966.
NOBLE, G. K. Contributions to the herpetology of the Belgian Congo based on the collection of the American Museum Congo Expedition, 1909-1915. Bull. American Mus. Nat. Hist., 49, 147-347; 1924.
MERTENS, R. Amphibien und Reptilien aus Angola. Senckenbergiana, 20, 425-443; 1938. (List)
LAURENT, R. F. Reptiles et amphibiens de l'Angola. III. Lisboa; 12-165; 1964.
ROSE, W. The reptiles and amphibians of southern Africa, 2nd ed. Cape Town; 1-494; 1962. (1st edition, 1-378, 1950)
POYNTON, J. C. Amphibia of southern Africa: a faunal study. Ann. Natal Mus., 17, 1-334; 1964.
van DIJK, D. E. Systematics and field keys to the families, genera and described species of southern African anuran tadpoles. Ann. Natal Mus., 18, 231-286; 1966.
WAGER, V. A. The frogs of South Africa. Cape Town; 1-242; 1965. (keys)
HEWITT, J. A guide to the vertebrate fauna of the Eastern Cape Province South Africa. Part II. Reptiles, amphibians, and freshwater fishes. Grahamstown; 1-141; 1937.

MANACAS, S. Batraquios de Mocambique. Ann. Junta Invest. Colon, 5, 181-197; 1950.
LOVERIDGE, A. East African reptiles and amphibians in the United States National Museum. United States Nat. Mus. Bull., 151, 1-135; 1929.
LOVERIDGE, A. A list of the Amphibia of the British territores in East Africa (Uganda, Kenya Colony, Tanganyika Territory and Zanzibar), together with keys . . . . Proc. Zool. Soc. London, 1930, 7-32; 1930.
LOVERIDGE, A. Check list of the reptiles and amphibians of East Africa (Uganda, Kenya, Tanganyika, Zanzibar). Bull. Mus. Comp. Zool., 117, 153-362; 1957.
SKELTON-BOURGEOIS, M. Reptiles et batraciens d'Afrique orientale. Rev. Zool. Bot. Africaines, 63, 309-336; 1961. (List)
LOVERIDGE, A. Amphibians from Nyasaland and Tete. Bull. Mus. Comp. Zool., 110, 325-406; 1953.
POYNTON, J. C. Amphibia of the Nyasa-Luangwa region of Africa. Senckenbergische Biol., 46, 193-225; 1964.
STEWART, M. M. Amphibians of Malawi. New York; 1-163; 1967. (List & keys)
SCORTECCI, G. Anfibi della Somalia Italiana. Atti Soc. Italiana Milano, 72, 5-70; 1933.
SCORTECCI, G. Rettili ed anfibi raccolti dal Prof. E. Zavattari in Eritrea. Atti Soc. Italiana Milano, 69, 193-217; 1930. (List)

Madagascar

[None noted]

Indian Ocean Islands

SMITH, M. A. The herpetology of the Andaman and Nicobar Islands. Proc. Linn. Soc. London, 153, 150-158; 1941. (List)
PARKER, H. W. Revised list of reptiles (excluding chelonians) and amphibians collected in the Seychelles. Trans. Linn. Soc. London, 19, 444-446; 1936.
VESEY-FITZGERALD, D. Reptiles and amphibians from the Seychelles Archipelago. Ann. Mag. Nat. Hist., (11), 14, 577-584; 1948.
GAYMER, R. Amphibians and reptiles of the Seychelles. British Journ. Herpet., 4, 24-28; 1968. (List)

India

BRUCK, G. Zur herpetofauna Afghanistans. Vist. Ceskoslovenska Spol. Zool., 32, 201-208; 1967.
MINTON, S. A., Jr. An annotated key to the amphibians and reptiles of Sind and Las Bela, West Pakistan. American Mus. Novit., 2081, 1-60; 1962.
MINTON, S. A, Jr. A contribution to the herpetology of West Pakistan. Bull. American Mus. Nat. Hist., 134, 27-184; 1966.
MERTENS, R. Die Amphibien und Reptilien West-Pakistans. Stuttgarter Beitr. Naturk., 197, 1-96; 1969.
BOULENGER, G. A. Reptilia and Batrachia. The fauna of British India including Ceylon and Burma. London; 1-541; 1890.
DANIEL, J. C. Field guide to the amphibians of western India. Part 1. Journ. Bombay Nat. Hist. Soc., 60, 415-438; 1963. Part 2. Journ., 60, 690-702; 1964.
de SILVA, P. H. D. H. A list of Amphibia recorded from Ceylon with a report on the Amphibia in the Colombo Museum. Spol. Zeylanica, 27, 243-250; 1955.
KIRTISINGHE, P. The Amphibia of Ceylon. Colombo; 1-112; 1957. (including keys)
SWAN, L. W. & LEVITON, A. E. The herpetology of Nepal: a history, check list, and zoogeographical analysis of the herpetofauna. Proc. California Acad. Sci., (4), 32, 103-147; 1962.

Indochina & Malay Peninsula

BOURRET, R. Les batraciens de l'Indochine. Mem. Inst. Oceanogr. Indochine, 6, 1-547; 1941.
SMITH, M. A. A list of the batrachians at present known to inhabit Siam. Journ. Nat. Hist. Soc. Siam, 2, 226-231, 256; 1917.
SUVATTI, C. Fauna of Thailand. Bangkok; 1-1100; 1950. (Amphibia, 447-463)
TAYLOR, E. H. & ELBEL, R. E. Contribution to the herpetology of Thailand. Univ. Kansas Sci. Bull., 38, 1033-1189; 1958.
TAYLOR, E. H. The amphibian fauna of Thailand. Univ. Kansas Sci. Bull., 43, 265-600; 1962.
BOULENGER, G. A. A vertebrate fauna of the Malay Peninsula from the Isthmus of Kra to Singapore . . adjacent islands. Reptilia and Batrachia. London; 1-294; 1912.
SMITH, M. A. The Reptilia and Amphibia of the Malay Peninsula . . . . Bull. Raffles Mus., 3, 1-149; 1930.

Indo-Australian Archipelago

van KAMPEN, P. N. The Amphibia of the Indo-Australian Archipelago. Leiden; 1-304; 1923.
MERTENS, R. Die Amphibien und Reptilien der Inseln Bali, Lombok, Sumbawa, und Flores. Abh. Senckenbergische Naturf. Ges., 42, 117-342; 1930.
ROBINSON, H. C. & KLOSS, C. B. A nominal list of the species of reptiles and batrachians occurring in Sumatra. Journ. Federated Malay States Mus., 8, 297-306; 1920.
MERTENS, R. Amphibien und Reptilien aus dem aussersten westen Javas und von benachbarten Eilanden. Treubia, 24, 83-105; 1958.

INGER, R. F. The systematics and zoogeography of the Amphibia of Borneo. Fieldiana, Zool., 52, 1-402; 1966.
BOULENGER, G. A. A catalogue of the reptiles and batrachians of Celebes . . . . Proc. Zool. Soc. London, 1897, 193-237; 1897. (Keys)
FORCART, L. Die Amphibien und Reptilien von Sumba. . . . Verh. Naturf. Ges. Basel, 64, 356-388; 1953. LOVERIDGE, A.
LOVERIDGE, A. New Guinean reptiles and amphibians in the Museum of Comparative Zoology and United States National Museum. Bull. Mus. Comp. Zool., 101, 305-430; 1948.

Australia

COGGER, H. C. The frogs of New South Wales. Sydney; 1-38; 1960.
MOORE, J. A. The frogs of eastern New South Wales. Bull. American Mus. Nat. Hist., 121, 149-386; 1961.
MARTIN, A. A. & LITTLEJOHN, M. J. The amphibian fauna of Victoria: Two new records and a check-list. Victoria Nat., 86, 169-172; 1969.
WAITE, E. R. The reptiles and amphibians of South Australia. Adelaide; 1-270; 1929.
MAIN, A. R. Key to frogs of south-western Australia. Western Australian Nat. Handb., 4, 114-124; 1954.
MAIN, A. R. Frogs of southern Western Australia. Western Australian Nat. Club Handb., 8, 1-73; 1965.

New Zealand

BARWICK, R. E. Illustrations of the New Zealand frog fauna. Tuatara, 8, 96-98; 1961.
STEPHENSON, E. M. New Zealand native frogs. Tuatara, 8, 99-106; 1961.

The Orient

TAYLOR, E. H. Philippine Amphibia. Philippine Journ. Sci., 16, 213-359; 1920.
TAYLOR, E. H. Amphibians and turtles of the Philippine Islands. Manila: 1-193; 1921.
INGER, R. F. Systematics and zoogeography of Philippine Amphibia. Fieldiana, Zool., 33, 181-531; 1954.
STEJNEGER, L. The batrachians and reptiles of Formosa. Proc. United States Nat. Mus., 38, 91-114; 1910.
SATO, I. The salamanders of Formosa. Trans. Nat. Hist. Soc. Formosa, 31, 114-124; 1941.
GEE, N. G. & BORING, A. M. A check list of Chinese Amphibia with notes . . . . Bull. Biol. Yenching Univ. Peiping, 1, 15-51; 1930.
CHANG, M. L. Y. Contribution a l'etude morphologique, biologique et systematique des Amphibiens urodeles de la Chine. Paris; 1-156; 1936.
POPE, C. H. & BORING, A. M. A survey of Chinese Amphibia. Peking Nat. Hist. Bull., 15, 13-74; 1940.
BORING, A. M. & LEROY, P. Chinese amphibians living and fossil forms. Publ. Inst. Geo. -Biol. Pekin, 13, 1-151; 1945.
CHANG, M. L. Y. Contribution a l'etude... des... urodeles de la Chine. Paris;1968. (Reprint of 1936)
BORING, A. M. The Amphibia of Hong Kong. Hongkong Nat., 5, 8-22, 95-107; 1934.
SOWERBY, A. D. Amphibians and reptiles recorded from or known to occur in the Shanghai area. Notes Herpet. Shanghai, 1, 1-16; 1943.
LIU, C. -C. Amphibians of western China. Fieldiana, Zool. Mem., 2, 1-400; 1950.
REEVES, C. D. Manual of the vertebrate animals of northeast and central China, exclusive of birds. Shanghai; 1-806; 1933.
BORING, A. M. et al. Handbook of North China Amphibia and reptiles, or herpetology of North China. Peiping; 1-64; 1932.
OKADA, Y. Amphibia of Jehol. Rep. 1st Sci. Exp. Manchoukuo, 1-47; 1935.
SHANNON, F. A. The reptiles and amphibians of Korea. Herpetologica, 12, 22-49; 1956.
INGER, R. F. Preliminary survey of the amphibians of the Riukiu Islands. Fieldiana, Zool., 32, 297-352; 1947.
JOHNSON, C. R. Herpetofauna of Okinawa, Ryu Kyu Islands. Herpetologica, 25, 206-210; 1969.
STEJNEGER, L. The herpetology of Japan and adjacent territory. United States Nat. Mus. Bull., 58, 1-577; 1907.
OKADA, Y. The tailless batrachians of the Japanese Empire. Tokyo; 1-310; 1931.
TAGO, K. The salamanders of Japan. Tokyo; 1-222; 1931. (In Japanese)
SATO, E. A general account of the Japanese tailed Amphibia. Osaka; 1-520; 1943. (In Japanese)
(anonymous). Illustrated encyclopedia of the fauna of Japan . . ., 2nd ed. Tokyo; 2 volumes; 1949.
SATO, I. & TAKASHIMA, H. A synopsis of the amphibians of Japan. Misc. Rep. Yamashine Inst. Ornith. Zool., 6, 260-268; 1955.
NAKAMURA, K. & UENO, S. -i. Japanese reptiles and amphibians in colour. Osaka; 1-214; 1963. (In Japanese)
OKADA, Y. Fauna Japonica. Anura (Amphibia). Tokyo; 1-234; 1966.

Oceania

BURT, C. E. & BURT, M. D. Herpetological results of the Whitney South Sea Expedition. VI. Pacific Island Amphibia and Reptilia in the collections of the American Museum of Natural History. Bull. American Mus. Nat. Hist., 63, 463-597; 1932. (Complete for western Pacific islands)

OLIVER, J. A. & SHAW, C. E. The amphibians and reptiles of the Hawaiian Islands. Zoologica, 38, 65-95; 1953.
KINGHORN, J. R. Herpetology of the Solomon Islands. Rec. Australian Mus., 16, 123-178; 1928.
BROWN, W. C. The amphibians of the Solomon Islands. Bull. Mus. Comp. Zool., 107, 1-64; 1952.

SPECIAL LISTS

Cave-dwelling. WOLF, B. Animalium cavernarum catalogus, pars 3, 700-704; 1936.

Salt-water. Neill, W. T. The occurrence of amphibians and reptiles in salt-water areas, and a bibliography. Bull. Mar. Sci. Gulf Caribbean, 8, 1-97; 1958.

Venomous. BOYS, F. & SMITH, H. M. Poisonous amphibians and reptiles. Recognition and bite treatment. Springfield, Illinois; 1-149; 1959.

MONOGRAPHIC, DESCRIPTIVE, REVISIONARY WORKS

Monographs

GADOW, H. Amphibia and reptiles. IN Cambridge Natural History, vol. VIII. London; 1-668; 1909.

Classifications

[See also works listed under Animalia]

von FEJERVARY, G. J. Kritische Bemerkungen zur Osteologie, Phylogenie, und Systematik der Anuren. Arch. Naturg., 37, (3), 1-30; 1921.
de MIRANDA-RIBEIRO, A. Notas para servirem ao estudo dos Gymnobatrachios (Anura) Brasileiros. Tomo primeiro. Arch. Mus. Rio de Janeiro, 27, 1-227; 1926. (World coverage)
PERRIER, E. Les Batraciens. Traite de Zool., 8, 2725-2882; 1928. (To families)
NOBLE, G. K. The biology of the Amphibia. New York; 1-577; 1944. (1st edition, 1931)
ROMER, A. S. Vertebrate paleontology, 2nd ed. Chicago; 1-687; 1945. (1st edition, 1933)
KUHN, O. Das System der fossilen und rezenten Amphibien und Reptilien. Ber. Naturf. Ges. Bamberg, 29, 49-67; 1946.
BLAIR, W. F. et al. Vertebrates of the United States, 2nd ed. New York; 1-616; 1968. (1st edition, 1957)
KUHN, O. Die Familien der rezenten und fossilen Amphibien und Reptilien. Bamberg; 1-79; 1961.
TERENT'EV, P. V. Herpetology. A study of amphibians and reptiles. Moskva; 1-336; 1961. (In Russian)
TERENT'EV, P. V. Herpetology. A manual of amphibians and reptiles. Jerusalem; 1-313; 1965. (Translation of 1961)
KUHN, O. Die Amphibien. System und Stammesgeschichte. Munchen; 1-102; 1965.
GOIN, C. J. & GOIN, O. B. Introduction to herpetology, 2nd ed. San Francisco; 1-353; 1971. (Pp. 331-332; to family)

Descriptive catalogs

BOULENGER, G. A. Catalogue of the Batrachia Gradientia S. Caudata and Batrachia Apoda in the collection of the British Museum, 2nd ed. London; 1-127; 1882.
BOULENGER, G. A. Catalogue of the Batrachia Salientia S. Ecaudata in the collection of the British Museum, 2nd ed. London; 1-503; 1882.

Synonymic catalogs

BRAME, A. H., Jr. A list of the world's Recent and fossil salamanders. Herpeton, 2, 1-26; 1967.
DONOSO-BARROS, R. Catalogo herpetologico chileno. Bol. Mus. Nac. Hist. Nat. Chile, 31, 42-124; 1970.
DUNN, E. R. Los generos de anfibios y reptiles de Colombia. Part I. Anfibios. Caldasia, 2, 497-529; 1944.
GORHAM, S. W. Liste der rezenten Amphibien und Reptilien. Gymnophiona. Das Tierreich, Lief. 78, 1-25; 1962.
RIEMER, W. J. et al. Catalogue of American amphibians and reptiles. Bethesda, Maryland; 1963 —

Revisionary works by families

Works cited by author

In order to save space in the list of works on individual families, the following studies, which are each referred to more than a few times, are cited by author and date and number only. Citations are otherwise given by title, if a book, or by the serial reference, if not a separate work. The titles of journal articles are not usually given, but the relevant limitations are shown by annotations.

Anderson, J., No. 1, 1898. Reptilia and Batrachia. Zoology of Egypt, vol. 1, 1-371.
Angel, F., No. 3, 1946. Reptiles et amphibiens. Faune de France, 45, 1-204.
Barbour, T. & Ramsden, C. T., No. 1, 1919. The herpetology of Cuba. Mem. Mus. Comp. Zool., 47, 69-213.
Bishop, S. C., No. 1, 1943. Handbook of salamanders. The salamanders of the United States, of Canada, and of Lower California. Ithaca, New York; 1-555. (Reprinted 1962 & 1967)
Boulenger, G. A., No. 1, 1890. Reptilia and Batrachia. The fauna of British India including Ceylon and Burma. London; 1-541.
Boulenger, G. A., No. 2, 1912. A vertebrate fauna of the Malay Peninsula from the Isthmus of Kra to Singapore . . . adjacent islands. Reptilia and Batrachia. London; 1-294.
Bourret, R., No. 2, 1942. Les batraciens de l'Indochine. Mem. Inst. Oceanogr. Indochine, 6, 1-547.
Cochran, D. D., No. 1, 1941. The herpetology of Hispaniola. United States Nat. Mus. Bull., 177, 1-398.
Cochran, D. M. & Goin, C. J., No. 1, 1970. Frogs of Colombia. United States Nat. Mus. Bull., 288, 1-655.
Cope, E. D. No. 1, 1889. The Batrachia of North America. United States Nat. Mus. Bull., 34, 1-525.
Fuhn, I. E., No. 1, 1960. Amphibia. Fauna Rep. Pop. Romane, 14, (1), 1-288.
Gunther, A. C. L. G., No. 1, 1885-1902. Reptilia and Batrachia. Biologia Centrali-Americana; 1-326.
Inger, R. F., No. 1, 1954. Systematics and zoogeography of Philippine Amphibia. Fieldiana, Zool., 33, 181-531.
Inger, R. F., No. 2, 1966. The systematics and zoogeography of the Amphibia of Borneo. Fieldiana, Zool., 52, 1-402.
van Kampen, P. N., No. 1, 1923. The Amphibia of the Indo-Australian Archipelago. Leiden; 1-304.
Kellogg, R., No. 1, 1932. Mexican tailless amphibians in the United States National Museum. United States Nat. Mus. Bull., 160, 1-224.
Mertens, R., No. 1, 1952. Die Amphibien und Reptilien von El Salvador, auf Grund der Reisen von R. Mertens und A. Zilch. Abh. Senckenbergische Naturf. Ges., 487, 1-83.
de Miranda-Ribeiro, A., No. 1, 1926. Notas para servirem ao estudo dos Gymnobatrachios (Anura) Brasileiros. Tomo premeiro. Arch. Mus. Rio de Janeiro, 27, 1-227.
Nieden, F., No. 1, 1923. Anura I, Subordo Aglossa und Phaneroglossa, Sectio I, Arcifera. Das Tierreich, Lief. 46, 1-584.
Nikolsky, A. M., No. 3, 1918. Amphibiens (Amphibia). Faune de la Russie, 1-309. (In Russian)
Nikol'skii, A. M., No. 6, 1962. Amphibians. Fauna of Russia and adjacent countries. Jerusalem; 1-225. (Translation of 1915)
Nikolskli, A. M., No. 7, 1962. Fauna of Russia and adjacent countries. Amphibians. Jerusalem; 1-225. (Translation of 1918)
Okada, Y., No. 1, 1931. The tailless Batrachians of the Japanese Empire. Tokyo; 1-210.
Pasteur, G. & Bons, J., No. 1, 1959. Les batraciens du Maroc. Trav. Inst. Sci. Cherifien, Zool., 17, 1-241.
Poynton, J. C., No. 1, 1964. Amphibia of southern Africa: A faunal study. Ann. Matal Mus., 17, 1-334.
Rivero, J. A., No. 1, 1961. Salientia of Venezuela. Bull. Mus. Comp. Zool., 126, 1-207.
Schreiber, E., No. 1, 1912. Herpetologia europaea. Eine systematische Bearbeitung der Amphibien und Reptilien. . . ., 2. Aufl. Jena; 1-960.
Smith, H. M., No. 2, 1956. Handbook of amphibians and reptiles of Kansas, ed. 2. Misc. Publ. Univ. Kansas Mus. Nat. Hist., 9, 1-356.
Smith, M. A., No. 1, 1930. The Reptilia and Amphibia of the Malay Peninsula. . . . Bull. Raffles Mus., 3, 1-149.
Stebbins, R. C., No. 1, 1951. Amphibians of western North America. Berkeley; 1-539.
Stebbins, R. C., No. 2, 1954. Amphibians and reptiles of western North America. New York; 1-528.
Stejneger, L., No. 1, 1907. The herpetology of Japan and adjacent territory. United States Nat. Mus. Bull., 58, 1-577.
Taylor, E. H., No. 10, 1921. Amphibians and turtles of the Philippine Islands. Manila; 1-193.
Taylor, E. H., No. 1, 1952. A review of the frogs and toads of Costa Rica. Univ. Kansas Sci. Bull., 35, 577-942.
Taylor, E. H., No. 3, 1962. The amphibian fauna of Thailand. Univ. Kansas Sci. Bull., 43, 265-600.
Wright, A. H. & Wright, A. A., No. 1, 1949. Handbook of frogs and toads of the United States, 3rd ed. Ithaca, New York; 1-640. (Reprinted 1965)

By families

Acanthostomatidae, as Zatrachyidae (only as fossils)
Adelogyrinidae (only as fossils)
Alytidae, as Discoglossidae
Amblystomatidae, as Ambystomatidae
Amblystomidae, as Ambystomatidae

Ambystomatidae (incl. Amblystomatidae, Amblystomidae, Ambystomidae). HILTON, W. A., 1946, Journ. Entom. Zool., 38, 29-36 (skeleton); COPE, E. D., No. 1, 1889 (North America); BISHOP, S. C., No. 1, 1962 (North America); SMITH, H. M., No. 2, 1956 (Kansas); STEBBINS, R. C., No. 1, 1951 (western North America); STEBBINS, R. C., No. 2, 1954 (western North America); NIKOLSKI, A. M., No. 3, 1918 (U. S. S. R.); NIKOL'SKII, A. M., No. 6, 1962 (Russia); STEJNEGER, L., No. 1, 1907 (Japan)
Ambystomidae, as Ambystomatidae
Amphibamidae (only as fossils)
Amphignathodontidae, as Hylidae
Amphiumidae. COPE, E. D., No. 1, 1889 (North America); BISHOP, S. C., No. 1, 1962 (North America)
Anthracosauridae (only as fossils)
Archegosauridae (only as fossils)
Ascalaphidae. GORHAM, S. W., 1966, Das Tierreich, 85, 1-122 (Liste der rezenten. . . .)
Ascaphidae (incl. Leiopelmidae, Leiopelmatidae, Liopelmidae). GORHAM, S. W., 1966, Das Tierreich, 85, 1-122 (Liste der rezenten . . . .); WRIGHT, A. H. & WRIGHT, A. A., No. 1, 1949 (North America); STEBBINS, R. C., No. 1, 1951 (North America, as Liopelmidae); STEBBINS, R. C., No. 2, 1954 (western North America)
Asterophryidae, as Microhylidae
Astrodactylidae. (Family not identified)
Atelopodidae. COCHRAN, D. M. & GOIN, C. J., No. 1, 1970 (Colombia); RIVERO, J. A., No. 1, 1961 (Venezuela); TAYLOR, E. H., No. 3, 1962 (Thailand)

Bactrachophrynidae, as Leptodactylidae
Benthosuchidae (only as fossils)
Bolitoglossidae, as Plethodontidae
Bombinatoridae, as Discoglossidae
Brachycephalidae (incl. Dendrobatidae, Dendrophryniscidae, Rhinodermatidae). GUNTHER, A. C. L. G., No. 1, 1885-1902 (Central America); SAVAGE, J. M., 1969, Copeia, 1968, 745-776 (Central America, as Dendrobatidae); TAYLOR, E. H., No. 1, 1952 (Costa Rica); RIVERO, J. A., No. 1, 1961 (Venezuela, as Dendrobatidae); de MIRANDA-RIBEIRO, A., No. 1, 1926 (Brasil); COCHRAN, D. M., 1955, United States Nat. Mus. Bull., 206, 1-423; GALLARDO, J. M., 1961, Reun. Trab. Comun. Cienc. Nat. Geog. Argentina, 1961, 205-212; CEI, J. M., N962, Batracios de Chile, Santiago, 1-228 (Chile)
Brachymeridae. (Family not identified)
Brachyopidae (only as fossils)
Brevicipitidae, as Microhylidae
Broilisauridae. (No revisionary references noted)
Bufavidae (only as fossils)
Bufondidae, as Bufonidae
Bufonidae (incl. Bufondidae, Hylodidae, Pseudidae, Telmatobiidae). NIEDER, F., No. 1, 1923 (World); GALLARDO, J. M., 1961, Bull. Mus. Comp. Zool., 125, 111-134 (World, as Pseudidae); COPE, E. D., No. 1, 1889 (North America); WRIGHT, A. H. & WRIGHT, A. A., No. 1, 1949 (North America); STEBBINS, R. C., No. 1, 1951 (western North America); STEBBINS, R. C., No. 2, 1954 (western North America); SMITH, H. M., No. 2, 1956 (Kansas); BARBOUR & RAMSDEN No. 1, 1919 (Cuba); COCHRAN, D. M., No. 1, 1941 (Hispaniola); SCHMIDT, K. P., No. 1, 1928 (Puerto Rico); GUNTHER, A. C. L. G., No. 1, 1885-1902 (Central America); KELLOGG, R., No. 1, 1932 (Mexico); MERTENS, R., No. 1, 1952 (El Salvador); TAYLOR, E. H., No. 1, 1952 (Costa Rica); COCHRAN, & GOIN, No. 1, 1970 (Colombia, also as Pseudidae); RIVERO, J. A., No. 1, 1961 (Venezuela, also as Pseudidae); de MIRANDA-RIBEIRO, A., No. 1, 1926 (Brasil, also as Hylodidae & Telmatobiidae); COCHRAN, D. M., 1955, United States Nat. Mus. Bull., 206, 1-423 (Brasil); CEI, J. M., 1962, Batracios de Chile, Santiago, 1-228 (Chile); SCHREIBER, E., No. 1, 1912 (Europe); ANGEL, F., No. 3, 1946 (France); FUHN, I. E., No. 1, 1960 (Romania); NIKOLSKI, A. M., No. 3, 1918 (U. S. S. R.); NIKOL'SLII, A. M., No. 6, 1962 (Russia); ANDERSON, J., No. 1, 1898 (Egypt); PASTEUR & BONS, No. 1, 1959 (Maroc); POYNTON, J. C., No. 1, 1964 (southern Africa); BOULENGER, G. A., No. 1, 1890 (British India); MERTENS, R., 1969, Stuttgarter Beitr. Naturk., 197, 1-96 (West Pakistan); ANNANDALE, N. & RAO, C. R. N., 1918, Rec. Indian Mus., 15, 25-40 (India); TAYLOR, E. H., No. 3, 1962 (Thailand); BOURRET, R., No. 2, 1942 (Indochina); BOULENGER, G. A., No. 2, 1912 (Malay Peninsula); SMITH, M. A., No. 1, 1930 (Malay Peninsula); van KAMPEN, P. N., No. 1, 1923 (Indo-Australian Arch.); INGER, R. F., No. 2, 1966 (Borneo); TAYLOR, E. H., No. 10, 1921 (Philippines); INGER, R. F., No. 1, 1954 (Philippines); OKADA, Y., 1935, Rep. 1st Sci. Exp. Manchoukuo, 1-47 (Jehol); STEJNEGER, L., No. 1, 1907 (Japan); OKADA, Y., No. 1, 1931 (Japan); OKADA, Y., 1966, Fauna Japonica, Anura (Amphibia), Tokyo, 1-234 (Japan); BROWN, W. C., 1952, Bull. Mus. Comp. Zool., 107, 1-64 (Solomon Is.)
Bystrowianidae. (No revisionary references noted)

Caecilidae, as Caeciliidae
Caeciliidae (incl. Caecilidae, Coecilidae, Coeciliidae). NIEDEN, F., 1913, Das Tierreich, 37, 1-31; TAYLOR, E. H., 1968, The Caecilians of the world, Lawrence, Kansas, 1-848 (monograph); DUNN, E. R., 1942, Bull. Mus. Comp. Zool., 91, 439-540 (America); MERTENS, R., No. 1, 1952 (El Salvador); TAYLOR, E. H., 1952, Univ. Kansas Sci. Bull., 34, 695-771 (Costa Rica); ROZE, J. A. & SOLANO, H., 1963, Acta Biol. Venezolana, 3, 287-300 (Venezuela); von IHERING, R., 1911, Rev. Mus. Paulista, 8, 89-111 (Brasil); PARKER, H. W., 1941, Ann. Mag. Nat.

Hist., (11), 7, 1-17 (Seychelles); PARKER, H. W., 1958, Copeia, 1958, 71-76 (Seychelles); BOULENGER, G. A., No. 1, 1890 (British India); DERANIYAGALA, P. E. P., 1933, Ceylon Journ. Sci., (B), 17, 231-235 (Ceylon); TAYLOR, E. H., No. 3, 1962 (Thailand); BOURRET, R., No. 2, 1942 (Indochina); BOULENGER, G. A., No. 2, 1912 (Malay Peninsula); SMITH, M. A., No. 1, 1930 (Malay Peninsula); van KAMPEN, P. N., No. 1, 1923 (Indo-Australian Arch.) INGER, R. F., No. 2, 1966 (Borneo); TAYLOR, E. H., No. 10, 1921 (Philippines); INGER, R. F. No. 1, 1954 (Philippines)

Capitosauridae (only as fossils)

Centrolenidae. TAYLOR, E. H., No. 1, 1952 (Costa Rica); COCHRAN & GOIN, No. 1, 1970 (Colombia); RIVERO, J. A., No. 1, 1961 (Venezuela); RIVERO, J. A., 1969, Mem. Soc. Cienc. Nat. La Salle, 28, 301-334 (Venezuela); TAYLOR, E. H. & COCHRAN, D., 1953, Univ. Kansas Sci. Bull., 35, 1625-1656 (Brasil); COCHRAN, D. M., 1955, United States Nat. Mus. Bull., 206, 1-423 (Brasil)

Ceratobatrachidae, as Ranidae

Ceratophrydidae, as Leptodactylidae

Ceratophryidae, as Leptodactylidae

Ceratophrynidae, as Leptodactylidae

Chenoprosopidae (only as fossils)

Coecilidae, as Caeciliidae

Coeciliidae, as Caeciliidae

Colosteidae. (No revisionary references noted)

Colostethidae. (Family not identified)

Cophylidae, as Microhylidae

Cricotidae. (Family not identified)

Cryptobranchidae. COPE, E. D., No. 1, 1889 (North America); BISHOP, S. C., No. 1, 1962 (North America); SMITH, H. M., No. 2, 1956 (Kansas); CHANG, M. L. Y., 1936, Contribution a l'etude morphologique, biologique et systematique des Amphibiens urodeles de la Chine, Paris, 1-156 (China); STEJNEGER, L., No. 1, 1907 (Japan)

Cystignathidae, as Leptodactylidae

Dactylethridae. (Family not identified)

Dendrerpetontidae (only as fossils)

Dendrobatidae, as Brachycephalidae

Dendrophryniscidae, as Brachycephalidae

Desmognathidae, as Plethodontidae

Diplovertebrontidae (Only as fossils)

Discoglossidae (incl. Alytidae, Bombinatoridae). GORHAM, S. W., 1966, Das Tierreich, 85, 1-122 (Liste der rezenten. . . .); NIEDEN, R., No. 1, 1923 (World); SCHREIBER, E., No. 1, 1912 (Europe); ANGEL, F., No. 3, 1946 (France); FUHN, I. E., No. 1, 1960 (Romania); NIKOLSKI, A. M., No. 3, 1918 (U. S. S. R.); NIKOL'SKII, A. M., No. 6, 1962 (Russia); PASTEUR, G. & BONS, J., No. 1, 1959 (Maroc); BOURRET, R., No. 2, 1942 (Indochina); INGER, R. F., No. 1, 1954 (Philippines); STEJNEGER, L., No. 1, 1907 (Japan); OKADA, Y., No. 1, 1931 (Japan); OKADA, Y., 1966, Fauna Japonia, Anura (Amphibia), Tokyo, 1-234 (Japan)

Dissorhophidae, as Dissorophidae (only as fossils)

Dissorophidae (incl. Dissorhophidae, Dissorphidae) (only as fossils)

Dissorphidae, as Dissorophidae

Dolichopareiidae (only as fossils)

Dolichosomidae (only as fossils)

Dvinosauridae (only as fossils)

Dyscophidae, as Microhylidae

Edopsidae (only as fossils)

Ellipsoglossidae (Family not identified)

Elosiidae, as Leptodactylidae

Elpistostegidae (only as fossils)

Engystomatidae, as Microhylidae

Engystomidae, as Microhylidae

Eogyrinidae (only as fossils)

Eoxenopoididae (only as fossils)

Eryopsidae (only as fossils)

Gymnarthridae (only as fossils)

Heleophrynidae. (No revisionary references noted)

Hemidactylidae, as Plethodontidae

Hemiphractidae, as Hylidae

Hemisidae, as Microhylidae

Hesperoherpetonidae. (No revisionary references noted)

Hylaebatrachidae (only as fossils)

Hylaedactylidae, as Microhylidae

Hylaplosidae. (Family not identified)

Hylidae (incl. Amphignathodontidae, Hemiphractidae, Phyllomedusidae). NIEDER, F., No. 1, 1923 (World); GOIN, C. J., 1961, Ann. Carnegie Mus., 36, 5-18 (synopsis of genera); DUELLMAN, W. E.,

1968, Publ. Univ. Kansas Mus. Nat. Hist., 18, 1-10 (phyllomedusine genera); COPE, E. D., No. 1, 1889 (North America); WRIGHT & WRIGHT, No. 1, 1949 (North America); SMITH, H. M., No. 2, 1956 (Kansas); STEBBINS, R. C., No. 1, 1951 (western North America); STEBBINS, R. C., No. 2, 1954 (western North America); LUTZ, B., 1968, Texas Mem. Mus. Pearce-Sellards Ser., 11, 3-26 (Neotropics); BARBOUR & RAMSDEN, No. 1, 1919 (Cuba); COCHRAN, D. M., No. 1, 1941 (Hispaniola); DUELLMAN, W. E., 1970, Monogr. Univ. Kansas Mus. Nat. Hist., 1, 2 volumes (Central America); KELLOGG, R., No. 1, 1932 (Mexico); MERTENS, R., No. 1, 1952 (El Salvador); TAYLOR, E. H., No. 1, 1952 (Costa Rica); COCHRAN & GOIN, No. 1, 1970 (Colombia); RIVERO, J. A., No. 1, 1961 (Venezuela); de MIRANDO-RIBEIRO, A., No. 1, 1926 (Brasil & classification); COCHRAN, D. M., 1955, United States Nat. Mus. Bull., 206, 1-423 (Brasil); SCHREIBER, E., No. 1, 1912 (Europe); ANGEL, F., No. 3, 1946 (France); FUHN, I. E., No. 1, 1960 (Romania); NIKOLSKI, A. M., No. 3, 1918 (U. S. S. R.); NIKOL'SKII, A. M., No. 6, 1962 (Russia); ANDERSON, J., No. 1, 1898 (Egypt); PASTEUR, & BONS, No. 1, 1959 (Maroc); BOULENGER, G. A., No. 1, 1890 (British India); TAYLOR, E. H., No. 3, 1962 (Thailand); BOURRET, R., No. 2, 1942 (Indochina); van KAMPEN, P. N., No. 1, 1923 (Indo-Australian Arch.); TYLER, M. J., 1968, Zool. Verhand., 96, 1-203 (Papuan Hyla); TYLER, M. J., 1963, Trans. Roy. Soc. South Australia, 86, 1-29, 105-130 (Central highlands of New Guinea); MOORE, J. A., 1961, Bull. American Mus. Nat. Hist., 121 149-386 (eastern New South Wales); INGER, R. F., No. 1, 1954 (Philippines); STEJNEGER, L., No. 1, 1907 (Japan); OKADA, Y., No. 1, 1931 (Japan); OKADA, Y., 1966, Fauna Japonica, Anura (Amphibia), Tokyo, 1-234 (Japan); BROWN, W. C., 1952, Bull. Mus. Comp. Zool., 107, 1-64 (Solomon Is.)

Hylodidae, as Bufonidae

Hylonomidae (only as fossils)

Hynobidae, as Hynobiidae

Hynobiidae (incl. Hynobidae). DUNN, E. R., 1923, Proc. American Acad. Arts Sci., 58, 445-523; HILTON, W. A., 1946, Journ. Entom. Zool., 38, 40-45 (skeleton); CHANG, M. L. Y., 1936, Contribution a l'etude morphologique, biologique et systematique des amphibiens urodeles de la Chine, Paris (China); SATO, I., 1937, Bull. Biogeogr. Soc. Tokyo, 7, (3), 31-45 (Japan)

Hyperolidae, as Hyperoliidae

Hyperoliidae (incl. Hyperolidae). LIEM, S. S., 1970, Fieldiana, Zool., 57, 1-145 (Old World)

Ichthyophidae. TAYLOR, E. H., 1968, The caecilians of the world, Lawrence, Kansas, 1-848 (new family)

Ichthyostegidae (only as fossils)

Intasuchidae (only as fossils)

Keraterpetontidae (incl. Scincosauridae). (No revisionary references noted)

Kotlassiidae (only as fossils)

Lasalidae (incl. Lasilidae) (Only as fossils)

Lasilidae, as Lasalidae (Only as fossils)

Leiopelmatidae, as Ascaphidae

Leiopelmidae, as Ascaphidae

Lepterpetontidae (only as fossils)

Leptodactylidae (incl. Bactrachophrynidae, Ceratophrydidae, Ceratophryidae, Ceratophrynidae, Elosiidae, Cystignathidae, Paludicolidae, Plectomantidae, Plectromantidae, Uperoliidae). GORHAM, S. W., 1966, Das Tierreich, 85, 1-122 (Liste der rezenten....); NIEDER, F., No. 1, 1923 (World, as Cystignathidae); COPE, E. D., No. 1, 1889 (North America, as Cystignathidae); WRIGHT & WRIGHT, No. 1, 1949 (North America); STEBBINS, R. C., No. 1, 1951, (western North America); STEBBINS, R. C., No. 2, 1954 (western North America); BARBOUR & RAMSDEN, No. 1, 1919 (Cuba); COCHRAN, D. M., No. 1, 1941 (Hispaniola); GUNTHER, A. C. L. G., No. 1, 1885-1902 (Central America, as Cystignathidae); KELLOGG, R., No. 1, 1932 (Mexico); LYNCH, J. D., 1968, Publ. Univ. Kansas Mus. Nat. Hist., 17, 503-515 (Mexican genera); MERTENS, R., No. 1, 1952 (El Salvador); TAYLOR, E. H., No. 1, 1952 (Costa Rica); COCHRAN & GOIN, No. 1, 1970 (Colombia); RIVERO, J. A., No. 1, 1961 (Venezuela); de MIRANDA-RIBEIRO, A., No. 1, 1926 (Brasil, also as Ceratophrydidae, Elosiidae, & Paludicolidae); COCHRAN, D. M., 1955, United States Nat. Mus. Bull., 206, 1-423 (Brasil); CEI, J. M., 1962, Batracios de Chile, Santiago de Chile, 1-228 (Chile); POYNTON, J. C., No. 1, 1964 (southern Africa); PARKER, H. W., 1940, Novit. Zool., 42, 1-206 (Australasia); van KAMPEN, P. N., No. 1, 1923 (Indo-Australian Arch., as Cystignathidae); MOORE, J. A., 1961, Bull. American Mus. Nat. Hist., 121, 149-386 (eastern New South Wales)

Limnerpetontidae (only as fossils)

Liopelmidae, as Ascaphidae

Loxommidae (only as fossils)

Lydekkerinidae (only as fossils)

Lysorophidae (only as fossils)

Metoposauridae (incl. Metropasauridae). (Only as fossils)

Metropasauridae, as Metoposauridae (Only as fossils)

Microbrachidae (only as fossils)

Microhylidae (incl. Asterophryidae, Brevicipitidae, Cophylidae, Dyscophidae, Engystomatidae, Engystomidae, Hemisidae, Hylaedactylidae, Xenorhinidae). PARKER, H. W., 1934, A monograph of the frogs of the family Microhylidae, London, 1-208 (monograph); NIEDEN, F., 1926, Das

Tierreich, 49, 1-110; de CARVALHO. A. L., 1954, Occ. Pap. Mus. Zool. Univ. Michigan, 555, 1-19 (synopsis American genera); COPE, E. D., No. 1, 1889 (North America, as Engystomidae); WRIGHT & WRIGHT, No. 1, 1949 (North America, as Brevicipitidae); SMITH, H. M., No. 2, 1956 (Kansas); STEBBINS, R. C., No. 1, 1951 (western North America); STEBBINS, R. C., No. 2, 1954 (western North America); KELLOGG, R., No. 1, 1932 (Mexico, as Brevicipitidae); TAYLOR, E. H., No. 1, 1952 (Costa Rica); COCHRAN & GOIN, No. 1, 1970 (Colombia); RIVERO, J. A., No. 1, 1961 (Venezuela); de MIRANDA-RIBEIRO, A., No. 1, 1926 (Brasil, as Engystomatidae); COCHRAN, D. M., 1955, United States Nat. Mus. Bull., 206, 1-423 (Brasil); POYNTON, J. C., No. 1, 1964 (southern Africa); NOBLE, G. K. & PARKER, H. W., 1926, American Mus. Novit., 232, 1-21 (Madagascar, as Brevicipitidae); BOULENGER, G. A., No. 1, 1890 (British India, as Dyscophidae & Engystomatidae); TAYLOR, E. H., No. 3, 1962 (Thailand); BOURRET, R., No. 2, 1942 (Indochina); BOULENGER, G. A., No. 2, 1912 (Malay Peninsula, as Engystomatidae); SMITH, M. A., No. 1, 1930 (Malay Peninsula, as Brevicipitidae); van KAMPEN, P. N., No. 1, 1923 (Indo-Australian Arch., as Brevicipitidae); INGER, R. F., No. 2, 1966 (Borneo); TYLER, M. J., 1963, Trans. Roy. Soc. South Australia, 86, 1-19, 105-130 (central highlands of New Guinea); ZWEIFEL, R. G., 1962, American Mus. Novit., 2113, 1-40 (Australia); TAYLOR, E. H., No. 10, 1921 (Philippines, as Brevicipitidae); INGER, R. F., No. 1, 1954 (Philippines); STEJNEGER, L., No. 1, 1907 (Japan, as Engystomidae); OKADA, Y., No. 1, 1931 (Japan, as Engystomidae); OKADA, Y., 1966, Fauna Japonica, Anura (Amphibia), Tokyo, 1-234 (Japan, as Microhylidae)

Micropholidae (only as fossils)
Molgidae. (Family not identified)
Montsechobatrachidae (only as fossils)

Necturidae. (No revisionary references noted)
Notobatrachidae. (No revisionary references noted)

Ophiderpatontidae, as Ophiderpetontidae. (Only as fossils)
Ophiderpetontidae (incl. Ophiderpatontidae). (Only as fossils)
Ostodolepidae (only as fossils)
Otocratiidae (only as fossils)

Palaeobatrachidae (only as fossils)
Palaeogyrinidae (only as fossils)
Paludicolidae, as Leptodactylidae
Pantylidae (only as fossils)
Peliontidae. (No revisionary references noted)
Pelobatidae. (Incl. Pelodytidae, Scaphiopodidae). NIEDEN, F., No. 1, 1923 (World); GORHAM, S. W., 1966, Das Tierreich, 85, 1-122 (Liste der rezenten....); COPE, E. D., No. 1, 1889 (North America, as Scaphiopodidae); WRIGHT & WRIGHT, No. 1, 1949 (North America, as Scaphiopodidae); SMITH, H. M., No. 2, 1956 (Kansas); STEBBINS, R. C., No. 1, 1951 (western North America); STEBBINS, R. C., No. 2, 1954 (western North America); KELLOGG, R., No. 1, 1932 (Mexico); SCHREIBER, E., No. 1, 1912 (Europe); ANGEL, F., No. 3, 1946 (France); FUHN, I., No. 1, 1960 (Romania); NIKOLSKI, A. M., No. 3, 1918 (U. S. S. R.); NIKOL'SKII, A. M., No. 6, 1962 (Russia); PASTEUR, G. & BONS, J., No. 1, 1959 (Maroc); BOULENGER, G. A., No. 1, 1890 (British India); TAYLOR, E. H., No. 3, 1962 (Thailand); BOURRET, R., No. 2, 1942 (Indochina); BOULENGER, G. A., No. 2, 1912 (Malay Peninsula); SMITH, M. A., No. 1, 1930 (Malay Peninsula); van KAMPEN, P. N., No. 1, 1923 (Indo-Australian Arch.); INGER, R. F., No. 2, 1966 (Borneo); TAYLOR, E. H., No. 10, 1921 (Philippines); INGER, R. F., No. 1, 1954 (Philippines)
Pelodryadidae. (Family not identified)
Pelodytidae, as Pelobatidae
Phlegethontiidae (only as fossils)
Pholidogasteridae (only as fossils)
Phryniscidae. COPE, E. D., No. 1, 1889 (North America)
Phrynomeridae. (No revisionary references noted)
Phrynosuchidae (only as fossils)
Phyllomedusidae, as Hylidae
Pididae, as Pipidae
Pipidae (incl. Pididae, Xenopidae). NIEDEN, F., No. 1, 1923 (World); GORHAM, S. W., 1966, Das Tierreich, 85, 1-122 (Liste der rezenten....); DUNN, E. R., 1948, American Mus. Novit., 1384, 1-13 (America); COCHRAN & GOIN, No. 1, 1970 (Colombia); GINES, H., 1958, Mem. Soc. Cienc. Nat. La Salle, 18, (49), 5-18 (Venezuela); RIVERO, J. A., No. 1, 1961 (Venezuela); de MIRANDA-RIBEIRO, A., No. 1, 1926 (Brasil); ARNOULT, J. & LAMOTTE, M., 1968, Bull. Inst. Fond. Afrique Noire, 30A, 270-306 (West Africa & Cameroun); POYNTON, J. C., No. 1, 1964 (southern Africa)
Plectomantidae, as Leptodactylidae
Plectromantidae, as Leptodactylidae
Plethodontidae (incl. Desmognathidae, Hemidactylidae, Typhlomolgidae). DUNN, E. R., 1926, The salamanders of the family Plethodontidae, Northampton, Massachusetts, 1-441 (monograph); also as: Smith Coll. 50th Ann. Publ., 7, 1-441, 1926; WAKE, D. B., 1966, Comparative osteology and evolution of the lungless salamanders, family Plethodontidae. Mem. California Acad. Sci., 4, 1-111; COPE, E. D., No. 1, 1889 (North America, also as Desmognathidae);

BISHOP, S. C., No. 1, 1962 (North America); SMITH, H. M., No. 2, 1956 (Kansas); STEBBINS, R. C., No. 1, 1951 (western North America); STEBBINS, R. C., No. 2, 1954 (western North America); HILTON, W. A., 1946, Journ. Entom. Zool., 38, 1-8 (Mexico & Central America); TAYLOR, E. H., 1944, Univ. Kansas Sci. Bull., 30, 189-232 (Mexico); MERTENS, R., No. 1, 1952 (El Salvador); TAYLOR, E. H., 1952, Univ. Kansas Sci. Bull., 34, 695-771 (Costa Rica); BRAME, A. H., Jr. & WAKE, D. B., 1963, Contr. Sci. Los Angeles, 69, 1-72 (South America); ANGEL, F., No. 3, 1946 (France);

Pleurodelidae, as Salamandridae

Polypedatidae, as Rhacophoridae

Prosirenidae (only as fossils)

Proteidae. COPE, E. D., No. 1, 1889 (North America); BISHOP, S. C., No. 1, 1962 (North America); HECHT, M. K., 1958, Proc. Staten Isl. Inst. Arts Sci., 21, 5-38 (eastern North America); SMITH, H. M., No. 2, 1956 (Kansas); SCHREIBER, E., No. 1, 1912 (Europe); NIKOL'SKII, A. M., No. 6, 1962 (Russia)

Protobatrachidae (only as fossils)

Protonopsidae. (Family not identified)

Pseudidae, as Bufonidae

Ranidae (incl. Ceratobatrachidae, Raniidae). DECKERT, K., 1938, Sitz. Ber. Ges. Naturf. Freunde, 1938, 127-184 (osteology); BOULENGER, G. A., 1919, Ann. Mag. Nat. Hist., (9), 3, 408-416 (America); BOULENGER, G. A., 1920, Proc. American Acad. Arts Sci., 55, 413-480 (America); COPE, E. D., No. 1, 1889 (North America); WRIGHT L& WRIGHT, No. 1, 1949 (North America); SMITH, H. M., No. 2, 1956 (Kansas); STEBBINS, R. C., No. 1, 1951 (western North America); STEBBINS, R. C., No. 2, 1954 (western North America); BARBOUR, T. & RAMSDEN, C. T., No. 1, 1919 (Cuba); KELLOGG, R., No. 1, 1932 (Mexico); MERTENS, R., No. 1, 1952 (El Salvador); TAYLOR, E. H., No. 1, 1952 (Costa Rica); COCHRAN & GOIN, No. 1, 1970 (Colombia); RIVERO, J. A., No. 1, 1961 (Venezuela); de MIRANDO-RIBEIRO, A., No. 1, 1926 (Brasil); SCHREIBER, E., No. 1, 1912 (Europe); ANGEL, F., No. 3, 1946 (France); FUHN, I. E., No. 1, 1960 (Romania); NIKOLSKI, A. M., No. 3, 1918 (Russia); NIKOL'SKII, A. M., No. 6 1962 (Russia); ANDERSON, J., No. 1, 1898 (Egypt); PASTEUR & BONS, No. 1, 1959 (Maroc); POYNTON, J. C., No. 1, 1964 (southern Africa); BOULENGER, G. A., 1920, Rec. Indian Mus., 20, 1-226 (Indo-Australian & Pacific Rana); BOULENGER, G. A., No. 1, 1890 (British India); MERTENS, R., 1969, Stuttgarter Beitr. Naturk., 197, 1-96 (West Pakistan); ANNANDALE, N. & RAO, C. R. N., 1918, Rec. Indian Mus., 15, 25-40 (India); TAYLOR, E. H., No. 3, 1962 (Thailand); BOURRET, R., No. 2, 1942 (Indochina); BOULENGER, G. A., No. 2, 1912 (Malay Peninsula); SMITH, M. A., No. 1, 1930 (Malay Peninsula); van KAMPEN, P. N., No. 1, 1923 (Indo-Australian Arch.); INGER, R. F., No. 2, 1966 (Borneo); TYLER, M. J., 1963, Trans. Roy. Soc. South Australia, 86, 1-29, 105-130 (central highlands of New Guinea); TAYLOR, E. H., No. 10, 1921 (Philippines); INGER, R. F., No. 1, 1954 (Philippines); OKADA, Y., 1935 Rep. 1st Sci. Exp. Manchoukuo, 1-47 (Jehol); STEJNEGER, L., No. 1, 1907 (Japan); OKADA, Y., No. 1, 1931 (Japan); OKADA, Y., 1966, Fauna Japonica, Anura(Amphibia), Tokyo, 1-234 (Japan); BROWN, W. C., 1952 (Bull. Mus. Comp. Zool., 107, 1-64 (Solomon Is.)

Raniidae, as Ranidae

Rhacophoridae. (incl. Polypedatidae). AHL. E, 1931, Das Tierreich, 55, 1-462; APSTEIN, C., 1931, Das Tierreich, 55, 476-477 (both as Polypedatidae); LIEM, S. S., 1970, Fieldiana, Zool., 57, 1-145 (Old World); SCHIØTZ, A., 1967, Spol. Zool. Mus. Hauni, 25, 1-346 (West Africa); POYNTON, J. C., 1968, Copeia, 1968, 653-654 (West Africa); TAYLOR, E. H., No. 3, 1962 (Thailand); BOURRET, R., No. 2, 1942 (Indochina); INGER, R. F., No. 2, 1966 (Borneo); INGER, R. F., No. 1, 1954 (Philippines)

Rhcophoridae, as Rhacophoridae

Rhinesuchidae (only as fossils)

Rhinodermatidae, as Brachycephalidae

Rhinophrynidae. GORHAM, S. W., 1966, Das Tierreich, 85, 1-122 (Liste der rezenten....); KELLOGG, R., No. 1, 1932 (Mexico)

Salamandridae (incl. Pleurodelidae). COPE, E. D., No. 1, 1889 (North America, as Pleurodelidae); BISHOP, S. C., No. 1, 1962 (North America); SMITH, H. M., No. 2, 1956 (Kansas); SCHREIBER E., No. 1, 1912 (Europe); ANGEL, F., No. 3, 1946 (France); FUHN, I. E., No. 1, 1960 (Romania); NIKOLSKI, A. M., No. 3, 1918 (U. S. S. R.); NIKOL'SKII, A. M., No. 6, 1962 (Russia); ANDERSON, J., No. 1, 1898 (Egypt); PASTEUR, G. & BONS, J., No. 1, 1959 (Maroc); BOULENGER, G. A., No. 1, 1890 (British India); TAYLOR, E. H., No. 3, 1962 (Thailand); BOURRET, R., No. 2, 1942 (Indochina); CHANG, T.-K. & BORING, A. M., 1935, Peking Nat. Hist. Bull., 9, 327-360 (China, key to genera); CHANG, M. L. Y., 1936, Contribution a l'etude morphologique, biologique et systematique des amphibiens urodeles de la Chine, Paris, 1936 (reprinted 1968) (China); STEJNEGER, L., No. 1, 1907 (Japan)

Scapherpetonidae (only as fossils)

Scaphiopodiae, as Pelobatidae

Scincosauridae[Amphibian]. (No revisionary references noted)

Scolecomorphidae. TAYLOR, E. H., 1969, Univ. Kansas Sci. Bull., 48, 297-305 (Africa, new family)

Seiranotidae (incl. Siranotidae). (Family not identified)

Seymouridae (only as fossils)

Siranotidae, as Seiranotidae

Sirenidae. COPE, E. D., No. 1, 1889 (North America); BISHOP, S. C., No. 1, 1962 (North America); STEBBINS, R. C., No. 1, 1951 (western North America); STEBBINS, R. C., No. 2, 1954 (western North America)
Sooglossidae. (No revisionary references noted)

Telmatobiidae, as Bufonidae
Thoridae, as Plethodontidae
Trematopsidae (only as fossils)
Trematosauridae (only as fossils)
Triadobatrachidae. (No revisionary references noted)
Trimerorhachidae (only as fossils)
Tritonidae, as Salamandridae
Tupilakosauridae. (No revisionary references noted)
Typhlomolgidae, as Plethodontidae
Typhlonectidae. TAYLOR, E. H., 1968, The caecilians of the world, Lawrence, Kansas, 1-848 (new family)

Uperoliidae, as Leptodactylidae
Urocordylidae (only as fossils)

Xenopidae, as Pipidae
Xenorhinidae, as Microhylidae

Yarengiidae. (No revisionary references noted)

Zatrachyidae (incl. Acanthostomatidae) (Only as fossils)

REPTILIA

GENERAL BOOKS

BARRETT, C. Reptiles of Australia. Crocodiles, snakes and lizards. Melbourne; 1-168; 1950.
BELLAIRS, A. A. Reptiles. London; 1-195; 1957.
BELLAIRS, A. A. The world of reptiles. New York; 1-153; 1966.
BELLAIRS, A. The life of reptiles. London; 2 volumes; 1969; New York; 1970.
BELLAIRS, A. & CARRINGTON, R. The world of reptiles. London; 1-153; 1966.
BOGERT, C. M. Amphibians and reptiles of the world. In Drimmer, The Animal Kingdom. Garden City, New York; 1189-1390; 1954.
BUHLER, W. Amphibien und reptilien. Aarau; 1-127; 1966.
DITMARS, R. L. The lizards. Bull. New York Zool. Soc., 31, 116-144; 1928.
DITMARS, R. L. Snakes of the world. New York; 1-207; 1931.
DITMARS, R. L. Reptiles of the world. London & New York; 1-321; 1933. (Previous ed. 1910 & 1922)
DOTTRENS, E. Batraciens et reptiles d'Europe. Neuchatel; 1-263; 1963.
FREIBERG, M. A. Vida de batracios y reptiles sudamaericanos. Buenos Aires; 1-192; 1954.
GANS, C. et al. Biology of the reptiles. London & New York; 3 volumes; 1969-1970.
GOIN, C. J. & GOIN, L. B. Introduction to herpetology, 2nd ed. San Francisco; 1-353; 1971. (1st edition 1962.
HVASS, H. Reptiles and amphibians of the world. London; 1-125; 1964.
KLAUBER, L. M. Rattlesnakes. Berkeley, California; 1-1476; 1956.
LOVERIDGE, A. Reptiles of the Pacific world. New York; 1-259; 1945.
MATTHEWS, L. H. The British Amphibia and Reptilia. London; 1-54; 1952.
MERTENS, R. The world of amphibians and reptiles. London; 1-207; 1960. (Translation)
OLIVER, J. A. The natural history of North American amphibians and reptiles. Princeton, New Jersey; 1-359; 1955.
PERRIER, E. Les Reptiles. Traite de Zool., 8, 2883-3118, 1928.
POPE, C. H. Snakes alive and how lthey live. New York; 1-238; 1937.
POPE, C. H. The reptile world. A natural history. London; 1-325; 1956.
PRITCHARD, P. C. H. Living turtles of the world. Jersey City, New Jersey; 1-288; 1967.
ROMER, A. S. Osteology of the reptiles. Chicago; 1-772; 1956.
ROSE, W. The reptiles and amphibians of southern Africa, 2nd ed. Cape Town; 1-494; 1962. (1st ed., 1950)
SMITH, M. A. The British amphibians and reptiles, 3rd ed. London; 1-322; 1964. (Previous editions, 1951 & 1954)
TERENT'EV, P. V. Herpetology. A study of amphibians and reptiles. Moskva; 1-336; 1961. (In Russian)
THOMSON, J. S. The anatomy of the tortoise. Sci. Proc. Roy. DublinSoc., 20, 359-461; 1932.
WERMUTH, H. & MERTENS, P. Schildkroten, Krokodile, Bruckenechsen. Jena; 1-422; 1961.
WERNER, F. Die Lurche und Kriechtiere von Alfred Brehm (neuarbeitet). Brehm's Tierleben, 4, 1-572, 5, 1-598; 1925.
WILLISTON, S. W. Water reptiles of the past and present. Chicago; 1-251; 1914.
WILLISTON, S. W. & GREGORY, W. K. The osteology of the reptiles. Cambridge, Massachusetts; 1-300; 1925.

BIBLIOGRAPHIC WORKS

Bibliographies, separate works

[See also: Zoological Record, under heading Bibliographies, in each volume]

(anonymous). Bibliography of amphibians and reptiles. New York; 1-11; 1958; and supplement; 1-4; 1958. (American Museum of Natural History)
ALLOUSE, B. G. A bibliography of the vertebrate fauna of Iraq and neighbouring countries. III. Reptiles and amphibians. Publ. Iraq Nat. Hist. Mus., 6, 1-23; 1955.
BANTA, B. H. An annotated chronological bibliography of the herpetology of the state of Nevada. Wasmann Journ. Biol., 23, 1-224; 1965.
DOWNES, M. C. A bibliography of the Recent crocodilians. New Guinea; 1-256, 1-37; 1970. (Document to be more formally published "later")
GANS, C. A bibliography of the herpetology of Japan. Bull. American Mus. Nat. Hist., 93, 391-496; 1949.
GANS, C. & PARSONS, T. S. Taxonomic literature on reptiles. In Biology of the reptiles, vol. 2; 1969.
PETERS, J. A. Supplemental list of titles of papers concerning the herpetology of Ecuador. In Carlos Larrea, Biblio. Cient. Ecuador, 5, 1067-1076; 1953.
REESE, A. M. Bibliography of the Crocodilia. Herpetologica, 4, 43-54; 1947.
RUSSELL, F. E. & SCHARFFENBERG, R. S. Bibliography of snake venoms and venomous snakes. Covina, California; 1-220; 1964.
SCHMIDT, K. P. Annotated bibliography of marine ecological relations of living amphibians and reptiles (except turtles). Marine Life, 1, 43-54; 1951.

Bibliographies in other works

[See also: References in most large works]

BERGER, L. et al. Catalogus faunae Poloniae. Amphibia et Reptilia. Warszawa; 1-73; 1969.
BESKOV, V. & BERON, P. Catalogue et bibliographie des amphibiens et des reptiles en Bulgarie. Sofia; 1-40; 1964.
BLAIR, W. F. et al. Vertebrates of the United States, 2nd ed. New York; 1-616; 1968. (Pp. 367-368) (1st edition; 1-819; 1957. Pp. 357-358)
BOURRET, R. Les tortues de l'Indochine. Notes Inst. Oceanogr. Indochine, 38, 1-235; 1941. (Pp. 31-56)
INGLE, R. M. & SMITH, F. G. W. Sea turtles and the turtle industry [of the Caribbean] with annotated bibliography. Miami, Florida; 1-107; 1949.
KLAUBER, L. M. Rattlesnakes. Berkeley, California; 1-1476; 1956. [New edition expected 1971]
LOVERIDGE, A. & WILLIAMS, E. E. Revision of the African tortoises and turtles of the suborder Cryptodira. Bull. Mus. Comp. Zool., 115, 163-557; 1957. (Pp. 503-541)
NEILL, W. T. The occurrence of amphibians and reptiles in saltwater areas, and a bibliography. Bull. Mar. Sci. Gulf Caribbean, 8, 1-97; 1958.
PENNEY, J. T. Distribution and bibliography of the amphibians and reptiles of South Carolina. Univ. South Carolina Publ., (III), Biol., 1, 3-28; 1952.
POPE, C. H. The reptiles of China. In Nat. Hist. Central Asia, X, 1-604; 1935.
PRITCHARD, P. C. H. Living turtles of the world. Jersey City, New Jersey; 1-288; 1967.
SMITH, H. M. Handbook of amphibians and reptiles of Kansas, 2nd ed. Misc. Publ. Univ. Kansas Mus. Nat. Hist., 9, 1-356; 1956. (Pp. 330-343)
TAYLOR, E. H. The turtles and crocodiles of Thailand and adjacent waters with a synoptic herpetological bibliography. Univ. Kansas Sci. Bull., 49, 87-179; 1970.
de WITTE, G. F. Genera des serpentes du Congo et du Ruanda-Urundi. Ann. Mus. Afrique Centrale, Zool., 104, 1-203; 1962. (Pp. 169-188)
WRIGHT, A. H. & WRIGHT, A. A. Handbook of snakes of the United States and Canada. Vol. III. Bibliography. Ithaca, New York; 1-179; 1962.

Indexes

CLARK, H. Synopsis of herpetological literature in the Proceedings of the Iowa Academy of Science, 1891-1944. Herpetologica, 3, 131-135; 1946.
MINTON, S. A. An annotated list of herpetological papers from the Proceedings of the Indiana Academy of Science. Herpetologica, 3, 195-201; 1947.
REED, C. F. Index to Copeia. 1913-1954. Part I. — Author index. Lancaster, Pennsylvania; 1-106; 1955.
REED, C. F. Index to Copeia. 1913-1954. Part II. — Subject index. Lancaster, Pennsylvania; 1-332; 1956.
REED, C. F. Index to Herpetologica 1936-1955. Lancaster, Pennsylvania; 1-132; 1956.
REED, C. F. Index to Copeia. Supplement 1955-1964. Lancaster, Pennsylvania; 1-329; 1965.

Personal bibliographies

[See under heading BIOGRAPHIES (Persons and Institutions) below]

Institutional bibliographies

[See also under ANIMALIA and VERTEBRATA]

PETERS, J. A. A list of the herpetological publications of the United States National Museum. 1853-1965. Washington, D. C.; 1-12; 1965.
PETERS, J. A. A list of the herpetological publications of the United States National Museum 1853-1968. Washington, D. C.; 1-14; 1968.

Periodical lists

[See lists under ANIMALIA]

PERSONS AND INSTITUTIONS

Biographies & Personal bibliographies

[See also: Zoological Record, in each volume, under heading: Obituaries]

WRIGHT, A. H. Scientific and popular writers on American snakes (1517-1944). A check list and short biography, by A. H. Wright. Herpetologica, 5, Suppl. 1, 1-55; 1949.
do AMARAL, A. Trabalhos de Afranio do Amaral. Siphilis, Atraves da Historia, 1969, 287-309; 1969.
BOULENGER, G. A. Liste des publications ichthyologiques et herpetologiques (1877-1920) de G. A. Boulenger. Ann. Soc. Roy. Zool. Malac. Belgique, 52, 11-88; 1921.
COCHRAN, D. M. Bibliography of Doris M. Cochran, 1922-1952; list of new forms described, by D. M. Cochran. Herpetologica, 8, 115-120; 1953.
DERANIYAGALA, P. E. P. [Bibliography of publications 1925-1960], by P. E. P. Deraniyagala. Spol. Zeylanica, 29, 5-143; 1960.

DUGES, A. Alfredo Duges' types of Mexican reptiles and amphibians, by H. M. Smith & W. L. Necker. Anal. Esc. Nac. Cienc. Biol. Mexico, 3, 179-189; 1943. Early foundations of Mexican herpetology. An annotated and indexed bibliography of the herpetological publications of Alfredo Duges, 1826-1910, by H. M. Smith & R. B. Smith. Urbana, Illinois; 1-85; 1969.

DUNN, E. R. Herpetological publications of Emmett Reid Dunn. 1894-1956, by A. M. Dunn. (In Contrib. Herpet. Colombia. 1943-1946; 1-296; privately published) (Pp. 275-283)

MEHELY, L. Die wissenschaftliche u. literarische Tatigkeit von Ludwig Mehely aus dem Gebiete der Zoologie, by O. G. Dely. Vert. Hung., 9, 21-64; 1967.

MYERS, G. S. Annotated bibliography of the publications of George Sprague Myers (to the end of 1969). Proc. California Acad. Sci., (4), 38, 19-52.

NOBLE, G. K. Gladwyn Kingsley Noble, 1894-1940. A herpetological bibliography, by W. L. Necker. Herpetologica, 2, 47-55.

POPE, C. H. Bibliography of Clifford H. Pope. Escondido, California; 1-8; 1967.

SCHMIDT, K. P. A bibliography and index of Karl P. Schmidt's papers on coral snakes, by J. A. Peters. Copeia, 1959, 192-196; 1959.

SMITH, M. A. Bibliography of the works of Malcolm Arthur Smith, by R. H. Ahrenfeldt. British Journ. Herpet., 2, 141-149; 1959.

STEJNEGER, L. A herpetological bibliography of Leonard Stejneger (1851-1943), by W. L. Necker. Herpetologica, 2, 86-92; 1943.

WALL, F. Bibliography of the herpetological papers of Frank Wall (1868-1950) 1898-1928, by S. M. Campdon-Main. Washington, D. C.; 1-7; 1969.

Directories

[See also: Directories listed under Animalia]

(anonymous). American Society of Ichthyologists and herpetologists. List of members. San Francisco; 1-42; 1967.

WRIGHT, A. H. Scientific and popular writers on American snakes (1517-1944). A check list and short bibliography. Herpetologica, 5, (supplement 1), 1-55; 1949.

Institutions

[See also under Animalia]

DOWLING, H. G. & GILBOA, I. Herpetological societies. Amateur and professional. Herpetological Rev., 2, (3), 6-9; 1970.

Museum lists

[See also: Zoological Record, each volume, under headings: Museums: Collections and Expeditions]

Museum of Comparative Zoology, Harvard University

CONSTABLE, J. D. Reptiles from the Indian Peninsula in the Museum of Comparative Zoology. Bull. Mus. Comp. Zool., 103, 59-160; 1949.

LOVERIDGE, A. New Guinean reptiles and amphibians in the Museum of Comparative Zoology and United States National Museum. Bull. Mus. Comp. Zool., 101, 305-430; 1948.

American Museum of Natural History

BURT, C. E. & BURT, M. D. South American lizards in the collection of the American Museum of Natural History. Bull. American Mus. Nat. Hist., 61, 227-395; 1931.

United States National Museum

YARROW, H. C. Check list of North American Reptilia and Batrachia, with catalogue of specimens in U.S.National Museum. United States Nat. Mus. Bull., 24, 1-249; 1882.

do AMARAL, A. South American snakes in the collection of the United States National Museum. Proc. United States Nat. Mus., 67, (24), 1-30; 1925.

LOVERIDGE, A. East African reptiles and amphibians in the United States National Museum. United States Nat. Mus. Bull., 151, 1-135; 1929.

LOVERIDGE, A. East African reptiles and amphibians in the Museum of Comparative Zoology and United States National Museum. Bull. Mus. Comp. Zool., 101, 305-430; 1948.

West Virginia Biological Survey

GREEN, N. B. The herpetological collections of the West Virginia Biological Survey. Proc. West Virginia Acad. Sci., 20, 57-64; 1949.

Carnegie Museum

GRIFFIN, L. E. A catalog of the Ophidia from South America at present (June, 1916) contained in the Carnegie Museum. Mem. Carnegie Mus., 7, 163-227; 1916.

GRIFFIN, L. E. A list of the South American lizards of the Carnegie Museum . . . . Ann. Carnegie Mus., 11, 304-320; 1917.

Field Museum of Natural History (Chicago Natural History Museum)

LOVERIDGE, A. African reptiles and amphibians in Field Museum of Natural History. Field Mus. Publ., Zool., 22 (1), 1-111; 1936.

Texas Cooperative Wildlife Collections

SMITH, H. M. & LAUFE, L. E. Mexican amphibians and reptiles in the Texas Cooperative Wildlife collections. Trans. Kansas Acad. Sci., 48, 325-354; 1945.

Brigham Young University

TANNER, W. W. & BANTA, B. H. A systematic review of the Great Basin reptiles in the collections of Brigham Young University and the University of Utah. Great Basin Nat., 26, 87-135; 1966.

University of Utah

TANNER, W. W. & BANTA, B. H. A systematic review of the Great Basin reptiles in the collections of Brigham Young University and the University of Utah. Great Basin Nat., 26, 87-135; 1966.

Stanford University Natural History Museum

BURT, C. E. & MYERS, G. S. Neotropical lizards in the collection of the Natural History Museum of Stanford University. Stanford Univ. Publ., Biol. Sci., 8, (2), 1-52; 1942.

Museu Paulista

LUEDERWALDT, H. Chave . . . com uma lista das especies (dos crocodilideos brasileiros) do Museu Paulista. Rev. Mus. Sao Paulo, 14, 387-392; 1926.

LUEDERWALDT, H. Os chelonios brasileiros, com a lista sas especias do Museu Paulista. Rev. Mus. Sao Paulo, 14, 405-469; 1926.

Musee Teyler de Haarlem

COCUDE-MICHEL, M. Revision des Rhynchocephales de la collection du Musee Teyler de Haarlem (Pays-Bas). 1-2. Proc. K. Ned. Akad. Wet., 70B, 538-555; 1967.

Musee Zoologique Strasbourg

ANGEL, F. Reptiles et amphibiens de Madagascar. . . du Musee Zoologique du Strasbourg. Bull. Mus. Nat. Hist. Nat., (2), 22, 553-558; 1950.

British Museum (Natural History)

BOULENGER, G. A. Catalogue of the chelonians, rhynchocephaliens and crocodiles in the British Museum (Natural History). Codicote, England; 1-311; 1966, reprint.

Museu Zoologica de Coimbra

ARMANDO THEMIDO, A. Repteis do Brasil. (Catalogo das coleccoes do Museu Zoologica de Coimbra). Mem. Mus. Zool. Univ. Coimbra, 168, 1-15; 1945.

Museo Civico di Storia Naturale Milano

LUGARO, G. Elenco sistematico dei rettili Italiani conservati nella collezione di studio esistente presso il Museo di Storia Naturali di Milano . . . . Atti Soc. Ital. Sci. Nat., 96, 20-36; 1957.

Institut Francaise d'Afrique Noire

VILLIERS, A. La collection de serpents de l'I. F. A. N. Dakar; 1-155; 1950.

VILLIERS, A. La collection de serpents de l'I. F. A. N. Bull. Inst. Francaise Afrique Noire, 13, 813-836; 1951.

Laboratoire des Sciences Naturelles de l'Universite . . . Hanoi

BOURRET, R. Notes herpetologique sur l'Indochine Francaise. XVI. Tortues de la collection du Laboratoire des Sciences Naturelles de l'Universite . . . . Bull. Gen. Instruc. Publ. Hanoi, 1939, 1-12; 1939.

Museum type lists

[See also: Zoological,Record, each volume, under heading: Museums and collections]

(various collections). BARBOUR, T. & LOVERIDGE, A. Typical reptiles and amphibians. Bull. Mus. Comp. Zool., 69, 205-360; 1929. (First supplement.... Bull. Mus. Comp. Zool., 96, 59-214; 1946)

TAYLOR, E. H. Present location of certain herpetological and other type specimens. Univ. Kansas Sci. Bull., 30, 117-187; 1944.

DOWLING, H. G. et al. A list of herpetological type lists. Herpet. Rev., 2, (3), 53-54; 1970.

United States National Museum

COCHRAN, D. M. Type specimens of reptiles and amphibians in the United States National Museum. United States Nat. Mus. Bull., 220, 1-291; 1961.

Carnegie Museum

McCOY, C. J. & RICHMOND, N. D. Herpetological type specimens in Carnegie Museum. Ann. Carnegie Mus., 38, 233-264; 1966.

University of Michigan Museum of Zoology

PETERS, J. A. Catalogue of type specimens in the herpetological collections of the University of Michigan Museum of Zoology. Occ. Pap. Univ. Mich. Mus. Zool., 539, 1-55; 1952.

Field Museum of Natural History (Chicago Natural History Museum)

MARX, H. Catalogue of type specimens of reptiles and amphibians in Chicago Natural History Museum. Fieldiana, Zool., 36, 409-496; 1958.

University of Illinois Museum of Natural History

SMITH. H. M. et al. Herpetological type-specimens in University of Illinois Museum of Natural History. Illinois Biol. Monogr., 32, 1-80; 1964.

University of Kansas Museum of Natural History

DUELLMAN, W. E. & BERG, B. Type specimens of amphibians and reptiles in the Museum of Natural History, the University of Kansas. Publ. Univ. Kansas Mus. Nat. Hist., 15, 183-204; 1962.

California Academy of Sciences

SLEVIN, J. R. & LEVITON, A. E. Holotype specimens of reptiles and amphibians in the collection of the California Academy of Sciences. Proc. California Acad. Sci., (4), 28, 529-560; 1956.956.

Museum of Vertebrate Zoology, University of California, Berkeley

CRIPPEN, R. G. Holotype specimens of amphibians and reptiles in the Museum of Vertebrate Zoology, University of California, Berkeley. Herpetologica, 18, 187-194; 1962.

Stanford University Natural History Museum

LEVITON, A. E. Catalogue of the amphibian and reptile types in the Natural History Museum of Stanford University. Herpetologica, 8, 121-132; 1953.
LEVITON, A. E. & BANTA, B. H. Supplement 1 [to above]. Herpetologica, 12, 213-219; 1956.

Zoological Museum, Amsterdam

DAAN, S. & HILLENIUS, D. Catalogue of the type specimens of amphibians and reptiles in the Zoological Museum, Amsterdam. Beaufortia, 13, 117-144: 1966.

Senckenbergischen Museums; Senckenberg Natur-Museums und Forschungs-Instituts

MERTENS, R. Verzeichnis der Type in der herpetologischen Sammlung des Senckenbergischen Museums. Senckenbergiana, 4, 162-183; 1922.
MERTENS, R. Die herpetologische Sektion des Natur-Museums und Forschungs-Institutes Senckenberg in Frankfurt a. M. nebst einem Verzeichnis ihrer Typen. Senckenberg. Biol., 48, 1-106; 1967.

Staatliches Museums fur Tierkunde zu Dresden

SCHUZ, E. Verzeichniss der Typen des Staatliches Museums . . . . I. Teil. Fische, Amphibien und Reptilien . . . . Abh. Mus. Dresden, 17, (2), 1-16; 1929.

Museum Nacional d'Histoire Naturelle

GUIBE, J. Catalogue des types de lizards du Museum National d'Histoire Naturelle. Paris; 1-119; 1954.

Macleay Museum, University of Sydney

GOLDMAN, J. et al. TYPE SPECIMENS IN THE Macleay Museum, University of Sydney. II. Amphibians and reptiles. Proc. Linn. Soc. New South Wales, 93, 427-438; 1969.

National Museum of Victoria

COVENTRY. A. J. Reptile and amphibian type specimens housed in the National Museum of Victoria. Mem. Nat. Mus. Victoria, 31, 115-124; 1970.

EXPLORATION AND LOCALITIES

[See this heading under Animalia]

Localities

[See also: Atlases, gazetteers, and general works listed under Animalia]

SMITH, H. M. & TAYLOR, E. H. Type localities of Mexican reptiles and amphibians. Univ. Kansas Sci. Bull., 33, 313-380; 1950.

PETERS, J. A. Herpetological type localities in Ecuador. Rev. Ecuatoriana Ent. Par., 2 - 3/4, 335-352; 1955.
STEJNEGER, L. The herpetology of Japan and adjacent territory. United States Nat. Mus. Bull., 58, 1-577; 1907.

METHODS

[See also: Zoological Record, each volume, under heading Techniques]

(anonymous). Instructions for collectors. No. 3, Reptiles, amphibians and fishes, 6th ed. London; 1-28; 1953. [British Museum (Natural History)]
BROWN. B. C. A simple method for collecting lizards. Herpetologica, 3, 75-76; 1946.
COOK, F. R. Collecting and preserving amphibians and reptiles. Bull. Nat. Mus. Canada, 69, 128-151; 1965.
DUELLMAN, W. E. Directions for preserving amphibians and reptiles. Misc. Publ. Univ. Kansas Mus. Nat. Hist., 30, 37-40; 1962.
GLOYD, H. K. Methods of preserving and labeling amphibians and reptiles for scientific study. Turtox News, 16, 49-53, 66-67; 1938.
MYERS, G. S. Manual of tropical herpetological collecting, 2nd ed. Circ. Stanford Univ. Nat. Hist. Mus., 4, 1-12; 1956. (1st edition, 1951)
MYERS, G. S. Brief directions for preserving and shipping specimens of fishes, amphibians and reptiles, 2nd ed. Circ. Stanford Univ. Nat. Hist. Mus., 5; 1956.
NEILL, W. T. How to preserve reptiles and amphibians for scientific study. Ross Allen Rept. Inst., Spec. Publ., 2, 1-15; 1950.
SMITH, A. G. Making a collection of amphibians and reptiles. Contr. Dep. Biol. Sci. Loyola Univ., 1, 1-5; 1951.
STEBBINS, R. C. Amphibians and reptiles of western North America. New York; 1-528; 1954. (Collecting and preserving, pp. 8-22)
STEJNEGER, L. Directions for collecting reptiles and batrachians. United States Nat. Mus. Bull., 39, (E), 1-13; 1891.
VOGEL, Z. Reptiles and Amphibia; their care and behaviour. London; 1-228; 1964.
ZWEIFEL, R. G. Guidelines for the care of a herpetological collection. Curator, 9, 24-35; 1966.

GLOSSARIES

[See also: Many of the general books listed above, as well as Glossaries listed under Animalia]

PETERS, J. A. Dictionary of herpetology. New York; 1-426; 1964.
PRITCHARD, P. C. H. Living turtles of the world. Jersey City, New Jersey; 1-288; 1967.
WRIGHT, A. H. & WRIGHT, A. A. Handbook of snakes of the United States and Canada. Ithaca, New York. 2 volumes, 1-1106; 1957.

NOMENCLATURAL STUDIES

Common name lists

[See also: Zoological Record, under heading Vernacular Names, in each volume]

(anonymous). Common names for North American amphibians and reptiles. Copeia, 1956, 172-185; 1956.
WRIGHT, A. H. Common names of the snakes of the United States. Herpetologica, 6, 141-186; 1950.
GRIFFITH, D. D. Snakes of California, scientific and common names. Herpeton, 1, 1-4; 1966.
NEMURAS, K. T. . . . . turtles of Pan America. Journ. Int. Turtle Tort. Soc., 1, 22-23; 1967.
do AMARAL, A. Nomes vulgares de ophidios do Brasil. Bol. Mus. Nac. Rio de Janeiro, 2, (2), 1-11; 1926.
BARRAN, E. F. & FREIBURG, M. A. Nombres vulgares de reptiles y batracios del Argentina. Physis, 20, (58), 303-319; 1951.
VAN RAMSHORST, J. D. Europese slangen. Het Aquarium, 37, 224-227; 1967. (English, Dutch, German, Grench, Italian, Polish, Czechoslovak)
STØP-BOWITZ, C. et al. Norske dyrenevn med tilhorende latinske navn. Fauna, Oslo, (14), 1 Bilag, 1-57; 1961.
MACLEOD, R. D. Key to the names of British fishes, mammals, amphibians and reptiles. London; 1-71; 1956.
FROMMHOLD, E. Wir bestimmen Lurche und Kriechtiere Mitteleuropas. Radebeul; 1-218; 1959. (In 8 languages)
HIRTZ, M. Dictionary of national [Croatian] zoological names. Part I. Zagreb; 1-197; 1928. (In Croatian)
VILLIERS, A. Tortues et crocodiles des l'Afrique Noire Francaise. Init. Africaines, 15, 1-354; 1958.
MINTON, S. A. A contribution to the herpetology of West Pakistan. Bull. American Mus. Nat. Hist., 134, 27-184.

PHILLIPS, W. W. A. A list of the more important animals, birds and reptiles of Ceylon with the Sinhalese and Tamil names of each. Loris, 2, 140-144; 1940.
HAILE, N. S. The snakes of Borneo . . . . Sarawak Mus. Journ., 8, 743-771; 1959.

Nomenclators

[See also: Nomenclators listed under Animalia]

KUHN, O. Amphibien und reptilien-Katalog der Subfamilien und hoheren Taxa mit Nachweis des ersten Auftretens. Stuttgart; 1-124; 1967.
ROMER, A. S. Vertebrate paleontology. Chicago; 1-687; 1945. (Generic names) (1st ed., 1933)

Generic name studies

DUNN, E. R. Los generos de anfibios y reptiles de Colombia. Parts II, III, IV. Caldasia, 3, 73-110, 155-224, 307-335; 1944-1945. (Part I is Amphibia) (Study of genera)
FOLLETT, W. I. & DEMPSTER, L. J. Ichthyological and herpetological names and works on which the International Commission on Zoological Nomenclature has ruled since the publication (in 1958) of the several Official Lists and Indexes. Copeia, 1965, 518-523; 1965.
KUHN, O. Amphibien und reptilien. Katalog der Subfamilien und hoheren Taxa mit Nachweis des ersten Auftretens. Stuttgart; 1-124; 1967.
LINDHOLM, W. A. Revidiertes Verzeichnis der Gattungen der rezenten Schildkroten nebst . . . . Zool. Anz., 81, 275-295; 1929.
MACLEOD, R. D. Key to the names of British fishes, mammals, amphibians and reptiles. London; 1-71; 1956. (Origin and meaning)

REGIONAL STUDIES, FAUNAS, LISTS, KEYS

Faunas, lists, keys

General

GANS, C. A check-list of Recent amphisbaenians.... Bull. American Mus. Nat. Hist., 135, 61-106; 1967.
PRITCHARD, P. C. H. Living turtles of the world. Jersey City, New Jersey; 1-288; 1967.
(anonymous). Poisonous snakes of the world..., 2nd ed. Washington, D. C.; 1-203; 1970.
MEDEM, F. & MARX, H. An artificial key to the New World species of crocodilians. Copeia, 1955, 1-2; 1955.

North America

TAYLOR, W. E. The box tortoises of North America. Proc. United States Nat. Mus., 17, 573-588; 1895.
COPE, E. D. The crocodilians, lizards, and snakes of North America. Rep. United States Nat. Mus., for 1898, 155-1270; 1900.
BLANCHARD, F. N. A key to the snakes of the United States, Canada and Lower California. Pap. Michigan Acad. Sci. Arts Lett., 4, (2), 1-65; 1924.
BURT, C. E. A key to the lizards of the United States and Canada. Trans. Kansas Acad. Sci., 38, 255-305 1935.
DITMARS, R. L. The reptiles of North America. . . . Garden City, New York; 1-476; 1936. (Keys)
DITMARS, R. L. A field book of North American snakes. New York; 1-305; 1939.
POPE, C. H. Turtles of the United States and Canada. New York; 1-337; 1939. (Keys)
SCHMIDT, K. P. & DAVIS, D. D. Field book of snakes of the United States and Canada. New York; 1-365; 1941.
STEJNEGER, L. H. & BARBOUR, T. A check list of North American amphibians and reptiles, 5th ed. Bull. Mus. Comp. Zool., 93, 1-260; 1943.
WRIGHT, A. H. & WRIGHT, A. A. A list of the snakes of the United States and Canada by states and provinces. American Midl. Nat., 48, 574-603; 1952.
SCHMIDT, K. P. A check list of North American amphibians and reptiles, 6th ed. Chicago; 1-280; 1953.
WRIGHT, A. H. & WRIGHT, A. A. Handbook of snakes of the United States and Canada. Ithaca, New York, 2 volumes, 1-1106; 1957.
BOULENGER, G. A. Contributions to American herpetology, collected papers. Reprints by Ohio Herpetological Society; since 1966, Society for the Study of Amphibians and Reptiles. Part 1, 1877-1881, 1965; 2, 1882-1883, 1966; 3, 1884-1885, 1966; 4. 1886-1887, 1967; 5, 1888, 1967; 6, 1889-1890, 1968; 7, 1891-1893, 1968; 8, 1894, 1969; 9, 1895-1896, 1969; 10, 1897, 1970; 11, 1898, 1970; [projected: 12 & 13, 1971; 14, 15, 7 16, 1972]
CONANT, R. A field guide to the reptiles and amphibians of the United States and Canada east of the 100th meridian. Boston; 1-366; 1958.
BRIMLEY, C. S. The turtles of North Carolina; with a key to the turtles of the eastern United States. Journ. Elisha Mitchell Sci. Soc., 36, 62-71; 1920.
POPE, C. H. Snakes alive and how they live. Appendix: How to identify the snakes of the United States. New York; 1-238; 1937.
PERKINS, C. B. A key to the snakes of the United States. Bull. Zool. Soc. San Diego, 24, 1-79; 1947.
ZIM, H. S. & SMITH, H. M. Reptiles and amphibians. A guide to familiar American species. New York; 1-157; 1953.

BLAIR, W. F. et al. Vertebrates of the United States, 2nd ed. New York; 1-616; 1968. (Keys to species pp. 213-268) (1st ed., 1957)
WHITAKER, J. O., Jr. Keys to the vertebrates of the eastern United States, exclusive of birds. Minneapolis, Minnesota; 1-256; 1969.
COCHRAN, D. M. & GOIN, C. J. The new field book of reptiles and amphibians. New York; 1-359; 1970. (United States)
LOGIER, E. B. S. & TONER, G. E. Checklist of the amphibians and reptiles of Canada and Alaska. Contr. Roy. Ontario Mus. Zool., 41, 1-88; 1955.
LOGIER, E. B. S. & TONER, G. E. Check list of the amphibians and reptiles of Canada and Alaska. Contr. Life Sci. Div. Roy. Ontario Mus., 53, 1-92; 1961.
MILLS, R. C. A check list of the reptiles and amphibians of Canada. Herpetologica, 4, (Supplement 2), 1-15; 1948.
BLEAKNEY, J. S. A zoogeographical study of the amphibians and reptiles of eastern Canada. Bull. Nat. Mus. Canada, 155, 1-119; 1958.
BLEAKNEY, S. The amphibians and reptiles of Nova Scotia. Canadian Field Nat., 66, 125-129; 1952.
LOGIER, E. B. S. The reptiles of Ontario. Handb. Roy. Ontario Mus. Zool., 4, 1-71; 1939.
LOGIER, E. B. S. The snakes of Ontario. Toronto; 1-94; 1958.
COWAN, I. M. A review of the reptiles and amphibians of British Columbia. Rep. Prov. Mus. Nat. Hist. Victoria, 1936, 16-25; 1937.
CARL, B. C. The reptiles of British Columbia. Handb. British Columbia Prov. Mus., 3, 1-60; 1944.
COOK. F. R. An analysis of the herpetofauna of Prince Edward Island. Bull. Nat. Mus. Canada, 212, 1-60; 1967.
DITMARS, R. L. Serpents of the eastern states. Bull. New York Zool. Soc., 32, 83-120; 1929. (Keys)
CONANT, R. & BRIDGES, W. What snake is that? A field guide to the snakes of the United States east of the Rocky Mountains. New York; 1-163; 1939.
DE SOLA, C. R. The turtles of the north-eastern states.... Bull. New York Zool. Soc., 34, 131-160; 1931.
(anonymous). Turtles of the northeastern states. Fauna, 7, 37-39; 1945. (Zool. Soc. Philadelphia)
POPE, C. H. Snakes of the northeastern United States. New York; 1-52; 1946. (Keys)
(anonymous). Snakes of the northeastern states. Fauna, 8, 48-52; 1946. (Zool. Soc. Philadelphia)
CONANT, R. Lizards of the northeastern states. Fauna, 9, 42-44; 1947.
CONANT, R. Reptiles and amphibians of the north-eastern states. Philadelphia; 1-40; 1947.
BABCOCK, H. L. The turtles of New England. Mem. Boston Soc. Nat. Hist., 8, 327-427; 1919.
BABCOCK, H. L. The snakes of New England. Boston Soc. Nat. Hist., Nat. Hist. Guides, 1, 1-30; 1929.
BABCOCK, H. L. Field guide to New England turtles. New England Mus. Nat. Hist., Nat. Hist. Guide, 2, 1-56; 1938.
TRAPIDO, H. The snakes of New Jersey. A guide. Newark, New Jersey; 1-60; 1937.
NETTING, M. G. The amphibians and reptiles of Pennsylvania. (Pennsylvania); 1-29; 1946. (Keys)
BURT, C. E. The lizards of the south-eastern United States. Trans. Kansas Acad. Sci., 40, 349-366; 1938.
MANSUETI, R. A descriptive catalogue of the amphibians and reptiles found in and around Baltimore City, Maryland.... Proc. Nat. Hist. Soc. Maryland, 7, 1-53; 1941.
McCAULEY, R. H. The reptiles of Maryland and the District of Columbia. Hagerstown, Maryland; 1-194; 1945.
DePOE, C. E. et al. The reptiles and amphibians of North Carolina; a preliminary check list and bibliography. Journ. Elisha Mitchell Sci. Soc., 77, 125-136; 1961.
BRIMLEY, C. S. The turtles of North Carolina; with a key to the turtles of the eastern United States. Journ. Elisha Mitchell Sci. Soc., 36, 62-71; 1920.
CHERMOCK, R. L. A key to the amphibians and reptiles of Alabama. Mus. Pap. Geol. Surv. Alabama, 33, 1-88; 1952.
CARR, A. & GOIN, C. J. Guide to the reptiles, amphibians and fresh-water fishes of Florida. Gainesville, Florida; 1-341; 1959.
KEISER, E. D., Jr. & WILSON, L. D. Checklist and key to the herpetofauna of Louisiana. Lafayette Nat. Hist. Mus. Tech. Bull., 1, 1-51; 1969.
CAHN, A. The turtles of Illinois. Illinois Biol. Monogr., 16, 1-218; 1937.
PARMALEE, P. W. Reptiles of Illinois. Springfield, Illinois; 1-88; 1955.
SMITH, P. W. The amphibians and reptiles of Illinois. Bull. Illinois Nat. Hist. Surv., 28, 1-198; 1961.
POPE, C. H. Amphibians and reptiles of the Chicago area. Chicago; 1-275;; 1944.
SMITH, A. G. Key to the amphibians and reptiles of the Chicago area. Contr. Dep. Biol. Sci. Loyola Univ., 7, 1-4; 1951.
DICKINSON, W. E. Field guide to the lizards and snakes of Wisconsin. Milwaukee Publ. Mus. Pop. Sci. Handb. Ser., 2, 1-72; 1949.
ANDERSON, P. The reptiles of Missouri. Columbia, Missouri; 1.330; 1965.5.
SMITH, H. M. Handbook of amphibians and reptiles of Kansas. Misc. Publ. Univ. Kansas Mus. Nat. Hist., 2, 1-336; 1950.
SMITH, H. M. Handbook of amphibians and reptiles of Kansas, 2nd ed. Misc. Publ. Univ. Kansas Mus. Nat. Hist., 9, 1-356; 1956.
HUDSON, G. E. The amphibians and reptiles of Nebraska. Nebraska Cons. Bull., 23, (4), 1-146; 1942.
MASLIN, T. P. An annotated check list of the amphibians and reptiles of Colorado. Univ. Colorado Stud., Biol., 6, 1-98; 1959.
JONES-BURDICK, W. H. A guide to the snakes of Colorado. Univ. Colorado Mus. Leaflet, 1, 1-11; 1939.
JONES-BURDICK, W. H. Guide to the snakes of Colorado. Univ. Colorado Mus. Leaflet, 5, 1-23; 1963.
BROWN, B. C. An annotated checklist of the reptiles and amphibians of Texas. Waco, Texas; 1-250; 1950.
VAN DENBURGH, J. The reptiles of western North America. San Francisco; 2 vols., 1-1028; 1922.

STEBBINS, R. C. Amphibians and reptiles of western North America. New York; 1-528; 1954.
STEBBINS, R. C. A field guide to western reptiles and amphibians.... Boston; 1-179; 1966.
JONES-BURDICK, W. H. Guide to the snakes of Colorado. Univ. Colorado Mus. Leaflet, 1, -11; 1939.
MASLIN, T. P. Guide to the lizards of Colorado. Univ. Colorado Mus. Leaflet 3, 1-14; 1947.
VAN DENBURGH, J. The reptiles of the Pacific Coast and Great Basin.... Occ. Pap. California Acad. Sci., 5, 1-236; 1897.
SLEVIN, J. R. A handbook of reptiles and amphibians of the Pacific States including certain eastern species. San Francisco; 1-71; 1934.
PICKWELL, G. Amphibians and reptiles of the Pacific States. Stanford, California; 1-236; 1947.
GORDON, K. The Amphibia and Reptilia of Oregon. Oregon State Monogr. Zool., 1, 1-82; 1939.
KESSEL, E. L. The snakes of California. Wasmann. Coll., 3, 19-24; 1938.
GRIFFITH, D. D. Snakes of California, scientific and common names. Herpeton, 1, 1-4; 1966

Neotropics in general

KLUGE, A. G. A revision of the South American gekkonid lizard genus Homonota Gray. American Mus. Novit., 2193, 1-41; 1964. (Key to New World genera)
PETERS, J. A. & DONOSO-BARROS, R. Catalogue of the neotropical Squamata. Part II. Lizards and amphisbaenians. United States Nat. Mus. Bull., 297, (II), 1-293; 1970. (Keys)
PETERS, J. A. & OREJAS-MIRANDA, B. Catalogue of the neotropical Squamata, Part I. Snakes. United States National Mus. Bull., 297, (I), 1-347; 1970. (keys)
do AMARAL, A. Lista remissiva dos ophidios da regiao neotropica. Mem. Inst. Butantan, 4, 129-271; 1929.

West Indies

BARBOUR, T. A contribution to the zoogeography of the West Indies, with especial reference to amphibians and reptiles. Mem. Mus. Comp. Zool., 44, 205-359; 1914.
BARBOUR, T. BARBOUR, T. Third list of Antillean reptiles and amphibians. Bull. Mus. Comp. Zool., 82, 77-166; 1937. Previous editions 1920, 1935)
UNDERWOOD, G. Reptiles of the eastern Caribbean. Caribbean Affairs, (n. s.), 1, 1-192; 1962.
COPE, E. D. List of the Batrachia and Reptilia of the Bahamas Islands. Proc. United States Nat. Mus., 10, 436-439; 1887.
BARBOUR, T. & RAMSDEN, C. T. The herpetology of Cuba. Mem. Mus. Comp. Zool., 47, 69-213; 1919.
BUIDE, M. S. Lista de los anfibios y reptiles de Cuba. Torreia, (n. s.), 1, 7-60; 1967.
GRANT, C. The herpetology of the Cayman Islands.... Bull. Inst. Jamaica, Sci., 2, 1-65; 1940.
BARBOUR, T. An annotated list of the Amphibia and Reptilia of Jamaica. Handb. Jamaica 1922; 1923.
GRANT, C. The herpetology of Jamaica. II. The reptiles. Bull. Inst. Jamaica, Sci., 1, 61-148; 1940.
UNDERWOOD, G. Introduction to the study of Jamaica reptiles, 4 parts. Nat. Hist. Notes, Nat. Hist. Soc. Jamaica, 41, 104-107; 42, 127-131; 43, 144-147; 44, 161-163; 1950.
COCHRAN, D. M. The herpetology of Hispaniola. United States Nat. Mus. Bull., 177, 1-298; 1941.
STEJNEGER, L. The herpetology of Porto Rico. Ann. Rep. United States Nat. Mus. for 1902, 542-727;1904.
SCHMIDT, K. P. Amphibians and land reptiles of Porto Rico. Sci. Serv. Porto Rico Virgin Islands, 10, (1), 1-160; 1928. (New York Academy of Sciences)
RUTHVEN, A. G. The reptiles of the Dutch Leeward Islands. Occ. Pap. Mus. Zool. Univ. Michigan, 143, 1-10; 1923. (list)
GRANT, C. Herpetology of Barbados, B. W. I. Herpetologica, 15, 97-101; 1959. (list)
MOLE, R. R. The Trinidad snakes. Proc. Zool. Soc. London, 1924, 235-278; 1924.
PARKER, H. W. The lizards of Trinidad. Trop. Agric., 12, 65-70; 1935.

Central America

GUNTHER, A. C. L. G. Reptilia and Batrachia. Biologia Centrali-Americana; 1-326; 1885-1902.
COPE, E. D. Catalogue of batrachians and reptiles of Central America and Mexico. United States Nat. Mus. Bull., 32, 1-98; 1887.
DUNN, E. R. Notes on Central American caecilians. Proc. New England Zool. Club, 10, 71-76; 1928. (keys)
SMITH, H. M. & TAYLOR, E. H. An annotated checklist and key to the snakes of Mexico. United States Nat. Mus. Bull., 187, 1-239; 1945.
SMITH, H. M. & TAYLOR, E. H. An annotated checklist and key to the reptiles of Mexico exclusive of the snakes. United States Nat. Mus. Bull., 199, 1-253; 1950.
SMITH, H. M. & TAYLOR, E. H. Herpetology of Mexico. Annotated checklists and keys to the amphibians and reptiles. Ashton, Maryland; 1966. (Reprint of U. S. N. M. Bull. 187, 194, 199)
SCHMIDT, K. P. Scientific results of the expedition to the Gulf of California . . . "Albatross" in 1911. VIII. The amphibians and reptiles of Lower California. Bull. American Mus. Nat. Hist., 16, 607-707; 192 2.
CLIFF, F. S. Snakes of the islands in the Gulf of California, Mexico. Trans. San Diego Soc. Nat. Hist., 12, 67-98; 1954.
ZWEIFEL, R. G. Results of the Puritan-American Museum of Natural History Expedition to Western Mexico. 9. Herpetology of the Tres Marias Islands. Bull. American Mus. Nat. Hist., 119, 77-128; 1960.
BOGERT, C. M. & OLIVER, J. A. A preliminary analysis of the herpetofauna of Sonora. Bull. American Mus. Nat. Hist., 83, 303-425; 1945.

DUELLMAN, W. E. A preliminary analysis of the herpetofauna of Colima, Mexico. Occ. Pap. Mus. Zool. Univ. Michigan, 589, 1-22; 1958. (list)
HARDY, L. M. & McDIARMID, R. W. The amphibians and reptiles of Sinaloa, Mexico. Publ. Univ. Kansas Mus. Nat. Hist., 18, 39-252; 1969.
PETERS, J. A. The amphibians and reptiles of the coast and coastal sierra of Michoacan, Mexico. Occ. Pap. Mus. Zool. Univ. Michigan, 554, 1-37; 1954. (list)
STUART, L. C. A checklist of the herpetofauna of Guatemala. Misc. Publ. Mus. Zool. Univ. Michigan, 122, 1-150; 1963.
STUART, L. C. The amphibians and reptiles of Alta Verapaz, Guatemala. Misc. Publ. Mus. Zool. Univ. Michigan, 69, 1-109; 1948. (list)
STUART, L. C. The herpetofauna of the Guatemalan plateau. Contr. Lab. Vert. Biol. Univ, Michigan, 49, 1-71; 1951.
STUART, L. C. Herpetofauna of the southeastern highlands of Guatemala. Contr. Lab. Vert. Biol. Univ. Michigan, 68, 1-65; 1954.
SCHMIDT, K. P. The amphibians and reptiles of British Honduras. Publ. Field Mus. Nat. Hist., 512, 475-510; 1941.
MERTENS, R. Die Amphibien und Reptilien von El Salvador, auf Grund der Reisen von R. Mertens und A. Zilch. Abh. Senckenbergische Naturf. Ges., 487, 1-83; 1952.
TAYLOR, E. H. A brief review of the snakes of Costa Rica. Univ. Kansas Sci. Bull., 34, 3-188; 1951.
TAYLOR. E. H. A review of the lizards of Costa Rica. Univ. Kansas Sci. Bull., 38, 1-322; 1956.
SMITH, H. M. Handlist of the snakes of Panama. Herpetologica, 14, 222-224; 1958.
MYERS, C. W. & RAND, A. S. Checklist of amphibians and reptiles of Barro Colorado Island, Panama, with comments on faunal change and sampling. Smithsonian Contr. Zool., 10, 1-21; 1969.

South America

BURT, C. E. & BURT, M. D. The South American lizards in the collection of the United States National Museum. Proc. United States Nat. Mus., 78, (6), 1-52; 1930.
BURT, C. E. & BURT, M. D. South American lizards in the collection of the American Museum of Natural History. Bull. American Mus. Nat. Hist., 61, 227-395; 1931. (keys)
BURT, C. E. & BURT, M. D. A preliminary check list of the lizards of South America. Trans. Acad. St. Louis, 28, 1-104; 1933.
KLUGE, A. G. A revision of the South American gekkonid lizard genus Homonota Gray. American Mus. Novit., 2193, 1-41; 1964. (Key to New World genera)
do AMARAL, A. Ophidia of Colombia. Bull. Antiven. Inst. America, 4, 89-94; 1931.
NICEFORO, M. Los ofidios de Colombia. Rev. Acad. Colombiana Cienc., 5, 84-101; 1942.
MARIA, H. N. Los ofidios de Colombia. Rev. Acad. Colombiana Cient., 5, 84-101; 1942. (List)
DANIEL, H. Las serpientes en Colombia. Rev. Fac. Nacion. Agron. Medellin, 10, 301-333; 1949.
PETERS, J. A. The snakes of Ecuador. A check list and key. Bull. Mus. Comp. Zool., 122, 491-541; 1960.
PETERS, J. A. The lizards of Ecuador. A check list and key. Proc. United States Nat. Mus., 119, (3545), 1-49; 1967.
GARMAN, S. The Galapagos tortoises. Mem. Mus. Comp. Zool., 30, 257-296; 1917.
de la ROCA, F. M. Introduccion al estudio de los ofidios de Venezuela. Bol. Soc. Venezolana Cienc. Nat., 10, 381-392; 1932. (list)
ROZE, J. A. La taxonomie y zoogeografia de los ofidios de Venezuela. Caracas; 1-498; 1966.
PARKER, H. W. The frogs, lizards and snakes of British Guiana. Proc. Zool. Soc. London, 1935, 505-530; 1935. (list)
PRITCHARD, P. C. H. Turtles of British Guiana. Journ. British Guiana Mus., 39, 19-45; 1964. (list)
LUEDERWALDT, H. Chave para a determinacao dos crocodilideos brasileiros, com uma lista das especies do Museu Paulista. Rev. Mus. Sao Paulo, 14, 387-392; 1926.
SIEBENROCK, F. Schildkroten von Brasilien. Denkschr. Akad. Wiss. Wien, 76, 1-28; 1926.
LUEDERWALDT, H. Os chelonios brasileiros, com a lista das especies do Museu Paulista. Rev. Mus. Sao Paulo, 14, 405-469; 1926.
do AMARAL, A. Lista remissiva dos ophidios do Brasil. Mem. Inst. Butantan, 4, 71-125; 1929.
do AMARAL, A. Estudos sobre Lacertilios Neotropicos. 4. Lista remissiva dos Lacertilios do Brasil. Mem. Inst. Butantan, 11, 167-204; 1938.
do AMARAL, A. Contribucao do conhecimento dos ophidios do Brasil. 11. Synopse das Crotalideas do Brasil. Mem. Inst. Butantan, 11, 217-229; 1938.
SANTOS, E., Anfibios e repteis do Brasil (Vida e costumes). Rio de Janeiro; 1-280; 1942. (keys)
PRADO. A. Serpentes do Brasil. Sao Paulo; 1-134; 1945. (keys)
SILVA JUNIOR, M. O ofidismo no Brasil. Rio de Janeiro; 1-346; 1956. (keys)
DEVINCENZI, G. F. Fauna erpetologica del Uruguay. Anal. Mus. Hist. Nat. Montevideo, (2), 2, 1-64; 1925.
KLAPPENBACH, M. A. & OREJAS-MIRANDA, B. Anfibios y reptiles. Nuestra Tiera, 11, 1-68, 1969.
SCHOUTEN, G. B. Contribuciones al conocimiento de la fauna herpetologica del Paraguay . . . . Rev. Soc. Cienc. Paraguay, 3, 5-32; 1931. (list)
SCHOUTEN, G. B. Fauna herpetologica del Paraguay. Novena Reun. Soc. Argentina Patol. Reg., 2, 1218-1237; 1937. (list)
HELLMICH, W. Die Sauria des Gran Chaco und seines Randgebiete. Abh. Bayerrischen Akad. Wiss., Mat.-Nat. Kl., (n. f.), 101, 1-131; 1960.
FREIBERG, A. Las tortugas Argentinas. Rev. Cent. Estud. Med. Vet. Buenos Aires, 1935, 1-5; 1935. (keys)
SERIE, P. Nueva enumeracion sistematica y distribucion geografica de los ofidios Argentinos. La Plata; 33-61; 1936. Obra Cincuentenario Mus. La Plata, 2, 33-61; 1937.

FREIBERG, M. A. Catalogo sistematico y descriptivo de las Tortugas Argentinas. Mem. Mus. Entre Rios Parana, 9, 1-25; 1938.
LIEBERMANN, J. Catalogo sistematico y zoogeografico de los lacertilios argentinos. Physis, 16, 61-82; 1939.
FREIBERG, M. A. El mundo de los ofidios. Buenos Aires; 1-43; 1970.
FREIBERG, M. El mundo de las tortugas. Buenos Aires; 1-143; 1971.
GIGOUX, E. E. Los ofidios Chilenos. Bol. Mus. Nac. Chile, 18, 5-7; 1940.
DONOSO-BARROS, R. Reptiles de Chile. Santiago; 1-458; 1966.
DONOSO-BARROS, R. Catalogo herpetologico chileno. Bol. Mus. Nac. Hist. Nat. Chile, 31, 42-124; 1970.

Europe

SCHREIBER, E. Herpetologia europaea. Eine systematische Bearbeitung der Amphibien und Reptilien... 2. Aufl. Jena; 1-960; 1912.
BOULENGER, G. A. The snakes of Europe. London; 1-269; 1913.
STEINHEIL, F. Die europaischen Schlangen. Jena; 9 parts; 1913-1931. (keys)
MERTENS, R. & MULLER, L. Liste der Amphibien und Reptilien Europas. Abh. Senckenbergische Naturf. Ges., 41, 1-62; 1928.
MERTENS, R. & MULLER, L. Die Amphibien und Reptilien Europas. Zweite Liste. Frankfurt-am-Main; 1940.
HELLMICH, W. Die Lurche und Kriechtiere Europas. Winters Naturwiss. Taschenb., 26, 1-266; 1956.
MERTENS, R. & WERMUTH, H. Die Amphibien und Reptilien Europas (Dritte Liste...). Frankfurt-am-Main; 1-264; 1960. (Previous editions, 1928, 1940; Mertens & Muller)
HELLMICH, W. Reptiles and amphibians of Europe. London; 1-160; 1962. (Translation of Die Lurche und Kriechtiere . . . , 1956)
STEWARD, J. W. The snakes of Europe. Newton Abbot, England; 1-238; 1971. (keys)
TAYLOR, R. H. R. The distribution of reptiles and amphibia in the British Isles . . . . British Journ. Herpet., 1, 1-38; 1948.
TAYLOR, R. H. R. The distribution of amphibians and reptiles in England and Wales, Scotland and Ireland and the Channel Isles. A revised survey. British Journ. Herpet., 3, 95-116; 1963.
APPLEBY, L. G. British snakes. (England); 1-36; 1971. (keys)
FITTER, R. S. R. A check-list of the mammals, reptiles and amphibia of the London area 1900-1949. Londo Natur., 28, 98-115; 1949. (list)
BAAL, H. J. The indigenous mammals, reptiles and amphibians of the Channel Islands. Bull. Soc. Jersiaise, 15, 101-110; 1949.
STØP-BOWITZ, C. et al., Norske dyrenavn med tilhorende latisske navn. Fauna, Oslo, (14), l Bilag, 1-57; 1961.
MERTENS, R. Amphibia, Reptilia. In Grimpe & Wagler, Die Tierweld der Nord- und Ostsee. 3 Teil. 1-20; 1926. (list)
MERTENS, R. Amphibia, Reptilia (Nachtrage). In Tierwelt der Nord- und Ostsee; Lief, 32, 1-2; 1938.
BRUUN, A. F. et al. List of Danish vertebrates. København; 1-180; 1960.
van KAMPEN, P. N. Fauna van Nederland. Amphibia en Reptilia. Leiden; 1-64; 1927.
de WITTE, G. F. Amphibians et reptiles. Faune de Belgique, 2nd edition. Bruxelles; 1-321; 1948.
FERRANT, V. Faune du Grand-Duche de Luxenbourg. Deuxieme partie: Amphibien et reptiles. Luxembourg; 1055; 1922.
MERTENS, R. Neue Liste der Amphibien und Reptilien Deutschlands. Bl. Aquar. Terrar., 12, 177-178; 1938.
WERNER, F. & HERTER, K. Fauna von Deutschland. Amphibia und Reptilia. Leipzig; 492-501; 1944.
WERMUTH, H. Taschenbuch der heimischen Amphibien und Reptilien. Leipzig/Jena; 1-106; 1957.
MERTENS, R. Welches Tier ist das? Kriechtiere und Lurche, 3rd ed. Stuttgart; 1-98; 1964. (Previous editions, 1952, 1960)
KUHN, O. Die deutschen Saurier. Munchen; 1-107; 1967.
STEMMLER, O. Die Reptilien der Schweiz. Veroff. Naturh. Mus. Basel, 5, 1-32; 1967.
ANGEL, F. Reptiles et amphibiens. Faune de France, 45, 1-204; 1946.
FERREIRA, J. B. & de SEABRA, A. F. Catalogue systematique des vertebres du Portugal. Reptiles et amphibiens. Bull. Soc. Portugu. Sci. Nat., 5, 97-128; 1913.
ARMANDO THEMIDO, A. Repteis e batraquios das Colonias Portuguesas. Mem. Mus. Zool. Univ. Coimbra, (1), 119, 1-31; 1941. (lists)
ARMANDO THEMIDO, A. Anfibios e repteis de Portugal. Mem. Mus. Zool. Univ. Coimbra, (1), 133, 1-49; 1941. (list)
MALUQUER, J. Les serpents de Catalunya. Barcelona Mus. Barcinonensis Sci. Nat. Op., Zool., 7, 1087; 1917.
MERTENS, R. Die Amphibien und Reptilien Corsikas. Senckenbergische Biol., 38, 175-192; 1958.
MERTENS, R. Reptilien van de Malta-eilanden. Lacerta, 27, 11-15; 1968.
HEDIGER, H. Die Schlangen Mitteleuropas. Basel; 1-54; 1937.
STERNFELD, R. Die Reptilien und Amphibien Mitteleuropas. Heidelberg; 1-94; 1952.
FROMMHOLD, E. Wir bestimmen Lurche und Kriechtiere Mitteleuropas. Radebeul; 1-219; 1959. (In 8 languages)
MŁYNARSKI, M. Keys for the identification of Polish vertebrates. Czesc III. Reptilia. Warszawa & Krakow; 1-49; 1960 (In Polish)
STEPANEK, O. Klic nasich obratlovcu. Praha; 1-250; 1950. (In Czech)
EISELT, J. Amphibia, Reptilia. Catalogue Faunae Austriacae, 21ab, 1-21; 1961. (In German)
MEHELY, A. L. Reptilia et Amphibia. Fauna Regni Hungariae, 1-12; 1918.

SOCHUREK, E. Liste der Lurche und Kriechtiere Karntens. Carinthia II, 67, 150-152; 1958.
FUHN, I. E. Broaste, serpi, sopirle. Bucuresti; 1-245; 1969.
RADOVANOVIC, M. Vodozemci i gmizavci nase zemlje. Belgrade; 1-249; 1951. (In Jugoslavian) (Amphibia & Reptilia Jugoslaviens)
POZZI, A. Geonomia e catalogo ragionato degli Anfibi e dei Rettili della Jugoslavia. Natura, 57, 5-55; 1966.
RADOVANOVITCH, M. & MARTINO, K. The snakes of the Balkan Peninsula. Nauehno-Popul. Spisi Srpska Acad. Nauka Beograd Kni, 1, 1-75; 1950. (In Jugoslavian)
KOPSTEIN, F. & WETTSTEIN, O. Reptilien und Amphibien aus Albanien. Verh. Zool. -Bot. Ges. Wien, 70, 387-457; 1921.
MERTENS, R. Die Amphibien und Reptilien der Insel Korfu. Senckenbergische Biol., 42, 1-29; 1961.
KIRITESCU, C. Cercetari asupra faunei herpetologice a Romaniei. Bucuresti; 1-117; 1930.
CALINESCU, R. I. Contributiuni sistematice si zoogeografice la studiul amphibulor si reptilelor din Romania. Mem. Soc. Stun. Acad. Romana Bucuresti, (3), 7, 119-172; 1931.
FUHN, I. E. & VANCEA, S. Reptilia (Testoase, Sopirie, Serpe). Fauna Republ. Popul. Romine, 14, (2), 1-352; 1962.
IONESCU, V. Vertebratele din Romania. Bucuresti; 1-496; 1968. (keys)
WETTSTEIN, O. Herpetologia aegaea. Sitz. Ber. Ost. Akad. Wiss., Abt. 1, 162, 651-833; 1953.
von WETTSTEIN, O. Herpetologie der Insel Kreta. Ann. Naturhist. Mus. Wien, 45, 159-172; 1931. (list)
BURESCH, I. & ZONKOW, J. Untersuchungen uber die Verbreitung der Reptilien und Amphibien in Bulgarien und der Balkanhalbinsel. II. Schlangen. Bull. Inst. Roy. Hist. Nat. Sofia, 7, 106-188; 1934. (In Bulgarian)
BESKOV, V. & BERON, P. Catalogue et bibliographie des amphibiens et des reptiles en Bulgarie. Sofia; 1-40; 1964.
BODENHEIMER, F. A. Introduction into the knowledge of the amphibia and reptilia of Turkey. Rev. Fac. Sci. Univ. Istanbul, 9B, 1-83; 1944.
MERTENS, R. Amphibien und Reptilien aus der Turkei. Rev. Fac. Sci. Univ. Istanbul, 17B, 41-75; 1952.
NIKOLSKY, A. M. Reptiles (Reptilia). Vol. 1. Chelonia et Sauria. Faune de la Russie, 1-532; 1915. (In Russian; English translation 1963)
NIKOLSKY, A. M. Reptiles (Reptilia). Vol. II. Ophidia. Faune de la Russie; 1-349; 1916. (In Russian; English translation 1964)
TERENTIEV, P. V. & CERNOV, S. A. Key to the Amphibia and Reptiles of the U. S. S. R. Moskva; 1-95; 1936. (In Russian)
TERENTIEV, P. V. & CHERNOV, S. A. A brief key of the reptiles and amphibia of the U. S. S. R. Leningrad; 1-184; 1940. (In Russian)
TERENTIEV, P. V. & CHERNOV, S. A. Guide to reptiles and amphibians. Moskva; 1-340; 1949. (In Russian)
TERENT'EV, P. V. Herpetology. A study of amphibians and reptiles. Moskva; 1-336; 1961. (In Russian; English translation 1965)
NIKOL'SKII, A. M. Chelonia and Sauria. Fauna of Russia and adjacent countries. Vol. 1. Jerusalem; 1963. (Translation of 1915)
NIKOL'SKII, A. M. Ophidia. Fauna of Russia and adjacent countries. Vol. 2. Jerusalem; 1-247; 1963. (Translation of 1916)
TERENT'EV, P. V. Herpetology. A manual of amphibians and reptiles. Jerusalem; 1-313; 1965. (Translation of 1961)
TERENT'EV, P. V. & CHERNOV, S. A. Key to amphibians and reptiles. Jerusalem; 1-315; 1965. (Translation of 1949)
TARASHCHUK, V. I. Amphibians and reptiles. Fauna Ukraini, 7, 1-246; 1959. (In Ukrainian)
CHERNOV, S. A. Systematic review of the animals of Crimea. III. Reptilia and Amphibia. Zhivot. Mir., 5, 73-78; 1958. (In Russian)
NIKOLSKI, A. M. Herpetologia caucasica. Tiflis; 1-272; 1913. (In Russian)
LANTZ, L. A. & CYREN, O. Les lezards sylvicole de la caucasie. Bull. Soc. Zool. France, 72, 169-191; 1947.
VERESHCHAGIN, N. K. Systematic review of the animals of the Caucasian isthmus. III. Reptilia. Zhivot. Mir., 5, 261-280; 1958. (In Russian)
CERNOV, S. A. On the herpetological fauna of Armenia and Nachichevan A. R. S. S. Zool. Pap. Bio. Inst. Erevan, 1, 77-194; 1939. (IN Russian)
BOGDANOV, O. P. Reptiles of Turkmenia. Izdatel'stvo Akad. Nauk Turkmenskoi S. S. R., 1962, 1-236; 1962. (In Russian)
YAKOVLEVA, I. D. Key to the reptiles of the Kirghiz. Frunze; 1-112; 1961. (In Russian)

Near East and Middle East

BARASH, A. & HOOFIEN, J. H. Reptiles of Israel. Israel; 1-180; 1956. (In Hebrew)
HOOFIEN, J. H. An alphabetical list of the reptiles of Israel. Tel-Aviv; 1-2; 1967.
AHARONI, I. Reptilia and Amphibia. In Luke & Keith-Roach, Handbook of Palestine and Trans-Jordan. London; 418-422; 1934.
CORKILL, N. L. The snakes of Iraq. Journ. Bombay Nat. Hist. Soc., 35, 552-572; 1932.
KHALAF, K. T. Reptiles of Iraq with some notes on the amphibians. Baghdad; 1-96; 1959. (keys)
LEVITON, A. E. Report on a collection of reptiles from Afghanistan. Proc. California Acad. Sci., (4), 29, 445-463; 1959.
BRUCK, G. Zur herpetofauna Afghanistans. Vest. Ceskoslovenske Spol. Zool., 32, 201-218; 1967.
SCHMIDT, K. P. Amphibians and reptiles of Yemen. Fieldiana, Zool., 34, 253-261; 1953.

CORKILL, N. L. & COCHRAN, J. A. The snakes of the Arabian Peninsula and Socotra. Journ. Bombay Nat. Hist. Soc., 62, 475-506; 1965.
CORKILL, N. L. The snakes of Aden Territory. Port of Aden Ann., 1961-62, 78-81; 1962. (list)
SCHMIDT, K. P. & MARX, H. The perpetology of Sinai. Fieldiana, Zool., 39, (4), 21-40; 1956.

North Africa

BOULENGER, G. A. A list of the snakes of North Africa. Proc. Zool. Soc. London, 1919, 299-307; 1919.
SAINT GIRONS, H. Les serpentes de l'Afrique palearctique. Compt. Rend. Soc. Biogeogr., 28, 99-102; 1952.
MARINKELLE, C. J. Slangen van Noor-Afrika. De Slangen van Marokko, Algerie, Tunisie, en Lybie. Lacerta, 21, 12-16; 1962. (list)
ANDERSON, J. Reptilia and Batrachia. Zoology of Egypt, I, 1-371; 1898.
FLOWER, S. S. Notes on the recent reptiles and amphibians of Egypt, with a list of lthe species recorded from that kingdom. Proc. Zool. Soc. London, 1933, 735-851; 1933.
KEIMER, L. Histoire de serpents dans l'Egypte ancienne et moderne. Mem. Inst. Egypt, 50, 1-110; 1947.
MARX, H. Keys to the lizards and snakes of Egypt. Cairo; 1-8; 1956. (U. S. NAMRU 3 Res. Rep.)
PELLEGRIN, J. Les reptiles et les batraciens de l'Afrique du Nord francaise. Compt. Rend. Assoc. Francaise Avanc. Sci., 51, 260-264; 1927. (list)
ZAVATTARI, E. Erpetologia della Cirenaica. Arch. Zool. Italiana Torino, 14, 253-287; 1930.
KRAMER, E. & SCHNURRENBERGER, H. Zur Systematik Libyscher Schlangen. Mitt. Naturf. Ges. Bern, (n. s.), 17, 1-17; 1959. (keys)
KRAMER, E. & SCHNURRENBERGER, H. Systematik, Verbreitung und Okologie der libyschen Schlangen. I Rev. Suisse Zool., 70, 453-568; 1963.
CHPAKOWSKY, N. & CHNEOUR, A. Les serpents de Tunisie. Bull. Soc. Hist. Nat. Tunis, 6, 125-146; 1954.
DOMERGUE, C. A. Liste des batraciens, cheloniens et sauriens. Bull. Soc. Sci. Nat. Tunis, 9-10, (1956-1966), 75-79; 1967.
DOMERGUE, C. A. Liste des Ophidiens de Tunisie, de l'Algerie et du Maroc. Arch. Inst. Pasteur Tunis, 36, 157-161; 1959.
AELLEN, V. Contribution a l'herpetologie du Maroc. Bull. Soc. Sci. Nat. Maroc, 31, 153-199; 1952.
SAINT GIRONS, H. Les serpents du Maroc. Arch. Sci. Inst. Chrifien, 8, 1-29; 1956.
SAINT GIRONS, H. Les serpents du Maroc. Var. Sci. Soc. Nat. Maroc, 8, 1-29; 1957.
PASTEUR, G. & BONS, J. Catalogue des reptiles actuels du Maroc. Trav. Inst. Sci. Cherifien, Zool., 21, 1-134; 1960. (list)
BONS, J. & GIROT, B. Cli illustree des reptiles du Maroc. Trav. Inst. Sci. Cherifien, Zool., 26, 1-62; 1962.

Atlantic Islands

SARMENTO, A. A. Repteis e batraquios de Arquipelago de Madeira. Funchal; 1-45; 1940.
SARMENTO, A. A. Vertebrados da Madeira. I (Mamiferos — Aves — Repteis — Batraquios). 2nd ed. Funchal; 1-317; 1948.
MERTENS, R. Die Eidechsen der Kapverden. Comment. Biol., 15, (5), 1-17; 1956.

Africa

HEWITT, J. Descriptions of some African tortoises. Ann. Natal Mus., 6, 461-505; 1931. (Keys to genera)
LOVERIDGE, A. & WILLIAMS, E. E. Revision of the African tortoises and turtles of the suborder Cryptodira. Bull. Mus. Comp. Zool., 115, 163-557; 1957.
ISEMONGER, R. M. Snakes of Africa. Southern — Central and East. Cape Town; 1-236; 1962. (keys)
BOULENGER, G. A. A list of the snakes of West Africa, from Mauritania to the French Congo. Proc. Zool. Soc. London, 1917, 267-298; 1919.
CANSDALE, G. S. West African snakes. London; 1-74; 1961. (keys)
VILLIERS, A. Initiations Africaines. II. Les serpents de l'ouest Africain, 2nd ed. Dakar; 1-190; 1963.
PAPENFUSS, T. J. Preliminary analysis of the reptiles of arid Central West Africa. Wasmann Journ. Biol., 27, 249-325; 1969.
ANGEL, F. Les serpents de l'Afrique occidentale francaise. Paris; 1-246; 1933
VILLIERS, A. Tortues et crocodiles de l'Afrique noire francaise. Init. Africaines, 15, 1-354; 1958.
MANACAS, S. Saurios e ofidios da Guine Portuguesa. Anais Junta Invest. Ultramar, 10, 189-213; 1955.
MENZIES, J. I. The snakes of Sierra Leone. Copeia, 1966, 169-179; 1966.
BARBOUR, T. & LOVERIDGE, A. Reptiles and amphibians from Liberia. In Strong, African Republic of Liberia and the Belgian Congo. Cambridge, Massachusetts; 769-785; 1930. (list)
MACGRAITH, B. G. The identification of the poisonous snakes of British West Africa. Ann. Trop. Med. Parasit., 38, 21-34, 119-138; 1944.
DOUCET, J. Les serpents de la Republique de Côte d'Ivoire. I, II. Acta Trop. Basel, 20, 201-259, 297-340; 1963.
LEESON, F. Identification of snakes of the Gold Coast. (? London); 1-142; 1950. (keys)
CANSDALE, G. S. Gold Coast snakes — a complete list. Niger Field, 19, 118-132; 1954.
HUGHES, B. & BARRY, D. H. The snakes of Ghana. A checklist and key. Bull. Inst. Francaise d'Afrique Noire, 31, (A), 1004-1041; 1969.
DUNGER, G. T. The lizards and snakes of Nigeria, 3 parts. Niger Field, 32, 53-74, 117-131, 170-178; 1967.

KNOEPFFLER, L.-P. Clef de determination des serpents actuellement connus de Gabon et ... Moyen-Congo, Rio Muni, Cabinda et Sud-Cameroun forestier). Biol. Gabon, 4, 183-194; 1968.
LAURENT, R. F. Contribution a l'herpetologie de la region des Grands Lacs de l'Afrique centrale. I. Generalites. II. Cheloniens. III. Ophidiens. Ann. Mus. Roy. Congo Belgique, 48, 1-390; 1956.
BOULENGER, G. A. A list of the snakes of the Belgian and Portuguese Congo, Northern Rhodesia and Angola. Proc. Zool. Soc. London, 1915, 193-223; 1915.
SCHMIDT, K. P. Contributions to the herpetology of the Belgian Congo based on the collection of the American Congo Expedition, 1909-1915. Part I. Turtles, crocodiles, lizards, and chameleons. Bull. American Mus. Nat. Hist., 39, 385-624; 1917.
SCHMIDT, K. P. Contributions to the herpetology of the Belgian Congo .... Part II. Snakes. Bull. American Museum Nat. Hist.; 49, 1-146; 1923.
de WITTE, G. F. Genera des serpents du Congo et du Ruanda-Urundi. Ann. Mus. Afrique Centrale, 104, 1-203; 1962. (keys to genera)
MERTENS, R. Amphibien und Reptilien aus Angola. Senckenbergiana, 20, 425-443; 1938.
LAURENT, R. F. Reptiles et amphibiens de l'Angola. Lisboa; 12-165; 1964.
FitzSIMONS, F. W. The snakes of South Africa, 2nd ed. Cape Town; 1-547; 1912.
FitzSIMONS, F. W. The lizards of South Africa. Pretoria; 1-528; 1948.
ROSE, W. The reptiles and amphibians of southern Africa, 2nd ed. Cape Town; 1-494; 1962. (1st edition, 1950) (keys)
FitzSIMONS, V. F. M. Snakes of southern Africa. London; 1-423; 1962.
FitzSIMONS, V. A check-list, with synoptic keys, to the snakes of southern Africa. Ann. Transvaal Mus., 25, 35-79; 1966.
HEWITT, J. A guide to the vertebrate fauna of the eastern Cape Province, South Africa. Part II. Reptilia . . . . Grahamstown; 1-141; 1937.
BOULENGER, G. A. A list of the snakes of East Africa, north of the Zambesi and south of the Soudan and Somaliland, and of Nyassaland. Proc. Zool. Soc. London, 1915, 611-640; 1915.
LOVERIDGE, A. A list of the lizards of British Territories in East Africa (Uganda, Kenya Colony, Tanganyika Territory and Zanzibar), with keys for the diagnosis of the species. Proc. Zool. Soc. London, 1923, 841-863; 1923.
LOVERIDGE, A. Notes on East African tortoises collected 1921-1923 . . . . Proc. Zool. Soc. London, 1923, 923-933; 1923. (list)
LOVERIDGE, A. Check list of the Reptilia recorded from the British Territories in East Africa. Journ. East Africa Uganda Nat. Hist. Soc., 1924, (spec. suppl. 3), 1-16; 1924.
LOVERIDGE, A. East African reptiles and amphibians in the United States National Museum. United States Nat. Mus. Bull., 151, 1-135; 1929.
LOVERIDGE, A. Check list of reptiles and amphibians of East Africa (Uganda, Kenya, Tanganyika, Zanzibar). Bull. Mus. Comp. Zool., 117, 153-362; 1957. (Also cited as 1958)
SKELTON-BOURGEOIS, M. Reptiles et batraciens d'Afrique orientale. Rev. Zool. Bot. Africaines, 63, 309-338; 1961. (list)
BROADLEY, D. G. A checklist of the snakes of Southern Rhodesia. Part I. Journ. Herpet. Assoc. Rhodesia, 3, 2-6; 1958.
BROADLEY, D. G. The herpetology of Southern Rhodesia. Part i. Snakes. Bull. Mus. Comp. Zool., 120, 1-100; 1959.
BROADLEY, D. G. Key to the snakes of Rhodesia and Nyasaland. Journ. Herpet. Assoc. Rhodesia, 4, 3-4; 1958.
SWEENEY, R. C. H. Snakes of Nyasaland. Zomba; Nyasaland; 1-200; 1961. (keys)
VESEY-FITZGERALD, D. F. The snakes of northern Rhodesia and the Tanganyika borderlands. Proc. Rhodesia Sci. Assoc., 46, 17-102; 1958.
IONIDES, C. J. P. Snakes of the southern Province. Tanganyika Notes Rec., 29, 98-108; 1950.
UTHMOLLER, W. Beitrag zur Kenntnis der Schlangenfauna des Kilimandscharo/Tanganyika-Territory, ehemaliges Deutsch-Ostafrica). Temminckia, 2, 97-134; 1937.
LOVERIDGE, A. A guide to the snakes of the Nairobi district. Journ. East Africa Nat. Hist. Soc., 18, 97-115; 1945.
PITMAN, C. R. S. A guide to the snakes of Uganda; 11 parts & addenda. Uganda Journ., 3-5; 1935-38.
BOULENGER, G. A. A list of the snakes of north-east Africa, from the Tropic to the Soudan and Somaliland, including Socotra. Proc. Zool. Soc. London, 1915, 641-658; 1915.
PARKER, H. W. The lizards of British Somaliland. Bull. Mus. Comp. Zool., 91, 1-101; 1942.
PARKER, H. W. The snakes of Somaliland and the Sokotra Islands. Zool. Verh., Leiden; 6, 1-115; 1949.
SCORTECCI, G. Rettili ed anfibi raccolti dal Prof. E. Zavattari in Eritra. Att. Soc. Italiana Milano, 69, 193-217; 1930. (list)
SCORTECCI, G. Gli ofidi velenosi dell'Africa Italiana. Milano; 1-292; 1939.

Madagascar and Indian Ocean

ANGEL, F. Les lezards de Madagascar. Mem. Acad. Malgache, 36, 1-293; 1942.
GUIBE, J. Les serpents des Madagascar. Mem. Inst. Sci. Madagascar, 12A, 189-260; 1958.
BOULENGER, G. A. A list of the snakes of Madagascar, Comoro, Mascarene and Seychelles. Proc. Zool. Soc. London, 1915, 369-382; 1915.
PARKER, H. W. Revised list of reptiles (excluding chelonians) and amphibians collected in the Seychelles. Trans. Linn. Soc., 19, 444-446; 1936.
RENDAHL, H. Zur Herpetologie der Seychellen. I. Reptilien. Zool. Jahrb., Syst., 72, 255-328; 1939.

VESEY-FITZGERALD, D. Reptiles and amphibians from the Seychelles Archipelago. Ann. Mag. Nat. Hist., (11), 14, 577-584; 1948.
GAYMER, R. Amphibians and reptiles of the Seychelles. British Journ. Herpet., 4, 24-28; 1968.
SMITH, M. A. The herpetology of the Andaman and Nicobar Islands. Proc. Linn. Soc., 153, 150-158; 1941.

India

BOULENGER, G. A. Reptilia and Batrachia. The fauna of British India. . . . London; 1-541; 1890.
SMITH, M. A. The fauna of British India, . . . Reptilia and Amphibia, vol. 1, Loricata, Testudines. London; 1-213; 1931.
SMITH, M. A., The fauna of British India, . . . Reptilia and Amphibia, vol. II, Sauria. London; 1-440; 1935.
SMITH, M. A. The fauna of British India, . . . Reptilia and Amphibia, vol. III, Serpentes. London; 1-583; 1943.
WALL, F. A hand-list of the snakes of the Indian Empire. Journ. Bombay Nat. Hist. Soc., 29, 345-361; 598-632; 864-878; 30, 12-24; 1923-1924.
MINTON, S. A., Jr. A contribution to the herpetology of West Pakistan. Bull. American Mus. Nat. Hist., 134, 27-184; 1966.
MINTON, S. A., Jr. An annotated key to the amphibians and reptiles of Sind and Las Bela, West Pakistan. American Mus. Novit., 2081, 1-60; 1962.
CAZALY, W. H. The common snakes of India and Burma and how to recognize them. Allahabad; 1-60; 1914.
WALL, F. How to identify the snakes of India. Karachi; 1-56; 1923.
HORA, S. L. The distribution of crocodiles and chelonians in Ceylon, India, Burma, and farther east. Proc. Nat. Inst. Sci. India, 14, 285-310; 1948.
SHAW, G. E. et al. The snakes of northern Bengal and Sikkim, 16 parts. Journ. Bengal Nat. Hist. Soc., 16, 57-67; 1941, etc.
SWAN, L. W. & LEVITON, A. E. The herpetology of Nepal. A history, check list.... Proc. California Acad. Sci., (4), 32, 103-147; 1962.
ABERCROMBY, A. F. The snakes of Ceylon. London; 1-89; 1910.
WALL, F. Ophidia Taprobanica or The snakes of Ceylon. Colombo; 1-582; 1921.
NICHOLLS, L. The identification of the land snakes of Ceylon. Ceylon Journ. Sci., 2, 91-142; 1929.
DERANIYAGALA, P. E. P. Some Ceylon lizards. Ceylon Journ. Sci., 16B, 139-180; 1931. (Complete review, except for Gekkonidae)
DERANIYAGALA, P. E. P. The tetrapod reptiles of Ceylon, vol. 1, Testudinates and crocodilians. Colombo; 1-412; 1939.
TAYLOR, E. H. A brief review of Ceylonese snakes. Univ. Kansas Sci. Bull., 33, 519-603; 1950.
TAYLOR, E. H. A review of the lizards of Ceylon. Kansas Univ. Sci. Bull., 35, 1525-1585; 1953.
DERANIYAGALA, P. E. P. A colored atlas of some vertebrates from Ceylon, vol. 2, Tetrapod Reptilia. Colombo; 1-101; 1953.
DERANIYAGALA, P. E. P. A colored atlas..., vol. 3, Serpentoid Reptilia. Colombo; 1-121; 1955.

Indochina and Malay Peninsula

SMITH, M. List of the snakes at present known to inhabit Siam. Journ. Nat. Hist. Soc. Siam, 1, 211-215; 1915.
SMITH, M. A. A list of the crocodiles, tortoises, turtles and lizards at present known to inhabit Siam. Journ. Nat. Hist. Soc. Siam, 2, 48-57; 1916.
SUVATTI, C. Fauna of Thailand. Bangkok; 1-1100; 1950. (Reptiles, 464-521)
TAYLOR, E. H. & ELBEL, R. E. Contributions to the herpetology of Thailand. Univ. Kansas Sci. Bull., 38, 1033-1189; 1958.
TAYLOR, E. H. The lizards of Thailand. Univ. Kansas Sci. Bull., 44, 687-1077; 1963.
TAYLOR, E. H. The serpents of Thailand and adjacent waters. Univ. Kansas Sci. Bull., 45, 609-1096; 1965.
TAYLOR, E. H. The turtles and crocodiles of Thailand and adjacent waters with a synoptic herpetological bibliography. Univ. Kansas Sci. Bull., 49, 87-179; 1970.
DEUVE, J. Liste annotee des serpents du Laos. Bull. Soc. Sci. Nat. Laos, 1, 5-32; 1961.
DEUVE, J. Serpents du Laos. Mem. ORSTOM, 39, 1-251; 1970.
BOURRET, R. Les serpents de l'Indochine. I. Etudes sur la fauna; 1-141; II. Catalogue systematique descriptif; 1-504; 1936. Toulouse.
BOURRET, R. Les tortues de l'Indochine. Notes Inst. Oceanogr. Indochine, 38, 1-235; 1941.
CAMPDEN-MAIN, S. M. A field guide to the snakes of South Vietnam. Washington, D, C,; 1-114; 1970.
BOURRET, R. Comment determiner un serpent d'Indochine. Hanoi; 1-28; 1935.
BOULENGER, G. A. A vertebrate fauna of the Malay Peninsula from the Isthmus of Kra to Singapore . . Reptilia and Batrachia. London; 1-294; 1912.
SMITH, M. A. The Reptilia and Amphibia of the Malay Peninsula. Bull. Raffles Mus., 3, 1-149; 1930.
TWEEDIE, M. W. F. The snakes of Malaya. Singapore; 1-139; 1953. (keys)
TWEEDIE, M. W. F. The snakes of Malaya. Singapore; 1-143; 1961. (keys)
SWORDER, G. H. A list of the snakes of Singapore Island. Singapore Nat., 2, 55-73; 1923.

Indo-Australian Archipelago

de ROOIJ, N. The reptiles of the Indo-Australian Archipelago. I. Lacertilia Chelonia Emydosauria. Leiden; 1-384; 1915.
de ROOIJ, N. The reptiles of the Indo-Australian Archipelago. II. Ophidia. Leiden; 1-334; 1917.
de HAAS, C. P. J. Checklist of the snakes of the Indo-Australian Archipelago. Treubia, 20, 511-625; 1950.
HAILE, N. S. The snakes of Borneo, with a key to the species. Sarawak Mus. Journ., 8, 743-771; 1959.
BOULENGER, G. A. A catalogue of the reptiles and batrachians of Celebes.... Proc. Zool. Soc. London, 1897, 193-237; 1897. (keys)
ROBINSON, H. C. & KLOSS, C. B. A nominal list of the species of reptiles and batrachians occurring in Sumatra. Journ. Federated Malay States Mus., 8, 297-306; 1920.
MERTENS, R. Amphibien und Reptilien aus dem aussersten westen Javas und von benachbarten Eilanden. Treubia, 24, 83-105; 1958. (list)
FORCART, L. Die Amphibien und Reptilien von Sumba.... Verh. Naturf. Ges. Basel, 64, 356-388; 1953.
MERTENS, R. Die Amphibien und Reptilien der Inseln Bali, Lombok, Sumbawa, und Flores. Abh. Senckenbergische Naturf. Ges., 42, 117-342; 1930.
LOVERIDGE, A. New Guinea reptiles and amphibians.... Bull. Mus. Comp. Zool., 101, 305-430; 1948.

Australia

GOODE, J. Freshwater tortoises of Australia and New Guinea. Melbourne; 1-154; 1967.
HUNT, R. A. A key to the identification of Australian snakes. Victoria Nat., 64, 153-167; 1947.
KINGHORN, J. R. The snakes of Australia, 2nd ed. Sydney; 1-198; 1956. (keys)
WORRELL, E. Reptiles of Australia — Crocodiles - turtles - tortoises - lizards - snakes. Sydney; 1-207; 1963.
WORRELL, E. Australian snakes, crocodiles, tortoises, turtles, and lizards. (Australia); 1-64; 1968. (keys)
BUSTARD, R. Australian lizards. (Australia); 1-162; 1970.
LITTLEJOHN, M. J. Zoology of the High Plains. Part I. Ichthyology and herpetology. Proc. Roy. Soc. Victoria, (n. s.), 75, 311-313; 1962.
WAITE, E. R. The reptiles and amphibians of South Australia. Adelaide; 1-270; 1929. (keys)
GLAUERT, L. The vertebrate fauna of Western Australia. Pt. 1. Reptilia (Testudinata, Ophidia). Journ. Roy. Soc. Western Australia, 14, 61-77; 1928. (keys)
GLAUERT, L. A handbook of the snakes of Western Australia, 3rd ed. Perth; 1-62; 1967. (Previous editions, 1st 1950; 2nd, 1957)
GLAUERT, L. A hand book of the lizards of Western Australia. Western Australian Nat. Club Handb., 6, 1-100; 1961.

New Zealand

McCANN, C. Key to the lizards of New Zealand. Tuatara, 6, 45-51; 1956.
SHARELL, R. The tuatara, lizards and frogs of New Zealand. London; 1-94; 1966.
McCANN, C. Key to the marine turtles and snakes occurring in New Zealand. Tuatara, 14, 73-81; 1966.
McCANN, C. The marine turtles and snakes occurring in New Zealand. Rec. Domin. Mus., 5, 201-205; 1966.

The Orient

TAYLOR, E. H. Amphibians and turtles of the Philippine Islands. Manila; 1-193; 1921.
TAYLOR, E. H. The lizards of the Philippine Islands. Manila; 1-269; 1922.
TAYLOR, E. H. The snakes of the Philippine Islands. Manila; 1-312; 1922.
STEJNEGER, L. The batrachians and reptiles of Formosa. Proc. United States Nat. Mus., 38, 91-114; 1910.
KUNTZ, R. E. Snakes of Taiwan. Taipei; 1-80; 1963. (U. S. NAMRU 2)
POPE, C. H. The reptiles of Hainan. Bull. American Mus. Nat. Hist., 54, 395-465; 1927.
GEE, N. G. A contribution toward a preliminary list of reptiles recorded from China. Bull. Dep. Biol. Yenching Univ. Peiping, 1, 53-84; 1930.
MELL, R. List of Chinese snakes. Lingnan Sci. Journ., 8, 199-219; 1931.
POPE, C. H. List of the Chinese turtles, crocodilians, and snakes, with keys. American Mus. Novit., 733, 1-29; 1934.
POPE, C. H. The reptiles of China. Turtles, crocodilians, snakes, lizards. Nat. Hist. Central Asia, X, 1-604; 1935.
ROMER, J. D. Annotated checklist with keys to the snakes of Hong Kong. Mem. Hong Kong Nat. Hist. Soc., 5, 1-14; 1961.
ROMER, J. D. Revised annotated checklist with keys to the snakes of Hong Kong. Mem. Hong Kong Nat. Hist. Soc., 8, 1-22; 1970.
SOWERBY, A. D. Amphibians and reptiles recorded from or known to occur in the Shanghai area. Notes Herpet. Shanghai, 1, 1-16; 1943. (list)
MASLIN, T. P. Snakes of the Kiukiang-Lushan area, Kiangi, China. Proc. California Acad. Sci., (4), 26, 419-466; 1950.
REEVES, C. D. Manual of the vertebrate animals of northeastern and Central China exclusive of birds. Shanghai; 1-806; 1933. (keys)

BORING, A. M. et al. Handbook of North China amphibia and reptiles, or herpetology of North China. Peiping; 1-64; 1932. (keys)
OKADA, Y. Reptilia of Jehol. Rep. 1st Sci. Exp. Manchoukuo, (V, II, II-2); 1-76; 1935. (In Japanese and English)
SHANNON, F. A. The reptiles and amphibians of Korea. Herpetologica, 12, 22-49; 1956.
OKADA, Y. Annotated list of animals of Okinawa Islands. Naha City, Japan; 1-384; 1959.
JOHNSON, C. R. Herpetofauna of Okinawa, Ryu Kyu Islands. Herpetologica, 25, 206-210; 1969. (list)
STEJNEGER, L. The herpetology of Japan and adjacent territory. United States Nat. Mus. Bull., 58, 1-577; 1907.
MAKI, M. Monograph of the snakes in Japan. Tokyo; 1-240; 1931. (In Japanese and English)
(anonymous). Illustrated encyclopedia of the fauna of Japan..., 2nd ed. Tokyo; 2 volumes; 1949.
TAKASHIMA, H. A synopsis of the reptiles of Japan. Misc. Rep. Yamashina Inst. Ornith. Zool., 12, 486-493; 1958. (In Japanese)
NAKAMURA, K. & UENO, S.-i. Japanese reptiles and amphibians in color. Osaka; 1-214; 1963. (In Japanese) (keys)
GORIS, R. C. The reptiles of Japan. (Japan); 1-127; 1966. (In Japanese)
MATSUI, T. The snakes of Japan. Snake, 1, 50-55; 1969. (In Japanese) (list)

Oceania

LOVERIDGE, A. Reptiles of the Pacific World. New York; 1-259; 1945. (keys)
BURT, C. E. & BURT, M. D. Herpetological results of the Whitney South Sea Expedition. VI Pacific Island Amphibia and Reptilia in the collection of the American Museum of Natural History. Bull. American Mus. Nat. Hist., 63, 463-597; 1932. (Complete for western Pacific Islands)
STEJNEGER, L. The land reptiles of the Hawaiian Islands. Proc. United States Nat. Mus., 21, 783-813; 1899.
OLIVER, J. A. & SHAW, C. E. The amphibians and reptiles of the Hawaiian Islands. Zoologica, 38, 65-95; 1953.
KINGHORN, J. R. Herpetology of the Solomon Islands. Rec. Australian Mus., 16, 123-178; 1928.

SPECIAL LISTS

Cavernicolous. WOLF, B. Animalium cavernarum catalogus, 3, 700-707; 1936.

Saltwater. NEILL, W. T. The occurrence of amphibians and reptiles in saltwater areas, and a bibliography. Bull. Mar. Sci. Gulf Caribbean, 8, 1-97; 1958.

Venomous. MAASS, T. A. Kennzeichen der sogenannten ungiftigen und verdachtigen Schlangen mit Gift-Apparat. Tab. Biol. Period., 3, 105-208; 1934.
COCHRAN, D. M. Poisonous reptiles of the world. A wartime handbook. Smithsonian Inst. War Backgr. Stud., 10, 1-137; 1943.
(anonymous). Poisonous snakes of the world. A manual for use by U. S. amphibious forces. Washington, D. C.; 1-168; 1962. (keys)
BOYS, F. & SMITH, H. M. Poisonous amphibians and reptiles. Recognition and bite treatment. Springfield, Illinois; 1-149; 1959.
MOORE, G. M. Poisonous snakes of the world. Washington, D. C.; 1-212; 1968.
(anonymous). Poisonous snakes of the world . . ., 2nd ed. Washington, D. C.; 1-203; 1970.
DITMARS, R. L. The poisonous serpents of the New World. Bull. New York Zool. Soc., 33, 79-99; 1930.
POPE, C. H. The poisonous snakes of the New World. New York; 1-56; 1944. (Also in: Bull. New York Zool. Soc.)
DA LIE, D. A. Poisonouss snakes of America. Journ. Forestry, 51, 243-248; 1953.
STICKEL, W. H. Venomous snakes of the United States and treatment of their bites. U. S. Dep. Interior, Wildl. Leafl., 339, 1-29; 1952.
WERLER, J. E. The poisonous snakes of Texas and the first aid treatment of their bites. Texas Game Fish Bull., 31, 1-40; 1950.
VELLARD, J. Serpents venimeux du Venezuela. Ann. Sci. Nat. Paris, 3, 193-225; 1941.
HOGE, A. R. & LANCINI, A. R. Sinopsis de las serpientes venenosas de Venezuela. Publ. Ocas. Mus. Cienc. Nat. Caracas, Zool., 1, 1-24; 1962.
KLEMMER, K. Classification and distribution of European, North African, and north and west Asiatic venomous snakes. In Venomous Animals and Their Venom. London & New York; 1, 309-325; 1968.
DUREN, A. Les serpents venimeux du Congo Belge. Mem. Inst. Roy. Colon. Belge, 15, (5), 3-45; 1946.
LEVITON, A. E. The venomous terrestrial snakes of East Asia, India, Malaya and Indonesia. In Bucherl, Venomous Animals and Their Venoms; 1, 529-576; 1968.
KOPSTEIN, F. Die Giftschlangen Java.... Zeitschr. Morph. Okol. Tiere, 19, 339-363; 1930.
WORRELL, E. Dangerous snakes of Australia and New Guinea. Sydney; 1-68; 1953/1961.

MONOGRAPHIC, DESCRIPTIVE, & REVISIONARY WORKS

Monographs

GADOW, H. Amphibia and Reptilia. In Cambridge Natural History, VIII. London, 1-668; 1909. (To families and genera)

SIEBENROCK, F. Synopsis der rezenten Schildkroten. Zool. Jahrb., Suppl. 10, Heft 3, 427-618; 1909. (To families, genera & species)

Descriptive catalogs

BOULENGER, G. A. Catalogue of the lizards in the British Museum (Natural History), 2nd ed. London; 3 volumes; 1885. (Reprinted 1965)

BOULENGER, G. A. Catalogue of the chelonians, rhynchocephalians, and crocodiles in the British Museum (Natural History), 2nd ed. London; 1-311; 1889. (Reprinted 1966)

BOULENGER, G. A. Catalogue of the snakes in the British Museum (Natural History). London; 3 vols.; 1893-1896. (Reprinted 1961)

RIEMER, W. J. et al. Catalogue of American amphibians and reptiles. Bethesda, Maryland. (Looseleaf; 1963 and continuation)

Synonymic catalogs

MERTENS, R. & WERMUTH, H. Die rezenten Schildkroten, Krokodile und Bruckenechsen. Eine kritische Liste der heute lebenden Arten und Rassen. Zool. Jahrb., 83, 323-440; 1955.

KIMWRA, W. Crocodiles of world. Japan; 1-127; 1966. (In Japanese)

PETERS, J. A. & OREJAS-MIRANDA, B. Catalogue of the neotropical Squamata. Part I. Snakes. United States Nat. Mus. Bull., 297, (I), 1-347; 1970.

PETERS, J. A. & DONOSO-BARROS, R. Catalogue of the neotropical Squamata. Part II. Lizards and amphisbaenians. United States Nat. Mus. Bull., 297, (II), 1-293; 1970.

DONOSO-BARROS, R. Catalogo herpetologico chileno. Bol. Mus. Nac. Hist. Nat. Chile, 31, 42-124; 1970.

BOULENGER, G. A. Catalogue of the snakes in the British Museum (Natural History). London; 3 vols.; 1893-1896. (Reprinted 1961)

BERGER, L. et al. Catalogus faunae Poloniae. Amphibia et Reptilia. Warszawa; 1-73; 1969.

Classifications

[See also: Works listed on CLASSIFICATION under Animalia, above]

COPE, E. D. The classification of the Ophidia. Trans. American Philos. Soc., 18, (II), 186-219; 1895.

COPE, E. D. The crocodilians, lizards, and snakes of North America. Rep. United States Nat. Mus. for 1898; 155-1270; 1900. (To family, for World)

GOODRICH, E. S. On the classification of the Reptilia. Proc. Roy. Soc., 89B, 261-276; 1916.

WILLISTON, S. W. The phylogenyyand classification of reptiles. Journ. Geol., 25, 411-421; 1917.

NOPCSA, F. Bemerkungen zur Systematik der Reptilien. Palaeont. Zeitschr., 5, 107-118; 1922.

CAMP, C. L. Classification of the lizards. Bull. American Mus. Nat. Hist., 48, 289-435; 1923.

NOPCSA, F. Die Familien der Reptilien. Fortschr. Geol. Palaeont. Berlin, 2, 1-210; 1923. (Family)

BROOM, R. On the classification of the reptiles. Bull. American Museum Nat. Hist., 51, 39-65; 1924.

WILLISTON, S. W. & GREGORY, W. K. The osteology of the reptiles. Cambridge, Massachusetts; 1-300; 1925.

DUNN, E. R. Notes on Central American caecilians. Proc. New England Zool. Club, 10, 71-76; 1928.

NOPCSA, F. The genera of reptiles. Palaeobiol. Wien, 1, 163-188; 1928. (To genera)

PERRIER, E. Les Reptiles. Traite de Zoologie, 8, 2883-3118; 1928. (To families)

ROMER, A. S. Vertebrate paleontology. Chicago; 1-687; 1933.

MOOK, C. C. The evolution and classification of the Crocodilia. Journ. Geol., 42, 295-304; 1934. (To families)

KALIN, J. A. Hispanochampsa mulleri nov. gen. nov sp., ein neuer Crocodilide aus dem unteren Oligocaen von Tarriga (Catalonien). Abhandl. Schweizerischen Palaeont. Ges., 58, (2), 1-39; 1936. (Includes classification of Crocodilia)

ROMER, A. S. Vertebrate paleontology. Chicago; 1-687; 1945.

KUHN, O. Das System der fossilen und rezenten Amphibien und Reptilien. Ber. Naturf. Ges. Bamberg, 29, 49-67; 1946.

WERMUTH, H. Systematik der rezenten Krokodile. Mitt. Zool. Mus. Berlin, 29, 375-574; 1953.

WATSON, D. M. S. On Bolosaurus and the origin and classification of reptiles. Bull. Mus. Comp. Zool., 111, 297-449.

BLAIR, W. F. et al. Vertebrates of the United States. New York; 1-819; 1957.7. (Pp. 273-358)

DOWLING, H. G. Classification of the Serpentes. A critical review. Copeia; 1959, 38-52; 1959.

DUELLMAN, W. E. A classification of the Recent Amphibia. Lawrence, Kansas; 1-108; 1961. (University of Kansas Zoology 182 syllabus)

KUHN, O. Die Familien der rezenten und fossilen Amphibien und Reptilien. Bamberg, 1-79; 1961.

TERENT'EV, P. V. Herpetology. A study of amphibians and reptiles. Moskva; 1-336; 1961. (In Russian)

TERENT'EV, P. V. Herpetology. Manual of Amphibia and Reptilia. Jerusalem; 1-313; 1965. (Translation of 1961)

UNDERWOOD, G. A contribution to the classification of snakes. London; 1-179; 1967.
BLAIR, W. F. et al. Vertebrates of the United States, 2nd ed. New York; 1-616; 1968. (To orders)
GOIN, G. J. & GOIN, O, B. Introduction to herpetology, 2nd ed. San Francisco; 1-353; 1971. (Pp. 333-335)

Revisionary works by groups

[These comprehensive works cover groups of families of various size. They are arranged here in alphabetical order. They include no revisions of the species or genera of families but studies of the families themselves in relation to other families.]

Chelonia. RUST, H. T. Systematische Liste der lebenden Schildkroten. Aquar. Terrar., 45; 1934.
RODECK, H. G. Guide to the turtles of Colorado. Univ. Colorado Mus. Leafl., 7, 1-9; 1950.
UNDERWOOD, G. Introduction to the study of Jamaican reptiles. Part V. Chelonia. Nat. Hist. Notes Nat. Hist. Soc. Jamaica, 46, 209-213; 49, 15-21; 51, 57-65; 1951.
CARR, A. Handbook of turtles. The turtles of the United States, Canada and Baja California. Ithaca, New York; 1-542; 1952.
WERMUTH, H. & MERTENS, R. Schildkroten, Krokodile, Bruckenechsen. Jena; 1-422; 1961.
OBST, F. J. & MEUSEL, W. Die Landschildkroten Europas. Wittenberg Lutherstadt; 1-52; 1963;
PRITCHARD, P. C. H. Living turtles of the world. Jersey City, New Jersey; 1-288; 1967. (key to families)
FRAIR, W. Turtle family relationships as determined by serological tests. Journ. Intern. Turtle Tort. Soc., 1, (5), 12-18.

Crocodilia. WERMUTH, H. & MERTENS, R. Schildkroten, Krokodile, Bruckenechsen. Jena;1-422;1961.

Lacertilia. SMITH, H. M. Handbook of lizards. Lizards of the United States and Canada. Ithaca, New York; 1-557; 1946. (With bibliography)
DUDA, P. L. A new simplified system of identification of lacertilian families. Kashmir Sci., 1, 70-72; 1964.

Ophidia. UNDERWOOD, G. A contribution to the classification of snakes. London; 1-179; 1967.

Rhynchocephalia. WERMUTH, H. & MERTENS, R. Schildkroten, Krokodile, Bruckenechsen. Jena; 1-422; 1961.

Sauria. COOMANS, R. Les lezards (1). Preliminaire — Clasification des Reptiles. Bull. Mens. Soc. Nat. Mons, 40, 28-34; 1957.
GLAUERT, L. Handbook of the lizards of Western Australia. (No reference available. 1961)

Squamata. TARLO, L. B. H. An outline classification of the Squamata. British Journ. Herpet., 4, 32-35; 1968.

Revisionary works by family

These lists of works were made up primarily from two sources: (1) listings in the Zoological Record for the years 1910 to 1968, which were not checked further if the title appeared to be explicit; and (2) library shelves and the office libraries of various specialists in each field; these were examined for appropriateness and, if revisionary, checked for included families. The works listed by families are therefore of two sorts, so far as examination by the compiler is concerned, those cited in the title as referring to one or a few named families (not usually examined or verified) and those directly examined and checked and listed as considered to be appropriate.

Works cited by author

In order to save space in the list of works on individual families, the following studies, which are each referred to more than a few times, are cited by author and date and number only. Citations are otherwise given by title, if a book, or by the serial reference, if not a separate work. The titles of the journal articles are not usually given, but the relevant limitations are shown by annotations.

ANDERSON, J., No. 1, 1898. Reptilia and Batrachia. Zoology of Egypt, vol. 1, 1-371.

ANGEL, F., No. 2, 1933. Les serpents de l'Afrique Occidentale Francaise. Paris; 1-246.

ANGEL, F., No. 1, 1942. Les lezards de Madagascar. Mem. Acad. Malgache, 36, 1-193.

ANGEL, F., No. 3, 1946. Reptiles et amphibiens. Faune de France, 45, 1-204.

BARBOUR, T. & RAMSDEN, C. T., No. 1, 1919. The herpetology of Cuba. Mem. Mus. Comp. Zool., 47, 69-213.

BOULENGER, G. A., No. 1, 1890. Reptilia and Batrachia. The fauna of British India including Ceylon and Burma. London; 1-541.

BOULENGER, G. A., No. 2, 1912. A vertebrate fauna of the Malay Peninsula from the Isthmus of Kra to Singapore.... Reptilia and Batrachia. London; 1-294.

BOURRET, R., No. 1, 1941. Les tortues de l'Indochine. Notes Inst. Oceanogr. Indochine, 38, 1-235.

BOURRET, R., No. 3, 1936. Les serpents de l'Indochine. Vol. I. Etudes sur la fauna. Vol. II. Catalogue systematique descriptif. Toulouse; 1-141; 1-504.

COCHRAN, D. M., No. 1, 1941. The herpetology of Hispaniola. United States Nat. Mus. Bull., 177, 1-398.

COPE, E. D., No. 2, 1900. The crocodilians, lizards, and snakes of North America. Rep. United States Nat. Mus. for 1898; 155-1270.

DERANIYAGALA, P. E. P., No. 1, 1931. Some Ceylon lizards. Ceylon Journ. Sci., 16B, 139-180.

DERANIYAGALA, P. E. P., No. 2, 1939. The tetrapod reptiles of Ceylon. Vol. 1. Testudinates and crocodilians. Colombo; 1-412.

DEVINCENZI, G. J., No. 1, 1925. Fauna erpetologica del Uruguay. Anal. Mus. Hist. Nat. Montevideo, (2), 2, 1-64.

DONOSA-BARROS, R., No. 1, 1966. Reptiles de Chile. Santiago; 1-458, i-clvi.

FitzSIMONS, V. F., No. 1, 1943. The lizards of South Africa. Pretoria; 1-528. (Also as Mem. Transvaal Mus., 1, 1-528.

FitzSIMONS, V. F. M., No. 2, 1962. Snakes of southern Africa. London; 1-423.

FUHN, I. E. & VANCEA, S., No. 2, 1962. Reptilia (Testoase, sopirle, serpi). Fauna Republicii Populare Romine, 14, (2), 1-352.

GUNTHER, A. C. L. G., No. 1, 1885-1902. Reptilia and Batrachia. Biologia Centrali-Americana; 1-326.

HELLMICH, W., No. 1, 1960. Die Sauria des Gran Chaco und seines Randgebiete. Abh. Bayerischen Akad. Wiss. Math. Nat. Kl., (n. f.), 101, 1-131.

KRAMER, E. & SCHNURRENBERGER, H., No. 1, 1963. Systematik, Verbreitung und Okologie der libyschen Schlangen. Rev. Suisse Zool., 70, 453-568.

LAURENT, R. F., No. 1, 1956. Contribution a l'herpetologie de la region des Grands Lacs de l-Afrique centrale. I. Generalitie. II. Cheloniens. III. Ophidiens. Ann. Mus. Roy. Congo Belge, 48, 1-390.

MAKE, M., No. 1, 1931. Monograph of the snakes in Japan. Tokyo; 1-240.

MERTENS, R., No. 1, 1952. Die Amphibien und Reptilien von El Salvador, auf Grund der Reisen von R. Mertens und A. Zilch. Abh. Senckenbergische Naturf. Ges., 487, 1-83.

MERTENS, R., No. 2, 1969. Die Amphibien und Reptilien West-Pakistans. Stuttgarter Beitr. Naturk., 197, 1-96.

NIKOLSKI, A. M., No. 1, 1915. Reptiles (Reptilia). Vol. Chelonia et Sauria. Faune de la Russie; 1-532.

NIKOLSKY, A. M., No. 2, 1916. Reptiles (Reptilia. Vol. II. Ophidia. Faune de la Russie; 1. 349.

NIKOL'SKII, A. M., No. 4, 1963. Chelonia and Sauria. Fauna of Russie and adjacent countries. Reptilia vol. I. Jerusalem. (Translation of No. 1)

NIKOL'SKII, A. M., No. 5, 1963. Ophidia. Fauna of Russia . . . ; 1-247.

OKADA, Y., No. 2, 1935. Reptilia of Jehol. Rep. 1st Sci. Exped. Manchoukuo; 1-76.

PETERS, J. A. & DONOSO-BARROS, R., No. 2, 1970. Catalogue of the neotropical Squamata. Part II. Lizards and amphisbaenians. United States Nat. Mus. Bull., 297, (II), 1-293.

PETERS, J. A. & OREJAS-MIRANDA, B., No. 1, 1970. Catalogue of the neotropical Squamata. Part I. Snakes. United States Nat. Mus. Bull., 297, (I), 1-347.

POPE, C. H., No. 1, 1935. The reptiles of China. Turtles, Crocodilians, Snakes, Lizards. Nat. Hist. Central Asia, X, 1-604.

de ROOIJ, N., No. 1, 1917. The reptiles of the Indo-Australian Archipelago. II. Ophidia. Leiden; 1-334.

de ROOIJ, N., No. 2, 1915. The reptiles of the Indo-Australian Archipelago. I. Lacertilia Chelonia Emydosauria. Leiden; 1-384.

ROZE, J. A., No. 1, 1966. La taxonomia y zoogeografia de los ofidios en Venezuela. Caracas; 1-498.
SCHMIDT, K. P. No. 1, 1928. Amphibians and land reptiles of Porto Rico. Sci. Surv. Porto Rico Virgin Islands, X, (1), 1-160.
SCHMIDT, K. P., No. 2, 1917. Contributions to the herpetology of the Belgian Congo based on the collection of the American Congo Expedition, 1909-1915. Part I. Turtles, crocodiles, lizards, and chameleons. Bull: American Mus. Nat. Hist., 39, 385-624.
SCHMIDT, K. P., No. 3, 1922. Scientific results of the expedition to the Gulf of California . . . "Albatross" in 1911. VIII. The amphibians and reptiles of Lower California. . . . Bull. American Mus. Nat. Hist., 46, 607-707.
SCHMIDT, K. P., No. 4, 1923. Contributions to the herpetology of the Belgian Congo based on the collec- of the American Museum Congo Expedition, 1909-1915. Part II. Snakes. Bull. American Mus. Nat. Hist., 49, 1-146.
SCHREIBER, E., No. 1, 1912. Herpetologia europaea. Eine systematische Bearbeitung der Amphibien und Reptilien. . ., 2nd ed. Jena; 1-960.
SIEBENROCK, F., No. 1, 1909. Synopsis der rezenten Schildkroten. . . . Zool. Jahrb., Suppl. 10, Heft 3, 427-618.
SMITH, H. M., No. 1, 1946. Handbook of lizards. Lizards of the United States and Canada. Ithaca, New York; 1-557. (Reprinted 1965)
SMITH, H. M., No. 2, 1956. Handbook of amphibians and reptiles of Kansas, 2nd ed. Misc. Publ. Univ.. Kansas Mus. Nat. Hist., 9, 1-356.
SMITH, M. A., No. 1, 1930. The Reptilia and Amphibia of lthe Malay Peninsula. . . . Bull. Raffles, Mus., 3, 1-149.
SMITH, M. A., No. 2, 1931. The fauna of British India, including Ceylon and Burma. Reptilia and Amphibia. Vol. 1. Loricata, Testudines. London; 1-213.
SMITH, M. A., No. 3, 1935. Reptilia and Amphibia. Vol. II. Sauria. Fauna of British India. London; 1-440.
SMITH, M. A., No. 4, 1943. Fauna of British India. Reptilia land Amphibia. III. Serpentes. London; 1-583.
STEBBINS, R. C., No. 2, 1954. Amphibians and reptiles of western North America. New York; 1-528.
STEJNEGER, L., No. 1, 1907. The herpetology of Japan and adjacent territory. United States Nat. Mus. Bull., 58, 1-577.
TAYLOR, E. H., No. 2, 1953. A review of the lizards of Ceylon. Kansas Univ. Sci. Bull., 35, 1525-1585.
TAYLOR, E. H., No. 4, 1951. A brief review of the snakes of Costa Rica. Univ. Kansas Sci. Bull., 34, 3-188.
TAYLOR, E. H., No. 5, 1965. The serpents of Thailand and adjacent waters. Univ. Kansas Sci. Bull., 45, 609-1096.
TAYLOR, E. H., No. 6, 1963. The lizards of Thailand. Univ. Kansas Sci. Bull., 44, 687-1077.
TAYLOR, E. H., No. 7, 1956. A review of the lizards of Costa Rica. Univ. Kansas Sci. Bull., 38, 1-322.
TAYLOR, E. H., No. 8, 1922. The lizards of the Philippine Islands. Manila; 1-269.
TAYLOR, E. H., No. 9, 1922. The snakes of the Philippine Islands. Manila; 1-312.
TAYLOR, E. H., No. 10, 1921. Amphibians and turtles of the Philippine Islands. Manila; 1-193.
TAYLOR, E. H., No. 11, 1970. The turtles and crocodiles of Thailand and adjacent areas with a synoptic herpetological bibliography. Univ. Kansas Sci. Bull., 49, 87-179.
VILLIERS, A., No. 1, 1958. Tortues et crocodiles de l'Afrique Noire Francaise. Init. Africaines, 15, 1-354.
WERMUTH. H., & MERTENS, R., No. 1, 1961. Schildkroten, Krokodile, Bruckenechsen. Jena; 1-422.
WRIGHT, A. H. & WRIGHT, A. A., No. 2, 1957. Handbook of snakes of the United States and Canada. Ithaca, New York; 2 volumes; 1-1106.

Families alphabetically

Acanthopholidae (only as fossils)
Acontiidae, as Scincidae
Acontiophidae, as Colubridae
Acrochordidae, as Colubridae
Aegyptosuchidae (only as fossils)
Aëtosauridae (only as fossils)
Agamidae (incl. Uromasticidae). WERMUTH, H., 1967, Das Tierreich, 86, 1-127 (list of Recent species); SCHREIBER, E., No. 1, 1912 (Europe); NIKOLSKY, A. M., No. 1, 1915 (U. S. S. R.); NIKOL'SKII, A. M., No. 4, 1963 (Russia); POPE, C. H., No. 1, 1935 (Central Asia); ANDERSON, J., No. 1, 1898 (Egypt); SCHMIDT, K. P., No. 2, 1917 (Belgian Congo); FitzSIMONS, V. F., No. 1, 1948 (South Africa); BOULENGER, G. A., No. 1, 1890 (British India); MERTENS, R., No. 2, 1969 (West Pakistan); DERANIYAGALA, P. E. P., No. 1, 1931 (Ceylon); TAYLOR, E. H., No. 2, 1953 (Ceylon); TAYLOR, E. H., No. 6, 1963 (Thailand); BOULENGER, G. A., No. 2, 1912 (Malay Peninsula); SMITH, M. S., No. 1, 1930 (Malay Peninsula); de ROOIJ, N., No. 2, 1915 (Indo-Australian Arch.); TAYLOR, E. H., No. 8, 1922 (Philippines); OKADA, Y., No. 2, 1935 (Jehol); STEJNEGER, L., No. 1, 1907 (Japan); OKADA, Y., 1937, Sci. Rep. Tokyo Bun Dais., 3B, 51, 83-94 (Japan)
Aigialosauridae (only as fossils)
Alligatoridae. WERNER, F., 1933, Reptilia, Loricata; Das Tierreich, 62, 1-40; WERMUTH, H., 1953, Mitt. Zool. Mus. Berlin, 29, 375-574; WERMUTH, H. & MERTENS, R., No. 1, 1961 (monograph); MERTENS, R., No. 1, 1952 (El Salvador); POPE, C. H., No. 1, 1935 (Central Asia)
Allosauridae (only as fossils)
Alopecopsidae (only as fossils)

Amblycephalidae. WERNER, F., 1922, Arch. Naturg., 88, (A8), 185-244 (synopsis); POPE, C. H., No. 1, 1935 (Central Asia); BOULENGER, G. A., No. 1, 1890 (British India); BOURRET, R., No. 3, 1936 (Indochina); BOULENGER, G. A., No. 2, 1912 (Malay Peninsula); SMITH, M. A. No. 1, 1930 (Malay Peninsula); de ROOIJ, N., No. 1, 1917 (Indo-Australian Arch.); TAYLOR, E. H., No. 9, 1922 (Philippines); MAKI, M., No. 1, 1931 (Japan)

Ammosauridae (only as fossils)

Amphisbaenidae (incl. Bipedidae, Chirotidae, Euchirotidae, Lepidosternidae). VANZOLINI, P. E., 1951, Herpetologia, 7, 113-123; (systematic arrangement); GANS, C., 1967, Bull. American Mus. Nat. Hist., 135, 63-105 (list of Recent); GANS, C., 1967; Bull. American Mus. Nat. Hist., 135, 61-105 (list of Recent); GANS, C., 1960; Bull. American Mus. Nat. Hist., 119, 129-204 (revision of Trogonophinae); COPE, E. D., No. 2, 1900 (North America, also as Euchirotidae) SMITH, H. M., No. 1, 1946 (North America); PETERS & DONOSO-BARROS, No. 2, 1970 (catalog & keys, neotropical); BARBOUR & RAMSDEN, No. 1, 1919 (Cuba); COCHRAN, D. M., No. 1, 1941 (Hispaniola); SCHMIDT, K. P. No. 1, 1928 (Puerto Rica); GUNTHER, A. C. L. G., No. 1, 1885-1902 (Central America); SCHMIDT, K. P., No. 3, 1922 (Baja California); DEVINCENZI, G. J., No. 1, 1925 (Uruguay); HELLMICH, W., No. 1, 1960 (Chaco); SCHREIBER, E., No. 1, 1912 (Europe); NIKOLSKY, A. M., No. 1, 1915 (U. S. S. R.); NIKOL'SKII, A. M., No. 4, 1963 (Russia); LOVERIDGE, A., 1941, Bull. Mus. Comp. Zool., 87, 353-451 (Africa); de WITTE, G. F. & LAURENT, R., 1942, Rev. Zool. Bot. Africaines, 36, 67-86 (Congo Belge); de WITTE, G. F., 1954, Vol. Jub. V. von Straelen, 2, 781-1010 (Congo Belge); FitzSIMONS, V. F., No. 1, 1948 (South Africa)

Anadiidae, as Teiidae

Anelytropidae (incl. Anelytropsidae). COPE, E. D., No. 2, 1900 (North America); GUNTHER, A. C. L. G., No. 1, 1885-1902 (Central America); SCHMIDT, K. P., No. 2, 1917 (Belgian Congo)

Anelytropsidae, as Anelytropidae

Angudidae, as Anguidae

Anguidae (incl. Angudidae, Anguididae, Gerrhonotidae). COPE, E. D., No. 2, 1900 (North America); SMITH, H. M., No. 1, 1946 (North America); SMITH, H. M., No. 2, 1956 (Kansas); STEBBINS, R. C., No. 2, 1954 (western North America); PETERS & DONOSO-BARROS, No. 2, 1970 (catalog & keys, neotropics); BARBOUR, T. & RAMSDEN, C. T., No. 1, 1919 (Cuba); COCHRAN, D. M., No. 1, 1941 (Hispaniola); GUNTHER, A. C. L. G., No. 1, 1885-1902 (Central America); SCHMIDT, K. P., No. 3, 1922 (Baja California); MERTENS, R., No. 1, 1952 (El Salvador); TAYLOR, E. H., No. 7, 1956 (Costa Rica); DEVINCENZI, G. J., No. 1, 1925 (Uruguay); HELLMICH, W., No. 1, 1960 (Chaco); SCHREIBER, E., No. 1, 1912 (Europe); ANGEL, F., No. 3, 1946 (France); FUHN & VANCEA, No. 2, 1962 (Romania); NIKOLSKY, A. M., No. 1, 1915 (U. S. S. R.); NIKOL'SKII, A. M., No. 4, 1963 (Russia); POPE, C. H., No. 1, 1935 (Central Asia); BOULENGER, G. A., No. 1, 1890 (British India); SMITH, M. A., No. 3, 1935 (India); de ROOIJ, N., No. 2, 1915 (Indo-Australian Archipelago)

Anguididae, as Anguidae

Anguinidae. TIHEN, J. A., 1949, American Midl. Nat., 41, 580-601 (genera)

Aniellidae, as Anniellidae

Anilidae (incl. Aniliidae, Ilysiidae, Loxocemidae, Rhinophidae, Tortricidae). PETERS & OREJAS-MIRANDA No. 1, 1970 (catalog & keys, Neotropics); ROZE, J. A., No. 1, 1966 (Venezuela); BOULENGER, G. A., No. 1, 1890 (British India, as Ilysiidae); SMITH, M. A., No. 4, 1943 (India); TAYLOR, E. H., No. 5, 1965 (Thailand); BOURRET, R., No. 3, 1936 (Indochina, as Ilysiidae); BOULENGER, G. A., No. 2, 1912 (Malay Peninsula); SMITH, M. A., No. 1, 1930 (Malay Peninsula); de ROOIJ, N., No. 1, 1917 (Indo-Australian Archipelago)

Aniliidae, as Anilidae

Anniellidae (incl. Aniellidae). COPE, E. D., No. 2, 1900 (North America); SMITH, H. M., No. 1, 1946 (North America); STEBBINS, R. C., No. 2, 1954 (western North America); SCHMIDT, K. P., No. 3, 1922 (Baja California)

Anolidae, as Iguanidae

Anomalepidae. TAYLOR, E. H., 1939, Proc. New England Zool. Club, 17, 87-96 (new family); TAYLOR, E. H., No. 4, 1951 (Costa Rica)

Anomalepididae. PETERS, & OREJAS-MIRANDA, No. 1, 1970 (catalog & keys, Neotropics)

Aprasiidae, as Pygopodidae

Araeoscelidae (only as fossils)

Ardeosuridae (only as fossils)

Argaliidae, as Teiidae

Arretosauridae (only as fossils)

Ascalabotidae, as Gekkonidae

Atoposauridae (only as fossils)

Atractaspididae, as Viperidae

Baënidae (only as fossils)

Bataguridae, as Testudinidae

Bauridae (only as fossils)

Bipedidae, as Amphisbaenidae

Boidae (incl. Charinidae, Erycidae, Ungualiidae). WERNER, F., 1921, Arch. Naturg., 87, (A7), 230-265 (synopsis); STULL, O. G., 1935, Proc. Boston Soc. Nat. Hist., 40, 387-408 (checklist); SINGH, L. et al., 1968 (chromosomes & classification); STIMSON, A. F., 1969, Das Tierreich, 89, 1-49 (synonymic catalog); COPE, E. D., No. 2, 1900 (North America, also as Charinidae); WRIGHT, & WRIGHT, No. 2, 1957 (North America); STEBBINS, R. C., No. 2,

1954 (western North America); PETERS & OREJAS-MIRANDA, No. 1, 1970 (catalog & keys, Neotropics); BARBOUR & RAMSDEN, No. 1, 1919 (Cuba); COCHRAN, D. M., No. 1, 1941 (Hispaniola); SCHMIDT, K. P., No. 1, 1928 (Puerto Rico); SCHMIDT, K. P., No. 3, 1922 (Baja California); MERTENS, R., No. 1, 1952 (El Salvador); TAYLOR, E. H., No. 4, 1951 (Costa Rica); ROZE, J. A., No. 1, 1966 (Venezuela); DEVINCENZI, G. J., No. 1, 1925 (Uruguay); SCHREIBER, E., No. 1, 1912 (Europe); BOULENGER, G. A., No. 3, The snakes of Europe, 1-269; FUHN & VANCEA, No. 2, 1962 (Romania); NIKOLSKY, A. M., No. 2, 1916 (U. S. S. R.); NIKOL'SKII, A. M., No. 5, 1963 (Russia); POPE, C. H., No. 1, 1935 (Central Asia); KRAMER & SCHNURRENBERGER, No. 1, 1963 (Libya); ANDERSON, J., No. 1, 1898 (Egypt); ANGEL, F., No. 2, 1933 (French West Africa); LAURENT, R. F., No. 1, 1956 (central Africa); SCHMIDT, K. P., No. 4, 1923 (Belgian Congo); GUIBE, J., 1949, Mem. Soc. Scient. Madagascar, 3A, 95-105 (Madagascar); GUIBE, J., 1958, Mem. Inst. Sci. Madagascar, 12A, 189-260 (Madagascar); HOFFSTETTER, R., 1946, Bull. Mus. Nat. Hist. Nat., (2), 18, 132-135 (Mascareignes); BOULENGER, G. A., No. 1, 1890 (British India); MERTENS, R., No. 2, 1969 (West Pakistan); SMITH, M. A., No. 4, 1943 (India); TAYLOR, E. H., No. 5, 1965 (Thailand); BOURRET, R., No. 3, 1936 (Indochina); BOULENGER, G. A., No. 2, 1912 (Malay Peninsula); SMITH, M. A., No. 1, 1930 (Malay Peninsula); de ROOIJ, N., No. 1, 1917 (Indo-Australian Arch.); WORREL, E., 1951, Proc. Roy. Zool. Soc. New South Wales, 1949-50, 20-25 (Australian); TAYLOR, E. H., No. 9, 1922 (Philippines)

Boigidae, as Colubridae
Bolosauridae (only as fossils)
Brachiosauridae (only as fossils)
Burnetiidae (only as fossils)

Calamaridae, as Colubridae
Calamariidae, as Colubridae
Camelionidae, as Chamaeleonidae
Caouanidae, as Cheloniidae
Captorhinidae (only as fossils)
Carettidae. ANGEL, F., No. 3, 1946 (France); DERANIYAGALA, P. E. P., 1933, Ceylon Journ. Sci., 18B, 61-72 (Ceylon); DERANIYAGALA, P. E. P., No. 3, 1939 (Ceylon)
Carettochelidae, Carettochelyidae
Carettochelydidae, as Carettochelyidae
Carettochelyidae (incl. Carettochelidae, Carettochelydidae). SIEBENROCK, F., No. 1, 1909 (World); WERMUTH & MERTENS, No. 1, 1961 (monograph); de ROOIJ, N., No. 2, 1915 (Indo-Australian Archipelago)
Caseidae (only as fossils)
Causidae, as Viperidae
Ceratopsidae (only as fossils)
Ceratosauridae (only as fossils)
Cercosauridae, as Teiidae
Cetiosauridae (only as fossils)
Chalcidae, as Scincidae
Chalcididae, as Scincidae
Chamadeontidae, as Chamaeleontidae
Chamaeleonidae, as Chamaeleontidae
Chamaeleontidae (incl. Chamadeontidae, Chamaeleonidae, Chamaelionidae, Chamaleonidae, Chameleontidae, Chamelionidae). WERNER, F., 1911, Das Tierreich, 27, 1-52; MERTENS, R., 1966, Das Tierreich, 83, 1-37 (Liste der rezenten....); BUSACK, S. D., 1969, Smithsonian Herpet. Inform. Serv., 17, 1-12 (bibliography 1864-1964); COPE, E. D., No. 2, 1900 (North America); SCHREIBER, E., No. 1, 1912 (Europe); NIKOLSKY, A. M., No. 1, 1915 (U. S. S. R.); NIKOL'SKII, A. M., No. 4, 1963 (Russia); ANDERSON, J., No. 1, 1898 (Egypt); SCHMIDT, K. P., No. 2, 1917 (Belgian Congo); FitzSIMONS, V. F., No. 1, 1948 (South Africa); ANGEL, F., No. 1, 1942 (Madagascar); BOULENGER, G. A., No. 1, 1890 (British India); MERTENS, R., No. 2, 1969 (West Pakistan); SMITH, M. A., No. 3, 1935 (India); DERANIYAGALA, P. E. P. No. 1, 1931 (Ceylon); TAYLOR, E. H., No. 3, 1953 (Ceylon)
Chamaelionidae, as Chamaeleontidae
Chamaesauridae, as Lacertidae
Chamalionidae, as Chamaeleontidae
Chameleontidae, as Chamaeleontidae
Chamelionidae, as Chamaeleontidae
Champsosauridae (only as fossils)
Charinidae, as Boidae
Chelidae (incl. Chelydidae, Chelyidae, Hydraspididae). SIEBENROCK, F., No. 1, 1909 (World); WERMUTH & MERTENS, No. 1, 1961 (monograph); de ROOIJ, N., No. 2, 1915 (Indo-Australian Arch.); MEHRTENS, J. M., 1967, Journ. Intern. Turtle Tort. Soc., 1, (6), 14-21 (Australasia); GOODE, J., 1967, Freshwater tortoises of Australia and New Guinea, Melbourne, 1-154 (biology & identification)
Chelonidae, as Cheloniidae
Cheloniidae (incl. Caouanidae, Chelonidae, Chelonioidae). SIEBENROCK, F., No. 1, 1909 (World); WERMUTH & MERTENS, No. 1, 1961 (monograph); STEBBINS, R. C., No. 2, 1954 (western North America); MERTENS, R., No. 1, 1952 (El Salvador); DONOSO-BARROS, R., No. 1, 1966 (Chile); SCHREIBER, E., No. 1, 1912 (Europe); DERANIYAGALA, P. E. P., 1952,

Herpetologica, 8, 57-58 (Carettinae of Europe); ANGEL, F., No. 3, 1946 (France); FUHN, I. E. & VANCEA, S., No. 2, 1962 (Romania); NIKOLSKY, A. M., No. 1, 1915 (U. S. S. R.); POPE, C. H., No. 1, 1935 (Central Asia); NIKOL'SKII, A. M., No. 4, 1963 (Russia); LOVERIDGE, A. & WILLIAMS, E. E., 1957, Bull. Mus. Comp. Zool., 115, 163-557 (African); VILLIERS, A., No. 1, 1958 (French West Africa); BOULENGER, G. A., No. 1, 1890 (British India); MERTENS, R., No. 2, 1969 (West Pakistan); SMITH, M. A., No. 2, 1931 (India); DERANIYAGALA, P. E. P., No. 3, 1939 (Ceylon); TAYLOR, E. H., No. 11, 1970 (Thailand); BOURRET, R., No. 1, 1941 (Indochina); BOULENGER, G. A., No. 2, 1912 (Malay Peninsula); SMITH, M. A., No. 1, 1930 (Malay Peninsula); de ROOIJ, N., No. 2, 1915 (Indo-Australian Arch.); TAYLOR, E. H., No. 10 1921 (Philippines); STEJNEGER, L., No. 1, 1907 (Japan)

Chelonioidae, as Cheloniidae

Chelydidae, as Chelidae

Chelydridae (incl. Chelydroidae, Cinosternidae, Staurotypidae). SIEBENROCK, F., No. 1, 1909 (World, also as Cinosternidae); WERMUTH, H. & MERTENS, R., No. 1, 1961 (monograph); CAHN, A., 1937 (Illinois); SMITH, H. M., No. 2, 1956 (Kansas); STEBBINS, R. C., No. 2, 1954 (western North America); de ROOIJ, N., No. 2, 1915 (Indo-Australian Archipelago)

Chelydroidae, as Chelydridae

Chelyidae, as Chelidae

Chelyridae. (Family not identified)

Chiniquodontidae (only as fossils)

Chirocolidae, as Teiidae

Chirotidae, as Amphisbaenidae

Chitridae, as Trionychidae

Cinosternidae, as Chelydridae

Cistudinidae, as Emydidae

Claraziidae (only as fossils)

Clemmydoidae, as Testudinidae

Cobridae, as Viperidae

Coeluridae (only as fossils)

Colubridae (incl. Acontiophidae, Acrochordidae, Boigidae, Calamaridae, Calamariidae, Coronellidae, Dendrophidae, Dipsadidae, Dipsadomorphidae, Dipsididae, Dryadidae, Homalopsidae, Lycodontidae, Oligodontidae, Potamophilidae, Probletorhinidae, Psammophidae, Rachiodontidae). WERNER, F., 1923, Arch. Naturg., 89, (A8), 138-199 (Übersicht der Gattungen und Arten); WERNER, F., 1925, Arch. Naturg., 90, (A12), 108-166 (No. II); KLAUBER, L. M., 1956, Rattlesnakes, Berkeley, California, 1-1476 (monograph, identification, glossary); PETERS, J. A., 1965, Das Tierreich, 81, 1-18 (list of Recent); GYI, K. K., 1970, Publ. Univ. Kansas Mus. Nat. Hist., 20, 47-723 (revision of Homalopsinae); WERNER, F., 1929, Zool. Jahrb., Syst., 57, 1-196 (No. III); PETERS, J. A., 1960, Misc. Publ. Univ. Michigan Mus. Zool., 114, 1-224 (Dipsodinae); PETERS, J. A., 1965, Das Tierreich, 81, 1-19 (subfamily Dipsadinae); SMITH, M. A., 1939, Ann. Mag. Nat. Hist., (11), 3, 393-395 (Acrochordinae); TAYLOR, E. H. & SMITH, H. M., 1943, Univ. Kansas Sci. Bull., 29, 301-336 (Sibynophine); DUNN, E. R., 1928, Bull. Antevenin Inst. America, 2, 18-24 (classification); COPE, E. D., No. 2, 1900 (North America, also as Dipsadidae); WRIGHT, & WRIGHT, No. 2, 1957 (United States & Canada); SMITH, H. M., No. 2, 1956 (Kansas); STEBBINS, R. C., No. 2, 1954 (western North America); PETERS & OREJAS-MIRANDA, B., No. 1, 1970 (catalog, keys, Neotropics); MAGLIO, V. J., 1970, Bull. Mus. Comp. Zool., 141, 1-53 (West Indies); COCHRAN, D. M., No. 1, 1941 (Hispaniola); SCHMIDT, K. P., No. 1, 1928 (Puerto Rico); GUNTHER, A. C. L. G., No. 1, 1885-1902 (Central America); SCHMIDT, K. P., No. 3, 1922 (Baja California); MERTENS, R., No. 1, 1952 (El Salvador); TAYLOR, E. H., No. 4, 1951 (Costa Rica); ROZE, J. A., No. 1, 1966 (Venezuela); DEVINCENZI, G. F., No. 1, 1925 (Uruguay); DONOSO-BARROS, R., No. 1, 1966 (Chile); SCHREIBER, E., No. 1, 1912 (Europe); BOULENGER, G. A., 1913, Snakes of Europe, London, 1-269; ANGEL, F., No. 3, 1946 (France); FUHN & VANCEA, No. 2, 1962 (Romania); POPE, C. H., No. 1, 1935 (Central Asia); BOGERT, C. M., 1940, Bull. American Mus. Nat. Hist., 77, 1-107 (classification, Africa); KRAMER & SCHNURRENBERGER, No. 1, 1963 (Libya); ANDERSON, J., No. 1, 1898 (Egypt); ANGEL, F., No. 2, 1933 (French West Africa); LAURENT, R. F., No. 1, 1956 (central Africa); SCHMIDT, K. P., No. 4, 1923 (Belgian Congo); de WITTE, G. R. & LAURENT, R., 1943, Rev. Zool. Bot. Africaines, 37, 157-189 (Boiginae, Congo Belge); FitzSIMONS, V. F. M., No. 2, 1962 (southern Africa); FitzSIMONS, F. W., The snakes of South Africa, Cape Town, 1-547; GUIBE, J., 1958, Mem. Inst. Sci. Madagascar, 12A, 189-260 (Madagascar); BOULENGER, G. A., No. 1, 1890 (British India); MERTENS, R., No. 2, 1969 (West Pakistan); SMITH, M. A., No. 4, 1943 (India); TAYLOR, E. H., No. 5, 1965 (Thailand); TAYLOR, E. H., No. 5, 1965 (Thailand, as Dipsadidae); BOURRET, R., No. 3, 1936 (Indochina); SMITH, M. A., No. 1, 1930 (Malay Peninsula); BOULENGER, G. A., No. 2, 1912 (Malay Peninsula); de ROOIJ, N., No. 1, 1917 (Indo-Australian Arch.); OKADA, Y., No. 2, 1935 (Jehol); MAKI, M., No. 1, 1931 (Japan)

Compsognathidae (only as fossils)

Cordylidae. WERMUTH, H., 1968, Das Tierreich, 87, 1-30 (list of Recent); LOVERIDGE, A., 1944, Bull. Mus. Comp. Zool., 95, 1-118 (Africa); FitzSIMONS, V. F., No. 1, 1948 (South Africa)

Coronellidae, as Colubridae

Crocodilidae, as Crocodylidae

Crocodylidae (incl. Crocodilidae). WERNER, F., 1933, Das Tierreich, 62, 1-40; WERMUTH, H., 1953, Mitt. Zool. Mus. Berlin, 29, 375-574 (Systematik der rezenten); WERMUTH & MERTENS,

No. 1, 1961 (monograph); COPE, E. D., No. 2, 1900 (North America); BARBOUR & RAMSDEN, No. 1, 1919 (Cuba); COCHRAN, D. M., No. 1, 1941 (Hispaniola); MERTENS, R., No. 1, 1952 (El Salvador); ANDERSON, J., No. 1, 1898 (Egypt); VILLIERS, A., No. 1, 1958 (French West Africa); SCHMIDT, K. P., No. 2, 1917 (Belgian Congo); BOULENGER, G. A., No. 1, 1890 (British India); MERTENS, R., No. 2, 1969 (West Pakistan); SMITH, M. A., No. 2, 1931 (India); DERANIYAGALA, P. E. P., No. 3, 1939 (Ceylon); TAYLOR, E. H., No. 11, 1970 (Thailand); BOULENGER, G. A., No. 2, 1912 (Malay Peninsula); SMITH, M. A., No. 1, 1930 (Malay Peninsula); de ROOIJ, N., No. 2, 1915 (Indo-Australian Archipelago)

Crotalidae. KLAUBER, L. M., 1936, Trans. San Diego Soc. Nat. Hist., 8, 185-276 (key to rattlesnakes); RUSSELL, F. E., 1969, Herpeton, 4, 1-8 (checklist, Western Hemisphere); COPE, E. D., No. 2, 1900 (North America); WRIGHT & WRIGHT, No. 2, 1957 (United States & Canada); SMITH, H. M., No. 2, 1956 (Kansas); STEBBINS, R. C., No. 2, 1954 (western North America); PETERS & OREJAS-MIRANDA, No. 1, 1970 (catalog, keys, Neotropics); TERRON, C. C., 1921, Mem. Rev. Soc. Cient. Antonio Alzate, 39, 173-194 (Mexico); KLAUBER, L. M., 1952, Bull. Zool. Soc. San Diego, 26, 1-143 (Crotalinae of Mexico); NIKOLSKY, A. M., No. 2, 1916 (U. S. S. R.); NIKOL'SKII, A. M., No. 5, 1963 (Russia); POPE, C. H., No. 1, 1935 (Central Asia); BOULENGER, G. A., No. 1, 1890 British India, as Crotalinae in Viperidae); MERTENS, R., No. 2, 1969 (West Pakistan); TAYLOR, E. H., No. 5, 1965 (Thailand); TAYLOR, E. H., No. 9, 1922 (Philippines); OKADA, Y., No. 2, 1935 (Jehol); STEJNEGER, L., No. 1, 1907 (Japan); MAKI, M., No. 1, 1931 (Japan)

Cyamodontidae (only as fossils)

Cyclanorbidae. DERANIYAGALA, P. E. P., No. 3, 1939 (Ceylon)

Cynognathidae (only as fossils)

Cynosauridae (only as fossils)

Dasypeltidae. SMITH, M. A., No. 4, 1943 (India)

Deinodontidae (only as fossils)

Deirochelyoidae, as Emydidae

Dendraspididae, as Elapidae

Dendrophidae, as Colubridae

Dermatemydidae (incl. Dermatemyidae). SIEBENROCK, F., No. 1, 1909 (World); WERMUTH & MERTENS, No. 1, 1961 (monograph)

Dermatemyidae, as Dermatemydidae

Dermatochelydae, as Dermochelidae

Dermochelidae (Dermatochelydae, Dermochelididae, Dermochelydidae, Dermochelyidae, Dermochelydae, Sphargidae). SIEBENROCK, F., No. 1, 1909 (World); WERMUTH & MERTENS, No. 1, 1961 (monograph); STEBBINS, R. C., No. 2, 1954 (western North America); DONOSO-BARROS, R., No. 1, 1966 (Chile); ANGEL, F., No. 3, 1946 (France); POPE, C. H., No. 1, 1935 (Central Asia); LOVERIDGE, A. & WILLIAMS, E. E., 1957, Bull. Mus. Comp. Zool., 115, 163-557 (Africa); VILLIERS, A., No. 1, 1958 (French West Africa); BOULENGER, G. A., No. 1, 1890 (British India, as Sphargidae); MERTENS, R., No. 2, 1969 (West Pakistan, as Dermochelyidae); SMITH, M. A., No. 2, 1931 (India, as Sphargidae); DERANIYAGALA, P. E. P., No. 3, 1939 (Ceylon); TAYLOR, E. H., No. 11, 1970 (Thailand, as Dermochelydidae); BOURRET, R., No. 1, 1941 (Indochina, as Sphargidae); BOULENGER, G. A., No. 2, 1912 (Malay Peninsula, as Sphargidae); SMITH, M. A., No. 1, 1930 (Malay Peninsula); de ROOIJ, N., No. 2, 1915 (Indo-Australian Arch.); TAYLOR, E. H., No. 10, 1921 (Philippines); STEJNEGER, L., No. 1, 1907 (Japan)

Dermochelididae, as Dermochelidae

Dermochelydae, as Dermochelidae

Dermochelydidae, as Dermochelidae

Dermochelyidae, as Dermochelidae

Desmognathidae. SOLER, E. I., 1950, Univ. Kansas Sci. Bull., 33, 459-480 (status of family)

Deuterosauridae (only as fossils)

Diadectidae (only as fossils)

Diademodontidae (only as fossils)

Dibamidae. BOULENGER, G. A., No. 1, 1890 (British India); SMITH, M. A., No. 3, 1935 (India); TAYLOR, E. H., No. 6, 1963 (Thailand); BOULENGER, G. A., No. 2, 1912 (Malay Peninsula); SMITH, M. A., No. 1, 1930 (Malay Peninsula); de ROOIJ, N., No. 2, 1915 (Indo-Australian Arch.); TAYLOR, E. H., No. 8, 1922 (Philippines)

Dicynodontidae (only as fossils)

Dimorphodontidae (only as fossils)

Dipsadidae, as Colubridae

Dipsadomorphidae, as Colubridae

Dipsididae, as Colubridae

Dolichosauridae (only as fossils)

Dromatheriidae (only as fossils)

Dryadidae, as Colubridae

Dryiophidae, as Iguanidae

Dryophilidae, as Iguanidae

Dyrosauridae (only as fossils)

Ecpleopidae, as Teiidae

Edaphosauridae (only as fossils)

Elapidae (incl. Dendraspididae, Elapsidae, Micruridae, Najidae). HOFFSTETTER, R., 1939, Arch. Mus. Hist. Nat., Lyon, 15, 1-78 (Recent & fossil); ROZE, J. A., 1967, American Mus. Novit., 2287, 1-60 (New World venomous); COPE, E. D., No. 2, 1900 (North America); WRIGHT & WRIGHT No. 2, 1957 (United States & Canada); STEBBINS, R. C., No. 2, 1954 (western North America); PETERS & OREJAS-MIRANDA, No. 1, 1970 (Catalog & keys, Neotropics); MERTENS, R., No. 1, 1952 (El Salvador); TAYLOR, E. H., No. 4, 1951 (Costa Rica); BURGER, W. L., 1955, Bol. Mus. Cienc. Nat. Venezuela, 1, 1-18 (key); ROZE, J. A., 1955, Acta Biol. Venezolana, 1, 453-500 (revision, Venezuela); ROZE, J. A., No. 1, 1966 (Venezuela); DONOSO-BARROS, No. 1, 1966 (Chile); NIKOLSKY, A. M., No. 2, 1916 (U. S. S. R.); NIKOL'SKII, A. M., No. 5, 1963 (Russia); POPE, C. H., No. 1, 1935 (Central Asia); KRAMER & SCHNURRENBERGER, H., No. 1, 1963 (Libya); LAURENT, R. F., No. 1, 1956 (central Africa); FitzSIMONS, V. F. M., No. 2, 1962 (southern Africa); MERTENS, R., No. 2, 1969 (West Pakistan); SMITH, M. A., No. 4, 1943 (India); TAYLOR, E. H., No. 5, 1965 (Thailand); TAYLOR, E. H., No. 9, 1922 (Philippines); STEJNEGER, L., No. 1, 1907 (Japan); MAKI, M., No. 1, 1931 (Japan); McDOWELL, S. B., 1970, Journ. Zool. Soc. London, 161, 145-190 (Solomon Islands)

Elapsidae, as Elapidae

Elasmosauridae (only as fossils)

Emydidae (incl. Cistudinidae, Deirochelyoidae, Emydinadae, Emydoidae, Emyidae, Pseudemydae). WERMUTH & MERTENS, No. 1, 1961 (monograph); SMITH, H. M., No. 2, 1956 (Kansas); SCHMIDT, K. P., No. 1, 1928 (Puerto Rico); MERTENS, R., No. 1, 1952 (El Salvador); FUHN & VANCEA, No. 2, 1962 (Romania); VILLIERS, A., No. 1, 1958 (French West Africa); MERTENS, R., No. 2, 1969 (West Pakistan); SMITH, M. A., No. 2, 1931 (India); DERANIYA-GALA, P. E. P., No. 3, 1939 (Ceylon); TAYLOR, E. H., No. 11, 1970 (Thailand); BOURRET, R., No. 1, 1941 (Indochina)

Emydinadae, as Emydidae

Emydoidae, as Emydidae

Emyidae, as Emydidae

Endothiodontidae (only as fossils)

Eothyrididae (only as fossils)

Erycidae, as Boidae

Erythrosuchidae (only as fossils)

Eublepharidae. WERNER, F., 1912, Das Tierreich, 33, 1-33; WELLBORN, V., 1933, Sitz. Ber. Ges. Naturf. Fr. Berlin, 1933, 126-199 (osteology); COPE, E. D., No. 2, 1900 (North America); SCHMIDT, K. P., No. 3, 1922 (Baja California); TAYLOR, E. H., No. 7, 1956 (Costa Rica); NIKOLSKY, A. M., No. 1, 1915 (U. S. S. R.); NIKOL'SKII, A. M., No. 4, 1963 (Russia); BOULENGER, G. A., No. 1, 1890 (British India); TAYLOR, E. H., No. 6, 1963 (Thailand)

Euchirotidae, as Amphisbaenidae

Eunotosauridae (only as fossils)

Euposauridae (only as fossils)

Evemydoidae. (Family not identified)

Feyliniidae. (No revisionary references noted)

Galeopsidae (only as fossils)

Galesauridae (only as fossils)

Gavialidae. WERNER, F., 1933, Das Tierreich, 62, 1-40; WERMUTH, H., 1953. Mitt. Zool. Mus. Berlin, 29, 375-574 (Systematik der rezenten); WERMUTH & MERTENS, No. 1, 1961 (monograph); MERTENS, R., No. 2, 1969 (West Pakistan)

Gecconidae, as Gekkonidae

Geckonidae, as Gekkonidae

Geckotidae, as Gekkonidae

Gekkonidae (incl. Ascalabotidae, Gecconidae, Geckonidae, Geckotidae, Sphaerodactylidae). WELLBORN, V., 1933, Sitz. Ber. Ges. Naturf. Fr. Berlin, 1933, 126-199 (osteology); UNDERWOOD, G., 1954, Proc. Zool. Soc. London, 124, 469-472 (classification & evolution); WERMUTH, H., 1965, Das Tierreich, 80, 1-246 (Liste der rezenten); KLUGE, A. G., 1967, Bull. American Mus. Nat. Hist., 135, 1-59 (classification); COPE, E. D., No. 2, 1900 (North America); SMITH, H. M., No. 1, 1946 (North America); STEBBINS, R. C., No. 2, 1954 (western North America); PETERS & DONOSO-BARROS, No. 2, 1970 (catalog & keys, Neotropics); BARBOUR & RAMSDEN, No. 1, 1919 (Cuba); UNDERWOOD, G., 1951, Nat. Hist. Notes Nat. Hist. Soc. Jamaica, 46, 209-213 (Jamaica); COCHRAN, D. M., No. 1, 1941 (Hispaniola); SCHMIDT, K. P., No. 1, 1928 (Puerto Rico); GUNTHER, A. C. L. G., No. 1, 1885-1902 (Central America); SCHMIDT, K. P., No. 3, 1922 (Baja California); MERTENS, R., No. 1, 1952 (El Salvador); TAYLOR, E. H., No. 7, 1956 (Costa Rica); TAYLOR, E. H., No. 7, 1956 (Costa Rica) VANZOLINI, P. E., 1968, Arch. Zool. Est. Sao Paulo, 17, 85-111 (distributioln in South America); MECHLER, B., 1968, Rev. Suisse Zool., 75, 305-371 (Colombia); VANZOLINI, P. E., 1968, Arch. Zool. Est. Sao Paulo, 17, 1-84 (Brasil); DEVINCENZI, G. J., No. 1, 1925 (Uruguay); HELLMICH, W., No. 1, 1960 (Chaco); DONOSO-BARROS, R., No. 1, 1966 (Chile); SCHREIBER, E., No. 1, 1912 (Europe); ANGEL, F., No. 3, 1946 (France); NIKOLSKY, A. M., No. 1, 1915 (U. S. S. R.); NIKOL'SKII, A. M., No. 4, 1963 (Russia); POPE, C. H., No. 1, 1935 (central Asia); LOVERIDGE, A., 1947, Bull. Mus. Comp. Zool., 98, 1-469 (Africa); ANDERSON, J., No. 1, 1898 (Egypt); BOOTH, A. H., 1956, Journ. West African Sci. Assoc., 2, 134-136 (Gold Coast); THYS VAN DEN AUDENAERDE, D. F. E., 1967, Rev. Zool. Bot. Africaines, 76, 163-172 (central Africa); SCHMIDT, K. P., No. 2, 1917 (Belgian Congo);

FitzSIMONS, V. F., No. 1, 1948 (South Africa); ANGEL, F., No. 1, 1942 (Madagascar); BOULENGER, G. A., No. 1, 1890 (British India); MERTENS, R., No. 2, 1969 (West Pakistan); SMITH, M. A., No. 3, 1935 (India); DERANIYAGALA, P. E. P., 1932, Ceylon Journ. Sci., 16, 291-310 (Ceylon); TAYLOR, E. H., No. 2, 1953 (Ceylon); TAYLOR, E. H., No. 6, 1963 (Thailand); BOULENGER, G. A., No. 2, 1912 (Malay Peninsula); SMITH, M. A., No. 1, 1930 (Malay Peninsula); de ROOIJ, N., No. 2, 1915 (Indo-Australian Arch.); GLAUERT, L., 1956, Western Australian Nat., 5, 49-56 (Western Australia?); McCANN, C., 1955, Dominion Mus. Bull., 17, 1-127 (New Zealand); TAYLOR, E. H., No. 8, 1922 (Philippines); ROMER, J. D., 1950, Copeia, 1950, 54-55 (Hong Kong); OKADA, Y., No. 2, 1935 (Jehol); STEJNEGER, L., No. 1, 1907 (Japan); OKADA, Y., 1936 (Japan); McCANN, C., 1953, Proc. 7th Pacific Sci. Congr., 4, 17-38 (Pacific)

Gerrhonotidae, as Anguidae

Gerrhosauridae. LOVERIDGE, A., 1942, Bull. Mus. Comp. Zool., 89, 483-543 (Africa); SCHMIDT, K. P., No. 2, 1917 (Belgian Congo); FitzSIMONS, V. F., No. 1, 1948 (South Africa); ANGEL, F., No. 1, 1942 (Madagascar)

Glauconiidae. WERNER, F., 1917, Jahr. Hamburg Wiss. Ant., 2 Beih., 34, 191-208 (synopsis); COPE, E. D. No. 2, 1900 (North America); ANDERSON, J., No. 1, 1898 (Egypt); BOULENGER, G. A., No. 1, 1890 (British India); BOURRET, R., No. 3, 1936 (Indochina)

Goniopholidae (only as fossils)

Gorgonopsidae (only as fossils)

Gryponychidae (only as fossils)

Gymnophthalmidae, as Teiidae

Hadrosauridae (incl. Trachodontidae). (Only as fossils)

Hatteriidae, as Sphenodontidae

Helodermatidae (incl. Helodermidae). WERMUTH, H., 1958, Das Tierreich, 72, 1-16; MERTENS, R., 1963, Das Tierreich, 79, 1-26 (Liste der rezenten); COPE, E. D., No. 2, 1900 (North America) SMITH, H. M., No. 1, 1946 (North America); STEBBINS, R. C., No. 2, 1954 (western North America); PETERS & DONOSO-BARROS, No. 2, 1970 (catalog & keys, Neotropics); de ROOIJ, N., No. 2, 1915 (Indo-Australian Archipelago)

Helodermidae, as Helodermatidae

Holaspidae, as Lacertidae

Homalopsidae, as Colubridae

Hydraspididae, as Chelidae

Hydridae. MAKI, M., No. 1, 1931 (Japan)

Hydrophidae (incl. Hydrophiidae). SMITH, M. A., 1926, Monograph of the sea-snakes (Hydrophiidae), London, 1-130; WRIGHT & WRIGHT, No. 2, 1957 (North America); PETERS & OREJAS-MIRANDA, No. 1, 1970 (catalog, keys, Neotropics);. TAYLOR, E. H., No. 4, 1951 (Costa Rica); FitzSIMONS, V. F. M., No. 2, 1962 (southern Africa); VOLSØE, H., 1939, Danish Sci. Invest. Iran, I, 1-45 (Iranian Gulf & Gulf of Oman); POPE, C. H., No. 1, 1935 (Central Asia); MERTENS, R., No. 2, 1969 (West Pakistan); SMITH, M. A., No. 4, 1943 (India); TAYLOR, E. H., No. 5, 1965 (Thailand); BOURRET, R., 1935, Inst. Oceanogr. Indochine, 1-69 (Indochine francaise); SMITH, M. A., 1920, Journ. Federated Malay States Mus., 10, 1-63 (Malay Peninsula, Siam, Cochin China); SMITH, M. A., No. 1, 1930 (Malay Peninsula)

Hydrophiidae, as Hydrophidae

Hylaeochampsidae (only as fossils)

Hypsilophodontidae (only as fossils)

Ichthyosauridae (only as fossils)

Iguanidae (incl. Anolidae, Dryiophidae, Dryophilidae). COPE, E. D., No. 2, 1900 (North America); SMITH, H. M., No. 1, 1946 (North America); SMITH, H. M., No. 2, 1956 (Kansas); STEBBINS, R. C., No. 2, 1954 (western North America); PETERS & DONOSO-BARROS, No. 2, 1970 (catalog & keys, Neotropics); BARBOUR & RAMSDEN, No. 1, 1919 (Cuba); UNDERWOOD, G., 1951, Nat. Hist. Notes Nat. Hist. Soc. Jamaica, 46, 209-213 (Jamaica); COCHRAN, D. M., No. 2, 1941 (Hispaniola); SCHMIDT, K. P., No. 1, 1928 (Puerto Rico); GUNTHER, A. C. L. G., No. 1, 1885-1902 (Central America); SCHMIDT, K. P., No. 3, 1922 (Baja California); MERTENS, R., No. 1, 1952 (El Salvador); TAYLOR, E. H., No. 7, 1956 (Costa Rica); DEVINCENZI, G. J., No. 1, 1925 (Uruguay); HELLMICH, W., No. 1, 1960 (Chaco); DONOSO-BARROS R., No. 1, 1966 (Chile); ANGEL, F., No. 1, 1942 (Madagascar)

Iguanodontidae (only as fossils)

Ilysiidae, as Anilidae

Kailokibotidae (only as fossils)

Kinosternidae. WERMUTH & MERTENS, No. 1, 1961 (Monograph); CAHN, A., 1937, Illinois Biol. Monogr., 16, 1-208 (Illinois); SMITH, H. M., No. 2, 1956 (Kansas); MERTENS, R., No. 1, 1952 (El Salvador)

Lacertidae (incl. Chamaesauridae, Holaspidae, Lacertinidae). BOULENGER, G. A., 1920-20, Monograph of the Lacertidae, London, 2 volumes; SMITH, H. M., No. 1, 1946 (North America); SCHREIBER, E., No. 1, 1912 (Europe); ANGEL, F., No. 3, 1946 (France); FUHN, I. E. & VANCEA, S., No. 2, 1962 (Romania); NIKOLSKY, A. M., No. 1, 1915 (U. S. S. R.); NIKOL'SKII, A. M., No. 4, 1963 (Russia); POPE, C. H., No. 1, 1935 (central Asia); ANDERSON, J., 1898, (Egypt); SCHMIDT, K. P., No. 2, 1917 (Belgian Congo); de WITTE, G. F. & LAURENT, R., 1942, Rev. Zool. Bot. Africaines, 36, 165-180 (Congo Belge); FitzSIMONS, V. F., No. 1, 1948

(South Africa); LOVERIDGE, A., 1959, Proc. Zool. Soc. London, 133, 29-44 (East African genera); BOULENGER, G. A., No. 1, 1890 (British India); MERTENS, R., No. 2, 1969 (West Pakistan); SMITH, M. A., No. 3, 1935 (India); DERANIYAGALA, P. E. P., No. 1, 1931 (Ceylon); TAYLOR, E. H., No. 2, 1953 (Ceylon); TAYLOR, E. H., No. 6, 1963 (Thailand); BOULENGER, G. A., No. 2, 1912 (Malay Peninsula); SMITH, M. A., No. 1, 1930 (Malay Peninsula); de ROOIJ, N., No. 2, 1915 (Indo-Australian Arch.); OKADA, Y., No. 2, 1935 (Jehol); STEJNEGER, L., No. 1, 1907 (Japan)

Lacertinidae, as Lacertidae

Lanthanotidae. MERTENS, R., 1963, Das Tierreich, 79, 1-26 (Liste der rezenten)

Lepidophymidae. GUNTHER, A. C. L. G., No. 1, 1885-1902 (Central America)

Lepidosternidae, as Amphisbaenidae

Leptotyphlopidae (incl. Stenostomatidae, Stenostomidae). LIST, J. C., 1966, Illinois Biol. Monogr., 36, 1-112 (comparative osteology); McDOWELL, S. B., 1967, Copeia, 1967, 686-692 (osteology); WRIGHT & WRIGHT, No. 2, 1957 (North America): SMITH, H. M., No. 2, 1956 (Kansas); STEBBINS, R. C., No. 2, 1954 (western North America); PETERS & OREJAS-MIRANDA, No. 1, 1970 (catalog & keys, Neotropics); GUNTHER, A. C. L. G., No. 1, 1885-1902 (Central America, as Stenostomatidae); SCHMIDT, K. P., No. 3, 1922 (Baja California); MERTENS, R., No. 1, 1952 (El Salvador); TAYLOR, E. H., No. 4, 1951 (Costa Rica); DUNN, E. R., 1944, Caldasia, 11, 47-55 (Colombia); ROZE, J. A., No. 1, 1966 (Venezuela); KRAMER & SCHNURRENBERGER, No. 1, 1963 (Libya); ANGEL, F., No. 2, 1933 (French West Africa); LAURENT, R. F., No. 1, 1956 (central Africa); SCHMIDT, K. P., No. 4, 1923 (Belgian Congo); FitzSIMONS, V. F. M., No. 2, 1962 (southern Africa); MERTENS, R., No. 2, 1969 (West Pakistan); SMITH, M. A., No. 4, 1943 (India)

Lialisidae, as Pygopodidae

Limnoscelidae (only as fossils)

Loxocemidae, as Aniliidae

Lupeosauridae (only as fossils)

Lycodontidae, as Colubridae

Lystrosauridae (only as fossils)

Malaclemmydae, as Testudinidae

Megalanidae. FEJERVARY, G. J., 1918, Ann. Hist. Nat. Mus. Hungarici, 16, 341-467 (monograph); also Ann., 29, 1-130.

Megalosauridae (only as fossils)

Meiolanidae (only as fossils)

Melanorosauridae (only as fossils)

Mesenosauridae (only as fossils)

Mesosauridae (only as fossils))

Metriorhynchidae (only as fossils)

Microcleptidae (only as fossils)

Micruridae, as Elapidae

Millerettidae (only as fossils)

Mixosauridae (only as fossils)

Monitoridae, as Varanidae

Mosasauridae (only as fossils)

Najidae, as Elapidae

Natricidae. BARBOUR, & RAMSDEN, No. 1, 1919 (Cuba); NIKOLSKY, A. M., No. 2, 1916 (U. S. S. R.); NIKOL'SKII, A. M., No. 5, 1963 (Russia); TAYLOR, E. H., No. 9, 1922 (Philippines); STEJNEGER, L., No. 1, 1907 (Japan)

Nectemydoidae, as Testudinidae

Nitosauridae (only as fossils)

Nodosauridae (only as fossils)

Nothosauridae (only as fossils)

Notochampsidae (only as fossils)

Notosuchidae (only as fossils)

Oligodontidae, as Colubridae

Omphalosauridae (only as fossils)

Ophiacodontidae (only as fossils)

Ophiomoridae, as Scincidae

Ophiosepidae, as Pygopodidae

Ornithocheiridae (only as fossils)

Ornithomimidae (only as fossils)

Ornithosuchidae (only as fossils)

Pachypleurosauridae (only as fossils)

Palaeophidae (only as fossils)

Palaeosauridae (only as fossils)

Pareiasauridae (only as fossils)

Pelomedusidae (incl. Podocnemididae, Sternothaeridae). SIEBENROCK, F., No. 1, 1909 (World); WERMUTH & MERTENS, No. 1, 1961 (monograph); LOVERIDGE, A., 1941, Bull. Mus. Comp. Zool., 88, 465-524 (Africa); VILLIERS, A., No. 1, 1958 (French West Africa); LAURENT, R. F., No. 1, 1956 (central Africa); SCHMIDT, K. P., No. 2, 1917 (Belgian Congo)

Pholidosauridae (only as fossils)
Phytosauridae (only as fossils)
Pistosauridae (only as fossils)
Placodontidae (only as fossils)
Plateosauravidae (only as fossils)
Plateosauridae (only as fossils)
Platysternidae. SIEBENROCK, F., No. 1, 1909 (World); WERMUTH & MERTENS, No. 1, 1961 (monograph); POPE, C. H., No. 1, 1935 (central Asia); BOULENGER, G. A., No. 1, 1890 (British India); SMITH, M. A., No. 2, 1931 (India); BOURRET, R., No. 1, 1941 (Indochina); TAYLOR, E. H., No. 11, 1970 (Thailand)
Plesiochelidae (only as fossils)
Plesiosauridae (only as fossils)
Pleurosauridae (only as fossils)
Pleurosternidae (only as fossils)
Pliosauridae (only as fossils)
Podocnemididae, as Pelomedusidae
Podokesauridae (only as fossils)
Polycotylidae (only as fossils)
Polyglyphanodontidae (only as fossils)
Potamophilidae, as Colubridae
Pristerognathidae (only as fossils)
Probletorhinidae, as Colubridae
Procolophonidae (only as fossils)
Procompsognathidae (only as fossils)
Procynosuchidae (only as fossils)
Prolacertidae (only as fossils)
Proterochersidae (only as fossils)
Protoceratopsidae (only as fossils)
Protorosauridae (only as fossils)
Protorothyrididae (only as fossils)
Protostegidae (only as fossils)
Psammophidae, as Colubridae
Pseudemydae, as Emydidae
Psittacosauridae (only as fossils)
Peranodontidae (only as fossils)
Pterodactylidae (only as fossils)
Pygopidae, as Pygopodidae
Pygopodidae (incl. Aprasiidae, Lialisidae, Ophiosepidae, Pygopidae). WERNER, F., 1912, Das Tierreich, 33, 1-33; KINGHORN, J. R., 1926, Rec. Australian Mus., 15, 40-64 (brief review); UNDERWOOD, G., 1957, Journ. Morph., 100, 207-268; WERMUTH, H., 1965, Das Tierreich, 80, 1-246 (Liste der rezenten); de ROOIJ, N., No. 2, 1915 (Indo-Australian Arch.)
Pythonidae. FitzSIMONS, V. F. M., No. 2, 1962 (southern Africa)

Rachiodontidae, as Colubridae
Rhamphorhynchidae (only as fossils)
Rhinophidae, as Anilidae
Rhizodontidae (only as fossils)
Rhodonidae, as Scincidae
Rhopalodontidae (only as fossils)
Rhynchosauridae (only as fossils)
Rhynchosauroidae (only as fossils)
Riamidae, as Teiidae
Rynchocephalidae, as Sphenodontidae

Sapheosauridae (only as fossils)
Scaloposauridae (only as fossils)
Scelidosauridae (only as fossils)
Scincidae (incl. Acontiidae, Chalcidae, Chalcididae, Ophiomoridae, Rhodonidae, Sepidae, Sepsidae, Sincidae). GREER, A. E., 1970, Bull. Mus. Comp. Zool., 139, 151-183 (subfamily classification); MITTLEMAN, M. B., 1952, Smithsonian Misc. Coll., 117, (17), 1-35 (generic synopsis); COPE, E. D., No. 2, 1900 (North America); SMITH, H. M., No. 1, 1946 (North America); SMITH, H. M., No. 2, 1956 (Kansas); STEBBINS, R. C., No. 2, 1954 (western North America); PETERS & DONOSO-BARROS, No. 2, 1970 (catalog & keys, Neotropics); COCHRAN, D. M., No. 1, 1941 (Hispaniola); SCHMIDT, K. P., No. 1, 1928 (Puerto Rico); GUNTHER, A. C. L. G., No. 1, 1885-1902 (Central America); SCHMIDT, K. P., No. 3, 1922 (Baja California); MERTENS, R., No. 1, 1952 (El Salvador); TAYLOR, E. H., No. 7, 1956 (Costa Rica); DEVINCENZI, G. J., No. 1, 1925 (Uruguay); HELLMICH, W., No. 1, 1960 (Chaco); DONOSO-BARROS No. 1, 1966 (Chile); SCHREIBER, E., No. 1, 1912 (Europe); ANGEL, F., No. 3, 1946 (France); FUHN & VANCEA, No. 2, 1962 (Romania); NIKOLSKY, A. M., No. 1, 1915 (U. S. S. R.); NIKOL'SKII, A. M., No. 4, 1963 (Russia); POPE, C. H., No. 1, 1935 (central Asia); GREER, 1971, Bull. Mus. Comp. Zool., 140, 1-23 (Scincinae, Africa, Seychelles, Mauritius); ANDERSON, J., No. 1, 1898 (Egypt); SCHMIDT, K. P., No. 2, 1917 (Belgian Congo); FitzSIMONS, V. F., No. 1, 1948 (South Africa); ANGEL, F., No. 1, 1942 (Madagascar);

BOULENGER, G. A., No. 1, 1890 (British India); MERTENS, R., No. 2, 1969 (West Pakistan); SMITH, M. A., No. 3, 1935 (India); DERANIYAGALA, P. E. P., No. 1, 1931 (Ceylon); TAYLOR E. H., 1950, Univ. Kansas Sci. Bull., 33, 481-518 (Ceylon); TAYLOR, E. H., No. 2, 1953 (Ceylon); TAYLOR, E. H., No. 6, 1963 (Thailand); BOULENGER, G. A., No. 2, 1912 (Malay Peninsula); SMITH, M. A., No. 1, 1930 (Malay Peninsula); de ROOIJ, M., No. 2, 1915 (Indo-Australian Arch.); McCANN, C., 1955, Dom. Mus. Bull., 17, 127 (New Zealand); TAYLOR, E. H., No. 8, 1922 (Philippines); OKADA, Y., No. 2, 1935 (Jehol); STEJNEGER, L., No. 1, 1907 (Japan); OKADA, Y., 1939, Sci. Rep. Tokyo Bun Daig, 4B, 159-214 (Japan)

Scytalidae, as Viperidae
Sebecidae (only as fossils)
Segisauridae (only as fossils)
Sepidae, as Scincidae
Sepsidae, as Scincidae
Shastasauridae (only as fossils)
Shinisauridae. POPE, C. H., No. 1, 1935 (central Asia)
Sincidae, as Scincidae
Solenodonsauridae (only as fossils)
Sphaerodactylidae, as Gekkonidae
Sphargidae, as Dermochelidae
Sphargididae, ass Dermochelidae
Sphenacodontidae (only as fossils)
Sphenodontidae (incl. Hatteriidae, Rynchocephalidae). WERMUTH & MERTENS, No. 1, 1961 (monograph); SHARELL, R., 1966, The tuatara, lizards and frogs of New Zealand, London, 1-94.
Sphenosuchidae (only as fossils)
Spinosauridae (only as fossils)
Stagonolepidae (only as fossils)
Staurotypidae, as Chelydridae
Stegosauridae (only as fossils)
Stellionidae, as Varanidae
Stenopterygidae (only as fossils)
Stenostomatidae, as Leptotyphlopidae
Stenostomidae, as Leptotyphlopidae
Sternothaeridae, as Pelomedusidae
Stomatosuchidae (only as fossils)

Tangasauridae (only as fossils)
Tanystropheidae (only as fossils)
Tapinocephalidae (only as fossils)
Teidae, as Teiidae
Teiidae (incl. Anadiidae, Argaliidae, Cercosauridae, Chirocolidae, Ecpleopidae, Gymnophthalmidae, Riamidae, Teidae, Tejidae, Tupinambidae). COPE, E. D., No. 2, 1900 (North America); SMITH, H. M., No. 1, 1946 (North America); SMITH, H. M., No. 2, 1956 (Kansas); STEBBINS, R. C., No. 2, 1954 (western North America); PETERS & DONOSO-BARROS, No. 2, 1970 (catalog & keys, Neotropics); BARBOUR & RAMSDEN, No. 1, 1919 (Cuba); COCHRAN, D. M., No. 1, 1941 (Hispaniola); SCHMIDT, K. P., No. 1, 1928 (Puerto Rico); GUNTHER, A. C. L. G., No. 1, 1885-1892 (Central America); SCHMIDT, K. P., No. 3, 1922 (Baja California); MERTENS, R., No. 1, 1952 (El Salvador); TAYLOR, E. H., No. 7, 1956 (Costa Rica); RUIBAL, R., 1952, Bull. Mus. Comp. Zool., 106, 477-529 (South American); DEVINCENZI, G. J., No. 1, 1925 (Uruguay); HELLMICH, W., No. 1, 1960 (Chaco); DONOSO-BARROS, R., No. 1, 1966 (Chile)
Tejidae, as Teiidae
Teleosauridae (only as fossils)
Teratosauridae (only as fossils)
Testudinidae (incl. Bataguridae, Clemmydoidae, Malademmydae, Nectemydoidae). SIEBENROCK, F., No. 1, 1909 (World); WERMUTH & MERTENS, No. 1, 1961 (monograph); McDOWELL, S. B., 1964, Proc. Zool. Soc. London, 143, 239-279 (taxonomic problems, aquatic); CAHN, A., 1937, Illinois Biol. Monogr., 16, 1-218 (Illinois); STEBBINS, R. C., No. 2, 1954 (western North America); BARBOUR & RAMSDEN, No. 1, 1919 (Cuba); COCHRAN, D. M., No. 2, 1941 (Hispaniola); GUNTHER, A. C. L. G., No. 1, 1885-1902 (Central America); SCHREIBER, E., No. 1, 1912 (Europe); ANGEL, F., No. 3, 1946 (France); FUHN & VANCEA, No. 2, 1962 (Romania); NIKOLSKY, A. M., No. 1, 1915 (U. S. S. R.); NIKOL'SKII, A. M., No. 4, 1963 (Russia); POPE, C. H., No. 1, 1935 (central Asia); LOVERIDGE, A. & WILLIAMS, E. E., 1957, Bull. Mus. Comp. Zool., 115, 163-557 (African); ANDERSON, J., No. 1, 1898 (Egypt); VILLIERS, A., No. 1, 1958 (French West Africa); LAURENT, R. F., No. 1, 1956 (central Africa); SCHMIDT, K. P., No. 1, 1917 (Belgian Congo); BOULENGER, G. A., No. 1, 1890 (British India); MERTENS, R., No. 2, 1969 (West Pakistan); SMITH, M. A., No. 2, 1931 (India); DERANIYAGALA, P. E. P., No. 3, 1939 (Ceylon); TAYLOR, E. H., No. 11, 1970 (Thailand); BOURRET, R., No. 1, 1941 (Indochina); BOULENGER, G. A., No. 2, 1912 (Malay Peninsula); SMITH, M. A., No. 1, 1930 (Malay Peninsula); de ROOIJ, N., No. 2, 1915 (Indo-Australian Arch.); TAYLOR, E. H., No. 10, 1921 (Philippines); STEJNEGER, L., No. 1, 1907 (Japan)
Thalassemydidae (only as fossils)
Thalattosauridae (only as fossils)

Thecodontosauridae (only as fossils)
Titanosuchidae (only as fossils)
Tortricidae, as Anilidaeidae
Toxochelidae (only as fossils)
Trachodontidae, as Hadrosauridae
Traversodontidae (only as fossils)
Triassochelidae (only as fossils)
Trilophosauridae (only as fossils)
Trionicidae, as Trionychidae
Trionychidae (incl. Chitridae, Trionicidae). SIEBENROCK, F., No. 1, 1909 (World); WERMUTH & MERTENS, No. 1, 1961 (monograph); WEBB, R. G., 1962, Publ. Univ. Kansas Mus. Nat. Hist., 13, 429-611 (North America); CAHN, A., 1937, Illinois Biol. Monogr., 16, 1-218 (Illinois); SMITH, H. M., No. 2, 1956 (Kansas); STEBBINS, R. C., No. 2, 1954 (western North America); NIKOLSKY, A. M., No. 1, 1915 (U. S. S. R.); NIKOL'SKII, A. M., No. 4, 1963 (Russia); POPE, C. H., No. 1, 1935 (central Asia); LOVERIDGE, A. & WILLIAMS, E. E.,, 1957, Bull. Mus. Comp. Zool., 115, 163-557 (Africa); ANDERSON, J., No. 1, 1898 (Egypt); VILLIERS, A., No. 1, 1958 (French West Africa); BOULENGER, G. A., No. 1, 1890 (British India); MERTENS, R., No. 2, 1969 (West Pakistan); SMITH, M. A., No. 2, 1931 (India); TAYLOR, E. H., No. 11, 1970 (Thailand); BOURRET, R., No. 1, 1941 (Indochina); BOULENGER, G. A., No. 2, 1912 (Malay Peninsula); SMITH, M. A., No. 1, 1930 (Malay Peninsula); de ROOIJ, N., No. 2, 1915 (Indo-Australian Arch.); TAYLOR, E. H., No. 10, 1921 (Philippines); OKADA, Y., No. 2, 1935 (Jehol); STEJNEGER, L., No. 1, 1907 (Japan)
Trithelodontidae (only as fossils)
Tritylodontidae (only as fossils)
Trogonophidae. GANS, C., 1967, Bull. American Mus. Nat. Hist., 135, 61-105 (checklist of Recent)
Troödontidae (only as fossils)
Tupinambidae, as Teiidae)
Typhlinidae, as Typhlopidae
Typhlopidae (incl. Typhlinidae, Typhlopsidae). WERNER, F., 1921, Arch. Naturg., 87, (A7), 266-330 (synopsis); LIST, J. C., 1966, Illinois Biol. Monogr., 36, 1-112 (osteology); McDOWELL, S. B., 1967, Copeia, 1967, 686-692 (osteology); PETERS & OREJAS-MIRANDA, No. 1, 1970 (catalog & keys, Neotropics); BARBOUR & RAMSDEN, No. 1, 1919 (Cuba); COCHRAN, D. M., No. 1, 1941 (Hispaniola); SCHMIDT, K. P., No. 1, 1928 (Puerto Rico); GUNTHER, A. C. L. G., No. 1, 1885-1902 (Central America); DUNN, E. R., 1944, Caldasia, 11, 47-55 (Colombia); ROZE, J. A., No. 1, 1966 (Venezuela); DEVINCENZI, G. J., No. 1, 1925 (Uruguay); SCHREIBER, E., No. 1, 1912 (Europe); BOULENGER, G. A., 1913, Snakes of Europe, London, 1-269; NIKOLSKY, A. M., No. 2, 1916 (U. S. S. R.); NIKOL'SKII, A. M., No. 5, 1963 (Russia); POPE, C. H., No. 1, 1935 (central Asia); ANGEL, F., No. 2, 1933 (French WEst Africa); LAURENT, R. F., No. 1, 1956 (central Africa); SCHMIDT, K. P., No. 4, 1923 (Belgian Congo); FitzSIMONS, V. F. M., No. 2, 1962 (southern Africa); GUIBE, J., 1958, Mem. Inst. Sci. Magagascar, 12A, 189-260. (Madagascar); BOULENGER, G. A, No. 1, 1890 (British India); MERTENS, R., No. 2, 1969 (West Pakistan); SMITH, M. A., No. 4, 1943 (India); TAYLOR, E. H. No. 5, 1965 (Thailand); BOURRET, R., No. 3, 1936 (Indochina); BOULENGER, G. A., No. 2, 1912 (Malay Peninsula); SMITH, M. A., No. 1, 1930 (Malay Peninsula); de ROOIJ, N., No. 1, 1917 (Indo-Australian Arch.); WAITE, E. R., 1918, Rec. South Australian Mus., 1, 1-34, 1918); (Australian); TAYLOR, E. H., No. 9, 1922 (Philippines); STEJNEGER, L., No. 1, 1907 (Japan); MAKI, M., No. 1, 1931 (Japan)
Typhlopsidae, as Typhlopidae

Uaranidae, as Varanidae
Ungualiidae, as Boidae
Uromasticidae, as Agamidae
Uropeltidae. GANS, C., 1966, Das Tierreich, 84, 1-29 (Liste der rezenten); BOULENGER, G. A., No. 1, 1890 (British India); SMITH, M. A., No. 4, 1943 (India); BOURRET, R., No. 3, 1936 (Indochina)
Uroplatidae. WERNER, F., 1912, Das Tierreich, 33, 1-33; WELLBORN, V., No. 1933 (osteology)

Varanidae (incl. Monitoridae, Stellionidae, Uaranidae). MERTENS, R., 1942, Abh. Senckenbergische Naturf. Ges., 462, 1-115, 119-233, 237-391; MERTENS, R., 1963, Das Tierreich, 79, 1-26 (Liste der rezenten); NIKOLSKY, A. M., No. 1, 1915 (U. S. S. R.); NIKOL'SKII, A. M., No. 4, 1963 (Russia); POPE, C. H., No. 1, 1935 (central Asia); ANDERSON, J., No. 1, 1898 (Egypt); SCHMIDT, K. P., No. 2, 1917 (Belgian Congo); FitzSIMONS, V. F., No. 1, 1948 (South Africa); BOULENGER, G. A., No. 1, 1890 (British India); MERTENS, R., No. 2, 1969 (West Pakistan); SMITH, M. A., No. 3, 1935 (India); DERANIYAGALA, P. E. P., No. 1, 1931 (Ceylon); TAYLOR, E. H., No. 2, 1953 (Ceylon); TAYLOR, E. H., No. 6, 1963 (Thailand); BOULENGER, G. A., No. 2, 1912 (Malay Peninsula); SMITH, M. A., No. 1, 1930 (Malay Peninsula); de ROOIJ, N., No. 2, 1915 (Indo-Australian Arch.); TAYLOR, E. H., No. 8, 1922 (Philippines)
Varanopsidae (only as fossils)
Venjukoviidae (only as fossils)

Viperidae (incl. Atractaspididae, Causidae, Cobridae, Scytalidae). WERNER, F., 1922, Arch. Naturg., 88, (A8), 185-244 (synopsis); PETERS & OREJAS-MIRANDA, No. 1, 1970 (catalog & keys, Neotropics); SCHMIDT, K. P., No. 3, 1922 (Baja California); MERTENS, R., No. 1, 1952 (El Salvador); TAYLOR, E. H., No. 4, 1951 (Costa Rica); ROZE, J. A., No. 1, 1966 (Venezuela); DEVINCENZI, G. J., No. 1, 1925 (Uruguay); SCHREIBER, E., No. 1, 1912 (Europe); BOULENGER, G. A., 1913, The snakes of Europe, 1-269; SCHWARZ, E., 1936, Behringwerkmitt. Marburg-Lahn, 7, 1-362 (Europe & Mediterranean); ANGEL, F., No. 3, 1946 (France); NIKOLSKY, A. M., No. 2, 1916 (U. S. S. R., as Cobridae); NIKOL'SKII, A. M., No. 5, 1963 (Russia, as Cobridae); POPE, C. H., No. 1, 1935 (central Asia); ANDERSON, J., No. 1, 1898 (Egypt); KRAMER & SCHNURRENBERGER, No. 1, 1963 (Libya); ANGEL, F., No. 2, 1933 (French West Africa); LAURENT, R. F., No. 1, 1956 (central Africa); SCHMIDT, K. P., No. 4, 1923 (Belgian Congo); FitzSIMONS, V. F. M., No. 2, 1962 (southern Africa); FitzSIMONS, F. W., 1912, The snakes of South Africa, 2nd ed., Cape Town, 1-547; BOULENGER, G. A., No. 1, 1890 (British India); MERTENS, R., No. 2, 1969 (West Pakistan); SMITH, M. A., No. 4, 1943 (India); TAYLOR, E. H., No. 5, 1965 (Thailand); BOURRET, R., No. 3, 1936 (Indochina); BOULENGER, G. A., No. 2, 1912 (Malay Peninsula); SMITH, M. A., No. 1, 1930 (Malay Peninsula); de ROOIJ, N., No. 1, 1917 (Indo-Australian Archipelago); STEJNEGER, L., No. 1, 1907 (Japan, as Cobridae); MAKI, M., No. 1, 1931 (Japan)

Weigeltisauridae (only as fossils)
Whaitsiidae (only as fossils)

Xantusidae, as Xantusiidae
Xantusiidae (incl. Xantusidae). WERMUTH, H., 1965, Das Tierreich, 80, 1-246 (Liste der rezenten); COPE, E. D., No. 2, 1900 (North America); SMITH, H. M., No. 1, 1946 (North America); STEBBINS, R. C., No. 2, 1954 (western North America); PETERS & DONOSO-BARROS, No. 2, 1970 (catalog & keys, Neotropics); BARBOUR & RAMSDEN, No. 1, 1919 (Cuba); SCHMIDT, K. P., No. 3, 1922 (Baja California); MERTENS, R., No. 1, 1952 (El Salvador); TAYLOR, E. H., No. 7, 1956 (Costa Rica)
Xenopeltidae. POPE, C. H., No. 1, 1935 (central Asia); BOULENGER, G. A., No. 1, 1890 (British India); SMITH, M. A., No. 4, 1943 (India); TAYLOR, E. H., No. 5, 1965 (Thailand); BOURRET, R., No. 3, 1936 (Indochina); BOULENGER, G. A., No. 2, 1912 (Malay Peninsula); SMITH, M. A., No. 1, 1930 (Malay Peninsula); de ROOIJ, N., No. 1, 1917 (Indo-Australian Archipelago); TAYLOR, E. H., No. 9, 1922 (Philippines)
Xenosauridae. PETERS & DONOSO-BARROS, No. 2, 1970 (catalog & keys, Neotropics)

Younginidae (only as fossils)

Zonuridae. (No revisionary references noted)

AVES

GENERAL BOOKS

ALDEN, P. 1969. Finding the birds in western Mexico.... Tucson, Arizona; 1-138.
ALI, S. & FUTEHALLY, L. 1967. Common birds. New Delhi; 1-113.
ALI, S. & FUTEHALLY, L. 1968. Common birds, a picture album. New Delhi; 1-51, 101 col. pl.
ALLEN, A. A. 1961. The book of bird life. A study of birds in their native haunts. Princeton, New Jersey; 1-396.
ARRIGONI degli ODDI, E. 1929. Ornitologia Italiana. Milano; 1-1046.
BEDDARD, F. E. 1898. The structure and classification of birds. London; 1-548.
BENT, A. C. 1919-1953. Life histories of North America birds. United States Nat. Mus. Bull., 1919-53. (Reprinted 1962-1968, by Dover)
BERNDT, R. & MEISE, W. 1961-62. Naturgeschichte der Vogel, Band 2, Lief. 22-28. Stuttgart.
DEMENTIEV, G. P. 1940. Textbook of zoology, vol. 6, Vertebrates, birds. Moskva; 1-856. (In Russian)
DEMENTIEV, G. P. 1962. The birds of this country, 2nd ed. Moskva; 1-168. (In Russian)
FISHER, J. & PETERSON, R. T. 1964. The world of birds, London; 1-288.
GEORGE, J. C. & BERGER, A. J. 1966. Avian myology. New York; 1-500.
GILLIARD, E. T. 1958. Living birds of the world. London; 1-400.
GOULD, J., 1967. Birds of Australia. London; 1-321.
GRASSE, P.-P. et al. 1950. Oiseaux. Traite de Zool., Zool., 15, 1-1164.
GROEBBELS, F. 1932. Der Vogel. Bau, Funktion, Lebenserscheinung, Einpassung, vol. 1, Berlin; 1-918.
GROSSMAN, M. L. & HAMLET, J. 1964. Birds of prey of the world. New York, 1-496.
GROSVENOR, G. & WETMORE, A. 1937. The book of birds. Washington, D. C.; 1-356.
HANZAK, J., 1967. The pictorial encyclopedia of birds. New York; 1-582. (Transl. from Czech)
HAUSMAN, L. A. 1944. The illustrated encyclopedia of American birds. New York; 1-541.
HVASS, H. 1961. Alverdens fugle. København; 1-214.
HVASS, H. 1963. Birds of the world. London; 1-210. (Transl. of 1961; also New York)
LAND, H. C. 1970. Birds of Guatemala. Wynnewood, Pennsylvania; 1-381.
LUCAS, A. M. & JAMROZ, C. 1961. Atlas of avian hematology. Agric. Monogr., United States Dep. Agric., 25, 1-271.
MARSHALL, A. J. et al. 1960-61. Biology and comparative physiology of birds. New York; 2 volumes.
MOREAU, R. E. 1966. The bird faunas of Africa and its islands. London; 1-424.
MUNRO, J. A. 1931. An introduction to bird study in British Columbia. Victoria, British Columbia; 1-99.
OLIVER, W. R. B. 1930. New Zealand birds. Wellington; 1-541.
PATTEN, C. J., 1928. The story of birds. A guide to the study of avian structure and habits. Sheffied; 1-478.
PERRIER, E. 1931. Les oiseaux. Traite de Zool., 9, 3117-3341.
PETTINGILL, O. S. 1962. A guide to bird finding east of the Mississippi. New York; 1-659.
PETTINGILL, O. S., Jr. 1953. A guide to bird finding west of the Mississippi. New York; 1-709.
PETTINGILL, O. S., Jr. 1970. Ornithology in laboratory and field, 4th ed. Minneapolis, Minnesota; 1-524.
PYCRAFT, W. P. 1934. Birds of Great Britain and their natural History. London; 1-206.
SCHONWETTER, M. 1962. Handbuch der Oologie, Lief. 6. Berlin; 321-384.
SKUTCH, A. F. 1954. Life histories of Central American birds. Pacific Coast Avifauna, 31.
STEAD, E. F. 1932. The life histories of New Zealand birds. London; 1-162.
STRESEMANN, E. 1927-34. Handbuch der Zoology, 7 (2), 1-899.
STRESEMANN, E. & STRESEMANN, V. 1966. Die Mauser der Vogel. Journ. Ornith., 107, Sonderheft, 1-445. (Molting)
STRESEMANN, E. et al. 1967. Atlas der Verbreitung palaearktischer Vogel. Second installment. Berlin; 16 maps.
STURKIE, P. D. 1965. Avian physiology, 2nd ed. Ithaca, New York; 1-766.
THOMSON, J. A. 1923. The biology of birds. London & New York; 1-436.
VAN TYNE, J. & BERGER, A. J. 1959. Fundamentals of ornithology. New York; 1-624.
VOOUS, K. H. 1960. Atlas of European birds. London; 1-284.
WALLACE, G. J. 1955. Introduction to ornithology. New York; 1-443.
WELTY, J. C. 1962. The life of birds. Philadelphia; 1-546.
WETMORE, A. 1931. Birds. Smithsonian Scientific Series, 9, (1), 1-166.
WILKINSON, E. S. Shanghai birds. Shanghai; 1-243. 1929.
YAMASHINA, Y. 1933-34. A natural history of Japanese birds, vol. 1. Tokyo; 1-524.
ZEMSKII, W. A. 1960. The animal kingdom of the Antarctic. Mammals and birds. Moskva; 1-180.

BIBLIOGRAPHIC WORKS

Bibliographies, separate works

[See also: Under Animalia above; also Zoological Record, in each volume, under heading Bibliographies; also below under each family]

(anonymous). Checklist of the Mathews ornithological collection of the National Library of Australia. Canberra; 1-309. 1966.
(anonymous). A list of books on ornithology contained in the Carnegie Library of Pittsburgh. Sewickley, Pennsylvania. 1929.
ALLOUSE, B. E. A bibliography of the vertebrate fauna of Iraq . . . , II. Birds. Publ. Iraq Nat. Hist. Mus., 5, 1954. (Includes Iran, Jordan, Palestine, Sinai, Syria, Lebanon, Turkey)
BURNS, F. L. A bibliography of scarce or out of print North American amateur and trade periodicals devoted more or less to ornithology. Oologist, 32, Suppl., 1-32. 1915.
COCKRUM, E. L. A check-list and bibliography of hybrid birds in North America north of Mexico. Wilson Bull., 64, 140-159. 1952.
COUES, E. Bibliography of ornithology, 4 parts. Bull. United States Geol. Geogr. Surv. Terr., 5, 521-1072. 1878-1880.
DEADERICK, W. H. An annotated bibliography of Arkansas ornithology. American Midl. Nat., 24, 490-496. 1940.
DEIGNAN, H. G. A bibliography of Thai ornithology, 1758-1939. Nat. Hist. Bull. Thailand Res. Soc., 13, 51-78. 1942.
FISHER, H. I. Bibliography of Hawaiian birds since 1890. Auk, 64, 78-97. 1947.
GAGINA, T. N. On history of study of ornithological fauna of Eastern Siberia. Izvest. Irkutsk sel'-skokhoz. Inst., 18, 259-299. 1960. (In Russian)
GALLATIN, F., Jr. A catalogue of a collection of books on ornithology in the library of Frederic Gallatin, Jr. New York. 1908.
GIEBEL, C. G. Thesaurus ornithologiae. Leipzig; 3 volumes. 1872-77.
GRINNELL, J. A bibliography of California ornithology. Pacific Coast Avifauna, 5, 16, 26. 1909-1939.
HACHISUKA, M. Bibliography of Chinese birds. Quart. Journ. Taiwan Mus., 5, 71-209. 1952.
HALLORAN, P. O. A bibliography of references to diseases of wild mammals and birds. American Journ. Veter. Res., 16, 1-465. 1955.
HOOGERSWERF, A. An ornithological bibliography having particular reference to the study of the birds of Java. Bull. Org. Sci. Res. Indonesia, 13-16, 1-168. 1953.
HOYS, S. F. A bibliography of New York state ornithology. Kingbird, 13, 90-95. 1963.
IRWIN, R. British bird books. An index to British ornithology, A. D. 1481-1948. London; 1-398. 1951.
KARAŠEK, J. Bibliographae ornithologicae republicae Bohemo-Slovenicae conspectus. Sborn. Klubu prir Brno, 11, 61-128. 1929.
LEGENDRE, M. Bibliographie des faunes ornithologique du nord de la France. Bull. Soc. Linn. Nord, 23, 42-48. 1927.
LEGENDRE, M. Bibliographie des faunes ornithologique des regions francaises. Rev. Francais Ornith., 19, 60-71, 153-161. 1927.
LEGENDRE, M. Bibliographie des faunes ornithologiques des regions francaises. Mem. Soc. Ornith. Mamm. France, 4, 1-127. 1936.
LOVELL, H. B. & SLACK, M. Bibliography of Kentucky ornithology. Occ. Pap. Kentucky Ornith. Soc., 1. 1949.
LOW, G. C. The literature of the Charadriiformes from 1894 to 1928, 2nd ed. London; 1-637. 1931. (1st edition, 1924)
MARELLI, C. A. Contribuciones al estudio de la fauna Argentina. Bibliographia relative a la ornitologia. Mem. Jardin Zool., 5, 37-106. 1934.
MATHEWS, G. M. Bibliography of the birds of Australia. London; 1-149. 1925.
MOBES, W. K. G. Bibliographie der Tauben. Halle; 1-268. 1945.
MULLENS, W. H. & SWANN, H. K. A bibliography of British ornithology from the earliest times to the end of 1912 . . . . London; 1-691. 1917.
MULLENS, W. H. et al. A geographical bibliography of British ornithology . . . to the end of 1918. London; 1-558. 1919-20.
NISSEN, C. Die illustrierten Vogelbucher. Stuttgart; 1-223. 1953.
OLENDORFF, R. R. & OLENDORFF, S. E. An extensive bibliography of falconry, eagles . . . , 3 parts. Fort Collins, Colorado; 1-244. 1969-70.
OLIVER, H. C. Annotated index to some early New Zealand bird literature. Wellington; 1-222. 1968.
OSBORN, H. Bibliography of Ohio zoology. Bull. Ohio Biol. Surv., 23. 1930.
PETTINGILL, O. S., Jr. Bibliography of life history studies. In Ornithology . . . . New York; 458-472. 1970.
PORTER, C. E. Bibliografia ornitolojica de Chile. Santiago. 1912.
REICHENOW, A. Bibliographia ornithologiae aethiopicae. Journ. Ornith., 42, 172-226. 1894. (Africa)
RILEY, J. H. & RICHMOND, C. W. Partial bibliography of Chinese birds. Journ. China R. Assoc. Soc. Shanghai, 53, 196-237. 1922.
RIPLEY, S. D. & SCRIBNER, L. L. Ornithological books in the Yale University Library . . . . New Haven, Connecticut; 1-338. 1961.
ROBERTS, B. A bibliography of Antarctic ornithology. Sci. Rep. British Grahamsland Exped., 1, 337-367. 1941.

ROHL, E. Apuntes para la historia y al bibliografica de la ornitologia Venezolana. Bol. Soc. Venezolana Cienc. Nat., 6, 201-248. 1932.
RONSIL, E. Bibliographie ornithologique francaise...1473-1944. Paris, 2 volumes. 1948-49. (French and Latin works published in France and its colonies)
SCHAANNING, H. T. L. Bibliotheca ornithologia Norvegica 1591-1924. Norsk Ornith. Tidsskr., (2), 5, 57-131. 1925.
SCHALOW, H. Beitrage zu einer ornithologischen Bibliographie des Atlas Gebietes. Journ. Ornith., 54, 100-143. 1906.
SCHLEGEL, R. Ornis taxonica. Ein Beitrag zur Bibliographie des Gebietes. Journ. Ornith., 73, 247-255. 1925.
SHAVER, J. M. A bibliography of Tennessee ornithology. Journ. Tennessee Acad. Sci., 6, 179-190; 1931.
STEPHENS, T. C. An annotated bibliography of South Dakota ornithology. Sioux City, Iowa. 1945.
STEPHENS, T. C. An annotated bibliography of North Dakota ornithology. Occ. Pap. Nebraska Orn. Union, 1, 1956.
STRONG, R. M. A bibliography of birds. Field Mus. Publ. Zool., 25, I, 1-464, 1939; II, 465-937, 1939; III, 1-528, 1946; IV, 1-186, 1950. (Largely non-taxonomic)
THAYER, E. & KEYES, V. Catalogue of a collection of books on ornithology in the library of John E. Thayer. Boston. 1913.
TOMIALOJC, L. Polisk ornithological bibliography. 1945-1960. Acta Ornith. Warszawa, 9, 1-76; 1965. (In Polish)
URBAN, E. K. Bibliography of the avifauna of Ethiopia. Addis Ababa; 1-28. 1970.
VAURIE, C. Russian ornithological literature. Auk, 81, 238-241. 1964.
WHITTELL, H. M. The literature of Australian birds: A history and bibliography .... Perth, 1-788;1954.
ZIMMER, J. T. Catalogue of the Edward E. Ayer ornithological library. I, II. Field Mus. Nat. Hist., Zool., 16, 1-706. 1926.

Bibliographies in other works

[See also: References in most large works]

CHAPIN, J. P. Birds of Belgian Congo. Part 4. Bull. American Mus. Nat. Hist., 75B, 739-809. 1954.
CHAPMAN, F. M. Handbook of birds of eastern North America. New York; 1-581. 1966. (Pp. 547-566; arranged by states)
GROSSMAN, M. L. & HAMLET, J. Birds of prey of the world. New York; 1-496. 1968.
HELLMAYR, C. E. The birds of Chile. Field Mus. Nat. Hist., Zool., 19, 1-472. 1932. (Pp. 429-458)
MURPHY, R. C. Oceanic birds of South America. New York; 2 volumes. 1936. (Pp. 1179-1210)
PETTINGILL, O. S., Jr. Ornithology.... Minneapolis, Minnesota; 1-524. 1970. (Pp. 455-497)
REICHENOW, A. Die Vogel Afrikas. Erster Band. Neudamm, 1-706. 1901. (Pp. xxxvii-lxxix)
TAKA-TSUKASA, N. The birds of Nippon, vol. 1, part 6. Bibliography. 1932.
VAN TYNE, J. & BERGER, A. J. Fundamentals of ornithology. New York; 1-624. 1959. (Pp. 553-558) (General works only)
WETMORE, A. & SWALES, B. H. The birds of Haiti and the Dominican Republic. United States Nat. Mus. Bull., 155, 1-483. 1931.

Personal bibliographies

[See under heading Biographies (Persons & Institutions) below]

Institutional bibliographies

[See under Animalia, above]

Periodical lists

[See also under Animalia and Vertebrata]

(anonymous). Biological serials, exclusive of botany, in the libraries of Philadelphia. Bull. Wistar Inst. 2. 1909.
(anonymous). Check list of periodicals and serials in biology and allied sciences available in libraries of the University of Minnesota and vicinity. Minneapolis, Minnesota. 1925.
PETTINGILL, O. S., Jr. Ornithology .... Minneapolis, Minnesota; 1-524. (Pp. 498-503)
STRONG, R. M. Bibliography of birds. Field Mus. Nat. Hist., Publ., Zool., 25, 1-937. (Pp. 17-82) 1939-1959. (Listed by abbreviation)

PERSONS AND INSTITUTIONS

Biographies & Personal bibliographies

[See also: Zoological Record, under heading Obituaries in each volume]

HUME, E. E. Ornithologists of the U. S. Army Medical Corps. Baltimore; 1-583. 1942.
PALMER, R. S. et al. Biographies of members of the American Ornithologists' Union. Washington, D. C. 1-630. 1954.

Directories

[See also: Directories listed under Animalia]

(anonymous). Membership list. American Ornithologists' Union. September 1967. Supplement to Auk, 84, c-lxxxiv. 1967.

Institutions

[See also under Animalia]

BRUNS, H. Ornithologische Organisationen in Grossbrittannien. Ornith. Mitt., 18, 177-181. 1966.
WALLACE, G. J. An introduction to ornithology. New York; 1-443. (Pp. 383-391)

Museum lists

United States. National Museum

RIDGWAY, R. List of species of Middle and South American birds not contained in the United States National Museum. Proc. United States Nat. Mus., 4, 165-203. 1881.
RIDGWAY, R. Catalogue of the Old World birds in the United States National Museum. Proc. United States Nat. Mus., 4, 317-333. 1882.

Estacion Biologica de los Llanos

AVELADO, H. R. Lista de las Aves colecciones en la Estacion Biologica de los Llanos. Bol. Soc. Venezolana Cienc. Nat., 22, 213-225. 1961.

British Museum (Natural History)

OATES, E. W. et al. Catalogue of the collection of bird's eggs in the British Museum (Natural History). London; 5 volumes. 1901-1912.

(various authors). Descriptive catalogs of most groups, published by the Museum.

Museum Alexander Koenig in Bonn a. Rhein

KOENIG, A. Katalog der Nido-Oologischen Sammlung(Vogeleier-sammlung) in Museum Alexander Koenig. . . . Bonn; 3 volumes. 1932.

Museum d'Histoire Naturelle de Nantes

MARCHAND, M. E. & KOWALSKI, M. J. Inventaire... du Museum d'Histoire Naturelle de Nantes. Bull. Soc. Sci. Nat. Ouest, (5), 4, 3-97. 1935.
KOWALSKI, M. J. Collection ornithologique regionale du Museum de Nantes. Index alphabetique (noms latins et francais). Bull. Soc. Sci. Nat. Ouest, (5), 8, 13-54. 1939.

R. Museo di Firenze

de GERMINY, G. Catalogo della collezione ornitologica generale del R. Museo di Firenze. Rasseg. Faunist. Roma, 3-4, 4 parts. 1937.

Wiener Naturhistorischen Museums

KEVE-KLEINER, A.. Einige systematische Bemerkungen uber das Ungarische ornithologische Material in der ... Wiener naturhistorischen museums. Aquila, 50, 307-310. 1943.

Indian Museum

ROONWAL, M. L. Catalogue of the birds in the Indian Museum, Calcutta, I, II. Rec. Indian Mus., 43, 281-360; 45, 57-731 1941, 1947.

Musee Hoangho Paiho de Tien Tsin

SEYS, G. & LICENT, E. La collection d'oiseaux du Musee Hoangho Paiho de Tien Tsin. Pub. Mus. Hoangho Pai-ho Tientsin, 19, 1-154; 31, 1-46. 1933.

Museum type lists

[See also: This heading under Animalia above; in each volume of Zoological Record, under heading Institutions & Collections]

Museum of Comparative Zoology

BANGS, O. Types of birds now in the Museum of Comparative Zoology. Bull. Mus. Comp. Zool., 70, 145-426. 1930.

United States National Museum

DEIGNAN, H. G. Type specimens of birds in the United States National Museum. United States Nat. Mus. Bull., 221, 1-718; 1961.

Carnegie Museum

TODD, W. E. C. List of types of birds in the collections of the Carnegie Museum on May 1, 1928. Ann. Carnegie Mus., 18, 329-385. 1928.

San Diego Natural History Museum

JEHL, J. R., Jr. Type specimens of birds in the San Diego Natural History Museum. Trans. San Diego Soc. Nat. Hist., 15, 133-139. 1968.

Museo Nacional de Historia Natural de Santiago

GIGOUX, E. E. & LOOSER, G. Los tipos de aves conservados en el Museo Nacional de Historia Natural de Santiago. Bul. Mus. Nac. Chile, 13, 5-36. 1936.

British Museum (Natural History)

WARREN, R. L. M. Type-specimens of birds in the British Museum (Natural History), vol. 1, Non-passerines. London; 1-320. 1966.

Tring Museum

HARTERT, E. Types of birds in the Tring Museum. (Many parts) Novit. Zool., 26-36. 1919-1931. (And others)

Zoologischen Sammlung Bayerischen Staates i Munchen

LAUBMANN, A. Die Types... der Zoologischen Sammlung der Bayerischen Staates i Munchen. I. Verh. Ornith. Ges. Bayern, 19, 532-541. 1932.

Staatlichen Museums fur Tierkunde in Dresden

MEISE, W. Verzeichnis der typen des Staatlichen Museums for Tierkunde in Dresden. Vogel, 1. Abh. Mus. Dresden, 17, (4), 1-22. 1929.

Museum National d'Histoire Naturelle, Paris

BERLIOZ, J. Catalogue systematique des types de la collection d'oiseaux du Museum. I Ratites. II. Palmipedes. Bull. Mus. Nat. Hist. Nat., (2), 1, 58-69. 1929.

Musee Polonaise d'Histoire Naturelle

SZTOLCMAN, J. & DOMANIEWSKI, J. Les types d'oiseaux au Musee Polonaise d'Histoire Naturelle. Ann. Mus. Polonaise, 6, 95-194. 1927.

Zoological Museum of the Moscow University

SUDILOVSKAYA, A. M. Types in the ornithological collection of the Zoological Museum of the Moscow University. Part I. Passeriformes. Ornitologia, 2, 81-88; 1959. Part II. Ornitologia, 5, 431-437; 1962. (In Russian)

Raffles Museum

GIBSON-HILL, C. A. Bird and mammal type specimens formerly in the Raffles Museum collections. Bull. Raffles Mus., 19, 133-198. 1949.

Australian Museum, Sydney

HINDWOOD, K. A. A list of types and paratypes of birds from Australian localities in the Australian Museum, Sydney, New South Wales. Rec. Australian Mus., 21, 386-393; 1946.

EXPLORATION AND LOCALITIES

[See also this heading under Animalia]

Localities

BEHLE, W. H. Types and type localities of birds described from Utah. Proc. Utah Acad. Sci., 38, 21-30. 1961.
HEIM de BALSAC, H. & MAYAUD, N. Les oiseaux du nord-ouest de l'Afrique.... Paris; 1-486. 1962. (localities of northwest Africa)
CHAPIN, J. P. The birds of the Belgian Congo, vol. 4. Bull. American Mus. Nat. Hist., 75B, 638-738. 1954. (localities of Belgian Congo)

METHODS

[See also: This heading under Animalia; and in each volume of Zoological Record, under heading Techniques]

(anonymous). Instructions for collectors: No. 2 — Birds and their eggs, 7th ed. London; 1-14. 1921.
ANDERSON, R. M. Instructions for preserving animal specimens for scientific purposes. Spec. Contr. Nat. Mus. Canada, 43, (2), 1-34. 1943.
BENDIRE, C. Directions for collecting, preparing, and preserving birds' eggs and nests. United States Nat. Mus. Bull., 39, part D, 1-10. 1891.
BERGER, A. J. Suggestions regarding alcoholic specimens and skeletons of birds. Auk, 72, 300-303; 1955.
BLAKE, E. R. Preserving birds for study. Fieldiana, Technique, 7, 1-38; 1949.

CLANCEY, P. A. Some temporary methods of preserving birds and mammals in the field — their uses and their failings. Bull. South African Mus. Assoc., 6, 206-209. 1856.
HUTSON, H. P. W. The ornithologist's guide. Especially for overseas. London; 1-287. 1956.
KENNARD, F. H. A method of blowing eggs. Auk, 45, 234-236. 1928.
MARTINO, V. E. & MATVEYEV, S. D. Ptitze Yugoslaviye. Beograd. 1947.
RIDGWAY, R. Directions for collecting birds. United States Nat. Mus. Bull., 39, part A, 1-27. 1891.
VAN TYNE, J. Principles and practices in collecting and taxonomic work. Auk, 69, 27-33. 1952.
WYTHE, M. W. Some procedures in caring for a research collection of birds. Auk, 46, 306-310. 1929.

GLOSSARIES

[See also many of the general books listed above, as well as Glossaries listed under Animalia]

MacLENNAN, J. M. Russian — English bird glossary. Ottawa; 1-94. 1958.
THOMSON, A. L. A new dictionary of birds. New York; 1-627. 1964.
VAN TYNE, J. & BERGER, A. J. Fundamentals of ornithology. New York; 1-624. 1959. (Pp. 559-586)

NOMENCLATURAL STUDIES

Miscellaneous studies

BOUBIER, M. Origine et etymologie des noms francais des oiseaux de l'Europe occidentale. Bull. Soc. Zool. Geneve, 3, (6), 5-29. 1927.
GLADSTONE, H. The meaning of the names of some British birds and their first use in British ornithology. Trans. Dumfries. Gall. Nat. Hist. Soc., (3), 23, 4-115. 1946.
GLADSTONE, H. British birds named after persons. Trans. Dunfries. Gall. Nat. Hist. Soc., (3), 23, 175-189. 1946.
McATEE, W. L. American bird names / Their histories and meanings. 1697 MS pages and 90 drawers of 2x5 cards; Cornell University Library, Ithaca, New York.
WYNNE, O. E. Biographical key / Names of birds of the world / To authors and those commemorated. Fordingbridge, England. 1-246. 1969.
WYNNE, O. E. Biographical key — Names of birds of the world. Sandleheath, England. 1-248. 1966.

Common names

North America

CHEESMAN, W. H. & OEHSER, R. H. The spelling of common names of birds. Auk, 1937, 333-340.
WETMORE, A. et al. Check-list of North American birds, 5th ed. Baltimore, Maryland; 1-691. 1957.
McATEE, W. L. Folk-names of Canadian birds. Bull. Nat. Mus. Canada, 149, 1-74. 1957.
(anonymous). Canadian bird names. French, English and scientific. Occ. Pap. Canadian Wildl, Serv., 2, 1-20. 1962.

South America

GARCIA, R. Nomes de Aves en lingua Tupi. Bol. Mus. Nac. Rio de Janeiro, 5, 1-54. 1929.
VIEIRA, C. O. C. Nomes vulgares de Aves do Brasil. Rev. Mus. Paulista, 20, 437-489. 1936.
PERGOLANI DE COSTA, M. J. I. Indice de los nombres vulgares de la Aves Argentinos. Idia, 64, 1-57. 1953.
BEHN, F. Nombres vulgares de Aves silvestres chilenas. Bol. Soc. Biol. Concepcion, 6, 117-122. 1942.

Europe

JØRGENSEN, H. I. Nomina avium Europaearum, 2nd ed. København; 1-283. 1958. (Prev. ed., 1941)
MERIKALLIO, E. Vogelnamen aus Kemijoki, Lappland. Ornis Fenn., 9, 65-68. 1932.
MERIKALLIO, E. Finnische Vogelnamen. Ann. Soc. Zool. Bot. Fenn., 2, 1-120. 1925. (In Finnish)
SCHIØLER, E. L. Danmarks fugle..., 3 volumes. København. 1925-31. (All Scandinavia)
SWANN, H. K. A dictionary of English and folk-names of British birds. London; 1-266. 1913. (Reprinted Detroit, Michigan, 1968)
FLOERICKE, C. Die deutschen Vogelnamen. Mitt. Vogeln Stuttgart, 29, 23-25, 52-54. 1930.
HOFFMANN, B. Die deutschen Vogelnamen. Anz. Ornith. Ges. Bayern, 2, 124-127. 1931.
MASAREY, A. Bemerkungen zur deutschen und deutsch-schweizerischen Benennung der Vogel. Ornith. Beobacht., 29, 133-138. 1932.
HOFFMANN, B. Vom Ursprung und Sinn deutschen Vogelnamen.... Bernberg; 1-106. 1937.
STRESEMANN, E. Einiges uber deutsche Vogelnamen. Journ. Ornith., Erg.-Band III, 65-104. 1941.
BEQUAERT, M. Bijdrage tot de kennis van de avifauna in het noroden van Brugge. Biol. Jaarb., 28, 37-65. 1960.
KOWALSKI, M. J. Collection ornithologique regionale du Museum de Nantes. Index alphabetique (noms latins et francaises). Bull. Soc. Sci. Nat. Ouest, (5), 8, 13-54. 1939.
SACARRAO, G. F. Nomes vernaculos das aves portugueses. Naturalia, 8, 39-58. 1963.

WHINNOM, K. A glossary of Spanish bird-names. London; 1-157. 1966.
MOLTONI. E. L'etimologia ed il significato dei nomi vulgari e scientifici degli uccelli Italiana. Riv. Italiana Ornith., 16, 33-50, 69-92. 1946.
THOMPSON, D. W. A glossary of Greek birds. Hildesheim; 1-342. 1966.
CSORNAI, R. Serbische und Kroatische Namen der Vogel der Bacska. Aquila, 50, 394-402. 1943.
SCHENK, J. V. Ungarische Beisvogelnamen. II. Abschliesender Teil. Aquila, 46-49, 5-145. 1939-42.
ZHORDANIYA, G. Terminological dictionary of the birds of Georgia (Latin, Georgian, Russian and German nomenclature). Tbilisi; 1-63. 1960.

Africa

ETCHECOPAR, R. D. & HUE, F. Les oiseaux du nord de l'Afrique. Paris; 1-606. 1964. (Arabic)
CHADWICK, J. M. K. Zulu names for birds. Ostrich, 18, 179-182. 1947.
SMITHERS, R. H. N. A checklist of the birds of Bechuanaland Protectorate and the Caprivi Strip . . . . Bulawayo; 1-188. 1964.
MOREAU, R. E. Bird-names used in coastal north-eastern Tanganyika Territory. Tanganyika Notes Rec., 10, 47-72. 1940.
STONEHAM, H. F. African bird-names in Dho-Luo. Bateleur, 2, 23-28. 1930.
BOWELL, D. W. African bird names in Ki-Swahili. Bateleur, 2, 68-70. 1930.

Indo-Malaya

KOUL, S. C. The vernacular names for Kashmir birds. Journ. Bonbay Nat. Hist. Soc., 34, 571-573. 1930.
MATTHEWS, W. H. Local bird-names in the Darjeeling district. Journ. Darjeeling Nat. Hist. Soc., 9, 156-162. 1935.
PHILLIPS, W. W. A. A list of ... animals, birds and reptiles of Ceylon with the Sinhalese and Tamil names of each. Loris, 2, 140-144. 1940.
EBBELS, D. L. English nomenclature of Ceylon avifauna. Loris, 9, 116-122. 1961.
DEUVE, M. Premiere liste de noms vernaculaires Lao des oiseaux de la vallee du Mekong de Pakse a Vientiane. Bull. Soc. Sci. Nat. Laos, 1, 49-55. 1961.
McCLURE, H. E. English vernacular names of the birds of the Malaysian subregion. Malay Nat. Journ., 17, 74-121. 1963.
BANKS, E. Notes on birds in Sarawak, with a list of native names. Sarawak Mus., 4, 267-325. 1935.

Australia

WEBB, T. T. Aboriginal bird names in East Arnheim Land. Emu, 33, 18-22. 1933.
LINDGREN, S. Natural history notes from Jigalong. III. The birds. V. Aboriginal flora and fauna names. Western Australian Nat., 7, 169-176, 195-201. 1961.

The Orient & Oceania

URAMOTO, M. Japanese names for orders and families of birds. Tori, 8, 202-211. 1967.
MERCER, R. A field guide to Fiji birds. Fiji Mus. Spec. Publ. Ser., 1, 1-39. 1965.

Nomenclators

[See also: Nomenclators listed under Animalia]

GIEBEL, C. G. Thesaurus ornithologiae. Leipzig; 3 volumes. 1872-77. (Pp. 253-824)
GRAY, G. R. Hand-list of genera and species of birds. London; 3 volumes. 1869-71. (Genera and species to about 1870)
LOW, G. C. The literature of Charadriiformes from 1894 to 1928. London, 2nd ed., 1-637. (1st ed., 1924) (Genera, species, and subspecies)
McGREGOR, R. C. Index to the genera of birds. Philippine Dep. Agric. Nat. Res. Publ., 14, 1-185. 1920.
OGILVIE-GRANT, W. R. General index to a Hand-list of the Genera and Species of Birds. . . . London; 1-200. 1912.
RICHMOND, C. W. List of generic terms proposed for birds during the years 1890 to 1900, inclusive, to which are added names omitted by Waterhouse in his "Index Generum Avium." Proc. United States Nat. Mus., 24, 663-730. 1902.
RICHMOND, C. W. Generic names applied to birds during the years 1901 to 1905, inclusive, with further additions to Waterhouse's "Index Generum Avium." Proc. United States Nat. Mus., 35, 583-655. 1908.
RICHMOND, C. W. Generic names applied to birds during the years 1906 to 1915, inclusive . . . . Proc. United States Nat. Mus., 53, 565-636. 1917.
RICHMOND, C. W. Generic names applied to birds during the years 1916 to 1922, inclusive . . . . Proc. United States Nat. Mus., 70, (Art. 15), 1-44. 1927.
SWANN, H. K. A bibliography of British ornithology . . . . Supplement: a chronological list of British birds. London; 1-42. 1923.
WATERHOUSE, F. H. Index generum avium. London; 1-240. 1889.
WATERHOUSE, F. H. Avium generum index alphabeticus; an alphabetical index to the genera adopted in the twenty-seven volumes of the catalogue of birds in the British Museum. Bull. British Ornith. Club, 9, 1-31. 1899.

Generic name studies

GRAY, G. R. The genera of birds. London; 3 volumes. 1844-49.
ALLEN, J. A. The types of the North American genera of birds. Bull. American Mus. Nat. Hist., 23, 279-384. 1907.
ALLEN, J. A. A list of the genera and subgenera of North American birds, with their types, according to Article 30 of the International Code of Zoological Nomenclature. Bull. American Mus. Nat. Hist., 24, 1-50. 1908.

Family name lists

BARDEN, A. A., Jr. Distribution of the families of birds. Auk, 58, 543-557. 1941.

REGIONAL STUDIES

Faunas, checklists, identification works

World

PETERS, J. L. Check-list of birds of the world. Cambridge, Massachusetts; 15 volumes. 1931 –
SWANN, H. K. & WETMORE, A. A monograph of the birds of prey (Accipitres), 16 parts. London; 2 volumes. 1924-45.

World oceans & oceanic islands

ALEXANDER, W. B. Birds of the ocean. A handbook...., 2nd ed. New York; 1-306. 1954 (1st edition, 1928.
ALEXANDER, W. B. Birds of the ocean. A handbook. . , 3rd ed. New York; 1963.
BOURNE, W. R. P. The breeding birds of Bermuda. Ibis, 99, 94-105. 1957.
MURPHY, R. C. Oceanic birds of South America.... New York; 2 volumes. 1936.
WATSON, G. E. Seabirds of the tropical Atlantic Ocean. Washington, D. C.; 1-120. 1966.
WATSON, G. E. et al. Preliminary field guide to the birds of the Indian Ocean. Washington, D. C.; 1-214. 1963.
BERLIOZ, J. Oiseaux de la Reunion. Faune Empire Francaise, 4, 1-81. 1946.
LOUSTAU-LALANNE, P. The birds of the Chagos Archipelago, Indian Ocean. Ibis, 104, 67-73. 1962.
RIDGWAY, R. Birds of the Galapagos Archipelago. Proc. United States Nat. Mus., 19, 459-670. 1897.
SWARTH, H. S. The avifauna of the Galapagos Islands. Occ. Paper California Acad. Sci., 18, 1-299. 1931.

New World

KELSO, L. & KELSO, E. H. A key to the species of American owls. With a list of the owls of the Americas. Biol. Leafl., 4, 1-101.
RIDGWAY, R. & FRIEDMANN, H. The birds of North and Middle America, 11 parts. United States Nat. Mus. Bull., 50. 1901-1950.

North America

RIDGWAY, R. A catalogue of the birds of North America. Proc. United States Nat. Mus., 3, 163-246. 1880.
COUES, E. Key to North American birds, 2nd ed. Boston; 1-863. 1884.
MAY, J. B. The hawks of North America, their field identification.... New York; 1-140. 1935.
WOLFE, L. R. A synopsis of North American birds of prey.... Bull. Chicago Acad. Sci., 5, 167-208; 1938. (list)
JAQUES, H. E. How to know the birds. Dubuque, Iowa; 1-196. 1947.
POUGH, R. H. Audubon land bird guide. Garden City, New York; 1-302. 1949.
WETMORE, A. et al. Check-list of North American birds, 5th ed. American Ornithologists' Union; 1957. (Previous editions 1886, 1895, 1910, 1931)
JAQUES, H. E. & OLLIVIER, R. How to know the water birds. Dubuque, Iowa; 1-172. 1960.
PETERSON, R. T. How to know the birds, 2nd ed. Dubuque, Iowa; 1-168. 1962.
PALMER, R. S. Handbook of North American birds. Volume 1. New Haven, Connecticut; 1-567. 1962.
REED, C. A. North American birds eggs, 2nd ed. New York; 1-372. 1964. (keys)
ROBBINS, C. S. et al. A guide to field identification. Birds of North America. New York; 1-340. 1966.
MATTHIESSEN, P. The shore birds of North America. New York; 1-270. 1967.
REILLY, E. M., Jr. The Audubon illustrated handbook of American birds. New York; 1-524. 1968.
SALOMONSEN, F. & GITZ -JOHANSEN, – Grønlands fugle. The birds of Greenland. København; 1-608. 1950.
SALOMONSEN, F. Fuglene på. Grønland. København, 1-340. 1967.
MACOUN, J. Catalogue of Canadian birds. Parts I - III, 2nd ed. Ottawa; 1-761. 1909. (1st ed., 1900-04)
TAVERNER, P. A. Birds of Canada. Nat. Mus. Canada Bull., 72, 1-445. 1934.
GODFREY, W. E. The birds of Canada. Bull. Nat. Mus. Canada, 203, 1-428. 1966.
TURNER, L. M. List of the birds of Labrador .... Proc. United States Nat. Mus., 8, 233-254. 1885.
TODD, W. E. C. Birds of the Labrador Peninsula .... Toronto; 1-819. 1963.
TAVERNER, P. A. Birds of eastern Canada. Geol. Surv. Canada Mem., 104, 1-297. 1919. (keys)

PETERS, H. S. & BURLEIGH, T. D. The birds of Newfoundland. St. Johns; 1-431. 1951. (keys)
TUFTS, R. W. The birds of Nova Scotia. Halifax; 1-481. 1961. (keys)
SQUIRES, W. A. Birds of New Brunswick. St. John; 1-164. 1952.
THOMPSON, E. E. The birds of Manitoba. Proc. United States Nat. Mus., 13, 457-643. 1891.
TAVERNER, P. A. Birds of western Canada. Victoria Mem. Mus. Bull., 41, 1-380. 1926. (keys)
GUIGNET, C. J. The birds of British Columbia, part 8. Handb. British Columbia Prov. Mus., 22, 1-66. 1964. (keys)
SNYDER, L. L. Arctic birds of Canada. Toronto; 1-310. 1957. (keys)
RAND, A. L. List of Yukon birds and those of the canol road. Bull. Nat. Mus. Canada, 105, 1-76. 1946.
GABRIELSON, I. N. & LINCOLN, F. C. The birds of Alaska. Harrisburg, Pennsylvania; 1-922. 1959.
MURIE, O. J. Fauna of the Aleutian Islands and Alaska Peninsula. North American Fauna, 61, 1-364. 1959. (Birds and mammals)
BLAIR, W. F. et al. Vertebrates of the United States, 2nd ed. New York; 1-616. 1968. (Previous edition, 1957. Pp. 269-451. Keys)
CHAPMAN, F. M. Handbook of the birds of eastern North America, 2nd ed. New York; 1-581. 1932. (keys)
PETERSON, R. T. A field guide to the birds... east of the Rockies. Boston; 1-167. 1939. (Prev. ed. 1934)
POUGH, R. H. Audubon bird guide. Eastern land birds. New York; 1-313. 1946.
HAUSMAN, L. A. Field book of eastern birds.... New York; 1-659. 1946.
PETERSON, R. T. A field guide to the birds. Eastern land and water birds, 2nd ed. Boston; 1-290. 1947. (keys)
POUGH, R. H. Audubon water bird guide... eastern and central United States. Garden City, New York; 1-352. 1951.
CHAPMAN, F. M. Handbook of birds of eastern North America. New York; 1-581. 1966.
PALMER, R. S. Maine birds. Bull. Mus. Comp. Zool., 102, 1-656. 1949.
RICHARDS, T. A list of the birds of New Hampshire. (New Hampshire); 1958.
FORTNER, H. C. et al. A list of Vermont birds. Bull. Vermont Dep. Agric., 41. 1933.
FORBUSH, E. H. Birds of Massachusetts, pts. I, II, III. Norwood, Massachusetts. 1927-29.
GRISCOM, L. & SNYDER, D. E. The birds of Massachusetts. An annotated and revised check-list. Salem, Massachusetts. 1955.
SAGE, J. H. et al. The birds of Connecticut. Bull. Connecticut State Geol. Nat. Hist. Surv., 20, 1-370; 1913.
EATON, E. H. Birds of New York, 2nd ed. Albany, New York; 2 volumes. 1923.
FABLES, D., Jr. Annotated list of New Jersey birds. Trenton, New Jersey. 1955.
POOLE, E. L. Pennsylvania birds. An annotated list. Wynnewood, Pennsylvania; 1-94. 1964.
STEWART, R. E. & ROBBINS, C. S. Birds of Maryland and the District of Columbia. North American Fauna, 62, 1-401. 1958.
MURRAY, J. J. A check-list of the birds of Virginia. Lexington, Virginia. 1952.
BROOKS, M. A check-list of West Virginia birds. Bull. West Virginia Univ. Agric. Exper. Sta., 316. 1944.
PEARSON, T. G. et al. Birds of North Carolina, 2nd ed. Raleigh, North Carolina; 1-434. 1959.
SPRUNT, A., Jr. & CHAMBERLAIN, E. B. South Carolina bird life. Columbia, South Carolina; 1949.
BURLEIGH, T. D. Georgia birds. Norman, Oklahoma; 1-746. 1958.
STEVENSON, H. M. A key to Florida birds. Tallahassee, Florida; 1960.
IMHOF, T. A. Alabama birds. University, Alabama. 1962.
LOWERY, G. H., Jr. Louisiana birds, 2nd ed. Baton Rouge, Louisiana; 1-567. 1960.
MENGEL, R. M. The birds of Kentucky. A. O. U. Ornith. Monogr., 3, 1965.
BORROR, D. J. A check-list of the birds of Ohio.... Ohio Journ. Sci., 50, 1-32. 1950.
WOOD, N. A. The birds of Michigan. Misc. Publ. Mus. Zool. Univ. Michigan, 75, 1-559. 1951.
ZIMMERMAN, D. A. & VAN TYNE, J. A distributional check-list of the birds of Michigan. Occ. Pap. Univ. Michigan Mus. Zool., 608. 1959.
GROMME, O. J. Birds of Wisconsin. Madison, Wisconsin. 1963.
BENNITT, R. Check-list of the birds of Missouri. Univ. Missouri Stud., 7, 1-81. 1932.
DuMONT, P. A. A revised list of the birds of Iowa. Iowa Stud. Nat. Hist., 15, 1-171. 1933.
ROBERTS, T. S. The birds of Minnesota. Minneapolis, Minnesota; 2 volumes. 1936.
OVER, W. H. & THOMAS, C. S. Birds of South Dakota, 2nd ed. Univ. South Dakota Mus. Nat. Hist. Stud., 1. 1946.
RAPP, W. F., Jr. et al. Revised check-list of Nebraska birds. Occ. Pap. Nebraska Ornith. Union, 5. 1958.
JOHNSTON, R. F. A directory of the birds of Kansas. Misc. Publ. Univ. Kansas Mus. Nat. Hist., 41, 1-67. 1965.
BAILEY, A. M. & NIEDRACH, R. J. Birds of Colorado. Denver, Colorado; 2 volumes. 1965.
SAUNDERS, A. A. A distributional list of the birds of Montana. Pacific Coast Avifauna, 14. 1921.
POUGH, R. H. Audubon western bird guide. Land, water, and game birds. Garden City, New York; 1-316. 1957. (keys)
PETERSON, R. T. A field guide to western birds, 2nd ed. Boston; 1-240. 1941. (1st ed., 1941)
WYMAN, L. E. & BURNELL, E. F. Field book of birds of the southwestern United States. Boston; 1-308. 1925.
WOLFE, L. R. Check-list of the birds of Texas. Lancaster, Pennsylvania; 1-89. 1956.
PETERSON, R. T. A field guide to the birds of Texas and adjoining states. Boston; 1-304. 1960.
BAILEY, F. M. Birds of New Mexico. Santa Fe, New Mexico; 1-807. 1928.
MONSON, G. & PHILLIPS, A. R. A checklist of the birds of Arizona. Tucson, Arizona; 1-74. 1964.
PHILLIPS, A. et al. The birds of Arizona. Tucson, Arizona; 1-212. 1964.
WOODBURY, A. M. et al. Annotated check-list of the birds of Utah. Bull. Univ. Utah, 39, (16), 1-40. 1949.
ARVEY, M. D. A check-list of the birds of Idaho. Publ. Univ. Kansas Mus Nat. Hist., 1, 193-216. 1947.

LINSDALE, J. M. The birds of Nevada. Pacific Coast Avif., 23, 1936. (Suppl.: Condor, 53, 228-249; 1936)
HOFFMAN, R. Birds of the Pacific states. Boston; 2nd ed.; 1-353. 1958. (Prev. ed., 1927)
JEWETT, S. G. et al. Birds of Washington State. Seattle. 1-767. 1953.
ALCORN, G. D. Checklist of birds of the State of Washington. Occ. Pap. Univ. Puget Sound, 1962, 17, 155-199. 1962.
GABRIELSON, I. N. & JEWETT, S. G. Birds of Oregon. Corvallis, Oregon; 1940.
GRINNELL, J. & MILLER, A. H. The distribution of the birds of California. Pacific Coast Avifauna, 27. 1944.

West Indies

ALLEN, R. P. Birds of the Caribbean. New York; 1-256; 1961. London; 1-256; 1962. (keys)
BOND, J. Field guide to the birds of the West Indies. New York; 1-257. 1947.
BOND, J. Check-list of the birds of the West Indies, 4th ed. Philadelphia; 1-214. Supplement 1-8. (Previous editions 1940, 1945, 1950)
BARBOUR, T. Cuban ornithology, 2nd ed. Mem. Nuttall Ornith. Club, 9, 1-142. 1943.
WETMORE, A. & SWALES, B. H. The birds of Haiti and the Dominican Republic. United States Nat. Mus. Bull., 155, 1-483. 1931.
WETMORE, A. The birds of Porto Rico and the Virgin Islands.... Sci. Surv. Porto Rico ..., New York Acad. Sci., 9, (3), 245-598. 1927.
LEOPOLD, N. F. Checklist of birds of Puerto Rico and the Virgin Islands. Rio Piedras, Puerto Rico. 1963.
LAWRENCE, G. N. A general catalogue of the birds noted from the islands of the Lesser Antilles.... Proc. United States Nat. Mus., 1, 486-488. 1879.
VOOUS, K. H. The birds of St. Martin, Saba, and St. Eustatius. Stud. Fauna Curacao, 6, 1-82. 1955.
VOOUS, K. H. & KOELERS, H. J. Check-list of the birds of St. Martin, Saba, and St. Eustatius. Ardea, 55, 115-137. 1967.
LAWRENCE, G. N. Catalogue of birds of Antigua and Barbuda.... Proc. United States Nat. Mus., 1, 232-242. 1878.
LAWRENCE, G. N. Catalogue of birds of Guadeloupe.... Proc. United States Nat. Mus., 1, 449-462; 1879.
LAWRENCE, G. N. Catalogue of birds of Dominica.... Proc. United States Nat. Mus., 1, 48-69. 1878.
LAWRENCE, G. N. Catalogue of birds of Martinique.... Proc. United States Nat. Mus., 1, 349-360. 1879.
LAWRENCE, G. N. Catalogue of birds of St. Vincent.... Proc. United States Nat. Mus., 1, 185-198. 1878.
LAWRENCE, G. N. Catalogue of birds of Grenada.... Proc. United States Nat. Mus., 1, 265-278. 1879.
WELLS, J. G. Catalogue of birds of Grenada. Proc. United States Nat. Mus., 9, 609-633. 1887.
JUNGE, G. C. A. & MEES, G. F. The avifauna of Trinidad and Tobago. Zool. Verh., 37, 1-172. 1958.
HERKLOTS, G. A. C. The birds of Trinidad and Tobago. London; 1-287. 1965. (Prev. ed., 1961)(keys)
VOOUS, K. H. De vogels van de nederlandse Antillen. Curacao; 1-205. 1955. (keys)
VOOUS, K. H. The birds of Aruba, Curacao and Bonaire. Natuur.-wet. Stud. Suriname, 14, 1-260; 1957.
VOOUS, K. H. Checklist of the birds of Aruba, Curacao, and Bonaire. Ardea, 53, 205-234; 1965.

Central America

SALVIN, O. & GODMAN, F. D. Biologia Centrali-Americana, Aves, vol. 1, 1-512. 1879-1904.
EISENMANN, E. The species of Middle American birds. Trans. Linn. Soc. New York, 7, 1-128. 1955.
FRIEDMANN, H. et al. Distributional check-list of the birds of Mexico, part 1. Pacific Coast Avifauna, 29, 1-202. 1950.
MILLER, A. H. et al. Distributional check-list of the birds of Mexico, part 2. Pacific Coast Avifauna, 33. 1957.
SUTTON, G. M. Mexican birds.... Norman, Oklahoma; 1-282. 1951.
BLAKE, E. R. Birds of Mexico. A guide for field identification. Chicago; 1-644. 1953.
EDWARDS, E. P. Finding birds in Mexico, 2nd ed. Sweet Briar, Pennsylvania; 1-282. 1968.
ALDEN, P. Finding the birds in western Mexico. Tucson, Arizona; 1-138. 1969.
GRANT, P. R. A systematic study of the terrestrial birds of the Tres Marias Islands, Mexico. Postilla, 90, 1-106. 1965.
SCHALDACH, W. J. The avifauna of Colima and adjacent Jalisco, Mexico. Proc. W. Fdn. Vert. Zool., 1, 1-100. 1963.
DAVIS, W. B. & RUSSELL, R. J. Aves y mamiferos des Estado de Morelas. Rev. Soc. Mex. Hist. Nat., 14, 77-147. 1953.
SMITHE, F. B. Las Aves de Tikal. Guatemala City; 1-372. 1968. (Transl. of the following) (keys)
SMITHE, F. B. The birds of Tikal. Garden City, New York; 1-350. 1966. (keys)
RUSSELL, S. M. A. A distributional study of the birds of British Honduras. Ann Arbor, Michigan; 1-195. 1964.
DICKEY, D. R. & VAN ROSSEM, A. J. The birds of El Salvador. Field Mus. Nat. Hist., Zool., 23, 1-609. 1938.
RAND, A. L. & TRAYLOR, M. A. Manual de las Aves de El Salvador. San Salvador; 1-308. 1954.
MONROE, B. L., Jr. A distributional survey of the birds of Honduras. American Ornithologists' Union, Ornith. Monogr., 7, 1-458. 1968.
ZELEDON, J. C. Catalogue of the birds of Costa Rica. Proc. United States Nat. Mus., 8, 104-118. 1885.
SLUD, P. The birds of Costa Rica. Bull. American Mus. Nat. Hist., 128, 1-430. 1964.
WETMORE, A. The birds of the Republic of Panama. Part I. Smithsonian Misc. Coll., 150, (1), 1-483. 1965. Part II. S. M. C., 150, (2), 1-605. 1968. Part III (to appear)
STURGIS, B. B. Field book of birds of the Panama Canal Zone. New York; 1-466. 1928.

EISENMANN, E. Annotated list of birds of Barro Colorado Island, Panama Canal Zone. Smithsonian Misc. Coll., 117, (5), 1-62. 1952.
WETMORE, A. The birds of San Jose and Pedro Gonzalez Islands, Republic of Panama. Smithsonian Misc. Coll., 106, 1-60. 1946.

South America

de SCHAUENSEE, R. M. The species of birds of South America.... Narberth, Pennsylvania; 1-577. 1966. (Also cited as Philadelphia [Academy of Natural Sciences])
OLROG, C. C. Las Aves sudamericanos. Una guia de campo, vol. 1. Tucuman; 1-507. 1968.
de SCHAUENSEE, R. M. A guide to the birds of South America. Wynnewood, Pennsylvania; 1-470. 1970.
HELLMAYR, C. E. A contribution to the ornithology of western Colombia. Proc. Zool. Soc. London, 1911, 1084-1213. 1911.
de SCHAUENSEE, R. M. The birds of the Republic of Colombia, Secunda entrega. Caldasia, 5, 251-379, 381-644, 645-871. 1948-50.
de SCHAUENSEE, R. M. The birds of Colombia and the adjacent areas of South and Central America. Narberth, Pennsylvania; 1-427. 1964. (keys)
KOEPCKE, M. Las Aves del Departamento de Lima. Lima; 1-128. 1964.
ROHL, E. Las Aves de rapina diurnes de Venezuela (Accipitres). Boll. Soc. Venezolana Cienc. Nat., 12, 33-86. 1933.
FERNANDEZ YEPES, A. Introduccion a la ornitologia Venezolana, 5 parts. Mem. Soc. Cienc. Nat. La Salle, 6, (16), 201-207. 1945-46. (List)
PHELPS, W. H. & PHELPS, W. H., Jr. Lista de las Aves de Venezuela. II. Passeriformes. Bol. Soc. Venezolana Cienc. Nat., 12, 1-427; 19, 1-317. 1950-58.
PHELPS, W. H. & PHELPS, W. H., Jr. Lista de las Aves de Venezuela. II. Passeriformes, 2nd ed. Bol. Soc. Venezolana Cienc. Nat., 24, 1-479. 1963.
CHERRIE, G. K. A contribution to the ornithology of the Orinoco Region. Brooklyn Mus. Inst. Arts Sci., 2, 133-374. 1916. (Keys)
PENARD, F. P. & PENARD, A. P. De vogels van Guyana (Suriname, Cayenne, en Demerara). Paramaribo; 2 volumes. 1908-10.
CHUBB, C. The birds of British Guiana. London; 2 volumes. 1916-21.
CLEARE, L. D. Birds. British Guiana Nature Studies. Georgetown; 1-65. 1943. (keys)
SNYDER, D. E. The birds of Guyana.... Salem, Massachusetts; 1-308. 1966. (list)
BLAKE, R. The birds of southern Suriname. Ardea, 51, 53-72. 1963.
HAVERSCHMIDT, F. Birds of Surinam. Edinburgh. 1968.
von IHERING, H. & von IHERING, R. Catalogus da fauna Brasileira..., vol. 1. Os Aves do Brazil. Sao Paulo; 1-485. 1907.
SANTOS, E. Passares do Brasil.... Rio de Janeira; 1-277. 1948.
FRISCH, S. & FRISCH, J. D. Aves brasileiros, Sao Paulo; 1-160. 1964.
PINTO, O. M. O. Ornitologia Brasiliense. Catalogo descritivo..., vol. 1. Sao Paulo; 1-181. 1964.
NAUMBERG, E. M. B. The birds of Matto Grosso, Brazil.... Bull. American Mus. Nat. Hist., 60, 1-432. 1930.
PODTIAGUIN, B. Catalogo sistematico de la Aves del Paraguay. Rev. Soc. Cienc. Paraguay, 5, (5), 3-107; 6, (3), 7-119; 6, (6), 63-80. 1944-45.
DEVINCENZI, G. J. Aves del Uruguay. Catalogo descriptivo. I, II, III, IV. Anal. Mus. Nat. Montevideo, (2), 3, 1-43. 1929.
CUELLO, J. & GERZENSTEIN, E. Las Aves del Uruguay. Lista sistematica, distribucion y notas. Comun. Zool. Mus. Nat. Hist. Montevideo, 6, 1-191. 1962.
CASARES, J. Palmipedos argentinos. El Hornero, 6, 1-21. 1935.
ZOTTA, A. R. & da FONSECA, S. Sinopsis de los Ciconiiformes argentinos. El Hornero, 6, 48-58; 1935. 6, 240-248; 1936. 6, 395-418; 1937.
STEULLET, A. B. & DEAUTIER, E. A. Catalogo sistematico de las Aves de la Republica Argentina. Obra Cincuent. Mus. La Plata, 1, 493-1006. 1939-46.
PLOTNICK, R. & PERGOLANI DE COSTA, M. J. I. Clave de las familias Passeriformes ... Argentina. Rev. Invest. Agric., 9, 65-88. 1955.
OLROG, C. C. Las Aves Argentinas: Una guia de campo. Tucuman. 1959.
OLROG, C. C. Lista y distribucion des las Aves Argentinas. Tucuman. 1963.
HUDSON, W. H. Birds of La Plata. London; 2 volumes. 1920. (Also New York)
HELLMAYR, C. E. The birds of Chile. Fieldl Mus. Nat. Hist., Zool., 19, 1-472. 1932.
HOUSSE, E. Les oiseaux de proie du Chili. Ann. Sci. Nat., 3, 1-96. 1941.
GOODALL, J. D. et al. Las Aves de Chile..., 2 volumes. Buenos Aires; 1, 1-358; 12, 1-445. 1946-51.
PHILIPPI, R. A. Catalogo de las Aves Chilenas.... Invest. Zool. Chilenas, 11, 1-179. 1964.
JOHNSON, A. W. The birds of Chile..., 2 volumes. Buenos Aires. 1965-67. (keys)
HUMPHREY, P. S. et al. Birds of Isla Grande (Tierra del Fuego). Washington, D. C.; 1-411. 1970.
BENNETT, A. G. A list of the birds of the Falkland Islands.... Ibis, (12), 2, 306-333. 1926. (Suppl.)

Europe

HARTERT, E. Die Vogel der palaarktischen Fauna. Berlin; 3 volumes; 1903-38. (Reprinted 1967)
MOLINEAUX, H. G. K. A catalogue of birds ... Palearctic Region. Eastbourne, England. 3 parts; 1-320. 1930.
VAURIE, C. The birds of the Palearctic fauna, a systematic reference, order Passeriformes. London; 1-762. 1959.

VAURIE, C. The birds of the Palearctic fauna. A systematic reference. Non-Passerines. London; 1-763. 1965. (list)
ENGELMANN, F. Die Raubvogel Europas.... Neudamm; 1-834. 1928-29.
EYKMAN, C. The European Anatidae.... (Nederland); 1-7. 1930. (keys)
PETERSON, R. et al. Die Vogel Europas.... Hamburg; 1-376. 1959. (Prev. ed., 1954)
GEROUDET, P. Les rapaces diurnes et nocturnes d'Europe. Neuchatel; 1-426. 1965.
PETERSON. R. T. et al. A field guide to the birds of Britain and Europe. London; 1-344; 1966. (Prev. edition, 1954)
PETERSON, R. et al. Guide des oiseaux d'Europe. Paris; 1-447. 1967.
TIMMERMANN, G. Die Vogel Islands. Visindafelag Islandinga, 21, 24, 28, 1-524. 1938-49.
HORTLING, I. Ornithologisk handbok..., 4 parts. Helsinki; 1-800. 1929-30. (keys)
MERIKALLIO, E. Uber regionali Verbreitung und Anzahl der Landvogel in sud- und mittlefinnland.... Ann. Zool. Soc. Zool. Bot. Fennica, 12, (1), 1-143. 1946.
MERIKALLIO, E. Finnish birds.... Fauna Fennica, V, 1-181. 1958. (Annotated list)
RENDALL, H. Fågelboken. Sveriges fågler i ord och bild. Stockholm; 1-546. 1935. (keys)
ROSIO, F. Svenska fåglar. Stockholm; 1-500. 1953.
LØVENSKIOLD, H. L. Håndbok over Norges fugler. Oslo; 1-887. 1947-50.
SALOMONSEN, F. Systematisk oversigt over Nordens fugle. København; 1-459. 1963.
SCHIØLER, E. L. Danmarks fugle.... København; 3 volumes. 1925-31.
HELMS, O. Danske fugle ved stranden. (Danish birds of the seashore). København; 1-97. 1927. (In Danish)
HEILMANN, G. & MANNICHE, A. L. V. Danmarks fugleliv. København; 3 volumes. 1928-39.
HØRRING, R. Fugle, III. Danmarks Fauna, 39, 1-309. 1934.
JESPERSON, P. The breeding birds of Denmark. København; 1-79. 1946. (list)
BRUUN, A. F. et al. List of Danish vertebrates. København; 1-180. 1950. (list)
SALOMONSEN, F.& RUDEBECK, G. Danmarks fugle. København; 2 volumes. 1961-62.
SANDERS, E. A bird book for the pocket. London; 1-246. 1927. (keys)
SAUNDERS, H. Manual of British birds, 3rd ed. London; 1-834. 1927. (keys)
HALL, C. A. A pocket book of British birds. London; 1-125. 1936. (keys)
BENSON, S. V. The observer's book of British birds. London; 1-223. 1937. (keys)
WITHERBY, H. F. et al. The handbook of British birds. London; 5 volumes. 1938-41.
COWARD, T. A. Birds of the British Isles, 7th ed. London; 3 volumes. 1950.
FITTER, R. S. R. & RICHARDSON, R. A. The pocket guide to British birds. London; 1-240. 1952.
BANNERMAN, D. A. & LODGE, G. E. Birds of the British Isles. Edinburgh; 12 volumes. 1953-63.
HOLLOM, P. A. D. The popular handbook of British birds, 2nd ed. London; 1-424. 1962. (Prev. ed. 1952)
FITTER, R. S. R. & CHARTERIS, G. The pocket guide to nests and eggs. London; 1-172. 1963. (keys)
FITTER, R. S. R. Collins pocket guide to British birds. London; 1-240. 1964. (keys)
KENNEDY, P. G. et al. The birds of Ireland.... Edinburgh; 1-437. 1954.
VENABLES L. S. V. & VENABLES, U. M. Birds and mammals of Shetland. Edinburgh; 1-391. 1955.
van OORT, E. D. De vogels van Nederland. I - V. 's Gravenhage. 1922-35.
EYKMAN, C. De nederlandsche Steltloopers (Gressores et Cursores). Levende Nat., 32, 243-268; 1927.
EYKMAN, C. et al. De Nederlandsche Vogels.... Uitgave; 3 parts. 1936.
EYKMAN, C. et al. De Nederlandsche vogels determineerlijst.... Wagenin; 1-381. 1937.
van IJZENDOORN, A. L. J. The breeding birds of the Netherlands. Leiden. 1950.
de VRIES, T. G. Aves Frisicae. Lyst fen Fryske fugelnammen. Ljouwert. Velde; 1-167. 1928. (list)
DUBOIS, A. Nouvelle revue des oiseaux observes en Belgique. Mem. Soc. Zool. Paris, 25, 162-209. 1913.
van HAVRE, G. C. M. Les oiseaux de la faune Belge.... Bruxelles; 1-497. 1928.
DUPOND, C. Faune de Belgique. Bruxelles; 10 parts. 1940-46.
LIPPENS, L. Les oiseaux d'eau de Belgique, 2nd ed. St. Andre-lez-Bruges; 1-306. 1954. (Prev. ed. 1941)
VERHEYEN, R. Les rapaces diurnes et nocturnes de Belgique, 2nd ed. Bruxelles; 1-248. 1944.
VERHEYEN, R. Les Passereaux de Belgique. Bruxelles; 2 volumes. 1946-57.
VERHEYEN, R. Les oiseaux d'eau de Belgique. ... Bruxelles; 1-182. 1951.
(anonymous). Avifauna de Belgique.... Gerfaut, 57, 273-465. (list)
HULTEN, M. & WASSENICH, V. Die Vogelfauna Luxembourgs. I, II. Arch. Inst. Gr. -Duc. Luxembourg Sec. Sci. Nat. Phys. Math., (n. s.), 27, 293-422; 1960; 28, 339-488; 1962.
DIETRICH, F. Hamburgs Vogelwelt. Hamburg; 1-398. 1928. (keys)
BRINKMANN, M. Die Vogelwelt Nordwestdeutschlands. Hildesheim; 1-227. 1933. (keys)
KLEINSCHMIDT, O. Die Raubvogel der Heimat. Leipzig; 1-100. 1934. (keys)
PFEIFER, S. Die Vogel unserer Heimat. Frankfurt-a-Main; 1-259. 1936. (keys)
FRIELING, H. Exkursionsbuch zum Bestimmen der Vogel..., 2nd ed. Berlin; 1-283. 1936. (keys)
NIETHAMMER, G. Handbuch der deutschen Vogelkunde. Band l. Passeres. Leipzig; 1-474. 1937. (keys)
VOIGT, A. & KLEINSCHMIDT, O. (Key to Aves) Fauna von Deutschland. Leipzig; 501-537. 1944.
PFEIFER, S. Taschenbuch der deutschen Vogelwelt. Senckenbergische Buch., 23, 1-352. 1949. (keys)
NIETHAMMER, G. et al. Die Vogel Deutschlands: Artenliste. Frankfurt a. Main; 1-138. 1964. (list)
FATIO, V. & STUDER, T. Catalogue des oiseaux de la Suisse.... Geneve; 17 livraisons. 1889-1940.
von BARG, G. & KNOPTLI, W. Les oiseaux de la Suisse. XVI. Geneve. 1930.
PARIS, P. Faune de France. 2. Oiseaux. Paris; 1-473. 1921.
MENEGAUX, A. Les oiseaux de France. Paris; vol. 1, 1-93; vol. II, 1-403. 1932-34.
MENEGAUXA & DIDIER, C. Les oiseaux de France. Encyclopedie Pratique Nat., 31, 1-267; 32, 1-402. 1937-39. (keys)
MAYAUD, N. et al. Inventaire des oiseaux de France. (France); 1-211. 1936. (list)
DELAPCHIER, L. Petit atlas des oiseaux. Part I - Passeres. Paris; 1-39. 1940. (keys)

BARRUEL, P. Notes pour servir a l'identification des oiseaux dans la nature. IV. Terre et la Vie, 94, 67-89. 1947.
MAYAUD, N. Liste des oiseaux de France. Alauda, 21, 1-63. 1953.
BERLIOZ, J. Les oiseaux. Paris; 1-128. 1962. (keys)
LLETGET, A. G. Sinopsis de las Aves de España y Portugal. Trab. Inst. Cienc. Nat. Madrid, Biol., 2, 1-346. 1945. (keys)
dos REIS, J. A., Jr. Aves de Portugal. I. Lariformes. Porto; 1-96. 1927.
ARMANDO THEMIDO, A. Aves de Portugal. Mem. Mus. Zool. Univ. Coimbra, 213, 1-241. 1952.
LLETGET, A. G. Sinopsis de las Charadriformes españoles. Reseñ. Cient. Soc. Españ. Hist. Nat., 8, 149-152. 1933.
PETERSON, R. et al. Guia de campo de las Aves de España y demas paises de Europe. Barcelona; 1-390. 1957.
KUNCKEL, P. Beitrag zur Avifauna Sardiniens. Vogelwelt, 84, 137-145. 1963. (list)
CATERINI, F. Gli uccelli del Pisano. Riv. Italiana Ornith., 10-13; 8 parts. 1940-43.
MOLTONI, E. Elenco degli uccelli Italiani con l'attuale nome scientifico e relativa pronuncia in riguardo all'accento. Riv. Italiana Ornith., 15, 33-78. 1945.
CATERINI, F. & UGOLINI, L. Manuale di ornitologia Italiana. III editio. Firenze; 1-708. 1953. (keys)
MARTORELLI, G. et al. Gli uccelli d'Italia, 3rd ed. Milano; 1-859. 1960. (Prev. ed. 1932)
ROBERT, E. L. The birds of Malta. Malta; 1954.
De LUCCA, C. A revised check-list of the birds of the Maltese Islands. London; 1-95. 1969.
HEINROTH, O. & HEINROTH, M. Die Vogel Mitteleuropas.... Berlin; 3 volumes. 1926-28.
FEHRINGER, O. Die Vogel Mitteleuropas. Band 3. Heidelberg; 1-111. 1931.
BAUER, K. M. & GLUTZ VON BLOTZHEIM, U. N. Handbuch der Vogel Mitteleuropas, Band I. Gaviiformes Phoenicopteriformes. Frankfurt a. M., 1-483. 1966.
HEINROTH, O. & HEINROTH, M. Die Vogel Mitteleuropas. Band 4. Nachtrag. Frankfurt & Zurich; 1-127; 1967.
IVANAUSKAS, T. L. Birds of Lithuania, 2nd ed. Vilmis; 3 volumes. 1957. (In Lithuanian)
DOMANIEWSKI, J. Uebersicht einheimischer Formen der Superordo Herodiones. Kosmos, Lwow, 58, 175-184. 1933.
DOMANIEWSKI, J. Ornitologia lowiecka .... Warszawa; 3 parts; 1-730. 1951.
SOKOLOWSKI, J. Ptäki ziem Polskich. Warszawa; I, 1-441; II, 1-569. 1958.
STEPANEK, O. Klic nasich obratlovcu. Praha; 1-250. 1950. (keys)
MATOUSEK, B. Faunisticky prehl'ad slovenkeho vtactva. Cast I. Acta Nat. Mus. Slov., 7, 3-609. 1961. 8, 68-139. 1962.
FERIANC, O. Stavovce Slovenska. II. Vtaky I, II. Bratislava; 1-598, 1-417. 1964, 1965. (Keys)
ROKITANSKY, C. Catalogus Faunae Austriae. Teil XXIb: Aves. Wien; 1-62. 1964. (List)
KEVE, A. Nomenclator avium Hungariae. Budapest. 1960. (list; in German & Hungarian)
LINTIA, D. Pasarile din R. P. R. (Pasarile Romaniei orbis Romaniae). Bukarest; 3 volumes. 1946-55.
IONESCU, V. Vertebratele din Romania. Bucuresti; 1-496. 1968. (keys)
MARTINO, V. E. & MATVEYEV, S. D. Ptitze Yugoslaviye. Beograd. 1947. (keys)
MATVEJEV, S. D. List of Yugoslav birds. Arch. Sci. Biol. Belgrade, 2, 146-158. 1950. (In Yugoslavian)
MARCETIC, M. & ANDREJEVIC, D. N. Ornitofauna kosova i metahije. Pristina, Yugoslavia; 1-116. 1960.
PATEFF, P. Birds of Bulgaria. Izv. Zool. Inst. Sofiya, 1, 1-364. 1951. (In Bulgarian)
LAMBERT, A. A specific check list of the birds of Greece. Ibis, 99, 43-68. 1957.
BAUER, W. et al. Catalogus faunae Graeciae, pars II, Aves. Thessaloniki; 1-213. 1969.
BIANCHI, V. L. Faune de la Russie. Oiseaux. Vol. I. St. Petersburg; 1-384. 1911.
BUTURLIN, S. A. & DEMENTIEV, G. P. Systema avium rossicarum. Paris; 3 parts. 1933-34. (list)
BUTURLIN, S. A. Complete description of the birds of U. S. S. R. Moskva; 2 volumes. 1934-35.
KOZLOVA, E. Tableaux analytiques de la fauna d'U. R. S. S. No. 17. L'ordre des Gruiformes. Leningrad; 1-40. 1935. (In Russian)
DEMENTIEV, G. P. Systema avium rossicarum. Catalogue critique oiseaux de l'U. R. S. S., vol. 1. Paris; 1-288. 1935.
STEGMANN, B. Faune de l'U. R. S. S. Oiseaux. I. 5. Falconiformes. Moskva; 1-294. 1937.
BUTURLIN, S. A. & DEMENTIEV, G. P. Birds of the U. R. S. S. Part IV, Passeres. Moskva; 10334. 1937. (In Russian)
TOUGARINOV. A. J. Fauna of U. S. S. R. Aves, I, 4, Anseriformes. Moskva; 1-383. 1941. (In Russian)
KOZLOVA, E. V. Faune de l'U. R. S. S. Oiseaux. I. 3. Colymbiformes, Procellariiformes. Moskva; 1-125. 1947. (In Russian)
TUGARINOV, A. J. Faune de l'U. R. S. S. Oiseaux. I. 3. Moskva; 125-316. 1947.
IVANOV, A. I. et al. The birds of U. S. S. R. Part I. Zool. Inst. Acad. Sci. Moscou, 39, 1-280. 1951. (In Russian)
DEMENTIEV, G. P. et al. Birds of the Soviet Union. Moskva; 6 volumes. 1951-54. (In Russian)
IVANOV, A. I. et al. The birds of U. S. S. R. Tabl. Anal. Faune U. R. S. S., 49, (2), 1-343. 1952. (In Russian)
PORTENKO, L. A. The birds of U. S. S. R. Tabl. Anal. Faune U. R. S. S., 54, 1-254. 1954. (In Russian)
KOZLOVA, E. V. Fauna U. R. S. S. Birds. II. 1. 2. Charadriiformes. II. 1. 3 (continued). Moskva; 1-500. 1961-62. (In Russian)
GLADKOV, N. A. et al. Key to the birds of the U. S. S. R. Moskva; 1-536. 1964. (In Russian)
IVANOV, A. I. & SHTEGMAN, B. K. A short guide to the birds of the U. S. S. R. Trud. Zool. Inst. Leningrad, 85, 1-528. 1964. (In Russian)
DEMENT'EV, G. P. & GLADKOV, N. A. Birds of the Soviet Union, vol. 1, Jerusalem; 1-704. 1966. (Translation of 1951)

FEYUSHIN, A. V. & DOLBIK, M. S. Birds of Belorussia. Minsk; 1-520. 1967. (In Russian)
JOHANSON, H. Die Vogelfauna Westsibiriens. Journ. Ornith., 91, 1-100; 92, 1-105; 102, 35-67, 237-269. 1943, 1944, 1961.
BERGMAN, S. Zur Kenntnis nordostasiatischer Vogel.... Stockholm; 1-268. 1935.
GAGINA, T. N. Birds of East Siberia. Trud. Barguzin. Zapov., 3, 99-123. 1961. (In Russian) (list)
VAS'KOVSKII, A. The list and geographical distribution of birds of the extreme north-east of the U. S. S. R. Kraeved. Zapiski Magadan Mus., 6, 84-124. 1966. (list)
VOINSTRENSKI, M. A. The birds of the steppe zone of the European part of the U. S. S. R. Kiev; 1-290. 1961. (In Russian)
KISTYAKYVSKI, O. B. The fauna of the Ukrainian S. S. R. Birds. Kiev Akad. Nauk Ukrain. S. S. R., 4, 1-432. (In Urainian; also as Fauna Ukraini, 4, Birds)
VOINSTVENSKII, M. A. & KISTYAKIVSKII, O. Key to the birds of the Ukraine, 2nd ed. Kiev; 1-371. 1962. (In Russian)
STRAUTMAN, F. I. The birds of the western provinces of the Ukrainian S. S. R. Lwow; 1-199. 1963. (In Russian)
ZHORDANIYA, R. G. The ornithological fauna of the Little Caucasus (in limits of the Georgian S. S. R.). Tbilisi; 1-288. 1962. (In Russian)
SALIKHBAEV, K. C. & BOGDANOV, A. N. Fauna Uzbekskoi S. S. R. Birds, part 3. Akad. Nauk. Uz. SSS. Tashkent, 2, 7-271. 1961. (In Russian)
DOLGUSHIN, I. A. Ptitsy Kazakhstana. Alma-Ata; 1-780. 1962. (In Russian)
GAVRIN, V. F. et al. The birds of Kazakhstan, vol. 2, Alma-Ata; 1-779. 1962. (In Russian)
YANUSHERICH, A. I. et al. The birds of Kirghizia, vol. II. Frunze; 1-361. 1961. (In Russian)
VOROB'EV, K. A. The birds of Yakutia. Moskva; 1-336. 1963. (In Russian)

Near East & Middle East

ERGENE, S. Fan fakultesi monografileri (Avifauna of Turkey). Istanbul; 1-361. 1945.
KASPARYAN, A. A preliminary systematic list of the birds of Turkey. Rev. Fac. Sci. Univ. Istanbul, 21B, 27-48. 1956.
KUMERLOEVE, H. A revised systematic list of the bird species of Turkey. Alauda, 34, 165-186. 1966. (In French & English)
BANNERMAN, D. A. & BANNERMAN, W. M. Birds of Cyprus. Edinburgh; 1-384. 1958.
KUMERLOEVE, H. Recherches sur l'avifauna de la Republique arabe syrienne. Alauda, 35, 243-266; 36, 1-26, 196-207; 37, 43-58, 114-134, 188-205. 1967-69.
ARNOLD, P. Birds of Israel. Haifa; 1-104. 1962. (In Hebrew & English) (keys)
HARDY, E. A handbook of the birds of Palestine. (?); 1-50. 1944. (keys)
BENSON, S. V. Birds of Lebanon and the Jordan area. London; 1-208. 1970.
HUE, F. & ETCHEKOPAR, R. D. Notes ornithologiques du Moyen-Orient. Oiseau, 36, 95-109, 233-251. 1966. (list)
ALLOUSE, B. E. Handlist of the birds of Iraq. Irak Nat. Hist. Mus. Publ., 2, 1-60. 1950.
ALLOUSE, B. E. Birds of Iraq. Baghdad; 3 volumes. 1960-62. (In Arabic)
WHISTLER, H. Materials for the ornithology of Afghanistan. Part 5. Journ. Bombay Nat. Hist. Soc., 45, 462-485. 1945 etc.
PALUDAN, K. On the birds of Afghanistan. Vidensk. Medd. Dansk Naturh. Foren., 122, 1-332. 1959.
MEINERTZHAGEN, R. Birds of Arabia. Edinburgh; 1-624. 1954.

Africa

REICHENOW, A. Die Vogel Afrikas. Neudamm; 3 volumes. 1901-1905.
MILLET -HORSIN, — Guide de l'amateur d'oiseaux debarquant sur la terre d'Afrique. Rev. Francaise Ornith., 13, 97-101, 134-137, 167-171, 180-182. 1929. (keys)
WHITE, C. M. N. A revised check list of African non-passerine birds. Lusaka; 1-299. 1965. (list)

North Africa

ETCHECOPAR, R. D. & HUE, F. Les oiseaux du nord de l'Afrique Paris; 1-606. 1964. (keys)
ETCHECOPAR, R. D. & HUE, F. The birds of North Africa from the Canary Islands to the Red Sea. Edinburgh; 1-612. 1967. (translation of 1964)
NICOLL, M. J. Handlist of the birds of Egypt. Cairo; 1-119. 1919.
MEINERTZHAGEN, R. Nicoll's Birds of Egypt. London; 2 volumes. 1930.
KOENIG, A. Die Schwimmvogel (Natatores) ... (Steganopodes) ... (Urinatores) Aegyptens. Journ. Ornith., 80, Sonderheft, 1-237. 1932.
CAVE, F. O. & MACDONALD, J. D. Birds of the Sudan, their identification . . . . Edinburgh; 1-444. 1955.
BLANCHET, A. Oiseaux de Tunisie, I, II. Mem. Soc. Sci. Nat. Tunis, 1, (1), 107; (2), 93-216. 1951-57.
HEIM DE BALSAC, H. & MAYAUD, N. Les oiseaux du nord-ouest de l'Afrique. Paris; 1-486. 1962.

Atlantic Islands

SARMENTO, A. A. Vertebrados da Madeira, 2nd ed. Funchal; 1-317. 1948.
BERNSTROM, J. Check-list of the breeding birds of the archipelago of Madeira. Bol. Mus. Nac. Funchal, 5, 64-82. 1951.

BANNERMAN, D. A. & BANNERMAN, W. M. Birds of the Atlantic Islands. Edinburgh; 4 volumes. 1963-68.
1: A history of the birds of the Canary Islands and of the Salvages. 1963.
2: A history of the birds of Madeira, the Desertes, and the Porto Santo Islands. 1965.
3. A history of the birds of the Azores. 1966.
4. History of the birds of the Cape Verde Islands. 1968.

Tropical & southern Africa

BATES, G. L. Birds of the southern Sahara and adjacent countries in West Africa, II, III, IV, V. Ibis, 1934, 61-79, 213-239, 439-446, 685-717. 1934.
VALVERDE, J. A. Aves del Sahara Español. Madrid. 1-487. 1957.
MALBRANT, R. Faune du Centre Africain Francaises. Encycl. Biol. Paris, 1-430. 1936.
DRAGESCO, J. Oiseaux des savanes d'Afrique equatoriale. Oiseau, 31, 179-192, 261-271. 1961.
BOUET, G. Oiseaux de l'Afrique tropicale. Part 1. Faune Union Francaise, 16, 1-412. 1955.
BOUET, G. Oiseaux de l'Afrique tropicale. Part 2. Faune Trop., 17, 421-798. 1961.
MACKWORTH-PRAED, C. W. & GRANT, C. H. B. Birds of west central and western Africa. London; 1-671. 1970. (African Hand Book of Birds, ser. 3, vol. 1)
BATES, G. L. Handbook of the birds of West Africa. London; 1-572. 1930.
BANNERMAN, D. A. Birds of tropical West Africa. London; 8 volumes. 1930-51.
ELGOOD, J. H. Birds of the West African town and garden. London; 1-66. 1960. (keys)
DEKEYSER, P. L. & DERIVOT, J. H. Les oiseaux de l'ouest Africain. Guide.... Dakar; 1-507. 1966.
BANNERMAN, D. A. The birds of west and equatorial Africa. Edinburgh & London; 2 volumes. 1953. (keys)
FRADE, F. & BACELAR, A. Catalog das Aves da Guine Portuguesa. I. Non-Passeres. Anal. Junta Inv. Ultramar, 10, 7-194. 1955.
ALLEN, G. M. The birds of Liberia. In Rep. Harvard-Afr. Exp. Liberia Belgian Congo. Cambridge, Massachusetts. Pp. 635-748. 1931.
BRUNEL, J. & THIOLLAY, J. M. Liste preliminaire des oiseaux de Côte d'Ivoire. Alauda, 37, 230-254, 315-337; 38, 72-73. 1969-70.
HUTSON, H. P. W. & BANNERMAN, D. A. The birds of Northern Nigeria. Ibis, (12), 6, 600-638; (13), 1, 18-43, 147-203. 1930-31. (list)
MALZY, P. La faune avienne du Mali (Bassin du Niger). Oiseaux, 32, special number, 1-81. 1962.
BASILIO, A. Aves de la Isla de Fernando Po. Madrid; 1-205. 1963.
GOOD, A. I. The birds of French Cameroon, I, II. Mem. Inst. Francaise Afrique Noire, Centre Cameroun Sci. Nat., 2, 1-203; 3, 1-269. 1952-53.
MALBRANT, P. Fauna du Centre Africain Francais (Mammiferes et oiseaux). Paris; 1-435. 1936.
CHAPIN, J. P. The birds of the Belgian Congo, 4 volumes. Bull. American Mus. Nat. Hist., 65, 75, 75A, 75B. 1932-54.
SCHOUTEDEN, H. De Vogels van Belgisch Congo en van Ruanda-Urundi, I, II. Ann. Mus. Congo, (4), 2, 1-199, 201-415. 1948-49.
SCHOUTEDEN, H. Faune du Congo Belge et du Ruanda-Urundi. Ann. Mus. Congo Belge, Zool., 89, 1-328. 1960.
MALBRANT, R. & MACLATCHY, A. Faune de l'Equateur Africain Francais. 1. Oiseaux. Paris; 1-460. 1949. (keys)
MACKWORTH-PRAED, C. W. & GRANT, C. H. B. Birds of the southern third of Africa. London; 1-688. 1962. (African Handbook of Birds, series 2, volume 1, 2) London; 1-747. 1963.
TRAYLOR, M. A. Check-list of Angolan birds. Publ. Cult. Cia Diament. Angola, 61, 1-250. 1963. (list)
DA ROSA PINTO, A. A. (A revised checklist of Angolan birds). Bonn Zool. Beitr., 19, 280-288. (In Portuguese) 1968.
WINTERBOTTOM, J. M. A preliminary list of the birds of South West Africa. Avifauna Ser. Fitzpatrick Inst. Afr. Ornith., 25, 1-38. 1965.
SMITHERS, R. H. N. A checklist of the birds of Bechuanaland Protectorate and the Caprivi Strip . . . . Bulawayo; 1-188. 1964.
STARK, A. C. & SCLATER. The birds of South Africa. London; 3 volumes. 1900-1903.
GILL, E. L. A first guide to South African birds. Cape Town; 1-223. 1936.
McLACHLIN, G. R. & LIVERSIDGE, R. Roberts Birds of South Africa. Cape Town. 1-504. 1957. (keys)
WHEELER, P. W. & WINTERBOTTOM, J. M. Key to South African ducks and geese. Bokmakierie, 11, 32-37. 1959.
CLANCEY, P. A. A catalogue of birds of the South African sub-region, 5 parts. Durban Mus. Novit., 7, etc. 1966.
WINTERBOTTOM, J. M. A checklist of the land and fresh-water birds of Western Cape Province. Ann. South African Mus., 53, 1-276. 1968.
CLANCEY, P. A. A preliminary list of the birds of Natal and Zululand. Durban; 1-55. 1953.
CLANCEY, P. A. The birds of Natal and Zululand. Edinburgh; 1-511. 1964. (keys)
FRADE, F. Catalogo das Aves de Mocambique. Ann. Junta Miss. Geogr. Lisboa, 6, (4, 4), 7-294. 1951.
PRIEST, C. D. A guide to the birds of Southern Rhodesia. London; 1-233. 1929.
PRIEST, C. D. The birds of Southern Rhodesia. London; 4 volumes. 1933-36.
SMITHERS, R. H. N. et al. A check list of the birds of Southern Rhodesia. Bulawayo; 1-175. 1957.
SMITHERS, R. H. Guide to the waterfowl of Southern Rhodesia. Bulawayo; 1-32. 1962.
WHITE, C. M. N. & WINTERBOTTOM, J. M. A check list of the birds of Northern Rhodesia. Lusaka; 1-168. 1949.
BENSON, C. W. & WHITE, C. M. N. Check list of the birds of Northern Rhodesia. Lusaka; 1-166. 1957.

BENSON, C. W. & IRWIN, M. P. S. A contribution to the ornithology of Zambia. Zambia Mus. Paper, 1. 1967.
WILLIAMS, J. G. A field guide to the birds of eastern and central Africa. London; 1963. Boston; 1-288. 1964.
MACKWORTH-PRAED, C. W. & GRANT, C. H. B. Birds of eastern and north-eastern Africa, 2 volumes. London; 1-846; 1-1100. (Vol. 1, 2nd ed.; keys) (African Handbook of Birds, ser. 1, vol. 1, 2)
WILLIAMS, J. G. A field guide to the natural parks of East Africa. Boston. 1968.
BELCHER, C. F. The birds of Nyasaland ... a classified list. London; 1-356. 1930.
BENSON, C. W. A check list of the birds of Nyasaland.... Blantyre & Lusaka; 1-118. 1953.
BAIRD, D. A. A check list of the birds of Zanzibar and Pemba. Tanganyika Notes, 58-59, 186-196. 1962.
van SOMEREN, V. G. L. The birds of Kenya and Uganda. Journ. East Africa Uganda Nat. Hist. Soc., 22-48 (9 parts). 1925-36.
JACKSON, F. J. The birds of Kenya Colony and the Uganda Protectorate. London; 3 volumes. 1938.
ARCHER, G. & GODMAN, E. M. The birds of British Somaliland and the Gulf of Aden.... London; 4 volumes. 1937-61.
FRIEDMANN, H. Birds collected by the Childs Frick Expedition to Ethiopia and Kenya Colony, 2 parts. United States Nat. Mus. Bull., 153, (1), 1-516; (2), 1-506. 1930, 1937.
SCLATER, W. L. Systema avium ethiopicarum.... London; 1-922. 1924-30. (list)
SMITH, K. D. An annotated checklist of the birds of Eritrea. Ibis, 99, 1-26. 1957.
MOLTONI, E. & RUSCONE, G. G. Gli uccelli dell-Africa orientale Italiana. Milano; 4 parts. 1940-44.

Madagascar and adjacent islands

RAND, A. L. The distribution and habits of Madagascar birds.... Bull. American Mus. Nat. Hist., 72, 143-499. 1936.

India

BLANFORD, W. T. Fauna of British India.... Birds. London; 4 volumes. 1895-98.
BAKER, E. C. S. Fauna of British India.... Birds, 2nd ed. London; 8 volumes. 1922-30.
BAKER, E. C. S. Handlist of genera and species of birds of the Indian Empire. Bombay; 1-240. 1924. (Reprint of 1921)
BAKER, E. C. S. Nidification of birds of the Indian Empire. London; 4 volumes. 1930-49.
BAKER, E. C. S. The waders and other semi-sporting birds of the Indian Empire. Parts XVI, XVII, XVIII. Journ. Bombay Nat. Hist. Soc., 35, 475-483; 36, 1-12. 1932.
RIPLEY, S. D. II. A synopsis of the birds of India and Pakistan. Bombay; 1-700. 1961. (list)
ALI, S. & RIPLEY, S. D. Handbook of the birds of India and Pakistan. London; 4 volumes. 1968-70.
WHISTLER, H. & KINNEAR, N. B. Popular handbook of Indian birds, 4th ed. London; 1-560. 1949. (Previous editions 1928, 1935, 19--)
BISWAS, B. A checklist of genera of Indian birds. Rec. Indian Mus., 50, 1-62. 1953.
ALI, S. The book of Indian birds, 7th ed. Bombay; 1-395. 1964. (Previous edition 1941)
BATES, R. S. P. & LOWTHER, E. H. N. Breeding birds of Kashmir. London; 1-369. 1952.
ALI, S. Indian hill birds. London; 1-188. 1949.
ALI, S. The birds of Sikkim. London; 1962.
BISWAS, B. The birds of Nepal, 12 parts. Journ. Bombay Nat. Hist. Soc., 57-60, etc. 1960-67.
DONALD, C. H. The birds of prey of the Punjab. Journ. Bombay Nat. Hist. Soc., 26-27, 6 parts. 1918-21.
ALI, S. The birds of Travancore and Cochin. Bombay; 1-436. 1953. (keys)
WAIT, W. E. Manual of the birds of Ceylon. Colombo; 1-492. 1925.
PHILLIPS, W. W. A. A (1952) revised checklist of the birds of Ceylon. Colombo; 1-132. 1953.
PHILLIPS, W. W. A. Birds of Ceylon. 3. Birds of our highlands. Colombo; 1-51. 1955. (keys)
HENRY, G. M. A guide to the birds of Ceylon. Oxford; 1-432. 1955. (keys)
NORRIS, C. E. Field guide to the birds of Ceylon. Loris, 10, 1-196. 1965.
BLYTH, E. Catalogue of mammals and birds of Burma. Journ. Asiatic Soc. Bengal, 44, (2), extra number, 1-167. 1875.
SMYTHIES, B. E. Birds of Burma, 1st ed. Rangoon; 1-589. 1940.
SMYTHIES, B. E. Birds of Burma, 2nd ed. Edinburgh & London; 1-668. 1953.

Indochina

AAGARD, C. J. The common birds of Bangkok. København; 1-239. 1930. (keys)
DEIGNAN, H. G. Birds of northern Thailand. United States Nat. Mus. Bull., 186, 1-616. 1945.
SUVATTI, C. Fauna of Thailand. Bangkok; 1-1100. (Birds, 603-654) 1950.
DEIGNAN, H. G. Checklist of the birds of Thailand. United States Nat. Mus. Bull., 226. 1963.
LEKAGUL, B. Bird guide of Thailand. Bangkok. 1968. (keys)
BOONSONG, L. Bird guide of Thailand. Bangkok; 1-272. 1968. (keys)
DELACOUR, J. & JABOUILLE, P. Les oiseaux de l'Indochine francaise. Paris; 4 volumes. 1931.
WILDASH, P. Birds of South Vietnam. Rutland, Vermont; 1-234. 1968. (keys)

Malay Peninsula

ROBINSON, H. C. & CHASEN, F. N. The birds of the Malay Peninsula. London; 4 volumes. 1927-39.
CHASEN, F. N. A handlist of Malasian birds. Bull. Raffles Mus., 11, 1-389. 1935.

GIBSON-HILL, C. A. An annotated check list of the birds of Malaya. Bull. Raffles Mus., 20, 1-300. 1949.
GLENISTER, A. G. The birds of the Malay Peninsula, Sumatra and Penang. London; 1-282. 1951.
MADOC, G. C. An introduction to Malayan birds, 2nd ed. Kuala Lumpur; 1-234. 1956. (keys)
DELACOUR, J. Birds of Malaysia. New York; 1-382. 1947. (keys)
ABDULALI, H. The birds of the Andaman and Nicobar Islands. Journ. Bombay Nat. Hist. Soc., 61, 483-571. 1965.
ABDULALI, H. The birds of the Nicobar Islands. Journ. Bombay Nat. Hist. Soc., 64, 139-190. 1967.
BUCKNILL, J. A. S. & CHASEN, F. N. The birds of Singapore Island. Singapore; 1-247. 1927. (keys)
GIBSON-HILL, C. A. A checklist of the birds of Singapore Island. Bull. Raffles Mus., 21, 132-183. 1950.

Indo-Australian Archipelago

RIPLEY, S. D., The bird fauna of the West Sumatra Islands. Bull. Mus. Comp. Zool., 94, 307-430. 1944.
KURODA, N. Birds of the island of Java. Tokyo; 2 volumes. 1933-36.
SMYTHIES, B. E. An annotated checklist of the birds of Europe. Sarawak Mus. Journ., 7, 523-818. 1957.
SMYTHIES, B. E. Birds of Borneo. Edinburgh; 1-561. 1960. (keys)
van BEMMEL, A. C. V. A faunal list of the birds of the Mollucan islands. Treubia, 19, 323-402. 1948.
MAYR, E. List of New Guinea birds. New York; 1-260. 1941.
IREDALE, T. Birds of New Guinea. Melbourne; 2 volumes. 1956.
RAND, A. L. & GILLIARD, E. T. Handbook of New Guinea birds. Garden City, New York; 1-628. 1968.

Australia

CAMPBELL, A. J. Nests and eggs of Australian birds.... Sheffield; 2 volumes. 1901.
MATHEWS, G. M. The birds of Australia. London; 12 volumes. 1910-27.
MATHEWS, G. M. A reference list to the birds of Australia. Nov. Zool., 18, 171-455. 1912.
MATHEWS, G. M. A list of the birds of Australia. London; 1-332. 1913.
MATHEWS, G. M. & IREDALE, T. A name-list of the birds of Australia. Australian Av. Rec., 4, 65-113. 1920.
MATHEWS, G. M. & IREDALE, T. A manual of the birds of Australia, vol. 1. London; 1-279. 1921.
(anonymous). Official checklist of the birds of Australia, 2nd ed. Melbourne; 1-212. 1926.
MATHEWS, G. M. Systema avium Australasianarum. London; 1-1047. 1927-30. (list)
MATTHEWS, G. M. A working list of Australian birds. Sydney; 1-184. 1946.
CAYLEY, N. W. What bird is that? A guide to the birds of Australia, 2nd ed. London; 1-319. 1948. (Previous edition 1931)
CAYLEY, N. W. What bird is that? A guide to the birds of Australia, 4th ed. Sydney; 1-344. 1966.
HILL, R. Australian birds. Melbourne. 1967.
WHEELER, W. R. A handlist of the birds of Victoria. (Victoria); 1-88. 1967.
McGILP, J. N. South Australian Ornith., 12, 225-258. 1934.
CONDON, H. T. A handlist of the birds of South Australia, 2nd ed. Adelaide; 1-141. 1968.
CONDON, H. T. A handlist of the birds of South Australia, 3rd ed. Adelaide. 1969.
SERVENTY, D. L. & WHITTELL, H. M. A handbook of the birds of Western Australia.... Perth; 1-384. 1951. (also as 1948)
SLATER, P. Western Australian birds. Kwinana, Western Australia; 1-466. 1962.
SERVENTY, D. L. & WHITTELL, H. M. Birds of Western Australia. Perth; 1-427. 1962. (keys)
SERVENTY, D. L. & WHITTELL, H. M. Birds of Western Australia, 4th ed. Perth. 1967. (keys)
LITTLER, F. M. A handbook of the birds of Tasmania. Launceston, Tasmania; 1-242. 1910. (keys)
SHARLAND, M. Tasmanian birds. A field guide.... Sydney; 1958.

New Zealand

BULLER, W. L. A history of the birds of New Zealand, 2nd ed. London; 2 volumes. 1888.
MATHEWS, G. M. & IREDALE, T. A reference list of the birds of New Zealand. Ibis, 1913, 201-263; 402-452. 1913.
MATHEWS, G. M. & IREDALE, T. A name list of the birds of New Zealand. Australian Avian Rev., 4, 49-64. 1920.
MONCRIEFF, P. New Zealand birds and how to identify them, 2nd ed. Auckland; 1-142. 1937.
(anonymous). Checklist of New Zealand birds. Wellington; 1-80. 1953.
OLIVER, W. R. B. New Zealand birds, 2nd ed. Wellington; 1-661. 1955.
FALLA, R. A. et al. A field guide to the birds of New Zealand and outlying islands. London; 1-254. 1966.
WESTERKOV, K. E. Know your New Zealand birds. Christchurch. 1-144. 1967.

The Orient

TAKA-TSUKASA, N. et al. Monograph of the birds of Greater East Asia. Tokyo; 1944.
McGREGOR, R. C. A manual of Philippine birds, I, II. Manila; 1-769. 1909.
HACHISUKA, M. The birds of the Philippine Islands, volume 1. London; 1931.
DELACOUR, J. & MAYR, E. Birds of the Philippines. New York; 1-309. 1946. (keys)
RAND, A. L. & RABOR, D. S. Birds of the Philippine Islands: Siquijor, Mount Malindang, Bohol, and Samar. Fieldiana, Zool., 35, 221-441. 1960.
GEE, N. G. et al. A tentative list of Chinese birds. Peiping; 1-370. 1948. (Previous editions: Bull. Peking Soc. Nat. Hist., 1926-27)
CALDWELL, H. R. & CALDWELL, J. C. South China birds. Shanghai; 1-447. 1931.

HUTSON, H. P. W. The birds of Hong Kong. Hong Kong; 8 parts or more. 1931.
HERKLOTS, G. A. C. The birds of Hong Kong. Field identification.... Hong Kong; 1-77. 1946.
MACFARLANE, A. M. & MACDONALD, A. D. An annotated check-list of the birds of Hong Kong. Hong Kong; 1-96. 1966. (Previous edition 1960)
CAUNTER, J. R. I. & MACFARLANE, A. M. An annotated check-list of the birds of Hong Kong. Hong Kong; 1-92. 1966.
LA TOUCHE, J. D. D. A handbook of the birds of eastern China. London; 2 volumes. 1925-32.
SOWERBY, A. D. Birds recorded from ... Shanghai area. Musee Heude Notes Ornit. Chang-hai, 1, 1-212. 1943.
CHENG, T.-H. et al. A preliminary survey of birds of Hunan Province. Part II. Passeriformes. Acta Zool. Sinica, 13, 97-121. 1961. (list)
KORELOV, M. N. The list of birds ... of the north Tian-Shan. Trud. Inst. Zool. Akad. Nauk Kazakh. SSR, 15. 55-103. 1961. (In Russian)
VAURIE, C. A survey of the birds of Mongolia. Bull. American Mus. Nat. Hist., 127, 103-143. 1964.
AUSTIN, O. L. H. Birds of Korea. Bull. Mus. Comp. Zool., 101, 1-301. 1948.
WON, P.-O. An annotated checklist of the birds of Korea. Seoul; 1-178. 1969.
STEJNEGER, L. Review of Japanese birds, 9 parts. Proc. United States Nat. Mus., 9-11. 1886-1889.
HACHISUKA, M. et al. A hand-list of the Japanese birds, 3rd ed. Tokyo; 1-238. 1942. (Prev. ed. 1932)
(anonymous). Illustrated encyclopedia of the fauna of Japan, 2nd ed. Tokyo; 2 volumes. 1949.
UCHIDA, S. A monograph of the Japanese birds, 3rd ed. Tokyo; 1-313. 1949. (In Japanese)
YAMASHINA, Y. Birds in Japan. A field guide.... Tokyo; 1-233. 1952.
AUSTIN, O. L. Jr. & KURODA, N. The birds of Japan.... Bull. Mus. Comp. Zool., 109, 277-637. 1953.
(anonymous). A hand-list of Japanese birds, 4th ed. Tokyo; 1-264. 1958.
TAKA-TSUKASA, N. The birds of Nippon. Tokyo; vol. 1, 1-701. 1967. (In English; prev. ed. 1943.)

Oceania

JOHANSEN, H. Revised list of the birds of the Commander Islands. Auk, 78, 44-56. 1961.
KING, W. B. Preliminary identification manual: Seabirds of the tropical Pacific Ocean. Washington, D. C.; 1-126. 1967.
BAILEY, A. M. Birds of Midway and Laysan Islands. Denver Mus. Nat. Hist. Mus. Pictorial, 12. 1956. (list)
MUNRO, G. C. Birds of Hawaii. Honolulu; 1-189. 1944.
MUNRO, G. C. Birds of Hawaii. Rutland, Vermont; 1-192. 1966.
BRYAN, E. H., Jr. & GREENWAY, J. C., Jr. Checklist of the birds of the Hawaiian Islands. Bull. Mus. Comp. Zool., 94, 92-140. 1944.
AMERSON, A. B. Ornithology of the Marshall and Gilbert Islands. Atoll Res. Bull., 127, 1-348. 1969.
MORRIS, R. O. The birds of the Gilbert Islands. Swallow, 16, 79-83. 1963. (list)
MAYR, E. Birds of the south-west Pacific. New York; 1-316. 1945. (keys)
BAKER, R. H. Avifauna of Micronesia. Publ. Univ. Kansas Mus. Nat. Hist., 3, 1-359. 1951. (No. 1)
TAKA-TSUKASA, N. The birds of Nippon. Tokyo; 1-701. 1967. (Reprint of 1943. Includes Bonin Island and Micronesia) (lists)
GALBRAITH, I. C. J. & GALBRAITH, E. H. Land birds of Guadalcanal and the San Cristoval group, eastern Solomon Islands. Bull. British Museum (Natural Hist.), 9, 1-80. 1962.
SARASIN, F. Die Vogel Neu-Caledoniens und der Loyalty-Inseln. In Sarasin, F. & Roux, J., Nova Caledonia, A. Zool., vol. 1, Heft 1. Wiesbaden. 1913.
DELACOUR, J. Guide des oiseaux de la Nouvelle-Caledonie et de ses dependances. Neuchatel; 1-172. 1966. (keys)
MERCER, R. A field guide to Fiji birds. Fiji Mus. Spec. Publ. Ser., 1, 1-39. 1965.
ARMSTRONG, J. S. Hand-list to the birds of Samoa. London; 1-91. 1932.
ASHMOLE, M. J. Guide to the birds of Samoa. Honolulu; 1-21. 1963.

Antarctica

HOLGERSEN, H. Antarctic and subantarctic birds. Sci. Res. Norweg. Antarct. Exped. 1927-28, 23, 1-100. 1945.

SPECIAL SUBJECT LISTS

Extinct. HAHN, P. Where is that vanished bird? An index to the known specimens of the extinct and near extinct North American species. Toronto; 1-347. 1963.
GREENWAY, J. C., Jr. Extinct and vanishing birds of the world. New York; 1-518. 1958. (American Committ. Intern. Wild Life Prot., Spec. Publ., 13)

Fossils. BRODKORB, P. Catalogue of fossil birds. Bull. Florida State Mus., Biol., 7, 179-293. 1963.

Game birds. DELACOUR, J. The waterfowl of the world. London; 4 volumes. 1954-64.
KORTRIGHT, F. H. The ducks, geese and swans of North America .... Washington, D. C. 1-476. 1942.
GRINNELL, J. et al. The game birds of California. Berkeley, California; 1-642. 1918.
OGILVIE-GRANT, W. R. et al. The gun at home and abroad. British game birds and wild fowl. London; 1-444. 1912.
MILLAIS, J. G. British diving ducks. London; 2 volumes. 1913.

HORSBRUGH, B. The game-birds and water-fowl of South Africa. London; 1-160. 1912.
FINN, F. The game birds of India and Asia. Calcutta; 1-177. 1911.
BAKER, E. C. S. The game birds of India, Burma, and Ceylon. Journ. Bombay Nat. Hist. Soc., 22, 1-12, 219-229, 427-433. 1913.
BAKER, E. C. S. The game-birds of India, Burma and Ceylon. London; 2 volumes. 1921.
WRIGHT, R. G. & DEWAR, D. The ducks of India.... London; 1-231. 1925.

Hybrids. COCKRUM, E. L. A check-list and bibliography of hybrid birds in North American north of Mexico. Wilson Bull., 64, 140-159. 1952.

Oceanic. KING, W. B. Preliminary identification manual. Seabirds of the tropical Pacific Ocean. Washington, D. C.; 1-126. 1967.
MURPHY, R. C. Oceanic birds of South America. New York; 2 volumes. 1936.

MONOGRAPHIC, DESCRIPTIVE, REVISIONARY WORKS

Monographs

EVANS, A. H. Birds. In Cambridge Natural History, IX, 1-635. 1899. (to family)
WYTSMAN, P. Genera avium. Bruxelles; 26 parts. 1905-1914.
REICHENOW, A. Die Vogel. Handbuch der systematischen Ornithologie. Stuttgart; 2 volumes. 1913-14.

Classifications

DUBOIS, A. Sur la classification des oiseaux. Rev. Francaise Ornith., 1913, 65-68, 81-83. 1913.
BOUBIER, M. Le monde des oiseaux. Systematique scientifique des ordres et des familles. Arch. Hist. Nat. Soc. Accl. France, 5, 1-212. 1930.
PERRIER, E. Les oiseaux. Traite de Zoologie, 9, 3117-3341. 1931. (to tribe)
WETMORE, A. A systematic classification for the birds of the world. Smithsonian Misc. Coll., 89, (13), 1-11. 1934.
ROMER, A. S. Vertebrate paleontology. Chicago; 1-687. 1945.
BERLIOZ, J. Oiseaux. Systematique. Traite de Zoologie, 15, 845-1055. 1950. (to family)
MAYR, E. & AMADON, D. A. A classification of Recent birds. American Mus. Novit., 1496, 1-42. 1951.
WALLACE, G. J. An introduction to ornithology. New York; 1-443. 1955. (Pp. 344-350)
MAYR, E. & GREENWAY, J. C., Jr. Sequence of passerine families (Aves). Breviora, 58, 1-11. 1956.
BLAIR, W. F. et al. Vertebrates of the United States. New York; 1-819. 1957. (Pp. 359-613)
VAN TYNE, J. & BERGER, A. J. Fundamentals of ornithology. New York; 1-624. 1959. (to family)
WETMORE, A. Classification for the birds of the world. Smithsonian Misc. Coll., 139, 1-37. 1960.
VERHEYEN, R. A new classification for the non-passerine birds of the world. Bull. Inst. Sci. Nat. Belge, 37, (27), 1-36. 1961.
RYDZEWSKI, W. On the classification for birds. Przegl. Zool., 6, 167-171. 1962.
THOMSON, A. L. et al. A new dictionary of birds. New York; 1-927. 1964.
BLAIR, W. F. et al. Vertebrates of the United States. New York; 1-616. 1968. (2nd ed., to orders)

Synonymic catalogs

(various authors). Catalogue of birds in the British Museum (Natural History). London; 27 vols. 1874-98.
SHARPE, R. B. A hand-list of the genera and species of birds. Nomenclator avium tum fossilium tum viventium. London; 4 volumes. 1899-1912.
PETERS, J. L. Checklist of the birds of the world. Cambridge, Massachusetts; 15 volumes. 1931
CORY, C. B., HELLMAYR, C. E. & CONOVER, B. Catalog of birds of the Americas and the adjacent islands. Field Mus. Nat. Hist. Publ. Zool., 13, 11 parts. 1918-1949.
CARRIKER, M. A., Jr. An annotated list of the birds of Costa Rica including Cocos Island. Ann. Carnegie Mus., 6, 314-915. 1910.
OLIVEIRO PINTO, O. M. Catalogo das Aves do Brasil. Sao Paulo; 2 parts; 1-566, 1-700. 1938-41.
STEULLET, A. B. & DEAUTIER, E. A. Catalogo sistematico de las Aves de la Republica Argentina. Obra Cincuentenarid Mus. La Plata, 1, 493-732, 733-932. 1939-45. 733-1006. 1946.
TROUESSART, E.-L. Catalogue des oiseaux d'Europe. Paris; 1-545. 1912.
DEMENTIEV, G. P. Systema avium rossicarum (Catalogue critique des oiseaux de l'U. R. S. S., vol. 1. Paris; 1-288. 1935.
SCLATER, W. L. Systema avium Ethiopicarum.... London; 1-922. 1924-30.
CLANCEY, P. A. A catalogue of birds of the South African sub-region. Durban; 1-633. 1965. Supplement 1: Durban Mus. Novit., 8, 275-324. 1969.
MATHEWS, G. M. A list of the birds of Australia. London; 1-332. 1913.
MATHEWS, G. M. Systema avium Australasianarum. London; 1-1047. 1927-30.

Descriptive catalogs

(various authors). Catalogue of birds in the British Museum (Natural History). London; 27 vols. 1874-98.
STUDER, T. & FATIO, V. Katalog der schweizerischen Vogel.... Bern; 17 Lieferungen, 6 volumes. 1889-1937.

Revisionary works by groups

Alciformes. VERHEYEN, R. Contribution a la systematique des Alciformes. Bull. Inst. Sci. Nat. Belg., 34, (45), 1-15. 1958.

Apterygiformes. Les kiwis (Apterygiformes) dans les systemes de classification. Bull. Soc. Zool. Anvers, 15, 1-11. 1960.

Charadriiformes. LOW, G. C. The literature of the Charadriiformes from 1894 to 1928 with a classification of the order ..., 2nd ed. London; 1-637. 1931. (1st edition, 1924)
VERHEYEN, R. Analyse du potentiel morphologique et projet d'une nouvelle classification de Charadriiformes. Bull. Inst. Sci. Nat. Belg., 34, (18), 1-35. 1958.
JUDIN, K. A. On mechanisms of the jaw in Charadriiformes.... Trud. Zool. Inst. Akad. Nauk S. S. S. R., 29, 257-302. 1961. (In Russian)
JUDIN, K. A. Phylogeny and classification of Charadriiformes. Fauna S. S. S. R., (n. s.), 91, 1-261. 1965. (In Russian)

Ciconiiformes. VERHEYEN, R. Contribution a l'anatomie a la systematique de base des Ciconiiformes (Parker 1868). Bull. Inst. Sci. Nat. Belg., 35, (24), 1-34. 1959.
VERHEYEN, R. What are Ciconiiformes. Proc. XII Intern. Ornith. Congr. Helsinki, 1958, -741-743. 1960.

Falconiformes. MENZBIER, M. A. Fauna de la Russie, vol. VI, livr. 1; Oiseaux (Aves Falconiformes). Petrograd; 1-344. 1916. (In Russian)

Galliformes. HOLMAN, J. A. Osteology of living and fossil New World quails (Aves, Galliformes). Bull. Florida State Mus., 6, 131-233. 1961.

Gruimorphae. LOWE, P. R. On the relations of the Gruimorphae.... Ibis, (13), 1, 491-534. 1931.

Humicolae. KOENIG, A. Die Saenger (Cantores) Aegyptens. Journ. Ornith., 72, Sonderheft, 1-277. 1924.

Lariformes. VERHEYEN, R. Note sur la systematique de base du Lariformes. Bull. Inst. Sci. Nat. Belg., 35, (9), 1-16. 1959.

Oscines. DELACOUR, J. & VAURIE, C. A classification of the Oscines. Contr. Sci. Los Angeles County Mus., 16, 1-6. 1957.

Passeriformes. ALESSANDRI, M. Proposte per una eventuale classificazione dei Passeriformi su base morfo-ecologia. Atti Accad. Ligure, 15, 326-339. 1959.

Pelecaniformes. LANHAM, U. N. Notes on the phylogeny of the Pelecaniformes. Auk, 64, 65-70. 1947.

Procellariiformes. MATHEWS, G. M. A check-list of the order Procellariiformes. Novit. Zool., 39, 151-206. 1934.
MURPHY, R. C. Oceanic birds of South America.... New York; 2 volumes. 1936.
MATHEWS, G. M. Key to the order Procellariiformes. Emu, 36, 40-48. 1936.
JUDIN, K. A. On mechanisms of the jaw in ... Procellariiformes.... Trud. Zool. Inst. Akad. Nauk S. S. S. R., 29, 257-302. 1961. (In Russian)
ALEXANDER, W. B. et al. The families of petrels and their names. Ibis, 107, 401-405. 1965.

Psittaciformes. VERHEYEN, R. Analyse du potential morphologique et project d'une nouvelle classification des Psittaciformes. Bull. Inst. Sci. Nat. Belg., 32, (55), 1-54. 1956.

Ratitae. ROTHSCHILD, W. On the forms and present distribution of the so-called Ratites.... Verh. V Intern. Ornith. Kongr. Berlin 1910, 144-174, 1000. 1911.
DUBOIS, A. Coup d'oeil sur les oiseaux Ratites. Bull. Soc. Zool. Paris, 37, 303-309; 38, 104-115. (Review; key to genera & species)

Sphenisci. MATHEWS, G. M. New genera and rearrangement of the families of the Sphenisci. Bull. Brit. Ornith. Club., 55, 74-75. 1934.

Telmatomorphae. LOWE, P. E. An anatomical review of ... Telmatomorphae.... Ibis, 1931, 712-771.

Tinami. von BOETTICHER, H. Beitrag zu einem phylogenetisch begrundeten naturlichen System ... auf Grund einer taxonomisch verwertbaren Charaktere. Jena Zeitschr. Naturw., 69, 169-192. 1934.

Tubinares. LOWE, P. R. On the classification of the Tubinares or petrels. Proc. Zool. Soc. London, 1925, 1433-1443. 1926.
von BOETTICHER, H. Die Gefiederfarbung der Sturmvogel.... Jena Zeitschr. Naturw., 67, 107-123. 1932.

Turniciformes. VERHEYEN, R. Contribution au demembrement de l'ordo artificiel des Gruiformes... IV. Les Turniciformes. Bull. Inst. Sci. Nat. Belg., 34, (2), 1-18. 1958.

Revisionary works by family

These lists of works were made up primarily from two sources: (1) listings in the Zoological Record for the years 1910 to 1967, which were not checked further if the title appeared to be explicit; and (2) library shelves and the office libraries of various specialists in each field; these were examined for appropriateness and, if revisionary, checked for included families. The works listed by families are therefore of two sorts, so far as examination by the compiler is concerned, those cited in the title as referring to one or a few named families (not usually examined or verified) and those directly examined and checked and listed as considered to be appropriate.

Works cited by author

In order to save space in the list of works on individual families, the following studies, which are each referred to more than a few times, are cited by author and date and number only. Citations are otherwise given by title, if a book, or by the serial reference, if not a separate work. The titles of the journal articles are not usually given, but the relevant limitations are shown by annotations.

ALI, S. & RIPLEY, S. D., No. 1, 1968. Handbook of the birds of India and Pakistan together with those of Nepal, Sikkim, Bhutan, and Ceylon. Vol. I. London.
ALI, S. & RIPLEY, S. D., No. 2, 1969. Same, vol. II.
ALI, S. & RIPLEY, S. D., No. 3, 1969. Same, vol. III.
ALI, S. & RIPLEY, S. D., No. 4, 1970. Same, vol. IV.
BAKER, E. C. S., No. 1, 1922. Fauna of British India: Birds, 2nd ed. London. Volume I.
BAKER, E. C. S., No. 2, 1924. Same, volume II.
BAKER, E. C. S., No. 3, 1926. Same, volume III.
BAKER, E. C. S., No. 4, 1927. Same, volume IV.
BAKER, E. C. S., No. 5, 1928. Same, volume V.
BAKER, E. C. S., No. 6, 1929. Same, volume VI.
Baker, R. H. No. 1, 1951. The avifauna of Micronesia.... Publ. Univ. Kansas Mus. Nat. Hist., 3, 1-359.
BANNERMAN, D. A., No. 1, 1930. The birds of tropical West Africa. London. Volume 1.
BANNERMAN, D. A., No. 2, 1931. Same, volume 2.
BANNERMAN, D. A., No. 3, 1933. Same, volume 3.
BANNERMAN, D. A., No. 4, 1936. Same, volume 4.
BANNERMAN, D. A., No. 5, 1939. Same, volume 5.
BANNERMAN, D. A., No. 6, 1948. Same, volume 6.
BANNERMAN, D. A., No. 7, 1949. Same, volume 7. (Volume 8 has no text)
BLANFORD, W. T., No. 3, 1895. The fauna of British India.... Birds, volume III. London.
BLANFORD, W. T., No. 4, 1898. Same, volume IV.
BOUCARD, A., No. 1, 1893-95. Genera of humming birds. Being also a complete monograph of these birds. London; 1-412.
BOUET, G., No. 1, 1955. Oiseaux de l'Afrique tropicale. Part i. Faune Union France, 16, 1-412.
CHAPIN, J. P., No. 1, 1932. The birds of the Belgian Congo, part I. Bull. American Mus. Nat. Hist., 65, 1-756.
CHAPIN, J. P., No. 2, 1939. Same, part II.
CHAPIN, J. P., No. 3, 1953. Same, part III.
CHAPIN, J. P., No. 4, 1954. Same, part IV.
CORY, C. B., No. 1, 1918. Catalogue of birds of the Americas, part II, number 1. Field Museum Nat. Hist., Zool., 13, 1-315.
CORY, C. B., No. 2, 1919. Same, number 2.
CORY, C. B. & HELLMAYR, C. E., No. 3, 1924. Same, part III. (See also Hellmayr et al.)
DEIGNAN, H. G., No. 1, 1945. The birds of northern Thailand. United States Nat. Mus. Bull., 186, 1-616.
DEMENTIEV, G. P. et al., No. 1, 1951. Birds of the Soviet Union. Moskva. (In Russian; sometimes also listed by individual authors)
DEMENTIEV, G. P. et al., No. 2, 1951. Same, vol. 2.
DEMENTIEV, G. P. et al., No. 3, 1951. Same, vol. 3.
DEMENTIEV, G. P. et al., No. 4, 1952. Same, vol. 44.
DEMENTIEV, G. P. et al., No. 5, 1952. Same, vol. 5.
DEMENTIEV, G. P. et al., No. 6, 1954. Same, vol. 6. (See also translations below)
DEMENT'EV, G. P. & GLADKOV, N. A., No. 7, 1966. Birds of the Soviet Union. Jerusalem, vol. I. Translation of No. 1, above)
DEMENT'EV, G. P. & GLADKOV, N. A., No. 8, 1968. Same, vol. II.
DEMENT'EV, G. P. & GLADKOV, N. A., No. 9, 1969. Same, vol. III.
DEMENT'EV, G. P. & GLADKOV, N. A., No. 10, 1967. Same, vol. IV.
DEMENT'EV, G. P. & GLADKOV, N. A., No. 11, 1970. Same, vol. V.
DEMENT'EV, G. P. & GLADKOV, N. A., No. 12, 1968. Same, vol. VI.
DEVINCENZI, G. L., No. 1, 1926. Aves del Uruguay. Catalogo descriptivo, I, Anal. Mus. Hist. Nat. Montevideo, (2), 2, 129-200.
DEVINCENZI, G. L., No. 2, 1927. Same, pages 215-264.
DEVINCENZI, G. L., No. 3, 1928. Same, pages 339-407.
FORBUSH, E. H., No. 1, 1927. Birds of Massachusetts ..., Part II. Norwood, Massachusetts; 1-461.
FORBUSH, E. H., No. 2, 1929. Same, part I; 1-481.
FORBUSH, E. H., No. 3, 1929. Same, part III; 1-466.

FRIEDMANN, H., No. 1, 1930. Birds collected by the Childs Frick Expedition to Ethiopia and Kenya Colony. Part 1. Non-Passeres. United States Nat. Mus. Bull., 153, (1), 1-516.
FRIEDMANN, H., No. 2, 1937. Same, Part 2. Passeres. Bull., 153, (2), 1-506.
FREIDMANN, H., No. 3, 1950. The birds of North and Middle America. Part XI. United States Nat. Mus. Bull., 50, (XI), 1-793. (See Ridgway, 1901-50)
GLADKOV, N. A. et al., No. 1, 1964. A study of the birds of the U. S. S. R. (Russia); 1-536. (In Russian)
HARTERT, E., No. 1, 1903-10. Die Vogel der palaarktischen Fauna, volume 1. Berlin; 1-817.
HARTERT, E., No. 2, 1912-21. Same, volume 2; 833-1764.
HARTERT, E., No. 3, 1921-22. Same; volume 3; 1765-2328.
HELLMAYR, C. E., No. 4, 1925. Catalogue of birds of the Americas, part IV; 1-390. (See also Cory and Hellmayr & Conover)
HELLMAYR, C. E., No. 5, 1927. Same, part V.
HELLMAYR, C. E., No. 6, 1929. Same, part VI.
HELLMAYR, C. E., No. 7, 1934. Same, part VII.
HELLMAYR, C. E., No. 8, 1935. Same, part VIII.
HELLMAYR, C. E., No. 9, 1936. Same, part IX.
HELLMAYR, C. E., No. 10, 1937. Same, part X.
HELLMAYR, C. E., No. 11, 1938. Same, part XI.
HELLMAYR, C. E. & CONOVER, B., No. 1, 1942. Same, part I, no. 1.
HELLMAYR, C. E. & CONOVER, B., No. 2, 1948. Same, part I, no. 2.
HELLMAYR, C. E. & CONOVER, B., No. 3, 1948. Same, part I, no. 3.
HELLMAYR, C. E. & CONOVER, B., No. 4, 1949. Same, part I, no. 4.
KURODA, N., No. 1, 1933. Birds of the island of Java. Tokyo; volume 1; 1-370. (Passeres)
KURODA, N., No. 2, 1936. Same, volume 2; 371-794. (Non-Passeres)
LINTIA, D., No. 1, 1946. Pasarile din R. P. R. Bucuresti; 1-298. Volume I.
LINTIA, D., No. 2, 1954. Same, volume II.
LINTIA, D., No. 3, 1955. Same, volume III.
LOW, G. C., No. 1, 1931. The literature of the Charadriiformes from 1894 to 1928, with a classification of the order, and lists of the genera, species, and subspecies, 2nd ed. London; 1-637.
McGREGOR, R. C., No. 1, 1909. A manual of Philippine birds. Part I. Galliformes to Eurylaemiformes. Manila; 1-412.
McGREGOR, R. C., No. 2, 1909. Same, part II. Passeriformes. Manila; 413-769.
MURPHY, R. C., No. 1, 1936. Oceanic birds of South America.... New York; 1-1245. (Two volumes)
PARIS, P., No. 1, 1921. Faune de France. 2. Oiseaux. Paris; 1-473.
PETERS, J. L., No. 1, 1931. Check-list of birds of the world. Cambridge, Massachusetts. Volume I.
PETERS. J. L., No. 2, 1963. Same, volume II.
PETERS, J. L., No. 3, 1937. Same, volume III.
PETERS, J. L., No. 4, 1940. Same, volume IV.
PETERS, J. L., No. 5, 1955. Same, volume V.
PETERS, J. L., No. 6, 1948. Same, volume VI.
PETERS, J. L., No. 7, 1951. Same, volume VII.
PETERS, J. L., No. 8. (Not yet issued)
PETERS, J. L., No. 9, 1960. Same, volume IX.
PETERS, J. L., No. 10, 1964. Same, volume X.
PETERS, J. L., No. 11. (Not yet issued)
PETERS, J. L., No. 12, 1967. Same, volume XII.
PETERS, J. L., No. 13, 1970. Same, volume XIII.
PETERS, J. L., No. 14, 1968. Same, volume XIV.
PETERS, J. L., No. 15, 1962. Same, volume XV.
REICHENOW, A., No. 1, 1901. Die Vogel Afrikas. Erster Band. Neudamm; 1-706.
REICHENOW, A., No. 2, 1903. Same, Zweiter Band.
REICHENOW, A., No. 3, 1905. Same, Dritter Band.
REICHENOW, A., No. 4, 1913. Same, Vierter Band.
REICHENOW, A., No. 5, 1914. Same, Funfter Band.
RIDGWAY, R., No. 1, 1901. The birds of North and Middle America.... Part 1. United States Nat. Mus. Bull., 50, (1), 1-715.
RIDGWAY, R., No. 2, 1902. Same, part 2.
RIDGWAY, R., No. 3, 1904. Same, part 3.
RIDGWAY, R., No. 4, 1907. Same, part 4.
RIDGWAY, R., No. 5, 1911. Same, part 5.
RIDGWAY, R., No. 6, 1914. Same, part 6.
RIDGWAY, R., No. 7, 1916. Same, part 7.
RIDGWAY, R., No. 8, 1919. Same, part 8.
RIDGWAY, R., No. 9, 1941. Same, part 9. (With H. Friedmann)
RIDGWAY, R., No. 10, 1946. Same, part 10. (With H. Friedmann)
SALVIN, O. & GODMAN, F. D., No. 1, 1879-1904. Biologia Centrali-Americana, Aves, 4 volumes. Vol. I, 1-512.
SALVIN, O. & GODMAN, F. D., No. 2, 1879-1904. Same, volume II.
SALVIN, O. & GODMAN, F. D., No. 3, 1879-1904. Same, volume III. (Volume IV is all plates)
van SOMEREN, V. G. L., No. 1. 1925-36. The birds of Kenya and Uganda. Part I. Guinea fowl. Part II. Phasianidae. Journ. East African Uganda Nat. Hist. Soc., 22, 1-21; 23, 95-104.
TUGARINOV, A. J., No. 1, 1947. Faune de l'U. R. S. S., Oiseaux, I, 3. Pelecaniformes, Ciconiiformes, Phoenicopteriformes. Moskva; 125-316.

WAIT, W. E., No. 1, 1920. The picarian birds and parrots of Ceylon. Spol. Zeylanica, 11, 197-272. 1920.
WAIT, W. E., No. 2, 1921. The owls and diurnal birds of prey found in Ceylon. Spol. Zeylanica, 11, 317-380.
WETMORE, A., No. 1, 1927. The birds of Porto Rico and the Virgin Islands. Colymbiformes to Columbiformes. Sci. Surv. Porto Rico..., vol. 9, (3), 245-598.
WETMORE, A., No. 2, 1965. The birds of the Republic of Panama. Part I. Smithsonian Misc. Coll., 150, part 1, 1-483.
WETMORE, A., No. 3, 1968. The birds of the Republic of Panama. Part II. Columbiformes (pigeons) to Picidae (woodpeckers). Smithsonian Misc. Coll., 150, (2), 1-605.

Families alphabetically

Acanthisittidae (incl. Traversiidae, Xenicidae, Xenicornithidae). (No revisionary references noted)
Acanthizidae, as Sylviidae
Accentoridae, as Prunellidae
Accipitridae (incl. Aegyptiidae, Aegypiidae, Aquilidae, Buteonidae). SWANN, H. K., 1919-22, A synopsis of the Accipitres..., 2nd ed.; London, 1-233, (as Aegypiidae & Aquilidae); SWANN, H. K. & WETMORE, A., A monograph of the birds of prey..., London, 2 volumes (as Aegypiidae); PETERS, J. L., No. 1, 1931 (synonymic catalog); BROWN, L. & AMADON, D., Eagles, hawks, and falcons of the world, Middlesex, England, 2 volumes, 1968; HELLMAYR & CONOVER, No. 4, 1949 (the Americas, catalog); FRIEDMANN, H., No. 3, 1950 (North & Middle America); FORBUSH, E. H., No. 1, 1927 (Massachusetts); WETMORE, A., No. 1, 1927 (Puerto Rico); WETMORE, A., No. 2, 1965 (Panama); DUGAND, A., 1941, Rev. Acad. Col., 4, 394-404 (Colombia, keys); HARTERT, E., No. 2, 1912-21 (Palearctic, as Aquilidae); LINTIA, D., No. 2, 1954 (Romania, as Aquilidae); STEGMANN, B., 1937, Faune de l'U. R. S. S., Oiseaux, I, 5, Moskva, 1-294 (as Aquilidae); DEMENTIEV & GLADKOV, No. 1, 1951 & No. 7, 1966 (Soviet Union); GLADKOV, N. A. et al., No. 1, 1964 (U. S. S. R.); BOUET, G., No. 1, 1955 (Afrique tropicale, as Aegypiidae & Aquilidae); BANNERMAN, No. 1, 1930 (tropical West Africa, as Aegyptiidae); CHAPIN, J. P., No. 1, 1932 (Belgian Congo, as Aegypiidae & Aquilidae); FRIEDMANN, H., No. 1, 1930 (Ethiopia & Kenya); ALI & RIPLEY, No. 1, 1968 (India & Pakistan); BAKER, E. C. S., No. 5, 1928 (British India, as Aegypiidae); DEIGNAN, H. G., No. 1, 1945 (Thailand); BAKER, R. H., No. 1, 1951 (Micronesia)
Aegialornithidae (only as fossils)
Aegithalidae. PETERS, J. L., No. 12, 1967 (world list)
Aegithinidae, as Irenidae
Aegothelidae. PETERS, J. L., No. 4, 1940 (world list)
Aegypiidae, as Accipitridae
Aegyptiidae, as Accipitridae
Aepyornithidae (only as fossils)
Aerocharidae, as Vangidae
Agnopteridae (only as fossils)
Aladinidae. BANNERMAN, D. A., No. 3, 1933 (tropical West Africa)
Alaudidae. REICHENOW, A., No. 5, 1914 (monograph); MEINERTZHAGEN, R., 1951, Proc. Zool. Soc. London, 121, 81-132 (revision); PETERS, J. L., No. 9, 1960 (world list); HELLMAYR, C. E., No. 8, 1935 (the Americas, catalog); RIDGWAY, R., No. 4, 1907 (North America); FORBUSH, E. H., No. 1, 1927 (Massachusetts); SALVIN, O. & GODMAN, F. D., No. 1, 1879-1904 (Central America); HARTERT, E., No. 1, 1903-10 (Palearctic); PARIS, P., No. 1, 1921 (France); LINTIA, D., No. 1, 1946 (Romania); DEMENTIEV & GLADKOV, No. 5, 1954 & No. 11, 1970 (Soviet Union); GLADKOV, N. A. et al., No. 1, 1964 (U. S. S. R.); VAURIE, C., 1951, Bull. American Mus. Nat. Hist., 97, 431-526 (Asiatic); REICHENOW, A., No. 3, 1905 (Africa); KOENIG, A., 1924, Journ. Ornith., 72, (Sonderheft), 1-277 (Egypt); BANNERMAN, D. A., No. 4, 1936 (tropical West Africa); CHAPIN, J. P., No. 3, 1953 (Belgian Congo); FRIEDMANN, H., No. 2, 1937 (Ethiopia & Kenya); BAKER, E. C. S., No. 3, 1926 (India); DEIGNAN, H. G., No. 1, 1945 (Thailand); KURODA, N., No. 1, 1933 (Java); McGREGOR, R. C., No. 2, 1909 (Philippines)
Alcedinidae (incl. Dacelonidae, Halcyonidae). SHARPE, R. B., 1868-71, Monograph of Alcedinidae..., London; MILLER, W. W., 1912, Bull. American Mus. Nat. Hist., 31, 239-311 (revision); REICHENOW, A., No. 5, 1914 (monograph); LAUBMANN, A., 1924, Verh. Ornith. Ges. Bayern, 16, 129-138 (conspectus generum); PETERS, J. L., No. 5, 1955 (world list); CORY, C. B., No. 1, 1918 (the Americas, catalog); RIDGWAY, R., No. 6, 1914 (North America); FORBUSH, E. H., No. 1, 1927 (Massachusetts); SALVIN, O. & GODMAN, F. D., No. 2, 1888-1904 (Central America); WETMORE, A., No. 3, 1968 (Panama); PEREYRA, J. A., 1932, El Hornero, 5, 51-53 (Argentina); HARTERT, E., No. 2, 1912-21 (Palearctic); PARIS, P., No. 1, 1921 (France); LINTIA, D., No. 2, 1954 (Romania); DEMENTIEV, & GLADKOV, No. 1, 1951 & No. 7, 1966 (Soviet Union); GLADKOV, N. A. et al., No. 1, 1964 (U. S. S. R., as Halcyonidae); REICHENOW, A., No. 2, 1903 (Africa); CHAPIN, J. P., No. 2, 1939 (Belgian Congo); FRIEDMANN, H., No. 1, 1930 (Ethiopia & Kenya); ALI, S. & RIPLEY, S. D., No. 4, 1970 (India & Pakistan); BLANFORD, W. T., No. 3, 1895 (British India); BAKER, E. C. S., No. 4, 1927 (British India); WAIT, W. E., No. 1, 1920 (Ceylon); DEIGNAN, H. G., No. 1, 1945 (Thailand); LAUBMANN, A., 1925, Arch. Naturg., 90, (A7), 55-153 (Sumatra); KURODA, N., No. 2, 1936 (Java); McGREGOR, R. C., No. 1, 1909 (Philippines); HUTSON, H. P. W., 1931, The birds of Hong Kong, VI, Hong Kong; BAKER, R. H., No. 1, 1951 (Micronesia)

Alcidae (incl. Uriidae). REICHENOW, A., No. 4, 1913 (monograph); DAWSON, W. L., 1920, Journ. Mus. Comp. oology Sta. Barbara, 1, 1-14 (oological revision); PETERS, J. L., No. 2, 1963 (world list); HELLMAYR & CONOVER, No. 3, 1948 (the Americas, catalog); RIDGWAY, R., No. 8, 1919 (North America); FORBUSH, E. H., No. 2, 1929 (Massachusetts); SALVIN, O. & GODMAN, F. D., No. 3, 1897-1904 (Central America); SALOMONSEN, F., 1944, K. Vet. O. Vitterh. Samb. Handl., F6, B, 3, (5), 1-138 (Atlantic); PARIS, P., No. 1, 1921 (France); DEMENTIEV & GLADKOV, No. 2, 1951 & No. 8, 1968 (Soviet Union); KOZLOVA, E. V., 1957, Fauna of U. S. S. R., Birds, II, 3 (in Russian); KOZLOVA, E. V., 1961, translation of preceding; GLADKOV, N. A. et al., No. 1, 1964 (U. S. S. R.); UDRARDY, M. D. F., 1963, 10th Pacific Sci. Congr., 1961, 85-111 (Pacific)

Aluconidae, as Tytonidae

Amaziliidae. BOUCARD, A., No. 1, 1893-95 (monograph)

Ampelidae, as Bombycillidae

Anatidae. REICHENOW, A., No. 4, 1913 (monograph); PHILLIPS, J. C., 1923-26, (A natural history of the ducks, London, 4 volumes (with bibliography); PHILLIPS, J. C., 1926, Bibliography of ... ducks, Cambridge, Massachusetts; von BOETTICHER, H., 1930, Anz. Ornith. Ges. Bayern, 2, 94-100 (Klassification); PETERS, J. L., No. 1, 1931 (world List); DELACOUR, J., 1936, Oiseaux, 6, 366-379 (classification); von BOETTICHER, H., 1939, Koczag, 9-11, 47-53 (Klassification); von BOETTICHER, H., 1942, Zool. Anz., 140, 37-48 (classification); KURODA, N., 1942, A bibliography of the duck tribe Anatidae, mostly 1926-1940..., Tokyo, 1-852; DELACOUR, J. & MAYR, E., 1945, Wilson Bull., 57, 1-55; YAMASHINA, Y., 1947, Zoo.. Mag. Tokyo Imp. Univ., 57, 12 (new classification); YAMASHINA, Y., 1952, Coord. Committ. Res. Genet., Yamashina Ornith. Inst., 3, 1-34 (classification on cytogenetics); JOHNSGARD, P. A., 1961, Ibis, 103a, 71-85 (behavioral taxonomy); SCOTT, P., 1965, Coloured key to wildfowl of world, 2nd ed., London, 1-91; HELLMAYR & CONOVER, No. 2, 1948 (the Americas, catalog); KORTRIGHT, F. H., 1960, The ducks, geese and swans of North America, Harrisburg, Pennsylvania, 1-476; FORBUSH, E. H., No. 2, 1929 (Massachusetts); WETMORE, A., No. 1, 1927 (Puerto Rico); SALVIN, O. & GODMAN, F. D., No. 3, 1897-1904 (Central America); WETMORE, A., No. 2, 1965 (Panama); MURPHY, R. C., No. 1, 1936 (South America); DEVINCENZI, G. J., No. 3, 1928 (Uruguay); HARTERT, E., No. 2, 1912-21 (Palearctic); VERHEYEN, R., 1944, Les Anatides de Belgique, 4th ed., Bruxelles, 1-140; von BOETTICHER, H., 1929, Anz. Ornith. Ges. Bayern, 2, 10-15; PARIS, P., No. 1, 1921 (France); DOMANIEWSKI, J., 1925, Spraw. Kom. Fisjogr. Polskiej Akad. Umiez, 1925, 113-127 (Poland, in Polish); BAUER, K. M. & GLUTZ VON BLATZHEIM, U. N., 1969, Handbuch der Vogel Mitteleuropas, Anseriformes, Frankfurt, 1-535; LINTIA, D., No. 3, 1955 (Romania); TOUGARINOV, A. J., 1941, Fauna of U. S. S. R., Aves, I, 4, Anseriformes, Moskva, 1-383 (In Russian); DEMENTIEV & GLADKOV, No. 4, 1952 & 10, 1967 (Soviet Union); GLADKOV, N. A. et al., No. 1, 1964 (U. S. S. R.); REICHENOW, A., No. 1, 1901 (Africa); BOUET, G., No. 1, 1955 (Afrique tropicale); BANNERMAN, D. A., No. 1, 1930 (tropical West Africa); SCHOUTEDEN, H., 1929, Bull. Cercle Zool. Congo, 5, 104-116 Belgian Congo, list); CHAPIN, J. P., No. 1, 1932 (Belgian Congo); van SOMEREN, V. G. L., No. 1, 1925-36 (Kenya & Uganda); FRIEDMANN, H., No. 1, 1930 (Ethiopia & Kenya); BLANFORD, W. T., No. 4, 1898 (British India); BAKER, E. C. S., No. 6, 1929 (British India); ALI & RIPLEY, No. 1, 1968 (India, Pakistan); BAKER, E. C. S., 1915, Journ. Bombay Nat. Hist. Soc., 23, 454-459 (swans of India); DEIGNAN, H. G., No. 1, 1945 (Thailand); KURODA, N., No. 2, 1936 (Java); McGREGOR, R. C., No. 1, 1909 (Philippines); BAKER, R. H., No. 1, 1951 (Micronesia)

Anhimidae (incl. Palamedeidae). REICHENOW, A., No. 4, 1913 (monograph); PETERS, J. L., No. 1, 1931 (world list); HELLMARY & CONOVER, No. 2, 1948 (the Americas, catalog); DEVINCENZI, G. L., No. 3, 1928 (Uruguay)

Anhingidae (incl. Plotidae). PETERS, J. L., No. 1, 1931 (world list); HELLMAYR & CONOVER, No. 2, 1948 (the Americas, catalog); SALVIN & GODMAN, No. 3, 1897-1904 (Central America, as Plotidae); WETMORE, A., No. 2, 1965 (Panama); BANNERMAN, D. A., No. 1, 1930 (tropical West Africa); CHAPIN, J. P., No. 1, 1932 (Belgian Congo); FRIEDMANN, H., No. 1, 1930 (Ethiopia & Kenya); DEIGNAN, H. G., No. 1, 1945 (Thailand); McGREGOR, R. C., No. 1, 1909 (Philippines)

Anomalopterygidae. ARCHEY, G., 1941, Bull. Auckland Inst. Mus., 1, 1-120 (the moa)

Anseridae. REICHENOW, A., No. 4, 1913 (monograph)

Apatornithidae (only as fossils)

Aphrizidae, as Scolopacidae

Apodidae (incl. Cypselidae, Micropidae, Micropodidae, Triopidae). PETERS, J. L., No. 4, 1940 (world list); LACK, D., 1956, Auk, 73, 1-32 (revision of genera of swifts); CORY, C. B., No. 1, 1918 (the Americas, catalog, as Cypselidae); RIDGWAY, R., No. 5, 1911 (North America, as Micropodidae); FORBUSH, E. H., No. 1, 1927 (Massachusetts, as Micropodidae); SALVIN & GODMAN, No. 2, 1888-1904 (Central America, as Cypselidae); WETMORE, A., No. 3, 1968 (Panama); DABBENE, R., 1917, El Hornero, 1, 3-8 (Argentina, as Cypselidae); HARTERT, E., No. 2, 1912-21 (Palearctic, as Cypselidae); PARIS, P., No. 1, 1921 (France, as Cypselidae); LINTIA, D., No. 2, 1954 (Romania, as Cypselidae); DEMENTIEV & GLADKOV, No. 1, 1951 & No. 7, 1966 (U. S. S. R.); GLADKOV, N. A. et al., No. 1, 1964 (U. S. S. R); BANNERMAN, D. A., No. 3, 1933 (tropical West Africa, as Micropodidae); CHAPIN, J. P., No. 2, 1939 (Belgian Congo, as Micropodidae); FRIEDMANN, H., No. 1, 1930 (Ethiopia & Kenya, as Micropodidae); MOLTONI, E., 1940, Riv. Italiana Ornith., 10, 143-162 (Africa Orientale Italiana, as Micropodidae); BLANFORD, W. T., No. 3, 1895 (British India, as Cypselidae); BAKER, E. C. S., No. 4, 1927 (British India, as Micropidae); ALI & RIPLEY, No. 4, 1970 (India &

Pakistan); WAIT, W. E., No. 1, 1920 (Ceylon, as Cypselidae); KURODA, N., No. 2, 1936 (Java, as Micropodidae); McGREGOR, R. E., No. 1, 1909 (Philippines, as Micropodidae); BAKER, R. H., No. 1, 1951 (Micronesia)

Aptenodytidae, as Spheniscidae

Apterygidae. REICHENOW, A., No. 4, 1913 (monograph); BRASIL, L., 1913, Genera Avium, 22, 1-4; PETERS, J. L., No. 1, 1931 (world list)

Aquilidae, as Accipitridae

Aramidae. REICHENOW, A., No. 4, 1913 (monograph); PETERS, J. L., No. 2, 1963 (world list); HELLMAYR & CONOVER, No. 1, 1942 (the Americas, catalog); WETMORE, A., No. 1, 1927 (Puerto Rico); SALVIN, & GODMAN, No. 3, 1897-1904 (Central America); WETMORE, A., No. 2, 1965 (Panama); DEVINCENZI, G. J., No. 3, 1928 (Uruguay)

Archaeopterygidae (only as fossils)

Archaeornithidae (only as fossils)

Ardeidae. REICHENOW, A., No. 4, 1913 (monograph); PETERS, J. L., No. 1, 1931 (world list); BOCK, W. J., 1956, American Mus. Novit., 1779, 1-49 (generic review); HELLMAYR & CONOVER, No. 2, 1948 (the Americas, catalog); PEARSON, T. G., 1924, Bull. Nat. Assoc. Audubon Soc., 5, 1-38 (herons of United States); FORBUSH, E. H., No. 2, 1929 (Massachusetts); WETMORE, A., No. 1, 1927 (Puerto Rico); SALVIN & GODMAN, No. 3, 1897-1904 (Central America); WETMORE, A., No. 2, 1965 (Panama); DEVINCENZI, G. J., No. 3, 1928 (Uruguay); HARTERT, E., No. 2, 1912-21 (Palearctic); PARIS, P., No. 1, 1921 (France); LINTIA, D., No. 3, 1955 (Romania); TUGARINOV, A. J., No. 1, 1947 (U. R. S. S.); DEMENTIEV & GLADKOV, No. 2, 1951 & No. 8, 1968 (Soviet Union); GLADKOV, N. A., et al., No. 1, 1964 (U. S. S. R.); REICHENOW, A., No. 1, 1901 (Africa); BOUET, G., No. 1, 1955 (Afrique tropicale); BANNERMAN, D. A., No. 1, 1930 (tropical West Africa); CHAPIN, J. P., No. 1, 1932 (Belgian Congo); SCHOUTEDEN, H., 1933, Bull. Cercle Zool. Congo, 9, 83-99 (Congo); FRIEDMANN, H., No. 1, 1930 (Ethiopia & Kenya); BLANFORD, W. T., No. 4, 1898 (British India); BAKER, E. C. S., No. 6, 1929 (British India); ALI & RIPLEY, No. 1, 1968 (India, Pakistan); DEIGNAN, H. G., No. 1, 1945 (Thailand); KURODA, N., No. 2, 1936 (Java); McGREGOR, R. C., No. 1, 1909 (Philippines); BAKER, R. H., No. 1, 1951 (Micronesia)

Arenariidae. RIDGWAY, R., No. 8, 1919 (North America)

Artamidae. REICHENOW, A., No. 5, 1914 (monograph); PETERS, J. L., No. 15, 1962 (world list); REICHENOW, A., No. 1, 1903 (Africa); BAKER, E. C. S., No. 2, 1924 (British India); DEIGNAN, H. G.,, No. 1, 1945 (Thailand); KURODA, N., No. 1, 1933 (Java); McGREGOR, No. 2, 1909 (Philippines); BAKER, R. H., No. 1, 1951 (Micronesia)

Asionidae, as Strigidae

Atrichiidae, as Atrichornithidae

Atrichornithidae (incl. Atrichiidae). (No revisionary references noted)

Attagidae, as Thinocoridae

Balaenicipidae, as Balaenicipitidae

Balaenicipitidae (incl. Balaenicipidae). REICHENOW, A., No. 4, 1913 (monograph); PETERS, J. L., No. 1, 1931 (world list); REICHENOW, A., No. 1, 1901 (Africa); BOUET, G., No. 1, 1955 (Afrique tropicale); BANNERMAN, D. A., No. 1, 1930 (tropical West Africa); CHAPIN, J. P., No. 1, 1932 (Belgian Congo)

Balearididae, as Gruidae

Baptornithidae (only as fossils)

Bathornithidae (only as fossils)

Bombycillidae (incl. Ampelidae, Hypocoliidae). PETERS, J. L., No. 9, 1960 (world list); HELLMAYR, C. E., No. 8, 1935 (the Americas, catalog); RIDGWAY, R., No. 3, 1904 (North America); FORBUSH, E. H., No. 3, 1929 (Massachusetts); SALVIN & GODMAN, No. 1, 1879-1904 (Central America, as Ampelidae); HARTERT, E., No. 1, 1903-10 (Palearctic, as Ampelidae); PARIS, P., No. 1, 1921 (France, as Ampelidae); LINTIA, D., No. 1, 1946 (Romania, as Ampelidae); DEMENTIEV & GLADKOV, No. 6, 1954 & No. 12, 1968 (Soviet Union); GLADKOV, N. A. et al., No. 1, 1964 (U. S. S. R.); BAKER, E. C. S., No. 3, 1926 (British India)

Brachypodidae, as Pycnonotidae

Brachypteraciidae, as Coraciidae

Brontornithidae (only as fossils)

Bubalornithidae, as Ploceidae

Bubonidae, as Strigidae

Bucconidae. SCLATER, P. L., 1879-82, A monograph of jacamas and puff-birds, London; REICHENOW, A., No. 5, 1914 (monograph); STEINBACHER, J., 1937, Arch. Naturg., 6, 417-515 (anatomy); PETERS, J. L., No. 6, 1948 (world list); CORY, C. B., No. 2, 1919 (the Americas, catalog); RIDGWAY, R., No. 6, 1914 (North America); SALVIN & GODMAN, No. 2, 1888-1904 (Central America); WETMORE, A., No. 3, 1968 (Panama)

Bucerotidae. ELLIOT, D. G., 1877-82, A monograph of the Bucerotidae, London; DUBOIS, A., 1911, Genera Avium, 13, 1-24; REICHENOW, A., No. 5, 1914 (monograph); PETERS, J. L., No. 5, 1955 (world list); SANFT, K., 1960, Das Tierreich, 76, 1-174; REICHENOW, A., No. 2, 1903 (Africa); MOREAU, R. E., 1937, Proc. Zool. Soc. London, 107A, 331-346 (comparative breeding biology); BANNERMAN, D. A., No. 3, 1933 (tropical West Africa); CHAPIN, J. P., No. 2, 1939 (Belgian Congo); VRYDAGH, J. M., 1941, Bull. Soc. Bot. Zool. Congolaise, 4, (2), 25-28 (Congo); FRIEDMANN, H., No. 1, 1930 (Ethiopia & Kenya); BLANFORD, W. T., No. 3, 1895 (British India); BAKER, E. C. S., No. 4, 1927 (British India); ALI & RIPLEY, No. 4, 1970 (India, Pakistan); WAIT, W. E., No. 1, 1920 (Ceylon); DEIGNAN, H. G., No. 1, 1945

(Thailand); KURODA, N., No. 2, 1936 (Java); McGREGOR, R. C., No. 1, 1909 (Philippines)
Buphagidae, as Sturnidae
Burhinidae (incl. Oedicnemidae). LOW, G. C., No. 1, 1931 (bibliography); LOWE, P. R., 1931, Ibis, (13), 1, 491-534 (relations of Gruimorphae); PETERS, J. L., No. 2, 1963 (world list); HELLMAYR & CONOVER, No. 3, 1948 (the Americas, catalog); RIDGWAY, R., No. 8, 1919 (North America, as Oedicnemidae); SALVIN & GODMAN, No. 3, 1897-1904 (Central America, as Oedicnemidae); HARTERT, E., No. 2, 1912-21 (Palearctic); PARIS, P., No. 1, 1921 (France, as Oedicnemidae); LINTIA, D., No. 3, 1955 (Romania); DEMENTIEF & GLADKOV, No. 3, 1951 & No. 9, 1969 (Soviet Union); GLADKOV, N. A. et al., No. 1, 1964 (U. S. S. R.); BANNERMAN, D. A., No. 2, 1931 (tropical West Africa); BOUET, G., No. 1, 1955 (Afrique tropicale); CHAPIN, J. P., No. 2, 1939 (Belgian Congo); van SOMEREN, V. G. L., No. 1, 1925-36 (Kenya & Uganda); FRIEDMANN, H., No. 1, 1930 (Ethiopia & Kenya, as Oedicnemidae); BLANFORD, W. T., No. 4, 1898 (British India, as Oedicnemidae); BAKER, E. C. S., 1929, Journ. Bombay Nat. Hist. Soc., 33, 745-752 (Indian Empire, as Oedicnemidae); BAKER, E. C. S., No. 6, 1929 (British India, as Oedicnemidae); ALI & RIPLEY, No. 3, 1969 (India, Pakistan); DEIGNAN, H. G., No. 1, 1945 (Thailand); KURODA, N., No. 2, 1936 (Java); McGREGOR, R. C., No. 1, 1909 (Philippines, as Oedicnemidae)
Buteonidae, as Accipitridae

Cacatuidae, as Psittacidae
Caenagnathidae (only as fossils)
Caerebidae, as Parulidae
Callaeatidae, as Callaeidae
Callaeidae (incl. Callaeatidae, Creadiontidae). PETERS, J. L., No. 15, 1962 (world list)
Calyptophilidae. (No revisionary references noted)
Campephagidae (incl. Campophagidae, Pericrocotidae). REICHENOW, A., No. 5, 1914 (monograph); PETERS, J. L., No. 9, 1960 (world list); HARTERT, E., No. 1, 1903-10 (Palearctic); DEMENTIEV & GLADKOV, No. 6, 1954 & No. 12, 1968 (Soviet Union); GLADKOV, N. A. et al., No. 1, 1964 (U. S. S. R.); REICHENOW, A., No. 2, 1903 (Africa); BANNERMAN, D. A., No. 5, 1939 (tropical West Africa); CHAPIN, J. P., No. 3, 1953 (Belgian Congo); FRIEDMANN, H., No. 2 1937 (Ethiopia & Kenya); BAKER, E. C. S., No. 2, 1924 (British India, as Pericrocotidae); DEIGNAN, H. G., No. 1, 1945 (Thailand); KURODA, N., No. 1, 1933 (Java); McGREGOR, R. C., No. 2, 1909 (Philippines); BAKER, R. H., No. 1, 1951 (Micronesia)
Campophagidae, as Campephagidae
Campylopteridae. BOUCARD, A., No. 1, 1893-95 (monograph)
Cancromidae, as Cochleariidae
Capitonidae (incl. Megalaemidae). MARSHALL, C. H. T. & MARSHALL, G. F. L., 1870-71, A monograph of the Capitonidae..., London, 73 plates with text; REICHENOW, A., No. 5, 1914 (monograph); PETERS, J. L., No. 6, 1948 (world list); CORY, C. B., No. 2, 1919 (the Americas, catalog); RIDGWAY, R., No. 6, 1914 (North America); BERLIOZ, J., 1937, Oiseau, 8, 221-239 (Neotropics); SALVIN & GODMAN, No. 2, 1888-1904 (Central America); WETMORE, A., No. 3, 1968 (Panama); REICHENOW, A., No. 2, 1903 (Africa); BANNERMAN, D. A., No. 3, 1933 (tropical West Africa); CHAPIN, J. P., No. 2, 1939 (Belgian Congo); FRIEDMANN, H., No. 1, 1930 (Ethiopia & Kenya); BLANFORD, W. T., No. 3, 1895 (British India); BAKER, E. C. S., No. 4, 1927 (British India); ALI & RIPLEY, No. 4, 1970 (India, Pakistan); MUKKERJEE, A. K., 1955, Rec. Indian Mus., 52, 157-175 (Indian Museum list); WAIT, W. E., No. 1, 1920 (Ceylon); DEIGNAN, H. G., No. 1, 1945 (Thailand); KURODA, N., No. 2, 1936 (Java); McGREGOR, R. C., No. 1, 1909 (Philippines); BERLIOZ, J., 1936, Oiseau, 6, 28-56 (Orient)
Caprimulgidae. HARTERT, E., 1897, Das Tierreich, 1, 1-98; REICHENOW, A., No. 5, 1914 (monograph); PETERS, J. L., No. 4, 1940 (world list); CORY, C. B., No. 1, 1918 (the Americas, catalog); RIDGWAY, R., No. 6, 1914 (North America); FORBUSH, E. H., No. 1, 1927 (Massachusetts); SALVIN & GODMAN, No. 2, 1888-1904 (Central America); WETMORE, A., No. 3, 1968 (Panama); ROHL, E., 1934, Bol. Soc. Venezolana Cienc. Nat., 19, 417-420 (Venezuela, catalog) PEREYRA, J. A., 1932, El Hornero, 5, 41-46 (Argentina); HARTERT, E., No. 2, 1912-21 (Palearctic); PARIS, P., No. 1, 1921 (France); LINTIA, D., No. 2, 1954 (Romania); DEMENTIEV & GLADKOV, No. 1, 1951 7 No. 7, 1966 (Soviet Union); GLADKOV, N. A. et al., No. 1, 1964 (U. S. S. R.); REICHENOW, A., No. 2, 1903 (Africa); KOENIG, A., 1919, Journ. Ornith., 67, 431-485 (Egypt); BANNERMAN, D. A., No. 3, 1933 (tropical West Africa); CHAPIN, J. P., No. 2, 1939 (Belgian Congo); FRIEDMANN, H., No. 1, 1930 (Ethiopia & Kenya); MOLTONI, E., 1946, Riv. Italiana Ornith., 10, 191-215 (Africa Orientale Italiana); BLANFORD, W. T., No. 3, 1895 (British India); BAKER, E. C. S., No. 4, 1927 (British India); ALI & RIPLEY, No. 4, 1970 (India, Pakistan); WAIT, W. E., No. 1, 1920 (Ceylon); DEIGNAN, H. G., No. 1, 1945 (Thailand); KURODA, N., No. 2, 1936 (Java); McGREGOR, R. C., No. 1, 1909 (Philippines); BAKER, R. H., No. 1, 1951 (Micronesia)
Carduelidae, as Ploceidae
Cariamidae (incl. Dicholophidae). PETERS, J. L., No. 2, 1963 (world list); HELLMAYR & CONOVER, No. 1, 1942 (the Americas, catalog); DEVINCENZI, G. J., No. 3, 1928 (Uruguay)
Carpophagidae. REICHENOW, A., No. 4, 1913 (monograph)
Casuariidae. ROTHSCHILD, W., 1900, Trans. Zool. Soc. London, 15, 109-148 (monograph); BRASIL, L., 1913, Genera Avium, 20, 1-10; REICHENOW, A., No. 4, 1913 (monograph); PETERS, J. L., No. 1, 1931 (world list)
Catamblyrhynchidae. HELLMAYR, C. E., No. 11, 1938 (the Americas, catalog)
Catharactidae, as Stercorariidae

Cathartidae (incl. Sarcoramphidae, Sarcorhamphidae). REICHENOW, A., No. 4, 1913 (monograph); SWANN, H. K., 1919-22, A synopsis of the Accipitres, 2nd ed., London, 1-233; SWANN, H. K. & WETMORE, A., 1924-45, A monograph of the birds of prey..., London, 2 volumes in 16 parts; PETERS, J. L., No. 1, 1931 (world list); BROWN, L. & AMADON, D., 1968, Eagles, hawks and falcons of the world, Middlesex, England, 2 volumes; HELLMAYR & CONOVER, No. 4, 1949 (the Americas, catalog); FRIEDMANN, H., No. 3, 1950 (North & Middle America); FORBUSH, E. H., No. 1, 1927 (Massachusetts); WETMORE, A., No. 1, 1927 (Puerto Rico); SALVIN & GODMAN, No. 3, 1897-1904 (Central America, as Sarcorhamphidae); WETMORE, A, No. 2, 1965 (Panama)

Cephalolepidae. BOUCARD, A., No. 1, 1893-95 (monograph)

Certhicidae. PETERS, J. L., No. 12, 1967 (world list)

Certhiidae (incl. Climacteridae). HELLMAYR, C. E., 1903, Das Tierreich, 18, 1-255; HELLMAYR, C. E. 1911, Genera Avium, 15, 1-16; REICHENOW, A., No. 5, 1914 (monograph); PETERS, J. L., No. 12, 1967 (world list, as Climacteridae); HELLMAYR, C. E., No. 7, 1934 (the Americas, catalog); RIDGWAY, R., No. 3, 1904 (North America); FORBUSH, E. H., No. 3, 1929 (Massachusetts); SALVIN & GODMAN, No. 1, 1879-1904 (Central America); HARTERT, E., No. 1, 1903-10 (Palearctic); PARIS, P., No. 1, 1921 (France); LINTIA, D., No. 1, 1946 (Romania); DEMENTIEV & GLADKOV, No. 5, 1954 & No. 11, 1970 (Soviet Union); GLADKOV, N. A. et al., No. 1, 1964 (U. S. S. R.); REICHENOW, A., No. 3, 1905 (Africa); BANNERMAN, D. A., No. 6, 1948 (tropical West Africa); CHAPIN, J. P., No. 4, 1954 (Belgian Congo); BAKER, E. C. S., No. 1, 1922 (British India); DEIGNAN, H. G., No. 1, 1945 (Thailand); McGREGOR, R. C., No. 2, 1909 (Philippines)

Chaeturidae. DEIGNAN, H. G., No. 1, 1945 (Thailand)

Chalcopariidae, as Nectariniidae

Chamaeidae, as Timaliidae

Charadriidae. REICHENOW, A., No. 4, 1913 (monograph); LOWE, P. R., 1922, Ibis, 1922, 475-495 (classification); SKINNER, K. L. et al., 1929, Ool. Rec., 9, 52-82 (eggs); LOW, G. C., No. 1, 1931 (bibliography); PETERS, J. L., No. 2, 1963 (world list); BOCK, W. J., 1958, Bull. Mus. Comp. Zool., 118, 27-97 (review of Charadriinae); HELLMAYR & CONOVER, No. 3, 1948 (the Americas, catalog); RIDGWAY, R., No. 8, 1919 (North America); FORBUSH, E. H., No. 2, 1929 (Massachusetts); WETMORE, A., No. 1, 1927 (Puerto Rico); SALVIN & GODMAN, No. 3, 1897-1904 (Central America); WETMORE, A., No. 2, 1965 (Panama); DEVINCENZI, G. J., No. 2, 1927 (Uruguay); HARTERT, E., No. 2, 1912-21 (Palearctic); PARIS, P., No. 1, 1921 (France); LINTIA, D., No. 3, 1955 (Romania); DEMENTIEV & GLADKOV, No. 3, 1951 & No. 9 1969 (Soviet Union); KOZLOVA, E. V., 1961-62, Fauna U. R. S. S. Birds, II, 1, 2; II, 1, 3. Moskva, 1-500 (in Russian); GLADKOV, N. A. et al., No. 1, 1964 (U. S. S. R.); REICHENOW, A., No. 1, 1901 (Africa); BOUET, G., No. 1, 1955 (Afrique tropicale); BANNERMAN, D. A., No. 2, 1931 (tropical West Africa); CHAPIN, J. P., No. 2, 1939 (Belgian Congo); van SOMEREN V. G. L., No. 1, 1925-36 (Kenya & Uganda); FRIEDMANN, H., No. 1, 1930 (Ethiopia & Kenya); BLANFORD, W. T., No. 4, 1898 (British India); BAKER, E. C. S., No. 6, 1929 (British India); ALI & RIPLEY, No. 2, 1969 (India, Pakistan); DEIGNAN, H. G., No. 1, 1945 (Thailand); KURODA, N., No. 2, 1936 (Java); McGREGOR, R. C., No. 1, 1909 (Philippines); UCHIDA, S., 1911, Dobuts. Z. Tokyo, 23, 243-248 (keys, Japan, in Japanese); TAKA-TSUKASA, N., No. 1, 1967 (Nippon); BAKER, R. H., No. 1, 1951 (Micronesia)

Chionidae, as Chionididae

Chionididae (incl. Chionidae). LOW, G. C., No. 1, 1931 (bibliography); PETERS, J. L., No. 2, 1963 (world list); HELLMAYR & CONOVER, No. 3, 1948 (the Americas, catalog); MURPHY, R. C., No. 1, 1936 (South America)

Chlorolampidae. BOUCARD, A., No. 1, 1893-95 (monograph)

Chloropseidae. (No revisionary references noted)

Ciconidae, as Ciconiidae

Ciconiidae (incl. Ciconidae). REICHENOW, A., No. 4, 1913 (monograph); PETERS, J. L., No. 1, 1931 (world list); HELLMAYR & CONOVER, No. 2, 1948 (the Americas, catalog); FORBUSH, E. H., No. 2, 1929 (Massachusetts); SALVIN & GODMAN, No. 3, 1897-1904 (Central America); WETMORE, A., No. 2, 1965 (Panama); DEVINCENZI, G. J., No. 3, 1928 (Uruguay); HARTERT, E., No. 2, 1912-21 (Palearctic); PARIS, P., No. 1, 1921 (France); LINTIA, D., No. 3, 1955 (Romania); TUGARINOV, A. J., No. 1, 1947 (U. R. S. S.); DEMENTIEV & GLADKOV, No. 2, 1951 & No. 8, 1968 (Soviet Union); GLADKOV, N. A. et al., No. 1, 1964 (U. S. S. R.); REICHENOW, A., No. 1, 1901(Africa); BOUET, G., No. 1, 1955 (Afrique tropicale); BANNERMAN, D. A. No. 1, 1930 (tropical West Africa); CHAPIN, J. P., No. 1, 1932 (Belgian Congo); FRIEDMANN, H., No. 1, 1930 (Ethiopia & Kenya); BLANFORD, W. T., No. 4, 1898 (British India); BAKER, E. C. S., No. 6, 1929 (British India); ALI & RIPLEY, No. 1, 1968 (India, Pakistan); DEIGNAN, H. G., No. 1, 1945 (Thailand); KURODA, N., No. 2, 1936 (Java); McGREGOR, R. C., No. 1, 1909 (Philippines)

Cinclidae. PETERS, J. L., No. 9, 1960 (world list); HELLMAYR, C. E., No. 7, 1934 (the Americas, catalog); RIDGWAY, R., No. 3, 1904 (North America); SALVIN & GODMAN, No. 1, 1879-1904 (Central America); PARIS, P., No. 1, 1921 (France); DEMENTIEV & GLADKOV, No. 6, 1954 & No. 12, 1968 (Soviet Union); GLADKOV, N. A. et al., No. 1, 1964 (U. S. S. R.); BAKER, E. C. S., No. 2, 1924 (British India); DEIGNAN, H. G., No. 1, 1945 (Thailand)

Cinclosomatidae, as Timaliidae

Cladornithidae (only as fossils)

Climacteridae, as Certhiidae

Cochleariidae (incl. Cancromidae). PETERS, J. L., No. 1, 1931 (world list); HELLMAYR & CONOVER, No. 2, 1948 (the Americas, catalog); SALVIN & GODMAN, No. 3, 1897-1904 (Central America as Cancromidae); WETMORE, A., No. 2, 1965 (Panama);

Coerebidae, as Parulidae

Coliidae. SCLATER, P. L., 1906, Genera Avium, 6, 1-6; REICHENOW, A., No. 5, 1914 (monograph); PETERS, J. L., No. 5, 1955 (world list); REICHENOW, A., No. 2, 1903 (Africa); BANNERMAN, D. A., No. 3, 1933 (tropical West Africa); CHAPIN, J. P., No. 2, 1939 (Belgian Congo); FRIEDMANN, H., No. 1, 1930 (Ethiopia & Kenya)

Columbidae (incl. Didunculidae, Gouridae, Peristeridae, Treronidae, Trerovidae, Turturidae). REICHENOW, A., No. 4, 1913 (monograph, also as Didunculidae); PETERS, J. L., 1936, Proc. 8th Intern. Congr. Ornith. 1934, 371-391 (generic limits, as Treronidae); PETERS, J. L., No. 3, 1937 (world list); MOBES, W. K. G., 1945, Bibliographie der Tauben, Halle, 1-268; GOODWIN, D., 1967, Pigeons and doves of the world, London, 1-446 (monograph); HELLMAYR & CONOVER, No. 1, 1942 (the Americas, catalog); RIDGWAY, R., No. 7, 1916 (North America); FORBUSH, E. H., No. 1, 1927 (Massachusetts); WETMORE, A., No. 1, 1927 (Puerto Rico); SALVIN & GODMAN, No. 3, 1897-1904 (Central America, also as Peristeridae); WETMORE, A., No. 3, 1968 (Panama); DEVINCENZI, G. J., No. 1, 1926 (Uruguay, also as Peristeridae); HARTERT, E., No. 2, 1912-21 (Palearctic); PARIS, P., No. 1, 1921 (France, also as Peristeridae); LINTIA, D., No. 3, 1955 (Romania); von BOETTICHER, H., 1930, Kocsag, 3, 16-27 (Bulgaria); DEMENTIEV & GLADKOV, No. 2, 1951 & No. 8, 1968 (Soviet Union); GLADKOV, N. A. et al., No. 1, 1964 (U. S. S. R.); REICHENOW, A., No. 1, 1901 (Africa); BANNERMAN, D. A., No. 2, 1931 (tropical West Africa); CHAPIN, J. P., No. 2, 1939 (Belgian Congo); van SOMEREN, V. G. L., No. 1, 1925-36 (Kenya & Uganda); FRIEDMANN, H., No. 1, 1930 (Ethiopia & Kenya); BLANFORD, W. T., No. 4, 1898 (British India); BAKER, E. C. S., 1913, Indian pigeons and doves, London, 1-260; BAKER, E. C. S., No. 5, 1928 (British India); ALI & RIPLEY, No. 3, 1969 (India, Pakistan); DEIGNAN, H. G., No. 1, 1945 (Thailand); DELACOUR, J. & JABOUILLE, P., 1927, Faune Colon. Francaise, 1, 369-461 (Annam); KURODA, N., No. 2, 1936 (Java); McGREGOR, R. C., No. 1, 1909 (Philippines, also as Peristeridae & Trerovidae); MANUEL, C. G., 1936, Philippine Journ. Sci., 59, 327-336 (Philippines); MANUEL, C. G., 1937, Philippine Journ. Sci., 63, 175-184 (Philippines); HUTSON, H. P. W., 1931, The birds of Hong Kong, VII, Hong Kong ; BAKER, R. H., No. 1, 1951 (Micronesia)

Colymbidae, as Podicipedidae

Compsothlypidae, as Parulidae

Conopophagidae. REICHENOW, A., No. 5, 1914 (monograph); PETERS, J. L., No. 7, 1951 (world list); CORY & HELLMAYR, No. 3, 1924 (the Americas, catalog)

Coracidae, as Coraciidae

Coraciidae (incl. Brachypteraceidae, Coracidae). DRESSER, H. E., 1893, A monograph of the Coraciidae ..., Farnborough, England, 1-111; PETERS, J. L., No. 5, 1955 (world list); HARTERT, No. 2, 1912-21 (Palearctic); PARIS, P., No. 1, 1921 (France); LINTIA, D., No. 2, 1954 (Romania); DEMENTIEV & GLADKOV, No. 1, 1951 & No. 7, 1966 (Soviet Union); GLADKOV, N. A. et al., No. 1, 1964 (U. S. S. R.); REICHENOW, A., No. 2, 1903 (Africa); BANNERMAN, D. A., No. 3, 1933 (tropical West Africa); CHAPIN, J. P., No. 2, 1939 (Belgian Congo); FRIEDMANN, H., No. 1, 1930 (Ethiopia & Kenya); BLANFORD, W. T., No. 3, 1895 (British India); BAKER, E. C. S., No. 4, 1927 (British India); ALI & RIPLEY, No. 4, 1970 (India, Pakistan); WAIT, W. E. No. 1, 1920 (Ceylon); DEIGNAN, H. G., No. 1, 1945 (Thailand); KURODA, N., No. 2, 1936 (Java); McGREGOR, R. C., No. 1, 1909 (Philippines); HUTSON, H. P. W., 1931, The birds of Hong Kong, VIII, Hong Kong; BAKER, R. H., No. 1, 1951 (Micronesia)

Corcoraciididae, as Grallinidae

Corvidae. REICHENOW, A., No. 5, 1914 (monograph); AMADON, D., 1944, American Mus. Novit., 1251, 1-21 (genera); PETERS, J. L., No. 15, 1962 (world list); HELLMAYR, C. E., No. 7, 1934 (the Americas, catalog); JOHNSTON, D. W., 1961, Biosystematics of American crows, Seattle, Washington, 1-119; HARDY, J. W., 1969, Condor, 71, 360-375 (New World jays); RIDGWAY, R., No. 3, 1904 (North America); ASHLEY, J. F., 1941, Condor, 43, 184-195 (key to North American); FORBUSH, E. H., No. 1, 1927 (Massachusetts); SALVIN & GODMAN, NO. 1, 1879-1904 (Central America); HARTERT, E., No. 1, 1903-10 (Palearctic); PARIS, P., No. 1, 1921 (France); LINTIA, D., No. 1, 1946 (Romania); STEGMANN, B., 1932, Les corbeaux, No. 6, Tableaux analytique de la faune de l'U. R. S. S., Leningrad, 1-32; DEMENTIEV & GLADKOV, No. 5, 1954 & No. 11, 1970 (Soviet Union); GLADKOV, N. A. et al., No. 1, 1964 (U. S. S. R.); REICHENOW, A., No. 2, 1903 (Africa); BANNERMAN, D. A., No. 6, 1948 (tropical West Africa); CHAPIN, J. P., No. 4, 1954 (Belgian Congo); FRIEDMANN, H., No. 2, 1937 (Ethiopia & Kenya); BAKER, E. C. S., No. 1, 1922 (British India); DEIGNAN, H. G., No. 1, 1945 (Thailand); KURODA, N., No. 1, 1933 (Java); McGREGOR, R. C., No. 2, 1909 (Philippines); BAKER, R. H., No. 1, 1951 (Micronesia)

Cotingidae (incl. Rupicolidae). REICHENOW, A., No. 5, 1914 (monograph); HELLMAYR, C. E., No. 6, 1929 (the Americas, catalog, also as Rupicolidae); RIDGWAY, R., No. 4, 1907 (North America); SALVIN & GODMAN, No. 2, 1888-1904 (Central America); VIEIRA, C. O. C., 1935, Rev. Mus. Paulista, 19, 327-397 (Brasil);

Cracidae. REICHENOW, A., No. 4, 1913 (monograph); PETERS, J. L., No. 2, 1963 (world list); VUILLEUMIER, F., 1965, Bull. Mus. Comp. Zool., 134-1-27 (relations); VAURIE, C., 1967, American Mus. Novit., 2296, 10 parts (systematics); VAURIE, C., 1968, Bull. American Mus. Nat. Hist. 138, 131-160 (taxonomy); HELLMAYR & CONOVER, No. 1, 1942 (the Americas, catalog); RIDGWAY, R. & FRIEDMANN, H., No. 10, 1946 (North America); SALVIN & GODMAN, No. 3,

1897-1904 (Central America); WETMORE, A., No. 2, 1965 (Panama); DENINCENZI, G. J., No. 1, 1926 (Uruguay)

Cracticidae (incl. Streperidae). PETERS, J. L., No. 15, 1962 (world list)

Crateropodidae. HARING, H. H., 1914, Journ. Bombay Nat. Hist. Soc., 23, 44-72, 311-340 (revision)

Creadiontidae, as Callaeidae

Crypturidae, as Tinamidae

Cuculidae. REICHENOW, A., No. 5, 1914 (monograph); PETERS, J. L., No. 4, 1940 (world list); CORY, C. B., No. 2, 1919 (the Americas, catalog); RIDGWAY, R., No. 7, 1916 (North America); FORBUSH, E. H., No. 1, 1927 (Massachusetts); SALVIN & GODMAN, No. 2, 1888-1904 (Central America); WETMORE, A., No. 3, 1968 (Panama); von IHERING, H., 1914, Rev. Mus. Sao Paulo, 9, 371-390 in Portuguese, 391-410 in German (biology & classification, Brasil); CUELLO, J. & ZORRILLA DE SAN MARTIN, J. C., 1960, Bol. Soc. Taguato, 1, 61-79 (Uruguay); HARTERT, E., No. 2, 1912-21 (Palearctic); VERHEYEN, R., Les pics et les coucous de Belgique, Bruxelles, 1-157; PARIS, P., No. 1, 1921 (France); LINTIA, D., No. 2, 1954 (Romania); DEMENTIEV & GLADKOV, No. 1, 1951 & No. 7, 1966 (Soviet Union); GLADKOV, N. A. et al., No. 1, 1964 (U. S. S. R.); REICHENOW, A., No. 2, 1903 (Africa); FRIEDMANN, H., 1948 (Africa); BANNERMAN, D. A., No. 3, 1933 (tropical West Africa); CHAPIN, J. P., No. 2, 1939 (Belgian Congo); FRIEDMANN, H., No. 1, 1930 (Ethiopia & Kenya); BLANFORD, W. T., No. 3, 1895 (British India); BAKER, E. C. S., No. 4, 1927 (British India); ALI & RIPLEY, No. 3, 1969 (India, Pakistan); WAIT, W. E., No. 1, 1920 (Ceylon); DEIGNAN, H. G., No. 1, 1945 (Thailand); KURODA, N., No. 2, 1936 (Java); MATHEWS, G. M., 1912, Australian Av. Rec., London, 1, 2-22 (revision, Australia); McGREGOR, R. C., No. 1, 1909 (Philippines); BAKER, R. H., No. 1, 1951 (Micronesia)

Cunampaiidae. (No revisionary references noted)

Cursoridae, as Glareolidae

Cursoriidae, as Glareolidae

Cyclarhidae (incl. Cyclorhidae). HELLMAYR, C. E., No. 8, 1935 (the Americas, catalog)

Cyclopsittacidae, as Psittacidae

Cyclorhidae, as Cyclarhidae

Cygnidae. REICHENOW, A., No. 4, 1913 (monograph)

Cyphornithidae (only as fossils)

Cypselidae, as Apodidae

Dacelonidae, as Alcedinidae

Dacnididae. REICHENOW, A., No. 5, 1914 (monograph)

Dendrochelidonidae, as Hemiprocnidae

Dendrocolaptidae. REICHENOW, A., No. 5, 1914 (monograph); von IHERING, H., 1915, Auk, 32, 145-153 (classification); CHUBB, C., 1919, Ann. Mag. Nat. Hist., (9), 3, 273-275 (notes); PETERS, J. L., No. 7, 1951 (world list); HELLMAYR, C. E., No. 4, 1925 (the Americas, catalog); RIDGWAY, R., No. 5, 1911 (North America); SALVIN & GODMAN, No. 2, 1888-1904 (Central America)

Diatrymidae (only as fossils)

Dicaeidae (incl. Paramythiidae). REICHENOW, A., No. 5, 1914 (monograph, also as Paramythiidae); MAYR, E. & AMADON, D., 1947, American Mus. Novit., 1360, 1-32 (review); PETERS, J. L., No. 12, 1967 (world list); BAKER, E. C. S., No. 3, 1926 (British India); INGLIS, C. M., 1963, Journ. Bengal Nat. Hist. Soc., (n. s.), 32, 1-9 (the Duars); DEIGNAN, H. G., No. 1, 1945 (Thailand); KURODA, N., No. 1, 1933 (Java); McGREGOR, R. C., No. 2, 1909 (Philippines)

Dicholophidae, as Cariamidae

Dicruridae (incl. Edoliidae). REICHENOW, A., No. 5, 1914 (monograph); VAURIE, C., 1949, Bull. American Mus. Nat. Hist., 93, 203-342 (revision); PETERS, J. L., No. 15, 1962 (world list); DEMENTIEV & GLADKOV, No. 5, 1954 & No. 11, 1970 (Soviet Union); GLADKOV, N. A. et al., No. 1, 1964 (U. S. S. R.); REICHENOW, A., No. 2, 1903 (Africa); BANNERMAN, D. A., No. 5, 1939 (tropical West Africa); CHAPIN, J. P., No. 4, 1954 (Belgian Congo); FRIEDMANN, H., No. 2, 1937 (Ethiopia & Kenya); BAKER, E. C. S., No. 2, 1924 (British India); DEIGNAN, H. G., No. 1, 1945 (Thailand); KURODA, N., No. 1, 1933 (Java); McGREGOR, No. 2, 1909 (Philippines) CHONG, L. T., 1932, Sinensia, 3, 161-172 (Kwangsi); BAKER, R. H., No. 1, 1951 (Micronesia)

Dididae, as Raphidae

Didunculidae, as Columbidae

Dinornithidae. ARCHEY, G., 1941, Bull. Auckland Inst. Mus., 1, 1-120 (the moa)

Diomedeidae (incl. Diomeidae). GODMAN, F. D., 1907-10, A monograph of the petrels..., London, 1-381; LOOMIS, L. M., 1918, Proc. California Acad. Sci., (4), 2, 1-187 (review); LOOMIS, L. M., 1923, Auk, 40, 596-602 (classification); von BOETTICHER, H., 1929, Anz. Ornith. Ges. Bayern, 2, 8-10 (genera); PETERS, J. L., No. 1, 1931 (world list); HELLMAYR & CONOVER, No. 2, 1948 (the Americas, catalog); SALVIN & GODMAN, No. 3, 1897-1904 (Central America); WETMORE, A., No. 2, 1965 (Panama); MURPHY, R. C., No. 1, 1936 (South America); DEVINCENZI, G. J., No. 1, 1926 (Uruguay); DABBENE, R., 1921-24, El Hornero, 2-3, 3 parts (petrels, South Atlantic); PARIS, P., No. 1, 1921 (France); GLADKOV, N. A. et al., No. 1, 1964 (U. S. S. R.); BOUET, G., No. 1, 1955 (Afrique tropicale); ALEXANDER, W. B., 1920, Emu, 20, 14-24, 66-74 (Australia); FRINGS, "H., M. &C.", 1966, Pacific Sci., 20, 312-317 (annotated bibliography of North Pacific albatrosses); BAKER, R. H., No. 1, 1951 (Micronesia)

Diomeidae, as Diomedeidae

Drepanidae, as Drepaniidae

Drepanididae, as Drepaniidae
Drepaniidae (incl. Drepanidae, Drepanididae). PERKINS, R. C. L., 1901, Ibis, 1901, 562-585; (anonymous) Bull. Zool. Nomencl., 18, 267-269 (put on official list); PETERS, J. L., No. 14, 1968 (world list); GREENWAY, J. C., Jr., 1944, Bull. Mus. Comp. Zool., 94, 140-142 (generic arrangement of Drepanidinae); AMADON, D., 1950, Bull. American Mus. Nat. Hist., 95, 151-262 (Hawaii)
Dromadidae. REICHENOW, A., No. 4, 1913 (monograph); LOW, G. C., No. 1, 1931 (bibliography); PETERS, J. L., No. 2, 1963 (world list); HARTERT, E., No. 2, 1912-21 (Palearctic); REICHENOW, A., No. 1, 1901 (Africa); BLANFORD, W. T., No. 4, 1898 (British India); BAKER, E. C. S., No. 6, 1929 (British India); ALI & RIPLEY, No. 2, 1969 (India, Pakistan)
Dromaeidae, as Dromiceiidae
Dromaiidae, as Dromiceiidae
Dromiceidae, as Dromiceiidae
Dromiceiidae (incl. Dromaeidae, Dromaiidae, Dromiceidae). REICHENOW, A., No. 4, 1913 (monograph); BRASIL, L., 1914, Genera Avium, 25, 1-5; PETERS, J. L., No. 1, 1931 (world list)
Dromornithidae (only as fossils)
Dulidae. PETERS, J. L., No. 9, 1960 (world list); HELLMAYR, C. E., No. 8, 1935 (the Americas, catalog); RIDGWAY, R., No. 3, 1904 (North America)

Edoliidae, as Dicruridae
Eleutherornithidae (only as fossils)
Elopterygidae (only as fossils)
Emberizidae. PETERS, J. L., No. 13, 1970 (world list); DEMENTIEV & GLADKOV, No. 5, 1954 & No. 11, 1970 (Soviet Union)
Enaliornithidae (only as fossils)
Enicuridae, as Turdidae
Eogruidae (only as fossils)
Epimachidae, as Paradisaeidae
Eremialectoridae, as Pteroclidae
Estrildidae, as Ploceidae
Eulabetidae, as Sturnidae
Eulebetidae, as Sturnidae
Eupetidae, as Timaliidae
Eurycerotidae, as Vangidae
Eurylaemidae, as Eurylaimidae
Eurylaimidae (incl. Eurylaemidae). HARTERT, E., 1905, Genera Avium, 1, 1-8; REICHENOW, A., No. 5, 1914 (monograph); PETERS, J. L., No. 7, 1951 (world list); BANNERMAN, D. A., No. 4, 1936 (tropical West Africa); CHAPIN, J. P., No. 3, 1953 (Belgian Congo); BLANFORD, W. T., No. 3, 1895 (British India); BAKER, E. C. S., No. 3, 1926 (British India); ALI & RIPLEY, No. 4, 1970 (India, Pakistan); DEIGNAN, H. G., No. 1, 1945 (Thailand); KURODA, N., No. 1, 1933 (Java); McGREGOR, R. C., No. 1, 1909 (Philippines)
Eurypigidae, as Eurypygidae
Eurypygidae (incl. Eurypigidae). REICHENOW, A., No. 4, 1913 (monograph); RIGGS, C. D., 1948, Wilson Bull., 60, 75-80 (review); PETERS, J. L., No. 2, 1963 (world list); HELLMAYR & CONOVER, No. 1, 1942 (the Americas, catalog); RIDGWAY & FRIEDMANN, No. 9, 1941 (North America); SALVIN & GODMAN, No. 3, 1897-1904 (Central America); WETMORE, A., No. 2, 1965 (Panama)

Falconidae (incl. Polyboridae). REICHENOW, A., No. 4, 1913 (monograph); SWANN, H. K. & WETMORE, A., 1924-45, A monograph of the birds of prey ..., London, 2 volumes; PETERS, J. L., No. 1, 1931 (world list); BROWN, L. & AMADON, D., 1968, Eagles, hawks and falcons of the world, Middlesex, England, 2 volumes; HELLMAYR & CONOVER, No. 4, 1949 (the Americas, catalog); FRIEDMANN, H., No. 3, 1950 (North & Middle America); FORBUSH, E. H., No. 1, 1927 (Massachusetts); WETMORE, A., No. 1, 1927 (Puerto Rico); SALVIN & GODMAN, No. 3, 1897-1904 (Central America); WETMORE, A., No. 2, 1965 (Panama); DUGAND, A., 1941, Rev. Acad. Colombiana, 4, 394-404 (keys, Colombia); LEHMANN, F. C., 1945, Rev. Univ. Cauca Popayan, 6, 81-124 (Buteoninae, Colombia); HARTERT. E., No. 2, 1912-21 (Palearctic); PARIS, P., No. 1, 1921 (France); LINTIA, D., No. 2, 1954 (Romania); STEGMANN, B., 1937, Faune de l'U. R. S. S., Oiseaux, I, 5, Falconiformes, Moskva, 1-294; DEMENTIEV & GLADKOV, No. 1, 1951 & No. 7, 1966 (Soviet Union); GLADKOV, N. A. et al., No. 1, 1964 (U. S. S. R.); REICHENOW, A., No. 1, 1901 (Africa); BOUET, G., No. 1, 1955 (Afrique tropicale); BANNERMAN, D. S., No. 1, 1930 (tropical West Africa); CHAPIN, J. P., No. 1, 1932 (Belgian Congo); FRIEDMANN, H., No. 1, 1930 (Ethiopia & Kenya); BLANFORD, W. T., No. 3, 1895 (British India); BAKER, E. C. S., No. 5, 1928 (British India); ALI & RIPLEY No. 1, 1968 (India, Pakistan); WAIT, W. E., No. 2, 1921 (Ceylon); DEIGNAN, H. G., No. 1, 1945 (Thailand); ENGELBACH, P., 1949, Terre et la Vie, 96, 1-14 (Indochine); KURODA, N., No. 2, 1936 (Java); McGREGOR, R. C., No. 1, 1909 (Philippines); BAKER, R. H., No. 1, 1951 (Micronesia)
Falculidae, as Vangidae
Falcunculidae, as Muscicapidae
Floricolidae. BOUCARD, A., No. 1, 1893-95 (monograph)
Formicariidae. REICHENOW, A., No. 5, 1914 (monograph); PETERS, J. L., No. 7, 1951 (world list); CORY & HELLMAYR, No. 3, 1924 (the Americas, catalog); RIDGWAY, R., No. 5, 1911 (North America) SALVIN & GODMAN, No. 2, 1888-1904 (Central America)

Fregatidae (incl. Tachypetidae). REICHENOW, A., No. 4, 1913 (monograph); PETERS, J. L., No. 1, 1931 (world list); HELLMAYR & CONOVER, No. 2, 1948 (the Americas, catalog); FORBUSH, E. H. No. 2, 1929 (Massachusetts); WETMORE, A., No. 1, 1927 (Puerto Rico); SALVIN & GODMAN, No. 3, 1897-1904 (Central America); WETMORE, A., No. 2, 1965 (Panama); MURPHY, R. C., No. 1, 1936 (South America); TUGARINOV, A. J., No. 1, 1947 (U. R. S. S.); DEMENTIEV & GLADKOV, No. 1951 & No. 7, 1966 (Soviet Union); BOUET, G., No. 1, 1955 (Afrique tropicale); BANNERMAN, D. A., No. 1, 1930 (tropical West Africa); CHAPIN, J. P., No. 1, 1932 (Belgian Congo); BLANFORD, W. T., No. 4, 1898 (British India); BAKER, E. C. S., No. 6, 1929 (British India); ALI & RIPLEY, No. 1, 1968 (India, Pakistan); KURODA, N., No. 2, 1936 (Java); McGREGOR, R. C., No. 1, 1909 (Philippines); BAKER, R. H., No. 1, 1951 (Micronesia)

Fringillidae (incl. Geospizidae). REICHENOW, A., No. 5, 1914 (monograph); SUSHKIN, P. P., 1924, Bull. British Ornith. Club, 45, 36-39 (morphology & classification); von BOETTICHER, H., 1931, Senckenbergiana, 13, 147-153 (Systematik); STALLCUP, W. B., 1954, Publ. Univ. Kansas Mus. Nat. Hist., 8, 157-211 (myology & serology); TORDOFF, H. B., 1954, Misc. Publ. Mus. Zool. Univ. Michigan, 81, 1-42 (skull structure); PETERS, J. L., No. 14, 1968 (world list); HELLMAYR, C. E., No. 11, 1938 (the Americas, catalog); RIDGWAY, R., No. 1, 1901 (North America); SUSHKIN, P. P., 1925, Auk, 42, 256-261 (America); FORBUSH, E. H., No. 3, 1929 (Massachusetts); SALVIN & GODMAN, No. 1, 1879-1904 (Central America); SWARTH, H. S., Occ. Pap. California Acad. Sci., 18, 1-299 (Galapagos Islands, as Geospizidae); LACK, D., 1945, Occ. Pap. California Acad. Sci., 21, 1-158 (Galapagos Islands); HARTERT, E., No. 1, 1903-10 (Palearctic); PARIS, P., No. 1, 1921 (France); LINTIA, D., No. 1, 1946 (Romania); DEMENTIEV & GLADKOV, No. 5, 1954 & No. 11, 1970 (Soviet Union); REICHENOW, A., No. 3, 1905 (Africa); BANNERMAN, D. A., No. 6, 1948 (tropical West Africa); CHAPIN, J. P., No. 4, 1954 (Belgian Congo); SKEAD, C. J. et al., 1960, The canaries, seedeaters and buntings of southern Africa, (South Africa), 1-152; WINTERBOTTOM, J. M., 1958, Ostrich, 29, 110-111 (Cape Province); FRIEDMANN, H., No. 2, 1937 (Ethiopia & Kenya); BAKER, E. C. S., No. 3, 1926 (British India); DEIGNAN, H. G., No. 1, 1945 (Thailand); KURODA, N., No. 1, 1933 (Java); McGREGOR, R. C., No. 2, 1909 (Philippines)

Fulicidae. BOUET, G., No. 1, 1955 (Afrique tropicale)

Fulmaridae, as Procellariidae

Furnaridae, as Furnariidae

Furnariidae (incl. Furnaridae). CHUBB, C., 1919, Ann. Mag. Nat. Hist., (9), 3, 273-275 (notes on family); PETERS, J. L., No. 7, 1951 (world list); HELLMAYR, C. E., No. 4, 1925 (the Americas, catalog); RIDGWAY, R., No. 5, 1911 (North America); GIGOUX, E. E., 1929, Bol. Mus. Nac. Chile, 12, 42-64 (Chile)

Galbulidae. SCLATER, P. L., 1879-82 (monograph); SCLATER, P. L., 1909, Genera Avium, 10, 1-7; REICHENOW, A., No. 5, 1914 (monograph); STEINBACHER, J., 1937, Arch. Naturg., 6, 417-515 (anatomy); PETERS, J. L., No. 6, 1948 (world list); CORY, C. B., No. 2, 1919 (the Americas, catalog); RIDGWAY, R., No. 6, 1914 (North America); SALVIN & GODMAN, No. 2, 1888-1904 (Central America); WETMORE, A., No. 3, 1968 (Panama)

Gallinuloididae (Only as fossils)

Gastornithidae (only as fossils)

Gaviidae (incl. Colymbidae of European authors, Urinatoridae). PETERS, J. L., No. 1, 1931 (world list); HELLMAYR & CONOVER, No. 2, 1948 (the Americas, catalog); FORBUSH, E. H., No. 2, 1929 (Massachusetts); HARTERT, E., No. 2, 1912-21 (Palearctic, as Urinatoridae); PARIS, P., No. 1, 1921 (France); LINTIA, D., No. 3, 1955 (Romania, as Urinatoridae); DEMENTIEV & GLADKOV, No. 2, 1951 & No. 8, 1968 (Soviet Union); GLADKOV, N. A. et al., No. 1, 1964 (U. S. S. R.); ALI & RIPLEY, No. 1, 1968 (India, Pakistan)

Geospizidae, as Fringillidae

Geotrygonidae. REICHENOW, A., No. 4, 1913 (monograph)

Geranoididae (only as fossils)

Glareolidae (incl. Cursoridae, Cursoriidae). LOW, G. C., No. 1, 1931 (bibliography); MEINERTZHAGEN, A. C., 1927, Ibis, 1927, 469-498 (review); LOW, G. C., No. 1, 1931 (bibliography, as Cursoridae); PETERS, J. L., No. 2, 1963 (world list); HARTERT, E., No. 2, 1912-21 (Palearctic, as Cursoridae); PARIS, P., No. 1, 1921 (France, as Cursoridae also); LINTIA, D., No. 3, 1955 (Romania, as Cursoridae); DEMENTIEV & GLADKOV, No. 3, 1951 & No. 9, 1969 (Soviet Union); GLADKOV, N. A. et al., No. 1, 1964 (U. S. S. R.); BOUET, G., No. 1, 1955 (Afrique tropicale); BANNERMAN, D. A., No. 2, 1931 (tropical West Africa); CHAPIN, J. P., No. 2, 1939 (Belgian Congo); FRIEDMANN, H., No. 1, 1930 (Ethiopia & Kenya); BLANFORD, W. T., No. 4, 1898 (British India); BAKER, E. C. S., No. 6, 1929 (British India); ALI & RIPLEY, No. 3, 1969 (India, Pakistan); DEIGNAN, H. G., No. 1, 1945 (Thailand); KURODA, N., No. 2, 1936 (Java); McGREGOR, R. C., No. 1, 1909 (Philippines); TAKA-TSUKASA, N., No. 1, 1967 (Nippon)

Gouridae, as Columbidae

Graculidae, part as Phalacrocoracidae, part as Sturnidae

Grallinidae (incl. Corcoraciididae, Struthiididae). PETERS, J. L., No. 15, 1962 (world list)

Gruidae (incl. Balearicidae, Megalornithidae, Psophiidae of Matthews). BLAAUW, F. E., 1897, (A monograph of the cranes, Leiden; REICHENOW, A., No. 4, 1913 (monograph); BRASIL, L., 1913, Genera Avium, 19, 1-9; PETERS, J. L., No. 2, 1963 (world list); HELLMAYR & CONOVER, No. 1, 1942 (the Americas, catalog); RIDGWAY, R. & FRIEDMANN, H., No. 9, 1941 (North America); FORBUSH, E. H., No. 2, 1929 (Massachusetts); SALVIN & GODMAN, No. 3, 1897-1904 (Central America); HARTERT, E., No. 3, 1921-22 (Palearctic); PARIS, P., No. 1, 1921

(France); LINTIA, D., No. 3, 1955 (Romania); DEMENTIEV & GLADKOV, No. 2, 1951 & No. 8 1968 (Soviet Union); GLADKOV, N. A. et al., No. 1, 1964 (U. S. S. R.); REICHENOW, A., No. 1, 1901 (Africa); BOUET, G., No. 1, 1955 (Afrique tropicale); BANNERMAN, D. A., No. 2, 1931 (tropical West Africa, as Balearicidae); CHAPIN, J. P., No. 2, 1939 (Belgian Congo); FRIEDMANN, H., No. 1, 1930 (Ethiopia & Kenya); MOLTONI, E., 1944, Riv. Italiana Ornith., 14, 77-89 (Africa Orientale, Italiana, as Balearicidae); BLANFORD, W. T., No. 4, 1898 (British India); BAKER, E. C. S., No. 6, 1929 (British India); ALI & RIPLEY, No. 2, 1919 (India, Pakistan); DEIGNAN, H. G., No. 1, 1945 (Thailand); McGREGOR, R. C., No. 1, 1909 (Philippines); HENG, L. T. & YUAN, L. H., 1932, Contr. Inst. Zool. Peiping, 1, (1), 1-22 (north China); TAKA-TSUKASA, N., No. 1, 1967 (Nippon)

Gypaetidae. PARIS, P., No. 1, 1921 (France); LINTIA, D., No. 2, 1954 (Romania)

Gypogeranidae, as Sagittariidae

Haematopidae, as Haematopodidae

Haematopodidae (incl. Haematopidae). PETERS, J. L., No. 2, 1963 (world list); HELLMAYR & CONOVER, No. 3, 1948 (the Americas, catalog); RIDGWAY, R., No. 8, 1919 (North America); FORBUSH, E. H., No. 2, 1929 (Massachusetts); WETMORE, A., No. 1, 1927 (Puerto Rico); WETMORE, A., No. 2, 1965 (Panama); MURPHY, R. C., No. 1, 1936 (South America); DEVINCENZI, G. J., No. 2, 1927 (Uruguay); ALI & RIPLEY, No. 2, 1969 (India, Pakistan)

Halcyonidae, as Alcedinidae

Heliangelidae. BOUCARD, A., No. 1, 1893-95 (monograph)

Heliantheidae. BOUCARD, A.,, No. 1, 1893-95 (monograph)

Heliodoxidae. BOUCARD, A., No. 1, 1893-95 (monograph)

Heliornithidae. PETERS, J. L., No. 2, 1963 (world list); HELLMAYR & CONOVER, No. 1, 1942 (the Americas, catalog); RIDGWAY & FRIEDMANN, No. 9, 1941 (North America); SALVIN & GODMAN, No. 3, 1897-1904 (Central America); WETMORE, A., No. 2, 1965 (Panama); BOUET, G., No. 1, 1955 (Afrique tropicale); BANNERMAN, D. A., No. 2, 1931 (tropical West Africa); CHAPIN, J. P., No. 2, 1939 (Belgian Congo); BLANFORD, W. T., No. 4, 1898 British India); BAKER, E. C. S., No. 6, 1929 (British India); ALI & RIPLEY, No. 2, 1969 (India, Pakistan); DEIGNAN, H. G., No. 1, 1945 (Thailand)

Hemiprocnidae (incl. Dendrochelidonidae, Macropterygidae). HARTERT, E., 1897, Das Tierreich, 1, 1-98; REICHENOW, A., No. 5, 1914 (monograph); PETERS, J. L., No. 4, 1940 (world list); REICHENOW, A., No. 2, 1903 (Africa, as Macropterygidae); DEIGNAN, H. G., No. 1, 1945 (Thailand); McGREGOR, R. C., No. 1, 1909 (Philippines)

Hemistephanidae. BOUCARD, A., No. 1, 1893-95 (monograph)

Hermosiornidae, as Hermosiornithidae

Hermosiornithidae (incl. Hermosiornidae). (No revisionary references noted)

Hesperornithidae (only as fossils)

Himantopodidae. DEVINCENZI, G. J., No. 2, 1927 (Uruguay)

Hirundinidae. SHARPE, R. B. & WYATT, C. W., 1885-1894, Monograph of Hirundinidae, London, 2 vol.; REICHENOW, A., No. 5, 1914 (monograph); MAYR, E. & RAND, J., 1943, Ibis, 85, 334-341 generic classification); MOREAU, R. E., 1944, Ibis, 1944, p. 101 (generic classification); PETERS, J. L., No. 9, 1960 (world list); HELLMAYR, C. E., No. 8, 1935 (world list); RIDGWAY, R., No. 3, 1904 (North America); FORBUSH, E. H., No. 3, 1929 (Massachusetts); SALVIN & GODMAN, No. 1, 1879-1904 (Central America); PARIS, P., No. 1, 1921 (France); LINTIA, D., No. 1, 1946 (Romania); DEMENTIEV & GLADKOV, No. 6, 1954 & No. 12, 1968 (Soviet Union); GLADKOV, N. A. et al., No. 1, 1964 (U. S. S. R.); REICHENOW, A., No. 2, 1903 (Africa); BANNERMAN, D. A., No. 5, 1939 (tropical West Africa); CHAPIN, J. P., No. 3, 1953 (Belgian Congo); FRIEDMANN, H., No. 2, 1937 (Ethiopia & Kenya); BAKER, E. C. S., No. 3, 1926 (British India); DEIGNAN, H. G., no. 1, 1945 (Thailand); KURODA, N., No. 1, 1933 (Java); McGREGOR, R. C., No. 2, 1909 (Philippines); BAKER, R. H., No. 1, 1951 (Micronesia)

Hydrobatidae (incl. Thalassidromidae); LOOMIS, L. M., 1918, Proc. California Acad. Sci., (4) 2, 1-187 (review); LOOMIS, L. M., 1923, Auk, 40, 596-602 (classification); PETERS, J. L., No. 1, 1931 (world list); MATHEWS, G. M., 1937, Emu, 37, 136-143 (keys); HELLMAYR & CONOVER, No. 2, 1948 (the Americas, catalog); LANGELIER, G., 1940, Nat. Canadienne, 67, 289-302; FORBUSH, E. H., No. 2, 1929 (Massachusetts); WETMORE, A., No. 1, 1927 (Puerto Rico); WETMORE, A., No. 2, 1965 (Panama); DEVINCENZI, G. J., No. 1, 1926 (Uruguay); DABBENE, R., El Hornero, 2, 157-179, 141-254, 3, 227-238 (South Atlantic); PARIS, P., No. 1, 1921 (France); GLADKOV, N. A. et al., No. 1, 1964 (U. S. S. R.); BOUET, G., No. 1, 1955 (Afrique tropicale); BANNERMAN, D. A., No. 1, 1930 (tropical West Africa); CHAPIN, J. P., No. 1, 1932 (Belgian Congo); ALI & RIPLEY, No. 1, 1968 (India, Pakistan); ALEXANDER, W. B., 1920, Emu, 20, 14-24, 66-74 (Australia)

Hylactidae, as Rhinocryptidae

Hypocoliidae, as Bombycillidae

Hyposittidae. HELLMAYR, C. E., 1913, Genera Avium, 24, 1-3.

Ibidae, as Threskiornithidae

Ibididae, as Threskiornithidae

Ichthyornithidae (only as fossils)

Icteridae. REICHENOW, A., No. 5, 1914 (monograph); FRIEDMANN, H., 1927, Auk, 44, 495-508 (revision); FRIEDMANN, H., 1929, The cowbirds, Springfield, Illinois, PETERS, J. L., No. 14, 1968 (world list); HELLMAYR, C. E., No. 10, 1937 (the American, catalog);

RIDGWAY, R., No. 2, 1902 (North America); FORBUSH, E. H., No. 1, 1927 (Massachusetts); SALVIN & GODMAN, No. 1, 1879-1904 (Central America); GLADKOV, N. A., et al., No. 1, 1964 (U. S. S. R.)

Illadopsidae, as Timaliidae

Indicatoridae. REICHENOW, A., No. 5, 1914 (monograph); PETERS, J. L., No. 6, 1948 (world list); FRIEDMANN, H., 1955, United States Nat. Mus. Bull., 208, 1-292 (the honey-guides); REICHENOW, A, No. 2, 1903 (Africa); BANNERMAN, D. A., No. 3, 1933 (tropical West Africa); CHAPIN, J. P., No. 2, 1939 (Belgian Congo); SCHOUTEDEN, H., 1953, Bull. Cercle Zool. Congolaise, 21, 30-36 (keys); FRIEDMANN, H., 1954, Ann. Mus. Congo Belge, (n. s.), Zool., 1, 21-27 (revision of classification); FRIEDMANN, H., No. 1, 1930 (Ethiopia & Kenya); MOLTONI, E., 1940, Riv. Italiana Ornith., 10, 1-10 (Africa orient. Italiana); BLANFORD, W. T., No. 3, 1895 (British India); BAKER, E. C. S., no. 4, 1927 (British India); ALI & RIPLEY, No. 4, 1970 (India, Pakistan)

Irenidae (incl. Aegithinidae, Phyllornithidae). PETERS, J. L., No. 9, 1960 (world list); BAKER, E. C. S., No. 3, 1926 (British India)

Irrisoridae, as Phoeniculidae

Jacanidae (incl. Parridae). REICHENOW, A., No. 4, 1913 (monograph); LOW, G. C., No. 1, 1931 (bibliography, as Parridae); LOWE, P. R., 1931 Ibis, (13), 1, 491-534 (relations of Gruimorphae); PETERS, J. L., No. 2, 1963 (world list); HELLMAYR & CONOVER, No. 3, 1948 (the Americas, catalog); RIDGWAY, R., No. 8, 1919 (North America); WETMORE, A., No. 1, 1927 (Puerto Rico); SALVIN & GODMAN, No. 3, 1897-1904 (Central America, as Parridae); WETMORE, A, No. 2, 1965 (Panama); DEVINCENZI, G. J., No. 2, 1927 (Uruguay); REICHENOW, A., No. 1, 1901 (Africa); BOUET, G., No. 1, 1955 (Afrique tropicale); BANNERMAN, D. A., No. 2, 1931 (tropical West Africa); CHAPIN, J. P., No. 2, 1939 (Belgian Congo); FRIEDMANN, H., No. 1, 1930 (Ethiopia & Kenya); MOLTONI, E., 1944, Riv. Italiana Ornith., 14, 77-89 (Africa Orientale Italiana); BLANFORD, W. T., No. 4, 1898 (British India, as Parridae); BAKER, E. C. S., No. 6, 1929 (British India); ALI & RIPLEY, No. 2, 1969 (India, Pakistan); DEIGNAN, H. G., No. 1, 1945 (Thailand); KURODA, N., No. 2, 1936 (Java); McGREGOR, R. C., No. 1, 1909 (Philippines)

Jyngidae (incl. Yungidae). CHAPIN, J. P., No. 2, 1939 (Belgian Congo)

Kakatoidae, as Psittacidae

Lampornidae. BOUCARD, A., No. 1, 1893-95 (monograph)

Laniariidae, as Laniidae

Laniidae (incl. Laniariidae). REICHENOW, A, No. 5, 1914 (monograph); OLIVIER, G., 1944, Monographie de pres-grieches du genre Lanius, Rouen, 1-326; PETERS, J. L., No. 9, 1960 (world list); HELLMAYR, C. E., No. 8, 1935 (the Americas, catalog); MILLER, A. H., 1931, Univ. California Publ. Zool., 38, 11-242 (systematic revision, America); RIDGWAY, R., No. 3, 1904 (North America); FORBUSH, E. H., No. 3, 1929 (Massachusetts); SALVIN & GODMAN, No. 1, 1879-1904 (Central America); HARTERT, E., No. 1, 1903-10 (Palestine); PARIS, P., No. 1, 1921 (France); LINTIA, D., No. 1, 1946 (Romania); DEMENTIEV & GLADKOV, No. 6, 1954 & No. 12, 1968 (Soviet Union); GLADKOV, N. A. et al., No. 1, 1964 (U. S. S. R.); REICHENOW, A, No. 2, 1903 (Africa); KOENIG, A., 1921, Journ. Ornith., 69, 426-456 (Aegyptens); BANNERMAN, D. A., No. 5, 1939 (tropical West Africa); CHAPIN, J. P., No. 4, 1954 (Belgian Congo); FRIEDMANN, H., No. 2, 1937 (Ethiopia & Congo); BAKER, E. C. S., No. 2, 1924 (British India); DEIGNAN, H. G., No. 1, 1945 (Thailand); KURODA, N., No. 1, 1933 (Java); KURODA, N., No. 1, 1933 (Java); McGREGOR, R. C., No. 2, 1909 (Philippines)

Laridae (incl. Sternidae). REICHENOW, A., No. 4, 1913 (monograph); DWIGHT, J., 1925, Bull. American Mus. Nat. Hist., 52, 63-401 (gulls of world); von BOETTICHER, H., 1939, Koczag, 9-11, 42-46 (Klassification); PETERS, J. L., No. 2, 1963 (world list); HELLMAYR & CONOVER, No. 3, 1948 (the Americas, catalog); RIDGWAY, R., No. 8, 1919 (North America, also as Sternidae); FORBUSH, E. H., No. 2, 1929 (Massachusetts); GRISCOM, L., Bull. Essex County Ornith. Club, 11, 13-26 (keys, Massachusetts gulls); WETMORE, A., No. 1, 1927 (Puerto Rico); SALVIN & GODMAN, No. 3, 1897-1904 (Central America); WETMORE, A., No. 2, 1965 (Panama); MURPHY, R. C., No. 1, 1936 (South America); DEVINCENZI, G. J., No. 1, 1926 (Uruguay, as Sternidae); DABBENE, R., 1918, El Hornero, 1, 49-56, 129-139 (Argentina); HARTERT, E., No. 2, 1912-21 (Palearctic); PARIS, P., No. 1, 1921 (France); GRANT, P. J. & SCOTT, R. E., 1967, British Birds, 60, 365-368 (keys, immature Mediterranean gulls); LINTIA, D., No. 3, 1955 (Romania); DEMENTIEV & GLADKOV, No. 3, 1951 & No. 9, 1969 (Soviet Union); GLADKOV, N. A., et al., No. 1, 1964 (U. S. S. R.); REICHENOW, A., No. 1, 1901 (Africa); BOUET, G., No. 1, 1955 (Afrique tropicale); BANNERMAN, D. A., No. 2, 1931 (tropical West Africa); CHAPIN, J. P., No. 2, 1939 (Belgian Congo); FRIEDMANN, H., No. 1, 1930 (Ethiopia & Kenya); BLANFORD, W. T., No. 4, 1898 (British India); BAKER, E. C. S., No. 6, 1929 (British India, also as Sternidae); ALI & RIPLEY, No. 3, 1969 (India, Pakistan); DEIGNAN, H. G., No. 1, 1945 (Thailand); KURODA, N., No. 2, 1936 (Java); FALLA, R. A., 1960, Tuatara, 8, 72-76 (New Zealand); McGREGOR, R. C., No. 1, 1909 (Philippines); BAKER, R. H., No. 1, 1951 (Micronesia)

Leiotrichidae, as Timaliidae

Leptosomatidae. PETERS, J. L., No. 5, 1955 (world list)

Lesbidae. BOUCARD, A., No. 1, 1893-1904 (monograph)

Liotrichidae, as Timaliidae

Lobivanellidae. DEVINCENZI, G. J., No. 2, 1927 (Uruguay)
Lonchodytidae (only as fossils)
Lophornidae. BOUCARD, A., No. 1, 1893-95 (monograph)
Loriidae, as Psittacidae

Macropterygidae, as Hemiprocnidae
Malimbidae, as Ploceidae
Maluridae, as Sylviidae
Megalaemidae, as Capitonidae
Megalornithidae, as Gruidae
Megapodidae, as Megapodiidae
Megapodiidae (incl. Megapodidae). REICHENOW, A., No. 4, 1913 (monograph); SHUFELDT, R. W., 1919, Emu, 19, 1-28, 107-127; PETERS, J. L., No. 2, 1963 (world list); BLANFORD, W. T., No. 4, (British India); BAKER, E. C. S., No. 5, 1928 (British India); ALI & RIPLEY, No. 2, 1969 (India, Pakistan); McGREGOR, R. C., No. 1, 1909 (Philippines); BAKER, R. H., No. 1, 1951 (Micronesia)
Meleagridae, as Meleagrididae
Meleagrididae (incl. Meleagridae). PETERS, J. L., No. 2, 1963 (world list); HELLMAYR & CONOVER, No. 1, 1942 (the Americas, catalog); RIDGWAY & FRIEDMANN, No. 10, 1946 (North America) SALVIN & GODMAN, No. 3, 1897-1904 (Central America); BAKER, R. H., No. 1, 1951 (Micronesia)
Meliphagidae (incl. Melithreptidae, Promeropidae). REICHENOW, A., No. 5, 1914 (monograph); PETERS, J. L., No. 12, 1967 (world list); OFFICER, H. R., 1964, Australian honeyeaters, Melbourne, 1-86, 1964.
Melithreptidae, as Meliphagidae
Menuridae. REICHENOW, A., No. 5, 1914 (monograph)
Mergidae. REICHENOW, A., No. 4, 1913 (monograph)
Meropidae. DRESSER, H. E., 1884-86, A monograph of the Meropidae..., London, 1-136; PARROT, C., 1911, Genera Avium, 14, 1-18; REICHENOW, A., No. 5, 1914 (monograph); von BOETTICHER, H., 1935, Koczag, 8, 33-44 (Systematik); PETERS, J. L., No. 5, 1955 (World list); HARTERT, E., No. 2, 1912-21 (Palearctic); PARIS, P., No. 1, 1921 (France); LINTIA, D., No. 2, 1954 (Romania); DEMENTIEV & GLADKOV, No. 1951 & No. 7, 1966 (Soviet Union); GLADKOV, N. A. et al., No. 1, 1964 (U. S. S. R.); REICHENOW, A., No. 2, 1903 (Africa); BANNERMAN, D. A., No. 3, 1933 (tropical West Africa); CHAPIN, J. P., No. 2, 1939 (Belgian Congo); FRIEDMANN, H., No. 1, 1930 (Ethiopia & Kenya); BLANFORD, W. T., No. 3, 1895 (British India); BAKER, E. C. S., No. 4, 1927 (British India); ALI & RIPLEY, No. 4, 1970 (India, Pakistan); WAIT, W. E., No. 1, 1920 (Ceylon); DEIGNAN, H. G., No. 1, 1945 (Thailand); KURODA, N., No. 2, 1936 (Java); McGREGOR, R. C., No. 1, 1909 (Philippines)
Mesitidae, as Mesoenatidae
Mesitornithidae, as Mesoenatidae
Mesoenatidae (incl. Mesitidae, Mesitornithidae, Mesoenidae). REICHENOW, A., No. 4, 1913 (monograph); BRASIL, L., 1914, Genera Avium, 26, 1-3, as Mesitidae); PETERS, J. L., No. 2, 1963 (world list)
Mesoenidae, as Mesoenatidae
Metalluridae. BOUCARD, A., No. 1, 1893-95 (monograph)
Micropidae, as Apodidae
Micropodidae, as Apodidae
Mimidae. PETERS, J. L., No. 9, 1960 (world list); HELLMAYR, C. E., No. 7, 1934 (the Americas, catalog); RIDGWAY, R., No. 4, 1907 (North America); FORBUSH, E. H., No. 3, 1929 (Massachusetts)
Mniotiltidae, as Parulidae
Momotidae (incl. Motmotidae); REICHENOW, A., No. 5, 1914 (monograph); CHAPMAN, F. M., 1923 (revision of genera & species); PETERS, J. L., No. 5, 1955 (world list); CORY, C. B., No. 1, 1918 (the Americas, catalog); RIDGWAY, R., No. 6, 1914 (North America); SALVIN & GODMAN, NO. 2, 1888-1904 (Central America); WETMORE, A., No. 3, 1968 (Panama); de MIRANDA RIBEIRO, A., 1931, Bol. Mus. Nac. Rio de Janeiro, 7, 73-91 (Brasil)
Motacillidae. REICHENOW, A.,, No. 5, 1914 (monograph); PETERS, J. L., No. 9, 1960 (world list); HELLMAYR, C. E., No. 8, 1935 (the Americas, catalog); RIDGWAY, R., No. 3, 1904 (North America); FORBUSH, E. H., No. 3, 1929 (Massachusetts); SALVIN & GODMAN, No. 1, 1879-1904 (Central America); HARTERT, E., No. 1, 1903-10 (Palearctic); PARIS, P., No. 1, 1921 (France); REIS, J. A., Jr., 1927, Mem. Est. Mus. Zool. Univ. Coimbra, (1), 13, 1-44 (Portugal); LINTIA, D., No. 1, 1946 (Romania); DEMENTIEV & GLADKOV, No. 5, 1954 & No. 11, 1970 (Soviet Union); GLADKOV, N. A. et al., No. 1, 1964 (U. S. S. R.); REICHENOW, A., No. 3, 1905 (Africa); KOENIG, A., 1924, Journ. Ornith., 72, (Sonderheft), 1-277 (Egypt); BANNERMAN, D. A., No. 4, 1936 (tropical West Africa); SCHOUTEDEN, H., 1940, Rev. Zool. Bot. Africaines, 33, 317-323 (Congo Belge); CHAPIN, J. P., No. 3, 1953 (Belgian Congo); FRIEDMANN, H., No. 2, 1937 (Ethiopia & Kenya); BAKER, E. C. S., No. 3, 1926 (British India); DEIGNAN, H. G., No. 1, 1945 (Thailand); KURODA, N., No. 1, 1933 (Java); McGREGOR, R. C., No. 2, 1909 (Philippines)
Motmotidae, as Momotidae
Muscicapidae (incl. Falcunculidae, Pachycephalidae, Turnagridae); REICHENOW, A., No. 5, 1914 (monograph); PETERS, J. L., No. 10 & 12, 1964 & 1967 (world list); VAURIE, C., 1953 (Bull. American Mus. Nat. Hist., 100, 453-638 (tribe Muscicapini); STORR, G. M., 1958, Emu, 58, 277-283 (classification, Old World); HARTERT, E., No. 1, 1903-10 (Palearctic);

PARIS, P., No. 1, 1921 (France); LINTIA, D., No. 1, 1946 (Romania); DEMENTIEV & GLADKOV, No. 6, 1954 & No. 12, 1968 (Soviet Union); GLADKOV, N. A. et al., No. 1, 1964 (U. S. S. R.) REICHENOW, A., No. 2, 1903 (Africa); BANNERMAN, D. A., No. 4, 1936 (tropical West Africa); CHAPIN, J. P., No. 3, 1953 (Belgian Congo); FRIEDMANN, H., No. 2, 1937 (Ethiopia & Kenya); WHITE, C. M. N., 1960-62, Occ. Pap. Nat. Mus. Southern Rhodesia, 3, (24B), 399-430, (26B, 653-738 (Ethiopian Sylviinae); BAKER, E. C. S, No. 2, 1924 (British India); DEIGNAN, H. G., No. 1, 1945 (Thailand); KURODA, N., No. 1, 1933 (Java); AMIET, L., 1960, Australian flycatchers of northern Queensland, Townsville Nat., 3, (13), 1-23; McGREGOR, R. C., No. 2, 1909 (Philippines); BAKER, R. H., No. 1, 1951 (Micronesia)

Musophagidae. DUBOIS, A., 1907, Genera Avium, 8, 1-9; REICHENOW, A., No. 5, 1914 (monograph); PETERS, J. L., No. 4, 1940 (world list); CLAY, T., 1947, Ibis, 89, 654-656 (systematic position, by mallophagan parasites); MOREAU, R. E., 1959, Publ. Syst. Assoc., 3, 113-119 (classification); REICHENOW, A., No. 2, 1903 (Africa); BANNERMAN, D. A., No. 3, 1933 (tropical West Africa); CHAPIN, J. P., No. 2, 1939 (Belgian Congo); FRIEDMANN, H., No. 1, 1930 (Ethiopia & Kenya)

Myiadestidae, as Turdidae

Nasiternidae. REICHENOW, A., No. 4, 1913 (monograph)

Nectarinidae, as Nectariniidae

Nectariniidae (incl. Chalcopariidae, Nectarinidae). SHELLEY, G. E., 1876-80, Monograph of Nectariniidae..., London, 1-393); REICHENOW, A., No. 5, 1914 (monograph); DELACOUR, J., 1944, Zoologica, 29, 17-38 (revision); PETERS, J. L., No. 12, 1967 (world list); HARTERT, E., No. 1, 1903-10 (Palearctic); REICHENOW, A., No. 3, 1905 (Africa); BANNERMAN, D. A., No. 6, 1948 (tropical West Africa); CHAPIN, J. P., No. 4, 1954 (Belgian Congo); SKEAD, C. J., 1967, The sunbirds of southern Africa, Cape Town, 1-352; FRIEDMANN, H., No. 2, 1937 (Ethiopia & Kenya); BAKER, E. C. S., No. 3, 1926 (British India, also as Chalcopariidae); BAKER, E. C. S., No. 3, 1926 (British India, as Chalcopariidae); DEIGNAN, H. G., No. 1, 1945 (Thailand); KURODA, N., No. 1, 1933 (Java); McGREGOR, R. C., No. 2, 1909 (Philippines)

Neocathartidae (only as fossils)

Neosittidae. (No revisionary references noted)

Nestoridae, as Psittacidae

Numididae. GHIGI, A., 1909-10, Mem. R. Accad. Sci. Inst. Bologna, Sci. Fis., (6), 7, 331-365; GHIGI, A., 1927, Pubbl. Staz. Sper. Pollicolt Rovigo, 2, 1-84 (monograph); PETERS, J. L., No. 2, 1963 (world list); HELLMAYR & CONOVER, No. 1, 1942 (the Americas, catalog); RIDGWAY & FRIEDMANN, No. 10, 1946 (North America); WETMORE, A., No. 1, 1927 (Puerto Rico); CHAPIN, J. P., No. 1, 1932 (Belgian Congo); van SOMEREN, V. G. L., 1925, Journ. East Africa Uganda Nat. Hist. Soc., 22, 1-21 (guinea fowl); FRIEDMANN, H., No. 1, 1930 (Ethiopia & Kenya)

Nyctibiidae. PETERS, J. L., No. 4, 1940 (world list); CORY, C. B., No. 1, 1918 (the Americas, catalog); RIDGWAY, R., No. 6, 1914 (North America); WETMORE, A., No. 3, 1968 (Panama)

Odobenidae. (No revisionary references noted)

Odontophoridae, as Phasianidae

Odontopterygidae (only as fossils)

Oedicnemidae, as Burhinidae

Opisthocomidae REICHENOW, A., No. 4, 1913 (monograph); PETERS, J. L., No. 2, 1963 (world list); HELLMAYR & CONOVER, No. 1, 1942 (the Americas, catalog)

Opisthodactylidae (only as fossils)

Oreotrochilidae. BOUCARD, No. 1, 1893-95 (monograph)

Oriolidae (incl. Sphecotheridae, Tylidae). REICHENOW, A., No. 5, 1914 (monograph); PETERS, J. L., No. 15, 1962 (world list); HARTERT, E., No. 1, 1903-10 (Palearctic); PARIS, R., No. 1, 1921 (France); LINTIA, D., No. 1, 1946 (Romania); DEMENTIEV & GLADKOV, No. 5, 1954 & No. 11, 1970 (Soviet Union); GLADKOV, N. A. et al., No. 1, 1964 (U. S. S. R.); REICHENOW, A., No. 2, 1903 (Africa); BANNERMAN, D. A., No. 5, 1939 (tropical West Africa); CHAPIN, J. P., No. 4, 1954 (Belgian Congo); FRIEDMANN, H., No. 2, 1937 (Ethiopia & Kenya); BAKER, E. C. S., No. 3, 1926 (British India); DEIGNAN, H. G., No. 1, 1945 (Thailand); KURODA, N., No. 1, 1933 (Java); McGREGOR, R. C., No. 2, 1909 (Philippines)

Orthocnemidae. (only as fossils)

Orthonycidae, as Timaliidae

Otidae, as Otididae

Otididae (incl. Otidae). REICHENOW, A., No. 4, 1913 (monograph); LOW, G. C., No. 1, 1931 (bibliography); PETERS, J. L., No. 2, 1963 (world list); HARTERT, E., No. 3, 1921-22 (Palearctic); PARIS, P., No. 1, 1921 (France); LINTIA, D., No. 3, 1955 (Romania); DEMENTIEV & GLADKOV, No. 2, 1951 & No. 8, 1968 (Soviet Union); GLADKOV, N. A. et al., No. 1, 1964 (U. S. S. R.); REICHENOW, A., No. 1, 1901 (Africa); BOUET, G., No. 1, 1955 (Afrique tropicale); BANNERMAN, D. A., No. 2, 1931 (tropical West Africa); CHAPIN, J. P., No. 2, 1939 (Belgian Congo); van SOMEREN, V. G. L., No. 1, 1925-36 (Kenya & Uganda); FRIEDMANN, H., No. 1, 1930 (Ethiopia & Kenya); BLANFORD, W. T., No. 4, 1898 (British India); BAKER, E. C. S., No. 6, 1929 (British India); ALI & RIPLEY, No. 2, 1969 (India, Pakistan); TAKA-TSUKASA, N., No. 1, 1967 (Nippon)

Oxyrhamphidae, as Oxyruncidae

Oxyruncidae (incl. Oxyrhamphidae). HELLMAYR, C. E., No. 6, 1929 (the Americas, catalog); RIDGWAY, R. No. 4, 1907 (North America); SALVIN & GODMAN, No. 2, 1888-1904 (Central America)

Pachycephalidae, as Muscicapidae
Paictidae, as Philepittidae
Palaelodidae. (No revisionary references noted)
Palaeoscinidae (only as fossils)
Palaeospizidae (only as fossils)
Palamedeidae, as Anhimidae
Pandionidae. BROWN, L. & AMADON, D., 1968, Eagles, hawks and falcons of the world, Middlesex, England, 2 volumes; HELLMAYR & CONOVER, No. 4, 1949 (the Americas, catalog); FRIEDMANN, H., No. 3, 1950 (North & Middle America); FORBUSH, E. H., No. 1, 1927 (massachusetts); SALVIN & GODMAN, No. 3, 1897-1904 (Central America); WETMORE, A., No. 2, 1965 (Panama); PARIS, P., No. 1, 1921 (France); BOUET, G., No. 1, 1955 (Afrique tropicale); BANNERMAN, D. A., No. 1, 1930 (tropical West Africa); BLANFORD, W. T., No. 3, 1895 (British India); BAKER, E. C. S., No. 5, 1928 (British India); WAIT, W. E., No. 2, 1921 (Ceylon); DEIGNAN, H. G., No. 1, 1945 (Thailand); KURODA, N., No. 2, 1936 (Java); McGREGOR, R. C., No. 1, 1909 (Philippines)
Panuridae, as Timaliidae
Paradisaeidae (incl. Epimachidae, Paradiseidae). SHARPE, R. B., 1891-98, Monograph of the Paradisaeidae and Ptilorhynchidae, London, 2 volumes; ROTHSCHILD, W., 1898, Das Tierreich, 2, 1-52; REICHENOW, A., No. 5, 1914 (monograph); STONOR, C. R., 1937, Proc. Zool. Soc. London, 1936, 1177-1185 (evolution & relationships); KURODA, N., 1944, Tori Tokyo, 11, 615-650 (list); IREDALE, T., 1948, Australian Zool., 11, 161-189 (checklist); IREDALE, T., 1950, Birds of paradise and bower birds, Melbourne, 1-239; PETERS, J. L., No. 15, 1962 (world list); GILLIARD, E. T., 1969, Birds of paradise and bower birds, London, 1-485.
Paradiseidae, as Paradisaeidae
Paradoxornithidae, as Timaliidae
Paramythiidae, as Dicaeidae
Paranyrocidae. (No revisionary references noted)
Paridae. HELLMAYR, C. E., 1903, Das Tierreich, 18, 1-255; HELLMAYR, C. E., 1911, Genera Avium, 18, 1-84; REICHENOW, A., No. 5, 1914 (monograph); PETERS, J. L., No. 12, 1967 (world list); HELLMAYR, C. E., No. 7, 1934 (the Americas, catalog); RIDGWAY, R., No. 3, 1904 (North America); FORBUSH, E. H., No. 3, 1929 (Massachusetts); SALVIN & GODMAN, No. 1, 1879-1904 (Central America); HARTERT, E., No. 1, 1903-10 (Palearctic); LEGENDRE, M., 1932, Monographie des mesanges d'Europe, Paris, 1-124; PARIS, P., No. 1, 1921 (France); LINTIA, D., No. 1, 1946 (Romania); DEMENTIEV & GLADKOV, No. 5, 1954 & No. 11, 1970 (Soviet Union); GLADKOV, N. A., et al., No. 1, 1964 (U. S. S. R.); REICHENOW, A., No. 3, 1905 (Africa); BANNERMAN, D. A., No. 6, 1948 (tropical West Africa); CHAPIN, J. P., No. 4, 1954 (Belgian Congo); FRIEDMANN, H., No. 2, 1937 (Ethiopia & Kenya); BAKER, E. C. S., No. 1, 1922 (British India); DEIGNAN, H. G., No. 1, 1945 (Thailand); KURODA, N., No. 1, 1933 (Java); McGREGOR, R. C., No. 2, 1909 (Philippines)
Parridae, as Jacanidae
Parulidae (incl. Caerebidae, Coerebidae, Compsothlypidae, Mniotiltidae, Sylvicolidae). REICHENOW, A., No. 5, 1914 (monograph); PETERS, J. L., No. 14, 1968 (world list); HELLMAYR, C. E., No. 8, 1935 (the Americas, catalog, as Coerebidae & Compsothlypidae); RIDGWAY, R., No. 2, 1902 (North America, as Coerebidae & Mniotiltidae); CHAPMAN, F. M., 1907, The warblers of North America, New York, 1-306, as Mniotiltidae; FORBUSH, E. H., No. 3, 1929 (Massachusetts, as Compsothlypidae); SALVIN & GODMAN, No. 1, 1879-1904 (Central America, as Coerebidae & Mniotiltidae); DEMENTIEV & GLADKOV, No. 5, 1954 & No. 11, 1970 (Soviet Union); GLADKOV, N. A., et al., No. 1, 1964 (U. S. S. R.)
Passeridae, as Ploceidae
Pedinomidae, as Pedionomidae
Pedionomidae (incl. Pedinomidae). PETERS, J. L., No. 2, 1963 (world list)
Pelagornithidae. (No revisionary references noted)
Pelecanidae. DUBOIS, A., 1907, Genera Avium, 7, 1-4; REICHENOW, A., No. 4, 1913 (monograph); PETERS, J. L., No. 1, 1931 (world list); HELLMAYR & CONOVER, No. 2, 1948 (the Americas catalog); FORBUSH, E. H., No. 2, 1929 (Massachusetts); WETMORE, A., No. 1, 1927 (Puerto Rico); SALVIN & GODMAN, No. 3, 1897-1904 (Central America); WETMORE, A., No. 2, 1965 (Panama); MURPHY, R. C., No. 1, 1936 (South America); HARTERT, E., No. 2, 1912-21 (Palearctic); PARIS, P., No. 1, 1921 (France); LINTIA, D., No. 3, 1955 (Romania); TUGARINOV, A. J., No. 1, 1947 (U. R. S. S.); DEMENTIEV & GLADKOV, No. 1, 1951 & No. 7, 1966 (Soviet Union); GLADKOV, N. A. et al., No. 1, 1964 (U. S. S. R.); REICHENOW, A., No. 1, 1901 (Africa); BOUET, G., No. 1, 1955 (Afrique tropicale); BANNERMAN, D. A., No. 1, 1930 (tropical West Africa); CHAPIN, J. P., No. 1, 1932 (Belgian Congo); FRIEDMANN, H., No. 1, 1930 (Ethiopia & Kenya); BLANFORD, W. T., No. 4, 1898 (British India); BAKER, E. C. S., No. 6, 1929 (British India); ALI & RIPLEY, No. 1, 1968 (India, Pakistan); DEIGNAN, H. G., No. 1, 1945 (Thailand); KURODA, N., No. 2, 1936 (Java); McGREGOR, R. C., No. 1, 1909 (Philippines)
Pelecanoidae, as Pelecanoididae
Pelecanoididae (incl. Pelecanoidae). GODMAN, F. C., 1907-10, A monograph of the petrels..., London, 1-381; LOOMIS, L. M., 1918, Proc. California Acad. Sci., (4), 2, 1-187 (review); MURPHY, R. C. & HARPER, F., 1921, Bull. American Mus. Nat. Hist., 44, 495-554 (review); LOOMIS, L. M., 1923, Auk, 40, 596-602 (classification); PETERS, J. L., No. 1, 1931 (world list); HELLMAYR & CONOVER, No. 2, 1948 (the Americas, catalog); LANGELIER, G.,

1940, Le Nat. Canadienne, 67, 289-302; MURPHY, R. C., No. 1, 1936 (South America); DABBENE, R., 1921, El Hornero, 2, 157-179, 141-254, 3, 227-238 (Atlantico austral); ALEXANDER, W. B., 1920, Emu, 20, 14-24, 66-74 (Australia)

Perdicidae, as Phasianidae

Pericrocotidae, as Campephagidae

Peristeridae, as Columbidae

Petasophoridae. BOUCARD, A., No. 1, 1893-95 (monograph)

Phaëthonidae, as Phaëthontidae

Phaethontidae, as Phaëthontidae

Phaëthontidae (incl. Phaëthonidae, Phaethontidae, Phaetonidae, Phaëtontidae). REICHENOW, A., No. 4, 1913 (monograph); PETERS, J. L., No. 1, 1931 (world list); HELLMAYR & CONOVER, No. 2, 1948 (the Americas, catalog); WETMORE, A., No. 1, 1927 (Puerto Rico); SALVIN & GODMAN, No. 3, 1897-1904 (Central America); WETMORE, A., No. 2, 1965 (Panama); MURPHY, R. C., No. 1, 1936 (South America); REICHENOW, A., No. 1, 1901 (Africa); BOUET, G., No. 1, 1955 (Afrique tropicale); BANNERMAN, D. A., No. 1, 1930 (tropical West Africa); BLANFORD, W. T., No. 4, 1898 (British India); BAKER, E. C. S., No. 6, 1929 (British India); ALI & RIPLEY, No. 1, 1968 (India, Pakistan); KURODA, N., No. 2, 1936 (Java); BAKER, R. H., No. 1, 1951 (Micronesia)

Phaethorniidae. BOUCARD, A., No. 1, 1893-95 (monograph)

Phaetonidae, as Phaëthontidae

Phaëtontidae, as Phaëthontidae

Phalacrocoracidae (incl. Graculidae part). REICHENOW, A., No. 4, 1913 (monograph); PETERS, J. L., No. 1, 1931 (world list); HELLMAYR & CONOVER, No. 2, 1948 (the Americas, catalog); FORBUSH, E. H., No. 2, 1929 (Massachusetts); SALVIN & GODMAN, No. 3, 1897-1904 (Central America); WETMORE, A., No. 2, 1965 (Panama); MURPHY, R. C., No. 1, 1936 (South America); HARTERT, E., No. 2, 1912-21 (Palearctic); PARIS, P., No. 1, 1921 (France); LINTIA, D., No. 3, 1955 (Romania); TOUGARINOV, A. J., No. 1, 1947 (U. R. S. S.); DEMENTIEV & GLADKOV, No. 1, 1951 & No. 7, 1966 (Soviet Union); GLADKOV, N. A. et al., No. 1, 1964 (U. S. S. R.); REICHENOW, A., No. 1, 1901 (Africa); BOUET, G., No. 1, 1955 (Afrique tropicale); BANNERMAN, D. A., No. 1, 1930 (tropical West Africa); CHAPIN, J. P., No. 1, 1932 (Belgian Congo); FRIEDMANN, H., No. 1, 1930 (Ethiopia & Kenya); BLANFORD, W. T., No. 4, 1898 (British India); BAKER, E. C. S., No. 6, 1929 (India); ALI & RIPLEY, No. 1, 1968 (India, Pakistan); DEIGNAN, H. G., No. 1, 1945 (Thailand); KURODA, N., No. 2, 1936 (Java); McGREGOR, R. C., No. 1, 1909 (Philippines); BAKER, R. H., No. 1, 1951 (Micronesia)

Phalaropidae, as Phalaropodidae

Phalaropodidae (incl. Phalaropidae). PETERS, J. L., No. 2, 1963 (world list); HELLMAYR & CONOVER, No. 3, 1948 (the Americas, catalog); RIDGWAY, R., No. 8, 1919 (North America); FORBUSH, E. H., No. 2, 1929 (Massachusetts); WETMORE, A., No. 2, 1965 (Panama); MURPHY, R. C., No. 1, 1936 (South America); BOUET, G., No. 1, 1955 (Afrique tropicale); CHAPIN, J. P., No. 2, 1939 (Belgian Congo); BAKER, R. H., No. 1, 1951 (Micronesia)

Phasianidae (incl. Odontophoridae, Perdicidae). REICHENOW, A., No. 4, 1913 (monograph); also as Odontophoridae); BEEBE, W., 1918-22, A monograph of the pheasants, London, 4 volumes; GHIGI, A., 1937, Monographia dei Fagiani, Bologna, 1-228; DELACOUR, J., 1951, The pheasants of the world, London, 1-347; PETERS, J. L., No. 2, 1963 (world list); HELLMAYR & CONOVER, No. 1, 1942 (the Americas, catalog); RIDGWAY & FRIEDMANN, No. 10, 1946 (North America); CRISPENS, C. G., Jr., 1960, Univ. Washington Publ. Biol., 20, 1-125 (North America & bibliography); FORBUSH, E. H., No. 1, 1927 (Massachusetts); WETMORE, A., No. 1, 1927 (Puerto Rico); SALVIN & GODMAN, No. 3, 1897-1904 (Central America); WETMORE, A., No. 2, 1965 (Panama); HARTERT, E., No. 3, 1921-22 (Palearctic); PARIS, P. No. 1, 1921 (France); LINTIA, D., No. 3, 1955 (Romania); DEMENTIEV & GLADKOV, No. 4, 1952 & No. 10, 1967 (Soviet Union); GLADKOV, N. A. et al., No. 1, 1964 (U. S. S. R.); REICHENOW, A., No. 1, 1901 (Africa); BOUET, G., No. 1, 1955 (Afrique tropicale); BANNERMAN, D. A., No. 1, 1930 (tropical West Africa); CHAPIN, J. P., No. 1, 1932 (Belgian Congo); van SOMEREN, V. G. L., 1925, Journ. East Africa Uganda Nat. Hist. Soc., 23, 95-104 (Kenya& Uganda); FRIEDMANN, H., No. 1, 1930 (Ethiopia & Kenya, also asPerdicidae); BLANFORD, W. T., No. 4, 1898 (British India); BAKER, E. C. S., No. 5, 1928 (British India); ALI & RIPLEY, No. 2, 1969 (India, Pakistan); BEEBE, C. W., 1910, Rec. Indian Mus., 5, 263-275 (Indian Museum list); DEIGNAN, H. G., No. 1, 1945 (Thailand); KURODA, N., No. 2, 1936 (Java); McGREGOR, R. C., No. 1, 1909 (Philippines); KURODA, N., 1926, A monograph of the pheasants of Japan, including Korea and Formosa, Tokyo, 1-40; TAKA-TSUKASA, N., 1944, Studies on the Galli of Nippon, Tokyo, 1-67; TAKA-TSUKASA, N., No. 1, 1967 (Nippon); BAKER, R. H., No. 1, 1951 (Micronesia)

Philepittidae (incl. Paictidae). REICHENOW, A., No. 5, 1914 (monograph)

Phoenicopteridae. REICHENOW, A., No. 4, 1913 (monograph); PETERS, J. L., No. 1, 1931 (world list); HELLMAYR & CONOVER, No. 2, 1948 (the Americas, catalog); WETMORE, A., No. 1, 1927 (Puerto Rico); SALVIN & GODMAN, No. 3, 1897-1904 (Central America); DEVINCENZI, G. J., No. 3, 1928 (Uruguay); PARIS, P., No. 1, 1921 (France); LINTIA, D., No. 3, 1955 (Romania); TUGARINOV, A. J., No. 1, 1947 (U. R. S. S.); GLADKOV, N. A. et al., No. 1, 1964 (U. S. S. R.); REICHENOW, A., No. 1, 1901 (Africa); BOUET, G., No. 1, 1955 (Afrique tropicale); BANNERMAN, D. A., No. 1, 1930 (tropical West Africa); CHAPIN, J. P., No. 1, 1932 (Belgian Congo); FRIEDMANN, H., No. 1, 1930 (Ethiopia & Kenya); BLANFORD, W. T., No. 4, 1898 (British India); BAKER, E. C. S., No. 6, 1929 (British India); ALI & RIPLEY, No. 1, 1968 (India, Pakistan)

Phoeniculidae (incl. Irrisoridae). PETERS, J. L., No. 5, 1955 (world list); BANNERMAN, D. A., No. 3, 1933 (tropical West Africa); FRIEDMANN, H., No. 1, 1930 (Ethiopia & Kenya); MOLTONI, E., 1940, Riv. Italiana Ornith., 10, 75-92 (Africa orientale Italiana)

Phororhacidae (only as fossils)

Phyllornithidae, as Irenidae

Phytotomidae. HELLMAYR, C. E., No. 6, 1929 (the Americas, catalog)

Picathartidae, as Timaliidae

Picidae. MALHERBE, A., 1859-62, Monographie des Picidees..., Metz, 4 volumes; HESSE, E., 1912, Mitt. Zool. Mus. Berlin, 6, 133-261 (Kritische Untersuchungen); REICHENOW, A., No. 5, 1914 (monograph); PETERS, J. L., No. 6, 1948 (world list); CORY, C. B., No. 2, 1919 (the Americas, catalog); RIDGWAY, R., No. 6, 1914 (North America); FORBUSH, E. H., No. 1, 1927 (Massachusetts); SALVIN & GODMAN, No. 2, 1888-1904 (Central America); WETMORE, A., No. 3, 1968 (Panama); PERGOLANI, N. J. I., 1940, El Hornero, 7, 382-395 (Argentina); 1942, El Hornero, 8, 155-170 (Argentina); HARTERT, E., No. 2, 1912-21 (Palearctic); VERHEYEN, R., 1946, Les pics et les coucous de Belgique, Bruxelles, 1-157; LEGENDRE, M., 1932, Gerfaut, 22, 8-19 (Belgium, France); PARIS, P., No. 1, 1921 (France); LEGENDRE, M., 1929, Oiseau, 10, 453-459 (list, France); DOMANIEWSKI, J., 1928, Spraw. Kom. Fizyogr. Krakow, 62, 135-143 (Poland); LINTIA, D., No. 2, 1954 (Romania); DEMENTIEV & GLADKOV, No. 1, 1951 & No. 7, 1966 (Soviet Union); GLADKOV, N. A. et al., No. 1, 1964 (U. S. S. R.); REICHENOW, A., No. 2, 1903 (Africa); BANNERMAN, D. A., No. 3, 1933 (tropical West Africa); CHAPIN, J. P., No. 2, 1939 (Belgian Congo); FRIEDMANN, H., No. 1, 1930 (Ethiopia & Kenya); BLANFORD, W. T., No. 3, 1895 (British India); BAKER, E. C. S., No. 4, 1927 (British India); ALI & RIPLEY, No. 4, 1970 (India, Pakistan); WAIT, W. E., No. 1, 1920 (Ceylon); DEIGNAN, H. G., No. 1, 1945 (Thailand); STRESEMANN, E., 1921, Arch. Naturg., 87, (A7), 64-120 (Sumatra); KURODA, N., No. 2, 1936 (Java); McGREGOR, R. C., No. 1, 1909 (Philippines)

Picumnidae. CHAPIN, J. P., No. 2, 1939 (Belgian Congo)

Pipridae. von PELZELN, A. & von. MADARASZ, J., 1887, Monographie der Pipridae ..., Budapest, 1-13; HELLMAYR, C. E., 1910, Genera Avium, 9, 1-31; HELLMAYR, C. E., No. 6, 1929 (the Americas, catalog); RIDGWAY, R., No. 4, 1907 (North America); SALVIN & GODMAN, No. 2, 1888-1904 (Central America)

Pittidae. ELLIOT, D. G., 1893-95, A monograph of the Pittidae, London; REICHENOW, A., No. 5, 1914 (monograph); REICHENOW, A., No. 2, 1903 (Africa); BANNERMAN, D. A., No. 4, 1936 (tropical West Africa); CHAPIN, J. P., No. 3, 1953 (Belgian Congo); BAKER, E. C. S., No. 3, 1926 (British India); ALI & RIPLEY, No. 4, 1970 (India, Pakistan); DEIGNAN, H. G., No. 1, 1945 (Thailand); KURODA, N., No. 1, 1933 (Java); McGREGOR, R. C., No. 2, 1909 (Philippines)

Pityriasidae, as Sturnidae

Plataleidae, as Threskiornithidae

Platycercidae. REICHENOW, A., No. 4, 1913 (monograph)

Plegadidae, as Threskiornithidae

Ploceidae (incl. Bubalornithidae, Carduelidae, Estrildidae, Malimbidae, Passeridae). REICHENOW, A., No. 5, 1914 (monograph); SUSHKIN, P. P., 1927, Bull. American Mus. Nat. Hist., 57, 1-32 (anatomy & classification); von BOETTICHER, H., 1931, Senckenbergiana, 13, 147-153 (Systematik); von BOETTICHER, H., 1941, Anz. Ornith. Ges. Bayern, 3, 126-130 (as subfamily of Passeridae); WOLTERS, H. E., 1954, Ann. Mus. Congo Belge, (n. s.), Zool., 1, 107-113 (genera); DELACOUR, J., 1943, Zoologica, 28, 69-86 (Estrildinae); WOLTERS, H. E., 1957, Bonn Zool. Beitra, 8, 90-129 (Klassification); PETERS, J. L., No. 15, 1962 & No. 14, 1968 (world list, as Estrildidae also); HELLMAYR, C. E., No. 11, 1938 (the Americas, catalog); RIDGWAY, R., No. 4, 1907 (North America); DEMENTIEV & GLADKOV, No. 5, 1954 & No. 11, 1970 (Soviet Union); GLADKOV, N. A. et al., No. 1, 1964 (U. S. S. R.); REICHENOW, No. 3, 1905 (Africa); BANNERMAN, D. A., No. 7, 1949 (tropical West Africa); BANNERMAN, D. A., No. 7, 1949 (tropical West Africa, as Bubalornithidae); LLETGET, A. G., 1935, Bol. Soc. Espanola Hist. Nat., 35, 69-82 (I, Congo belge); CHAPIN, J. P., No. 4 1954, (Belgian Congo); FRIEDMANN, H., No. 2, 1937 (Ethiopia & Kenya); BAKER, E. C. S., No. 3, 1926 (British India); DEIGNAN, H. G., No. 1, 1945 (Thailand); KURODA, N., No. 1, 1933 (Java); CAYLEY, N. W., 1932, Australian finches in bush and aviary, Sydney, 1-256 (review of weaver-finches); McGREGOR, R. C., No. 2, 1909 (Philippines); BAKER, R. H., No. 1, 1951 (Micronesia)

Plotidae, as Anhingidae

Podargidae. HARTERT, E., 1897, Das Tierreich, 1, 1-98; PETERS, J. L., No. 4, (1940 (World list); BLANFORD, W. T., No. 3, 1895 (British India); BAKER, E. C. S., No. 4, 1927 (British India); ALI & RIPLEY, No. 4, 1970 (India, Pakistan); WAIT, W. E., No. 1, 1920 (Ceylon); DEIGNAN, H. G., No. 1, 1945 (Thailand); KURODA, N., No. 2, 1936 (Java); McGREGOR, R. C., No. 1, 1909 (Philippines)

Podicepidae, as Podicipedidae

Podicipedidae (incl. Colymbidae, Podicepidae, Podicipidae, Podicipitidae). REICHENOW, A., No. 4, 1913 (monograph, as Colymbidae); PETERS, J. L., No. 1, 1931 (world list, as Colymbidae); HELLMAYR & CONOVER, No. 2, 1948 (the Americas, catalog, as Colymbidae); FORBUSH, E. H., No. 2, 1929 (Massachusetts, as Colymbidae); WETMORE, A., No. 1, 1917 (Puerto Rico) SALVIN & GODMAN, No. 3, 1897-1904 (Central America, also as Colymbidae); WETMORE, A., No. 2, 1965 (Panama); DEVINCENZI, G. J., No. 1, 1926 (Uruguay); HARTERT, E., No. 2, 1912-21 (Palearctic); PARIS, P., No. 1, 1921 (France, as Colymbidae); LINTIA, D., No. 3, 1955 (Romania); KOZLOVA, E. V., 1947, Faune l'U. R. S. S., Oiseaux, I, 3, Moskva, 1-125

(U. S. S. R., also as Colymbidae, in Russian); DEMENTIEV & GLADKOV, No. 2, 1951 & No. 8, 1968 (Soviet Union, as Colymbidae); GLADKOV, N. A., et al., No. 1, 1964 (U. S. S. R.); REICHENOW, A., No. 1, 1901 (Africa, as Colymbidae); BOUET, G., No. 1, 1955 (Afrique tropicale); BANNERMAN, D. A., No. 1, 1930 (tropical West Africa); CHAPIN, J. P., No. 1, 1932 (Belgian Congo); VOOUS, K. H. & PAYNE, H. A. W., Ardea, 53, 9-31 (The grebes of Madagascar); BLANFORD, W. T., No. 4, 1898 (British India); BAKER, E. C. S., No. 6, 1929 (British India, also as Colymbidae); ALI & RIPLEY, No. 1, 1968 (India, Pakistan); DEIGNAN, H. G., No. 1, 1945 (Thailand); KURODA, N., No. 2, 1936 (Java); McGREGOR, R. C., No. 1, 1909 (Philippines, as Colymbidae)

Podicipidae, as Podicipedidae

Podicipitidae, as Podicipedidae

Polioptilidae, as Sylviidae

Polyboridae, as Falconidae

Presbyornithidae (only as fossils)

Prionopidae. BANNERMAN, D. A., No. 5, 1939 (tropical West Africa); CHAPIN, J. P., No. 4, 1954 (Belgian Congo); FRIEDMANN, H., No. 2, 1937 (Ethiopia & Kenya)

Procellaridae, as Procellariidae

Procellariidae (incl. Fulmaridae, Procellaridae, Puffinidae). GODMAN, F. C., 1907-10, A monograph of the petrels..., London, 1-381 (also as Puffinidae); REICHENOW, A., No. 4, 1913; PETERS, H. L., No. 1, 1931 (world list); HELLMAYR & CONOVER, No. 2, 1948 (the Americas, catalog); SALVIN & GODMAN, No. 3, 1897-1904 (Central America, also as Puffinidae); WETMORE, A., No. 2, 1965 (Panama); MURPHY, R. C., No. 1, 1936 (South America); DEVINCENZI, G. J., No. 1, 1926 (Uruguay); PARIS, P., No. 1, 1921 (France, Puffinidae); LINTIA, D., No. 3, 1955 (Romania); KOZLOVA, E. V., 1947, Faune de l'U. R. S. S., Oiseaux, I, 3, Moskva, 1-125 (in Russian); DEMENTIEV & GLADKOV, No. 2, 1951 & No. 8, 1968 (Soviet Union); GLADKOV, N. A., No. 1, 1964 (U. S. S. R.); REICHENOW, A., No. 1, 1901 (Africa); BOUET, G., No. 1, 1955 (Afrique tropicale); BANNERMAN, D. A., No. 1, 1930 (tropical West Africa); BLANFORD, W. T., No. 4, 1898 (British India); BAKER, E. C. S., No. 6, 1929 (British India); ALI & RIPLEY No. 1, 1968 (India, Pakistan); KURODA, N., No. 2, 1936 (Java); ALEXANDER, W. B., 1920, Emu, 20, 14-24, 66-74 (Australia); McGREGOR, R. C., No. 1, 1909 (Philippines, also as Puffinidae); BAKER, R. H., No. 1, 1951 (Micronesia)

Procniatidae, as Tersinidae

Promeropidae, as Meliphagidae

Prosphiidae. (No revisionary references noted)

Protostrigidae (only as fossils)

Prunellidae (incl. Accentoridae). PETERS, J. L., No. 10, 1964 (world list); HELLMAYR, C. E., No. 7, 1934 (the Americas, catalog); PARIS, P., No. 1, 1921 (France, as Accentoridae); LINTIA, D., No. 1, 1946 (Romania, as Accentoridae); DEMENTIEV & GLADKOV, No. 6, 1954 & No. 12, 1968 (Soviet Union); GLADKOV, N. A., et al., No. 1, 1964 (U. S. S. R.)

Pseudodontornithidae (only as fossils)

Psilopteridae (only as fossils)

Psittacidae (incl. Cacatuidae, Cyclopsittacidae, Kakatoeidae, Loriidae, Nestoridae, Strigopidae, Stringopidae). MIVART, S., 1896, Monograph of the ... Loriidae, London, 1-193; SALVADORI, T., 1905, Genera Avium, 3, 1-2; (also as Nestoidae, Cacatuidae, Cyclopsittacidae, Loriidae); REICHENOW, A., No. 4, 1913 (monograph, also as Cacatuidae, Cyclopsittacidae, Loriidae, Nestoridae, Strigopidae); SETH-SMITH, D., 1926, Parakeets. A handbook of the imported species, 2nd ed., London, 1-295 (also as Cacatuidae & Loriidae); von BOETTICHER, H., 1929, Anz. Ornith. Ges. Bayern, 2, 15-16 (Systematik, as Kakatoeidae); SNOUCKAERT VAN SCHAUBERG, R., 1929, Org. Cl. Nederl. Vogelk. Zutphen, 1, Extra No., 3-68 (synopsis, as Kakatoeidae); PETERS, J. L., No. 3, 1937 (world list); CORY, C. B., No. 1, 1918 (the Americas, catalog); RIDGWAY, R., No. 7, 1916 (North America); SALVIN & GODMAN, No. 2, 1888-1904 (Central America); WETMORE, A., No. 3, 1968 (Panama); de MIRANDA-RIBEIRO, A., 1920, Rev. Mus. Paulista, 12, (2), 1-82 (Brasil); ORFILA, R. N., 1936, El Hornero, 6, 197-225 (Argentina); ORFILA, R. N., 1937, El Hornero, 6, 367-381 (Argentina); ORFILA, R. N., 1938, El Hornero, 7, 2-21 (Argentina); STIGAND, O. P., 1932, Riv. Italiana Ornith., 2, 56-62 (? Italy); REICHENOW, A., No. 2, 1903 (Africa); STIGAND, O. P., 1935, Riv. Italiana Ornith., 5, 36-47 (Africa); BANNERMAN, D. A., No. 2, 1931 (tropical West Africa); CHAPIN, J. P., No. 2, 1939 (Belgian Congo); FRIEDMANN, H., No. 1, 1930 (Ethiopia & Kenya); BLANFORD, NO. 3, 1895 (British India); BAKER, E. C. S., No. 4, 1927 (British India); ALI & RIPLEY, No. 3, 1969 (India, Pakistan); WAIT, W. E., No. 1, 1920 (Ceylon); DEIGNAN, H. G., No. 1, 1945 (Thailand); KURODA, N., No. 2, 1936 (Java); EASTMAN, W. R., Jr. & HUNT, A. C., 1966, The parrots of Australia, Narberth, Pennsylvania, 1-194; FORSHAW, J. M., 1969, Australian parrots, Melbourne, 1-306; CONDON, H. T., 1941, Rec. South Australian Mus., 7, 117-144 (subf. Platycercinae, Australia); McGREGOR, R. C., No. 1, 1909 (Philippines); also as Cacatuidae & Loriidae); HERKLOTS, G. A. C., 1940, Hong Kong Nat., 10, 1-4, 5-10, 75-78 (Birds of Hong Kong part 33); BAKER, R. H., No. 1, 1951 (Micronesia); BERLIOZ, J., 1945, Oiseau, 15, 1-9 (New Caledonia)

Psophiidae. PETERS, J. L., No. 2, 1963 (world list); HELLMAYR & CONOVER, No. 1, 1942 (the Americas, catalog)

Pterocletidae, as Pteroclidae

Pteroclidae (incl. Eremialectoridae, Pterocletidae, Pteroclididae, Syrrhaptidae). REICHENOW, A., No. 4, 1913 (monograph); BOWEN, W. W., 1927, American Mus. Novit., 273, 1-12 (Classification); PETERS, J. L., No. 3, 1937 (world list); MACLEAN, G. L., 1967, Journ. Ornith., 108, 203-217 (systematic position); PARIS, P., No. 1, 1921 (France); LINTIA, D., No. 3, 1955 (Romania); DEMENTIEV & GLADKOV, No. 2, 1951 & No. 8, 1968 (Soviet Union); GLADKOV, N. A. et al., No. 1, 1964 (U. S. S. R.); REICHENOW, A., No. 1, 1901 (Africa); BOWEN, W. W., 1928, Sudan Notes, 11, 69-82 (Sudan); BANNERMAN, D. A., No. 2, 1931 (tropical West Africa); CHAPIN, J. P., No. 2, 1939 (Belgian Congo); FRIEDMANN, H., No. 1, 1930 (Ethiopia & Kenya); MOLTONI, E., 1945, Riv. Italiana Ornith., 15, 1-18 (Africa Orientale Italiana); BLANFORD, W. T., No. 4, 1898 (British India); BAKER, E. C. S., No. 5, 1928 (British India); ALI & RIPLEY, No. 3, 1969 (India, Pakistan)

Pteroclididae, as Pteroclidae

Pteroptochidae, as Rhinocryptidae

Ptilogonatidae. HELLMAYR, C. E., No. 8, 1935 (the Americas, catalog); RIDGWAY, R., No. 3, 1904 (North America)

Ptilonorhynchidae (incl. Pteorhynchidae). SHARPE, R. B., 1891-98, Monograph of the Paradiseidae ... and the Ptilorhynchidae, London, 2 volumes; STONOR, C. R., 1937, Proc. Zool. Soc. London, 107B, 475-490 (systematic position); KURODA, N., 1944, Tori Tokyo, 11, 615-650 (list); IREDALE, T., 1948, Australian Zool., 11, 161-189 (checklist); IREDALE, T., 1950, Birds of paradise and bower birds, Melbourne, 1-239; MARSHALL, A. J., 1954, Biol. Rev., 29, 1-45 (bower-birds); PETERS, J. L., No. 15, 1962 (world list); GILLIARD, E. T., 1969, Birds of paradise and bower birds, London, 1-485)

Ptilorhynchidae, as Ptilonorhynchidae

Puffinidae, as Procellariidae

Pycnonotidae (incl. Brachypodidae). REICHENOW, A., No. 5, 1914 (monograph); DELACOUR, J., 1943, Zoologica, 28, 17-28 (revision); PETERS, J. L., No. 9, 1960 (world list); HARTERT, E., No. 1, 1903-10 (Palearctic, as Bradypodidae); DEMENTIEV & GLADKOV, No. 6, 1954 & No. 12, 1968 (Soviet Union); GLADKOV, N. A. et al., No. 1, 1964 (U. S. S. R.); REICHENOW, A., No. 3, 1905 (Africa); BANNERMAN, D. A., 1924, Rev. Zool. Afric., 12, 17-38 (Africa); BANNERMAN, D. A., No. 4, 1936 (tropical West Africa); CHAPIN, J. P., No. 3, 1953 (Belgian Congo) FRIEDMANN, H., No. 2, 1937 (Ethiopia & Kenya); BAKER, E. C. S., No. 1, 1922 (British India); SNOUCKAERT VAN SCHAUBERG, R. C. E. G. L. Org. Cl. Nederl. Vogelk., 9, 102-122 (5 parts, Asia & Indian Archipelago); DEIGNAN, H. G., No. 1, 1945 (Thailand); DELACOUR, J., 1929, Oiseau, 10, 709-728 (Indochine francaise); KURODA, N., No. 1, 1933 (Java); McGREGOR, R. C., No. 2, 1909 (Philippines)

Rallidae. REICHENOW, A., No. 4, 1913 (monograph); PETERS, J. L., No. 2, 1963 (world list); HELLMAYR & CONOVER, No. 1, 1942 (the Americas, catalog); RIDGWAY & FRIEDMANN, No. 9, 1941 (North America); FORBUSH, E. H., No. 2, 1929 (Massachusetts); WETMORE, A., No. 1, 1927 (Puerto Rico); SALVIN & GODMAN, No. 3, 1897-1904 (Central America); WETMORE, A., No. 2, 1965 (Panama); DEVINCENZI, G. J., No. 1, 1926 (Uruguay); HARTERT, E., No. 3, 1921-22 (Palearctic); PARIS, P., No. 1, 1921 (France); LINTIA, D., No. 3, 1955 (Romania); DEMENTIEV & GLADKOV, No. 3, 1951 & No. 9, 1969 (Soviet Union); GLADKOV, N. A. et al., No. 1, 1964 (U. S. S. R.); REICHENOW, A., No. 1, 1901 (Africa); BOUET, G., No. 1, 1955 (Afrique tropicale); BANNERMAN, D. A., No. 2, 1931 (tropical West Africa); CHAPIN, J. P. No. 2, 1939 (Belgian Congo); FRIEDMANN, H., No. 1, 1930 (Ethiopia & Kenya); BLANFORD, W. T., No. 4, 1898 (British India); BAKER, E. C. S., No. 6, 1929 (British India); ALI & RIPLEY No. 2, 1969 (India, Pakistan); DEIGNAN, H. G., No. 1, 1945 (Thailand); KURODA, N., No. 2, 1936 (Java); McGREGOR, R. C., No. 1, 1909 (Philippines); HERKLOTS, G. A. C., 1940, Hong Kong Nat., 10, 1-4, 5-10, 75-78 (Hong Kong); TAKA-TSUKASA, N., No. 1, 1967 (Nippon); BAKER, R. H., No. 1, 1951 (Micronesia)

Ramphastidae (incl. Rhamphastidae). GOULD, J., 1852-55, Monograph of the Ramphastidae, 2nd ed., London; LANGELIER, G., 1937, Le Nat. Canadienne, 64, 40-63 (revision with keys); PETERS, J. L., No. 6, 1948 (world list); CORY, C. B., No. 2, 1919 (the Americas, catalog); RIDGWAY, R., No. 6, 1914 (North America); SALVIN & GODMAN, No. 2, 1888-1904 (Central America); WETMORE, A., No. 3, 1968 (Panama); DUGAND, A., 1941, Rev. Acad. Colombiana, 4, 356-362 (Colombia); DABBENE, R., 1929, El Hornero, 4, 265-271 (Argentina)

Raphidae (incl. Dididae). (Only as subfossils). PETERS, J. L., No. 3, 1937 (world list)

Recurvirostridae. PETERS, J. L., No. 2, 1963 (world list); HELLMAYR & CONOVER, No. 3, 1948 (the Americas, catalog); RIDGWAY, R., No. 8, 1919 (North America); FORBUSH, E. H., No. 2, 1929 (Massachusetts); WETMORE, A., No. 1, 1927 (Puerto Rico); WETMORE, A., No. 2, 1965 (Panama); FRIEDMANN, H., No. 1, 1930 (Ethiopia & Kenya); ALI & RIPLEY, No. 2, 1969 (India, Pakistan); DEIGNAN, H. G., No. 1, 1945 (Thailand); KURODA, No. 2, 1936 (Java)

Regulidae, as Sylviidae

Remizidae. PETERS, J. L., No. 12, 1967 (world list)

Rhabdornithidae. PETERS, J. L., No. 12, 1967 (world list)

Rhamphastidae, as Ramphastidae

Rhegminornithidae. (No revisionary references noted)

Rheidae. REICHENOW, A., No. 4, 1913 (monograph); CRANDALL, L. S., 1929, Bull. New York Zool. Soc., 32, 193-212; PETERS, J. L., No. 1, 1931 (world list); HELLMAYR & CONOVER, No. 1, 1942 (the Americas, catalog); DEVINCENZI, G. J., No. 1, 1926 (Uruguay); DABBENE, R., 1920, El Hornero, 2, 81-84 (Argentina)

Rhinochetidae, as Rhynochetidae

Rhinocryptidae (incl. Hylactidae, Pteroptochidae). REICHENOW, A., No. 5, 1914 (monograph); SICK, H., 1960, Journ. Ornith., 101, 141-174 (Systematik); PETERS, J. L., No. 7, 1951 & No. 10, 1964 (world list); CORY & HELLMAYR, No. 3, 1924 (the Americas, catalog, as Pteroptochidae); RIDGWAY, R., No. 5, 1911 (North America, as Pteroptochidae); SALVIN & GODMAN, No. 2, 1888-1904 (Central America, as Pteroptochidae);

Rhynchopidae, as Rynchopidae

Rhyncopidae, as Rynchopidae

Rhynochetidae (incl. Rhinochetidae). BRASIL, L., 1913, Genera Avium, 21, 1-3; PTERS, J. L., No. 2, 1963 (world list)

Rostratulidae. LOWE, P. R., 1931, Ibis, (13), 1, 491-534 (relations); LOWE, P. R., Ibis, 1932, 507-530 (anatomical review); PETERS, J. L., No. 2, 1963 (world list); HELLMAYR & CONOVER, No. 3, 1948 (the Americas, catalog); KOZLOVA, E. V., 1961-62, Fauna U. R. S. S., Oiseaux, II, 1, 2 & 3, Moskva, 1-500 (Russia, in Russian); DEMENTIEV & GLADKOV, No. 3, 1951 & No. 9, 1969 (Soviet Union); BOUET, G., No. 1, 1955 (Afrique tropicale); BANNERMAN, D. A., No. 2, 1931 (tropical West Africa); CHAPIN, J. P., No. 2, 1939 (Belgian Congo); BAKER, E. C. S., No. 6, 1929 (British India); ALI & RIPLEY, No. 2, 1969 (India, Pakistan); DEIGNAN, H. G., No. 1, 1945 (Thailand)

Rupicolidae, as Cotingidae

Rynchopidae (incl. Rhynchopidae, Rhyncopidae). PETERS, J. L., No. 2, 1963 (world list); HELLMAYR & CONOVER, No. 3, 1948 (the Americas, catalog); RIDGWAY, R., No. 8, 1919 (North America); FORBUSH, E. H., No. 2, 1929 (Massachusetts); WETMORE, A., No. 2, 1965 (Panama); MURPHY, R. C., No. 1, 1936 (South America); DEVINCENZI, G. J., No. 1, 1926 (Uruguay); DABBENE, R., 1918, El Hornero, 1, 49-56, 129-139 (Argentina, I, II); BAKER, E. C. S., No. 6, 1929 (British India);

Sagittariidae (incl. Gypogeranidae, Serpentariidae). REICHENOW, A., No. 4, 1913 (monograph, as Serpentariidae); SWANN, H. K., 1919-22 A synopsis of the Accipitres ..., 2nd ed., London, 1-233 (3 parts, as Serpentariidae); SWANN & WETMORE, 1924-45, A monograph of the birds of prey, London, 2 volumes; PETERS, J. L., No. 1, 1931 (world list); BROWN, L. & AMADON, D., 1968, Eagles, hawks and falcons of the world, Middlesex, England, 2 volumes; BOUET, G., No. 1, 1955 (Afrique tropicale); BANNERMAN, D. A., No. 1, 1930 (tropical West Africa); CHAPIN, J. P., No. 1, 1932 (Belgian Congo); FRIEDMANN, H., No. 1, 1930 (Ethiopia & Kenya)

Sarcoramphidae, as Cathartidae

Sarcorhamphidae, as Cathartidae

Scaniornithidae (only as fossils)

Scolopacidae (incl. Aphrizidae). REICHENOW, A., No. 4, 1913 (monograph); MATHEWS, G. M. & IREDALE DALE, T., 1920, Australian Avian Rec., 4, 123-129 (a rearrangement); LOW, G. C., No. 1, 1931 (bibliography); PETERS, J. L., No. 2, 1963 (world list); MEINERTZHAGEN, A. C., 1926, Ibis, 1926, 477-521 (review of Scolopacinae); HELLMAYR & CONOVER, No. 3, 1948 (the Americas, catalog); RIDGWAY, R., No. 8, 1919 (North America, also as Aphrizidae); FORBUSH, E. H., No. 2, 1929 (Massachusetts, also as Aphrizidae); WETMORE, A., No. 1, 1927 (Puerto Rico); WETMORE, A., No. 2, 1965 (Panama); DEVINCENZI, G. J., No. 2, 1927 (Uruguay); REICHENOW, A., No. 1, 1901 (Africa); BANNERMAN, D. A., No. 2, 1931 (tropical West Africa); FRIEDMANN, H., No. 1, 1930 (Ethiopia & Kenya); BAKER, E. C. S., No. 6, 1929 (British India); DEIGNAN, H. G., No. 1, 1945 (Thailand); KURODA, N., No. 2, 1936 (Java); TAKA-TSUKASA, N., No. 1, 1967 (Nippon); BAKER, R. H., No. 1, 1951 (Micronesia)

Scopidae. REICHENOW, A., No. 4, 1913 (monograph); PETERS, J. L., No. 1, 1931 (world list); REICHENOW, A., No. 1, 1901 (Africa); BOUET, G., No. 1, 1955 (Afrique tropicale); BANNERMAN, D. A., No. 1, 1930 (tropical West Africa); CHAPIN, J. P., No. 1, 1932 (Belgian Congo); FRIEDMANN, H., No. 1, 1930 (Ethiopia & Kenya)

Serpentariidae, as Sagittariidae

Sittidae. HELLMAYR, C. E., 1903, Das Tierreich, 18, 1-255; HELLMAYR, C. E., 1911, Genera Avium, 16, 1-18; PETERS, J. L., No. 12, 1967 (world list); HELLMAYR, C. E., No. 7, 1934 (the Americas, catalog); RIDGWAY, R., No. 3, 1904 (North America); FORBUSH, E. H., No. 3, 1929 (Massachusetts); SALVIN & GODMAN, No. 1, 1879-1904 (Central America); HARTERT, E., No. 1, 1903-10 (Palearctic); PARIS, P., No. 1, 1921 (France); DUNAJEWSKI, A., 1934, Acta Ornith. Mus. Zool. Polonici, 1, 181-251 (Poland); LINTIA, D., No. 1, 1946 (Romania); DEMENTIEV & GLADKOV, No. 5, 1954 & No. 11, 1970 (Soviet Union); GLADKOV, N. A. et al., No. 1, 1964 (U. S. S. R.); BAKER, E. C. S., No. 1, 1922 (British India); DEIGNAN, H. G., No. 1, 1945 (Thailand); KURODA, N., No. 1, 1933 (Java); McGREGOR, R. C., No. 2, 1909 (Philippines)

Spermestidae, STEINER, H., 1960, Journ. Ornith., 101, 92-112 (Klassification)

Sphecotheridae, as Oriolidae

Spheniscidae (incl. Aptenodytidae). REICHENOW, A., No. 4, 1913 (monograph); STUART-SUTHERLAND, R., 1920, Emu, 20, 24-30, 74-81; PETERS, J. L., No. 1, 1931 (world list); HELLMAYR & CONOVER, No. 2, 1948 (the Americas, catalog); MURPHY, R. C., No. 1, 1936 (South America); DEVINCENZI, G. J., No. 1, 1926 (Uruguay); DABBENE, R., 1920, El Hornero, 2, 1-9 (Argentina); REICHENOW, A., No. 1, 1901 (Africa); STUART-SUTHERLAND, R., 1922, Emu, 21, 198-201 (types of Australasian genera)

Steatornithidae. PETERS, J. L., No. 4, 1940 (world list); CORY, C. B., No. 1, 1918 (the Americas, catalog); WETMORE, A., No. 3, 1968 (Panama);

Stercorariidae (incl. Catharactidae). PETERS, J. L., No. 2, 1963 (worldl list); HELLMAYR & CONOVER, No. 3, 1948 (the Americas, catalog); RIDGWAY, R., No. 8, 1919 (North America); FORBUSH, E. H., No. 2, 1929 (Massachusetts); WETMORE, A., No. 2, 1965 (Panama); MURPHY, R. C., No. 1, 1936 (South America); DABBENE, R., 1918, El Hornero, 1, 49-56, 129-139 (Argentina); PARIS, P., No. 1, 1921 (France); BANNERMAN, D. A., No. 2, 1931 (tropical West Africa);

BLANFORD, W. T, , No. 4, 1898 (British India); BAKER, E. C. S., No. 6, 1929 (British India); ALI & RIPLEY, No. 3, 1969 (India, Pakistan)

Sternidae, as Laridae

Streperidae, as Cracticidae

Strigidae (incl. Asionidae, Bubonidae). REICHENOW, A., No. 4, 1913 (monograph); PETERS, J. L., No. 4, 1940 (world list); CORY, C. B., No. 1, 1918 (the Americas, catalog, as Bubonidae); RIDGWAY, R., No. 6, 1914 (North America, as Bubonidae); FORBUSH, E. H., No. 1, 1927 (Massachusetts); SALVIN & GODMAN, No. 3, 187-1904 (Central America, also as Asionidae); WETMORE, A., No. 3, 1968 (Panama); PEREYRA, J. A., 1931, El Hornero, 4, 392-397 (Argentina); HARTERT, E., No. 2, 1912-21 (Palearctic); PARIS, P., No. 1, 1921 (France); LINTIA, D., No. 2, 1954 (Romania); DEMENTIEV & GLADKOW, No. 1, 1951 & No. 7, 1966 (Soviet Union); GLADKOV, N. A. et al., NO. 1, 1964 (U. S. S. R.); REICHENOW, A., No. 1, 1901 (Africa); KOENIG, A., 1917, Journ. Ornith., 65, (II), 129-160 (Egypt); BANNERMAN, D. A., No. 3, 1933 (tropical West Africa); CHAPIN, J. P., No. 2, 1939 (Belgian Congo); FRIEDMANN, H., No. 1, 1930 (Ethiopia & Kenya); BLANFORD, W. T., No. 3, 1895 (British India, also as Asionidae); BAKER, E. C. S., No. 4, 1927 (British India, as Asionidae); ALI & RIPLEY, No. 3, 1969 (India, Pakistan); WAIT, W. E., No. 2, 1921 (Ceylon, also as Asionidae); DEIGNAN, H. G., No. 1, 1945 (Thailand); KURODA, N., No. 2, 1936 (Java); McGREGOR, R. C., No. 1, 1909 (Philippines) BAKER, R. H., No. 1, 1951 (Micronesia)

Strigopidae, as Psittacidae

Stringopidae, as Psittacidae

Struthiididae, as Grallinidae

Struthionidae. BEEBE, C. W., 1905, Ninth Ann. Rep. New York Zool. Soc., (ostriches & allies); REICHENOW, A., No. 4, 1913 (monograph); POISSON, H., 1926, Encyclo. Ornith., 2, 1-203; CRANDALL, L. S., 1929, Bull. New York Zool. Soc., 32, 193-212 (struthious birds, II); PETERS, J. L., No. 1, 1931 (world list); HARTERT, E., No. 3, 1921-22 (Palearctic); REICHENOW, A., No. 1, 1901 (Africa); BOUET, G., 1955, Faune Union Francaise, 16, 1-412 (Afrique tropicale, part 1); BANNERMAN, D. A., No. 1, 1930 (tropical West Africa); FRIEDMANN, H., No. 1, 1930 (Ethiopia & Kenya)

Sturnidae (incl. Buphagidae, Eulabetidae, Eulebetidae, Graculidae part, Pityriasidae). REICHENOW, A., No. 5, 1914 (monograph); von BOETTICHER, H., 1931, Senckenbergiana, 13, 147-153 (Systematik); AMADON, D., 1943, American Mus. Novit., 1247, 1-16 (genera of Starlings); PETERS, J. L., No. 15, 1962 (world list); HELLMAYR, C. E., No. 8, 1935 (the Americas, catalog); RIDGWAY, R., No. 4, 1907 (North America); FORBUSH, E. H., No. 1, 1927 (Massachusetts); HARTERT, E., No. 1, 1903-10 (Palearctic); PARIS, P., No. 1, 1921 (France); LINTIA, D., No. 1, 1946 (Romania); DEMENTIEV & GLADKOV, No. 5, 1954 & No. 11, 1970 (Soviet Union); GLADKOV, N. A. et al., No. 1, 1964 (U. S. S. R.); REICHENOW, A., No. 2, 1903 (Africa) BANNERMAN, D. A., No. 6, 1948 (tropical West Africa); CHAPIN, J. P., No. 4, 1954 (Belgian Congo); FRIEDMANN, H., No. 2, 1937 (Ethiopia & Kenya); BAKER, E. C. S., No. 3, 1926 (British India, also as Eulabetidae); DEIGNAN, H. G., No. 1, 1945 (Thailand); GERMAIN, R., 1912, Rev. Francaise Ornith., 2, 302-305, 337-338 (Cochinchine francaise); KURODA, N., No. 1, 1933 (Java); McGREGOR, R. C., No. 2, 1909 (Philippines); BAKER, R. H., No. 1, 1951 (Micronesia)

Sulidae. REICHENOW, A., No. 4, 1913 (monograph); PETERS, J. L., No. 1, 1931 (world list); HELLMAYR & CONOVER, No. 2, 1948 (the Americas, catalog); FORBUSH, E. H., No. 2, 1929 (Massachusetts); WETMORE, A., No. 1, 1927 (Puerto Rico); SALVIN & GODMAN, No. 3, 1897-1904 (Central America); WETMORE, A., No. 2, 1965 (Panama); MURPHY, R. C., No. 1, 1936 (South America); HARTERT, E., No. 2, 1912-21 (Palearctic); PARIS, P., No. 1, 1921 (France) TUGARINOV, A. J., No. 1, 1947 (U. R. S. S.); DEMENTIEV & GLADKOV, No. 1, 1951 & No. 7, 1966 (Soviet Union); REICHENOW, A., No. 1, 1901 (Africa); BOUET, G., No. 1, 1955 (Afrique tropicale); BANNERMAN, D. A., No. 1, 1930 (tropical West Africa); CHAPIN, J. P., No. 1, 1932 (Belgian Congo); BLANFORD, W. T., No. 4, 1898 (British India); BAKER, E. C. S., No. 6, 1929 (British India); ALI & RIPLEY, No. 1, 1968 (India, Pakistan); KURODA, N., No. 2, 1936 (Java); McGREGOR, R. C., No. 1, 1909 (Philippines); BAKER, R. H., No. 1, 1951 (Micronesia)

Sylvicolidae, as Parulidae

Sylvidae, as Sylviidae

Sylviidae (incl. Acanthizidae, Maluridae, Polioptilidae, Regulidae, Sylvidae). HELLMAYR, C. E., 1911, Genera Avium, 17, 1-17 (as Regulidae); REICHENOW, A., No. 5, 1914 (monograph); CORTI, U. A., 1927, Ornith. Beob., 25, 6-8 (Systematik, as Regulidae); DELACOUR, J., 1942, Ibis, 1942, 509-515 (bush-warblers); HELLMAYR, C. E., No. 7, 1934 (the Americas, catalog); RIDGWAY, R., No. 3, 1904 (North America); FORBUSH, E. H., No. 3, 1929 (Massachusetts); SALVIN & GODMAN, No. 1, 1879-1904 (Central America); HOWARD, H. E., 1907-15, The British warblers, London, 2 volumes: PARIS, P., No. 1, 1921 (France, also as Regulidae); DEMENTIEV & GLADKOV, No. 6, 1954 & No. 12, 1968 (Soviet Union, also as Regulidae); GLADKOV, N. A. et al., No. 1, 1964 (U. S. S. R.); REICHENOW, A., No. 3, 1905 (Africa); KOENIG, A., 1924, Journ. Ornith., 72, (Sonderheft), 1-277 (Egypt); BANNERMAN, D. A., No. 5, 1939 (tropical West Africa); CHAPIN, J. P., No. 3, 1953 (Belgian Congo); FRIEDMANN, H., No. 2, 1937 (Ethiopia & Kenya); BAKER, E. C. S., No. 1, 1924 (British India, also as Regulidae); DEIGNAN, H. G., No. 1, 1945 (Thailand); KURODA, N., No. 1, 1933 (Java); McGREGOR, R. C., No. 2, 1909 (Philippines); BAKER, R. H., No. 1, 1951 (Micronesia)

Synallaxidae. CHUBB, C., 1919, Ann. Mag. Nat. Hist., (9), 3, 273-275.

Syrrhaptidae, as Pteroclidae

Tachypetidae, as Fregatidae
Tanagridae, as Thraupidae
Tangaridae, as Thraupidae
Tantalidae, as Threskiornithidae
Teratornithidae (only as fossils)
Tersinidae (incl. Procniatidae). HELLMAYR, C. E., No. 9, 1936 (the Americas, catalog)
Tetraonidae. REICHENOW, A., No. 4, 1913 (monograph); PETERS, J. L., No. 2, 1963 (world list); SHORT, L. L., Jr., 1967, American Mus. Novit., 2289, 1-39 (generic review); HELLMAYR & CONOVER, No. 1, 1942 (the Americas, catalog); RIDGWAY & FRIEDMANN, No. 10, 1946 (North America); FORBUSH, E. H., No. 1, 1927 (Massachusetts); HARTERT, E., No. 3, 1921-22 (Palearctic); PARIS, P., No. 1, 1921 (France); LINTIA, D., No. 3, 1955 (Romania); KLEINSCHMIDT, O., 1949, Beitr. Taxon. Zool., 1, 101-122 (Classification; Russia); DEMENTIEV & GLADKOV, No. 4, 1952 & No. 10, 1967 (Soviet Union); GLADKOV, N. A. et al., No. 1, 1964 (U. S. S. R.); TAKA-TSUKASA, N., 1932, The birds of Nippon, volume I, part 6, Tokyo, 291-326); TAKA-TSUKASA, N., No. 1, 1967 (Nippon)
Thalassidromidae, as Hydrobatidae
Thaluranidae. BOUCARD, A., No. 1, 1893-95 (monograph)
Thinocoridae (incl. Attagidae, Thinocorythidae). REICHENOW, A., No. 4, 1913 (monograph); LOW, G. C., No. 1, 1931 (bibliography, as Attagidae); PETERS, J. L., No. 2, 1963 (world list); HELLMAYR & CONOVER, No. 3, 1948 (the Americas, catalog)
Thinocorythidae, as Thinocoridae
Thraupidae (incl. Tanagridae, Tangaridae). von BERLEPSCH, H., 1911, Verh. V. Intern. Ornith. Kongr. Berlin, 1910 (1911), 1001-1161 (revision, as Tanagridae); REICHENOW, A., No. 5, 1914 (monograph, as Tanagridae); HELLMAYR, C. E., No. 9, 1936 (the Americas, catalog); RIDGWAY, R., No. 2, 1902 (North America, as Tanagridae); FORBUSH, E. H., No. 3, 1929 (Massachusetts, as Tangaridae); SALVIN & GODMAN, No. 1, 1879-1904 (Central America, as Tanagridae)
Threskiornithidae (incl. Ibidae, Ibididae, Plataleidae, Plegadidae, Tantalidae). REICHENOW, A., No. 4, 1913 (monograph, as Ibididae); PETERS, J. L., No. 1, 1931 (world List); HELLMAYR & CONOVER, No. 2, 1948 (the Americas, catalog); FORBUSH, E. H., No. 2, 1929 (Massachusetts); WETMORE, A., No. 1, 1927 (Puerto Rico); SALVIN & GODMAN, No. 3, 1897-1904 (Central America, as Ibididae & Plataleidae); WETMORE, A., No. 2, 1965 (Panama); DEVINCENZI, G. J., No. 3, 1928 (Uruguay, as Ibididae & Plataleidae); HARTERT, E., No. 2, 1912-21 (Palearctic, as Ibididae); PARIS, P., No. 1, 1921 (France, as Ibididae & Plataleidae); LINTIA, D., No. 3, 1955 (Romania, as Ibididae); TUGARINOV, A. J., No. 1, 1947 (U. R. S. S.); DEMENTIEV & GLADKOV, No. 2, 1951 & No. 8, 1968 (Soviet Union, as Ibididae); GLADKOV, N. A. et al., No. 1, 1964 (U. S. S. R., as Ibididae); REICHENOW, A., No. 1, 1901 (Africa, as Ibidae); BOUET, G., No. 1, 1955 (Afrique tropicale, as Plataleidae); BANNERMAN, D. A., No. 1, 1930 (tropical West Africa, as Plataleidae & Plegadidae); CHAPIN, J. P., No. 1, 1932 (Belgian Congo, as Plataleidae); FRIEDMANN, H., No. 1, 1930 (Ethiopia & Kenya); BLANFORD, W. T., No. 4, 1898 (British India, as Ibididae & Plataleidae); BAKER, E. C. S., No. 6, 1929 (British India, as Ibididae & Plataleidae); ALI & RIPLEY, No. 1, 1968 (India, Pakistan); DEIGNAN, H. G., No. 1, 1945 (Thailand); KURODA, N., No. 2, 1936 (Java, as Plegadidae); McGREGOR, R. C., No. 1, 1909 (Philippines, as Ibididae & Plataleidae)
Timaliidae (incl. Chamaeidae, Cinclosomatidae, Eupetidae, Illadopsidae, Leiotrichidae, Liotrichidae, Orthonycidae, Panuridae, Paradoxornithidae, Picathartidae, Timeliidae, Turnoididae). HELLMAYR, C. E., 1912, Genera Avium, 23, 1-3 (as Chamaeidae); DELACOUR, J., 1946, Oiseau, 16, 7-36 (classification); HELLMAYR, C. E., No. 7, 1934 (the Americas, catalog, as Chamaeidae); RIDGWAY, R., No. 3, 1904 (North America, as Chamaeidae); DEMENTIEV & GLADKOV, No. 5, 1954 & No. 11, 1970 (Soviet Union) as Paradoxornithidae also); GLADKOV, N. A. et al., No. 1, 1964 (U. S. S. R., also as Paradoxornithidae); BANNERMAN, D. A., No. 4, 1936 (tropical West Africa); CHAPIN, J. P., No. 3, 1953 (Belgian Congo); FRIEDMANN, H., No. 2, 1937 (Ethiopia & Kenya); BAKER, E. C. S., No. 1, 1922 (British India, also as Paradoxornithidae); HARINGTON, H. H., 1914-15, Journ. Bombay Nat. Hist. Soc., 23, 44-72, 311-340, 417-453, 614-657, etc. (revision, also as Paradoxornithidae); WHISTLER, H., 1941, Ibis, 1941, 172-173 (eastern Himalayas); DEIGNAN, H. G., No. 1, 1945 (Thailand, also as Paradoxornithidae); KURODA, N., No. 1, 1933 (Java); McGREGOR, R. C., No. 2, 1909 (Philippines)
Timeliidae, as Timaliidae
Tinamidae (incl. Crypturidae). REICHENOW, A., No. 4, 1913 (monograph); TODD, W. E. C., 1941, Ann. Carnegie Mus., 29, 1-29 (Carnegie Museum list); PETERS, J. L., No. 1, 1931 (world list); HELLMAYR, & CONOVER, No. 1, 1942 (the Americas, catalog); SALVIN & GODMAN, No. 3, 1897-1904 (Central America); WETMORE, A., No. 2, 1965 (Panama); de MIRANDA-RIBEIRO, A., 1938 Rev. Mus. Paulista, 23, 667-788 (keys to Brasilian species); DEVINCENZI, G. J., No. 1, 1926 (Uruguay); LIEBERMANN, J., 1936, Monografia de los Tinamiformes argentinas ..., Buenos Aires, 1-103.
Todidae. WYTSMAN, P., 1905, Genera Avium, 2, 1-4; REICHENOW, A., No. 5, 1914 (monograph); PETERS, J. L., No. 5, 1955 (world list); CORY, C. B., No. 1, 1918 (the Americas, catalog); RIDGWAY, R., No. 6, 1914 (North America)
Totanidae. DEVINCENZI, G. J., No. 2, 1927 (Uruguay)
Traversiidae, as Acanthisittidae
Treronidae, as Columbidae
Trerovidae, as Columbidae
Triopidae, as Apodidae

Trochilidae. GOULD, J., 1849-87, Monograph of the Trochilidae, London, 1849-61 & supplement 1880-1887; RIDGWAY, R., 1880, Proc. United States Nat. Mus., 3, 308-320 (US. N. M. catalog); BOUCARD, A., No. 1, 1893-95 (monograph); HARTERT, E., 1900, Das Tierreich, 9, 1-254; REICHENOW, A., No. 5, 1914 (monograph); SIMON, E., 1921, Histoire naturelle des Trochilidae (Synopsis et catalogue), Paris, 1-416; TODD, W. E. C., 1942, Ann. Carnegie Mus., 29, 271-370 (Carnegie Museum list); PETERS, J. L., No. 5, 1955 (world list); CORY, C. B., No. 1, 1918 (the Americas, catalog); RIDGWAY, R., No. 5, 1911 (North America); FORBUSH, E. H., No. 1, 1927 (Massachusetts); JOUANIN, C., 1946, Oiseau, 16, 103-112 (collections commerciales de Trinidad); SALVIN & GODMAN, No. 2, 1888-1904 (Central America); ALVARADO, R., 1915, Bol. Div. Est. Biol. Mexico, 1, 43-95 (Mexico); BERLIOZ, J., 1932, Oiseau, 2, 120-132 (Mexico); WETMORE, A., No. 3, 1968 (Panama); DABBENE, R., 1929, Rev. Chilena Hist. Nat., 33, 489-503 (Chile)

Troginidae, as Trogonidae

Troglodytidae. PETERS, J. L., No. 9, 1960 (world list); HELLMAYR, C. E., No. 7, 1934 (the Americas, catalog); RIDGWAY, R., No. 3, 1904 (North America); FORBUSH, E. H., No. 3, 1929 (Massachusetts); SALVIN & GODMAN, No. 1, 1879-1904 (Central America); PARIS, P., No. 1, 1921 (France); LINTIA, D., No. 1, 1946 (Romania); DEMENTIEV & GLADKOV, No. 6, 1954 & No. 12, 1968 (Soviet Union); GLADKOV, N. A. et al., No. 1, 1964 (U. S. S. R.); BAKER, E. C. S., No. 1, 1922 (British India); KURODA, N., No. 1, 1933 (Java)

Trogonidae (incl. Troginidae). GOULD, J., 1858-75, A monograph of the Trogonidae..., London; REICHENOW, A., No. 5, 1914 (monograph); PETERS, J. L., No. 5, 1955 (world list); CORY, C. B., No. 2, 1919 (the Americas, catalog); RIDGWAY, R., No. 5, 1911 (North America); SALVIN & GODMAN, No. 2, 1888-1904 (Central America); WETMORE, A., No. 3, 1968 (Panama); DUGAND, A., 1942, Rev. Acad. Colombiana, 5, 69-75 (Colombia); PINTO, O., 1950, Pap. Avuls. Dep. Zool. Sao Paulo, 9, 89-136 (monograph, Brasil); REICHENOW, A., No. 2, 1903 (Africa); BANNERMAN, D. A., No. 3, 1933 (tropical West Africa); CHAPIN, J. P., No. 2, 1939 (Belgian Congo); FRIEDMANN, H., No. 1, 1930 (Ethiopia & Kenya); BLANFORD, W. T., No. 3, 1895 (British India); BAKER, E. C. S., No. 4, 1927 (British India); ALI & RIPLEY, No. 4, 1970 (India, Pakistan); WAIT, W. E., No. 1, 1920 (Ceylon); DEIGNAN, H. G., No. 1, 1945 (Thailand); KURODA, N., No. 2, 1936 (Java); McGREGOR, R. C., No. 1, 1909 (Philippines)

Turdidae (incl. Enicuridae, Myiadestidae). SEEBOHM, H. & SHARPE, R. B., 1898-1902, Monograph of Turdidae..., London, 2 volumes; RIPLEY, S. D., 1952, Postilla, 13, 1-48 (revision); HELLMAYR, C. E., No. 7, 1934 (the Americas, catalog); RIDGWAY, R., No. 4, 1907 (North America); FORBUSH, E. H., No. 3, 1929 (Massachusetts); SALVIN & GODMAN, No. 1, 1879-1904 (Central America); BIANCHI, V., 1921, Bull. Acad. Sci. Russ., (6), 15, 509-584 (in Russian, keys to palearctic & himalochinese); PARIS, P., No. 1, 1921 (France); DUNAJEWSKI, A., 1934, Acta Ornith. Mus. Zool. Polon., 1, 275-301 (Poland); DEMENTIEV & GLADKOV, No. 6, 1954 & No. 12, 1968 (Soviet Union); GLADKOV, N. A. et al., No. 1, 1964 (U. S. S. R.); KOENIG, A., 1924, Journ. Ornith., 72, (Sonderheft), 1-277 (Egypt); BANNERMAN, D. A., No. 4, 1936 (tropical West Africa); CHAPIN, J. P., No. 3, 1953 (Belgian Congo); FRIEDMANN, H., No. 2, 1937 (Ethiopia & Kenya); BAKER, E. C. S., No. 2 (British India); DEIGNAN, H. G., No. 1, 1945 (Thailand); KURODA, N., No. 1, 1933 (Java); McGREGOR, R. C., No. 2, 1909 (Philippines); HUTSON, H. P. W., 1931, The birds of Hong Kong, V, Hong Kong; BAKER, R. H., No. 1, 1951 (Micronesia)

Turnagridae, as Muscicapidae

Turnicidae. REICHENOW, A., No. 4, 1913 (monograph); PETERS, J. L., No. 2, 1963 (world list); HARTERT E., No. 3, 1921-22 (Palearctic); PARIS, P., No. 1, 1921 (France); DEMENTIEV & GLADKOV, No. 2, 1951 & No. 8, 1968 (Soviet Union); GLADKOV, N. A. et al., No. 1, 1964 (U. S. S. R.); REICHENOW, A., No. 1, 1901 (Africa); BOUET, G., No. 1, 1955 (Afrique tropicale); BANNERMAN, D. A., No. 2, 1931 (tropical West Africa); CHAPIN, J. P., No. 1, 1932 (Belgian Congo); BLANFORD, W. T., No. 4, 1898 (British India); BAKER, E. C. S., No. 5, 1928 (British India); ALI & RIPLEY, No. 2, 1969 (India, Pakistan); DEIGNAN, H. G., No. 1, 1945 (Thailand); DELACOUR, J. & JAFOUILLE, P., 1927, Faune Colon. Francaise, 1, 369-461 (Annam); KURODA, N., No. 2, 1936 (Java); McGREGOR, R. C., No. 1, 1909 (Philippines); TAKA-TSUKASA, N., No. 1, 1967 (Nippon)

Turnoididae, as Timaliidae

Turturidae, as Columbidae

Tylidae, as Oriolidae

Tyrannidae. von IHERING, H., 1904, Auk, 21, 313-322 (systematic arrangement); REICHENOW, A., No. 5, 1914 (monograph); HELLMAYR, C. E., No. 5, 1927 (the Americas, catalog); RIDGWAY, R., No. 4, 1907 (North America); FORBUSH, E. H., No. 1, 1927 (Massachusetts); SALVIN & GODMAN, No. 2, 1888-1904 (Central America); GIGOUX, E. E., 1929, Bol. Mus. Nac. Chile, 12, 42-64 (Chile)

Tytidae, as Tytonidae

Tytonidae (incl. Aluconidae, Tytidae). PETERS, J. L., No. 4, 1940 (world list); CORY, C. B., No. 1, 1918 (the Americas, catalog); RIDGWAY, R., No. 6, 1914 (North America); FORBUSH, E. H., No. 1, 1927 (Massachusetts, as Aluconidae); WETMORE, A., No. 3, 1968 (Panama); PARIS, P., No. 1, 1921 (France); DEMENTIEV & GLADKOV, No. 1, 1951 & No. 7, 1966 (Soviet Union); GLADKOV, N. A. et al., No. 1, 1964 (U. S. S. R.); BANNERMAN, D. A., No. 3, 1933 (tropical West Africa); CHAPIN, J. P., No. 2, 1939 (Belgian Congo); FRIEDMANN, H., No. 1, 1930 (Ethiopia & Kenya); BAKER, E. C. S., No. 4, 1927 (British India); DEIGNAN, H. G., No. 1, 1945 (Thailand); McGREGOR, R. C., No. 1, 1909 (Philippines, as Alluconidae)

Upupidae. PETERS, J. L., No. 5, 1955 (world list); HARTERT, E., No. 2, 1912-21 (Palearctic); PARIS, P., No. 1, 1921 (France); LINTIA, D., No. 2, 1954 (Romania); DEMENTIEV & GLADKOV, No. 1, 1951 & No. 7, 1966 (Soviet Union); GLADKOV, N. A. et al., No. 1, 1964 (U. S. S. R.); REICHENOW, A., No. 2, 1903 (Africa); BANNERMAN, D. A., No. 3, 1933 (tropical West Africa); CHAPIN, J. P., No. 2, 1939 (Belgian Congo); FRIEDMANN, H., No. 1, 1930 (Ethiopia & Kenya); MOLTONI, E., 1940, Riv. Italiana Ornith., 10, 75-92 (Africa orientale Italiana); BLANFORD, W. T., No. 3, 1895 (British India); BAKER, E. C. S., No. 4, 1927 (British India); ALI & RIPLEY, No. 4, 1970 (India, Pakistan); WAIT, W. E., 1920, Spol. Zeylanica, 11, 197-272 (Ceylon); DEIGNAN, H. G., No. 1, 1945 (Thailand)

Uriidae, as Alcidae

Urinatoridae, as Gaviidae

Vangidae (incl. Aerocharidae, Eurycerotidae, Falculidae). PETERS, J. L., No. 9, 1960 (world list)

Vireolaniidae. HELLMAYR, C. E., No. 8, 1935 (the Americas, catalog)

Vireonidae. PETERS, J. L., No. 14, 1968 (world list); HELLMAYR, C. E., No. 8, 1935 (the Americas, catalog); RIDGWAY, R., No. 3, 1904 (North America); FORBUSH, E. H., No. 3, 1929 (Massachusetts); SALVIN & GODMAN, No. 1, 1879-1904 (Central America)

Vulturidae. REICHENOW, A., No. 4, 1913 (monograph); PARIS, P., No. 1, 1921 (France); REICHENOW, A., No. 1, 1901 (Africa); BLANFORD, W. T., No. 3, 1895 (British India); WAIT, W. E., No. 2, 1921 (Ceylon)

Xenicidae, as Acanthisittidae

Xenicornithidae, as Acanthisittidae

Xenopidae. CHUBB, C., 1919, Ann. Mag. Nat. Hist., (9), 3, 273-275 (under Dendrocolaptidae)

Yungidae, as Jyngidae

Zeledoniidae. HELLMAYR, C. E., No. 7, 1934 (the Americas, catalog); RIDGWAY, R., No. 4, 1907 (North America)

Zosteropidae. FINSCH, O., 1901, Das Tierreich, 15, 1-55; REICHENOW, A., No. 5, 1914 (monograph); PETERS, J. L., No. 12, 1967 (world list); HARTERT, E., No. 1, 1903-10 (Palearctic); DEMENTIEV & GLADKOV, No. 5, 1954 & No. 11, 1970 (Soviet Union); GLADKOV, N. A. et al., No. 1, 1964 (U. S. S. R.); REICHENOW, A., No. 3, 1905 (Africa); BANNERMAN, D. A., No. 6, 1948 (tropical West Africa); CHAPIN, J. P., No. 4, 1954 (Belgian Congo); FRIEDMANN, H., No. 2, 1937 (Ethiopia & Kenya); BAKER, E. C. S., No. 3, 1926 (British India); DEIGNAN, H. G., No. 1, 1945 (Thailand); MEES, G. F., 1957, Zool. Verhandel, 35, 1-204 (part I, Indo-Australia), Zool. Verhandel, 50, 1-168 (part II); KURODA, N., No. 1, 1933 (Java); McGREGOR, R. C., No. 2, 1909 (Philippines); BAKER, R. H., No. 1, 1951 (Micronesia)

MAMMALIA

GENERAL BOOKS

(anonymous). 1935. The wild animals of the Indian Empire, 4 parts. Journ. Bombay Nat. Hist. Soc., vols. 37, 38, etc.
ALLEN, G. M. 1942. Extinct and vanishing mammals of the Western Hemisphere. Lancaster, Pennsylvania; 1-620.
ANDERSEN, H. T. 1969. The biology of marine mammals. New York; 1-511.
ANTHONY, H. E. 1947. Mammals of America. Garden City, N. Y.; 1-354. (Previous eds. 1917, 1937)
BANKS, E. 1931. A popular account of the mammals of Borneo. Journ. Malayan Branch Asiatic Soc., 9, (2), 1-139.
BANKS, E. 1949. Bornean mammals. Kuching, Sarawak; 1-83.
BEDDARD, F. E. 1902. Mammalia. London; 1-605.
BLACKMORE, M. 1948. Mammals in Britain. London; 1-128.
BOURLIERE, F. 1954. Le monde des mammiferes. Paris; 1-224.
BOURLIERE, F. 1955. The natural history of mammals.... London; 1-362. (Translation of 1954)
BOURLIERE, F. 1956. The natural history of mammals.... New York; 1-364. (Second edition)
BOURLIERE, F. 1964. The natural history of mammals, 3rd ed. New York; 1-361.
BOURLIERE, F. 1955. Mammals of the world, their life and habits. London; 1-223. (Also New York)
BRANDER, A. A. D. 1923. Wild animals in central India. London; 1-296.
BURTON. M. 1962. Systematic dictionary of mammals of the world. London; 1-307. (Not a dictionary)
BURTON, M. 1968. University dictionary of mammals of the world. New York; 1-307. (Not a dictionary)
CABRERA, A. 1922. Manual de mastozoologia. Madrid-Barcelona; 1-440.
CABRERA, A. 1940, Historia natural ediar. Mamiferos Sud-americanos. Buenos Aires; 1-370.
CAHALANE, V. H. 1947. Mammals of North America. New York; 1-682.
CLARK, W. E. L. 1950. History of the Primates. London; 1-117.
COCKRUM, E. L. 1962. Introduction to mammalogy. New York; 1-455.
CORBET, G. B. 1966. The terrestrial mammals of western Europe. Philadelphia; 1-264.
da CRUZ LIMA, E. 1945. Mammals of Amazonia, vol. 1, General introduction and Primates. Belem; Brasil; 10274.
DAVIS, D. E. & GOLLEY, F. B. 1963. Principles in mammalogy. New York & London; 1-335.
EDWARDS, D. L. 1941. All about elephants. New York; 1-88.
FEHRINGER, O. 1953. Die West der Saugetiere. Munchen; 1-431.
FitzSIMONS, F. W. 1919 -. The natural history of Southern Africa. London; 5 volumes.
FLOWER, W. H. 1885. An introduction to the osteology of the mammals. London; 1-373.
FLOWER, W. H. & LYDEKKER, R. 1891. An introduction to the study of mammals living and extinct. London; 1-763.
GEE, E. P. The wild life of India. London; 1-192.
GOODWIN, G. G. In Drimmer, The Animal Kingdom.... New York; 5 volumes. 1954.
GOODWIN, G. C. 1964. Mammals. London & Melbourne.
GRASSE, P-P. 1955. Traite de Zoologie, 17, 1-2300.
GRAY, A. P. 1954. Mammalian hybrids. Tech. Comm. Commonw. Bur. Anim. Breed. Genet., 10, 1-144. (Edinburgh)
GREGORY, W. K. 1951. Evolution emerging. New York; 2 volumes.
GRZIMEK, H. C. B. et al. 1967-1968. Grzimeks Tierleben. Enzyklopaedie des Tierreichs - Saugetaire. Zurich; 4 volumes.
HAMILTON, W. J., Jr. 1939. American mammals.... New York; 434 pp.
HAMILTON, W. J., Jr. Mammals of eastern United States.... Ithaca, New York; 1-432.
HARPER, F. Extinct ... animals of the Old World. New York; 1-850.
HARRISON, R. J. & KING, J. E. Marine mammals. London; 1-192.
HEDIGER, H. 1952. La vie des animaux sauvages d'Europe. Paris; 1-170.
HOFER, H. et al. 1956. Primatologica. Handbook of Primatology.... Basel & New York; 1-1063. (Also 1962.)
HOWELL, A. B. 1930. Aquatic mammals. Springfield, Illinois & London; 2-338.
HVASS, H. 1961. Mammals of the world. London; 1-212.
KRUMBIEGEL, I. 1953. Biologie der Saugetiere. Ein Handbuch. Krefeld; 1-144. (Continued 1954 and 1955 to 844 pp.)
LEACH, W. J. 1946. Functional anatomy of the mammals. New York; 1-231.
LEOPOLD, A. S. 1959. Wildlife in Mexico. Berkeley, California; 1-568.
Le SOUET, A. S. et al. 1926. The wild animals of Australasia. London; 1-388.
LYDEKKER, R. 1894. The Royal Natural History. London; 6 volumes. (Re-issued 1922)
LYNE, A. G. Marsupials and monotremes of Australia. New York; 1-72.
MARSHALL, P. 1968. Wild mammals of Hong Kong. Oxford; 1-64.
MATSUURA, Y. 1943. Marine mammals. Tokyo; 1-298. (In Japanese)
MATTHEWS, L. H. 1952. British Mammals. London; 1-410.
MATTHEWS, L. H. 1969-71. The life of mammals, 2 volumes. London (1969-71), New York (1970-1971)
MAXWELL, G. 1967. Seals of the world .... London; 1-153.

MOCHI, U. & CARTER, T. D. 1953. Hoofed mammals of the world. New York; pl. i-xl.
MOHR, E. 1954. Die freilebenden Nagetiere Deutschlands und der Nachbarlander, 3rd edition. Jena; 1-212.
MORRIS, D. 1965. The mammals. A guide to living species. London; 1-448.
NAORA, N. 1942-43. History of Japanese mammals. Nat. Sci. Mus. Tokyo, 13-14, 5 parts. (In Japanese)
NAPIER, J. R. & NAPIER, P. H. 1967. A handbook of living primates. New York; 1-456.
OGNEV, S. I. 1959. Saugetiere und ihre Welt. Berlin; 1-362. (Translated from 1951 Russian)
PAVLOVSKII, E. N. 1965. Marine mammals. Moskva; 1-320. (In Russian)
PERRIER, E. 1932. Les mammiferes. Traite de Zoologie, 10, 3344-3610.
PRATER, S. H. 1948. The book of Indian mammals. Bombay; 1-263.
PRATER, S. H. 1965. The book of Indian mammals, 2nd ed. Bombay; 1-323.
RODE, P. 2942-44. Petit atlas des mammiferes. Paris; 4 fascicules.
ROWLANDS, I. W. 1966. Comparative biology of reproduction in mammals. Zool. Soc. London Symposium 15, 1-559.
SANDERSON, I. T. 1955. Living mammals of the world. Garden City, New York; 1-303.
SCOTT, W. B. 1937. A history of land mammals in the Western Hemisphere, 2nd ed. New York; 1-786. (Previous edition 1913)
SHORTRIDGE, G. C. 1934. The mammals of South Africa. A biological account.... London; 2 volumes.
SIIVONEN, L. 1956. Suuri nisakaskirja. Helsinki, Finland. ("Great book of mammals...")
SOUTHERN, H. N. 1963. The handbook of British mammals. Oxford; 1-465.
STEVENSON-HAMILTON, J. 1947. Wild life in South Africa. London; 1-364.
TATE, G. H. H. 1947. Mammals of eastern Asia. New York; 1-366.
THEVENIN, R. 1952. Les petits carnivores d'Europe. Paris; 1-293.
TROUGHTON, E., 1941. Furred animals of Australia. Sydney; 1-374.
TROUGHTON, E. 1946. Furred animals of Australia, 3rd ed. Sydney; 1-376. (New York; 1-374)
VERESHCHAGIN, N. K. 1967. The mammals of the Caucasus .... Jerusalem; 1-810. (Translation of 1959 Russian)
WALKER, E. P. 1964. Mammals of the world. Baltimore, Maryland; 3 volumes.
WARD, A. E. 1925-26. The mammals and birds of Kashmir .... V. Cervidae. Vi. Carnivora. Journ. Bombay Nat. Hist. Soc., 30, 719-724, 31, 1-11, etc.
WEBER, M. W. C. 1928. Die Saugetiere, 2nd ed. Jena; 2 volumes.
WENDER, L. 1948. Animal encyclopaedia. Mammals. London; 1-266.
YOUNG, J. Z. 1957. The life of mammals. London; 1-820.
ZEMSKII, W. A. 1960. The animal kingdom of the Antarctic. Mammals and birds. Moskva; 1-180.

BIBLIOGRAPHIC WORKS

Bibliographies, separate works

[See also: Under General and Animalia above; see also Zoological Record, in each volume, under heading Bibliographies]

ALLEN, J. A. 1881. Preliminary list of works and papers relating to the mammalian orders of Cete and Sirenia. Bull. United States Geol. Surv., 6, 399-562.
ALLOUSE, B. E. 1954. A bibliography on the vertebrate fauna of Iraq.... I. Mammals. Publ. Iraq Nat. Hist. Mus., 4, 1-30.
FREUND, L. 1951. A bibliography of the mammalian order Sirenia. Mem. Soc. Zool. Tcheosl., 14, 161-181.
HALLORAN, P. O. 1955. A bibliography of references to diseases of wild mammals and birds. American Journ. Veter. Res., 16, 1-465.
HAYWARD, C. L. 1936. A bibliography of Utah mammalogy.... Proc. Utah Acad. Sci., 13, 121-146.
HAYWARD, C. L. 1941. A bibliography of Utah mammalogy.... Great Basin Nat., 2, 125-136.
KURODA, N. 1952. Mammalogical history of Formosa, with zoogeography and bibliography. Quart. Journ. Taiwan Mus., 5, 267-301.
MARELLI, C. A. 1936. Bibliographia euristica de los mammiferos, de caza e caza maritima. Contribuciones al estudio de la fauna Argentina. La Plata; 1-189.
RAUN, G. G. 1962. A bibliography of the Recent mammals of Texas. Bull. Texas Mem. Mus., 3, 1-81.
SHERMAN, H. B. 1952. A history and bibliography of the mammals of Florida, living and extinct. Quart. Journ. Florida Acad. Sci., 15, 86-126.
VOSS, H. 1955. Bibliographie der Menschenaffen (Schimpanse, Orang, Gorilla). Jena; 1-163.
WALKER, E. P. 1904. Mammals of the world, volume 3. Baltimore, Maryland; 1-769.

Bibliographies in other works

[See also: References, in most large works]

ANDERSON, S. & JONES, J. K., Jr. Recent mammals of the world. Synopsis of families. New York; 1-453. (Pp. 407-433)
BARBOUR, R. W. & DAVIS, W. H. Bats of America. Lexington, Kentucky; 1-286. (Pp. 256-283)
BLAIR, W. F. et al. 1957. Vertebrates of the United States. New York; 1-819. (Pp. 773-774)
GILL, T. & COUES, E. 1877. Material for a bibliography of North American mammals. United States Geol. Surv. Terr., XI, Appendix B, 951-1081. (Chronological)
GREGORY, W. K. 1910. The orders of mammals. Bull. American Mus. Nat. Hist., 27, 1-524. (Pp. 469-513)

HALL, E. R. & KELSON, K. R. 1959. The mammals of North America. New York; 2 volumes. (In volume II, pp. 1049-1078)
HERSHKOVITZ, P. 1966. Catalog of living whales. United States Nat. Mus. Bull., 246, 1-259. (Pp. 222-226)
NAPIER, J. R. & NAPIER, P. H. 1967. A handbook of living primates. New York; 1-456. (Pp. 417-446)
NOVIKOV, G. A. Mammals of the U. S. S. R., vol. 2. Moskva. (In Russian)
ROMER, A. S. 1945. Vertebrate paleontology. Chicago; 1-687. (Principally fossils)
SIMPSON, G. G. 1945. Principles of classification and a classification of mammals. Bull. American Mus. Nat. Hist., 85, 1-350. (Pp. 273-307)
TOMICH, P. Q. 1969. Mammals in Hawaii. A synopsis and notational bibliography. Honolulu; 1-238. (Pp. 140-225)
VERESHCHAGIN, N. K. 1959. Mammals of Caucasia. Moskva; 1-703. (In Russian; pp. 583-628)
VERESHCHAGIN, N. K. 1967. The mammals of the Caucasus .... Jerusalem; 1-810. (Translation of 1959)
WALKER, E. P. 1964. Mammmals of the world, vol. I. Baltimore, Maryland; 1-644. (Selected bibliography pp. 508-622. See also volume III, above)

Periodical lists

[See also under Animalia and Vertebrata]

VERESHCHAGIN, N. K. 1967. The mammals of the Caucasus .... Jerusalem; 1-816. (Translation of 1959)

Indexes

SCHANTZ, V. S. 1945. Twenty-year index to Journal of Mammalogy Volumes 1-20 inclusive 1919-1939. (No city); 1-219. (Supplements for 1940-1949 and 1950-1959)

Personal bibliographies

[See under heading Biographies (Persons & Institutions) below]

PERSONS AND INSTITUTIONS

Biographies & personal bibliographies

[See also: Zoological Record, in each volume, under heading Obituaries]

ALLEN, G. M. Lawrence, B., Bibliography of publications of Glover Morrill Allen. Proc. New England Zool. Club, 24, 1-81. 1947.
ALLEN, J. A. Allen, J. A., 1916, Autobiographical notes and a bibliography of the scientific publications of J. A. Allen. New York; 1-215.
AMEGHINO, C. & AMEGHINO, F. Simpson, G. G., 1948, The beginning of the age of mammals in South America, part I. Bull. American Mus. Nat. Hist., 91, 1-232. (anonymous). 1957, Ameghiniana. Rev. Asoc. Paleont. Argentina, 1.
MATTHEW, W. D. Abel, O., 1931, William Diller Matthew 1871-1930. Palaeobiol., Wien, 4, 1-24. Osborn, H. F., 1931, Memorial of William Diller Matthew. Bull. Geol. Soc. America, 42, 55-95. Camp, C. L. & Vanderhoof, V. L., 1939, Annotated bibliography of William Diller Matthew. Spec. Publ. New York Acad. Sci., 1, 201-223.
MERRIAM, C. H. Grinnell, H. W., 1943, Bibliography of Clinton Hart Merrian [1855-1942]. Journ. Mammalogy., 24, 436-457.
MILLER, G. S., Jr. Miller, G. S., Jr. & Johnson, D. H., 1954, Bibliography of Gerrit Smith Miller, Jr. Journ. Mammalogy, 35, 329-344.
de MIRANDA RIBEIRO, A. Kretz, D. J., 1942, Alipio de Miranda Ribeiro (Conferencia biografica). Sao Paulo, 1-24. (anonymous), 1955, Bibliografia de Alipio de Miranda Ribeiro. Arqu. Mus. Nac. Rio de Janeiro, 42, xix-xxxvi.
OGNEV, G. P. Dementiev, G. P., 1952, Scientific activities of S. I. Ognev. Bull. Soc. Nat. Moscou, Biol., 57, (2), 12-39.
RUSCONI, C. Rusconi, C., 1948, Resumenes de publicaciones de Carlos Rusconi. Rev. Mus. Hist. Nat. Mendoza, 2, 255-302. Rusconi, C., 1949, Lista de familias, generos y especies fundados por Carlos Rusconi desde 1927 a 1948. Rev. Mus. Hist. Nat. Mendoza, 3, 138-156. Alos, J. M., 1967, Biography of the scientific work of Carlos Rusconi. Rev. Mus. Hist. Nat. Mendoza, 19, 23-139 (in Spanish).
SCHOUTEDEN, H. (Anonymous), 1054, Liste des publications du Professeur Dr. H. Schouteden. Ann. Mus. Roy. Congo Belge, (n. s.), Sci. Zool., 1, xiii-xxviii.

Directories

[See: Directories under Animalia]

Institutions & Collections

[See also under Animalia]

HOWELL, A. B., 1923, The mammal collections of North America. Journ. Mamm., 4, 113-120.
DOUTT, J. K. et al., 1945, The mammal collections of North America. Journ. Mamm., 26, 231-272.
ANDERSON, S. et al., 1963, Collections of mammals in North America. Journ. Mamm., 44, 471-500.
WALKER, E. P., 1964, Mammals of the world, vol. III. Baltimore, Maryland, 1-769. (Lists of museums, catalogs, collections, pp. 737-739)

Museum lists

(Various museums)

KROGMAN, W. M. & SCHULTE, A. H. 1938. Anthropoid ape materials in American collections. American Journ. Physic. Anthrop., 24, 199-234.

IMAIZUMI, Y. 1958. Collections of marine mammals in Japanese museums. Nat. Sci. Mus. Tokyo, 25, (5-6), 1-9.

United States National Museum

MILLER, G. S., Jr. 1912. List of North American land mammals in the United States National Museum. United States Nat. Mus. Bull., 79, 1-455.

HOLLISTER, N. 1918-1924. East African mammals in the United States National Museum, 3 parts. United States National Mus. Bull., 99, 194 & 184 & 164 pp.

Carnegie Museum

SANBORN, C. C. 1932. Neotropical bats in the Carnegie Museum. Ann. Carnegie Mus., 21, 171-183.

Field Columbian Museum (Field Museum of Natural History)

ELLIOT, D. G. 1907. A catalogue of the collection of mammals in the Field Columbian Museum. Field Col. Mus., Zool. Ser., 8, 1-694.

Museu Paulista

LIMA, J. L. 1926. Os morcegos da collecao do Museu Paulista. Rev. Mus. Paulista, 14, 43-127.

British Museum (Natural History)

LYDEKKER, R. 1913-16. Catalogue of the ungulate mammals in the British Museum (Natural History). London; 5 volumes.

Royal Scottish Museum

HILL, W. C. O. 1952. Primates in the Royal Scottish Museum. I. Strepsirhini & Tarsioidea. Proc. Roy. Phys. Soc. Edinburgh, 23, 155-164.

HILL, W. C. O. 1955. (Same) II. Platyrrhini. Proc., 24, 49-62, etc.

Anatomical Museum of the University of Edinburgh

TURNER, W. 1912. The marine mammals in the Anatomical Museum of the University of Edinburgh. London; 1-207.

Musee Royal d'Histoire Naturelle de Belgique

SLIJPUR, E. J. 1938. Die Sammlung rezenten Cetacea des Musee Royal d'Histoire Naturelle de Belgique. Bull. Mus. Hist. Nat. Belgique, 14, (10), 1-33.

Jardin Zoologique de la Societe Royale de Zoologie d'Anvers

GIJZEN, A. 1960. Liste des mammiferes ayant figure au figurant dans le collections du Jardin . . . . 1843-1960 . . . . Bull. Soc. Roy. Zool. Antwerp, 16, 1-18.

Musee Zoologique de Strasbourg

AELLEN, V. 1957. Les Chiropteres africains du Musee Zoologique de Strasbourg. Rev. Suisse Zool., 64, 189-214.

Musee de l'Emperi a Salon de Provence

LAURENT, P. 1945. Les chauves-souris du Musee de l'Emperi a Salon de Provence. Bull. Mus. Hist. Nat. Marseille, 5, (4), 141-151.

Musee Oceanographique de Monaco

RODE, P. 1939. La collection de Cetaces du Musee Oceanographique de Monaco. Notes osteometriques. Bull. Inst. Oceanogr. Monaco, 780, 1-20.

Museu Zoologico de Coimbra

ARMANDO THEMIDO, A. 1928. Catalog des Primates (& rongeurs & carnivores). . . . Mem. Est. Mus. Zool. Univ. Coimbra, (1), 19-30, etc.

ARMANDO THEMIDO, A. 1946. Mamiferos das colonias Portuguesas. (Catalogo das coleccoes do Museu Zoologico de Coimbra). Mem. Mus. Zool. Univ. Coimbra, 174, 1-52.

Museo de Ciencias Naturales de Madrid

CABRERA, A. 1912. Catalogo metodoco de las colecciones de mamiferos del Museo de Ciencias Naturales de Madrid. Trab. Mus. Cienc. Nat. Zool., 11, 1-447.

Museo di Anatomia comparata della R. Universita de Pisa

REPETTI, V. 1924. Catalogo dei cetacei del Museo di Anatomia comparata della R. Universita de Pisa. Proc. Verb. Att. Soc. Tosc., Sci. Nat., 33, 46-55.

Museo Civico de Storia Naturale "Giacomo Doria" in Genoa

Van der FEEN, P. J. 1962. Catalogue of the Marsupialia from New Guinea, the Moluccas and Celebes in the Museo Civico de Storia Naturale "Giacomo Doria" in Genoa. Ann. Mus. Civ. Stor. Nat. Genova, 73, ?-17.

Museo Civico di Milano

de BEAUX, O. 1923. Mammiferi della Somalia Italiana ... nel Museo Civico di Milano. Att. Soc. Italiana Sci. Nat., 62, 247-316.

Museo di Storia Naturale della Venezia Tridentina in Trento

BONOMI, L. 1932. La raccolta di micromammiferi del Museo di Storia Naturale della Venezia Tridentina in Trento. Studi Trentini, 13, 120-125.

Indian Museum

DOBSON, G. E. 1876. Monograph of the Asiatic Chiroptera and catalogue of the species of bats in the collection of the Indian Museum, Calcutta. London; 1-228.

ANDERSON, J. 1881, 1891. Catalogue of Mammalia in the Indian Museum, Calcutta. Calcutta; 1-223, 1-375.

Zoological Survey of India

KHAJURIA, H. 1953. Catalogue of mammals in the Zoological Survey of India. I. Primates: Hominoidea. Rec. Indian Mus., 50, 129-145. (Also II, III, IV)

Laboratoire de Zoologie de l'Ecole Superieure des Sciences (Hanoi)

BOURRET, R. 1941. Les mammiferes de la collection du Laboratoire de Zoologie de l' Ecole Superieure des Sciences. Notes Trav. Ecole Sup. Sci. Hanoi, 1, 1-44.

Musee Ho-angho Paiho de Tien Tsin

JAKOVLEFF, B. P. 1935. Collection des mammiferes du Musee Ho-angho Paiho de Tien Tsin. Ungulata.... Publ. Mus. Hoang ho Pai ho, Tien Tsin, 35, 1-31.

Collection of Marquis Yamashina

KURODA, N. 1939. Mammal fauna of Manchoukuo preserved in the collection of Marquis Yamashina. Bull. Biogeogr. Soc. Tokyo, 9, 1-50.

Museum type lists

(Lists of such lists)

WALKER, E. P. 1964. Mammals of the world, vol. III. Baltimore; 1-769. (List of museum type lists, pp. 737-739)

Museum of Comparative Zoology (Harvard University)

ALLEN, G. M. 1931. Type specimens of mammals in the Museum of Comparative Zoology. Bull. Mus. Comp. Zool., 74, 229-289.

American Museum of Natural History

GOODWIN, G. G. 1953. Catalogue of type specimens of Recent mammals in the American Museum .... Bull. American Mus. Nat. Hist., 102, 209-411.

United States National Museum

LYON, M. W. & OSGOOD, W. H. 1909. Catalogue of the type specimens of mammals in the United States National Museum. United States Nat. Mus. Bull., 62, 1-325.

POOLE, A. J. & SCHANTZ, V. S. 1942. Catalog of the type specimens of mammals in the United States National Museum. United States Nat. Mus. Bull., 178, 1-705.

Carnegie Museum

DOUTT, J. K. 1934. List of types of mammals in the collection of the Carnegie Museum on April 2, 1934. Ann. Carnegie Mus., 22, 317-321.

Chicago Natural History Museum (Field Museum of Natural History)

SANBORN, C. C. 1947. Catalogue of type specimens of mammals in the Chicago Natural History Museum. Fieldiana, 32, 209-293.

British Columbia Provincial Museum

COWAN, I. M. 1936. Notes on some mammals in the British Columbia Provincial Museum with a list of the type specimens .... Canadian Field Nat., 50, 145-148.

Senckenbergischen Museums

MERTENS, R. 1925. Verzeichnis der Saugetier-Typen des Senckenbergischen Museums. Senckenbergiana, 7, 18-37.

Zoologischen Museum Halle an der Saale

MOHR, E. 1941. Saugetiertypen im Zoologischen Museum Halle an der Saale. Zeitschr. Naturwissensch., 94, 215-226.

Museum National d'Histoire Naturelle

RODE, P. 1945. Catalogue des types de mammiferes du Museum .... Ordre des Rongeurs (suite). Bull. Mus. Nat. Hist. Nat., (2), 17, 201-208, 292-300.
RODE, P. 1942. Catalogue des types de mammiferes du Museum .... III. Ordre des Insectivores. Bull. Mus. Nat. Hist. Nat., (2), 14, 307-314, 382-387.
RODE, P. 1941. Catalogue des types des mammiferes du Museum .... II. Ordre des Chiropteres. Bull. Mus. Nat. Hist. Nat., (2), 13, 227-252.
de BEAUFORT, F. 1964. Catalogue des types d'ongules du Museum National d'Histoire Naturelle, Paris. Bull. Mus. Nat. Hist. Nat., 35, 551-579.
de BEAUFORT, F. 1966. Catalogue des types de mammiferes du Museum .... 6. Monotremata. 7. Marsupialia. Bull. Bus. Nat. Hist. Nat., (2), 38, 509-553.

Raffles Museum

GIBSON-HILL, C. A. 1949. Bird and mammal type specimens formerly in Raffles Museum. Bull. Raffles Mus., 19, 133-198.

Western Australian Museum

(Anonymous). 1961. Type specimens in the Western Australian Museum. Rep. Western Australian Mus., 1959-60, 28-31.

EXPLORATION AND LOCALITIES

[See also this heading under Animalia]

WALKER, E. P. 1964. Mammals of the world. III. A classified bibliography. Baltimore, Maryland. (Pp. 712-713, expeditions)

Localities

GOODWIN, G. G. 1946. Mammals of Costa Rica. Bull. American Mus. Nat. Hist., 87, 271-473. (Pp. 454-458, collecting localities)
MOREAU, R. E. et al. 1946. The type localities of some African mammals. Proc. Zool. Soc. London, 115, 387-447.
HILL, J. E. & CARTER, T. D. 1941. The mammals of Angola, Africa. Bull. American Mus. Nat. Hist., 78, 1-211. (Pp. 5-9, localities)

METHODS

[See also: Zoological Record, each volume, under heading Technique; see also this heading under Animalia]

(anonymous). 1968. Instructors for collectors. No. 1, Mammals (non-marine), 6th ed. London; 1-55.
ANTHONY, H. E. 1925. The capture and preservation of small mammals for study. American Museum Natural History Guide Leaflet, 61, 1-53.
ANTHONY, H. E. 1925. The capture and preservation of small mammals for study. New York; 1-53.
ANTHONY, H. E. 1945. The capture and preservation of small mammals for study. American Museum Natural History, Science Guide, 61, 1-54.
BURT, W. H. 1957. Mammals of the Great Lakes region. Ann Arbor, Michigan; 1-246. (Pp. 160-178)
CLARK, J. L. 1937. The preservation of mammal skins in the field. Journ. Mamm., 18, 89-92.
COCKRUM, E. L. 1962. Laboratory and field manual for Introduction to Mammalogy, 2nd ed. New York; 1-121.
ELTON, C. 1938. A convenient method of mounting, and storing skins of small mammals. Journ. Mamm., 19, 244-245.
GAST, R. 1935. Sammlung und Behandlung von tierkundlichen Material zur naturalisten Aufstellung. Kleintier und Pelztier, 11, 191-222.
HALL, E. R. 1955. Handbook of mammals of Kansas. Publ. Univ. Kansas Mus. Nat. Hist., 7, 1-303. (Pp. 256-287)
HALL, E. R. & KELSON, K. R. 1959. The mammals of North America. New York; 2 volumes. (Methods pp. 1036-1047)

HUDSON, G. E. 1935. A practical method of degreasing study skins. Journ. Mamm., 16, 329-330.
KRITZINGER, C. C. 1945. Preservation of skins for museum purposes. Bull. South African Mus. Assoc., 3, 351-352.
MILLER, G. S. 1932. Directions for preparing specimens of mammals, 6th ed. United States Nat. Mus. Bull., 39, part N.
NORRIS, K. S. 1960. Standardized methods for measuring and recording data on the smaller cetaceans. Journ. Mamm., 42, 471-476.
PALMER, R. S. 1946. Plastic vials for storing small skulls. Journ. Mamm., 27, 276-277.
SCHULTZ, A. H. 1924. Preparation and preservation of anatomical and embryological material in the field. Journ. Mamm., 5, 16-24.

GLOSSARIES

[See many of the general books listed above, as well as Glossaries listed under Animalia]

NOMENCLATURAL STUDIES

Common names

HARRINGTON, J. P. 1947. The word for seal ... in various languages. Journ. Washington Acad. Sci., 37, 109-111.
NAPIER, J. R. & NAPIER, P. H. 1967. A handbook of living Primates. New York; 1-456.
SIMPSON, G. G. 1945. Principles of classification and a classification of mammals. Bull. American Mus. Nat. Hist., 85, 1-350. (World, 339-350)
ELLIOT, D. G. 1905. A check list of the mammals of the North American continent .... Field Col. Mus., Zool. Ser., 6, 1-761. (Supplement, New York, 1-192, 1917)
HALL, E. R. et al. 1957. Vernacular names for North American mammals north of Mexico. Misc. Publ. Univ. Kansas Mus. Nat. Hist., 14, 1-16.
HALL, E. R. & KELSON, K. R. 1959. The mammals of North America. New York; 2 volumes.
HALL, E. R. 1965. Names of species of North American mammals north of Mexico. Misc. Publ. Univ. Kansas Mus. Nat. Hist., 43, 1-6.
SIMPSON, G. G. 1941. Some Caribbean indian mammal names. American Mus. Novit., 1119, 1-10.
SIMPSON, G. G. 1941. Vernacular names of South American mammals. Journ. Mamm., 22, 1-17.
DENNLER, J. G. 1939. Los nombres indigenas en guarani de los mamiferos de la Argentina . . . . Physis, 16, 225-244.
ELLERMAN, J. R. & MORRISON-SCOTT, T. C. S. 1966. Checklist of Palaearctic and Indian mammals 1758-1946, 2nd ed. London; 1-810. (Prev. ed., 1951)
MOHR. E. 1961. Glossarium Europae Mammalium Terrestrium. Wittenberg Lutherstadt; 1-72. (In 17 languages)
van den BRINK, F. H. 1957. Die Saugetiere Europas .... Hamburg; 1-225.
CONSTANT, P. 1960. Lexique des noms communs de Chiropteres. Sous le Plancher, 1960, 35-39. (In French, English, German)
DEGERBOL, M.I. 1950. Mammals, in List of Danish Vertebrates. København.
ALSTON, C. H. 1913. A list of the Gaelic names of the British mammals. Scott. Nat., 1913, 145-153.
MACLEOD, R. D. 1956. Key to the names of British fishes, mammals, amphibians and reptiles. London; 1-71.
EBLE, H. et al. 1954. Beitrag zur deutschsprachigen Nomenklatur der rezenten Mammalia. Wissensch. Zeitschr. Martin-Luther-Univ., 4.
EBLE. H. et al. 1955. Beitrag zur deutschsprachigen Nomenklatur der rezenten Mammalia. Wissensch. Zeitschr. Univ. Halle-Wittenberg, Math.-Nat. R., 4, (1), 169-172.
FREYE, H.-A. et al. 1956. Die deutschen Namen der deutschen Saugetiere. Saugetierk. Mitt., 4, 171-174.
GAFFREY, G. 1955. Uber die deutschsprachige Nomenklatur der Saugetiere, insbesondere der deutschen. Abh. Mus. Tierk. Dresden, 22, 185-205.
HALTENORTH, T. 1955. Mammalia, in Doderlein, Bestimmungsbuch dur deutsche Land- und Suss-wassertiere. Munchen.
ZIMMERMANN, K. et al. 1958. Zur Liste der deutschen Namen fur deutsche Saugetiere. Saugetierk. Mitt., 6, 124-126.
MEYLAN, A. 1966. Liste de mammiferes de Suisse. Bull. Soc. Vaud Sci. Nat., 69, 233-246. (In French, German, Italian)
DORST, J. et al. 1957. Liste des noms francaise et espagnols des mammiferes d'Europe. Mammalia, 21, 258-266.
CABRERA, A. 1914. Fauna Iberica. Mamiferos. Madrid; 1-441.
VERESHCHAGIN, N. K. 1967. The mammals of the Caucasus.... Jerusalem; 1-816. (Transl. of 1959)
TOKTOSUNOV, A. 1958. Rodents of the Kirghiz. Frunze; 1-172.
HATT, R. T. 1959. The mammals of Iraq. Misc. Publ. Univ. Michigan Mus. Zool., 106, 1-113.
de BEAUFORT, F. & PUJOL, R. 1966. Noms vernaculaires des mammiferes de l'Afrique Equatoriale Occidentale. Cah. La Maboke, 4, 151-157.
ROSEVEAR, D. R. 1951. Nigerian mammals. Lagos; 1-87.
PETTER, F. & PUJOL, R. 1963. Noms vernaculaires lissongo des mammiferes de la region de la Maboke. Cah. La Maboke, 1, 119-122. (Central African Republic)
MISONNE, X. 1959. Les noms vernaculaires des rongeurs en Kilendu et en Kinande. Zooleo, 53, 111-113. (Congo)

SHORTRIDGE, G. C. 1934. The mammals of Southwest Africa. London; 2 volumes.
ROBERTS, A. 1951. The mammals of South Africa. (South Africa), 1-700.
WHITE, C. M. N. & ANSELL, W. F. H. A list of Luvale and Lunde mammal names. Puku, 4, 181-185. (Northern Rhodesia)
SWYNNERTON, G. H. 1946. Vernacular names for ... mammals in ... Tanganyika Territory. Tanganyika Notes Rec., 21, 21-38.
van ROSEN, B. 1953. Game animals of Ethiopia. Addis Ababa; 1-92.
VIRA, R. et al. 1953. Indian scientific nomenclature of the mammals in India, Burma and Ceylon. Sarawak Vihara Series, 30, 1-186.
MATTHEWS, W. H. 1940. Nepali names of mammals. Journ. Bengal Nat. Hist. Soc., 14, 153.
PHILLIPS, W. W. A. 1940. A list of ... animals, birds and reptiles of Ceylon with the Sinhalese and Tamil names of each. Loris, 2, 140-144.
HARRISON, J. L. 1953. Vernacular names of mammals. Malayan Nature Journ., 8, 37-40.
HARRISON, J. et al. 1955. Aboriginal names of mammals. Malayan Hist. Journ., 2, 53-57.
HARRISON, J. L. & QUAH, S. K. 1964. Vernacular names of Malaysian mammals. Malay Nat. Journ., 18, 70-72.
DENTAN, R. K. 1967. The mammalian taxonomy of the Senoi Semai. Malay Nat., 20, 100-106.
HARRISON, J. L. & TRAUB, K. Names of mammals in the Dusun language of North Borneo. Malayan Nature Journ., 8, 120-128.
HARRISON, J. 1964. An introduction to the mammals of Sabah. Jesselton, Malaysia; 1-244. (Borneo)
SODY, H. J. V. 1938. Lijst van de zoogdieren van Java (met hollandsche en Inlandsche namen). Tectona Buitenzorg, 31, 741-764.
de RAADT, O. L. E. 1933. De namen der in Nederlandsch-Indie voorkomende Ratten. Zool. Medd., 16, 31-32.
JOHNSTON, T. H. 1943. Aboriginal names and utilization of the fauna of the Eyrean region. Trans. Roy. Soc. South Australia, 67, 244-311.
TAKASHIMA, H. 1930. Standard Japanese names for the mammals of Japan. I. Trans. Nat. Hist. Soc. Formosa, 20, 195-217. (In Japanese)
SIMPSON, G. G. 1938. Mongolian mammal names. American Mus. Novit., 980, 1-26.
WON, P. O. & WOO, H. C. 1958. A distributional list of the Korean birds and mammals. Korea, 1-10, 1-96, 1-12.

Generic name studies

HERSHKOVITZ, P. 1948. Names of mammals dated from Frisch, 1775, and Zimmermann, 1777. Journ. Mamm., 29, 272-277.
MACLEOD, R. D. 1956. Key to the names of British fishes, mammals, amphibians and reptiles. London; 1-71. (Origin and meaning)
CONISBEE, L. R. 1964. New proposed genera, 1957-1961. Journ. Mamm., 45, 474-475.

Family name lists

PALMER, T. S. 1904. Index generum mammalium. A list of the genera and families of mammals. North American Fauna, 23, 1-984. (Reprinted 1967)
ROMER, A. S. 1933. Vertebrate paleontology. Chicago; 1-687. (Recent & fossil, 610-627)
SIMPSON, G. G. 1945. The principles of classification and a classification of mammals. Bull. American Mus. Nat. Hist., 85, 1-350. (Recent & fossil, 39-162)
ANDERSON, S. & JONES, J. K., Jr. 1967. Recent mammals of the world: A synopsis of families. New York; 1-453.

Nomenclators

[See also: Nomenclators listed under Animalia]

WATERHOUSE, F. H. 1894. Index generum mammalium. (Original manuscript, 1-158)
PALMER, T. S. 1904. Index generum mammalium: A list of the genera and families of mammals. North American Fauna, 23, 1-984.
HAYMAN, R. W. & HOLD, G. W. C. 1940. With a list of named forms (1758 - 1936). In Ellerman, The families and genera of living rodents, volume 1. London.
SIMPSON, G. G. 1945. Principles of classification and a classification of mammals. Bull. American Mus. Nat. Hist., 85, 1-350. (Genera and higher taxa, 309-338, in index)
CONISBEE, L. R. 1953. List of the names proposed for genera and subgenera of Recent mammals from the publication of T. S. Palmer's Index Generum Mammalium 1904 to the end of 1951. London; 1-109.
MACLEOD, R. D. 1956. Key to names of British ... mammals. London; 1-71. (Meaning & origin)

REGIONAL STUDIES

Faunas, checklists, identification works

General

(anonymous). Guide to the whales, porpoises, and dolphins (order Cetacea). London; 1-47; 1909.
BERTRAM, G. C. L. The seals of the Empire. Journ. Soc. Preserv. Fauna Emp., 40, 19-28; 1940.
KING, J. E. Seals of the world. London; 1-154; 1964.
NAPIER, J. R. & NAPIER, P. H. A handbook of living Primates. New York; 1-456; 1967.

Arctic

MANNICHE, A. L. V. The terrestrial mammals and birds of north-eastern Greenland. Medd. Gronl., 45, 1-200; 1910.
ADLERBERG, G. P. et al. Arctic animals. Introduction to the study of Ungulata, Carnivora, Pinnipedia, Cetacea, Insectivora, and Chiroptera. Leningrad; 1-579; 1935. (In Russian)
SAEMUNDSSON, B. The zoology of Iceland. Reykjavik; IV, (76), 1-52.
DUNBAR, M. J. The Pinnipedia of the Arctic and Subarctic. Bull. Fish. Res. Board Canada, 85, 1-22; 1949.
BEE, J. W. & HALL, E. R. Mammals of northern Alaska on the Arctic Slope. Misc. Publ. Univ. Kansas Mus. Nat. Hist., 8, 1-309; 1956.

North America

ELLIOT, D. G. A synopsis of the mammals of North America .... Field Columbian Mus., Zool. Series, 2, 1-471; 1901.
ELLIOT, D. G. A check list of the mammals of the North American continent.... Field Columbian Mus., Zool. Series, 6, 1-761; 1905. (Supplement, New York, 1-192; 1917)
MILLER, G. S., Jr. List of North American Recent mammals. United States Nat. Mus. Bull., 128, 1-673; 1924.
ANTHONY, H. E. Field book of North American mammals.... New York; 1-625; 1928.
BOOTH, E. S. How to know the mammals, 2nd ed. Dubuque, Iowa; 1-203; 1961. (Prev. ed., 1950) (Keys)
GLASS, B. P. A key to the skulls of North American mammals. Minneapolis, Minnesota; 1-54; 1951.
SANDERSON, I. T. How to know the American mammals. New York; 1-164; 1951. (Keys)
BURT, W. H. Field guide to the mammals. Boston, 1-200; 1952.
PALMER, R. S. The mammal guide. New York, 1-384; 1954. (Keys)
PALMER, T. S. Palmer's Fieldbook of Mammals. New York; 1-321; 1957.
HALL, E. R. & KELSON, K. R. The mammals of North America. New York; 2 volumes; 1959. (Keys)
BURT, W. H. & GROSSENHEIDER, R. P. A field guide to the mammals, 2nd ed. Boston; 1-284; 1964.
COUES, E. & ALLEN, J. A. Monographs of North American Rodentia. Washington, D. C.; 1-1091; 1877.
ALLEN, J. A. History of North American pinnipeds.... United States Geol. Geogr. Surv. Terr., Misc. Publ., 12, 1-785; 1880.
HALL, E. R. A synopsis of the North American Lagomorpha. Publ. Univ. Kansas Mus. Nat. Hist., 5, 119-202; 1951.
BARBOUR, R. W. & DAVIS, W. H. Bats of America. Lexington, Kentucky; 1-286; 1969. (Keys)

Alaska & Yukon

MURIE, O. J. Fauna of the Aleutian Islands and Alaska Peninsula. North American Fauna, , 1-364; 1959.
RAND, A. L. Mammals of Yukon, Canada. Bull. Nat. Mus. Canada, 100, 1-93; 1945.

Canada

ANDERSON, R. M. Catalogue of Canadian Recent mammals. Bull. Nat. Mus. Canada, 102, 1-238; 1947.
CAMERON, A. W. Canadian mammals. Ottawa; 1-81; 1958.
CAMERON, A. W. Les mammiferes du Canada. Ottawa; 1-89; 1960.
CAMERON, A. W. A guide to eastern Canadian mammals. Ottawa; 1-72; 1956. (Keys)
SERGEANT, D. E. & FISHER, H. D. The smaller Cetacea of eastern Canadian waters. Journ. Fish. Res. Board Canada, 14, 83-115; 1957.
PETERSON, R. L. The mammals of eastern Canada. Toronto; 1-465; 1966. (Keys)
SMITH, R. W. Land mammals of Nova Scotia. American Midl. Nat., 24, 213-241; 1940.
ANDERSON, R. M. Mammals of the Province of Quebec. Rep. Provancher Soc. Nat. Hist., 1938, 50-114.
DOWNING, S. C. A provisional check-list of the mammals of Ontario. Misc. Publ. Roy. Ontario Mus., 2, 1-11; 1948.
SOPER, J. D. The mammals of Manitoba. Canadian Field Nat., 75, 171-219; 1961. (Also in: Wildl. Manag. Bull., (1), 17, 1-51; 1961)
ANDERSON, R. M. & RAND, A. L. A synopsis of the rodents of the southern parts of the prairie provinces of Canada. Spec. Contr. Nat. Mus. Canada, 43, (1), 1-46.
RAND, A. L. Mammals of the eastern Rockies and western plains of Canada. Nat. Mus. Canada, 108, 1-237; 1948. (Keys)
BECK, W. M. A guide to Saskatchewan mammals. Regina; 1-52; 1958.

COWAN, I. M. & GUIGET, C. J. The mammals of British Columbia, Victoria, B. C. Handb. British Columbia Prov. Mus., 11, 1-413; 1956. (Keys)

United States

BLAIR, W. F. et al. Vertebrates of the United States, 2nd ed. New York; 1-616; 1968. (Prev. ed., 1957) (Keys, 453-562)
RUE, L. L. Pictorial guide to the mammals of North America. New York; 1967. (Keys)
SANBORN, C. C. Bats of the United States. Publ. Health Rep., United States Publ. Health Serv., 69, 17-28; 1954.
HAMILTON, W. J., Jr. The mammals of eastern United States. Ithaca, New York; 1-432; 1943. (Reprinted, New York, 1963)
WHITAKER, J. O., Jr. Keys to the vertebrates of the eastern United States, excluding birds. Minneapolis, Minnesota; 1-256; 1969.
MILLER, G. S., Jr. Key to the land mammals of northeastern North America. Bull. New York State Mus., 38, (8), 59-160; 1900.
ALLEN, G. M. The whalebone whales of New England. Mem. Boston Soc. Nat. Hist., 8, 107-322; 1916.
GRAYCE, R. L. Checklist of New England mammals. Bull. Massachusetts Audubon Soc., 41, 15-24; 1957.
GRAYCE, R. L. Key to the marine mammals of New England. Bull. Massachusetts Audubon Soc., 41, 417-419; 1957.
GAUGHRAN, G. R. I. A key to genera of rodents found in Pennsylvania based on skull and mandible characteristics. Proc. Pennsylvania Acad. Sci., 19, 49-57; 1945.
DOUTT, J. K. et al. Mammals of Pennsylvania. Harrisburg, Pennsylvania; 1-281; 1967.
BAILEY, J. W. The mammals of Virginia. Richmond, Virginia; 1-416; 1946.
HANDLEY, C. O., Jr. & PATTON, C. P. Wild animals of Virginia. Richmond, Virginia; 1-220; 1947.
GOLLEY, F. B. Mammals of Georgia .... Athens, Georgia; 1-218; 1962. (Keys, list)
SHERMAN, H. B. A list and bibliography of the mammals of Florida, living and extinct. Quart. Journ. Florida Acad. Sci., 15, 86-126; 1952.
LYON, M. W. Mammals of Indiana. American Midl. Nat., 17, 1-384; 1936.
BURT, W. H. Mammals of the Great Lakes Region. Ann Arbor, Michigan; 1-246; 1957.
WOOD, N. A. An annotated checklist of Michigan mammals. Occ. Pap. Mus. Zool. Univ. Michigan, 4, 1-13; 1914.
BURT, W. H. The mammals of Michigan. (Michigan), 1-288; 1948.
HOFFMEISTER, D. F. & MOHR, C. O. Fieldbook of Illinois mammals. Illinois Nat. Hist. Surv. Manual, 4, 1-233; 1957. (Keys)
CORY, C. B. Mammals of Illinois and Wisconsin. Publ. Field Museum, 153, 1-505; 1912.
HOLLISTER, N. A check-list of Wisconsin mammals. Bull. Wisconsin Nat. Hist. Soc., (n. s.), 8, 21-31; 1910.
SWANSON, G. et al. The mammals of Minnesota. Tech. Bull. Div. Game Fish Minnesota Dept. Conserv., 2, 1-108; 1945.
GUNDERSON, H. L. & BEER, J. R. The mammals of Minnesota. Occ. Pap. Minnesota Mus. Nat. Hist., 6, 1-190; 1953.
VAN HYNING, T. & PELLETT, F. C. An annotated catalogue of the Recent mammals of Iowa. Proc. Iowa Acad. Sci., 17, 211-218; 1910.
POLDER, E. A check list of mammals of present occurrence in Iowa .... Proc. Iowa Acad. Sci., 60, 716-724; 1954.
SCHWARTZ, C. W. & SCHWARTZ, E. R. The wild mammals of Missouri. Columbia, Missouri; 1-341; 1964.
SEALANDER, J. A., Jr. A provisional check-list and key to the mammals of Arkansas.... American Midl. Nat., 56, 257-296; 1956.
DAVIS, W. B. The mammals of Texas. Austin, Texas; 1-252; 1960.
HIBBARD, C. W. A checklist of Kansas mammals, 1943. Trans. Kansas Acad. Sci., 47, 61-88; 1945.
HALL, E. R. Handbook of mammals of Kansas. Misc. Publ. Univ. Kansas Mus. Nat. Hist., 7, 1-303; 1955.
WARREN, E. R. The mammals of Colorado .... New York; 1-300; 1910. (Keys)
RODECK, H. G. Guide to the mammals of Colorado. Leafl. Univ. Colorado Mus., 10, 1-72; 1952.
RODECK, H. G. Guide to the mammals of Colorado, 2nd ed. Leafl. Univ. Colorado Mus., 10, 1-72; 1966.
LONG, C. A. The mammals of Wyoming. Publ. Univ. Kansas Mus. Nat. Hist., 14, 493-758; 1965. (Keys)
KRUTZSCH, P. H. A key to the bats of western North America .... Chicago Acad. Sci., Nat. Hist. Misc., 133, 1-6; 1954.
DURRANT, S. D. Mammals of Utah. Publ. Univ. Kansas Mus. Nat. Hist., 6, 1-549; 1952.
COCKRUM, E. L. The Recent mammals of Arizona .... Tucson, Arizona; 1-276; 1960.
HALL, E. R. The mammals of Nevada. Berkeley, California; 1-710; 1946. (Keys & list)
DAVIS, W. B. The Recent mammals of Idaho. Caldwell, Idaho; 1-400; 1939.
SCHEFFER, V. B. A list of marine mammals on the West Coast of North America. Murrelet, 23, 42-47; 1942.
ORR, R. T. Cetacean records from the Pacific Coast of North America. Wasmann Journ. Biol., 9, 147-148; 1951.
INGLES, L. G. Mammals of the Pacific States. California, Oregon, and Washington. Stanford, California; 1-506; 1965.
SCHEFFER, V. B. Whales and dolphins of Washington State, and key to cetaceans of west coast of North America. American Midl. Nat., 39, 257-337; 1948.
DALQUEST, W. W. Mammals of Washington. Publ. Univ. Kansas Mus. Nat. Hist., 2, 1-444; 1948.

SCHEFFER, V. B. & SLIPP, J. W. The whales and dolphins of Washington State, with a key to the cetaceans of the West Coast of North America. American Midl. Nat., 39, 257-337; 1948.
BAILEY, V. The mammals and life zones of Oregon. North American Fauna, 55, 1-416; 1936.
GRINNELL, J. A distributional list of the mammals of California. Proc. Califom ia Acad. Sci., (4), 3, 265-390; 1913.
GRINNELL, H. W. A synopsis of the bats of California. Univ. California Publ. Zool., 17, 223-404; 1918.
GRINNELL, J. Review of the Recent mammal fauna of California. Univ. California Publ. Zool., 40, 71-234; 1933.
INGLES, L. G. Mammals of California. Stanford, California; 1-258; 1947.
INGLES, L. G. Mammals of California and its coastal waters, 2nd ed. Stanford, California; 1-396; 1954.

Neotropics

ELLIOT, D. G. The land and sea mammals of Middle America and the West Indies. Publ. Field Columbian Mus., Zool., 4, (2), 441-850; 1904.
HERSHKOVITZ, P. A geographic classification of neotropical mammals. Fieldiana, Zool., 36, 579-620; 1958. (List & distribution)
TATE, G. H. H. The taxonomy of the genera of neotropical hystricoid rodents. Bull. American Mus. Nat. Hist., 68, 295-447; 1935.

West Indies

VAN GELDER, R. G. & WINGATE, D. B. The taxonomy and status of bats in Bermuda. American Mus. Novit., 2029, 1-9; 1961.
KOOPMAN, K. F. et al. Notes on the mammals of the Bahamas. Journ. Mamm., 38, 164-174; 1957.(List)
GUNTER, G. Mammals of the Gulf of Mexico. United States Fish Wildl. Serv. Fish. Bull., 55, 543-551; 1954. (List)
AGUAGYO, C. G. & RIVERO, L. H. Sinopsis de los mamiferos cubanos. Circ. Mus. Malac. Habana, 1954, 1283-1324; 1954.
LLYN, W. G. The bats of Jamaica. Nat. Hist. Notes, 1, (11), 15-19; 1943.
ANTHONY, H. E. The indigenous land mammals of Porto Rico, living and extinct. Mem. American Mus. Nat. Hist., (n. s.), 2, 332-435.
STARRETT, A. The bats of Puerto Rico and the Virgin Islands, with a checklist and keys for identification. Caribbean Journ. Sci., 2, 1-7; 1962.
ANTHONY, H. E. Mammals of Porto Rico, living and extinct — Rodentia and Edentata. New York Acad. Sci., Surv. Porto Rico & Virgin Is., 9, (2), 97-241; 1925-26.
BEATTY, H. A. The mammals of St. Croix, Virgin Islands. Journ. Agric. Univ. Puerto Rico, 28, 181-185; 1944. (List)
GREENHALL, A. M. Bats of Trinidad. Journ. Trinidad Field Nat. Club, 1956, 16-18; 1957.
GREENHALL, A. M. Checklist of the bats of Trinidad and Tobago. Caribbean Med. Journ., 21, 169-171; 1959.
GOODWIN, G. G. & GREENHALL, A. M. A review of the bats of Trinidad and Tobago. .... Bull. American Mus. Nat. Hist., 122, 189-301; 1961. (Keys)

Central America

VILLA, B. & BIAGI, F. F. Clave para familias de mamiferos Mexicanos. Acta Zool. Mexicana, 3, (5-6), 1-8; 1959. (Keys)
WRIGHT, N. P. A guide to Mexican mammals. Mexico, D. F.; 1-96; 1965.
VILLA-R., B. Los murcielagos de Mexico. Mexico, D. F.; 1-491; 1966. (Keys)
MARTINEZ, L. & RAMIREZ, B. V. Contribuciones al conocimiento de los murcielagos de Mexico. Anal. Inst. Biol., 9-12, 3 parts & continuacion; 1938-41.
HUEY, L. M. The mammals of Baja California. Trans. San Diego Soc. Nat. Hist., 13, 85-168; 1964.
BAKER, R. H. Mammals of Coahuila, Mexico. Publ. Univ. Kansas Mus. Nat. Hist., 9, 125-335; 1956.
BAKER, R. H. & GREER, J. K. Mammals of the Mexican state of Durango. Publ. Mus. Michigan State Univ., Biol., 2, 25-154; 1962.
DALQUEST, W. W. Mammals of the Mexican state of San Luis Potosi. Louisiana State Univ. Studies, 1, 1-233; 1953.
HALL, E. R. & VILLA R., B. An annotated check list of the mammals of Michoacan, Mexico. Publ. Univ. Kansas Mus. Nat. Hist., 1, 431-471; 1949.
HALL, E. R. & VILLA R., B. Lista anotada de los mamiferos de Michoacan, Mexico. Anal. Inst. Biol. Mexicana, 21, 159-214; 1950.
MARTINEZ, L. & VILLA R., B. Segunda contribucion al conocimiento de los murcielagos mexicanos. II. Estado de Guerrero. Anal. Inst. Biol. Mexico, 11, 291-361; 1940.
LUKENS, P. W. & DAVIS, W. B. Bats of the Mexican state of Guerrero. Journ. Mamm., 38, 1-14; 1957.
DAVIS, W. B. & LUKENS, P. W. Mammals of the Mexican state of Guerrero, exclusive of Chiroptera and Rodentia. Journ. Mamm., 39, 347-367; 1959.
DAVIS, W. B. & RUSSELL, R. J. Aves y mamiferos del Estado de Morelos. Rev. Soc. Mexicana Hist. Nat., 14, 77-147; 1953.
DAVIS, W. B. & RUSSELL, R. J. Mammals of the Mexican state of Morelos. Journ. Mamm., 35, 63-80; 1954.
GAUMER, G. F. Monografia de los mamiferos de Yucatan. Mexico; 1-331; 1917.
GOODWIN, G. G. Mammals of Honduras. Bull. American Mus. Nat. Hist., 79, 107-195; 1942.

FELTEN, H. Fledermause ... aus El Salvador. Senckenbergiana Biol., 36, 271-285; 37, 69-86, 179-212, 342-367; 38, 1-22; 1955-57.
FELTEN, H. Nagetiere (Mammalia, Rodentia) aus El Salvador. Senckenbergiana Biol., 38, 145-155; 1957.
BURT, W. H. & STIRTON, R. A. The mammals of El Salvador. Misc. Publ. Mus. Zool. Univ. Michigan, 117, 1-69; 1961.
GOODWIN, G. G. Mammals of Costa Rica. Bull. American Mus. Nat. Hist., 87, 271-473; 1946.
GOLDMAN, E. A. Mammals of Panama. Smithsonian Misc. Coll., 69, (5), 1-309; 1920. (List)
WENZEL, R. L. & TIPTON, V. J. Ectoparasites of Panama. Chicago; 1-861; 1966. (Checklist of the mammals, 753-795)

South America

CABRERA, A. & YEPES, J. Mamiferos sud-americanos. Buenos Aires; 1-370; 1940.
CABRERA, A. Catalogo de los mamiferos der America del Sur. I, II. Rev. Mus. Argentino Cienc. Nat. "Bernardino Rivadavia", Zool., 4, 1-732; 1957-61.
MIRANDA RIBEIRO, A. Didelphia ou Mammalia — ovovivipara Marsupiaes, Didelphos, Pedimanos ou Metatherios. Rev. Mus. Paulista, 20, 245-427; 1936. (South America)
SIMPSON, G. G. The development of marsupials in South America. Physis, 14, 373-398; 1939.
PEARSON, O. P. Mammals in the highlands of southern Peru. Bull. Mus. Comp. Zool., 106, 117-174; 1951.
ORTIZ DE LA PUENTE, J. Estudio monografico de los quiropteros de Lima y alrededores. Publ. Mus. Hist. Nat. "Javier Prado", (A, Zool.), 7, 1-48; 1951.
CEBULLOS, B. I. Clave dicotomica para ordenes de mamiferos des Cuzco. Rev. Univ. Cuzco, 41, 120-125; 1952.
SOUKOP, J. Materiales para el catalogo de los mamiferos Peruanos. Biota, 3, (26), 240-276, 1960; 3, (27), 277-324, 1961; 3, (28), 325-331, 1961.
ROHL, E. Fauna descriptiva de Venezuela, 3rd ed. Madrid; 1-516; 1956.
HUMMELINCK, P. W. Studies on the fauna of Curacao, Aruba, Bonaire and the Venezuelan Islands. Utrecht; 1-130; 1940.
TATE, G. H. H. The mammals of the guiana region. Bull. AMerican Mus. Nat. Hist., 76, 151-229; 1939.
ROTH, V. Notes and observations on animal life in British Guiana. A popular guide to colonial Mammalia. Georgetown; 1-164; 1941.
SANDERSON, I. T. A brief review of the mammals of Suriname .... Proc. Zool. Soc. London, 119, 755-789; 1949.
HUSSON, A. M. The bats of Suriname. Leiden; 1-282; 1962.
HUSSON, A. M. The bats of Suriname. Zool. Verh., 58, 1-282; 1962.
BROSSET, A. & DUBOST, G. Chiropteres de la Guyane francaise. Mammalia, 31, 583-594; 1967.
(anonymous). Provisional list of the fauna of Guyana. Journ. Guyana Mus. Zool, 42, 58-60; 1967.
VIEIRA, C. C. Lista remissiva dos mamiferos do Brasil. Arqu. Zool. Sao Paulo, 8, 341-474; 1955.
da CUNHA VIEIRA, C. O. Ensaio monografico os quiropteros do Brasil. Arqu. Zool. Est. Sao Paulo, 3, 219-471; 1942.
MOOJEN, J. Os roedores do Brasil. Rio de Janeiro; 1-214; 1952.
da CRUZ LIMA, E. Mammals of Amazonia. I. Rio de Janeiro; 1-274; 1945.
von IHERING, H. Os mamiferos do Brazil meridional. Rev. Mus. Paulista, 8, 148-272; 1910.
VIEIRA, C. Roedores e lagomorfos do Estado de Sao Paulo. Arqu. Zool. Sao Paulo, 8, 129-168; 1953.
PODTIAGUIN, B. Contribuciones al conocimiento de los murcielagos del Paraguay. Rev. Soc. Cient. Paraguay, 6, 25-62; 1944.
ACOSTA y LARA, E. F. Quiropteros del Uruguay. Comun. Zool. Mus. Hist. Nat. Montevideo, 3, (58), 1-71; 1950.
YEPES, J. Epitome de la sistematica de los roedores argentinos. Rev. Inst. Bact. Buenos Aires, 7, 213-269; 1935.
YEPES, J. Los "Edentata" Argentinas sistematica y distribucion. Rev. Univ. Buenos Aires, (2), (5), 1, 461-515; 1928. (Also as vol. 25)
OSGOOD, W. H. The mammals of Chile. Field Mus. Nat. Hist., Zool. Ser., 30, 1-268; 1943.
YANEZ, A. P. Vertebrados marinos chilenos. I. Mamiferos. Rev. Biol. Marin Valparaiso, 1, 103-123; 1948.
MANN, G. Clave de determinacion para les especies de mamiferos Silvestres de Chile. Invest. Zool. Chile, 4, 89-126; 1957.
CABRERA, A. Sinopsis de los Quiropteros chilenos. Rev. Chilena Hist. Nat., 7, 278-308; 1903.
SCHNEIDER, O. C. Catalogo de los mamiferos de la Provincia de Concepcion. Bol. Soc. Biol. Concepcion, 21, 67-83; 1946.
HAMILTON, J. E. Cetacea of the Falkland Islands. Comun. Zool. Mus. Montevideo, 4, (66), 1-6; 1952.

Palearctic

ELLERMAN, J. R. & MORRISON-SCOTT, T. C. S. Checklist of Palearctic and Indian mammals — amendments. Journ. Mamm., 34, 516-518; 1953.

Europe, western

TROUESSART, E.-L. Faune des mammiferes d'Europe. Berlin; 1-266; 1910.
HAINARD, R. Les mammiferes sauvages d'Europe, I, II. Paris; 1-268, 1-274; 1948-49.
HAINARD, R. Mammiferes sauvages d'Europe, 2nd ed. Neuchatel; 2 volumes; 1961.

van den BRINK, F. H. A field guide to the mammals of Britain and Europe. London; 1-221; 1967. (New York; 1-221; 1968)
CORBET, G. B. The terrestrial mammals of western Europe. Philadelphia; 1-264; 1966. (List & keys)
SIIVONEN, L. Pohjolan nisakkaat. Helsinki; 1-181; 1967. (Mammals of northern Europe)
CURRY-LINDAHL, K. Djureni farg. Daggdjur — Kraldjur — Groddjur. Stockholm; 1-172; 1955. (Keys)
COLLETT, R. Norges hvirveldyr. I. Kristiania; 1-744; 1911-12.
MANNICHE, A. L. V. et al. Danmarks pattedyr. København; 1-480; 1935.
BRUUN, A. F. et al. List of Danish vertebrates. København; 1-180; 1950.
BARRETT-HAMILTON, G. E. H. A history of British Mammals. London; 3 volumes; 1910-21.
CORY, H. The mammals of the British Islands. London; 1-292; 1941. (Keys)
MORRISON-SCOTT, T. C. S. A list of British mammals. London; 1-24; 1952.
SOUTHERN, H. N. Key to identification of small mammal skulls in owl pellets. Bull. Mamm. Soc. British Isles, 2, 14; 1954.
PAGE, F. J. T. Field guide to British deer. London; 1-80; 1957.
FORREST, H. E. British bats: How to identify them. Trans. Caradoc Field Club, 11, 256-259; 1945.
TURNER, J. F. British bats. Proc. South London Ento. Nat. Hist. Soc., 1946-47, 79-84; 1947.
MORRISON-SCOTT, T. C. S. A key to the British bats. The Naturalist, 985, 33-36; 1939.
SOUTHERN, H. N. The handbook of British mammals. Oxford; 1-465; 1963.
CORBET, G. B. The identification of British mammals. London; 1-46; 1964.
SAVAGE, R. J. G. Key to the identification of British bat remains. Proc. Speleol. Soc., 10, 116-118; 1964.
SOUTHERN, H. N. The handbook of British mammals. Oxford; 1-465; 1964.
LAWRENCE, M. J. & BROWN, R. W. Mammals of Britain. Their tracks, trails and signs. London; 1-233; 1967.
FITTER, R. S. R. Check-list of mammals, reptiles and amphibians of London .... London Nat., 28, 98-115; 1949.
VENABLES, L. S. V. & VENABLES, U. M. Birds and mammals of Shetland. Edinburgh; 1-391; 1955.
MOFFAT, C. B. The mammals of Ireland. Proc. Roy. Irish Acad., 44B, 61-128; 1938.
BAAL, H. J. The indigenous mammals, reptiles and amphibians of the Channel Islands. Bull. Soc. Jersiaise, xv, 101-110; 1949.
HUSSON, A. M. The identification of skull remains from pellets of owls in the Netherlands, Belgium and Luxembourg. Zool. Bijdr., 5, 1-63; 1962. (In Dutch)
van DEINSE, A. B. Het determineeren van de Nederlandsche Cetacea. Nat. Mus. Rotterdam, 1, 11-22; 1932.
IJSSELING, M. A. & SCHEYGROND, A. De zoogdieren van Nederland. Zutphen; 2 volumes; 1943.
van den BRINK, F. H. Zoogdierengids van Europe ten westen van 30° oosterlengte. Amsterdam; 1-231; 1955. (Keys)
IJSSELING, M. A. & SCHEYGROND, A. Wat is dat voor dier? Zutphen; 1-96; 1957. (N. w. Europe)
VANDEN EECKHOUDT, J.-P. Faune elementaire des mammiferes de Belgique. Nat. Belges, 35, 3-50; 1954.
FRECHKOP, S. Faune de Belgique. Mammiferes. Bruxelles; 1-545; 1958.
FREUND, L. Cetacea. In Grimpe, Die Tierwelt der Nord- und Ostsee. Leipzig; Lief. 22, Teil 12K, 1-64; 1932.
MOHR, E. Die freilebenden Nagetiere Deutschlands. Jena; 1-112; 1938.
HALTENORTH, T. Mammalia, in Doderlein, Bestimmungsbuch fur deutsche Land- und Susswassertiere. Munchen; 1955.
MOHR. E. Die freilebenden Nagetiere Deutschlands. Jena; 1-152; 1950.
DODERLEIN, L. Bestimmungsbuch fur deutsch Land- und Susswassertiere. Wirbeltiere. 2. Munchen; 1-304; 1955. (See Haltenorth, above)
GERBER, W. Die wildlebende Raubtiere Deutschlands. Wittenberg Lutherstadt; 1-96; 1960.
NATUSCHKA, G. Heimische Fledermause. Wittenberg Lutherstadt; 1-146; 1960.
MEHL, S. Kleine Saugetiere der Heimat in naturlicher Frosse. l. Lief. Munchen; 1-36; 1960.
NATUSCHKE, G. Heimische Fledermause. Wittenberg Lutherstadt; 1-146; 1964.
BAUMANN, F. Die freilebenden Saugetiere der Schweiz. Bern; 1-492; 1949.
MEYLAN, A. Liste de mammiferes de Suisse. Bull. Soc. Vaud Sci. Nat., 69, 233-246; 1966.
CHAPPELLIER, A. Les rongeurs de France .... Arch. Hist. Nat., 9, 1-138; 1932.
DIDIER, R. & RODE, P. Les mammiferes de France. Paris; 1-398; 1935. (Keys)
RODE, P. & DIDIER, R. Atlas des mammiferes de France. Paris; 1-219; 1946. (Keys)
RODE, P. Les chauves-souris de France. Atlas Vert., 6, 1-70; 1947.
von LEHMANN, E. Die Saugetiere des Furstentums Liechtenstein. Jahrb. Hist. Ver. Fusste Liechtenstein, 62, 159-362; 1963.
CABRERA, A. Fauna Iberica. Mamiferos. Madrid; 1-441; 1914.
de SEABRA, A. F. Catalogue systematique des Vertebres du Portugal. Bull. Soc. Port. Sci. Nat., 4, 91-114; 1910.
NOBRE, A. Descricao dos mamiferos marinhos de Portugal. In Fauna Marinha do Portugal, I, Vertebrados. Porto; 1-21; 1935.
da GAMA, M. M. Mamiferos de Portugal (Chaves para a sua determinacao). Mem. Mus. Zool. Univ. coimbra, 246, 1-246; 1957. (Keys)
CABRERA, A. Los roedores de España. Madrid; 1-63; 1910.
GULINO, G. & DAL PIAZ, G. I Chirotteri Italiani. .... Boll. Mus. Zool. Anat. Comp. Torino, (3), 47, (91), 1-43; 1939. (List)
TOSCHI, A. & LANZA, B. Fauna d'Italia. Bologna; 1-485; 1959. (Chiroptera) (Insectivora)
ZANGHERI, P. Fauna di Romagna, Mammiferi. Boll. Zool., 24, 17-38; 1957. (List)
TOSCHI, A. Fauna d'Italia, Mammalia. Bologna; 1-647; 1965.

TORTONESE, E. Pesci e cetacei del mar Ligure. Genova; 1-216; 1965.

Europe, central & eastern

BROHMER, P. Saugetiere. In Tierwelt Mitteleuropas. Leipzig; Lief. 7, 1-61; 1927.
BIEGER, W. & WAHLSTROM, A. Die wildlebenden Saugetiere Mitteleuropas. Heidelberg; 1-88; 1938.
GAFFREY, G. Merkmale de wildlebenden Saugetiere Mitteleuropas. Leipzig; 1-284; 1961. (Keys)
LING, H. & PAAVER, K. List of mammals of Estonia. Abiks. Loodus., Tartu, 7, 1-18; 1952. (In Estonian)
VANAUSKAS, T. et al. Vadovas Lietuvos zinduoliams pazinti. Vilnius; 1-340; 1964. (In Lett) (Keys)
KOWALSKIEGO, K. Klucze do osnaczania kregowcow Polski. Czesc V. Szaki — Mammalia. Warszawa; 1-280; 1964.
STEPANEK, O. Klic nasich obratlovcu. Praha; 1-256; 1950.
FERIANCOVA-MASAROVA, Z. & HANAH, V. Stavovce Slovenska. IV. Cicavce. Bratislava; 1-333; 1965.
GAISLER, J. et al. Netopyri Ceskoslovenska. Urcovaci klice, popisy a poznamky o vyskytu. Acta Univ. Carolinae (Prague); 1957, (Biol.), 1-65; 1957.
REBEL, H. Die freilebenden Saugetiere Osterreichs. Wien; 1-119; ? date. (Keys)
WETTSTEIN-WESTERSHEIMB, O. Mammalia. Cat. Faunae Austriae, 21C, 1-16; 1955. Wien.
IONESCU, V. Vertebratele din Romania. Bucuresti; 1-496; 1968. (Keys)
MARTINO, V. Kluic za odrieivanie glodara. Glasn. Minist. Polopr. Belgrade, 29, 1-24; 1930.
PESHEV, T. & BOEV, N. Vertebrates. In Fauna of Bulgaria. A short classification key. (Russia); 1-520; 1962.
OGNEV, S. I. The mammals of the eastern Europe and of the northern Asia. I - VII. Moskva; 1928-50. (In Russian)
OGNEV, S. I. Mammals of eastern Europe and northern Asia. Jerusalem; 7 volumes; 1962-1966. (Translation of 1928-50)
BIANCHI, V. Table synoptique ... Chiropteres de la Russie europeenne. Ann. Mus. Zool. Petrograd, 22, 200-217; 1922. (In Russian)
VINOGRADOV, B. Tableau analytique de la faune de l'URSS. Les Rongeurs.... Moskva; 1-87; 1933. (In Russian)
OGNEV, S. T. A synopsis of the Russian bats. Journ. Mamm., 8, 140-157; 1927.
VINOGRADOV, B. S. Key to rodents. Moskva; 1-241; 1941. (In Russian)
BOBRINSKOY, N. et al. Mammals of USSR. No. 1, Moskva; 1-440; 1944. (In Russian)
BOBRINSKII, N. A. et al. Key to the mammals of the U. S. S. R. Moskva; 1-368; 1944. (In Russian)
GROMOVA, V. Key to the identification of the mammals of USSR based on skeletal bones. Part I. Trav. Comm. Etud. Quatern. Acad. Sci. USSR, 9, 1-240; 1950. (In Russian)
KUZYAKIN, A. P. Bats. Moskva; 1-443; 1950. (Review of USSR species. In Russian)
NOVIKOV, G. A. Carnivorous mammals of the fauna of the USSR. (Russia; 1-293; 1956. (In Russian)
NOVIKOV, G. A. Carnivorous mammals of the USSR. In Keys to the fauna of the USSR ... 62. Jerusalem; 1-284; 1962. (Translation of 1956)
TOMILIN, A. G. Mammals of USSR..., XI. Moskva; 1-756; 1957. (In Russian) (Cetacea)
TOMILIN, A. G. Mammals of the USSR. IX. Jerusalem; 1-717; 1967. (Cetacea. Translation of 1957)
OGNEV, S. I. Contribution a la classification of the mammalian Insectivores of Russia. Yearb. Zool. Mus. Russian Acad. Sci., 1921, 311-350; 1922.
ARGYROPULO, A. I. Faune de l'URSS. Mammiferes. Tableaux des rongeurs. Inst. Zool. Acad. Sci. URSS, (n. s.), 29, 1-241; 1941. (In Russian)
KUZNETZOV, B. A. Key to the mammals of the USSR (Rodentia). Moskva; 262-368; 1944. (In Russian)
VINOGRADOV, B. & GROMOV, I. M. Rodents of the USSR. Tableau Analytique Fauna USSR, 48, 1-296; 1952. (In Russian)
VINOGRADOV, B. Short list of the rodents of the USSR. Moskva; 1956. (In Russian)
VINOGRADOV, B. S. & GROMOV, I. M. Rodents of the USSR. Tableau Analytique Fauna USSR, 48, 1-296; 1952. (In Russian)
NOVIKOV, G. A. Predaceous mammals of the USSR. Fauna USSR, 62, 1-294; 1956. (In Russian)
VINOGRADOV, B. S. & GHROMOV, I. M. Short guide to the rodent fauna of the USSR. Moskva; 1-120; 1956. (In Russian)
SOKOLOV, V. E. Fauna SSSR.,, Moskva; 1-639; 1959. (In Russian)
GROMOVA, V. Key to the identification of the mammals of the USSR based on skeletal bones. Part 2. Trav. Comm. Quatern. Acad. Sci. URSS, 16, 1-116; 1960. (In Russian)
HEPTNER, V. G. The mammals of the Soviet Union. Volume I. (Russia); 1961. (In Russian)
NOVIKOV, G. B. Keys to the fauna of the USSR, No. 62. Carnivorous mammals.... Jerusalem; 1962. (Translation of 1956)
TOMILIN, A. G. Cetacean fauna of the seas of the USSR. Opred. Faune SSSR, 79, 1-212; 1962. (In Russian)
CHAMSKY, K. K. Mammals of the USSR. Vol. 2. Cetacea and Pinnipedia. Moskva; 2 volumes; 1963. (In Russian)
GOUREIEV, A. A. Mammals of the USSR. I. Insectivores and lagomorphs. Moskva; 1963. (In Russian)
GROMOV, J. M. Mammals of the USSR. Vol. I, Rodents. Moskva; 1963. (In Russian)
GROMOV, I. M. et al. Mammals of the fauna of the USSR. Vol. 8. Moskva; 1-2000; 1963. (In Russian)
NOVIKOV, G. A. Mammals of the USSR. Vol. 2. Carnivora and bibliography. Moskva; 2 volumes; 1963. (In Russian)
SOKOLOV, I. I. Mammals of the USSR. Moskva; 2 volumes; 1963. (In Russian)
GUREYEV, A. A. Fauna SSSR. Leporines. Moskva; 1-276; 1964. (In Russian)
BOBRINSKOY, N. et al. Opredelitel' mlekopitaiushchikh SSSR. Moskva; 1-382; 1965. (In Russian)(Keys)

HEPTNER, V. G. et al. Die Saugetiere der Sowjetunion. Band I. Jena; 1-939; 1966. (Translation of 1961 Russian)
HEPTNER, V. G. & NAUMOV, N. P. Mammals of the Soviet Union, vol. 2, Sirenia and Carnivora. Moskva; 5-1004; 1967. (In Russian)
VINOGRADOV, B. S. & ARGIROPULO, A. I. Fauna of the USSR. Mammals. Key to rodents. Jerusalem; 1-241; 1 1968. (Translation of 1941)
SERZKANIN, I. N. Mammalia of the Belorussian S. S. R. Minsk; 1-311; 1955.
VINOGRADOV, R. S. et al. Rodents of central Asiatic part of USSR. Moskva; 1-228; 1936. (In Russian)
STROGANOV, S. U. Carnivorous mammals of Siberia. Jerusalem; 1-522; 1969.
STROGANOV, S. U. Insectivores of Siberia. Moskva; 1-267; 1957. (In Russian)
VINOGRADOV, B. S. Directions for the determination of the rodents of Central Asia. Samarkand; 1-50; 1930.
BOBRINSKOY, N. Bats of central Asia. Ann. Mus. Zool. Acad. Sci. URSS, 1929, 217-249; 1929.
STROGANOV, S. U. Animals of Siberia. Carnivora. Moskva; 1-458; 1962.
de POUSARGUES, E. Etude sur les ruminants de l'Asie centrale. Mem. Soc. Zool. France, 11, 126-224; 1898. (Artiodactyla)
MIGULIN, O. O. Mammals of the Ukrainian S. S. R. Kiev; 1-426; 1938. (In Russian)
POPOV, B. A. Mlekopitaiushchie Volzhsko-Kamskogo Kraia. Kazan; 1-468; 1960. (In Russian)
DAL, S. K. Systematic review of the animals of the Crimea. I. Mammalia. Zhivot. Mir. USSR, 5, 42-55; 1958. (In Russian; list)
ELLERMAN, J. R. Key to the rodents of southwest Asia in the British Museum collection. Proc. Zool. Soc. London, 118, 765-816; 1948.
KUZNETZOV, N. Y. Rodents of the northern Caucasian steppes. (Russia); 1-230; 1935. (In Russian)
ARGYROPULO, A. I. Catalogue of the rodents of the Caucasus. Trans. Azerbaidjan Sect. Trans-Caucas. Filial Acad. Sci. USSR, 20, 45-70; 1937. (In Russian)
HEPTNER, W. G. & FORMOZOV, A. N. Les mammiferes du Daghestan (Caucase oriental). Arch. Mus. Zool. Moscou, 6, 3-74; 1941. (In Russian)
VERESHCHAGIN, N. K. Systematic review of the animals of the Caucasian isthmus. I. Mammals. Zhivat. Mir. USSR, 5, 180-219; 1958.
VERESHCHAGIN, N. K. Mammals of Caucasia. Noskva; 1-703; 1959. (In Russian)
KUZJAKIN, A. P. The bats from Tashkent and ... Caucasus, Bucharia and Turkmenia. Bull. Soc. Nat. Moscou, (Biol.), 40, 316-330; 1934. (In Russian)
ISHUNIN, G. I. Fauna of the Uzbek SSR. III. Mammals (Carnivores and Ungulates). Taschkent; 1-230; 1961. (In Russian)
TOKTOSUNOV, A. Rodents of the Kirghiz. Frunze; 1-172; 1958. (In Russian)
BELJAEV, A. M. Les rongeurs du Kazakstan. Rodentia. Trav. Inst. Zool. Acad. Sci. Leningrad, 2, 27-40; 1934. (In Russian)
AFANASER, A. V. et al. The animals of Kazakhstan. Alma-Ata; 1-536; 1953. (In Russian)
PRAKASH, I. A list of the mammals of the Rajasthan Desert. Journ. Bengal Nat. Hist. Soc., 28, 1-7;1965.
RASORENOVA, A. Materialen zur Kenntnis der Nagetiere des Hochgebirges von Altai. Bull. Soc. Nat. Moscou, (Biol.), 42, 78-84. (Mongolia)

Near East & Middle East

LAY, D. M. A study of the mammals of Iran .... Fieldiana, Zool., 54, 1-282; 1967.
NIETHAMMER, J. Die Saugetiere Afghanistans (Teil 2). Insectivora, Lagomorpha and Rodentia. Kabul; 19-41; 1965.
BOBRINSKOY, N. A. Materials on the bat fauna of Turkestan.... Bull. Soc. Nat. Moscou, 34, 330-374; 1925. (In Russian)
HATT, R. T. The mammals of Iraq. Misc. Publ. Univ. Michigan Mus. Zool., 106, 1-113; 1959.
LEWIS, R. W. et al. A review of Lebanese mammals. Lagomorpha and Rodentia. Journ. Zool., 153, 45-70; 1967. (Keys)
HARRISON, D. L. A key to the identification of the bats (Chiroptera) of the Arabian Peninsula. Proc. Zool. Soc. London, 127, 447-452; 1956.
HARRISON, D. L. The mammals of Arabia, I - III. London; 1964 —
WASSIF, K. & HOOGSTRAAL, H. The mammals of southern Sinai, Egypt. Proc. Egyptian Acad. Sci., 9, 63-79; 1954.

Africa altogether

JOLEAUD, L. Les ruminants cervicornes d'Afrique. Mem. Inst. Egypte, 27, 1-85; 1935.
RODE, P. Les Primates de l'Afrique. Paris; 1-222; 1937.
ALLEN, G. M. A checklist of African mammals. Bull. Mus. Comp. Zool., 83, 1-763; 1939.
SIDNEY, J. The past & present distribution of some African ungulates. Trans. Zool. Soc. London, 30, 1-397; 1965.
MEESTER, J. Preliminary identification manual for African mammals. Washington, D. C.; 1966.
DORST, J. A field guide to the larger mammals of Africa. Boston; 1-287; 1970.

North Africa

HOOGSTRAAL, H. A brief review of the contemporaneous land mammals of Egypt ... 1. Journ. Egypt. Publ. Health Assoc., 37, 143-162; 38, 1-35; 1962-63.

HOOGSTRAAL, H. A brief review of the contemporaneous land mammals of Egypt (including Sinai). 3. Carnivora, Hyracoidea, Perissodactyla and Artiodactyla. Journ. Egypt. Publ. Health Assoc., 39, 205-239; 1964.
SANBORN, C. C. & HOOGSTRAAL, H. The identification of Egyptian bats. Journ. Egyptian Publ. Health Assoc., 30, 103-121; 1955.
LAURENT, P. Essai d'une clef dichotomique des Cheiropteres de la Barbarie. Mammalia, 1, 133-159; 1937.
TOSCHI, A. Mammiferi della Libia. Suppl. Ric. Zool. Applic. Caccia Bologna, 2, 137-177; 1951.
SETZER, H. W. A review of the Libyan mammals. Journ. Egyptian Publ. Health Assoc., 32, 41-82; 1957.
RANCK, G. L. The rodents of Libya .... United States Nat. Mus. Bull., 275, 1-264; 1968.
CABRERA, A. Los mamiferos de Marruecos. Trab. Mus. Cienc. Madrid, 57, 1-361; 1932.
LAURENT, P. Essai d'une etude de la repartition geographique des rongeurs du Maroc. Bull. Inst. Hygiene Maroc, 1-2, 1-40; 1936.
PANOUSE, J. B. Les mammiferes du Maroc. Primates, Carnivores, Pinnipedes, Artiodactyles. Trav. Inst. Sci. Cherifien, Zool., 5, 1-206; 1957.
SAINT GERONS, M. C. & PETTER, F. Les rongeurs du Maroc. Trav. Inst. Scien. Cherifien, Zool., 31, 1-55; 1965.

Atlantic Islands

SARMENTO, A. A. Vertebrados da Madeira. I. Mamiferos - Aves - Repteis - Batraquios, 2nd ed. Funchal; 1-317; 1948.

West Africa

DEKEYSER, P. L. Les mammiferes de l'Afrique Noire Francaise. Dakar; 1-61; 1948.
DEKEYSER, P. L. Les mammiferes de l'Afrique Noire Francaise, I, 2nd ed. Dakar; 1-428; 1955.
ROSEVEAR, D. R. The bats of West Africa. London; 1-418; 1965.
URBAIN, A. Faune de l'equateur africain francais. II. Encycl. Biol., 1-323; 1949.
DEKEYSER, P. L. Les mammiferes de l'Afrique Noire Francaise, ed. 2 Init. Africaines, I, 1-426; 1955.
WEBB, G. C. A guide to West African mammals. Ibadan; 1-40; 1957.
BOOTH, A. H. Small mammals of West Africa. London; 1-68; 1960.
ROSEVEAR, D. R. The bats of West Africa. London; 1-418; 1965.
ROSEVEAR, D. R. The rodents of West Africa. London; 1-604; 1969.
MONARD, A. Resultats de la mission scientifique du Dr. Monard en Guinee Portugaise 1937-1938, Primates, Ongules. Arqu. Mus. Bocage, 9, 121-149, 150-196; 1938. (Lists)
BASILIO, A. La vida animal en la Guinea Espanola. Madrid; 1-190; 1962.
CANSDALE, G. S. Provisional check list of Gold Coast mammals. Accra; 1-16; 1948.
ROSEVEAR, D. R. The flesh-eating mammals of Nigeria. Nigerian Field, 4, 64-69; 1935.
ROSEVEAR, D. R. The insect-eaters of Nigeria. Nigerian Field, 10, 98-103; 1941.
ROSEVEAR, D. R. Rodents of Nigeria, I - IV. Nigerian Field, 14 & 15; 1949-50.
ROSEVEAR, D. R. Nigerian mammals. Lagos; 1-87; 1951. (Keys)
ROSEVEAR, D. R. Checklist and atlas of Nigerian mammals. Lagos; 1-131; 1953.
SANDERSON, I. T. The mammals of the north Cameroon forest area. Trans. Zool. Soc. London, 24, 623-725; 1940.
AELLEN, V. Contribution a l'etude des chiropteres du Cameroun. Mem. Soc. Neuchatel Sci. Nat., 8, (1), 1-121; 1952.

Africa, Central

MALBRANT, P. Fauna du Centre Africain Francais (Mammiferes et oiseaux). Paris; 1-435; 1936. (Also as: Encycl. Biol., Paris, 1-430)
URBAIN, A. Faune l'equateur africain francais. II. Encycl. Biol., 1-312; 1949.
MALBRANT, R. & MACLATCHY, A. Faune de l'equateur Africain Francais. II. Mammiferes. Encycl. Biol., 36, 1-323; 1949.
(anonymous). Catalogo sistematico de la fauna de las posesiones espanoles del Golfo de Guinea. Mem. Soc. Española Hist. Nat., 1, 545-593; 1910.
EISENTRAUT, M. La faune de chiropteres de Fernando-Po. Mammalia, 28, 529-552; 1964.
ALLEN, J. A. Carnivora collected by the American Museum Congo Expedition. Bull. American Mus. Nat. Hist., 47, 73-281; 1924.
ALLEN, J. A. Primates collected by the American Museum Congo Expedition. Bull. American Mus. Nat. Hist., 47, 283-499; 1925.
SCHWARZ, E. Die Sammlung africanischer Affen im Congo-Museum. Rev. Zool. Bot. Africaines, 16, 105-152; 1928.
SCHOUTEDEN, H. Catalogue des mammiferes du Congo Belge et du Ruanda-Urundi. Rev. Zool. Bot. Africaines, 37, 102-125; 1943.
SCHOUTEDEN, H. De zoogdieren van Belgisch Congo en van Ruanda-Urundi. I, II. Ann. Mus. Congo, (II), 3, 1-168, 169-332; 1944. II. Ungulata (2), Rodentia. Ann., (II), 3, 333-576; 1946.
SCHOUTEDEN, H. Faune du Congo Belge et du Ruanda-Urundi. I. Mammiferes. Ann. Mus. Congo Belge, Sci. Zool., 1, 1-331; 1948. (Translation . Keys)
HAYMAN, R. W. et al. The bats of the Congo and of Rwanda and Burundi. Ann. Rep. Afr. Cent., 154, 1-105; 1967.
FRECHKOP, S. Mammiferes en: Explor. Parc Nat. Albert Mission G. F. de Witte (1933-1935). Fasc. 10, 1-103; 1938. (Revision of bats of Belgian Congo)

Southern Africa

MONARD, A. Contribution a la mammalogie d'Angola. Arqu. Mus. Bocage, 6, 1-314; 1935.
HILL, J. E. & CARTER, T. D. The mammals of Angola, Africa. Bull. American Mus. Nat. Hist., 78, 1-211; 1941. (Keys)
SHORTRIDGE, G. C. The mammals of South West Africa. London; 2 volumes; 1934.
SCLATER, W. L. The mammals of South Africa. Vol. I, Primates, Carnivora and Ungulata. London; 1-324; 1900. Vol. 2, Rodentia, Chiroptera, Insectivora, Cetacea and Edentata. London; 1-242; 1901.
HAAGNER, A. South African mammals. A short manual .... London; 1-248; 1920.
ROBERTS, A. The mammals of South Africa. (South Africa); 1-700; 1951. (Keys)
ELLERMAN, J. R. et al. Southern African mammals. 1758 - 1951: a reclassification. London; 1-363; 1953. (List)
ELLERMAN, J. R. Die taksonomie van die soogdiere van die Unie van Suid-Afrika. Ann. Univ. Stellenbosch, 30A, 1-125; 1954. (Keys)
MEESTER, J. et al. An interim classification of southern Africa mammals. (South Africa); unpaged. 1964. (Keys)
SCLATER, W. L. Descriptive list of the rodents of South Africa. Ann. South African Mus., 1, (2), 1-181; 1899.
BARNARD, K. H. Guide book to South African whales and dolphins. Guide South African Mus., 4, 1-3; 1954.
SMITHERS, R. H. N. A check list and atlas of the mammals of Botswana. Salisbury; 1-169; 1968.
SMITHERS, R. H. N. The mammals of Rhodesia, Zambia, and Malawi. London; 1-159; 1966. (Keys)
SIMPSON, C. D. A key to the identification of carnivore skulls in Rhodesia and Zambia. Arnoldia, 2, (18), 1-32; 1966.
LANCASTER, D. G. A check list of the mammals of Northern Rhodesia. Lusaka; 1-56; 1953.
ANSELL, W. F. H. Mammals of Northern Rhodesia. Lusaka; 1-155; 1960.
SWEENEY, R. C. H. A preliminary annotated checklist of the mammals of Nyasaland. Blantyre; 1-71; 1959.

East Africa & Sudan

HOLLISTER, N. East African mammals in the United States National Museum, 3 parts. United States Nat. Mus. Bull., 99, 1-194, 1-184, 1-164; 1918, 1919, 1924.
HESSE, P. R. The identification of the spoor and dung of East African mammals. Journ. East Africa Uganda Nat. Hist. Soc., 22, 107-110; 1954. (Keys)
HESSE, P. R. Identification of the spoor and dung of East African mammals. Part II. The carnivores, ant-bear and hyrax. African Wild Life, 11, 317-322; 1957.
FOSTER, J. B. & DUFF-MACKAY, A. Keys to the genera of Insectivora, Chiroptera and Rodentia of East Africa. Journ. East Africa Uganda Nat. Hist. Soc., 25, 189-204; 1966.
SWYNNERTON, G. H. & HAYMAN, R. W. A check list of the land mammals of the Tanganyika Territory and the Zanzibar Protectorate. Journ. East Africa Uganda Nat. Hist. Soc., 20, 274-392; 1951.
COPLEY, H. Small mammals of Kenya. Nairobi; 1-96; 1950. (Keys)
HARRISON, D. L. A checklist of the bats ... of Kenya Colony. Journ. East Africa Uganda Nat. Hist. Soc., 23, 286-295; 1961.
BERE, R. M. The wild mammals of Uganda. London; 1-148; 1962.
FUNAIOLI, U. & SIMONETTA, A. M. The mammalian fauna of the Somali Republic .... Monitore Zool. Ital. (Suppl), 74, 285-347; 1966.
van ROSEN, B. Game animals of Ethiopia. Addis Ababa; 1-92; 1953. (Keys)
OWEN, J. S. A field key to the genera of Sudan rodents. Sudan Notes Rec., 34, 104-113; 1953.
McKENNA, P. Z. Catalogue of wild animals of the Sudan .... Artiodactyla and Perissodactyla. Publ. Sudan Mus. (Nat. Hist.), 4, 1-21; 1954.
SETZER, H. W. Mammals of the Anglo-Egyptian Sudan. Proc. United States Nat. Mus., 106, 447-587; 1956.

Madagascar

DORST, J. Essai d'une clef de determination des chauves-souris malgaches. Mem. Inst. Sci. Madagascar, (A), 1, 81-88; 1947.

India

BLANFORD, W. T. The fauna of British India .... Mammalia. London; 2 volumes; 1888-91.
FINN, F. Sterndale's Mammalia of India. Calcutta; 1-347; 1929.
POCOCK, R. I. The fauna of British India .... Mammalia. London; 2 volumes; 1939-41.
ELLERMAN, J. R. A key to the Rodentia inhabiting India, Ceylon and Burma. Journ. Mamm., 28, 357-387; 1947.
ELLERMAN, J. R. The fauna of British India..... Mammalia, 2nd ed. Vol. 3, Rodentia, part 1 & 2. Calcutta; 1-884; 1961.
AKHTAR, S. A. The rodents of West Pakistan. Pakistan Journ. Sci., 10, 269-292; 1958. 10, 5-18, 79-90; 1959. 12, 17-37; 1960.
SIDDIQI, M. S. Checklist of mammals of Pakistan.... Biologia, Lahore, 7, 93-225; 1961.
BROSSET, A. The bats of central and western India, part II, III, IV. Journ. Bombay Nat. Hist. Soc., 59, 582-624; 707-746; 60, 337-355; 1962-63.

PHILLIPS, W. W. A. A guide to the mammals of Ceylon, 8 parts. Spol. Zeylanica, 13-14; 1924-28.
PHILLIPS, W. W. A. Manual of mammals of Ceylon. Ceylon; 1-373; 1935. (Keys)
HILL, W. C. O. A revised check-list of the mammals of Ceylon. Spol. Zeylanica, 21, 139-184; 1939.

Malay Peninsula

GYLDENSTOLPE, N. A list of the mammals at present known to inhabit Siam. Journ. Nat. Hist. Soc. Siam, 3, 127-175; 1919.
SUVATTI, C. Fauna of Thailand. Bangkok; 1-1100; 1950. (Mammals pp. 522-602)
DELACOUR, J. Liste provisoire des mammiferes de l'Indochine francaise (suite). Mammalia, 4, 46-58; 1940.
VAN PEENEN, P. F. D. Preliminary identification manual for mammals of South Vietnam. Washington, D. C.: 1-310; 1969.
KLOSS, C. B. A list of the bats occurring in the Peninsular Region with a key to the genera. Journ. Fed. Malay States Mus., 2, 151-161; 1908.
CHASEN, F. N. Handlist of malaysian mammals .... Bull. Raffles Mus., 15, 1-209; 1940.
GIBSON-HILL, C. A. The whales, porpoises and dolphins known in Malayan waters. Malayan Nat. Journ., 4, 44-61; 1949.
ELLERMAN, J. R. & MORRISON-SCOTT, T. C. S. Supplement to Chasen (1940), A Handlist of Malaysian Mammals, containing a generic synonymy and a complete index. London; 1-66; 1955.
MEDWAY, — Identification of Malaysian cave bats. Malay Nat. Journ., 19, 88-107; 1965.
HARRISON, J. An introduction to mammals of Singapore and Malaya. Singapore; 1-340; 1966. (Keys)

Indo-Australian Archipelago

TATE, G. H. H. Geographical distribution of the bats in the Australasian Archipelago. American Mus. Novit., 1323, 1-21; 1946.
SODY, H. J. V. Naamlijst van de Vleermuizen van Java. Natuurk. Tijdschr., 89, 28-66; 1929.
DAMMERMAN, K. W. The mammals of Java. I. Rodentia. Treubia, 13, 429-470; 1931.
DAVIS, D. Mammals of the lowland rain-forest of North Borneo. Bull. Nat. Mus. Singapore, 31, 1-129; 1962.
HARRISON, J. An introduction to the mammals of Sabah. Jesselton, Malaysia; 1-244; 1964.
SHAMEL, H. H. The insectivorous bats collected by H. C. Raven in Celebes. Journ. Mamm., 21, 352-354; 1940. (List)
LAURIE, E. M. O. & HILL, J. E. List of land mammals of New Guinea, Celebes and adjacent islands. 1758-1952. London; 1-175; 1954.
MATSCHIE, P. Die Verbreitung der Beuteltiere auf Neuguinea mit einiger Bemerkungen uber ihre Einteilung in Untergattungen. Mitt. Zool. Mus. Berlin, 8, 259-308; 1916.
HUSSON, A. M. Tabel voor het determineren van de landzoogdieren van Nederlands Nieuw-Guinea. Zool. Bijdr., 1, 1-35; 1955. (Keys)

Australia

TROUGHTON, E. L. The bats of Australia and New Guinea. In The Wild Animals of Australasia, by Le Souef & Burwell. London; 1-388. (Pp. 21-88)
OGILBY, J. D. Catalogue of Australian mammals .... Australian Mus. Cat., 16, 1-142; 1892.
LONGMAN, H. A. Notes on classification of common rodents with list of Australian species. Commonw. Australia Quart. Serv. Publ., 8, 1-27; 1916. (Keys & list)
IREDALE, T. & TROUGHTON, E. L. Check-list of the mammals recorded from Australia. Pinnipedia. Mem. Australian Mus., 6, 1-122; 1934.
TATE, G. H. H. Results of the Archbold Expeditions. No. 65. The rodents of Australia and New Guinea. Bull. American Mus. Nat. Hist., 97, 185-430; 1951.
MACKERRAS, M. J. Catalogue of Australian mammals and their recorded parasites. I - IV. Proc. Linn. Soc. New South Wales, 83, 101-160; 1958.
MARLOW, B. J. Marsupials of Australia. Brisbane; 1-141; 1962. (Field guide)
RIDE, W. D. L. A guide to the native mammals of Australia. Melbourne; 1-263; 1970.
LONGMAN, H. A. The marsupials of Queensland. Mem. Queensland Mus., 10, 55-64; 1930. (List)
FINLAYSON, H. H. On eastern Australian mammals. Part IV. Rec. Australian Mus., 14, 141-191; 1961.
MARLOW, B. J. A survey of the marsupials of New South Wales. CSIRO Wildl. Res., 3, 71-114; 1958.
JONES, F. W. The mammals of South Australia.... Adelaide; 1-458; 1923-25.
LORD, C. Existing Tasmanian marsupials. Pap. Proc. Roy. Soc. Tasmania, 1927, 17-24; 1928. (List)

New Zealand

OLIVER, W. B. R. A review of the Cetacea of the New Zealand seas. I. Proc. Zool. Soc. London, 1922, 557-583; 1922.
TATE, G. H. H. Results of the Archbold Expeditions. No. 65. The rodents of Australia and New Guinea. Bull. American Mus. Nat. Hist., 97, 185-430; 1951.
McCANN, C. Key to the seals (Pinnipedia) of New Zealand. Tuatara, 12, 40-49; 1964.

The Orient

HOLLISTER, N. A list of the mammals of the Philippine Islands, exclusive of the Cetacea. Philippine Journ. Sci., 7, 1-64; 1912.
de ELERA, M. L. P. C. Contribucion a la fauna Filipina. Manila; 1-284; 1915.
TAYLOR, E. H. Philippine land mammals. Philippine Bur. Sci. Monogr., 30, 1-548; 1934.
AOKI, B. A hand-list of Japanese and Formosan mammals. Annot. Zool. Japan, 8, 261-353; 1913.
TANAKA, R. On the insectivorous fauna in Taiwan, Japan. Trans. Nat. Hist. Soc. Formosa, 26, 310-313; 1936. (In Japanese)
KURODA, N. Mammals of the island of Hainan. Bull. Biogeogr. Soc. Tokyo, 10, 139-161; 1940.
SOWERBY, A. D. The insectivores off China and neighboring regions. Pars I and II. China Journ., 33, 116-125, 156-166; 1940.
REEVES, C. D. Manual of the vertebrate animals of southeastern and central China exclusive of birds. Shanghai; 1-806; 1933.
ALLEN, G. M. The mammals of China and Mongolia, 2 parts. Nat. Hist. of Central China, vol. 10, 1-1350; 1938-40. (New York)
HSIA, W.-P. et al. Illustrated description of animals in China. Mammals. (China); 1-104; 1964. (Keys)
BANNIKOV, A. G. List of the mammals of the Mongolian People's Republic. Rep. Mongol. Comm. 51, Acad. Sci. USSR, 1953, 1-111; 1953. (In Russian)
BANNIKOV, A. G. Mammals of the Mongolian People's Republic. Rep. Mongol. Comm. 53, Acad. Sci. USSR, 1954, 1-669. (In Russian)
MORI, T. Mammalia of Jehol and the district north of it. Rep. 1st Sci. Exped. Manchoukuo, V, II, 4, 1-84. (In Japanese)
TATE, G. H. H. Mammals of eastern Asia. New York; 1-366; 1947.
WON, P. O. & WOO, H. C. A distributional list of the Korean birds and mammals. (Korea); 1-10, 1-96, 1-12; 1958.
JONES, J. K. & JOHNSON, D. H. Review of the insectivores of Korea. Publ. Univ. Kansas Mus. Nat. Hist., 9, 549-578; 1960.
JONES, J. K. & JOHNSON, D. H. Synopsis of the lagomorphs and rodents of Korea. Publ. Univ. Kansas Mus. Nat. Hist., 16, 357-407; 1965.
MOSKOVSKOGO, U. Z. Pinnipedia of the Okhotsk Sea .... Gosud. Pedag. Inst., 24, 19-74; 1941. (In Russian)
AOKI, B. A hand-list of Japanese and Formosan mammals. Annot. Zool. Japan, 8, 261-353; 1913.
KISHIDA, H. Monograph of Japanese mammals. Tokyo; 1924. (In Japanese)
TAKASHIMA, H. A handlist of Japanese marine mammals. Trans. Nat. Hist. Soc. Formosa, 23, 249-258; 1933.
OGAWA, T. List of the Odontoceti in Japan. Ann. Rep. Saito Ho-on Kai, Sendai, 12, 63-65; 1937.
KURODA, N. A list of the Japanese mammals. Tokyo; 1-122; 1938. (In Japanese & English)
KURODA, N. A monograph of the Japanese mammals.... Tokyo & Osaka; 1-311; 1940. (In Japanese)
(anonymous). Illustrated encyclopedia of the fauna of Japan ..., 2nd ed. Tokyo; 2 volumes; 1949.
IMAIZUMI, Y. Coloured illustrations of the mammals of Japan. Osaka; 1-196; 1960.
IMAIZUMI, Y. The handbook of Japanese land mammals. 1970. (Not seen)

Oceania

CARTER, T. D. et al. Mammals of the Pacific world. New York; 1-227; 1945. (Keys)
HIRASAKA, K. On the distribution of the Sirenians in the Pacific. Proc. Fifth Pan-Pacific Sci. Congr. Canada, 5, 4221-4222; 1934.
TOMICH, P. Q. Mammals in Hawaii .... Honolulu; 1-238; 1969. (List)
TATE, G. H. H. A list of the mammals of the Japanese War Area. New York; 4 parts; 1944. (Includes: New Guinea, Greater Sunda Islands, Lesser Sunda, Moluccas, Celebes, Borneo, China Sea)
NICHOLSON, A. J. & WARNER, D. W. The rodents of New Caledonia. Journ. Mamm., 34, 168-179; 1953.
BROSSET, A. Mammiferes des iles Galapagos. .... Mammalia, 27, 323-338; 1963.
HONEGGER, R. E. Beobachtungen aus eingefuhrten Saugetieren auf Galapagos. Natur. Mus. Frankfurt, 96, 20-27; 1966.

Antarctic Seas

SAPIN-JALOUSTRE, J. L'identification des Cetaces antarctiques a la mer. Mammalia, 17, 221-259; 1953.

SPECIAL SUBJECT LISTS

Anatomy. BRAUER, K. & SCHOBER, W. Katalog der Saugetiergehirne. Jena; 1-164; 1970. (Brains)
Extinct. ALLEN, G. M. Extinct and vanishing mammals of the Western Hemisphere with the marine species of all the oceans. Lankaster, Pennsylvania; 1-620; 1942.
HARPER, F. Extinct ... animals of the Old World. New York; 1-850; 1945.
Marine. SCHEFFER, V. B. & RICE, D. W. A list of the marine mammals of the world. United States Fish Wildl. Serv. Sci. Rep. - Fisheries, 431, 1-12; 1963.
Subfossil. LAMBERTON, C. Contribution a la connaissance de la fauna subfossile de Madagascar. Lemuriens et Cryptoproctes. Mem. Acad. Malgache, 27, 1-203; 1939.

MONOGRAPHIC, DESCRIPTIVE, REVISIONARY

Monographs

DOBSON, G. E. Monograph of the Asiatic Chiroptera and catalogue of the species of bats in the collection of the Indian Museum, Calcutta. London; 1-228; 1876.
DOBSON, G. C. Catalogue of the Chiroptera in the collection of the British Museum. London;1-567; 1878. (Reprinted 1966)
DOBSON, G. E. A monograph of the Insectivora .... London, 1-172; 1882-83.
SCHEFFER, V. B. Seals ..., a review of the Pinnipedia. Stanford, California; 1-179; 1958.
FORBES, H. O. A hand-book to the Primates. London; 2 volumes; 1896-97.
HILL, W. C. O. Primates. A monograph. Edinburgh & New York; 7 volumes & continuation; 1953
OSBORN, H. F. Proboscidea. A monograph of the discovery, evolution, migration and extinction of the mastodonts and elephants of the world. New York; 2 volumes; 1936.

Classifications

GILL, T. Arangement of the families of mammals .... Smithsonian Misc. Coll., 11, 1-98; 1872.
FLOWER, W. H. On the arrangement of the orders and families of existing Mammalia. Proc. Zool. Soc. London, 1883, 178-186; 1883.
PALMER, T. S. Index generum mammalium: A list of the genera and families of mammals. North American Fauna, 23, 1-984; 1904. (Reprinted 1967)
BEDDARD, F. E. Mammalia. In Cambridge Natural History, vol. X, 1-605; 1909.
GREGORY, W. The orders of mammals. Bull. American Mus. Nat. Hist., 27, 1-524; 1910.
SHUFELDT, R. W. Arrangement of the families and the higher groups of the Mammalia. Nyt Mag. Naturvidensk., 49, 65-80; 1911.
WINGE, H. Pattedyr-Slaegter. Kjøbenhavn; 3 volumes; 1923.
SIMPSON, G. G. New classification of mammals. Bull. American Mus. Nat. Hist., 59, 259-293; 1931.
PERRIER, E. Les mammiferes. Traite de Zoologie, 10, 3344-3610; 1932.
SLIJPER, E. J. Die Cetaceen .... Den Haag; 1-590; 1936. (To family)
WINGE, H. The interrelationships of the mammalian genera. Kjøbenhavn; 2 volumes; 1941.
LAMEERE, A. Precis de zoologie. Tome VII. Chpitre XXIII. Les mammiferes. Rec. Inst. Zool. Torley-Rousseau, 9, 188-203.
ROMER, A. S. Vertebrate paleontology. Chicago; 1-687; 1945.
SIMPSON, G. G. The principles of classification and a classification of mammals. Bull. American Mus. Nat. Hist., 85, 1-350; 1945. (Pp. 39-162.)
BLAIR, W. F. et al. Vertebrates of the United States. New York; 1-819; 1957. (Pp. 615-774)
BURTON, M. Systematic dictionary of mammals of the world. London; 1-307; 1962. (To family)
BUETTNER-JANUSCH, J. Evolutionary and genetic biology of Primates. New York; 2 volumes; 1963.
JOLLY, C. J. Introduction to the Cercopithecoidea .... Symp. Zool. Soc. London, 17, 427-457; 1966.
ANDERSON, S. & JONES, J. K., Jr. Recent mammals of the world. A synopsis of families. New York; 1-453; 1967.
BLAIR, W. F. et al. Vertebrates of the United States, 2nd ed. New York; 1-616; 1968. (Pp. 453-562)

Descriptive catalogs

THOMAS, O. Catalogue of the Marsupialia and Monotremata in the collection of the British Museum (Natural History). London; 1-401; 1888.
ANDERSEN, K. Catalogue of the Chiroptera in the collection of the British Museum, 2nd ed. Volume I: Megachiroptera. London; 1-854; 1912. (Prev. ed., 1878, by G. E. Dobson.)
LYDEKKER, R. Catalogue of the ungulate mammals in the British Museum (Natural History); London; 5 volumes; 1913-1916. (Reprinted 1966)
MILLER, G. S., Jr. Catalogue of the mammals of western Europe ... in the collection of the British Museum. London; 1-1019; 1912. (Reprinted 1966)
CABRERA, A. Catalogo descriptiva de los mammiferos de la Guinea Espanola. Mem. R. Soc. Española Hist. Nat., Mem. i, vol. 16, 5-121; 1929.
BLYTH, E. Catalogue of mammals and birds of Burma. Journ. Asiatic Soc. Bengal, 44, (2), extra number, 1-167; 1875.

Synonymic catalogs

JENTINK, F. A. Catalogue systematique des Mammiferes. (Rongeurs, Insectivores, Cheiropteres, Edentes et Marsupiaux). Mus. Hist. Nat. Pay-Bas, vol. 12; 1888.
SCHULZE, E. Catalogus mammalium europaeorum. Zeitschr. Naturwiss., 73, 187-224; 1900.
TROUESSART, E. L. Catalogus mammalium, 2nd ed. Berlin; 1-929; 1904-05.
LYDEKKER, R. Catalogue of the ungulate mammals in the British Museum (Natural History). London; 5 volumes; 1913-16.
ANDERSON, R. M. Catalogue of Canadian Recent mammals. Nat. Mus. Canada Bull., 102, 1-238; 1947.
ELLIOT, D. G. A check list of the mammals of the North American continent .... Field Columbian Mus., Zool. Series, 6, 1-761; 1905. (Supplement, New York 1-192; 1917)
MILLER, G. S., Jr. List of North American Recent mammals. 1923. United States Nat. Mus. Bull., 128, 1-673; 1924.

MILLER, G. S., Jr. & KELLOGG, R. List of North American Recent mammals. United States Nat. Mus. Bull., 205, 1-954; 1955.

CABRERA, A. Catalogo de los mamiferos de America del Sur. I, II. Rev. Mus. Argentino Cienc. Nat. "Bernardino Rivadavia", Zool., 4, 1-732; 1957-1961.

ELLERMAN, J. R. & MORRISON-SCOTT, T. C. S. Checklist of Palaearctic and Indian mammals 1758 to 1946, 2nd ed. London; 1-810; 1966. (1st edition, 1951)

MILLER, G. S., Jr. Catalogue of the mammals of western Europe (Europe exclusive of Russia) in the collection of the British Museum. London; 1-1019; 1912.

ELLERMAN, J. R. et al. Southern African mammals 1758-1951: a reclassification. London; 1-363; 1953.

ALLEN, G. M. A checklist of African mammals, 2nd ed. Boston; 1-763; 1954. (1st edition as Bull. Mus. Comp. Zool., 83, 1-763; 1939)

LAURIE, E. M. O. & HILL, J. E. List of land mammals of New Guinea, Celebes, and adjacent islands. 1758-1954. London; 1-175; 1954.

Revisionary works by groups

[These comprehensive works cover groups of families of various size. They are arranged here in alphabetical order. They include no revisions of the species or genera of families but studies of the families themselves in relation to other families.]

Aeluroidea. KRETZOI, N. Materialien zur phylogenetischen Klassifikation der Aeluroideen. Tenth Int. Congr. Zool. Budapest, pt. 2, 1293-1355; 1929.

Artiodactyla. MATTHEW, W. D. Reclassification of the artiodactyle families. Bull. Geol. Soc. America, 40, 40, 403-408; 1929.

DUTILLY, A. A bibliography of reindeer, caribou, and musk-ox. United States Dep. Army, Off. Quarterm. Gen., Envir. Prc. Sec., Rep. 129, 1-462; 1949.

HALTENORTH, T. Die Klassifikation der Saugetiere: Artiodactyla. Handb. der Zoologie, 8, 1-167; 1963.

Carnivora. LYDEKKER, R. A hand-book to the Carnivora, Part I. London; 1-312; 1896.

LEONE, C. A. & WIENS, A. L. Comparative serology of carnivores. Journ. Mamm., 37, 11-23; 1956.

Cercopithecoidea. NAPIER, J. R. & NAPIER, P. H. Old World monkeys . . . . New York; 1-660; 1970.

Cetacea. TRUE, F. W. On the classification of the Cetacea. Proc. American Philos. Soc., 47, 385-391; 1908.

WINGE, H. A review of the interrelationships of the Cetacea. Smithsonian Misc. Coll., 72, (8), 1-97; 1921.

KUKENTHAL, W. Zur Stammesgeschichte der Wale. Sitz. Ber. Akad. Wiss. Berlin, 1922, 72-87; 1922.

KELLOGG, A. R. The history of whales — their adaptation to life in the water. Quart. Rev. Biol., 3, 29-76, 174-208; 1928.

JENKINS, J. T. Bibliography of whaling. Journ. Soc. Biblio. Nat. Hist., 2, 71-166; 1948.

BOYDEN, A. & GEMEROY, D. The relative position of the Cetacea among the orders of mammals as indicated by precipitin tests. Zoologica, 35, 145-151; 1950.

SLIJPER, E. J. Whales. New York; 1-475; 1962.

NORRIS, K. S. Whales, dolphins, and porpoises. Los Angeles; 1-789; 1966.

HERSHKOVITZ, P. Catalogue of living whales. United States Nat. Mus. Bull., 246, 1-194; 1966.

Chiroptera. ALLEN, J. A. The families and genera of bats. American Nat., 41, 671-672; 1907.

MILLER, G. S., Jr. The families and genera of bats. United States Nat. Mus. Bull., 57, 1-282; 1907.

ALLEN, G. M. Bats. Cambridge, Massachusetts; 1-368; 1939.

Desmostylia. REINHART, R. H. A review of the Sirenia and Desmostylia. Univ. California Pubs. Geol. Sci., 36, 1-145; 1959.

Edentata. SIMPSON, G. G. Metachiromys and the Edentata. Bull. American Mus. Nat. Hist., 59, 295-381; 1931.

Glires. ALSTON, E. R. On the classification of the order Glires. Proc. Zool. Soc. London, 1876, 61-98; 1876.

Hominoidea. LEAKEY, L. S. B. East African fossil Hominoidea and the classification within this super-family. Publ. Anthr. Viking Fund, 37, 32-49; 1964.

Hyracoidea. HALL, R. T. Annotated catalogue of the Hyracoidea in the American Museum of Natural History.... American Mus. Novit., 594, 1-13; 1933.

Insectivora. GILL, T. N. Synopsis of insectivorous mammals. Bull. United States Geol. Geogr. Surv. Terr., 2, 91-120; 1875.

GILL, T. N. On the classification of insectivorous mammals. Bull. Philos. Soc. Washington, 5, 118-120; 1882.

SABAN, R. Phylogenie des Insectivores. Bull. Mus. Nat. Hist. Nat., 26, (2), 419-432; 1954.

BUTLER, P. M. The skull of Ictops and the classification of the Insectivora. Proc. Zool. Soc. London, 126, 453-481; 1956.

McDOWELL, S. B., Jr. The Greater Antillean insectivores. Bull. American Mus. Nat. Hist., 115, 115-214; 1958. (To family)

Lagomorpha. GIDLEY, J. W. The lagomorphs an independent order. Science, (n. s.), 36, 285-286; 1912.
STOHL, G. Uber die Stellung der Lagomorpha im System der Saugetiere. Ann. Inst. Biol. Tihany Hungaricae Acad. Sci., 24, 51-57; 1957.

Lemuroidea. GREGORY, W. K. On the classification and phylogeny of the Lemuroidea. Bull. Geol. Soc. America, 26, 426-446; 1915.

Marsupialia. LYDEKKER, R. A handbook of the Marsupialia and Monotremata. London; 1-302; 1894.
ABBIE, A. A. Some observations on the major subdivisions of the Marsupialia with special reference to the position of the Peramelida and Coenolestidae. Journ. Anat., 71, 429-435; 1937.
GILMORE, R. M. Classification of South American marsupial genera. American Journ. Trop. Med., 21, 314-319; 1941.
HALTENORTH, T. Klassification der Saugetiere, I (Monotremata, Marsupialia). Handb. der Zoologie, 8, (16), 1-40; 1958.
RIDE, W. D. L. A review of Australian fossil marsupials. Journ. Roy. Soc. Western Australia, 47, 97-131; 1964. (Classification)

Monotremata. LYDEKKER, R. A handbook of the Marsupialia and Monotremata. London; 1-302; 1894.
GREGORY, W. K. The monotremes and the palimpsest theory. Bull. American Mus. Nat. Hist., 88, 1-52; 1947.
HALTENORTH, T. Klassification der Saugetiere, I (Monotremata, Marsupialia). Handb. der Zoologie, 8, (16), 1-40; 1958.

Pinnipedia. MAXWELL, G. Seals of the world .... London; 1-153; 1967. (World list)

Primates. ELLIOT, D. G. A review of the Primates. New York; 3 volumes; 1912. (Also as 1913)
SMITH, G. E. The subdivisions of the order Primates. Nature, 124, 876-877; 1929.
ANTHONY, R. & COUPIN, F. Tableaux resume d'une classification generique des Primates fossiles et actuels. Bull. Mus. Nat. Hist. Nat., (2), 3, 566-569; 1931.
ARAMBOURG, C. La classification des Primates et particulierement des Hominiens. Mammalia, 12, 123-135; 1948.
HOFER, H. et al. Primatologia. Handbook of Primatology. Basel; 1-1063; 1956.
WARWICK, J. W. The jaws and teeth of Primates. London; 1-328; 1960.
BUETTNER, J. Evolutionary and genetic biology of Primates. New York; 2 volumes; 1963.

Proboscidea. WATSON, D. M. S. The evolution of the Proboscidea. Biol. Rev., 21, 15-29; 1946.

Prototheria. GRIFFITHS, M. Echidnas. London; 1-282; 1968.

Rhinocerotoidea. RADINSKY, L. B. The families of the Rhinocerotoidea.... Journ. Mammalogy, 47, 631-639; 1966.

Rodentia. PALMER, T. S. A list of the generic and family names of rodents. Proc. Biol. Soc. Washington, 11, 241-270; 1897.
THOMAS, O. On the genera of rodents .... Proc. Zool. Soc. London, 1896, 1012-1028; 1897.
TULLBERG, T. Uber das System der Nagethiere: eine phylogenetische Studie. Upsala, 1-514; 1899.
ALLEN, J. A. Generic and family names of rodents. American Nat., 33, 70-72; 1899.
MILLER, G. S., Jr. & GIDLEY, J. W. Synopsis of the supergeneric groups of rodents. Journ. Washington Acad. Sci., 8, 431-448; 1918.
ST. LEGER, J. A key to the families and genera of African rodents. Proc. Zool. Soc. London, 1931, 957-997; 1931.
ELLERMAN, J. R. et al. The families and genera of living rodents. London; 3 volumes; 1940-49. (Reprinted 1966)
WOOD, A. E. Comments on the classification of rodents. Breviora, 41, 1-9; 1954.
WOOD, A. E. A revised classification of the rodents. Journ. Mamm., 36, 165-187; 1955.
LANDRY, S. O. The interrelationships of the New and Old World histricomorph rodents. Publ. Univ. California Zool., 56, 1-118; 1957.

Sirenia. FREUND, L. Bibliography of the mammalian order Sirenia. Vetn. Ceskol. Zool. Spol., 14, 161-181; 1950.
REINHART, R. H. A review of the Sirenia and Desmostylia. Univ. California Publ. Geol. Sci., 36, 1-145; 1959.

Tubulidentata. FRICK, H. Zur Taxonomie der Tubulidentata. Saugetierk. Mitt., 4, 15-17; 1956.

Ungulata. FRECHKOP, S. Sur la classification des Ongules. Mammalia, 1, 39-48; 1936.
FRECHKOP, S. La classification des ongules actuels. Compt. Rend. Congr. Intern. Zool. Paris, 1948, 379-383; 1949.
LAVOCAT, R. Classification des Ongules d'apres leur origine et leur evolution. Mammalia, 22, 28-40; 1958.

Revisionary works by family

These lists of works were made up primarily from two sources: (1) listings in the Zoological Record for the years 1910 to 1968, which were not checked further if the title appeared to be explicit; and (2) library shelves and the office libraries of various specialists in each field; these were examined for appropriateness and, if revisionary, checked for included families. The works listed by families are therefore of two sorts, so far as examination by the compiler is concerned, those cited in the title as referring to one or a few named families (not usually examined or verified) and those directly examined and checked and listed as considered to be appropriate.

Works cited by author

In order to save space in the list of works on individual families, the following studies, which are each referred to more than a few times, are cited by author and date and number only. Citations are otherwise given by title, if a book or by the serial reference, if not a separate work. The titles of the journal articles are not usually given, but the relevant limitations are shown by annotations.

ALLEN, G. M., No. 1, 1938. The mammals of China and Mongolia, Part 1 of 2 parts. Nat. Hist. Central Asia, XI, (2), 1-620.

ALLEN, G. M., No. 2, 1940. Same; part 2; 621-1350.

ALLEN, G. M., No. 3, 1939. A checklist of African mammals. Bull. Mus. Comp. Zool., 83, 1-763. (Second printing, Boston; 1-763; 1954)

ALLEN, J. A., No. 1, 1880. History of North American pinnipeds. A monograph of the walruses, sea-lions, sea-bears and seals of North America. United States Geol. Geogr. Surv. Terr. Misc. Publ., 12, 1-785.

ANDERSON, S. & JONES, J. K., Jr. Recent mammals of the world: a synopsis of families. New York; 1-453.

ANTHONY, H. E., No. 1, 1918. The indigenous land mammals of Porto Rico, living and extinct. Mem. American Mus. Nat. Hist., (n. s.), 2, (2), 333-435.

ANTHONY, H. E., No. 2, 1925-26. Mammals of Porto Rico, living and extinct, 2 parts. Sci. Surv. Porto Rico Virgin Is., 1-238.

BLANFORD, W. T., No. 1, 1888-91. Mammalia, The fauna of British India; London; 1-617.

BOBRINSKOY, N. et al., No. 1, 1944. Mammals of U. S. S. R. Moskva; 1-440. (In Russian)

BOBRINSKOY, N. et al., No. 2, 1944. Same; English translation.

BUETTNER-JANUSCH, J., No. 1, 1963-64. Evolutionary and genetic biology of Primates. New York; 2 volumes.

CABRERA, A., No. 1, 1919. Genera mammalium. Monotremata Marsupialia. Madrid; 1-177.

CABRERA, A., No. 2, 1925. Genera mammalium. Insectivora Galeopithecia. Madrid; 1-232.

CABRERA, A., No. 3, 1914. Fauna Iberica. Mamiferos. Madrid; 1-441.

da CUNHA VIEIRA, C. O., No. 1, 1942. Ensaio monografico sobre os quiropteros do Brasil. Arqu. Zool. Est. Sao Paulo, 3, 219-471. (Also as: Rev. Mus. Paulista, 26, 219-471)

DOBSON, G. E., No. 1, 1876. Monograph of the Asiatic Chiroptera and catalogue of the species of bats in the collection of the Indian Museum Calcutta. London; 1-228.

DOBSON, G. E., No. 2, 1878. Catalogue of the Chiroptera in the collection of the British Museum. London; 1-567.

DOBSON, G. E., No. 3, 1882-83. A monograph of the Insectivora, systematic and anatomical. London; 3 parts; 1-172.

ELLERMAN, J. R., No. 1, 1961. The fauna of British India.... Mammalia, 2nd ed., vol. 3, Rodentia, part 1. Calcutta; 1-482; part 2, 483-884.

ELLERMAN, J. R. et al., No. 1, 1940-49. The families and genera of living rodents. London; 3 volumes; reprinted 1966.

ELLERMAN, J. R. et al., No. 2, 1953. Southern African mammals. 1758-1951: a reclassification. London; 1-363.

ELLIOT, D. G., No. 1, 1904. The land and sea mammals of Middle America and the West Indies. Field Columbian Mus., Zool., IV, 1-850.

ELLIOT, D. G., No. 2, 1912. A review of the Primates. New York; 3 volumes.

ELLIOT, D. G., No. 3, 1913. A review of the Primates. New York; 3 volumes.

GOLDMAN, E. A., No. 1, 1920. Mammals of Panama. Smithsonian Misc. Coll., 69, (5), 1-309.

GOODWIN, G. G., No. 1, 1942. Mammals of Honduras. Bull. American Mus. Nat. Hist., 79, 107-195.

GOODWIN, G. G., No. 2, 1946. Mammals of Costa Rica. Bull. American Mus. Nat. Hist., 87, 271-473.

GOODWIN, G. G. & GREENHALL, A. M., No. 1, 1961. A review of the bats of Trinidad and Tobago. Descriptions, rabies infection, and ecology. Bull. American Mus. Nat. Hist., 122, 189-301.

HALL, E. R. & KELSON, K. R. The mammals of North America. New York; 2 volumes. (Includes Central America)

HARRISON, D. L., No. 1, 1964. The mammals of Arabia. Vol. 1. Introduction. Insectivora — Chiroptera — Primates. London; 1-192.

HARRISON, D. L., No. 2, 1968. Same; volume 2.

HERSHKOVITZ, P., No. 1, 1966. Catalog of living whales. United States Nat. Mus. Bull., 246, 1-259.

HILL, J. E. & CARTER, T. D., No. 1, 1941. The mammals of Angola, Africa. Bull. American Mus. Nat. Hist., 78, (1), 1-211.

HILL, W. C. O., No. 1, 1953. Primates. Comparative anatomy and taxonomy. 1 — Strepsirhini. A monograph. Edinburgh; 1-798.

HOLLISTER, N., No. 1, 1918. East African mammals in the United States National Museum. United States Nat. Mus. Bull., 99, 1-194.
HOLLISTER, N., No. 2, 1919. Same; part 2; 1-184.
HOLLISTER, N., No. 3, 1924. Same; part 3; 1-164.
HUSSON, A. M., No. 1, 1962. The bats of Suriname. Leiden; 1-282.
JONES, F. W., No. 1, 1923-25. The mammals of South Australia. Adelaide; 3 parts; 1-458.
MILLER, G. S., No. 1, 1912. Catalogue of the mammals of western Europe (Europe exclusive of Russia) in the collection of the British Museum. London; 1-1019. (Reprinted 1966)
NOVIKOV, G. A., No. 1, 1956. Carnivorous mammals of the fauna of the USSR. Moskva; 1-293.(In Russian)
NOVIKOV, G. A., No. 2, 1962. Carnivorous mammals of the fauna of the USSR. In Keys to the fauna of the USSR..., 62. Jerusalem; 1-284. (Translation of 1956)
OGNEV, S. I., No. 1, 1928. The mammals of the eastern Europe and of the northern Asia, vol. 1. Insectivora and Chiroptera. Moskva; 1-631. (In Russian)
OGNEV, S. I., No. 2, 1931. Same; volume 2. Moskva; 1-776. (In Russian)
OGNEV, S. I., No. 3, 1935. Same; volume 3. Moskva; 1-752. (In Russian) (Fissipedia & Pinnipedia)
OGNEV, S. I., No. 4, 1940. Same; volume 4. Moskva; 1-615. (In Russian) (Rodents)
OGNEV, S. I., No. 5, 1947. Same; volume 5, Rodentia continued. Moskva; 1-809. (In Russian)
OGNEV, S. I., No. 6, 1948. Same; volume 6, Rodentia continued. Moskva; 1-559. (In Russian)
OGNEV, S. I., No. 7, 1950. Same; volume 7, Rodentia continued. Moskva; 1-706. (In Russian)
OGNEV, S. I., No. 8, 1962. Same; volume 1. Jerusalem; 1-487. (Translation of 1928)
OGNEV, S. I., No. 9, 1962. Same; volume 2. Jerusalem; 1-590. (Translation of 1931)
OGNEV, S. I., No. 10, 1962. Same; volume 3. Jerusalem; 1-641. (Translation of 1935.
OGNEV, S. I., No. 11, 1966. Same; volume 4. Jerusalem; 1-429. (Translation of 1940)
OGNEV, S. I., No. 12, 1963. Same; volume 5. Jerusalem; 1-662. (Translation of 1963)
OGNEV, S. I., No. 13, 1963. Same; volume 6. Jerusalem; 1-508. (Translation of 1948)
OGNEV, S. I., No. 14, 1964. Same; volume 7. Jerusalem; 1-626. (Translation of 1950)
PHILLIPS, W. W. A., No. 1, 1935. Manual of mammals of Ceylon. Ceylon; 1-373.
POCOCK, R. I., No. 1, 1939. The fauna of British India ..., Mammalia. - Volume I, Primates and Carnivora (in part), families Felidae and Viverridae. London; 1-463.
POCOCK, R. I., No. 2, 1941. The fauna of British India ..., Mammalia. — Volume II, Carnivora cont. London; 1-503.
RANCK, G. L., No. 1, 1968. The rodents of Libya .... United States Nat. Mus. Bull., 275, 1-264.
ROSEVEAR, D. R., No. 1, 1965. The bats of West Africa. London; 1-418.
SETZER, H. W. Mammals of the Anglo-Egyptian Sudan. Proc. United States Nat. Mus., 106, 447-587.
SIMPSON, G. G., No. 1, 1945. The principles of classification and a classification of mammals. Bull. American Mus. Nat. Hist., 85, 1-350.
SOKOLOV, V. E., No. 1, 1959. Fauna of the SSSR. Perissodactyla and Artiodactyla. Moskva; 1-639. (In Russian)
TATE, G. H. H., No. 1, 1939. The mammals of the Guiana region. Bull. American Mus. Nat. Hist., 76, 151-229.
TAYLOR, E. H., No. 1, 1934. Philippine land mammals. Philippine Bur. Sci. Monogr., 30, 1-548.
TOMILIN, A. G., No. 2, 1957. Mammals of the USSR. IX. Cetacea. Moskva; 1-756. (In Russian)
TOMILIN, A. G., No. 1, 1967. Mammals of the USSR ..., volume IX, Cetacea. Jerusalem; 1-717. (Translation of 1957.)
TOSCHI, A., No. 2, 1965. Fauna d'Italia, 7, Mammalia. Bologna; 1-647.
TOSCHI, A. & LANZA, B., No. 1, 1959. Fauna d'Italia. Mammalia. Generalite — Insectivora — Chiroptera. Bologna; 1-485.
TROUGHTON, E. L., No. 1, 1926. The bats of Australia and New Guinea. In The Wild Animals of Australasia, by LeSouef & Burwell. London; 1-388. (Pp. 21-88)
VILLA-R., B., No. 1, 1966. Los murcielagos de Mexico. Mexico; 1-491.

Families alphabetically

Abderitesidae, as Caenolestidae. (Only as fossils)
Abrocomidae. SIMPSON, G. G., No. 1, 1945 (classification); ANDERSON & JONES, No. 1, 1967 (families)
Acanthionidae, as Erinaceidae
Achaenodontidae, as Entelodontidae. (Only as fossils)
Acoelodidae, as Oldfieldthomasiidae. (Only as fossils)
Acotherulidae, as Cebochoeridae. (Only as fossils)
Acrodelphidae. (Only as fossils)
Acrotherulidae, as Cebochoeridae. (Only as fossils)
Acyonidae. (Only as fossils)
Adapidae (incl. Notharctidae). (Only as fossils)
Adapisoricidae, as Leptictidae. (Only as fossils)
Adianthidae, as Macraucheniidae. (Only as fossils)
Adiantidae, as Macraucheniidae. (Only as fossils)
Adiastaltidae. (Only as fossils)
Adjidaumidae, as Eomyidae. (Only as fossils)
Aegosceridae, as Oegosceridae
Aepycerotidae, as Bovidae
Agaphelidae. (Family not identified)
Agorophiidae. (Only as fossils)

Agoutidae, as Dasyproctidae
Agriochaeridae, as Agriochoeridae. (Only as fossils)
Agriochoeridae (incl. Agriochaeridae). (Only as fossils)
Ailuridae, as Procyonidae
Ailuromachairodontidae, as Felidae. (Only as fossils)
Ailuropodidae, as Procyonidae
Albertogaudryidae, as Astrapotheriidae. (Only as fossils)
Alcelaphidae, as Bovidae
Allodesmidae, as Otariidae
Allodontidae. (Only as fossils)
Allomyidae, as Aplodontiidae. (Only as fossils)
Alouattidae, as Cebidae
Ambloctonidae (incl. Amblyctonidae). (Only as fossils)
Amblotheridae, as Dryolestidae. (Only as fossils)
Amblotheriidae, as Dryolestidae. (Only as fossils)
Amblyctonidae, as Ambloctonidae. (Only as fossils)
Amblytheriidae, as Dryolestidae. (Only as fossils)
Ameghinotheriidae. (Only as fossils)
Amphictidae. (Only as fossils)
Amphicyonidae, as Canidae. (Only as fossils)
Amphidontidae. (Only as fossils)
Amphilemuridae. (Only as fossils)
Amphilestidae. (Only as fossils)
Amphimerycidae. (Only as fossils)
Amphiproviverridae. (Only as fossils)
Amphitheriidae. (Only as fossils)
Amynodontidae. (Only as fossils)
Anagalidae. (Only as fossils)
Anaptomorphidae (incl. Microchoeridae, Necrolemuridae, Tetoniidae). (Only as fossils)
Anathitidae. (Only as fossils)
Anchippodontidae, as Tillotheriidae. (Only as fossils)
Anchitheridae, as Equidae. (Only as fossils)
Ancodontidae, as Anthracotheriidae. (Only as fossils)
Ancylotheridae. (Only as fossils)
Anomalomyidae, as Cricetidae. (Only as fossils)
Anomaluridae (incl. Idiuridae). ELLERMAN, J. R. et al., No. 1, 1940-49. (genera); SIMPSON, G. G., No. 1, 1945 (classification); ANDERSON & JONES, No. 1, 1967 (families); RUEMMLER, H., 1934, Sitz. Ber. Gesell. Naturf. Freunde, 1933, 389-391 (Africa); ALLEN, G. M., No. 3, 1954 (catalog, Africa); HILL & CARTER, No. 1, 1941 (Angola); ELLERMAN, J . R. et al., No. 2, 1953 (southern Africa)
Anoplotheridae, as Anoplotheriidae. (Only as fossils)
Anoplotheriidae (incl. Anoplotheridae) & Tapirulidae). (Only as fossils)
Antelopidae, as Bovidae
Anthracotheridae, as Anthracotheriidae. (Only as fossils)
Anthracotheriidae (incl. Ancodontidae, Anthracotheridae, Hyopotamidae, Merycopotamidae). (only as fossils)
Anthropidae, as Pongidae
Anthropoidae, as Pongidae
Anthropomorphidae. (Family not identified)
Antilocapridae (incl. Merycodontidae). LYDEKKER, R., 1914, Catalogue of the Ungulate mammals in the British Museum (Natural History). London; 3 volumes; SIMPSON, G. G., No. 1, 1945 (classification); ANDERSON & JONES, No. 1, 1967 (families); FRICK, C., 1937, Bull. American Mus. Nat. Hist., 69, 1-669 (North America); HALL & KELSON, No. 1, 1959 (North America); ELLIOT, D. G., No. 1, 1904 (Central America & West Indies)
Antilopidae, as Bovidae
Aotidae, as Cebidae
Apatemyidae. (Only as fossils)
Apheliscidae. (Only as fossils)
Aplodontidae, as Aplodontiidae
Aplodontiidae (incl. Allomyidae, Aplodontidae, Haplodontidae-Alston, Haploodontidae). TAYLOR, W. P., 1918, Univ. California Publ. Zool., 17, 435-504. (revision of Aplddontia); McGREW, P. O., 1941, Field Mus. Publ. Zool., 9, (1), 1-30 (classification); ELLERMAN, J. R. et al., No. 1, 1940-1949 (genera); SIMPSON, G. G., No. 1, 1945 (classification); ANDERSON & JONES, No. 1, 1967 (families); HALL & KELSON, No. 1, 1959 (North America)
Archaeohyracidae. (Only as fossils)
Archaeomyidae, as Theridomyidae. (Only as fossils)
Archaeonycteridae. (Only as fossils)
Archaeopithecidae. (Only as fossils)
Arctictidae, as Viverridae
Arctocyonidae (incl. Chriacidae, Oxyclaenidae, Triisodontidae). (Only as fossils)
Arctogalidae, as Viverridae
Arctostylopidae. (Only as fossils)
Armadillidae, as Dasypodidae
Arminiheringiidae, as Borhyaenidae. (Only as fossils)

Arsinoitheriidae. (Only as fossils)
Artionychidae. (Family not identified)
Arvicolidae, as Cricetidae
Aspalacidae. (Family not identified)
Astrapotheridae, as Astrapotheriidae
Astrapotheriidae (incl. Albertogaudryidae, Astrapotheridae). (Only as fossils)
Atelidae, as Cebidae
Athrodontidae. (Only as fossils)
Atryptheridae. (Only as fossils)
Axeidae, as Cervidae
Axidae, as Cervidae

Balaenidae (incl. Neobalaenidae). SIMPSON, G. G., No. 1, 1945 (classification); HERSHKOVITZ, P., No. 1, 1966 (synonymic catalog); ANDERSON & JONES, No. 1, 1967 (families); HALL & KELSON, No. 1, 1959 (North America); ELLIOT, D. G., No. 1, 1904 (Central America & West Indies); CABRERA, A., No. 3, 1914 (Iberian Peninsula); TOSCHI, A., No. 2, 1965 (Italia); BOBRINSKOY, N. et al., No. 1, 1944 (U. S. S. R.); TOMILIN, A. G., No. 2, 1957 (U. S. S. R.); BOBRINSKOY, N. et al., No. 2, 1965 (U. S. S. R.); TOMILIN, A. G., No. 1, 1967 (U. S. S. R.); ALLEN, G. M., No. 3, 1954 (catalog, Africa); ELLERMAN, J. R. et al., No. 2, 1953 (southern Africa); BLANFORD, W. T., No. 1, 1888-91 (British India)
Balaenopteridae (incl. Megapteridae). SIMPSON, G. G., No. 1, 1945 (classification); HERSHKOVITZ, P., No. 1, 1966 (synonymic catalog); ANDERSON & JONES, No. 1, 1967 (families); HALL & KELSON, No. 1, 1959 (North America); CABRERA, A., No. 3, 1914 (Iberian Peninsula); TOSCHI, A., No. 2, 1965 (Italia); TOMILIN, A. G., No. 2, 1957 (U. S. S. R.); TOMILIN, A. G., No. 1, 1967 (U. S. S. R.); ALLEN, G. M., No. 3, 1954 (catalog, Africa); ELLERMAN, J. R. et al., No. 2, 1953 (southern Africa)
Barylambdidae. (Only as fossils)
Barytheriidae. (Only as fossils)
Basilosauridae (incl. Zeuglodontidae). (Only as fossils)
Bassaricyonidae, as Procyonidae
Bassaridae (including Bassarididae). No revisionary references noted)
Bassarididae, as Bassaridae
Bassariscidae, as Procyonidae
Bathmodontidae. (Only as fossils)
Bathyergidae (incl. Georychidae). ELLERMAN, J. R. et al., No. 1, 1940-49 (genera); SIMPSON, G. G., No. 1, 1945 (classification); ANDERSON & JONES, No. 1, 1967 (families); ALLEN, G. M., No. 3, 1954 (catalog, Africa); HILL & CARTER, No. 1, 1941 (Angola); ELLERMAN, J. R. et al., No. 2, 1953 (southern Africa); HOLLISTER, N., No. 2, 1919 (East Africa); SETZER, H. W., No. 1, 1956 (Anglo-Egyptian Sudan)
Bathyopsidae, as Uintatheriidae. (Only as fossils)
Belugidae, as Monodontidae
Bolodontidae, as Plagiaulacidae. (Only as fossils)
Borhyaenidae (incl. Arminiheringiidae, Proborhyaenidae, Sparassodontidae). (Only as fossils)
Bovidae (incl. Aepycerotidae, Alcelaphidae, Antelopidae, Antilopidae, Capridae, Cephalophidae, Cephalophoridae, Cervicapridae, Connochoetidae, Damalidae, Hippotragidae, Nesotragidae, Orygidae, Ovibovidae, Ovidae, Pantholopidae, Saigiidae, Strepsicerotidae, Tetracerocidae, Tragelaphidae). SCLATER, P. L. & THOMAS, O., 1894-1900, The book of antelopes, London, 4 volumes; LYDEKKER, R., 1898, Wild oxen, sheep, and goats of all lands, London, 1-318; LYDEKKER, R., 1913-14, Catalogue of the ungulate mammals in the British Museum (Natural History), London, 3 volumes; SIMPSON, G. G., No. 1, 1945 (classification); SOKOLOV, I. I., 1954, Trav. Inst. Zool. Acad. Sci. URSS, 14, 3-295 (classification, in Russian); ANDERSON & JONES, No. 1, 1967 (families); FRICK, C., 1937, Bull. American Mus. Nat. Hist., 69, 1-669 (North America); HALL & KELSON, No. 1, 1959 (North America); ELLIOT, D. G., No. 1, 1904 (Central America & West Indies, also as Brachypodidae); NASSONOV, N., 1923, Distribution geographique des moutons sauvages du Monde ancien, Petrograd, 1-255 (In Russian); SUCHKIN, P. P., 1925, Journ. Mamm., 6, 145-157 (wild sheep, Old World); MILLER, G. S., Jr., No. 1, 1912 (western Europe); CABRERA, A., No. 3, 1914 (Iberian Peninsula); TOSCHI, A., No. 2, 1965 (Italia); SERAFINSKI, W., 1957, Przegl. Zool., 1, 39-47 (taxonomy, in Polish); BOBRINSKOY, N. et al., No. 1, 1944 (U. S. S. R.); SOKOLOV, V. E., No. 1, 1959 (S. S. S. R.); BOBRINSKOY, N. et al., No. 2, 1965 (U. S. S. R.); HARRISON, D. L., No. 2, 1968 (Arabia); ALLEN, G. M., No. 3, 1954 (catalog, Africa); BOUET, G., 1934, Comm. Proc. Verb. Acad. Sci. Coloniales, 12, 1-46 (Afrique francaise); RODE, P., 1943, Faune de l'Empire Francaise, II, 1, Paris, 1-122 (Afrique Noire); ROSEVEAR, D. R., 1938, The antelopes of Nigeria, Nigèrian Field, 7. (Bovinae, Neotraginae); HILL & CARTER, No. 1, 1941 (Angola); ELLERMAN, J. R. et al., No. 2, 1953 (southern Africa); HOLLISTER, N., No. 3, 1924 (East Africa); SETZER, H. W., No. 1, 1956 (Anglo-Egyptian Sudan); BLANFORD, W. T., No. 1, 1888-91 (British India); PHILLIPS, W. W. A., No. 1, 1935 (Ceylon); TAYLOR, E. H., No. 1, 1934 (Philippines); ALLEN, G. M., No. 2, 1940 (China)
Brachypodidae. (Family not identified)
Brachyuridae. (Family not identified)
Bradypidae, as Bradypodidae
Bradypodidae (incl. Bradypidae, Choloepodidae). SIMPSON, G. G., No. 1, 1945 (classification); ANDERSON & JONES, No. 1, 1967 (families); HALL & KELSON, No. 1, 1959 (North America); GOODWIN, G. G.

G. G., No. 2, 1946 (Costa Rica, also as Choloepodidae); GOLDMAN, E. A., No. 1, 1920 (Panama, also as Choloepodidae); TATE, G. H. H., No. 1, 1939 (Guiana); VIEIRA, C., 1950, Arqu. Zool. Sao Paulo, 7, 325-362 (Estado de Sao Paulo)

Brontotheridae, as Brontotheriidae

Brontotheriidae (incl. Brontotheridae, Lambdotheriidae, Palaeosyopidae). (Only as fossils)

Bunodontheridae. (Only as fossils)

Bunolitopternidae, as Didolodontidae. (Only as fossils)

Bunomastodontidae, as Gomphotheriidae. (Only as fossils)

Bunotheriidae. (Family not identified)

Caenolestidae (incl. Abderitesidae, Coenolestidae, Epanorthidae, Garzonidae, Palaeothentidae). CABRERA, A., No. 1, 1919 (classification); OSGOOD, W. H., 1924, Field Mus. Nat. Hist. Publ., Zool., 14, 165-172 (world review); SIMPSON, G. G., No. 1, 1945 (classification); ANDERSON & JONES, No. 1, 1967 (families); OSGOOD, W. H., 1921, Field Mus. Nat. Hist. Publ., Zool. 14, 1-162 (monograph)

Caenopidae, as Rhinocerotidae. (Only as fossils)

Caenotheriidae, as Cainotheriidae. (Only as fossils)

Cainotheriidae (incl. Caenotheriidae). (Only as fossils)

Calamodontidae, as Stylinodontidae. (Only as fossils)

Callimiconidae, as Cebidae

Callithricidae (incl. Callitrichidae, Callitricidae, Hapalidae, Harpaladae, Mididae). ELLIOTT, D. G., No. 3, 1913 (monograph, as Hapalidae); POCOCK, R. I., 1917, Ann. Mag. Nat. Hist., (8), 20, 147-258 (genera, as Hapalidae); THOMAS, O., 1922, Ann. Mag. Nat. Hist., (9), 9, 196-199 (systematic arrangement); SIMPSON, G. G., No. 1, 1945 (classification); HILL, W. C. O., 1957, Primates ..., III, Edinburgh, 1-354 (monograph, as Hapalidae); BUETTNER-JANUSCH, J., No. 1, 1963 (world list, as Hapalidae); ANDERSON & JNOES, No. 1, 1967 (families); ELLIOT, D. G., No. 2, 1912 (World); HALL & KELSON, No. 1, 1959 (North America); ELLIOT, D. G., No. 1, 1904 (Central America & West Indies); GOLDMAN, E. A., No. 1, 1920 (Panama); TATE, G. H. H., No. 1, 1939 (Guiana); da CRUZ LIMA, E., 1945, Mammals of Amazonia, I, Rio de Janeiro, 1-274.

Callitrichidae, as Callithricidae

Callitricidae, as Callithricidae

Cameleopardalidae, as Giraffidae

Camelidae (incl. Poebrotheriidae, Protolabidae, Protolabididae). LYDEKKER, R., 1913-16, Catalogue of the ungulate mammals in the British Museum (Natural History), London, volumes 1-4; SIMPSON, G. G., No. 1, 1945 (classification); ANDERSON & JONES, No. 1, 1967 (families); FRICK, C., 1937, Bull. American Mus. Nat. Hist., 69, 1-669 (North America); WEBB, S. D., 1965, Bull. Los Angeles City Mus. Sci., 1, 1-54 (Camelinae, North America); CABRERA, A., 1932, Rev. Mus. La Plata, 33, 89-117 (South America); LADERA TUMIALAN, A. G., 1967, I cameli di Sud Americain. Riv. Zootec. Num. Spec., 1967, 178-182; LEON, J. A., 1939, Bol. Mus. Hist. Nat., "Javier Prado", 3, (11), 95-105 (Andes); SOKOLOV, V. E., No. 1, 1959 (S. S. S. R.); ALLEN, G. M., No. 3, 1954 (catalog, Africa)

Camelopardalidae, as Giraffidae

Canidae (incl. Amphicyonidae, Lycaonidae, Megalotidae, Otocyonidae, Simocyonidae, Vulpidae). COPE, E. D., 1879, Proc. Acad. Nat. Sci. Philedelphia, 13, 1-27 (genera); MIVART, S., 1890, A monograph of the Canidae, London, 1-216; KLATT, B., 1928, Zool. Jahrb., Abt. Allg. Zool. Phys. Tiere, 45, 217-292; JAKOVLEFF, B. P., 1933, Publ. Mus. Hoang ho Paiho Tien Tsin, 26, 1-24 (museum list); SIMPSON, G. G., No. 1, 1945 (classification); ANDERSON & JONES, No. 1, 1967 (families); YOUNG, S. P. & GOLDMAN, E. A., 1944, The wolves of North America, Washington, D. C., 1-636 (North America); HALL & KELSON, No. 1, 1959 (North America); ELLIOT, D. G., No. 1, 1904 (Central America & West Indies); GOODWIN, G. G., No. 1, 1942 (Honduras); GOODWIN, G. G., No. 2, 1942 (Costa Rica); GOLDMAN, E. A., No. 1, 1920 (Panama); KRAGIEVICH, L., 1930, Physis, 10, 35-73 (South America); OSGOOD, W. H., 1934, Journ. Mamm., 15, 45-50 (genera, South America); KUHLHORN, F., 1938, Arch. Naturg., 7, 29-45 (South America); HERSHKOVITZ, P., 1957, Noved. Colombianes, 3, 157-161 (Colombia); TATE, G. H. H., No. 1, 1939 (Guiana); CABRERA, A., 1932, Rev. Centro. Ing. Agronomos y Centro E. Agronomia, 145, 489-501 (Argentina); MILLER, G. S., Jr., No. 1, 1912 (western Europe); CABRERA, A., No. 3, 1914 (Iberian Peninsula); TOSCHI, A., No. 2, 1965 (Italia); NOVIKOV, G. A., No. 1, 1956 (U. S. S. R.); NOVIKOV, G. A., No. 2, 1962 (USSR); OGNEV, S. I., No. 2, 1931 (eastern Europe & northern Asia); OGNEV, S. I., No. 9, 1962 (eastern Europe & northern Asia); BOBRINSKOY, N. et al., No. 1, 1944 (U. S. S. R.); BOBRINSKOY, N. et al., No. 2, 1965 (U. S. S. R.); STROGANOV, S. U., 1969, Carnivorous mammals of Siberia, Jerusalem, 1-522; HARRISON, DLL., No. 2, 1968 (Arabia); ALLEN, G. M., No. 3, 1954 (catalog, Africa); SETZER, H. W., 1961, Journ. Egyptian Publ. Health Assoc., 36, 113-118 (Egypt); ALLEN, J. A., 1924, Bull. American Mus. Nat. Hist., 47, 73-281 (Belgian Congo); HILL & CARTER, No. 1, 1941 (Angola); ELLERMAN, J. R. et al., No. 2, 1953 (southern Africa); HOLLISTER, N., No. 1, 1918 (East Africa); SETZER, H. W., No. 1, 1956 (Anglo-Egyptian Sudan); BLANFORD, W. T., No. 1, 1888-91 (British India); POCOCK, R. I., No. 2, 1941 (British India); PHILLIPS, W. W. A., No. 1, 1935 (Ceylon); JONES, F. W., No. 1, 1923-25 (South Australia); TAYLOR, E. H., No. 1, 1934 (Philippines); ALLEN, G. M., No. 1, 1938 (China)

Capreolidae, as Cervidae

Capridae, as Bovidae

Capromyidae. SIMPSON, G. G., No. 1, 1945 (classification); ANDERSON & JONES, No. 1, 1967 (families); HALL & KELSON, No. 1, 1959 (North America); BOBRINSKOY, N. et al., No. 1, 1944 (USSR); BOBRINSKOY, N. et al., No. 2, 1965 (U. S. S. R.)
Caroloameghiniidae. (Only as fossils)
Carolozittelidae, as Pyrotheriidae. (Only as fossils)
Carpolestidae. (Only as fossils)
Castoridae (incl. Castoroididae, Chalicomyidae). ELLERMAN, J. R. et al., No. 1, 1941-49 (genera); SIMPSON, G. G., No. 1, 1945 (classification); FREYE, H. -A., 1960, Mitt. Zool. Mus. Berlin, 36, 105-122 (Systematik); ANDERSON & JONES, No. 1, 1967 (families); HALL & KELSON, No. 1, 1959 (North America); ELLIOT, D. G., No. 1, 1904 (Central America & West Indies); SEREBRENNIKOV, M., 1929, Compt. Rend. Acad. Sci. USSR, 1929, 271-276 (Palearctic); MILLER, G. S., Jr., No. 1, 1912 (western Europe); OGNEV, S. I., No. 5, 1947 (U. S. S. R.); OGNEV, S. I., No. 12, 1963 (U. S. S. R.); BOBRINSKOY, N. et al., No. 1, 1944 (U. S. S. R.); BOBRINSKOY, N. et al., No. 2, 1965 (U. S. S. R.); ALLEN, G. M., No. 2, 1940 (China)
Castoroididae, as Castoridae. (Only as fossils)
Catodontidae. (Family not identified)
Cavidae, as Caviidae
Caviidae (incl. Cavidae, Coelogenyidae, Galeidae). THOMAS, O., 1916, Ann. Mag. Nat. Hist., (8), 18, 301-303 (classification); KRAGLIEVICH, L., 1931, Anal. Mus. Nac. Buenos Aires, 36, 59-96 (classification & genera); ELLERMAN, J. R. et al., No. 1, 1941-49 (genera); SIMPSON, G. G., No. 1, 1945 (classification); ANDERSON & JONES, No. 1, 1967 (families); GOLDMAN, E. A., No. 1, 1920 (Panama); TATE, G. H. H., No. 1, 1939 (Guiana); CABRERA, A., 1954, Publ. Esc. Vet. Fac. Agron. Vet. Univ. Buenos Aires, 6, 1-93 (Argentina)
Cavicornidae. (Family not identified)
Cebidae (incl. Aotidae, Atelidae, Alouattidae, Callimiconidae, Homonculidae, Saimiridae). ELLIOTT, D. G., No. 3, 1913 (monograph); KELLOGG, R. & GOLDMAN, E. A., 1944, Proc. United States Nat. Mus., 96, 1-45 (review of spider monkeys); SIMPSON, G. G., No. 1, 1945 (classification); HILL, W. C. O., 1957, Primates ..., III, Edinburgh, 1-354 (monograph, as Callimiconidae); HILL, W. C. O., 1960-62, Primates ..., IV, V, Edinburgh, 1-523, 1-537 (monograph); BUETTNER-JANUSCH, J., No. 1, 1963 (world list); ANDERSON & JONES, No. 1, 1967 (families); ELLIOT, D. G., No. 2, 1912 (world); ANTHONY, J., 1947, Bull. Mus. Nat. Hist. Nat., (2), 19, 47-50 (key to American, head bones); HALL & KELSON, No. 1, 1959 (North America); ELLIOT, D. G., No. 1, 1904 (Central America & West Indies); GOODWIN, G. G., No. 1, 1942 (Honduras, also as Alouattidae & Atelidae); GOODWIN, G. G., No. 2, 1946 (Costa Rica, also as Alouattidae & Saimiridae); GOLDMAN, E. A., No. 1, 1920 (Panama, also as Alouattidae, Aotidae, Atelidae & Saimiridae); TATE, G. H. H., No. 1, 1939 (Guiana); da CRUZ LIMA, E., 1945, Mammals of Amazonia, I, Rio de Janeiro, 1-274; CABRERA, A., 1939, Los monos de la Argentina, Physis, 16.
Cebochoeridae (incl. Acotherulidae, Acrotherulidae, Mixtotheriidae, Mixtotheriodontidae). (Only as fossils)
Ceciliolemuridae. (Only as fossils)
Centetidae, as Tenrecidae
Cephalomyidae. (Only as fossils)
Cephalophidae, as Bovidae
Cephalophoridae, as Bovidae
Cephalotidae, as Pteropidae
Cercolabidae. (Family not identified)
Cercoleptidae, as Procyonidae
Cercoleptididae, as Procyonidae
Cercopithecidae (incl. Colobidae, Cynocephalidae of Ameghino, Cynopithecidae, Lasiopygidae, Macacidae, Papionidae, Pithecidae, Semnopithecidae). ELLIOT, D. G., No. 2, 1912 (world, as Lasiopygidae); ELLIOT, D. G., No. 3, 1913 (monograph); CABRERA, A., 1914, Rev. Real Acad. Cienc. Exact. Fis. Nat. Madrid, 1914, 1-7 (classification & nomenclature, as Lasiopygidae); HILL, W. C. O., 1939, Spol. Zeylanica, 21, 277-305 (systematic list); WASHBURN, S. L., 1944, Journ. Mamm., 25, 289-294 (genera of langurs); SIMPSON, G. G., No. 1, 1945 (classification); BUETTNER-JANUSCH, J., No. 1, 1963 (world list); HILL, W. C. O., 1966, Primates ..., VI, Edinburgh, 1-757; ANDERSON & JONES, No. 1, 1967 (families); HALL & KELSON, No. 1, 1959 (North America); NAPIER, J. R. & NAPIER, P. H., 1970, Old World monkeys ..., New York, 1-660; CABRERA, A., No. 3, 1914 (Iberian peninsula); HARRISON, D. L., No. 1, 1964 (Arabia); RODE, P., 1937, Les Primates de l'Afrique, Paris, 1-222 (also as Semnopithecidae); ALLEN, G. M., No. 3, 1954 (catalog, Africa, also as Colobidae); BOOTH, A. H., 1956, Ann. Mag. Nat. Hist., (12), 9, 476-480 (Gold & Ivory coasts); ROSEVEAR, D. R., 1934, Nigerian Field, 3, (4), 138-151 (monkeys & baboons, Nigeria); ALLEN, J. A., 1925, Bull. American Mus. Nat. Hist., 47, 283-499 (Belgian Congo, as Lasiopygidae); HILL & CARTER, No. 1, 1941 (Angola); ELLERMAN, J. R. et al., No. 2, 1953 (southern Africa); HOLLISTER, N., No. 3, 1924 (East Africa, as Colobidae & Lasiopygidae); SETZER, H. W., No. 1, 1956 (Anglo-Egyptian Sudan); KHAJURIA, H., 1955, Rec. Indian Mus., 52, 101-127 (museum list, also as Colobidae); BLANFORD, W. T., No. 1, 1888-91 (British India); POCOCK, R. I., No. 1, 1939 (British India, also as Colobidae); PHILLIPS, W. W. A., No. 1, 1935 (Ceylon, also as Colobidae); TAYLOR, E. H., No. 1, 1934 (Philippines); ALLEN, G. M., No. 1, 1938 (China, also as Colobidae)
Cervicapridae, as Bovidae
Cervidae (incl. Axeidae, Axidae, Capreolidae, Cervulidae, Elaphidae, Moschidae, Palaeomerycidae, Rangiferidae, Rangiferinidae). BROOKE, W., 1878, Proc. Zool. Zoc. London, 1878, 883-928

(classification & synopsis of species); LYDEKKER, R., 1898, The deer of all lands ..., London, 1-329; LYDEKKER, R., 1913-16, Catalogue of the ungulate mammals in the British Museum (Nat. Hist), London, volumes 1-4; LOOMIS, F. B., 1928, American Journ. Sci., 16, 531-542 (phylogeny of deer); SIMPSON, G. G., No. 1, 1945 (classification); ANDERSON & JONES, No. 1, 1967 (families); FRICK, C., 1937, Horned ruminants of North America, Bull. American Mus. Nat. Hist., 69, 1-669 (North America); HALL & KELSON, No. 1, 1959 (North America); SHELDON, H. H., 1933, Occ. Pap. Santa Barbara Mus., 3, 1-71 (deer of California) ELLIOT, D. G., No. 1, 1904 (Central America & West Indies); GOODWIN, G. G., No. 1, 1942 GOODWIN, G. G., No. 2, 1942 (Costa Rica); GOLDMAN, E. A., No. 1, 1920 (Panama); TATE, G. H. H., No. 1, 1939 (Guiana); de MIRANDA RIBEIRO, A., 1919, Rev. Mus. Paulista, 11, 209-307 (Brasil); CASTELLANOS, A., 1924, Riv. Univ. Nac. Cordoba, 11, (4-6), 1-26 (Argentina); MILLER, G. S., Jr., No. 1, 1912 (western Europe); CABRERA, A., No. 3, 1914 (Iberian Peninsula); TOSCHI, A., No. 2, 1965 (Italia); BOBRINSKOY, N. et al., No. 1, 1944 (U. S. S. R.); HEPTNER, V. G. & TZALKIN, V. I., 1947, Deer of the U. S. S. R., Moskva, 1-174 (in Russian); FLEROV, K. K., 1952, Zool. Inst. Akad. Nauk SSSR, (n. s.), 55, 1-255 (Fauna SSR..., also as Moschidae); SOKOLOV, V. E., No. 1, 1959 (S. S. S. R., also as Moschidae); FLEROV, K. K., 1960, Fauna of U. S. S. R., Mammalia, volume 1, no. 2, Jerusalem, 1-257 (translation of 1952, also as Moschidae); BOBRINSKOY, N. et al., No. 2, 1965 (U. S. S. R.); HARRISON, D. L., No. 2, 1968 (Arabia); ALLEN, G. M., No. 3, 1954 (catalog, Africa); BLANFORD, W. T., No. 1, 1888-91 (British India); PHILLIPS, W. W. A., No. 1, 1935 (Ceylon); TAYLOR, E. H., No. 1, 1934 (Philippines); ALLEN, G. M., No. 2, 1940 (China)

Cervulidae, as Cervidae

Cetotheriidae. (Only as fossils)

Chalicomyidae, as Castoridae. (Only as fossils)

Chalicotheriidae (incl. Macrotheriidae, Moropodidae). (Only as fossils)

Cheiromydae, as Daubentoniidae

Chinchillidae (incl. Lagostomidae, Viscaceidae). PRELL, H., 1934, Zool. Anz., 108, 97-104 (monograph of Chinchilla); ELLERMAN, J. R., et al., No. 1, 1941-49 (genera); SIMPSON, G. G., No. 1, 1945 (classification); ANDERSON & JONES, No. 1, 1967 (families); ANTHONY, H. E., No. 1, 1918 (Puerto Rico)

Chirogidae, as Ptilodontidae. (Only as fossils)

Chiromidae, as Chiromyidae

Chiromyidae (incl. Chiromidae, Chyromysidae). (No revisionary references noted)

Chironectidae, as Didelphidae

Chlamydophoridae, as Dasypodidae

Chlamydotheridae, as Dasypodidae. (Only as fossils)

Choeropotamidae (incl. Helohyidae). (Only as fossils)

Choloepodidae, as Bradypodidae

Chriacidae, as Arctocyonidae. (Only as fossils)

Chrysochloridae. DOBSON, G. E., No. 3, 1882-83 (monograph); CABRERA, A., No. 2, 1925 (classification); FORCART, L., 1942, Rev. Suisse Zool., 49, 1-6; SIMPSON, G. G., No. 1, 1945 (classification); ANDERSON & JONES, No. 1, 1967 (families); ALLEN, G. M., No. 3, 1954 (catalog, Africa); HILL & CARTER, No. 1, 1941 (Angola); ELLERMAN, J. R. et al., No. 2, 1953 (southern Africa)

Chyromysidae, as Chiromyidae

Cimolestidae. (Only as fossils)

Cimolodontidae, as Ptilodontidae. (Only as fossils)

Cimolomidae. (Only as fossils)

Coelogenyidae (part as Caviidae, part as Dasyproctidae)

Coendidae, as Erethizontidae

Coenolestidae, as Caenolestidae

Colobidae, as Cercopithecidae

Colpodontidae, as Leontiniidae. (Only as fossils)

Connochetidae, as Bovidae

Conoryctidae, as Stylinodontidae. (Only as fossils)

Coryphodontidae (incl. Pantolambdidae). (Only as fossils)

Cotylopidae. (Only as fossils)

Creotarsidae. (Only as fossils)

Cricetidae (incl. Anomalomyidae, Arvicolidae, Cricetodontidae, Cricetopidae, Gerbillidae, Hesperomyidae, Lophiomyidae, Melissiodontidae, Merionidae, Microtidae, Mystromyidae, Nesomyidae). MILLER, G. S., Jr., 1896, North American Fauna, 12, 1-85 (genera of voles & lemmings); HEPTNER, W. G., 1933 etc., Zool. Anz., 102, 107-112 (Notizen, as Gerbillidae); ELLERMAN, J. R. et al., No. 1, 1941-49 (genera, as Lophiomyidae); SIMPSON, G. G., No. 1, 1945 (classification); ANDERSON & JONES, No. 1, 1967 (families); HALL & KELSON, No. 1, 1959 (North America); HOOPER, E. T. & HART, B. S., 1962, Misc. Publ. Univ. Michigan Mus. Zool., 120, 1-68 (synopsis of North American microtine); GYKDENSTOLPE, N., 1932, Kungl. Svenska Vetenskap. Handl., (3), 11, (3), 1-164 (neotropics); GOODWIN, G. G., No. 1, 1942 (Honduras); GOODWIN, G. G., No. 2, 1946 (Costa Rica); TATE, G. H. H., No. 1, 1939 (Guiana); ARGYROPULO, A. I., 1933, Trav. Inst. Zool. Acad. Sci. Leningrad, 1, 239-248 also in Zeitschr. Saugetierk., 3, 129-149, 1933, Palearctic, Cricetinae); SCHREUDER, A., 1933, Verh. Akad. Wet. Amsterdam, (II), 30, (1), 1-37 (Microtinae, Netherlands); HEPTNER, W. G., 1940, Nouv. Mem. Soc. Nat. Moscou, 20, 5-71 (Persia, as Gerbillidae); PETTER, F., 1957, Mammalia, 21, 241-257 (Liste, Palestine, as Gerbillidae); ALLEN, G. M., No. 3,

1954 (catalog, Africa); SETZER, H. W., 1958, Journ. Egyptian Publ. Health Assoc., 33, 205-227 (gerbils of Egypt); RANCK, G. L., No. 1, 1968 (Libya); HOLLISTER, N., No. 2, 1918 (East Africa); SETZER, H. W., No. 1, 1956 (Anglo-Egyptian Sudan); ALLEN, G. M., No. 2, 1940 (China)
Cricetodontidae, as Cricetidae. (Only as fossils)
Cricetopidae, as Cricetidae. (Only as fossils)
Cryptoproctidae, as Viverridae
Ctenodactylidae. ELLERMAN, J. R. et al., No. 1, 1941-49 (genera); SIMPSON, G. G., No. 1, 1945 (classification); ANDERSON & JONES, No. 1, 1967 (families); ALLEN, G. M., No. 3, 1954 (catalog, Africa); RANCK, G. L., No. 1, 1968 (Libya)
Ctenomyidae. SIMPSON, G. G., No. 1, 1945 (classification); ANDERSON & JONES, No. 1, 1967 (families)
Cuniculidae, as Dasyproctidae
Curtognathidae, as Deinotheriidae. (Only as fossils)
Cylindrodontidae, as Ischyromyidae. (Only as fossils)
Cynarctidae. (Only as fossils)
Cynictidae, as Viverridae
Cynidae. (Family not identified)
Cynocephalidae of Ameghino, as Cercopithecidae
Cynocephalidae of Simpson, as Galeopithecidae
Cynogalidae, as Viverridae
Cynopithecidae, as Cercopithecidae
Cynorchidae. (Only as fossils)
Cyomorphidae. (Only as fossils)
Cyrtodontidae, as Dryolestidae
Cystophoridae, as Phocidae

Damalidae, as Bovidae
Dasipidae, as Dasypodidae
Dasipodidae, as Dasypodidae
Dasypidae, as Dasypodidae
Dasypodidae (incl. Armadillidae, Chlamydophoridae, Chlamydotheridae, Dasipidae, Dasipodidae, Dasypidae, Praopidae, Scleropleuridae, Stegotheridae, Tatusidae, Tatusiidae, Tolypeutidae). SIMPSON, G. G., No. 1, 1945 (classification); ANDERSON & JONES, No. 1, 1967 (families); HALL & KELSON, No. 1, 1959 (North America); ELLIOT, D. G., No. 1, 1904 (Central America & West Indies); GOODWIN, G. G., No. 1, 1942 (Honduras); GOODWIN, G. G., No. 2, 1946 (Costa Rica); GOLDMAN, E. A., No. 1, 1920 (Panama); TATE, G. H. H., No. 1, 1939 (Guiana); VIEIRA, C., 1950, Arqu. Zool. Sao Paulo, 7, 325-362.
Dasyproctidae (incl. Agoutidae, Coelogenyidae part, Cuniculidae). ELLERMAN, J. R. et al., No. 1, 1941-49 (genera, also as Cuniculidae); SIMPSON, G. G., No. 1, 1945 (classification); ANDERSON & JONES, No. 1, 1967 (families); HALL & KELSON, No. 1, 1959 (North America); ELLIOT, D. G., No. 1, 1904 (Central America & West Indies, as Agoutidae); ANTHONY, H. E., No. 1, 1918 (Puerto Rico); GOODWIN, G. G., No. 1, 1942 (Honduras); GOODWIN, G. G., No. 2, 1942 (Costa Rica, also as Cuniculidae); GOLDMAN, E. A., No. 1, 1920 (Panama)
Dasyuridae (incl. Myrmecobiidae, Thylacinidae, Thylaxinidae). CABRERA, A., No. 1, 1919 (classification, also as Myrmecobiidae & Thylaxinidae); SIMPSON, G. G., No. 1, 1945 (classification); TATE, G. H. H., Bull. American Mus. Nat. Hist., 88, 97-156 (anatomy & classification); TATE, G. H. H., 1951, American Mus. Novit., 1521, 1-8; ANDERSON & JONES, No. 1, 1967 (families); JONES, F. W., No. 1, 1923-25 (South Australia, as Myrmecobiidae & Thylacinidae also); FINLAYSON, H. H., 1933, Trans. Proc. Roy. Soc. South Australia, 57, 195-202 (Lake Eyre basin)
Daubentoniidae (incl. Cheiromydae). SIMPSON, G. G., No. 1, 1945 (classification); HILL, W. C. O., No. 1, 1953 (monograph); BUETTNER-JANUSCH, J., No. 1, 1963 (world list); ANDERSON & JONES, No. 1, 1967 (families); ELLIOT, D. G., No. 2, 1912 (world review); ALLEN, G. M., No. 3, 1954 (catalog, Africa)
Decastidae. (Only as fossils)
Deinotheridae, as Deinotheriidae
Deinotheriidae (incl. Curtognathidae, Deinotheridae). (Only as fossils)
Delphinapteridae, as Monodontidae
Delphinidae (incl. Delphinoidae). GRUE, F. W., 1889, United States Nat. Mus. Bull., 36, 1-191 (review); BARABASH, I. I., 1938, Zool. Journ. (Moscou); 17, 1091-1104 (methods of investigation, in Russian); SIMPSON, G. G., No. 1, 1945 (classification); NISHIWAKI, M., 1963-64, Sci. Rep. Whales Res. Inst., 17, 93-103, 18, 171-172 (on genera); HERSHKOVITZ, P., No. 1, 1966 (synonymic catalog); ANDERSON & JONES, No. 1, 1967 (families); HALL & KELSON, No. 1, 1959 (North America); ELLIOT, D. G., No. 1, 1904 (Central America & West Indies); CABRERA, A., No. 3, 1914 (Iberian Peninsula); TOSCHI, A., No. 2, 1965 (Italia); TOMILIN, A. G., No. 2, 1957 (U. S. S. R.); TOMILIN, A. G., No. 1, 1967 (U. S. S. R.); BOBRINSKOY, N. et al., No. 1, 1944 (U. S. S. R.); BOBRINSKOY, N. et al., NO. 2, 1965 (U. S. S. R.); ALLEN, G. M., No. 3, 1954 (catalog, Africa); ELLERMAN, J. R. et al., No. 2, 1953 (southern Africa); BLANFORD, W. T., No. 1, 1888-91 (British India); DAMMERMAN, K. W., 1924, Treubia, 5, 340-352 (Indo-Australian Arch.); JONES, F. W., No. 1, 1923-25 (South Australia); ALLEN, G. M., No. 1, 1938 (China); OGAWA, T., 1932, Saito Ho-on Kai Jiho, 69-70, 1-57 (Japan); OKADA, Y. & HANAOKA, T., 1940, Sci. Rep. Tokyo Bun. Daig., 4B, 77, 285-306 (Japan)
Delphinoidae, as Delphinidae

Delphinorhynchidae. (Family not identified)
Deltatheridiidae. (Only as fossils)
Desmatophocidae, as Otariidae
Desmodidae, as Desmodontidae
Desmodontidae (incl. Desmodidae). SIMPSON, G. G., No. 1, 1945 (classification); ANDERSON & JONES, No. 1, 1967 (families); HALL & KELSON, No. 1, 1959 (North America); GOODWIN & GREENHALL, No. 1, 1961 (Trinidad & Tobago); VILLA-R., B., No. 1, 1966 (Mexico); GOODWIN, G. G., No. 1, 1942 (Honduras); GOODWIN, G. G., No. 2, 1946 (Costa Rica); GOLDMAN, E. A., No. 1, 1920 (Panama); HUSSON, A. M., No. 1, 1962 (Surinam); da CUNHA VIEIRA, C. O., No. 1, 1942 (Brasil);
Desmostylidae. (Only as fossils)
Dibunodontidae, as Gomphoteriidae. (Only as fossils)
Dichobunidae. (Only as fossils)
Dichodontidae. (Only as fossils)
Dicotylidae, as Tayussuidae
Dicrocynodontidae, as Docodontidae. (Only as fossils)
Dideilotheridae. (Only as fossils)
Didelphidae (incl. Chironectidae, Didelphididae, Didelphiidae, Didelphyidae, Microbiotheridae, Microbiotheriidae, Thlaeodontidae). CABRERA, A., No. 1, 1919 (classification); SIMPSON, G. G., 1935, Journ. Mamm., 16, 134-137 (classification); SIMPSON, G. G., No. 1, 1945 (classification); ANDERSON & JONES, No. 1, 1967 (families); ALLEN, J. A., 1901, Bull. American Mus. Nat. Hist., 14, 149-188 (North American); HALL & KELSON, No. 1, 1959 (North America); ELLIOT, D. G., No. 1, 1904 (Central America & West Indies); GOODWIN, G. G., No. 1, 1942 (Honduras); GOODWIN, G. G., No. 2, 1946 (Costa Rica); GOLDMAN, E. A., No. 1, 1920 (Panama); TATE, G. H. H., No. 1, 1939 (Guiana); VIEIRA, C., 1950, Arqu. Zool. Sao Paulo, 7, 325-362 (Estado de Sao Paulo); JONES, F. W., No. 1, 1923-25 (South Australia)
Didelphididae, as Didelphidae
Didelphiidae, as Didelphidae
Didelphyidae, as Didelphidae
Didolodidae, as Didolodontidae. (Only as fossils)
Didolodontidae (incl. Bunolitopternidae, Didolodidae). (Only as fossils)
Dimylidae. (Only as fossils)
Dinoceratidae, as Uintatheriidae. (Only as fossils)
Dinomyidae. ELLERMAN, J. R. et al., No. 1, 1941-49 (genera); SIMPSON, G. G., No. 1, 1945 (classification); ANDERSON & JONES, No. 1, 1967 (families); ANTHONY, H. E., No. 2, 1925-26 (Puerto Rico)
Dinotheridae, as Dinotheriidae. (Only as fossils)
Dinotheriidae (incl. Dinotheridae). (Only as fossils)
Diplocynodontidae, as Docodontidae. (Only as fossils)
Diplopidae. (Only as fossils)
Diplopodidae. (Only as fossils)
Dipodidae (incl. Dipsidae, Gerboidae, Ierboidae, Jaculidae). VINOGRADOV, B. S., 1930, Bull. Acad. Sci. USSR, 4, 331-350, 5, 453-466 (classification); ELLERMAN, J. R. et al., No. 1, 1941-49 (genera); SIMPSON, G. G., No. 1, 1945 (classification); ANDERSON & JONES, No. 1, 1967 (families); VINOGRADOV, B. S., 1937, Fauna of USSR, Mammals, vol. 3, No. 4, 1-197 (U. S. S. R., In Russian); BOBRINSKOY, N. et al., No. 1, 1944 (U. S. S. R., as Jaculidae); BOBRINSKOY, N. et al., No. 2, 1965 (U. S. S. R., as Jaculidae); OGNEV, S. I., No. 6, 1948 (U. S. S. R.); OGNEV, S. I., No. 13, 1963 (U. S. S. R.); SETZER, H. W., 1957, Journ. Egyptian Publ. Health Assoc., 32, 265-271 (jerboas of Egypt); SETZER, H. W., 1958, Journ. Egyptian Publ. Health Assoc., 33, 87-94 (Egypt); RANCK, G. L., No. 1, 1968 (Libya); SETZER, H. W., No. 1, 1956 (Anglo-Egyptian Sudan); BLANFORD, W. T., No. 1, 1888-91 (British India); ELLERMAN, J. R., No. 1, 1961 (India); ALLEN, G. M., No. 2, 1940 (China); BANNIKOV, A. G., 1947, Bull. Soc. Nat. Moscou, Biol., (n. s.), 52, (4), 25-43 (Mongolia, in Russian)
Dipriodontidae. (Only as fossils)
Diprotodontidae (incl. Nototheriidae). (Only as fossils)
Dipsidae, as Dipodidae
Docodontidae (incl. Diplocynodontidae, Dicrocynodontidae). (Only as fossils)
Doedicuridae, as Glyptodontidae. (Only as fossils)
Dorudontidae. (Only as fossils)
Dromatheridae, as Dromatheriidae. (Only as fossils)
Dromatheriidae (incl. Dromatheridae). (Only as fossils)
Dryolestidae (incl. Amblotheridae, Amblotheriidae, Amblytheriidae, Cyrtodontidae, Kurtodontidae, Stylacodontidae). (Only as fossils)
Dugongidae (incl. Halicoridae, Halitheriidae, Hydrodamalidae, Rhytinidae, Rytinadae). SIMPSON, G. G., No. 1, 1945 (classification); ANDERSON & JONES, No. 1, 1967 (families); ALLEN, G. M., No. 3, 1954 (catalog, Africa); ELLERMAN, J. R. et al., No. 2, 1953 (southern Africa); ALLEN, G. M., No. 2, 1940 (China)

Echidnidae, as Tachyglossidae
Echimyidae (incl. Echinomyidae, Echymidae, Echymyidae, Loncheridae). ELLERMAN, J. R. et al., No. 1, 1941-49 (genera); SIMPSON, G. G., No. 1, 1945 (classification); ANDERSON & JONES, No. 1, 1967 (families); HALL & KELSON, No. 1, 1959 (North America); ANTHONY, H. E., No. 2, 1925-26 (fossil, Puerto Rico); GOODWIN, G. G., No. 1, 1942 (Honduras);

GOODWIN, G. G., No. 2, 1942 (Costa Rica); HERSHKOVITZ, P., 1948, Proc. United States Nat. Mus., 97, 125-140 (northern Colombia); TATE, G. H. H., No. 1, 1939 (Guiana)

Echingidae. (Family not identified)

Echinomyidae, as Echimyidae

Echymidae, as Echimyidae

Echymyidae, as Echimyidae

Ectoganidae, as Stylinodontidae. (Only as fossils)

Elaphidae, as Cervidae

Elasmotheriidae, as Rhinocerotidae. (Only as fossils)

Elephantidae (incl. Stagodontidae, Stegodontidae). LYDEKKER, R., 1916, Catalogue of the ungulate mammals in the British Museum (Natural History), London, volume 5; OSBORN, H. F., 1936-42, Proboscidea, A monograph ... of the mastodonts and elephants of the world, New York, 2 volumes (monograph & bibliography); SIMPSON, G. G., No. 1, 1945 (classification); ANDERSON & JONES, No. 1, 1967 (families); DUBROVA, I. A., 1957, Vertebrata Palasiatica, 1, 223-232 (genera of Elephantinae, In Russian); ALLEN, G. M., No. 3, 1954 (catalog, Africa); GRANDIDIER, G., 1932, La Terre et la Vie, 2, 130-134 (Mauritanie); RODE, P., 1944, Faune de l'Empire Francaise, II, Paris, 125-210 (Afrique Noire); FRADE, F., 1933, Bull. Soc. Portugaise Sci. Nat., 11, 319-333 (Angola); HILL & CARTER, No. 1, 1941 (Angola); ELLERMAN, J. R. et al., No. 2, 1953 (southern Africa); SETZER, H. W., No. 1, 1956 (Anglo-Egyptian Sudan); HILL, W. C. O., 1953, The elephant in East Africa, A monograph, London, 1-150 (East Africa); HOLLISTER, N., No. 3, 1924 (East Africa); FRADE, F., 1933, Bull. Soc. Portugaise Sci. Nat., 11, 307-318 (Mozambique); BLANFORD, W. T., No. 1, 1888-91 (British India); PHILLIPS, W. W. A., No. 1, 1935 (Ceylon); DERANIYAGALA, P. E. P., 1944, Ann. Mag. Nat. Hist., (11), 11, 61-64 (Ceylon)

Elotheriidae, as Entelodontidae. (Only as fossils)

Emballonuridae (incl. Taphozoidae). DOBSON, G. E., No. 2, 1878 (descriptive catalog); SIMPSON, G. G., No. 1, 1945 (classification); ANDERSON & JONES, No. 1, 1967 (families); SANBORN, C. C., 1937, Field Mus. Nat. Hist., Zool., 20, 321-354 (America, subf. Emballonurinae); HALL & KELSON, No. 1, 1959 (North America); VILLA-R., B., No. 1, 1966 (Mexico); GOODWIN, G. G., No. 1, 1942 (Honduras); GOODWIN, G. G., No. 2, 1946 (Costa Rica); GOLDMAN, E. A., No. 1, 1920 (Panama); GOODWIN, G. G. & GREENHALL, A. M., No. 1, 1961 (Trinidad and Tobago); HUSSON, A. M., No. 1, 1962 (Surinam); da CUNHA VIEIRA, C. O., No. 1, 1942 (Brasil); CABRERA, A., 1904, Mem. Soc. Espanola Hist. Nat., 2, 249-287 (España); DOBSON, G. E., No. 1, 1876 (Asia); HARRISON, D. L., No. 1, 1964 (Arabia); ALLEN, G. M., No. 3, 1954 (catalog, Africa); ROSEVEAR, D. R., No. 1, 1965 (West Africa); HILL & CARTER, No. 1, 1941 (Angola); ELLERMAN, J. R. et al., No. 2, 1953 (southern Africa); HOLLISTER, N., No. 1, 1915 (East Africa); BLANFORD, W. T., No. 1, 1888-91 (British India); PHILLIPS, W. W. A., No. 1, 1935 (Ceylon); TROUGHTON, E. L., No. 1, 1926 (Australia, New Guinea); JONES, F. W., No. 1, 1923-25 (South Australia); TAYLOR, E. H., No. 1, 1934 (Philippines)

Enhydridae, as Mustelidae

Entelodontidae (incl. Achaenodontidae, Elotheriidae). (Only as fossils)

Entelopsidae. (Only as fossils)

Eobasileidae, as Uintatheriidae. (Only as fossils)

Eocardidae, as Eocardiidae. (Only as fossils)

Eocardiidae (incl. Eocardidae). (Only as fossils)

Eohyidae. (Only as fossils)

Eomericidae. (Only as fossils)

Eomyidae (incl. Adjidaumidae). (Only as fossils)

Epanorthidae, as Caenolestidae. (Only as fossils)

Epoicotheriidae. (Only as fossils)

Epiodontidae. (Family not identified)

Equidae (incl. Anchitheridae, Hyracotheridae). LYDEKKER, R., 1916, Catalogue of the ungulate mammals in the British Museum (Natural History), London, volume IV; d'ANDRADE, R., 1937, Bull. Soc. Portugaise Sci. Nat., 12, 267-293 (classification); ANTONIUS, O., 1938, Proc. Zool. Soc. London, 107B, 557-564 (distribution past and present); SIMPSON, G. G., No. 1, 1945 (classification); BOURDELLE, E. & FRECHKOP, S., 1950, Mammalia, 14, 126-139 (classification); SIMPSON, G. G., 1951, Horses, New York, 1-247; ANDERSON & JONES, No. 1, 1967 (families); CABRERA, A., Caballos de America, Buenos Aires, 1-405, 1945; DEPERET, C., 1901, Bull. Soc. Geol. France, (4), 1, 199-225 (Europe, as Hyracotheridae); BOBRINSKOY, N. et al., No. 1, 1944 (U. S. S. R.); BOBRINSKOY, N. et al., No. 2, 1965 (U. S. S. R.); SOKOLOV, V. E., No. 1, 1959 (S. S. S. R.); BOURDELLE, E., 1935, Arch. Mus. Nat. Hist. Nat., (6), 12, 475-483 (Asia); ALLEN, G. M., No. 3, 1939 (checklist, Africa); HILL & CARTER, No. 1, 1941 (Angola); ELLERMAN, J. R. et al., No. 2, 1953 (southern Africa); HOLLISTER, N., No. 3, 1924 (East Africa); SETZER, H. W., No. 1, 1956 (Anglo-Egyptian Sudan); BLANFORD, W. T., No. 1, 1888-91 (British India); ALLEN, G. M., No. 2, 1940 (China); JAKOVLEFF, B. P., 1932, Publ. Mus. Hoang ho Pai Ho Tien Tsin, 10, 1-10 (museum list)

Erethizontidae (incl. Coendidae). ELLERMAN, J. R. et al., No. 1, 1941-49 (genera); SIMPSON, G. G., No. 1, 1945 (classification); ANDERSON & JONES, No. 1, 1967 (families); HALL & KELSON, No. 1, 1959 (North America); ELLIOT, D. G., No. 1, 1904 (Central America & West Indies); GOODWIN, G. G., No. 1, 1942 (Honduras); GOODWIN, G. G., No. 2, 1946 (Costa Rica); GOLDMAN, E. A., No. 1, 1920 (Panama); TATE, G. H. H., No. 1, 1939 (Guiana)

Erinaceidae (incl. Acanthionidae, Erinacidae, Gymnuridae, Hylomidae). DOBSON, G. E., No. 3, 1882-83 (monograph); THOMAS, O., 1918, Ann. Mag. Nat. Hist., (9), 1, 193-196 (generic division);

CABRERA, A., No. 2, 1925 (classification); FRIANT, M., 1935, Bull. Soc. Zool. France, 59, 508-516 (classification); FRIANT, M., 1943, Catalogue raisonne ... des collections d'osteologique ... du Museum ..., Mammiferes, II, Paris, 1-56; SIMPSON, G. G., No. 1, 1945 (classification); HERTER, C., 1965, Hedgehogs, London, 1-69 (generic classification); ANDERSON & JONES, No. 1, 1967 (families); OGNEV, S. I., No. 1, 1928 (eastern Europe & northern Asia); OGNEV, S. I., No. 8, 1962, same as 1928; MILLER, G. S., Jr., No. 1, 1912 (western Europe); CABRERA, A., No. 3, 1914 (Iberian Peninsula); TOSCHI, A. & LANZA, B., No. 1, 1959 (Italia); BOBRINSKOY, N. et al., No. 1, 1944 (U. S. S. R.); BOBRINSKOY, N. et al., same as 1944; STROGANOV, S. U., 1957, Insectivores of Siberia, Moskva, 1-267; HARRISON, D. L., No. 1, 1964 (Arabia); ALLEN, G. M., No. 3, 1954 (catalog, Africa); HILL & CARTER, No. 1, 1941 (Angola); ELLERMAN, J. R. et al., No. 2, 1953 (southern Africa); SETZER, H. W., No. 1, 1956 (Anglo-Egyptian Sudan); HOLLISTER, N., No. 1, 1918 (East Africa); BLANFORD, W. T., No. 1, 1888-91 (British India); TAYLOR, E. H., No. 1, 1934 (Philippines); ALLEN, G. M., No. 1, 1938 (China)

Erinacidae, as Erinaceidae

Eriomyidae. (Family not identified)

Eschatiidae. (Only as fossils)

Eschrichtidae, as Eschrichtiidae

Eschrichtiidae (incl. Eschrichtidae, Rhachianectidae). SIMPSON, G. G., No. 1, 1945 (classification, as Rhachianectidae); HERSHKOVITZ, P., No. 1, 1966 (synonymic catalog); ANDERSON & JONES, No. 1, 1967 (families); HALL & KELSON, No. 1, 1959 (North America); TOMILIN, A. G., No. 2, 1957 (U. S. S. R.); TOMILIN, A. G., No. 1, 1967 (U. S. S. R.)

Esthonychidae, as Tillotheriidae. (Only as fossils)

Eumegamyidae, as Leptaxodontidae. (Only as fossils)

Eupleridae, as Viverridae

Eurhinodelphidae. (Only as fossils)

Eurymylidae. (Only as fossils)

Eurytheriidae. (Only as fossils)

Eutrachytheriidae, as Mesotheriidae. (Only as fossils)

Eutypomyidae. (Only as fossils)

Felidae (incl. Ailuromachairodontidae, Guepardidae, Lyncidae, Machaerodontidae, Megantereontidae, Nimravidae). COPE, E. D., 1879, Proc. Acad. Nat. Sci. Philadelphia, 13, 1-27 (genera); ELLIOT, D. G., 1883, Monograph of the Felidae, London, 1-108; MATTHEW, W. D., 1910, Bull. American Mus. Nat. Hist., 28, 289-316 (phylogeny); POCOCK, R. I., 1917, Ann. Mag. Nat. Hist., (8), 20, 329-350 (classification); MATTHEW, W. D., 1930, Journ. Mamm., 11, 117-138 (phylogeny of dogs); HALTENORTH, T., 1937, Zeitschr. Saugetierk., 12, 98-240 (verwandschaftliche Stellung); SIMPSON, G. G., No. 1, 1945 (classification); POCOCK, R. I., 1951, Catalogue of the genus Felis); DENIS, A., 1964, Cats of the world, London, 1-119; de BEAUMONT, G., 1964, Eclog. Geol. Helv., 57, 837-845 (classification); ANDERSON & JONES, No. 1, 1967 (families); HALL & KELSON, No. 1, 1959 (North America); ELLIOT, D. G., No. 1, 1904 (Central America & West Indies); GOODWIN, G. G., No. 1, 1942 (Honduras); GOODWIN, G. G., No. 2, 1942 (Costa Rica); GOLDMAN, E. A., No. 1, 1920 (Panama); CABRERA, A., 1911, Rev. Chilena Hist. Nat., 15, 40-54 (synonymic catalog, South America); DUBUC, H. L., 1942, Mem. Soc. Cienc. Nat. LaSalle, 2, (3), 37-39, 2, (5), 26-27; TATE, G. H. H. No. 1, 1939 (Guiana); CABRERA, A., 1961, Rev. Mus. Argentino Cienc. Nat. Bern. Rivadavia, 6, 161-247 (Argentina); MILLER, G. S., Jr., No. 1, 1912 (western Europe); CABRERA, A., No. 3, 1914 (Iberian Peninsula); TOSCHI, A., No. 2, 1965 (Italia); NOVIKOV, G. A., No. 1, 1956 (U. S. S. R.); NOVIKOV, G. A., No. 2, 1962 (U. S. S. R.); BOBRINSKOY, N. et al., No. 1, BOBRINSKOY, N. et al., No. 1, 1944 (U. S. S. R.); BOBRINSKOY, N. et al., No. 2, 1965 (U. S. S. R.); OGNEV, S. I., No. 3, 1933 (U. S. S. R.); OGNEV, S. I., No. 10, 1962; STROGANOV, S. U., 1969, Carnivorous mammals of Siberia, Jerusalem, 1-522; HARRISON, D. A., No. 2, 1968 (Arabia); GLOVER, G. M., No. 3, 1954 (catalog, Africa); BOURKE, D. O., 1966, Sci. Nat. 1966, Sci. Nat. (Paris), 74, 9-15 (Nigeria); ALLEN, J. A., 1924, Bull. American Mus. Nat. Hist., 47, 73-281 (Belgian Congo); HILL & CARTER, No. 1, 1941 (Angola); ELLERMAN, J. R. et al., No. 2, 1953 (southern Africa); HOLLISTER, N., No. 1, 1918 (East Africa); SETZER, H. W., No. 1, 1956 (Anglo-Egyptian Sudan); BLANFORD, W. T., No. 1, 1888-91, (British India); POCOCK, R. I., No. 1, 1939 (British India); PHILLIPS, W. W. A., No. 1, 1935 (Ceylon); JONES, F. W., No. 1, 1923-25 (South Australia); TAYLOR, E. H., No. 1, 1934 (Philippines); ALLEN, G. M., No. 1, 1938 (China); JAKOVLEFF, B. P., 1932, Publ. Mus. Hoang ho Pai ho Tien Tsin, 9, 1-19 (museum list); TEILHARD DE CHARDIN, P. & LEROY, P., 1945, Publ. Inst. Geo.-Biol. Pekin, 11, 1-58 (China)

Furipteridae. SIMPSON, G. G., No. 1, 1945 (classification); ANDERSON & JONES, No. 1, 1967 (families); GOODWIN, & GREENHALL, No. 1, 1961 (Trinidad & Tobago); HUSSON, A. M., No. 1, 1962 (Surinam); da CUNHA VIEIRA, C. O., No. 1, 1942 (Brasil)

Galagidae, as Lorisidae

Galaginidae, as Lorisidae

Galechinidae. (Family not identified)

Galeidae, as Caviidae

Galeopithecidae (incl. Cynocephalidae of Simpson, Galeopteridae). SHUFELDT, R. W., 1911, Philippine Journ. Sci., 6, 139-211 (skeletons); CABRERA, A., No. 2, 1925 (classification); SIMPSON, G. G., No. 1, 1945 (classification); ANDERSON & JONES, No. 1, 1967 (families); BLANFORD,

W. T., No. 1, 1888-91 (India); TAYLOR, E. H., No. 1, 1934 (Philippines, as Galeopteridae)

Galeopteridae, as Galeopithecidae

Garzonidae, as Caenolestidae. (Only as fossils)

Gelocidae. (Only as fossils)

Genettidae, as Viverridae

Geniohyidae (incl. Titanohyracidae). (Only as fossils)

Geogalidae, as Tenrecidae

Geomyidae. MERRIAM, C. H., 1895, North American Fauna, 8, 1-258 (monograph); ELLERMAN, J. R. et al., No. 1, 1941-49 (genera); SIMPSON, G. G., No. 1, 1945 (classification); ANDERSON, S. & JONES, J. K., Jr., No. 1, 1967 (families); HALL & KELSON, No. 1, 1959 (North America); BAILEY, V., 1895, Bull. Div. Ornith. Mamm. United States Dep. Agric., 5, 1047 (pocket gophers of United States); ELLIOTT, D. G., No. 1, 1904 (Central America & West Indies); GOODWIN, G. G., No. 1, 1942 (Honduras); GOODWIN, G. G., No. 2, 1946 (Costa Rica); GOLDMAN, E. A., No. 1, 1920 (Panama)

Georychidae, as Bathyergidae

Gerbillidae, as Cricetidae

Gerboidae, as Dipodidae

Giraffidae (incl. Cameleopardalidae, Camelopardalidae, Helladotheridae, Helladotheriidae, Sivatheriidae). LYDEKKER, R., Catalogue of the ungulate mammals in the British Museum (Natural History), London, vol. 2, 3; SIMPSON, G. G,, No. 1, 1945 (classification); ANDERSON & JONES, No. 1, 1967 (families); ALLEN, G. M., No. 3, 1954 (catalog, Africa); RODE, P., 1944, Faune de l'Empire Francaise, II, Paris, 125-210 (Afrique noire); HILL & CARTER, No. 1, 1941 (Angola); ELLERMAN, J. R. et al., No. 2, 1953 (southern Africa); HOLLISTER, N., No. 3, 1924 (East Africa); SETZER, H. W., No. 1, 1956 (Anglo-Egyptian Sudan); BOHLIN, B., 1926, Palaeont. Sinica, (C), 4, 5-178 (China)

Gliridae (incl. Graphiuridae, Muscardinidae, Myosidae, Myoxidae). ELLERMAN, J. R. et al., No. 1, 1941-49 (genera, as Muscardinidae); SIMPSON, G. G., No. 1, 1945 (classification); ANDERSON & JONES, No. 1, 1967 (families); MILLER, G. S., Jr., No. 1, 1912 (western Europe, as Muscardinidae); STAUDENMAYER, T., 1967, Saugetierk. Mitt., 15, 119-126 (common names in German); CABRERA, A., No. 3, 1914 (Iberian Peninsula, as Muscardinidae); TOSCHI, A., No. 2, 1965 (Italia, as Muscardinidae); OGNEV, S. I., No. 5, 1947 (U. S. S. R., as Myoxidae); OGNEV, S. I., No. 12, 1963 (U. S. S. R., as Myoxidae); BOBRINSKOY, N. et al., No. 1, 1944 (U. S. S. R., as Myoxidae); BOBRINSKOY, N. et al., No. 2, 1965 (U. S. S. R., as Myoxidae); ALLEN, G. M., No. 3, 1954 (Catalog, Africa, as Muscardinidae); RANCK, G. L., No. 1, 1968 (Libya); HILL & CARTER, No. 1, 1941 (Angola, as Muscardinidae); ELLERMAN, J. R. et al., No. 2, 1953 (southern Africa, as Muscardinidae); HOLLISTER, N., No. 2, 1919 (East Africa, as Graphiuridae); SETZER, H. W., No. 1, 1956 (Anglo-Egyptian Sudan); ELLERMAN, J. R., No. 1, 1961 (India, as Muscardinidae)

Globiocephalidae. (Family not identified)

Glyptodontidae (incl. Doedicuridae, Hoplophoridae, Palaeopeltidae, Propalaeohoplophoridae). (Only as fossils)

Gomphotheriidae (incl. Bunomastodontidae, Dibunodontidae, Humboldtidae, Palaeomastodontidae, Serridentidae, Trilophodontidae). (Only as fossils)

Grampidae, as Delphinidae

Graphiuridae, as Gliridae

Guepardidae, as Felidae

Gymnorhinidae, as Squalodontidae. (Only as fossils)

Gymnuridae, as Erinaceidae

Halamydae, as Pedetidae

Halicoridae, as Dugongidae

Halitheriidae, as Dugongidae

Halmaturidae, as Macropodidae

Hapalidae, as Callithricidae

Haplodontheriidae, as Toxodontidae. (Only as fossils)

Haplodontidae, as Aplodontiidae

Haplodontidae, as Toxodontidae. (Only as fossils)

Haploodontidae, as Aplodontidae

Harpaladae, as Callithricidae

Harpyidae, as Pteropidae

Hathlyacynidae. (Only as fossils)

Hegetotheridae, as Hegetotheriidae. (Only as fossils)

Hegetotheriidae (incl. Hegetotheridae, Pachyrucidae). (Only as fossils)

Helaletidae. (Only as fossils)

Heleotragidae. (Family not identified)

Helladotheridae, as Giraffidae

Helladotheriidae, as Giraffidae

Helohyidae, as Choeropotamidae. (Only as fossils)

Hemicyonidae, as Ursidae

Hemisyntrachelidae. (Only as fossils)

Henricosborniidae (incl. Henricosbornidae, Pantostylopidae, Selenoconidae). (Only as fossils)

Henricosbornidae, as Henricosborniidae

Heptaxodontidae (incl. Eumegamyidae, Megamyidae, Neoepiblemidae, Potamarchidae). ANDERSON & JONES, No. 1, 1967 (families); HALL & KELSON, No. 1, 1959 (North America); ANTHONY, H. E., No. 2, 1925-26 (Puerto Rico, fossils)
Herpestidae, as Viverridae
Hesperomyidae, as Cricetidae
Heterodontidae. (Family not identified)
Heteromyidae (incl. Saccomyidae). WOOD, A. E., 1930, American Mus. Novit., 501, 1-19 (review of genera); ELLERMAN, J. R. et al., No. 1, 1941-49 (genera); SIMPSON, G. G., No. 1, 1945 (classification); ANDERSON & JONES, No. 1, 1967 (families); COUES, E., 1875, Proc. Acad. Nat. Sci. Philadelphia, 27, 272-327 (review of North American Saccomyidae); MERRIAM, C. H., 1889, North American Fauna, 1, 1-29 (North America); HALL & KELSON, No. 1, 1959 (North America); ELLIOT, D. G., No. 1, 1904 (Central America & West Indies); GOODWIN, G. G., No. 1, 1942 (Honduras); GOODWIN, G. G., No. 2, 1946 (Costa Rica); GOLDMAN, E. A., No. 1, 1920 (Panama)
Hippidae. (Family not identified)
Hippopotamidae. LYDEKKER, R., 1915, Catalogue of the ungulate mammals in the British Museum (Natural History), London, volume IV; SIMPSON, G. G., No. 1, 1945 (classification); ANDERSON & JONES, No. 1, 1967 (families); ALLEN, G. M., No. 3, 1954 (Catalog, Africa); RODE, P., 1944, Faune de l'Empire Francaise, II, Paris, 125-210 (Afrique noire); HILL & CARTER, No. 1, 1941 (Angola); ELLERMAN, J. R. et al., No. 2, 1953 Southern Africa); HOLLISTER, N., No. 3, 1924 (East Africa); SETZER, H. W., No. 1, 1956 (Anglo-Egyptian Sudan)
Hippotragidae, as Bovidae
Hipposideridae (incl. Phyllorhinidae). SIMPSON, G. G., No. 1, 1945 (classification); HARRISON, D. L., No. 1, 1964 (Arabia); ALLEN, G. M., No. 3, 1954 (catalog, Africa); ROSEVEAR, D. R., No. 1, 1965 (West Africa); HOLLISTER, N., No. 1, 1918 (East Africa); TATE, G. H. H., 1941, Bull. American Mus. Nat. Hist., 78, 353-393 (Indo-Australia); TROUGHTON, E. L., No. 1, 1926 (Australia, New Guinea); JONES, F. W., No. 1, 1923-25 (South Australia); TAYLOR, E. H., No. 1, 1934 (Philippines); ALLEN, G. M., No. 1, 1938 (China)
Hircidae. (Family not identified)
Histricidae, as Hystricidae
Histridae, as Hystricidae
Holoodontidae. (Family not identified)
Homacodontidae. (Only as fossils)
Homalodontotheridae, as Homalodotheriidae. (Only as fossils)
Homalodotheriidae (incl. Homalodontotheridae). (Only as fossils)
Hominidae. ELLIOT, D. G., No. 3, 1913 (monograph); SERGI, G., 1927, Boll. Soc. Adriat. Sci. Nat. Trieste, 29, 1-48 (classification); HILL, W. C. O., Nature, 146, 402-403 (classification); SIMPSON, G. G., No. 1, 1945 (classification); ARAMBOURG, C., 1948, Mammalia, 12, 123-135 (classification; also as 1949); WOO, J. K., 1964, Trud. Mosk. Obshch. Isp. Prir., 14, 161-168 (systematics); BUETTNER-JANUSCH, J., No. 1, 1963 (world list); CAMPBELL, B. G., 1965, Occ. Pap. R. Anthrop. Soc., 22, 1-23 (world list); ANDERSON & JONES, No. 1, 1967 (families); ROBINSON, J. T., 1954, American Journ. Phys. Anthrop., 12, 181-200 (genera & species of Australopithecinae); HALL & KELSON, No. 1, 1959 (North America); ALLEN, G. M., No. 3, 1954 (catalog, Africa)
Homonculidae, as Cebidae. (Only as fossils)
Hoplophoridae, as Glyptodontidae. (Only as fossils)
Humboldtidae, as Gomphotheriidae. (Only as fossils)
Hyaenidae (incl. Proteleidae, Protelidae). SIMPSON, G. G., No. 1, 1945 (classification); ANDERSON & JONES, No. 1, 1967 (families); OGNEV, S. I., No. 2, 1931 (eastern Europe & northern Asia); OGNEV, S. I., No. 9, 1962 (eastern Europe & northern Asia); BOBRINSKOY, N. et al., No. 1, 1944 (U. S. S. R.); BOBRINSKOY, N. et al., No. 2, 1965 (U. S. S. R.); NOVIKOV, G. A., No. 1, 1956 (U. S. S. R.); NOVIKOV, G. A., No. 2, 1962 (U. S. S. R.); HARRISON, D. L., No. 2, 1968 (Arabia); ALLEN, G. M., No. 3, 1954 (catalog, Africa, as Protelidae also); ALLEN, J. A., 1924, Bull. American Mus. Nat. Hist., 47, 73-281 (Belgian Congo); HILL & CARTER, No. 1, 1941 (Angola, also as Protelidae); ELLERMAN, J. R. et al., No. 2, 1953 (southern Africa, also as Protelidae); HOLLISTER, N., No. 1, 1918 (East Africa, also as Protelidae); SETZER, H. W., No. 1, 1956 (Anglo-Egyptian Sudan); BLANFORD, W. T., No. 1, 1888-1891 (British India); POCOCK, R. I., No. 2, 1941 (British India)
Hyaenodontidae (incl. Proviverridae). (Only as fossils)
Hydrarchidae. (Family not identified)
Hydrochaeridae, as Hydrochoeridae
Hydrochoeridae (incl. Hydrochaeridae). SIMPSON, G. G., No. 1, 1945 (classification); ANDERSON & JONES, No. 1, 1967 (families); HALL & KELSON, No. 1, 1959 (North America)
Hydrodamalidae, as Dugongidae
Hydrosoridae. (Family not identified)
Hyemoschidae, as Tragulidae
Hylobatidae, as Pongidae
Hylomidae, as Erinaceidae
Hyopotamidae, as Anthracotheriidae. (Only as fossils)
Hyopsodidae. (Only as fossils)
Hyopsodontidae (incl. Mioclaenidae). (Only as fossils)
Hyperoodontidae. HERSHKOVITZ, P., No. 1, 1966 (synonymic catalog); MOORE, J. C., 1968, Fieldiana, Zool., 53, 205-298 (key to genera)

Hypertragulidae (incl. Leptomerycidae). (Only as fossils)
Hypognathodontidae. (Family not identified)
Hypsiprimnodontidae, as Macropodidae
Hypsiprimnidae, as Macropodidae
Hypsiprymnodontidae, as Macropodidae
Hyrachyidae. (Only as fossils)
Hyracidae, as Procaviidae
Hyracodontidae (incl. Triplopidae, Triplopodidae). (Only as fossils)
Hyracotheridae, as Equidae
Hystrichidae, as Hystricidae
Hystricidae (incl. Histricidae, Histridae, Hystrichidae). ELLERMAN, J. R. et al., No. 1, 1941-49 (genera); SIMPSON, G. G., No. 1, 1945 (classification); ANDERSON & JONES, No. 1, 1967 (families); BOBRINSKOY, N. et al., No. 1, 1944 (U. S. S. R.); BOBRINSKOY, N. et al., No. 2, 1965 (U. S. S. R.); MOHR, E., 1965, Altweltliche Stachelsschweine, Wittenberg Lutherstadt, 1-164; MILLER, G. S., Jr., No. 1, 1912 (western Europe); De GERMINY, G., 1929, Rev. Hist. Nat., 1, A, Mammiferes, 10, (11), 386-391 (Italia); TOSCHI, A., No. 2, 1965 (Italia); ALLEN, G. M., No. 3, 1954 (catalog, Africa); RANCK, G. L., No. 1, 1968 (Libya); HILL & CARTER, No. 1, 1941 (Angola); ELLERMAN, J. R. et al., No. 2, 1953 (southern Africa); HOLLISTER, N., No. 2, 1919 (East Africa); SETZER, H. W., No. 1, 1956 (Anglo-Egyptian Sudan); BLANFORD, W. T., No. 1, 1888-91 (British India); ELLERMAN, J. R., No. 1, 1961 (India); PHILLIPS, W. W. A., No. 1, 1935 (Ceylon); DAMMERMAN, K. W., 1931, Trebbia, 13, 419-470 (Java); TAYLOR, E. H., No. 1, 1934 (Philippines); ALLEN, G. M., No. 2, 1940 (China)

Ictopsidae, as Leptictidae. (Only as fossils)
Idiuridae, as Anomaluridae
Ierboidae, as Dipodidae
Indridae (incl. Indriidae, Indrisidae, Lichanotidae). SIMPSON, G. G., No. 1, 1945 (classification); HILL, W. C. O., No. 1, 1953 (monograph); BUETTNER-JANUSCH, J., No. 1, 1963 (world list); ANDERSON & JONES, No. 1, 1967 (families); ALLEN, G. M., No. 3, 1954 (catalog, Africa)
Indriidae, as Indridae
Indrisidae, as Indridae
Iniidae, as Platanistidae
Interatheridae, as Interatheriidae. (Only as fossils)
Interatheriidae (incl. Interatheridae, Notopithecidae, Protypotheridae). (only as fossils)
Isacidae. (Only as fossils)
Ischyromyidae (incl. Cylindrodontidae, Paramyidae, Sciuravidae). (Only as fossils)
Isectolophidae. (Only as fossils)
Isotemnidae. (Only as fossils)

Jaculidae, as Dipodidae

Kangeroidae. (Family not identified)
Koalidae, as Phalangeridae
Kogiidae, as Physeteridae
Kurtodontidae, as Dryolestidae. (Only as fossils)

Lagidae. (Family not identified)
Lagomerycidae. (Only as fossils)
Lagomyidae, as Ochotonidae
Lagostomidae, as Chinchillidae
Lambdotheriidae, as Brontotheriidae. (Only as fossils)
Lasiopygidae, as Cercopithecidae
Leithiidae. (Only as fossils)
Lemuravidae. (Only as fossils)
Lemuridae (incl. Megaladapidae, Nesopithecidae). ELLIOT, D. G., No. 3, 1913 (monograph); SIMPSON, G. G., No. 1, 1945 (classification); FRIANT, M., 1948, Acta Anat. Basel, 6, (1-2), 151-174 (generic classification); HILL, W. C. O., No. 1, 1953 (monograph, also as Mega.); BUETTNER-JANUSCH, J., No. 1, 1963 (world list); ANDERSON & JONES, No. 1, 1967 (families); ALLEN, G. M., No. 3, 1954 (catalog, Africa); ROSEVEAR, D. R., 1935, Nigerian Field, 4, 16-21 (lemurs of Nigeria); SCHOUTEDEN, H., 1940, Cercle Zool. Congolais, 16, 120-126 (Congo); SCHWARZ, E., 1931, Proc. Zool. Soc. London, 1931, 399-428 (Madagascar); BLANFORD, W. T., No. 1, 1888-91 (British India)
Leontiniidae (incl. Colpodontidae). (Only as fossils)
Leporidae (incl. Lepusidae, Palaeolagidae). LYON, M. W., Jr., 1903, Smithsonian Misc. Coll., 45, 321-447 (classification); DICE, L. R., 1929, Journ. Mamm., 10, 340-344 (phylogeny); SIMPSON, G. G., No. 1, 1945 (classification); ANDERSON & JONES, No. 1, 1967 (families); NELSON, E. W., 1909, North American Fauna, 29, 1-314 (North America); HALL, E. R., 1951, Publ. Univ. Kans. Mus. Nat. Hist., 5, 119-202 (North America); HALL & KELSON, No. 1, 1959 (North America); HOWELL, A. H., 1936, Journ. Mamm., 17, 315-337 (revision American arctic Hares); PALMER, T. S., 1896-97, United States Dep. Agric., Div. Biol. Surv., Bull., 8, 5-88 (United States); ORR, R. T., 1940, Occ. Pap. California Acad. Sci., 19, 1-227 (California); ELLIOT, D. G., No. 1, 1904 (Central America & West Indies); GOODWIN, G. G., No. 1, 1942 (Honduras); GOODWIN, G. G., No. 2, 1946 (Costa Rica); GOLDMAN, E. A., No. 1, 1920

(Panama); HERSHKOVITZ, P., 1950, Proc. United States Nat. Mus., 100, 327-375 (South America); TATE, G. H. H., No. 1, 1939 (Guiana); MILLER, G. S., Jr., No. 1, 1912 (western Europe); CABRERA, A., No. 3, 1914 (Iberian Peninsula); TOSCHI, A., No. 2, 1965 (Italia); OGNEV, S. I., 1922, Yearbook Zool. Mus. Russ. Acad. Sci., 23, 474-496 (Russia, in Russian); OGNEV, S. I., 1929, Zool. Anz., 84, 69-83 (Russia, in German); OGNEV, S. I., No. 4, 1940 (U. S. S. R.); OGNEV, S. I., No. 11, 1966 (U. S. S. R.); BOBRINSKOY, N. et al., No. 1, 1944 (U. S. S. R.); BOBRINSKOY, N. et al., No. 2, 1965 (U. S. S. R.); GUREYEV, A. A., 1964, Fauna S. S. S. R. Leporines, Moskva, 1-276 (S. S. S. R., as Paleolagidae also); ALLEN, G. M., No. 3, 1954 (catalog, Africa); SETZER, H. W., 1958, Journ. Egyptian Publ. Health Assoc., 33, 145-151 (hares, Egypt); HILL & CARTER, No. 1, 1941 (Angola); ELLERMAN, J. R. et al., No. 2, 1953 (southern Africa); HOLLISTER, N., No. 2, 1919 (East Africa); SETZER, H. W., No. 1, 1956 (Anglo-Egyptian Sudan); BLANFORD, W. T., No. 1, 1888-91 (British India); PHILLIPS, W. W. A., No. 1, 1935 (Ceylon); DAMMERMAN, K. W., 1931, Treubia, 13, 429-470 (Java); JONES, F. W., No. 1, 1923-25 (South Australia); ALLEN, G. M., No. 1, 1938 (China); ABE, Y., 1931, Journ. Sci. Hiroshima Univ., (B1), 1, 45-63 (Japan)

Leptictidae (incl. Adapisoricidae, Ictopsidae). (Only as fossils)
Leptochoeridae. (Only as fossils)
Leptomerycidae, as Hypertragulidae
Lepusidae, as Leporidae
Lestodontidae. (Only as fossils)
Lichanotidae, as Indridae
Limnocyonidae. (Only as fossils)
Limnohyidae. (Only as fossils)
Limnotheridae. (Only as fossils)
Listriodontidae, as Suidae. (Only as fossils)
Lobostomidae, as Phyllostomatidae
Loncheridae, as Echimyidae
Lophiodontidae. (Only as fossils)
Lophiomyidae, as Cricetidae
Loridae, as Lorisidae

Lorisidae (incl. Galagidae, Galaginidae, Loridae, Nycticebidae, Nycticibidae, Perodicticinidae). ELLIOT, D. G., No. 2, 1912 (world); ELLIOT, D. G., No. 3, 1913 (monograph, as Nycticebidae); SIMPSON, G. G., No. 1, 1945 (classification); HILL, W. C. O., No. 1, 1953 (monograph, also as Galagidae); BUETTNER-JANUSCH, J., No. 1, 1963 (world list); ANDERSON & JONES, No. 1, 1967 (families); ALLEN, G. M., No. 3, 1954 (catalog, Africa); RODE, P., 1937, Les Primates de l'Afrique, Paris, 1-222; ALLEN, J. A., 1925, Bull. American Mus. Nat. Hist., 47, 283-499 (Belgian Congo); SCHOUTEDEN, H., 1940, Rev. Zool. Bot. Africaines, 33, 289-292 (Belgian Congo); HILL & CARTER, No. 1, 1941 (Angola); ELLERMAN, J. R. et al., No. 2, 1953 (southern Africa); HOLLISTER, N., No. 3, 1924 (East Africa, also as Galagidae); SETZER, H. W., No. 1, 1956 (Anglo-Egyptian Sudan); POCOCK, R. I., No. 1, 1939 (British India); PHILLIPS, W. W. A., No. 1, 1935 (Ceylon); TAYLOR, E. H., No. 1, 1934 (Philippines, as Nycticebidae); ALLEN, G. M., No. 1, 1938 (China)

Lutridae, as Mustelidae
Lycaonidae, as Canidae
Lyncidae, as Felidae

Macacidae, as Cercopithecidae
Machaerodontidae, as Felidae. (Only as fossils)
Macraucheniidae (incl. Adianthidae, Adiantidae). (Only as fossils)
Macropidae, as Macropodidae

Macropodidae (incl. Halmaturidae, Hypsiprimnodontidae, Hypsiprymnodontidae, Hypsiprymnidae, Macropidae, Potoridae). CABRERA, A., No. 1, 1919 (classification); SIMPSON, G. G., No. 1, 1945 (classification); TATE, G. H. H., 1948, Bull. American Mus. Nat. Hist., 91, 233-352 (anatomy & phylogeny); ANDERSON & JONES, No. 1, 1967 (families); FINLAYSON, H. H., 1958, Rec. South Australian Mus., 13, 235-302 (Potoroinae, central Australia); JONES, F. W., No. 1, 1923-25 (South Australia)

Macropristidae. (Only as fossils)
Macroscelidae, as Macroscelididae

Macroscelididae (incl. Macroscelidae, Macroscelidoidae, Rhynchocyonidae). DOBSON, G. E., No. 3, 1882-83 (monograph); CARLSSON, A., 1909, Zool. Jahrb., Syst. Okol. Geogr. Tiere, 28, 349-400 (classification); CABRERA, A., No. 2, 1925 (classification); EVANS, F. G., 1942, Bull. American Mus. Nat. Hist., 80, 85-125 (osteology & relationships); SIMPSON, G. G., No. 1, 1945 (classification); ANDERSON & JONES, No. 1, 1967 (families); CORBET & HANKS, 1968, Bull. British Museum (Nat. Hist.), 16, 47-111 (revision); ALLEN, G. M., No. 3, 1954 (catalog, Africa; earlier ed. as Bull. Mus. Comp. Zool., 83, 1939); HILL & CARTER, No. 1, 1941 (Angola); ELLERMAN, J. R. et al., No. 2, 1953 (southern Africa); HOLLISTER, N., No. 1, 1918 (East Africa); SETZER, H. W., No. 1, 1956 (Anglo-Egyptian Sudan)

Macroscelidoidae, as Macroscelididae
Macrotheriidae, as Chalicotheriidae. (Only as fossils)
Mammutidae (incl. Mastodontidae). (Only as fossils)
Manatidae, as Trichechidae

Manidae (incl. Manididae). JENTINCK, F. A., 1882, Notes Leyden Mus., 4, 193-209 (revision); POCOCK, R. I., 1924, Proc. Zool. Soc. London, 1924, 707-723 (classification); FRECHKOP, S., 1931,

Bull. Mus. Roy. Hist. Nat. Belgique, 7, (22), 1-14 (classification); SIMPSON, G. G., No. 1, 1945 (classification); ANDERSON & JONES, No. 1, 1967 (families); ALLEN, G. M., No. 3, 1954 (catalog, Africa); ROSEVEAR, D. R., 1937, Nigerian Field, 6, 11-14 (Nigeria); PAGES, E., 1965, Biol. Gabon, 1, 209-238 (Gabon); HILL & CARTER, No. 1, 1941 (Angola); ELLERMAN, J. R. et al., No. 2, 1953 (southern Africa); SETZER, H. W., No. 1, 1956 (Anglo-Egyptian Sudan); BLANFORD, W. T., No. 1, 1888-91 (British India); PHILLIPS, W. W. A., No. 1, 1935 (Ceylon); TAYLOR, E. H., No. 1, 1934 (Philippines); ALLEN, G. M., No. 1, 1938 (China)

Manididae, as Manidae

Marsupidae. (Family not identified)

Martidae, as Mustelidae

Mastodontidae, as Mammutidae. (Only as fossils)

Megadermatidae (incl. Megadermidae). SIMPSON, G. G., No. 1, 1945 (classification); ANDERSON & JONES No. 1, 1967 (families); ALLEN, G. M., No. 3, 1954 (catalog, Africa); ROSEVEAR, D. R., No. 1, 1965 (West Africa); ELLERMAN, J. R. et al., No. 2, 1953 (southern Africa); HOLLISTER, N., No. 1, 1918 (East Africa); PHILLIPS, W. W. A., No. 1, 1935 (Ceylon); TROUGHTON, E. L., No. 1, 1926 (Australia & New Guinea); JONES, F. W., No. 1, 1923-25 (South Australia); TAYLOR, E. H., No. 1, 1934 (Philippines)

Megadermidae, as Megadermatidae

Megaladapidae, as Lemuridae. (Only as fossils)

Megalonichidae, as Megalonychidae

Megalonychidae (incl. Megalonichidae, Megalonycidae, Orophodontidae, Ortotheridae). (Only as fossils)

Megalonycidae, as Megalonychidae. (Only as fossils)

Megalotheriidae. (Family not identified)

Megalotidae, as Canidae

Megamyidae, as Heptaxodontidae. (Only as fossils)

Megantereontidae, as Felidae. (Only as fossils)

Megapteridae, as Balaenopteridae

Megatheriidae (incl. Planopsidae, Prepotheridae). (Only as fossils)

Melidae, as Mustelidae

Melididae, as Mustelidae

Melinidae, as Mustelidae

Melissiodontidae, as Cricetidae. (Only as fossils)

Mellivoridae, as Mustelidae

Meniscotheriidae (incl. Pleuraspidotheridae). (Only as fossils)

Menodontidae. (Only as fossils)

Mephitidae, as Mustelidae

Merionidae, as Cricetidae

Merycodontidae, as Antilocapridae. (Only as fossils)

Merycoidodontidae (incl. Oreodontidae, Protoreodontidae). (Only as fossils)

Merycopotamidae, as Anthracotheriidae. (Only as fossils)

Mesonychidae. (Only as fossils)

Mesorhinidae. (Only as fossils)

Mesotheridae, as Mesotheriidae. (Only as fossils)

Mesotheriidae (incl. Eutrachytheriidae, Mesotheridae, Trachytheridae, Typotheriidae). (Only as fossils)

Metacheironyidae. (Only as fossils)

Miacidae (incl. Uintacyonidae, Viverravidae). (Only as fossils)

Microbiotheridae, as Didelphidae. (Only as fossils)

Microbiotheriidae, as Didelphidae. (Only as fossils)

Microchoeridae, as Anaptomorphidae. (Only as fossils)

Microcleptidae (incl. Microlestidae). (Only as fossils)

Microlestidae, as Microcleptidae. (only as fossils)

Microsyopsidae, as Mixodectidae. (Only as fossils)

Microtidae, as Cricetidae

Mididae, as Callithricidae

Mioclaenidae, as Hyopsodontidae. (Only as fossils)

Mixodectidae (incl. Microsyopsidae). (Only as fossils)

Mixtotheriidae, as Cebochoeridae. (Only as fossils)

Mixtotheriodontidae, as Cebochoeridae. (Only as fossils)

Moeritheriidae. (Only as fossils)

Molosidae, as Molossidae

Molossidae (incl. Molosidae). SIMPSON, G. G., No. 1, 1945 (classification); ANDERSON & JONES, No. 1, 1967 (families); PETERSON, R. L., 1965, Contr. Roy. Ontario Mus. Zool., 64, 3-32 (South America & Africa); BARBOUR, R. W. & DAVIS, W. A., 1969, Bats of America, Lexington, Kentucky, 1-286; HALL & KELSON, No. 1, 1959 (North America); ELLIOT, D. G., No. 1, 1904 (Central America & West Indies); ANTHONY, H. E., 1918, Mem. American Mus. Nat. Hist., (n. s.), 2, (2), 335-435 (Puerto Rico); ANTHONY, H. E., No. 2, 1925-26 (Puerto Rico); VILLA-R., B., No. 1, 1966 (Mexico); GOODWIN, G. G., No. 1, 1942 (Honduras); GOODWIN, G. G., No. 2, 1946 (Costa Rica); GOLDMAN, E. A., No. 1, 1920 (Panama); GOODWIN & GREENHALL, No. 1, 1961 (Trinidad & Tobago); HUSSON, A. M., No. 1, 1962 (Surinam); da CUNHA VIEIRA, C. O.. No. 1, 1942 (Brasil); OGNEV, S. I., No. 1, 1928 (eastern Euorpe & northern Asia); OGNEV, S. I., No. 8, 1962 (eastern Europe & northern Asia); MILLER, G. S., Jr., No. 1, 1912 (western Europe); CABRERA, A., No. 3, 1914 (Iberian Peninsula); TOSCHI & LANZA, No. 1, 1959 (Italia); KUZJAKIN, A. P., 1950, Letuchie myshi, Moskva, 1-443

(U. S. S. R.); HARRISON, D. L., No. 1, 1964 (Arabia); ALLEN, G. M., No. 3, 1954 (catalog, Africa); ROSEVEAR, D. R., No. 1, 1965 (West Africa); HILL & CARTER, No. 1, 1941 (Angola); ELLERMAN, J. R. et al., No. 2, 1953 (southern Africa); HOLLISTER, N., No. 1, 1918 (East Africa); PHILLIPS, W. W. A., No. 1, 1935 (Ceylon); TROUGHTON, E. L., No. 1, 1926 (Australia, New Guinea); JONES, F. W., No. 1, 1923-25 (South Australia); TAYLOR, E. H., No. 1, 1934 (Philippines); ALLEN, G. M., No. 1, 1938 (China)

Monodontidae (incl. Belugidae, Delphinapteridae). SIMPSON, G. G., No. 1, 1945 (classification); HERSHKOVITZ, P., No. 1, 1966 (synonymic catalog); ANDERSON & JONES, No. 1, 1967 (families); HALL & KELSON, No. 1, 1959 (North America)

Mormopidae, as Phyllostomatidae

Moropodidae, as Chalicotheriidae. (Only as fossils)

Moschidae, as Cervidae

Mungotidae, as Viverridae

Muridae (incl. Rattidae). HINTON, M. A. C., 1926, Monograph of the ... Microtinae, I, London, 1-488 (subfamily Microtinae); ELLERMAN, J. R. et al., No. 1, 1941-49 (genera); SIMPSON, G. G., No. 1, 1945 (classification); HERSHKOVITZ, P., 1962, Fieldiana, Zool., 46, 1-524 (phylogeny) ANDERSON & JONES, No. 1, 1967 (families); THOMAS, O., 1918, Ann. Mag. Nat. Hist., (9), 2, 203-211 (classification, Otomyinae); HALL, E. R. & COCKRUM, E. L., 1953, Publ. Univ. Kansas Mus. Nat. Hist., 5, (27), 373-498 (Microtinae, North America); HALL & KELSON, No. 1, 1959 (North America); ELLIOT, D. G., No. 1, 1904 (Central America & West Indies); THOMAS, O., 1898, Ann. Mag. Nat. Hist., (7), 1, 176-180 (West Indies); GOODWIN, G. G., No. 1, 1942 (Honduras); GOODWIN, G. G., No. 2, 1946 (Costa Rica); GOLDMAN, E. A., No. 1, 1920 (Panama); PHILIPPI, R. A., 1900, Anal. Mus. Nac. Chile, 14a, 5-70 (Chile); VASVARI, N., 1923, Zool. Palearc. Dresden, 1, 23-32 (Europe); MILLER, G. S., Jr., No. 1, 1912 (western Europe); DIDIER, R. & RODE, P., 1942, Mammalia, 6, 36-45 (Murinae, France); DIDIER, R. & RODE, P., 1944, Mammalia, 8, 47-66 (France); CABRERA, A., No. 3, 1914 (Iberian Peninsula); TOSCHI, A., No. 2, 1965 (Italia); LAPIN, I., 1956, Vestn. Latv. P. S. R. zin Akad., 2, (103), 87-92 (Latvia, in Lett); ARGYROPULO, A. I., 1940, Faune de l'URSS, Mammiferes, III, 5, Muridae, Moskva, 1-170; BOBRINSKOY, N. et al., No. 1, 1944 (U. S. S. R.); BOBRINSKOY, N. et al., No. 2, 1965 (U. S. S. R.); OGNEV, S. I., No. 6, 1948 (U. S. S. R.); OGNEV, S. I., No. 13, 1963 (U. S. S. R.); OGNEV, S. I., No. 7, 1950 (U. S. S. R.); OGNEV, S. I., No. 14, 1964 (U. S. S. R.); NEUHAUSER, G., 1936, Journ. Saugetierk., 11, 163-236 (Asia Minor); AKARONI, B., 1932, Zeitschr. Saugetierk., 7, 166-240 (Palestine & Syria); ALLEN, G. M., No. 3, 1954 (catalog, Africa); DAVIS, D. H. S., 1965, Zool. Afric., 1, 121-145 (Africa); RANCK, G. L., No. 1, 1968 (Libya); CABRERA, A., 1921, Roy. Soc. Est. Hist. Nat., 50, 42-58 (Morocco); HATT, R. T., 1940, Bull. American Mus. Nat. Hist., 76, 457-604 (Belgian Congo); HILL & CARTER, No. 1, 1941 (Angola); ELLERMAN, J. R. et al., No. 2, 1953 (southern Africa); HOLLISTER, N., No. 2, 1919 (East Africa); DELANY M. J. & NEAL, B. R., 1966, Bull. British Museum (Nat. Hist.), Zool., 13, 294-355 (Uganda); SETZER, H. W., No. 1, 1956 (Anglo-Egyptian Sudan); TATE, G. H. H., 1936, Bull. American Mus. Nat. Hist., 72, 501-728 (review of Indo-Australian Murinae); BLANFORD, W. T., No. 1, 1888-91 British India); ELLERMAN, J. R., No. 1, 1961 (India); WROUGHTON, R. C., 1913, Journ. Bombay Nat. Hist. Soc., 22, 19-21 (generic classification, Murinae, India); PHILLIPS W. W. A., No. 1, 1935 (Ceylon); MARSHALL, J. T. & SOMSAK PANTUWATANA, 1966, Identification of rats in Thailand, Bangkok, 1-22; HARRISON, J. L., 1948, Malayan Nature Journ., 3, 130-141 (Malaya); HARRISON, J. L., 1949, Med. Journ. Malaya, 4, 96-105 (domestic rats of Malaya); TATE, G. H. H., 1950, Bull. Raffles Mus., 22, 271-277 (Cocos & Keeling Is.); SODY, H. J. V., 1930, Zool. Meded., 13, 100-140 (Java); RUEMMLER, H., 1938, Mitt. Zool. Mus. Berlin, 23, (1), 1-197 (New Guinea); LONGMAN, H. A., 1916, Mem. Queensland Mus., 5, 23-45 (list, Australia & Australo-Pacific); BRAZENOR, C. W., 1936, Mem. Nat. Mus. M Melbourne, 10, 62-85 (Victoria); JONES, F. W., No. 1, 1923-25 (South Australia); TAYLOR, E. H., No. 1, 1934 (Philippines); AOKI, B. & TANAKA, R., 1941, Mem. Fac. Sci. Agric. Taihoku Imp. Univ., 23, 121-191 (Formosa); TANAKA, R., 1944, A study of Muridae of Hainan I., Taihoku, 129-147; BONHOTE, J. L., 1906, Proc. Zool. Soc. London, 1905, 384-397 (subfamily Murinae, China); ALLEN, G. M., No. 2, 1940 (China); TOKUDA, M., 1941, Biogeographica, 4, 1-155 (Japan, Korea, Manchukuo); TOKUDA, M., 1932, Trans. Nat. Hist. Soc. Sapporo, 12, 206-211(museum list, Univ. Mus. Nat. Hist. Sapporo); TOKUDA, M., 1934, Bot. & Zool., Tokyo, 2, 1995-2008 (keys to Muridae of Nippon, in Japanese); NICHOLSON, A. J., 1953, Journ. Mamm., 34, 168-179 (New Caledonia)

Muriformidae. (Family not identified)

Muscardinidae, as Gliridae

Mustelidae (incl. Enhydridae, Lutridae, Martidae, Melidae, Melididae, Melinidae, Mellivoridae, Mephitidae, Zorillidae). SATUNIN, K. A., 1911, Mitt. Kaukas. Mus. Tiflis, 5, 243-280 (systematics); SIMPSON, G. G., No. 1, 1945 (classification); GOETHE, F., 1964, Handb. der Zool., 37, 1-80 (monograph); ANDERSON & JONES, No. 1, 1967 (families); POCOCK, R. I., 1922, Proc. Zool. Soc. London, 1921, 803-837 (external characters & classification); ALLEN, J. A., 1901, Bull. American Mus. Nat. Hist., 14, 325-334 (generic names of Mephitinae); POHLE, H., 1919, Arch. Naturg., 185, (A9), 1-247 (subfamily Lutrinae); HARRIS, C. J., 1968, Otters. A study of the Recent Lutrinae, London, 1-389; HALL, E. R., 1944, Proc. California Acad. Acad. Sci., (4), 23, 555-560 (classification of ermines); HALL, E. R., 1951, Publ. Univ. Kansas Mus. Nat. Hist., 4, 1-466 (America); HALL & KELSON, No. 1, 1959 (North America) ELLIOT, D. G., No. 1, 1904 (Central America & West Indies); GOODWIN, G. G., No. 1, 1942 (Honduras); GOODWIN, G. G., No. 2, 1942 (Costa Rica); GOLDMAN, E. A., No. 1, 1920

(Panama); TATE, G. H. H., No. 1, 1939 (Guiana); HOUSSE, R. P. R., 1951, Rev. Univ. Santiago, 36, 105-117, 145-165 (Chile); MILLER, G. S., Jr., No. 1, 1912 (western Europe); CABRERA, A., No. 3, 1914 (Iberian Peninsula); TOSCHI, A., No. 2, 1965 (Italia); CALINESCU, R. I., 1931, Trav. Inst. Geogr. Univ. Cluj, 4, 61-77 (Romania, in Romanian); CALINESCU, R. I., 1933, Verh. Siebenburg. Ver. Naturw., 81-82, 16-34 (Romania); OGNEV, S. I., No. 3, 1935 (U. S. S. R.) OGNEV, S. I., No. 10, 1962 (U. S. S. R.); BOBRINSKOY, N. et al., No. 1, 1944 (U. S. S. R.); BOBRINSKOY, N. et al., No. 2, 1965 (U. S. S. R.); NOVIKOV, G. A., No. 1, 1956 (U. S. S. R.); NOVIKOV, G. A., No. 2, 1962 (U. S. S. R.); OGNEV, S. I., No. 2, 1931 (eastern Europe & northern Asia); OGNEV, S. I., No. 9, 1962 (eastern Europe & Northern Asia); STROGANOV, S. U., 1969 (Carnivorous mammals of Siberia, Jerusalem, 1-522; HARRISON, D. L., No. 2, 1968 (Arabia); ALLEN, G. M., No. 3, 1954 (catalog, Africa); SETZER, H. W., 1958, Journ. Egyptian Publ. Health Assoc., 33, 199-204 (Egypt); ALLEN, J. A., 1924, Bull. American Mus. Nat. Hist., 47, 73-281 (Belgian Congo); SCHOUTEDEN, H., 1940, Rev. Zool. Bot. Africaines, 35, 412-416 (Belgian Congo, as Lutridae); HILL & CARTER, No. 1, 1941 (Angola); ELLERMAN, J. R. et al., No. 2, 1953 (southern Africa); HOLLISTER, N., No. 1, 1918 (East Africa); SETZER, H. W., No. 1, 1956 (Anglo-Egyptian Sudan); BLANFORD, W. T., No. 1, 1888-91 (India); POCOCK, R. I., No. 2, 1941 (British India); PHILLIPS, W. W. A., No. 1, 1935 (Ceylon); TAYLOR, E. H., No. 1, 1934 (Philippines); ALLEN, G. M., No. 1, 1938 (China); TEILHARD de CHARDIN, P. & LEROY, P., 1945, Publ. Inst. Geo-Biol. Pekin, 12, 1-56 (China); JAKOVLEFF, B. P., 1934, Publ. Mus. Hoangho Paiho Tien Tsin, 28, 1-30 (museum list); BANNIKOV, A. G., 1952, Bull. Soc. Nat. Moscou, Biol., 57, 30-44 (Mongolia)

Myaladae, as Talpidae
Mylagaulidae. (Only as fossils)
Mylodontidae (incl. Scelidotheridae). (Only as fossils)
Myocastoridae, as Octodontidae
Myogalidae, as Talpidae
Myohyracidae. (Only as fossils)
Myosidae, as Gliridae
Myoxidae, as Gliridae
Myrmecobiidae, as Dasyuridae

Myrmecophagidae. SIMPSON, G. G., No. 1, 1945 (classification); ANDERSON & JONES, No. 1, 1967 (families); HALL & KELSON, No. 1, 1959 (North America); GOODWIN, G. G., No. 2, 1946 (Costa Rica); GOODWIN, G. G., No. 1, 1942 (Honduras); GOLDMAN, E. A., No. 1, 1920 (Panama); TATE, G. H. H., No. 1, 1939 (Guiana); VIEIRA, C., 1950, Arqu. Zool. Sao Paulo, 7, 325-362 (Estado de Sao Paulo)

Mystacinidae (incl. Mystacopidae). SIMPSON, G. G., No. 1, 1945 (classification); ANDERSON & JONES, No. 1, 1967 (families); ELLIOT, D. G., No. 1, 1904 (Central America & West Indies)

Mystacopidae, as Mystacinidae
Mystromyidae, as Cricetidae
Mythomyidae. (Family not identified)

Myzopodidae. SIMPSON, G. G., No. 1, 1945 (classification); ANDERSON & JONES, No. 1, 1967 (families); ALLEN, G. M., No. 3, 1954 (catalog, Africa)

Narvallidae. (Family not identified)
Nasuidae, as Procyonidae

Natalidae. SIMPSON, G. G., No. 1, 1945 (classification); DALQUEST, W. W., 1950, Journ. Mamm., 31, 436-443 (genera); ANDERSON & JONES, No. 1, 1967 (families); HALL & KELSON, No. 1, 1959 (North America); ELLIOT, D. G., No. 1, 1904 (Central America & West Indies); VILLA-R., B., No. 1, 1966 (Mexico); GOODWIN, G. G., No. 1, 1942 (Honduras); GOODWIN, G. G., No. 2, 1946 (Costa Rica); GOLDMAN, E. A., No. 1, 1920 (Panama); GOODWIN & GREENHALL, No. 1, 1961 (Trinidad & Tobago); da CUNHA VIEIRA, C. O., No. 1, 1942 (Brasil)

Necrolemuridae, as Anaptomorphidae. (Only as fossils)
Necrolestidae. (Only as fossils)
Nematheridae. (Only as fossils)
Neobalaenidae, as Balaenidae
Neoepiblemidae, as Heptaxodontidae. (Only as fossils)
Neoplagiaulacidae, as Ptilodontidae. (Only as fossils)
Nesodontidae, as Toxodontidae. (Only as fossils)
Nesokerodontidae, as Theridonyidae. (Only as fossils)
Nesomyidae, as Cricetidae

Nesophontidae (as sub-fossils). ANTHONY, H. E., 1916, Bull. American Mus. Nat. Hist., 35, 725-728 (new family); CABRERA, A., No. 2, 1925 (classification); HALL & KELSON, No. 1, 1959 (North America); ANTHONY, H. E., 1918, Mem. American Mus. Nat. Hist., (n. s.), 2, (2), 335-435 (Puerto Rico); ANTHONY, H. E., No. 2, 1925-26 (Puerto Rico)

Nesopithecidae, as Lemuridae. (Only as fossils)
Nesotragidae, as Bovidae
Nimravidae, as Felidae

Noctilionidae. SIMPSON, G. G., No. 1, 1945 (classification); ANDERSON & JONES, No. 1, 1967 (families); HALL & KELSON, No. 1, 1959 (North America); ANTHONY, H. E., 1918, Mem. American Mus. Nat. Hist., (n. s.), 2, (2), 333-435 (Puerto Rico); ANTHONY, H. E., No. 2, 1925-26 (Puerto Rico); ELLIOT, D. G., No. 1, 1904 (Central America & West Indies); VILLA-R., B., No. 1, 1966 (Mexico); GOODWIN, G. G., No. 1, 1942 (Honduras); GOODWIN, G. G., No. 2, 1946

(Costa Rica); GOLDMAN, E. A., No. 1, 1920 (Panama); GOODWIN & GREENHALL, No. 1, 1961 (Trinidad & Tobago); HUSSON, A. M., No. 1, 1962 (Surinam); da CUNHA VIEIRA, C. O., No. 1, 1942 (Brasil)

Notharctidae, as Adapidae. (Only as fossils)

Notohippidae (incl. Rhynchippidae). (Only as fossils)

Notopithecidae, as Interatheriidae. (Only as fossils)

Notoryctidae. CABRERA, A., No. 1, 1919 (classification); SIMPSON, G. G., No. 1, 1945 (classification); ANDERSON & JONES, No. 1, 1967 (families); JONES, F. W., No. 1, 1923-25 (South Australia)

Notostylopidae. (Only as fossils)

Nototheriidae, as Diprotodontidae. (Only as fossils)

Nycteridae (incl. Petaliidae). DOBSON, G. E., No. 2, 1878-(British Museum catalogue, reprinted in 1966); SIMPSON, G. G., No. 1, 1945 (classification); ANDERSON & JONES, No. 1, 1967 (families); TOSCHI & LANZA, No. 1, 1959 (Italia); DOBSON, G. E., No. 1, 1876 (Asia); HARRISON, D. L., No. 1, 1964 (Arabia); ALLEN, G. M., No. 3, 1954 (catalog, Africa); ROSEVEAR, D. R., No. 1, 1965 (West Africa); HILL & CARTER, No. 1, 1941 (Angola); ELLERMAN, J. R. et al., No. 2, 1953 (southern Africa); HOLLISTER, N., No. 1, 1918 (East Africa, as Petaliidae); BLANFORD, W. T., No. 1, 1888-91 (British India)

Nycticebidae, as Lorisidae

Nycticibidae, as Lorisidae

Nyctitheriidae. (Only as fossils)

Ochotonidae (incl. Lagomyidae). SIMPSON, G. G., No. 1, 1945 (classification); ANDERSON & JONES, No. 1, 1967 (families); HOWELL, A. H., 1924, North American Fauna, 47, 1-57 (America); HALL, E. R., 1951, Publ. Univ. Kansas Mus. Nat. Hist., 5, 119-202 (North America); HALL & KELSON, No. 1, 1959 (North America); OGNEV, S. I., No. 4, 1940 (U. S. S. R.); OGNEV, S. I., No. 11, 1966 (U. S. S. R.); BOBRINSKOY, N. et al., No. 1, 1944 (U. S. S. R.); BOBRINSKOY, N. et al., No. 2, 1965 (U. S. S. R.); GUREYEV, A, A., 1964, Fauna SSSR, Leporines, Moskva, 1-276; BLANFORD, W. T., No. 1, 1888-91 (British India); ALLEN, G. M., No. 1, 1938 (China); LOUKASHKIN, A. S., 1940, Journ. Mamm., 21, 402-405 (Manchuria)

Octodontidae (incl. Myocastoridae, Spalacopodidae). SIMPSON, G. G., No. 1, 1945 (classification); ANDERSON & JONES, No. 1, 1967 (families, also as Myocastoridae); FISCHER, G. M., 1970, Bol. Mus. Nac. Hist. Nat. Santiago, 18, 103-124 (anatomy); ELLIOT, D. G., No. 1, 1904 (Central American & West Indies); ANTHONY, H. E., No. 1, 1918 (Puerto Rico); GOLDMAN, E. A., No. 1, 1920 (Panama); ELLERMAN, J. R. et al., No. 2, 1953 (southern Africa)

Odobaenidae, as Odobenidae

Odobenidae (incl. Odobaenidae, Rosmaridae, Trichecidae). SIMPSON, G. G., No. 1, 1945 (classification); CHARTERS, E. M., 1947, Wildlife Leaflet, 293, 1-14 (Bibliography on walrus); SCHEFFER, V. B., 1958, Seals ... review of the Pinnipedia, Stanford, California, 1-179 (world); KING, J. E., 1964, Seals of the world, London, 1-154 (monograph); MAXWELL, G., 1967, Seals of the world, London, 1-153 (world list); ANDERSON & JONES, No. 1, 1967 (families); ALLEN, J. A., 1880, United States Geol. Geogr. Surv. Terr., Misc. Publ., 12, 1-785 (North America); HALL & KELSON, No. 1, 1959 (North America); OGNEV, S. I., No. 3, 1933 (U. S. S. R.); OGNEV, S. I., No. 10, 1962 (U. S. S. R.); BOBRINSKOY, N. et al., No. 1, 1944 (U. S. S. R.); BOBRINSKOY, N. et al., No. 2, 1965 (U. S. S. R.)

Odontomysopidae. (Only as fossils)

Oegosceridae (incl. Aegosceridae). (No revisionary references Noted)

Oldfieldthomasiidae (incl. Acoelodidae). (Only as fossils)

Oreodontidae, as Merycoidodontidae. (Only as fossils)

Ornithorhynchidae. CABRERA, A., No. 1, 1919 (classification); BURRELL, H., 1927 The platypus ..., Sydney, 1-227; SIMPSON, G. G., No. 1, 1945 (classification); ANDERSON & JONES, No. 1, 1967 (families); JONES, F. W., No. 1, 1923-25 (South Australia)

Orophodontidae, as Megalonychidae

Ortholophodontidae. (Family not identified)

Ortotheridae, as Megalonychidae. (ONly as fossils)

Orycteropidae, as Orycteropodidae

Orycteropodidae (incl. Orycteropidae). SIMPSON, G. G., No. 1, 1945 (classification); FRICK, H., 1956, Saugetierk. Mitt., 4, 15-17 (Taxonomie); ANDERSON & JONES, No. 1, 1967 (families); ALLEN, G. M., No. 3, 1954 (catalog, Africa); HILL & CARTER, No. 1, 1941 (Angola); ELLERMAN, J. R. et al., No. 2, 1953 (southern Africa); HOLLISTER, N., No. 2, 1919 (East Africa); SETZER, H. W., No. 1, 1956 (Anglo-Egyptian Sudan)

Orygidae, as Bovidae

Oryzoryctidae, as Tenrecidae

Otariidae (incl. Allodesmidae, Desmatophocidae). SIMPSON, G. G., No. 1, 1945 (classification); SIVERTSEN, E., 1956, 14th Intern. Congr. Zool. 1953, 522-523 (review); SCHEFFER, V. B., 1958, Seals ... review of the Pinnipedia, Stanford, California, 1-179; KING, J. E., 1964, Seals of the world, London, 1-154 (monograph); MAXWELL, G., 1967, Seals of the world ..., London, 1-153 (world list); ANDERSON & JONES, No. 1, 1967 (families); ALLEN, J. A., 1880, United States Geol. Geogr. Surv. Terr., Misc. Publ., 12, 1-785 (North America); HALL & KELSON, No. 1, 1959 (North America); KING, J. E., 1954, Bull. British Mus. (Nat. Hist.), Zool., 2, 309-337 (Pacific Coast of America); ELLIOT, D. G., No. 1, 1904 (Central America & West Indies); OGNEV, S. I., No. 3, 1935 (U. S. S. R.); OGNEV, S. I., No. 10, 1962 (U. S. S. R.); BOBRINSKOY, N. et al., No. 1, 1944 (U. S. S. R.); BOBRINSKOY, N. et al., No. 2, 1965 (U. S. S. R.); ALLEN, G. M., No. 3, 1954 (catalog, Africa); ELLERMAN, J. R. et al.,

No. 2, 1953 (southern Africa); JONES, F. W., No. 1, 1923-25 (South Australia); ALLEN, G. M., No. 1, 1938 (China); EIBL-EIBESFELDT, I., 1957, Journ. Mamm., 37, 549 (Galapagos Is.)
Otocyonidae, as Canidae
Ouistitidae. (Family not identified)
Ovibovidae, as Bovidae
Ovidae, as Bovidae
Oxyaenidae. (Only as fossils)
Oxyclaenidae, as Arctocyonidae. (Only as fossils)

Pachylemuridae. (Only as fossils)
Pachynolophidae. (Only as fossils)
Pachyrucidae, as Hegetotheriidae. (Only as fossils)
Palaeocetidae. (Only as fossils)
Palaeochiropterygidae. (Only as fossils)
Palaeolagidae, as Leporidae. (Only as fossils)
Palaeomastodontidae, as Gomphotheriidae. (Only as fossils)
Palaeomerycidae, as Cervidae. (Only as fossils)
Palaeonictidae. (Only as fossils)
Palaeopeltidae. (Only as fossils)
Palaeoryctidae. (Only as fossils)
Palaeosyopidae, as Brontotheriidae. (Only as fossils)
Palaeothentidae, as Caenolestidae. (Only as fossils)
Palaeotheridae, as Palaeotheriidae. (Only as fossils)
Palaeotheriidae (incl. Palaeotheridae). (Only as fossils)
Pantholopidae, as Bovidae
Pantolambdidae, as Coryphodontidae. (Only as fossils)
Pantolambdodontidae. (Only as fossils)
Pantolestidae. (Only as fossils)
Pantostylopidae, as Henricosborniidae. (Only as fossils)
Papionidae, as Cercopithecidae
Paradoxuridae, as Viverridae
Paramyidae, as Ischyromyidae. (Only as fossils)
Parapithecidae. (Only as fossils)
Parasoricidae. (Only as fossils)
Paurodontidae. (Only as fossils)
Pectinatoridae, as Ctenodactylidae
Pedetidae (incl. Halamydae). ELLERMAN, J. R. et al., No. 1, 1941-49 (genera); SIMPSON, G. G., No. 1, 1945 (classification); ANDERSON & JONES, No. 1, 1967 (families); ALLEN, G. M., No. 3, 1954 (African, catalog); HILL & CARTER, No. 1, 1941 (Angola); ELLERMAN, J. R. et al., No. 2, 1953 (southern Africa); HOLLISTER, N., No. 2, 1919 (East Africa)
Peltephilidae. (Only as fossils)
Peralestidae. (Only as fossils)
Peramelidae. WROUGHTON, R. C., 1908, Journ. Bombay Nat. Hist. Soc., 18, 736-752 (classification); CABRERA, A., No. 1, 1919 (classification); SIMPSON, G. G., No. 1, 1945 (classification); TATE, G. H. H., 1948, Bull. American Mus. Nat. Hist., 92, 313-346 ("studies"); ANDERSON & JONES, No. 1, 1967 (families); JONES, F. W., No. 1, 1923-25 (South Australia); FINLAYSON, 1935, Trans. Proc. Roy. Soc. South Australia, 59, 227-236 (Lake Eyre Basin)
Periptychidae. (Only as fossils)
Perodicticinidae, as Lorisidae
Petaliidae, as Nycteridae
Petauristidae, as Sciuridae
Petromyidae. SIMPSON, G. G., No. 1, 1945 (classification); ANDERSON & JONES, No. 1, 1967 (families)
Phacochoeridae, as Suidae
Phalangeridae (incl. Koalidae, Phalangistidae, Phascolarctidae, Tarsipedidae). CABRERA, A., No. 1, 1919 (classification); SONNTAG, C. F., 1923, Proc. Zool. Soc. London, 1922, 863-869 (myology & classification); SIMPSON, G. G., No. 1, 1945 (classification); ANDERSON & JONES, No. 1, 1967 (families); JONES, F. W., No. 1, 1923-25 (South Australia)
Phalangistidae, as Phalangeridae
Phascolarctidae, as Phalangeridae
Phascolomidae (incl. Phascolomyidae, Vombatidae). CABRERA, A., No. 1, 1919 (classification); SONNTAG, C. F., 1923, Proc. Zool. Soc. London, 1922, 863-869 (myology & classification); SIMPSON, G. G., No. 1, 1945 (classification); TATE, G. H. H., American Mus. Novit., 1925, 1-18; ANDERSON & JONES, No. 1, 1967 (families); ANDERSON & JONES, No. 1, 1967 (families, as Vombatidae); SPENCER, B., 1910, The existing species of the genus Phascolomys, Melbourne; JONES, F. W., No. 1, 1923-25 (South Australia)
Phascolomyidae, as Phascolomidae
Phascolotheridae. (Only as fossils)
Phascolotheriidae, as Triconodontidae. (Only as fossils)
Phenacodidae, as Phenacodontidae
Phenacodontidae (incl. Phenacodidae). (Only as fossils)
Phenacolemuridae. (Only as fossils)
Phocaenidae (incl. Phocaenoidae). SIMPSON, G. G., No. 1, 1945 (classification); ANDERSON & JONES, No. 1, 1967 (families); TOSCHI, A., No. 2, 1965 (Italia)

Phocaenoidae, as Phocaenidae
Phocidae (incl. Cystophoridae). SIMPSON, G. G., No. 1, 1945 (classification); SCHEFFER, V. B., 1958, Seals ... review of the Pinnipedia, Stanford, California, 1-179; KING, J. E., 1964, Seals of the world, London, 1-154 (monograph); MAXWELL, G., 1967, Seals of the world ..., London, 1-153 (world list); ANDERSON & JONES, No. 1, 1967 (families); CHAPSKII, K. K., 1955, Transl. Ser. Fish. Res. Board Canada, 114, 1-57 (subfamily Phocinae); ALLEN, J. A., United States Geol. Geogr. Surv. Terr. Misc. Publ., 12, 1-785, 1880, (North America); HALL & KELSON, No. 1, 1959 (North America); ELLIOT, D. G., No. 1, 1904 (Central America & West Indies); GOODWIN, G. G., No. 1, 1942 (Honduras); GOODWIN, G. G., No. 2, 1942 (Costa Rica); MOHR, E., 1952, Monogr. Wildsauget., 12, 1-283 (Deutschland); CABRERA, A., No. 3, 1914 (Iberian Peninsula); TOSCHI, A., No. 2, 1965 (Italia); OGNEV, S. I., No. 3, 1935 (U. S. S. R.); OGNEV, S. I., No. 10, 1962 (U. S. S. R.); BOBRINSKOY, N. et al., No. 1, 1944 (U. S. S. R.); BOBRINSKOY, N. et al., No. 2, 1965 (U. S. S. R.); CHAPSKII, K. K., 1955, Trudi Zool. Inst. Akad. Nauk SSSR, 17, 160-199 (subfamily Phocinae, U. S. S. R.); ELLERMAN, J. R. et al., No. 2, 1953 (southern Africa); JONES, F. W., No. 1, 1923-25 (South Australia) ALLEN, G. M., No. 1, 1938 (China); ALLEN, G. A., 1902, Bull. American Mus. Nat. Hist., 16, 459-499 (North Pacific Ocean & Bering Sea)
Phyllorhinidae, as Hipposideridae
Phyllostomatidae (incl. Lobostomidae, Mormopidae, Phyllostomidae, Stenodermatidae, Vampyridae). DOBSON, G. E., No. 2, 1878 (British Museum catalog); SIMPSON, G. G., No. 1, 1945 (classification); ANDERSON & JONES, No. 1, 1967 (families); BAKER, R. J., 1967, Southwestern Nat., 12, 407-428 (karyotypes); BARBOUR, R. W. & DAVIS, W. A., 1969, Bats of America, Lexington, Kentucky, 1-286; ALLEN, H., 1893, United States Nat. Mus. Bull., 43, 1-198 (North America); HALL & KELSON, No. 1, 1959 (North America); ELLIOT, D. G., No. 1, 1904 (Central America & West Indies); ANTHONY, H. E., 1918, Mem. American Mus. Nat. Hist., (n. s.), 2, (2), 335-435 (Puerto Rico); ANTHONY, H. E., No. 2, 1925-26 (Puerto Rico); VILLA-R., B., No. 1, 1966 (Mexico); GOODWIN, G. G. G., No. 1, 1942 (Honduras); GOODWIN, G. G., No. 2, 1946 (Costa Rica); GOLDMAN, E. A., No. 1, 1920 (Panama); GOODWIN, G. G. & GREENHALL, A. M., No. 1, 1961 (Trinidad & Tobago); HUSSON, A. M., No. 1, 1962 (Surinam); da CUNHA VIEIRA, C. O., No. 1, 1942 (Brasil); DOBSON, G. E., No. 1, 1876 (Asia)
Phyllostomidae, as Phyllostomatidae
Physalinidae. (Family not identified)
Physalidae. (Family not identified)
Physeteridae (incl. Kogiidae, Physodontidae). SIMPSON, G. G., No. 1, 1945 (classification); HERSHKOVITZ, P., No. 1, 1966 (synonymic catalog); ANDERSON & JONES, No. 1, 1967 (families); HALL & KELSON, No. 1, 1959 (North America, also as Kogiidae); ELLIOT, D. G., No. 1, 1904 (Central America & West Indies); CABRERA, A., No. 3, 1914 (Iberian Peninsula); TOSCHI, A., No. 2, 1965 (Italia); BOBRINSKOY, N. et al., No. 1, 1944 (U. S. S. R.); BOBRINSKOY, N. et al., No. 2, 1965 (U. S. S. R.); TOMILIN, A. G., No. 2, 1957 (U. S. S. R.); TOMILIN, A. G., No. 1, 1967 (U. S. S. R.); ALLEN, G. M., No. 3, 1954 (Africa, catalog, also as Kogiidae); ELLERMAN, J. R. et al., No. 2, 1953 (southern Africa); BLANFORD, W. T., No. 1, 1888-91 (British India); JONES, F. W., No. 1, 1923-25 (South Australia); ALLEN, D. G., No. 1, 1938 (China)
Physodontidae, as Physeteridae
Picrodontidae. (Only as fossils)
Pithecanthropidae. (Only as fossils)
Pithecidae, as Cercopithecidae
Plagiaulacidae (incl. Bolodontidae). (Only as fossils)
Plagiomenidae. (Only as fossils)
Planopsidae, as Megatheriidae. (Only as fossils)
Platacanthomyidae. SIMPSON, G. G., No. 1, 1945 (classification); ANDERSON & JONES, No. 1, 1967 (families); ALLEN, G. M., No. 2, 1940 (China)
Platanistidae (incl. Iniidae, Susuidae). SIMPSON, G. G., No. 1, 1945 (classification); HERSHKOVITZ, P., No. 1, 1966 (synonymic catalog, as Susuidae); ANDERSON & JONES, No. 1, 1967 (families); BLANFORD, W. T., No. 1, 1888-91 (British India); ALLEN, G. M., No. 1, 1938 (China)
Platycerinidae. (Family not identified)
Platychoeropidae. (Only as fossils)
Pleopodidae. (Family not identified)
Plesiadapidae. (Only as fossils)
Pleuraspidotheridae, as Meniscotheriidae. (Only as fossils)
Pleuropteridae. (Family not identified)
Pliohyracidae, as Procaviidae. (Only as fossils)
Pliolophidae. (Family not identified)
Poebrotheriidae, as Camelidae. (Only as fossils)
Polydolopidae. (Only as fossils)
Polymastodontidae, as Taeniolabididae. (Only as fossils)
Pongidae (incl. Anthropidae, Anthropoidae, Hylobatidae, Simidae, Simiidae). ELLIOT, D. G., No. 3, 1913 (monograph, as Simiidae); MILLER, G. S., 1933, Journ. Mamm., 14, 158-159 (classification of gibbons); SIMPSON, G. G., No. 1, 1945 (classification); BUETTNER-JANUSCH, J., No. 1, 1963 (world list); SIMONS, E. L. & PILBEAM, D. R., 1965, Folia Primat., 3, 81-152 (preliminary revision); ANDERSON & JONES, No. 1, 1967 (families); ELLIOT, D. G., No. 2, 1912 (World, also as Hylobatidae); RODE, P., 1937, Les Primates de l'Afrique, Paris, 1-222 (as Anthropoidae); ALLEN, G. M., No. 3, 1954 (Africa, catalog); ALLEN, J. A., 1925,

Bull. American Mus. Nat. Hist., 47, 283-499 (Belgian Congo); HOLLISTER, N., No. 3, 1924 (East Africa); BLANFORD, W. T., No. 1, 1888-91 (British India, as Simiidae); POCOCK, R. I., No. 1, 1939 (British India); BOURRET, L., 1946, Les mammiferes d'Indochine, Les gibbons, (Indochina), 1-41; ALLEN, G. M., No. 1, 1938 (China)

Pontoplanodidae. (Only as fossils)
Porcidae. (Family not identified)
Potamarchidae, as Heptaxodontidae
Potamogalidae. DOBSON, G. E., No. 3, 1882-83 (monograph); SIMPSON, G. G., No. 1, 1945 (classification); ALLEN, G. M., No. 3, 1954 (Africa, catalog); HILL & CARTER, No. 1, 1941 (Angola); ELLERMAN, J. R. et al., No. 2, 1953 (southern Africa)
Potidae, as Procyonidae
Potoridae, as Macropodidae
Praopidae, as Dasypodidae
Prepotheridae, as Megatheriidae. (Only as fossils)
Proborhyaenidae, as Borhyaenidae. (Only as fossils)
Proboscidae. (Family not identified)
Procaviidae (incl. Hyracidae, Pliohyracidae, Saghatheriidae). LYDEKKER, R., 1916, Catalogue of the ungulate mammals in the British Museum (Natural History), London, volume V; THOMAS, O., 1916, Ann. Mag. Nat. Hist., (8), 18, 301-303 (classification); HAHN, H., 1935, Die Familie der Procaviidae, Jeitschr. Saugetierk., 9, 207-358; SIMPSON, G. G., No. 1, 1945 (classification); SALE, J. B., 1960, Journ. East African Nat. Hist. Soc., 23, 185-188 (systematic position & biology); ANDERSON & JONES, No. 1, 1967 (families); ALLEN, G. M., No. 3, 1954 (Africa, catalog); RODE, P., 1944, Faune de l'Empire francaise, II, Paris, 125-210 (Afrique noire); HATT, R. T., 1936, Bull. American Mus. Nat. Hist., 72, 117-141 (Belgian Congo); HILL & CARTER, No. 1, 1941 (Angola); ELLERMAN, J. R. et al., No. 2, 1953 (southern Africa) HOLLISTER, N., No. 3, 1924 (East Africa); SETZER, H. W., No. 1, 1956 (Anglo-Egyptian Sudan); HARRISON, D. L., No. 2, 1968 (Arabia)
Procyonidae (incl. Ailuridae, Ailuropodidae, Bassaricyonidae. Cercoleptidae, Cercoleptididae, Bassariscidae, Nasuidae, Potidae). HOLLISTER, N., 1915, Proc. United States Nat. Mus., 49, 143-150 (genera & subgenera); POCOCK, R. I., 1921, Proc. Zool. Soc. London, 1921, 389-422 (classification); KLATT, B., 1928, Zool. Jahrb., Abt. Allg. Zool. Phys. Tiere, 45, 217-292; SIMPSON, G. G., No. 1, 1945 (classification); ANDERSON & JONES, No. 1, 1967 (families); HALL & KELSON, No. 1, 1959 (North America); ELLIOT, D. G., No. 1, 1904 (Central America & West Indies); GOODWIN, G. G., No. 1, 1942 (Honduras, also as Bassariscidae); GOODWIN, G. G., No. 2, 1942 (Costa Rica, as Bassariscidae also); GOLDMAN, E. A., No. 1, 1920 (Panama); TATE, G. H. H., No. 1, 1939 (Guiana); BOBRINSKOY, N. et al., No. 1, 1944 (U. S. S. R.); NOVIKOV, G. A., No. 1, 1956 (U. S. S. R.); NOVIKOV, G. A., No. 2, 1962 (U. S. S. R.)) BOBRINSKOY, N. et al., No. 2, 1965 (U. S. S. R.); BLANFORD, W. T., No. 1, 1888-91 (British India); POCOCK, R. I., No. 2, 1941 (British India, as Ailuridae & Ailuropodidae); ALLEN, G. M., No. 1, 1938 (China, as Ailuropodidae also)
Promysopidae. (Only as fossils)
Propalaeohoplophoridae, as Glyptodontidae. (Only as fossils)
Prorastomidae. (Only as fossils)
Proteleidae, as Hyaenidae
Protelidae, as Hyaenidae
Protemnodontidae. (Only as fossils)
Protequidae. (Only as fossils)
Proterocetidae. (Only as fossils)
Proterotheridae, as Proterotheriidae. (Only as fossils)
Proterotheriidae (incl. Proterotheridae). (Only as fossils)
Prothylacynidae. (Only as fossils)
Protoceratidae. (Only as fossils)
Protocetidae. (Only as fossils)
Protolabidae, as Camelidae. (Only as fossils)
Protolabididae, as Camelidae. (Only as fossils)
Protomyidae. (Only as fossils)
Protoptychidae. (Only as fossils)
Protoreodontidae, as Merycoidodontidae. (Only as fossils)
Protosirenidae. (Only as fossils)
Protoxodontidae. (Only as fossils)
Protypotheridae, as Interatheriidae. (Only as fossils)
Proviverridae, as Hyaenodontidae. (Only as fossils)
Psammoryctidae. (Family not identified)
Pseudolemuridae. (Only as fossils)
Pseudosciuridae. (Only as fossils)
Pseudostomidae. (Family not identified)
Pteromidae, as Sciuridae
Pteromyidae, as Sciuridae
Pteropidae (incl. Cephalotidae, Harpyidae, Pteropodidae, Pteropusidae). DOBSON, G. E., No. 2, 1878 (British Museum catalog); ANDERSEN, K., 1912, Catalogue of the Chiroptera in the ... British Museum, ed. 2, London, vol. 1 (reprinted 1966); MENDEZ, R. L., 1937, Rev. Acad. Cienc. Zaragoza, 20, 57-66 (museum List); SIMPSON, G. G., No. 1, 1945 (classification); ANDERSON & JONES, No. 1, 1967 (families); ALLEN, G. M., No. 3, 1954 (Africa,

catalog); ROSEVEAR, D. R., No. 1, 1965 (West Africa); HILL & CARTER, No. 1, 1941 (Angola); ELLERMAN, J. R. et al., No. 2, 1953 (southern Africa); HOLLISTER, N., No. 1, 1918 (East Africa); HARRISON, D. L., No. 1, 1964 (Arabia); DOBSON, G. E., No. 1, 1876 (Asia); BLANFORD, W. T., No. 1, 1888-91 (India); PHILLIPS, W. W. A., No. 1, 1935 (Ceylon); TROUGHTON, E. L., No. 1, 1926 (Australia & New Guinea); TAYLOR, E. H., No. 1, 1934 (Philippines); ALLEN, G. M., No. 1. 1938 (China)

Pteropodidae, as Pteropidae

Pteropusidae, as Pteropidae

Ptilodontidae (incl. Chirogidae, Cimolodontidae, Neoplagiaulacidae). (Only as fossils)

Pyrotheridae, as Pyrotheriidae. (Only as fossils)

Pyrotheriidae (incl. Carolozittelidae, Pyrotheridae). (Only as fossils)

Rangiferidae, as Cervidae

Rangiferinidae, as Cervidae

Rattidae, as Muridae

Rhabdosteidae. (Only as fossils)

Rhachianectidae, as Eschrichtiidae

Rhinocerosidae, as Rhinocerotidae

Rhinocerotidae (incl. Caenopidae, Elasmotheriidae, Rhinocerosidae, Rhynocerotidae). LYDEKKER, R., 1916, Catalogue of the ungulate mammals in the British Museum (Natural History), London, vol. V; MATTHEW, W. D., 1931, Univ. California Publ. Geol., 20, 1-9 (classification); MATTHEW, W. D., 1932, Bull. Dep. Geol. Sci. Univ. California, 20, 411-480 (review); SIMPSON, G. G., No. 1, 1945 (classification); ANDERSON & JONES, No. 1, 1967 (families); DERANIYAGALA, P. E. P., 1953, Spol. Zeylanica, 27, 13-14 (Africa); ALLEN, G. M., No. 3, 1954 (Africa, catalog); RODE, P., 1944, Faune de l'Empire Francaise, II, Paris, 125-210 (Afrique Noire); HILL & CARTER, No. 1, 1941 (Angola); ELLERMAN, J. R. et al., No. 2, 1953 (southern Africa); HOLLISTER, N., No. 3, 1924 (East Africa); SETZER, H. W., No. 1, 1956 (Anglo-Egyptian Sudan); BLANFORD, W. T., No. 1, 1888-91 (British India)

Rhinogalidae, as Viverridae

Rhinolophidae. DOBSON, G. E., No. 2, 1878 (British Museum catalog); SIMPSON, G. G., No. 1, 1945 (classification); FRIANT, M., 1963, Acta Zool., 44, 161-178 (Tertiary & classification); ANDERSON & JONES, No. 1, 1967 (families); MILLER, G. S., Jr., No. 1, 1912 (western Europe); CABRERA, A., No. 3, 1914 (Iberian Peninsula); CABRERA, A., 1904, Mem. Soc. Espanola Hist. Nat., 2, 249-287 (España); LAURENT, —, Bull. Mus. Nat. Hist. Nat., 13, 28-31 (Corsica) TOSCHI & LANZA, No. 1, 1959 (Italia); OGNEV, S. I., No. 1, 1928 (U. S. S. R.); OGNEV, S. I., No. 8, 1962 (U. S. S. R.); KUZJAKIN, A. P., 1950, Letuchie myshi, Moskva, 1-443 (U. S. S. R.); BOBRINSKOY, N. et al., No. 2, 1965 (U. S. S. R.); DOBSON, G. E., No. 1, 1876 (Asia); ALLEN, G. M., No. 3, 1954 (Africa, catalog); ROSEVEAR, D. R., No. 1, 1965 (West Africa); HILL & CARTER, No. 1, 1941 (Angola); ELLERMAN, J. R. et al., No. 2, 1953 (southern Africa); HOLLISTER, N., No. 1, 1918 (East Africa); HARRISON, D. L., No. 1, 1964 (Arabia); PHILLIPS, W. W. A., No. 1, 1935 (Ceylon); TROUGHTON, E. L., No. 1, 1926 (Australia & New Guinea); JONES, F. W., No. 1, 1923-25 (South Australia); TAYLOR, E. H., No. 1, 1934 (Philippines) BLANFORD, W. T., No. 1, 1888-91 (British India)

Rhinopomatidae (incl. Rhinopomidae). SIMPSON, G. G., No. 1, 1945 (classification); ANDERSON & JONES, No. 1, 1967 (families); ALLEN, G. M., No. 3, 1954 (Africa, catalog); ROSEVEAR, D. R., No. 1, 1965 (West Africa); HOLLISTER, N., No. 1, 1918 (East Africa); HARRISON, D. L., No. 1, 1964 (Arabia)

Rhinopomidae, as Rhinopomatidae

Rhizomyidae. ELLERMAN, J. R. et al., No. 1, 1941-49 (genera); SIMPSON, G. G., No. 1, 1945 (classification); ANDERSON & JONES, No. 1, 1967 (families); BOBRINSKOY, N. et al., No. 1, 1944 (U. S. S. R.); ALLEN, G. M., No. 3, 1954 (Africa, catalog); HOLLISTER, N., No. 2, 1919 (East Africa); ELLERMAN, J. R., No. 1, 1961 (India); ALLEN, G. M., No. 2, 1940 (China)

Rhynchippidae, as Notohippidae. (Only as fossils)

Rhynchocyonidae, as Macroscelididae

Rhynocerotidae, as Rhinocerotidae

Rhytinidae, as Dugongidae

Rosmaridae, as Odobenidae

Rytinadae, as Dugongidae

Saccomyidae, as Heteromyidae

Saghatheriidae, as Procaviidae. (Only as fossils)

Saigiidae, as Bovidae

Saimiridae, as Cebidae

Sariguidae. (Family not identified)

Saurocetidae. (Only as fossils)

Scalopidae, as Talpidae

Scansoridae. (Family not identified)

Scelidotheridae, as Mylodontidae. (Only as fossils)

Schismotheridae. (Only as fossils)

Sciuravidae, as Ischyromyidae. (Only as fossils)

Sciuridae (incl. Petauristidae, Pteromidae, Pteromyidae). ELLERMAN, J. R. et al., No. 1, 1941-49 (genera); POCOCK, R. I., 1923, Proc. Zool. Soc. London, 1923, 209-246 (classification); HARRIS, W. P., Jr., 1944, Occ. Pap. Iniv. Michigan Mus. Zool., 484, 1-21 (additions &

corrections to Ellerman); SIMPSON, G. G., No. 1, 1945 (classification); FLYGER, V. F., 1951, Quart. Rep. Pennsylvania Coop. Wildl. Res. Unit, 13, 19-38 (bibliography); ANDERSON, & JONES, No. 1, 1967 (families); HOWELL, A. H., 1918, North American Fauna, 44, 1-64 (America); ALLEN, G. A., 1899, American Nat., 33, 635-642 (arboreal, North America); HOWELL, A. H., 1938, North American Fauna, 56, 1-256 (North America); BRYANT, M. D., 1945, American Midland Nat., 33, 157-390 (phylogeny, Nearctic); HALL & KELSON, No. 1, 1959 (North American); MacCLINTOCK, D., 1970, Squirrels of North America, New York, 1-184; HOWELL, A. H., 1915, North American Fauna, 37, 1-80 (revision, American marmots); BANGS, O., 1896, Proc. Biol. Soc. Washington, 10, 145-167 (eastern North America); ELLIOT, D. G., No. 1, 1904 (Central America & West Indies); THOMAS, O., 1909, Ann. Mag. Nat. Hist., (8), 3, 467-475 (West Indies); NELSON, E. W., 1899, Proc. Washington Acad. Sci., 1, 15-106 (Mexico & Central America); GOODWIN, G. G., No. 1, 1942 (Honduras); GOODWIN, G. G., No. 2, 1946 (Costa Rica); GOLDMAN, E. A., No. 1, 1920 (Panama); ALLEN, J. A., 1915, Bull. American Mus. Nat. Hist., 34, 147-309 (South American); HERSHKOVITZ, P., 1947, Proc. United States Nat. Mus., 97, 1-46 (Colombia); TATE, G. H. H., No. 1, 1939 (Guiana); PINTO, O. M. O., 1931, Rev. Mus. Paulista, 17, 263-321 (Brasil); MILLER, G. S., Jr., No. 1, 1912 (western Europe, also as Petauristidae); CABRERO, A., No. 3, 1914 (Iberian Peninsula); TOSCHI, A., No. 2, 1965 (Italia); OGNEV, S. I., No. 4, 1940 (U. S. S. R., also as Pteromyidae); BOBRINSKOY, N. et al., No. 1, 1944 (U. S. S. R., also as Petauristidae); OGNEV, S. I., No. 5, 1947 (U. S. S. R.); OGNEV, S. I., No. 12, 1963 (U. S. S. R.); BOBRINSKOY, N. et al., No. 2, 1965 (U. S. S. R., also as Petauristidae); OGNEV, S. I., No. 11, 1966 (U.S.S.R., also as Pteromyidae); GROMOV, I. M. et al., 1965, Nazemnye belichi (Marmotinae), Moskva, 1-467; JENTINCK, F. A., 1882, Notes Leyden Mus., 4, 1-53 (Africa & Leyden Museum list); ALLEN, G. M., No. 3, 1954 (Africa, catalog); HILL & CARTER, No. 1, 1941 (Angola); ELLERMAN, J. R. et al., No. 2, 1953 (southern Africa); HOLLISTER, N., No. 2, 1919 (East Africa); SETZER, H. W., No. 1, 1956 (Anglo-Egyptian Sudan); MOORE, J. C. & TATE, G. H. H., 1965, Fieldiana, Zool., 48, 1-351 (Sciurinae, India & Indochina); BLANFORD, W. T., No. 1, 1888-91 (British India); ELLERMAN, J. R., No. 1, 1961 (India); PHILLIPS, W. W. A., No. 1, 1935 (Ceylon); HARRISON, J. D., 1948, Malayan Nat. Journ., 3, 201-209 (Malaya); DAMMERMAN, K. W., 1931, Treubia, 13, 429-470 (Java); ROBINSON, H. C., & KLOSS, C. B., 1918, Rec. Indian Mus., 15, 171-254 (Oriental region, list); TAYLOR, E. H., No. 1, 1934 (Philippines, also as Petauristidae); ALLEN, G. M., No. 2, 1940 (China, also as Petauristidae)

Scleropleuridae, as Dasypodidae

Scoteopsidae. (Only as fossils)

Selenoconidae, as Henricosborniidae. (Only as fossils)

Selenolophodontidae. (Only as fossils)

Seleviniidae. SIMPSON, G. G., No. 1, 1945 (classification); ANDERSON & JONES, No. 1, 1967 (families); BOBRINSKOY, N. et al., No. 1, 1944 (U. S. S. R.); BOBRINSKOY, N. et al., No. 2, 1965 (USSR); BASHANOV, B. S. & BELOSLUDOV, B. A., 1941, Journ. Mamm., 22, 311-315 (Kazakhstan)

Semantoridae. (Only as fossils)

Semnopithecidae, as Cercopithecidae

Serridentidae, as Gomphotheriidae. (Only as fossils)

Sicistidae, as Zapodidae

Simidae, as Pongidae

Simiidae, as Pongidae

Simocyonidae, as Canidae

Sivatheriidae, as Giraffidae. (Only as fossils)

Sminthidae, as Zapodidae

Solenodontidae. DOBSON, G. E., No. 3, 1882-83 (monograph); CABRERA, A., No. 2, 1925 (classification); SIMPSON, G. G., No. 1, 1945 (classification); ANDERSON & JONES, No. 1, 1967 (families); HALL & KELSON, No. 1, 1959 (North America); ELLIOT, D. G., No. 1, 1904 (Central America & West Indies)

Soricidae. DOBSON, G. E., No. 3, 1882-83 (monograph); DOBSON, G. E., 1890, Proc. Zool. Soc. London, 1890, (4), 49-51 (synopsis of genera); KOLLER, O., 1920, Treubia, 11, 313-324 (museum list, Buitenzorg); CABRERA, A., No. 2, 1925 (classification); KOLLER, O., 1930, Treubia, 11, 313-324 (museum list, Buitenzorg); STEHLIN, H. G., 1940, Eclog. Geol. Helvet., 33, 298-306 (phylogeny); SIMPSON, G. G., No. 1, 1945 (classification); PRUITT, W. O., 1957, Saugetierk. Mitt., 5, 18-27 (survey of families); ANDERSON & JONES, No. 1, 1967 (families) MERRIAM, C. H., 1895, North American Fauna, 10, 1-100 (America); JACKSON, H. H. T., 1928, North American Fauna, 51, 1-238 (America); HALL & KELSON, No. 1, 1959 (North America); GUILDAY, J. E., 1962, Ann. Carnegie Mus., 36, 87-122 (key to eastern North American, based on mandibles); ELLIOT, D. G., No. 1, 1904 (Central America & West Indies) GOODWIN, G. G., No. 1, 1942 (Honduras); GOODWIN, G. G., No. 2, 1946 (Costa Rica); GOLDMAN, E. A., No. 1, 1920 (Panama); TATE, G. H. H., Journ. Mamm., 13, 223-228 (South America); OGNEV, S. I., 1933, Zool. Anz., 105, 77-85 (Palearctic, in German); MILLER, G. S., Jr., No. 1, 1912 (western Europe); CABRERA, A., No. 3, 1914 (Iberian Peninsula); TOSCHI & LANZA, No. 1, 1959 (Italia); FRIANT, M., 1948, Ann. Soc. Geol. Nord Lille, 67, 222-269 (western Europe); OGNEV, S. I., No. 1, 1928 (U. S. S. R.); OGNEV, S. I., 1933, Zool. Journ., Moscou, 12, (4), 8-16; BOBRINSKOY, N. et al., No. 1, 1944 (U. S. S. R.); BOBRINSKOY, N. et al., No. 2, 1965 (U. S. S. R.); OGNEV, S. I., No. 8, 1962 (U. S. S. R.); STROGANOV, S. U., 1957, Insectivores of Siberia, Moskva, 1-267 (Siberia); MEESTER, J., 1953, Ann. Transvaal Mus., 22, 205-213 (genera, Africa); ALLEN, G. M., No. 3, 1954 (Africa, catalog); HILL & CARTER, No. 1, 1941 (Angola); ELLERMAN, J. R. et al., No. 2, 1953 (southern

Africa); HOLLISTER, N., No. 1, 1918 (East Africa); HEIM de BALSAC, H., 1957, Zool. Anz., 158, 143-153 (East Africa); SETZER, H. W., No. 1, 1956 (Anglo-Egyptian Sudan); HARRISON, D. L., No. 1, 1964 (Arabia); BLANFORD, W. T., No. 1, 1888-91 (British India); PHILLIPS, W. W. A., 1928, Spol. Zeylanica, 14, 295-332 (Ceylon); PHILLIPS, W. W. A., No. 1, 1935 (Ceylon); TAYLOR, E. H., No. 1, 1934 (Philippines); ALLEN, G. M., No. 1, 1938 (China); CHEN, J. T. F., 1948, Quart. Journ. Taiwan Mus., 1, (1), 36-46 (China, in Chinese)

Spalacidae. ELLERMAN, J. R. et al., No. 1, 1941-49 (genera); SIMPSON, G. G., No. 1, 1945 (classification); ANDERSON & JONES, No. 1, 1967 (families); MILLER, G. S., Jr., No. 1, 1912 (western Europe); BOBRINSKOY, N. et al., No. 1, 1944 (U. S. S. R.); OGNEV, S. I., No. 5, 1947 (U. S. S. R.) OGNEV, S. I., No. 12, 1963 (U. S. S. R.); BOBRINSKOY, N. et al., No. 2, 1965 (U. S. S. R.); RESHNETNIK, E. G., 1939, Acad. Sci. Ukrainian SSR, Rep. Zool. Mus., 23, 3-21 (Ukrainian SSR); ALLEN, G. M., No. 3, 1954 (Africa, catalog); RANCK, G. L., No. 1, 1968 (Libya); BLANFORD, W. T., No. 1, 1888-91 (India); ALLEN, G. M., No. 2, 1940 (China)

Spalacogalidae. (Family not identified)

Spalacopodidae, as Octodontidae

Spalacotheridae, as Spalacotheriidae. (Only as fossils)

Spalacotheriidae (incl. Spalacotheridae, Tinodontidae). (Only as fossils)

Sparassodontidae, as Borhyaenidae. (Only as fossils)

Squalodelphidae, as Ziphiidae

Squalodontidae (incl. Gymnorhinidae). (Only as fossils)

Stagodontidae, as Elephantidae. (Only as fossils)

Stegodontidae, as Elephantidae. (Only as fossils)

Stegorhinidae. (Only as fossils)

Stegotheridae, as Dasypodidae. (Only as fossils)

Stenidae. ANDERSON & JONES, No. 1, 1967 (families)

Stenodermatidae, as Phyllostomatidae

Stentoridae. (Family not identified)

Stereognathidae. (Family not identified)

Strepsicerotidae, as Bovidae

Stylacodontidae, as Dryolestidae. (Only as fossils)

Stylinodontidae (incl. Calamodontidae, Conoryctidae, Ectoganidae). (Only as fossils)

Stylocerinidae. (Family not identified)

Stylodontidae. (Only as fossils)

Subulidae. (Family not identified)

Suidae (incl. Listriodontidae, Phacochoeridae, Syidae, Tetraconodontidae). LYDEKKER, R., 1915, Catalogue of the ungulate mammals in the British Museum (Natural History), London, volume IV; SIMPSON, G. G., No. 1, 1945 (classification); ANDERSON & JONES, No. 1, 1967 (families); MILLER, G. S., Jr., No. 1, 1912 (western Europe); CABRERA, A., No. 3, 1914 (Iberian Peninsula); TOSCHI, A., No. 2, 1965 (Italia); ADLERBURG, G., 1930, Compt. Rend. Acad. Sci. URSS, 1930, 91-96 (Russian & Mongolia); BOBRINSKOY, N. et al., No. 1, 1944 (U. S. S. R.); SOKOLOV, V. E., No. 1, 1959 (S. S. S. R.); BOBRINSKOY, N. et al., No. 2, 1965 (U. S. S. R.); ALLEN, G. M., No. 3, 1954 (Africa, catalog); RODE, P., 1944, Faune de l'Empire Francaise, II, Paris, 125-210 (Afrique Noire); HILL & CARTER, No. 1, 1941 (Angola); ELLERMAN J. R., et al., No. 2, 1953 (southern Africa); HOLLISTER, N., No. 3, 1924 (East Africa); SETZER, H. W., No. 1, 1956 (Anglo-Egyptian Sudan); HARRISON, D. L., No. 2, 1968 (Arabia); BLANFORD, W. T., No. 1, 1888-91 (British India); PHILLIPS, W. W. A., No. 1, 1935 (Ceylon); TAYLOR, E. H., No. 1, 1934 (Philippines); ALLEN, G. M., No. 2, 1940 (China)

Suricatidae, as Viverridae

Susuidae, as Platanistidae

Syidae, as Suidae

Systemodontidae. (Only as fossils)

Tachyglossidae (incl. Echidnidae). CABRERA, A., No. 1. 1919 (classification); SIMPSON, G. G., No. 1, 1945 (classification); ANDERSON & JONES, No. 1, 1967 (families); GRIFFITHS, M., 1968, Echidnas, London, 1-282 (key to species); JONES, F. W., No. 1, 1923-25 (South Australia, as Echidnidae)

Tachynicidae. (Family not identified)

Taeniolabididae (incl. Polymastodontidae). (Only as fossils)

Tagussuidae, as Tayassuidae

Talpidae (incl. Myaladae, Myogalidae, Scalopidae). DOBSON, G. E., No. 3, 1882-83 (monograph); CABRERA, A., No. 2, 1925 (classification); SIMPSON, G. G., No. 1, 1945 (classification); ANDERSON & JONES, No. 1, 1967 (families); TRUE, F. W., 1896, Proc. United States Nat. Mus., 19, 1-112 (American); JACKSON, H. H. T., 1915, North American Fauna, 38, 1-100 (America); HALL & KELSON, No. 1, 1959 (North America); ELLIOT, D. G., No. 1, 1904 (Central America & West Indies); MILLER, G. S., Jr., No. 1, 1912 (western Europe); CABRERA, A., No. 3, 1914 (Iberian Peninsula); TOSCHI & LANZA, No. 1, 1959 (Italia); OGNEV, S. I., No. 1, 1928 (U. S. S. R.); BOBRINSKOY, N. et al., No. 1, 1944 (U. S. S. R.); STROGANOV, S. V., 1948, Trud. Zool. Inst. Acad. Sci. USSR, 8, (2), 286-405 (systematics); OGNEV, S. I., No. 8, 1962 (U. S. S. R.); BOBRINSKOY, N. et al., No. 2, 1965 (U. S. S. R.); STROGANOV, S. U., 1957, Insectivores of Siberia, Moskva, 1-267; BLANFORD, W. T., No. 1, 1888-91 (British India); ALLEN, G. M., No. 1, 1938 (China)

Taperidae, as Tapiridae

Taphozoidae, as Emballonuridae

Tapiridae (incl. Taperidae). HATCHER, J. B., 1896, American Journ. Sci., (4), 1, 161-180 (Recent and fossil); LYDEKKER, R., 1916, Catalogue of the ungulate mammals in the British Museum (Natural History), London, vol. V; SIMPSON, G. G., No. 1, 1945 (classification); ANDERSON & JONES, No. 1, 1967 (families); HALL & KELSON, No. 1, 1959 (North America); ELLIOT, D. G., No. 1, 1904 (Central America & West Indies); GOODWIN, G. G., No. 1, 1942 (Honduras); GOODWIN, G. G., No. 2, 1942 (Costa Rica); GOLDMAN, E. A., No. 1, 1920 (Panama); HERSHKOVITZ, P., 1954, Proc. United States Nat. Mus., 103, 465-496 (northern Colombia); TATE, G. H. H., No. 1, 1939 (Guiana); BLANFORD, W. T., No. 1, 1888-91 (India)
Tapirulidae, as Anoplotheriidae. (Only as fossils)
Tarsidae, as Tarsiidae
Tarsiidae (incl. Tarsidae). ELLIOT, D. G., No. 2, 1912 (World); SIMPSON, G. G., No. 1, 1945 (classification); HILL, W. C. O., 1955, Primates..., 2, New York & Edinburgh, 1-347 (monograph); BUETTNER-JANUSCH, J., No. 1, 1963 (world list); ANDERSON & JONES, No. 1, 1967 (families); TAYLOR, E. H., No. 1, 1934 (Philippines)
Tarsipedidae, as Phalangeridae
Tatusidae, as Dasypodidae
Tatusiidae, as Dasypodidae
Tayassuidae (incl. Dicotylidae, Tagussuidae, Tayussuidae). SIMPSON, G. G., No. 1, 1945 (classification); ANDERSON & JONES, No. 1, 1967 (families); HALL & KELSON, No. 1, 1959 (North America); ELLIOT, D. G., No. 1, 1904 (Central America & West Indies); GOODWIN, G. G., No. 1, 1942 (Honduras); GOODWIN, G. G., No. 2, 1942 (Costa Rica); GOLDMAN, E. A., No. 1, 1920 (Panama); TATE, G. H. H., No. 1, 1939 (Guiana)
Tayussuidae, as Tayassuidae
Tembotheridae. (Only as fossils)
Tenrecidae (incl. Centetidae, Geogalidae, Oryzoryctidae). DOBSON, G. E., No. 3, 1882-83 (monograph, as Centetidae); CABRERA, A., No. 2, 1925 (classification); SIMPSON, G. G., No. 1, 1945 (classification); GOULD, E. & EISENBERG, J. F., 1966, Journ. Mamm., 47, 660-686 (biology); ANDERSON & JONES, No. 1, 1967 (families); ALLEN, G. M., No. 3, 1954 (Africa, catalog)
Tetoniidae, as Anaptomorphidae. (Only as fossils)
Tetracerocidae, as Bovidae
Tetraconodontidae, as Suidae. (Only as fossils)
Theridomyidae (incl. Archaeomyidae, Nesokerodontidae). (Only as fossils)
Thlaeodontidae, as Didelphidae. (Only as fossils)
Thrynomyidae, as Thryonomyidae
Thryonomyidae (incl. Thrynomyidae). SIMPSON, G. G., No. 1, 1945 (classification); ANDERSON & JONES, No. 1, 1967 (families); ALLEN, G. M., No. 3, 1954 (Africa, catalog); HILL & CARTER, No. 1, 1941 (Angola); HOLLISTER, N., No. 2, 1919 (East Africa); SETZER, H. W., No. 1, 1956 (Anglo-Egyptian Sudan)
Thylacinidae, as Dasyuridae
Thylacoleonidae (incl. Thylacoleontidae). (Only as fossils)
Thylacoleontidae, as Thylacoleonidae. (Only as fossils)
Thylaxinidae, as Dasyuridae
Thyropteridae. SIMPSON, G. G., No. 1, 1945 (classification); ANDERSON & JONES, No. 1, 1967 (families); HALL & KELSON, No. 1, 1959 (North America); VILLA-R., B., No. 1, 1966 (Mexico); GOODWIN, G. G., No. 1, 1942 (Honduras); GOODWIN, G. G., No. 2, 1946 (Costa Rica); GOODWIN & GREENHALL, No. 1, 1961 (Trinidad & Tobago); HUSSON, A. M., No. 1, 1962 (Surinam); da CUNHA VIEIRA, C. O., No. 1, 1942 (Brasil)
Tillotheridae, as Tillotheriidae. (Only as fossils)
Tillotheriidae (incl. Anchippodontidae, Esthonychidae, Tillotheridae). (Only as fossils)
Tinoceridae, as Uintatheriidae. (Only as fossils)
Tinoceratidae, as Uintatheriidae. (Only as fossils)
Tinodontidae, as Spalacotheriidae. (Only as fossils)
Titanohyracidae, as Geniohyidae. (Only as fossils)
Titanotheridae, as Titanotheriidae. (Only as fossils)
Titanotheriidae (incl. Titanotheridae). (Only as fossils)
Titidae. (Family not identified)
Tolypeutidae, as Dasypodidae
Toxodontidae (incl. Haplodontheriidae, Haplodontidae Ameghino, Nesodontidae, Xotodontidae). (Only as fossils)
Trachynichidae. (Family not identified)
Trachytheridae, as Mesotheriidae. (Only as fossils)
Tragelaphidae, as Bovidae
Tragulidae (incl. Hyemoschidae). LYDEKKER, R., 1915, Catalogue of the ungulate mammals in the British Museum (Natural History), London, vol. IV; THOMAS, O., 1916, Ann. Mag. Nat. Hist., (8), 18, 72-73 (generic names); FLEROV, K. K., 1931, Compt. Rend. Acad. Sci. URSS, 1931, 75-79 (generic characters); SIMPSON, G. G., No. 1, 1945 (classification); ANDERSON & JONES, No. 1, 1967 (families); ALLEN, G. M., No. 3, 1954 (Africa, catalog); BOUET, G., 1934, Comm. Proc. Verb. Acad. Sci. Coloniales, 12, 1-46 (Afrique francaise); RODE, P., 1944, Faune de l'Empire Francaise, II, Paris, 125-210 (Afrique Noire); BLANFORD, W. T., No. 1, 1888-91 (British India); PHILLIPS, W. W. A., No. 1, 1935 (Ceylon); TAYLOR, E. H., No. 1, 1934 (Philippines)
Trichechidae [Gill 1872] (incl. Manatidae). SIMPSON, G. G., No. 1, 1945 (classification); ANDERSON & JONES, No. 1, 1967 (families); HALL & KELSON, No. 1, 1959 (North America);

ELLIOT, D. G., No. 1, 1904 (Central America & West Indies); GOODWIN, G. G., No. 1, 1942 (Honduras); GOODWIN, G. G., No. 2, 1942 (Costa Rica); GOLDMAN, E. A., No. 1, 1920 (Panama); ALLEN, G. M., No. 3, 1954 (Africa, catalog); ELLERMAN, J. R. et al., No. 2, 1953 (southern Africa); BLANFORD, W. T., No. 1, 1888-91 (British India)

Trichecidae [Gray 1825], as Odobenidae

Triconodontidae (incl. Phascolotheriidae). (Only as fossils)

Tricuspiodontidae. (Only as fossils)

Trigonostylopidae. (Only as fossils)

Triisodontidae, as Arctocyonidae. (Only as fossils)

Trilophodontidae, as Gomphotheriidae. (Only as fossils)

Triplopidae, as Hyracodontidae. (Only as fossils)

Triplopodidae, as Hyracodontidae. (Only as fossils)

Tripriodontidae. (Only as fossils)

Tritylodontidae. (Only as fossils)

Tupaiidae (incl. Tupajidae, Tupayidae). DOBSON, G. E., No. 3, 1882-83 (monograph); LYON, M. W., Jr., 1913, Proc. United States Nat. Mus., 45, 1-188 (monograph); CARLSSON, A., 1922, Acta Zool., 3, 227-270; CABRERA, A., No. 2, 1925 (classification); SIMPSON, G. G., No. 1, 1945 (classification); BUETTNER, -JANUSCH, J., No. 1, 1963 (world list); ANDERSON & JONES, No. 1, 1967 (families); FRIANT, M., 1947, Bull. Mus. Nat. Hist. Nat., (2), 19, 257-260 (genera of Chirogalinae); BLANFORD, W. T., No. 1, 1888-91 (British India); TAYLOR, E. H., No. 1, 1934 (Philippines); ALLEN, G. M., No. 1, 1938 (China)

Tupajidae, as Tupaiidae

Tupayidae, as Tupaiidae

Tylopodidae. (Family not identified)

Typotheriidae, as Mesotheriidae. (Only as fossils)

Uintacyonidae, as Miacidae. (Only as fossils)

Uintatheriidae (incl. Bathopsidae, Dinoceratidae, Eobasileidae, Tinoceratidae, Tinoceridae). (Only as fossils)

Ulacodidae. (Family not identified)

Ursidae (incl. Hemicyonidae, Ursinidae). JAKOVLEFF, B. P., 1934, Publ. Mus. Hoangho Paiho Tien Tsin, 28, 1-30 (museum list); SIMPSON, G. G., No. 1, 1945 (classification); ANDERSON & Jones, No. 1, 1967 (families); HALL & KELSON, No. 1, 1959 (North America); ELLIOT, D. G., No. 1, 1904 (Central America & West Indies); ERDBRINK, D. P., 1953, A review of the fossil & Recent bears of the Old World . . ., I, II, Deventer, 1-597; HURZELER, J., 1944, Verh. Naturf. Ges. Basel, 55, 131-157 (revision, Europe); MILLER, G. S., Jr., No. 1, 1912 (western Europe); CABRERA, A., No. 3, 1914 (Iberian Peninsula); TOSCHI, A., No. 2, 1965 (Italia); OGNEV, S. I., No. 2, 1931 (U. S. S. R.); BOBRINSKOY, N. et al., No. 1, 1944 (U. S. S. R.); NOVIKOV, G. A., No. 1, 1956 (U. S. S. R.); OGNEV, S. I., No. 9, 1962 (U. S. S. R.); NOVIKOV, G. A., No. 1, 1962 (U. S. S. R.); BOBRINSKOY, N. et al., No. 2, 1965 (U. S. S. R.); STROGANOV, S. U., 1969, Carnivorous mammals of Siberia, Jerusalem, 1-522; ALLEN, G. M., No. 3, 1954 (Africa, catalog); ENNOUCHI, E., 1958, Bull. Soc. Sci. Nat. Maroc, 37, 201-223 (Maroc); HARRISON, D. L., No. 2, 1968 (Arabia); BLANFORD, W. T., No. 1, 1888-91 (British India); POCOCK, R. I., No. 2, 1941 (British India); PHILLIPS, W. W. A., No. 1, 1935 (Ceylon); ALLEN, G. M., No. 1, 1938 (China)

Ursinidae, as Ursidae

Vampyridae, as Phyllostomatidae

Vespertilionidae. DOBSON, G. E., No. 2, 1878 (British Museum catalog); ALLEN, H., 1893, A monograph of the bats of North America, United States Nat. Mus. Bull., 43, 1-198; SIMPSON, G. G., No. 1, 1945 (classification); ANDERSON & JONES, No. 1, 1967 (families); TROUGHTON, E. G., 1929, Rec. Australian Mus., 17, 85-99 (genera of Kerivoulinae); BARBOUR, R. W. & DAVIS, W. A., 1969, Bats of America, Lexington, Kentucky, 1-286; ALLEN, H., 1893, Proc. United States Nat. Mus., 16, 1-31 (North America); ALLEN, H., 1893, United States Nat. Mus. Bull., 43, 1-198 (monograph, North America); MILLER, G. S., Jr., 1897, North American Fauna, 13, 1-135 (North America); HALL & KELSON, No. 1, 1959 (North America); ELLIOT, D. G., No. 1, 1904 (Central America & West Indies); ANTHONY, H. E., No. 1, 1918, Mem. American Mus. Nat. Hist., (n. s.), 2, (2), 335-435 (Puerto Rico); ANTHONY, H. E., No. 2, 1925-26 (Puerto Rico); VILLA-R., B., No. 1, 1966 (Mexico); GOODWIN, G. G., No. 1, 1942 (Honduras); GOODWIN, G. G., No. 2, 1946 (Costa Rica); GOLDMAN, E. A., No. 1, 1920 (Panama); GOODWIN & GREENHALL, No. 1, 1961 (Trinidad & Tobago); HUSSON, A. M., No. 1, 1962 (Surinam); da CUNHA VIEIRA, C. O., No. 1, 1942 (Brasil); MILLER, G. S., Jr., No. 1, 1912 (western Europe); CABRERA, A., No. 3, 1914 (Iberian Peninsula); CABRERA, A., 1904, Mem. Soc. Española Hist. Nat., 2, 249-287 (España); TOSCHI & LANZA, No. 1, 1959 (Italia); OGNEV, S. I., No. 1, 1928 (U. S. S. R.); BOBRINSKOY, N. et al., No. 1, 1944 (U. S. S. R.) KUZJAKIN, A. P., 1950, Letuchie myshi, Moskva, 1-443 (U. S. S. R.); OGNEV, S. I., No. 8, 1962 (U. S. S. R.); BOBRINSKOY, N. et al., No. 2, 1965 (U. S. S. R.); DOBSON, G. E., No. 1, 1876 (Asia); ALLEN, G. M., No. 3, 1954 (Africa, catalog); ROSEVEAR, D. R., No. 1, 1965 (West Africa); HILL & CARTER, No. 1, 1941 (Angola); ELLERMAN, J. R. et al., No. 2, 1953 (southern Africa); HOLLISTER, N., No. 1, 1918 (East Africa); HARRISON, D. L., No. 1, 1964 (Arabia); BLANFORD, W. T., No. 1, 1888-91 (British India); PHILLIPS, W. W. A., No. 1, 1935 (Ceylon); TROUGHTON, E. L., No. 1, 1926 (Australia & New Guinea); JONES, F. W., No. 1, 1923-25 (South Australia); TAYLOR, E. H., No. 1, 1934 (Philippines); ALLEN, G. M., No. 1, 1938 (China)

Viscaccidae, as Chinchillidae
Viveridae, as Viverridae
Viverravidae, as Miacidae. (Only as fossils)
Viverridae (incl. Arctictidae, Arctogalidae, Cryptoproctidae, Cynictidae, Cynogalidae, Eupleridae, Genettidae, Herpestidae, Mungotidae, Paradoxuridae, Rhinogalidae, Viveridae, Suricatidae). POCOCK, R. I., 1919, Ann. Mag. Nat. Hist., (9), 3, 515-524 (classification of mongooses); JAKOVLEFF, B. P., 1933, Publ. Mus. Hoangho Paiho Tien Tsin, 26, 1-24 (museum list); GREGORY, W. K. & HELLMAN, M., 1939, Proc. American Philos. Soc., 81, 309-392 (classification); SIMPSON, G. G., No. 1, 1945 (classification); ANDERSON & JONES, No. 1, 1967 (families); HINTON, H. E. & DUNN, A. M. S., 1967, Mongooses ..., Edinburgh (Herpestinae, synonymic catalog, common names); HALL & KELSON, No. 1, 1959 (North America); ELLIOT, D. G., No. 1, 1904 (Central America & West Indies); MILLER, G. S., Jr., No. 1, 1912 (western Europe); CABRERA, A., No. 3, 1914 (Iberian Peninsula); POCOCK, R. I., 1933, Journ. Bombay Nat. Hist. Soc., 36, 423-449 (Asia, civet cats); ALLEN, G. M., No. 3, 1954 (Africa, catalog); ROSEVEAR, D. R., 1935, Nigerian Field, 4, 114-122 (Nigeria, civets); ROSEVEAR, D. R., 1935, Nigerian Field, 4, 165-170 (Nigeria, mongooses); ALLEN, J. A., 1924, Bull. American Mus. Nat. Hist., 47, 73-281 (Belgian Congo); HILL & CARTER, No. 1, 1941 (Angola); ELLERMAN, J. R. et al., No. 2, 1953 (southern Africa); HOLLISTER, N., No. 1, 1918 (East Africa); SETZER, H. W., No. 1, 1956 (Anglo-Egyptian Sudan); HARRISON, D. L., No. 2, 1968 (Arabia); BLANFORD, W. T., No. 1, 1888-91 (British India); POCOCK, R. I., 1937, Journ. Bombay Nat. Hist. Soc., 39, 211-245 (British India, mongooses); POCOCK, R. I., No. 1, 1939 (British India); POCOCK, R. I., No. 2, 1941 (British India, as Herpestidae); PHILLIPS, W. W. A., No. 1, 1935 (Ceylon); POCOCK, R. I., 1933, Proc. Zool. Soc. London, 1933, 969-1035 (rarer genera, Orient); TAYLOR, E. H., No. 1, 1934 (Philippines); HERKLOTS, G. A. C., 1934, Hong Kong Nat., 5, 91-94 (civets of Hong Kong); ALLEN, G. M., No. 1, 1938 (China)
Vombatidae, as Phascolomidae
Vulpidae, as Canidae

Xiphidae, as Ziphiidae
Xiphiidae, as Ziphiidae
Xiphodontidae. (Only as fossils)
Xotodontidae, as Toxodontidae. (Only as fossils)

Zalambdalestidae. (Only as fossils)
Zapodidae (incl. Sicistidae, Sminthidae). SIMPSON, G. G., No. 1, 1945 (classification); ANDERSON & JONES, No. 1, 1967 (families); HALL & KELSON, No. 1, 1959 (North America); KRUTZSCH, P. H., 1954, Publ. Univ. Kansas Mus. Nat. Hist., 7, 349-472 (North American Zapus); MILLER, G. S., Jr., No. 1, 1912 (western Europe); ALLEN, G. M., No. 2, 1940 (China)
Zeuglodontidae, as Basilosauridae. (Only as fossils)
Ziphidae, as Ziphiidae
Ziphiidae (incl. Squalodelphidae, Xiphidae, Xiphiidae, Ziphidae). TRUE, F. W., 1910, United States Nat. Mus. Bull., 73, 1-89 (U. S. N. M. museum list); SIMPSON, G. G., No. 1, 1945 (classification); McCANN, C., 1962, Tuatara, 10, 13-18 (keys, as Xiphiidae); ANDERSON & JONES, No. 1, 1967 (families); HALL & KELSON, No. 1, 1959 (North America); ULMER, F. A., 1941, Proc. Acad. Nat. Sci. Philadelphia, 93, 107-122 (Atlantic coast of North America, as Xiphiidae); CABRERA, A., No. 3, 1914 (Iberian Peninsula); TOSCHI, A., No. 2, 1965 (Italia); BOBRINSKOY, N. et al., No. 1, 1944 (U. S. S. R.); TOMILIN, A. G., No. 2, 1957 (U. S. S. R.); BOBRINSKOY, N. et al., No. 2, 1965 (U. S. S. R.); TOMILIN, A. G., No. 1, 1967 (U. S. S. R.); ALLEN, G. M., No. 3, 1954 (catalog, Africa); ELLERMAN, J. R. et al., No. 2, 1953 (southern Africa); JONES, F. W., No. 1, 1923-25 (South Australia)
Zorillidae, as Mustelidae

ADDENDA

PISCES

GENERAL BOOKS

FONTAINE, M. Traite de Zoologie. XIII. Agnathes et Poissons. Paris; 3 volumes. 1958.

BIBLIOGRAPHIES

CHAUHAN, B. S. A list of references relating to Indian Zoology (... fishes ...) published during the years 1838-1950. Rec. Indian Mus., 51, 427-480. 1954.
CORWIN, G. A bibliography of the tunas. Fish. Bull. Sacramento, 22, 1-103. 1930.
MOROVIC, D. Contributo alla bibliografia della pesa nell'Adriatico (1864-1949). Posebna Izd. Inst. Oceanogr. Jugoslav, 1, 1-143. 1950.
MURTY, V. S. et al.. Bibliography of marine fishes ... of the Indian Ocean 1962-1967. Bull. Cent. Mar. Fish. Res. Inst., 1, 1-208. 1968.

NOMENCLATURE

MACLEOD, R. D. Key to the names of British fishes, (etc.). London; 1-71. 1956. (Common names)
REDEKE, H. C. Fauna van Nederland. 10. Leiden; 1-331. 1941. (Common names)

FAUNAS & IDENTIFICATION WORKS

REDEKE, H. C. Fauna van Nederland. 10. Leiden; 1-331. 1941.
La MONTE, F. Marine game fishes of the world. Garden City, New York; 1-190. 1952.
WHITAKER, J. O., Jr. Keys to the vertebrates of the eastern United States, excluding birds. Minneapolis, Minnesota; 1-256. 1969.
GOSLINE, W. A. et al. Fishes of the western North Atlantic, order Iniomi. Mem. Sears Found. Mar. Res. 1, (5), 1-18. 1966.
TORTONESE, E. & LANZA, B. Piccola fauna Italiana. Pesci, anfibi e rettili. Milano; 1-185. 1968.
GASOWSKIEJ, M. Klucze do oznaczama kregowcow Polski. I. Cyclostomi et Pisces. Warszawa; 1-240. 1962.
IONESCU, V. Vertebratele din Romania. Bucuresti; 1-496. 1968.
WHITEHEAD, P. J. P. A review of the elopoid and clupeoid fishes of the Red Sea .... Bull. British Museum (Nat. Hist.), 12, 225-281. 1965.
SUVATTI, C. Fauna of Thailand. Bangkok; 1-1100. 1950. (Pp. 180-446)
(anonymous). Illustrated encyclopedia of the fauna of Japan ..., 2nd ed. Tokyo; 2 volumes. 1949.

MONOGRAPHS

SCHNAKENBECK, W. Acrania ... — Cyclostoma — Pisces. Handbuch der Zoologie, VI, (1), 3 parts. 1930-1960.
HOLLY, M. Cyclostomata. Das Tierreich, 59, 61, 63, 67. 1933-1936. (Classification)
GOSLINE, W. A. The morphology and systematic position of alepocephaloid fishes. Bull. British Mus. (Nat. Hist.), 18, 185-218. 1969.

CLASSIFICATION

PERRIER, E. Poissons. Traite de Zoologie, 6, 2355-2726. 1903. (To family)
FONTAINE, M. Agnathes et Poissons. Traite de Zoologie, XIII. Paris; 3 volumes. 1958. (To family)
ARAMBOURG, C. & BERTIN, L. Sous-classe des selaciens. Traite de Zoologie, XIII, fasc. 3, 2016-2056. 1958. (To family)
GREENWOOD, P. H. et al. Phyletic studies of teleostean fishes, with a provisional classification of living forms. Bull. American Mus. Nat. Hist., 131, 339-456. 1966.

FAMILIES

Albulidae. WHITEHEAD, P. J. P., 1965, Bull. British Mus. (Nat. Hist.), 12, 225-281 (Red Sea)

Chirocentridae. WHITEHEAD, P. J. P., 1965, Bull. British Mus. (Nat. Hist.), 12, 225-281 (Red Sea)

Cichlidae. REGAN, C. T., 1905-06, Proc. Zool. Soc. London, 1905 & Ann. Mag. Nat. Hist., (7), 15, 16. 17. (8 papers; revision of American Cichlidae)

Elopidae. WHITEHEAD, P. J. P., 1965. Bull. British Mus. (Nat. Hist.), 12, 225-281 (Red Sea)

Fitzroyididae, as Jenynsiidae

Characidae (incl. also Gasteropelecidae). FRASER-BRUNNER, A., 1950, Ann. Mag. Nat. Hist., (12), 3, 959-970. (revision)

Rhinodontidae. de BUEN, F., 1958, Invest. Zool. Chile, 4, 201-271 (Chile)

AMPHIBIA

FAUNAS & IDENTIFICATION WORKS

TORTONESE, E. & LANZA, B. Piccola fauna Italiana. Pesci, anfibi e rettili. Milano; 1-185. 1968.
TAYLOR, E. H. Herpetology of the Philippine Islands. 1. Amphibians and turtles .... Amsterdam; 1-193. 1966.
BRAME, A. H., Jr. List of Chinese salamanders. In Chang, Amphibiens Urodeles de la Chine. Ann Arbor, Michigan. 1968.

MONOGRAPHS

WERNER, F. Apoda — Gymnophiona. Schleichenlurche. Handbuch der Zoologie, 6, (2, 2), 143-208. 1931.

REPTILIA

GENERAL BOOKS

BELLAIRS, A. A. & PARSONS, T. S. Biology of the Reptilia, vol. 1, morphology. London. 1969.

BIBLIOGRAPHIES

MEDEM, F. Bibliografia comentada de reptiles Colombianes. Rev. Acad. Colombo Cienc. Exact. Fis. Nat. 12, 299-346. 1965.

FAUNAS & IDENTIFICATION

WEBB, R. G. Reptiles of Oklahoma. Norman, Oklahoma; 1-370. 1970.
SAVAGE, J. M. An illustrated key to the lizards, snakes and turtles of the West, 2nd ed. San Martin, California; 1-36. 1959.
MEDEM, F. El desarrolla de la herpetologia en Colombia. Rev. Acad. Colombiana Cienc. Exact. Fis. Nat., 13, 149-199. 1968.
DONOSO-BARROS, R. D. Contribucion al conocimiento de los cocodrilos de Venezuela. Physis, 26, 15, 32, 263-274, and other parts. 1966.
TORTONESE, E. & LANZA, B. Piccola fauna Italiana. Pesci, anfibi e rettili. Milano; 1-185. 1968.
FitzSIMONS, V. F. M. A field guide to the snakes of southern Africa. London; 1-221. 1970.
TAYLOR, E. H. Herpetology of the Philippine Islands. 1. Amphibians and turtles .... Amsterdam; 1-193. 1966.
TAYLOR, E. H. Herpetology of the Philippine Islands. 2. The lizards .... Amsterdam; 1-269. 1966.

MONOGRAPHS

von WETTSTEIN, O. Crocodilia. Handbuch der Zoologie, 7, (1, 3), 236-424. 1937-1954.
WERNER, F. Stegocephali - Panzerlurche. Handbuch der Zoologie, 7, (2, 1-2), 93-142. 1930-31)

AVES

[for de Schauensee, R. M. read Meyer de Schauensee, R.]

GENERAL BOOKS

PEARSON, R. G. The avian brain. London; 1-500. 1971.

BIBLIOGRAPHIES

ANDERSON, A. H. A bibliography of Arizona ornithology. Tucson, Arizona; 1-272. 1971.

FAUNAS & IDENTIFICATION WORKS

MANNICHE, A. L. V. The terrestrial mammals and birds of north-eastern Greenland. Medd. Gronl., 45; 1-200. 1910.
(anonymous). Lista sistematico de las Aves Argentinas. El Hornero, 6-8, several parts. 1938 —.
HENRY, G. M. A guide to the birds of Ceylon, 2nd ed. London; 1-500 —. 1971.
WON, P. O. & WOO, H. C. A distributional list of the Korean birds and mammals. Korea, 1-10, 1-96, 1-12. 1958.

NOMENCLATURE

WON, P. O. & WOO, H. C. A distributional list of the Korean birds and mammals. Korea, 1-10, 1-96, 1-12. 1958. (Common names)

MAMMALIA

[for da Costa Lima, E. read da Cruz Lima, E.]

BIBLIOGRAPHIES

(anonymous). Bibliography of F. S. Bodenheimer (1897-1957). Jerusalem; 119-145. 1957.

REVISIONS BY GROUPS

CHIARELLI, B. Taxonomic atlas of living Primates. London; 1-200. 1971.